Aegean International Textile and Advanced Engineering Conference (AITAE 2018)

IOP Conference Series: Materials Science and Engineering Volume 459

Lesvos, Greece
5 – 7 September 2018

ISBN: 978-1-5108-8005-4
ISSN: 1757-8981

Printed from e-media with permission by:

Curran Associates, Inc.
57 Morehouse Lane
Red Hook, NY 12571

Some format issues inherent in the e-media version may also appear in this print version.

Printed by Curran Associates, Inc. (2019)

For permission requests, please contact the Institute of Physics at the address below.

Institute of Physics
Dirac House, Temple Back
Bristol BS1 6BE UK

Phone: 44 1 17 929 7481
Fax: 44 1 17 920 0979

techtracking@iop.org

Additional copies of this publication are available from:

Curran Associates, Inc.
57 Morehouse Lane
Red Hook, NY 12571 USA
Phone: 845-758-0400
Fax: 845-758-2633
Email: curran@proceedings.com
Web: www.proceedings.com

TABLE OF CONTENTS

TEXTILE ENGINEERING

GARMENT ENGINEERING

MARKETING, BRANDING AND FASHION CONSUMERS

Preface

Scientists from all over the world participated in the conference and they have contributed with the communication of their valuable research work. Representatives from the industry and important decision makers have been together with the academic researchers and the creation and exchange of ideas has formed a fertile community, active from the first until the last moment of the AITAE 2018 Conference.

The main goal of the AITAE 2018 conference was to bring together academic scientists and researchers, industrialists and industry engineers, policy and administration specialists to exchange ideas and to share their experiences and results. After the successful organization of the AITAE 2018 Conference, the feedback received has confirmed that the goal of the AITAE 2018 has been fully achieved.

We would like to thank all the members of the Organizing Committee and the Scientific Committee for their continuous support during the time of the preparation and even after the Conference as well as for their contribution to maintain the high level of the AITAE 2018 Conference.

Sincere thanks to the members of the International Scientific Committee for their huge efforts in the abstracts review period and then during the full text review of the articles. The big number of the contributions imposed a very high work load in the phase of the reviews, however the members of the International Scientific Committee with the high degree of responsibility have carried out successfully the difficult tasks.

The AITAE 2018 conference couldn't be successful without the support of many members of the academic and administrative community of the University of West Attica. We truly thank all those who made this event happen and among them, Dr. Konstantinos - Stefanos Nikas, Mr. Evangelos Louris, Mr. Ioannis Chronis, Mr. Anastasios Tzerachoglou, and Mr. Aristidis Repoulias for the time spent and the insistence to overcome the minor and major difficulties with their hard work. In addition to them, the At-holidays travel agency personnel had guaranteed the quality of the travel and accommodation of the majority of our guests.

Last but not least we thank our sponsors Ouzo of Plomari, CreThiDev, IGFA for their real support and care for the strengthening of the and AITAE 2018 Conference.

After a very efficient cooperation with the IOP Publishing personnel, we have the pleasure to present to the participants and to the textile academic, research and industrial community the proceedings of the AITAE 2018 Conference. Thanks once again for your participation and active contribution to the success of AITAE 2018 Conference.

Prof. Dr. Savvas Vassiliadis

Dr. Georgios Priniotakis

IOP Conf. Series: Materials Science and Engineering **459** (2019) 011002 doi:10.1088/1757-899X/459/1/011002

Conference Committees

Organising Committee Chairs
Dr. Georgios Priniotakis
Prof. Dr. Kadoglou Huseyin
Scientific Committee Chairs
Prof. Dr. Savvas Vassiliadis
Prof. Dr. Marmarali Arzu

International Scientific Committee

Blaga Mirela	Technical University of Iasi, Faculty of Textiles and Leather Engineering	Romania
Boussu Francois	Ecole Nationale Supérieure des Arts et Industries Textiles (ENSAIT)	France
Demsar Andrej	University of Ljubljana, Department of Textiles	Slovenia
Dinis Marques Antonio	University of Minho, Department of Textile Engineering	Portugal
Garcia Pablo Diaz	Universitat Politecnica de Valencia	Spain
Kaldelis Ioannis	University of West Attica	Greece
Kazlacheva Zlatina	Trakia University	Bulgaria
Kiekens Paul	Universiteit Gent, Department of Textiles	Belgium
Kolgjini Blerina	Polytechnic University of Tirana	Albania
Nierstrasz Vincent	University of Boras, Department of Textile Technology	Sweden
Nikas Stefanos Konstantinos	University of West Attica	Greece
Nunes Ferreira Fernando	University of Minho, Department of Textile Engineering	Portugal
Papageorgas Panagiotis	University of West Attica	Greece
Patrikakis Charalampos	University of West Attica	Greece
Pyromalis Dimitrios	University of West Attica	Greece
Radu Cezar Doru	University of IASI	Romania
Rosace Giuseppe	University of Bergamo, Department of Engineering and Applied Sciences	Italy
Roshan Paul	University of Beira Interior	Portugal
Ryklin Dzmitry	Vitebsk State Technological University	Belarus
Schwarz-Pfeifer Anne	Niederrhein University of Applied Sciences	Germany
Silberman Jeffrey	Fashion Institute of Technology (FIT), New York	USA
Skrifvars Mikael	University of Boras, Department of Textile Technology	Sweden
Stjepanovic Zoran	University of Maribor, Department of Textile Engineering	Slovenia
Van Langenhove Lieva	Universiteit Gent, Department of Textiles	Belgium
Voyiatzis Georgios	Foundation for Research and Technology, Hellas (FORTH), Institute of Chemical Engineering Sciences (ICE/HT)	Greece

IOP Conf. Series: Materials Science and Engineering **459** (2019) 011003 doi:10.1088/1757-899X/459/1/011003

Peer review statement

All papers published in this volume of *IOP Conference Series: Materials Science and Engineering* have been peer reviewed through processes administered by the proceedings Editors. Reviews were conducted by expert referees to the professional and scientific standards expected of a proceedings journal published by IOP Publishing.

IOP Conf. Series: Materials Science and Engineering **459** (2019) 012001 doi:10.1088/1757-899X/459/1/012001

Talent assessment in software development firms

L Alawneh

Software Engineering Department
Jordan University of Science & Technology, Irbid - Jordan

lmalawneh@just.edu.jo

Abstract. Developing talents in a firm starts by the acquisition of the right candidates. The firm should have a well-defined process to acquire highly talented people. Software development companies constantly look for fresh graduates for strategic and economic reasons. Fresh graduates usually require intensive training to pick up on the pace of the new job requirements. Several software development companies started conducting in-house training for carefully selected candidates who passed a comprehensive placement test followed by a technical interview. At the end of this training, only highly qualified trainees will be hired for full time positions. In this paper, we present a number of methods for talent assessment in software development companies followed by a questionnaire completed by 11 software engineers in a software company in Jordan.

1. Introduction

Though talent acquisition is a big problem to the companies, the development of their talents [1] is at a higher concern to the decision makers. Employees should sustain a competitive level of knowledge and expertise. This opens new horizons to the firm by discovering new markets based on the skill set of its employees. Talent development means that the employees should continue to advance their careers by gaining new skills and knowledge and be up-to-date with the latest technologies in the market. In order to be competitive and capable of handling new challenges, we always need to have the required talents available. The lack of talented resources will have a direct impact on the turnover rates as resources with obsolete skills shall be forced to leave if they cannot keep up with the learning pace [2]. On the other hand, scarce innovative and skilled employees will look for new workplaces especially when their current employer is incapable of sufficing their needs as they look for more challenging opportunities. Therefore, continuous talent development is becoming a necessity for each company in order to stay and grow in the market. The rest of the paper is organized as follows. In Section 2, we present our talent development plan followed by a questionnaire in Section 3. Finally, we conclude our paper in Section 4.

2. Talent Development Plan

A talent development plan should be aligned with the company's objectives. Therefore, building the right plan should be carefully studied [3]. In the following, we present a number of methods that could

IOP Conf. Series: Materials Science and Engineering **459** (2019) 012001 doi:10.1088/1757-899X/459/1/012001

be integrated in software development companies' processes for developing the talents of their current employees [4]. We present these methods as bullets where each one could be applied separately or in conjunction with others.

- The company should avail all needed sources of information and learning material – Thanks to Google for this, but, we think the company should also have the latest useful material and books available when necessary. This can be achieved by building an online library at the firm. The employees themselves will be part of developing this library by adding the material and books they found useful. Moreover, this library should contain the list of pilot projects, developed by the employees in their learning process, to be set as a preference for other employees.
- The normal culture in the company should be 'I need to have a mentor'. Communication between mentors and mentees should be available and facilitated. Employees undergoing talent development should be mentored by the senior level employees in order to encourage, motivate, and assess their learning progress.
- At the beginning of each year, the employee should have a one-on-one formal meeting with his manager (or senior as some companies tend not to have a manager title) where they will discuss the weaknesses and areas of improvement (skills required by the company) and the new desired skills by the employee. The mentor should advise the employee where to focus in order to be aligned with the company's strategies and objectives.
- The company should also conduct a survey periodically (annually, semi-annually or quarterly) where the employees will be asked about their preferences on the skills they would like to gain and the courses they are interested in. The outcome of the survey should not be applied immediately but should be used for future planning (following year). People tend to forget what they were thinking of if it was not documented. Therefore, we think such a survey should be very useful in recording all the potential ideas, needed skills, etc.
- The learning process could be improved by having employees coach each other. Senior technical resources can always guide in the process of talent development and can provide the right feedback on each employee's technical abilities.
- The best way a person can gain new skills is by 'getting his hands dirty'. Therefore, asking the employees to come up with new ideas where they can apply and implement the new knowledge is an excellent strategy to make sure that they are ready to drive when new projects rise in the horizon.
- Another way is by applying Stretch Tasks. These kinds of tasks will force the employees to acquire the knowledge they are lacking much faster. Stretch tasks could be part of an ongoing project. However, this should be handled carefully with the severity of the project.
- Employees with best ideas translated into tangible pilot projects will be asked to lead their projects and build up their own teams. This will encourage the employees to innovate and excel as it will reflect directly on their career paths and will make them more engaged and loyal to the company.
- Knowledge sharing. The culture of sharing the knowledge should be the norm at the company. Employees who just delivered a major milestone should share their experience with the rest of the team members and the other teams in the company (daily stand-up meeting may be insufficient). In every project there is something new to be learned. Spreading the new

knowledge among the other teams is substantial in building the talent across the company. Moreover, knowledge gained from reading a new book, training course or any other experience should be presented to the interested teams in the company.

- Cross-team collaboration is also important in developing the talents at the company.
- Competition. Setting up a monthly/quarterly challenge where the best pilot project will be awarded. This will motivate the employees to innovate.
- Learning new skills shall be integrated in the annual review process. Employees who gain more knowledge and translate this knowledge in their work should be rewarded.

Finally, losing some of the company's most talented resources is inevitable. However, by applying certain measures, the company can reduce this number by providing the right environment to its employees. Talent retention through talent development is currently the trending approach in major software firms and proved its validity. Outstanding employees always look for competitive remuneration packages and a very challenging and highly skilled environment. Thus, companies should take talent development more seriously as it will have its benefits on all the stakeholders.

3. Employee Questionnaire Results

The objective of this survey is to measure the level of knowledge and expertise in the team and to identify opportunities for skills development. The survey was answered by 11 developers (Table 1).

Table 1. Target Employees

Employee Category	Number of Employees
Junior (0-2 years)	8
Intermediate (2-5 years)	1
Senior (5-8 years)	0
Principal (more than 8 years)	2

The topics included in the survey are core java skills, requirements engineering skills, configuration management and quality assurance skills. Developers were categorized based on their level of experience in four categories.

Table 2. Java Related Responses (average per developer group)

Question	Average per Category		
	Junior	**Intermediate**	**Principal**
Do you apply 'Exception Handling' in your code properly?	3.375	4	3.5
Rate your understanding of 'Generics'	3.375	4	3.5
Rate your skills in writing multithreaded programs	2.75	4	3
Rate your skills in using Reflection in Java?	2.875	3	2.5
Rate your skills in using Dynamic Class Loading features?	2.5	3	3
Are you familiar with Java's Dynamic Language Support?	1.875	1	2.5
Are you familiar with using Lambda Expressions in Java?	2.375	4	2.5
Rate your skills in using Annotations?	2.875	4	3
Are you familiar with Java Native Interfaces?	2.125	2	2
Rate your understanding of session management	2.625	3	3
Average	**2.73**	**3.4**	**3**

Table 2 provides a summary of the responses for the Java related questions. Each value represents the average for each answer based on each developer category (Junior, Intermediate, and Principal). The last row provides the averages of all the responses for each category. The results for junior engineers seem to be normal and aligned with their level of expertise. However, for advanced Java topics, the intermediate and principal engineers show similar knowledge compared to the junior engineers.

When asked about which requirements (functional and non-functional) they find more difficult to develop, the responses were as follows:

1. Two (2) responses were: Neither of them.
2. Four (4) responses were: Non-Functional Requirements.
3. Five (5) responses were: Depends on the task's size.

Being part of a highly dynamic and agile environment means that requirements may change at any time. Thus, the developers need to be familiar with the best practices for requirements gathering, analysis, and specifications. One important aspect is requirements traceability. The usage of Atlassian Suite at the company facilitates the traceability of requirements especially with the availability of tools such as Jira, Confluence and BitBucket.

Table 3 provides a summary of the responses related to requirements engineering. The responses indicate that most of the developers in the team need to improve on this important aspect. Requirements could come from business or from technical teams. In both ways, the engineer needs to be well acquainted with Requirements Engineering skills.

Table 3. Requirements Engineering Responses (average per developer group)

Question	Average		
	Junior	**Intermediate**	**Principal**
Are you able to elicit business/technical requirements properly?	2.875	3	3.5
Are you able to eliminate conflicts from elicited requirements?	2.875	3	3
Are you able to prioritize requirements correctly?	2.875	3	3
Do you prefer to provide a visual description of the specifications?	3.375	3	3.5
Are you able to specify the requirements in a unified and consistent manner?	2.875	3	3
Do your colleagues understand your specifications?	3	3	3
Are you following the convention at PS in writing the specifications?	3.25	3	3
Are you always capable of writing your test cases right after the requirements specifications are complete?	3.125	4	3.5
Average	**3.03**	**3.125**	**3.19**

When asked about which requirements the developers find more difficult to elicit, the responses were as follows:

1. Two (2) responses were: Depends on the stakeholders.
2. Three (3) responses were: Non-Functional Requirements.
3. Six (6) responses were: Depends on the requirements complexity.

Table 4 provides a summary of the responses regarding Configuration Management, DevOps and Software Quality Assurance. The engineers seem to have good knowledge in this area. However, the junior engineers still need to improve on some aspects.

Table 4. CM, DevOps and Quality Assurance Responses (average per developer group)

Question	Average		
	Junior	**Intermediate**	**Principal**
Rate your acquaintance level with Continuous Delivery?	2.75	3	3
Do you apply best practices when using Version Control Systems?	3	3	3
Your checked-in fixes work from the first time	3.5	5	3
Your test cases cover all scenarios	3.25	4	3
Rate your level of comfort when applying Continuous Delivery	3	5	3
Rate your level of satisfaction with applying Continuous Delivery	2.75	3	3.5
Rate your level of experience with Bamboo	2.875	3	4
Do you always provide a proper description of your changes in Jira?	3.625	2	4
Are you comfortable in taking the role of the scrum master?	3.125	3	3
Average	**3.1**	**3.44**	**3.28**

Table 5 provides a summary of the averages per section per developer. Averages below 3 were highlighted to show that these developers need to improve on the required skill.

Table 5. Average for each section per developer (J: Junior, I: Intermediate, P: Principal)

Question	J	J	J	J	J	J	J	J	I	P	P
How many years have you been at the Company? *1 year, 2 years, more than 2 years (> 2)*	1	1	1	2	2	2	1	2	> 2	> 2	> 2
How do you rate yourself in Java? *B-Beginner, I-Intermediate, A-Advanced*	I	A	I	B	I	B	I	A	A	A	B
Rate your acquaintance level with Object Oriented Principles? *B-Beginner, I-Intermediate, E-Expert*	I	I	I	E	I	I	I	E	E	E	I
Core Java Skills (average) *1 – Weak, 5 – Strong*	*2*	3.5	3.2	*2.5*	*2.8*	*2.2*	*2.6*	3.1	3.4	3.8	*2.2*
Requirements Engineering (average) *1 – Weak, 5 – Strong*	3.5	3.1	*2.1*	*2.7*	3	3.6	3	3.1	3.1	3.1	3.3
Configuration Management and Quality Assurance (average) *1 – Weak, 5 – Strong*	3.3	*2.2*	3.1	*2.8*	3.1	3.8	3.2	3.2	3.4	3.3	3.2

Table 6 summarizes the list of references the developers use to enrich their skills. As we can see, the developers depend mainly on well-known books and available blogs.

Table 6. References used by the team members

Title	Type
Design Patterns: Elements of Reusable Object-Oriented Software	Book
Object-oriented and classical software engineering	Book
Java in a Nutshell	Book
Competitive Programming 3rd Edition	Book
OAuth 2 in Action	Book
Clean Code	Book
Cucumber Cookbook	Book
Java Testing with Spock	Book
Head First Java	Book
Effective java	Book
Functional programming, reactive programming, design principles	Blogs

4. Conclusion

In this paper we proposed different methods that can be used by the talent development teams in software development companies. Additionally, we presented the results of a survey conducted by 11 developers in a well-known software development company in Jordan. The survey shows that in most cases, the knowledge of the developers is mainly related to the number of years of experience and the number of years spent at the company. In the future, we intend to present an assessment of the developers' skills after applying the proposed talent development methods and their impact on their productivity.

References

[1] A. H. Church, C. T. Rotolo, N. M. Ginther, R. Levine, "How are top companies designing and managing their high-potential programs? A follow-up talent management benchmark study," Consulting Psychology Journal: Practice and Research, Vol. 67(1), Mar 2015, 17-47

[2] E. Mone, M. London, "Employee Engagement Through Effective Performance Management," New York: Routledge, 2018.

[3] J. A Gruman., A. M. Saks, "Performance management and employee engagement," Human Resource Management Review, 2012, Vol. 21(2), 123-136.

[4] A. H. Mehdiabadi, J. Li, "Understanding talent development and implications for human resource Development: An integrative literature review. Human Resource Development Review," 2016, Vol. 15(3), 263–294.

IOP Conf. Series: Materials Science and Engineering **459** (2019) 012002 doi:10.1088/1757-899X/459/1/012002

Data Envelopment Analysis for Cities Efficiency

F Ali, H Njadat and Q Yaseen
Jordan University of Science and Technology
Irbid, Jordan 22110
{faali14, qmyaseen, najadat}@just.edu.jo

Abstract. This paper discusses the envelopment analysis for selected cities. It uses the DEA (Data Envelopment Analysis) for both input and output oriented methods using a data set of 176 cities, and tests several classifiers in classifying the dataset using both cross-validation and percentage split.

1. Introduction

Data Envelopment Analysis (DEA) is used to estimate the production frontiers in the fields of economics and operations research. Empirically, it measures the decision making units DMUs according to the productive efficiency. Generally speaking, data envelopment analysis (DEA) is a data-driven way of measuring efficiency for each decision making unit (DMU) [1][2]. In addition to using DEA in production theory, it is an important tool for benchmarking in operations research.

DEA is used in many applications such as the electric utilities sector, where DEA is used as a tool to determine and specify individualized regulatory rates for firms given their comparative efficiency [3]. Moreover, DEA is used to assess the efficiency of various non-profit or public organizations such as hospitals [4], police forces [5], etc.

Measuring the cities efficiency is crucial for sustainable development of cities. Minimizing the resources consumptions and maintaining and improving the delivered services and the construction activities level is mandatory for cities efficiency. The aforementioned measuring process is performed using the DEA tool.

This work discusses the cities efficiency and uses the DEA for this purpose. It calculates the efficiency and shows how correctly the classifiers work to classify the cities in the used data set using many of the DEA models [6]. The rest of work is organized as follows. Next section discusses some related work. Section 3 introduces the approach and the results. Finally, section 4 concludes the work.

2. Related work

Some research about using DEA for cities efficiency in China were performed by Fang et. al [7] and Bian and Yang [8]. The authors investigated the DEA for the China's Urban Agglomerations UAs using various sizes of populations and locations of about 35 central cities. They discussed the relationship among the elements the decomposed model and compares their efficiency performance. Moreover, the came up with important recommendations about improving the UAs efficiency performance.

Kuosmanen and Kortelainen [9] and Loikkanen and Susiluoto [10] used the DEA for cities efficiency in Finland. The authors used a large dataset of regional input and output variables and regional characteristics of about 83 NUTs 4-level regions in Finland during 1988-1999. The used the DEA to get the annual efficiency scores of all regions, and then, they explained the differences in inefficiency among regions using econometric approaches (logistic regression, Tobit Models).

Many other researchers investigated the using of DEA in many countries such as Raab and Lichty [11] who used it in USA cities, Bal and Orkcu [12] who used DEA in Turkey, and Suzuki et al. [13] who used the DEA to measure cities efficiency in Italy.

3. The Approach

The approach starts by obtaining the dataset that contains 176 DMU [14]. Each DMU represents a country with 14 attributes as input and 5 attributes as output. The attributes are { Real per capita GDP(Gross Domestic Product), Real per capita GDP growth, Population, Growth in population, Secondary school enrollment rates, IMF indicator of capital account restrictions, Institutional Investor Risk Ratings, Trade openness (exports + imports)/GDP, Financial openness (financial account inflows + financial account outflows)/GDP, Capital inflows (Inward FDI + inward portfolio + inward other)/GDP, Correlation of annual growth rates with OECD (Organization for Economic Cooperation and Development) average annual growth rates, Average annual growth rate of GDP deflator, Proportion of population speaking English, Proportion of population speaking major European language, Total ODA from all sources, Real per capita GNP, Oil export, Non-Oil Commodity Exporters, Ethnolinguistic fractionalization } and delete these fields from the dataset { Investment to GDP at PPP, Terms of trade shock (Growth in export deflator minus growth in import deflator, weighted by current LCU shares of exports and imports in GDP), Unweighted terms of trade shock }. The work is divided into three main parts, which are:

A. *Data preprocessing:* The dataset is not ready for DEA. Some processing has to be performed on the data to prepare it. First, some attributes are removed, while others are normalized to remove negative numbers since DEA does not deal with negative numbers. Moreover, the dataset contains missing values. In this case, it is filled by the attribute average. This preprocessing task prepares the dataset for efficiency calculation.

B. *Efficiency calculation:* The efficiency is calculated for the cities using input and output oriented and both CRS and VRS using EMS (Efficiency Measuring System). Figure 1 and Figure 2 show the calculation of input orient for both CRS and VRS. Figure 3 and Figure 4 show the efficiency calculation of output oriented for both CRS and VRS.

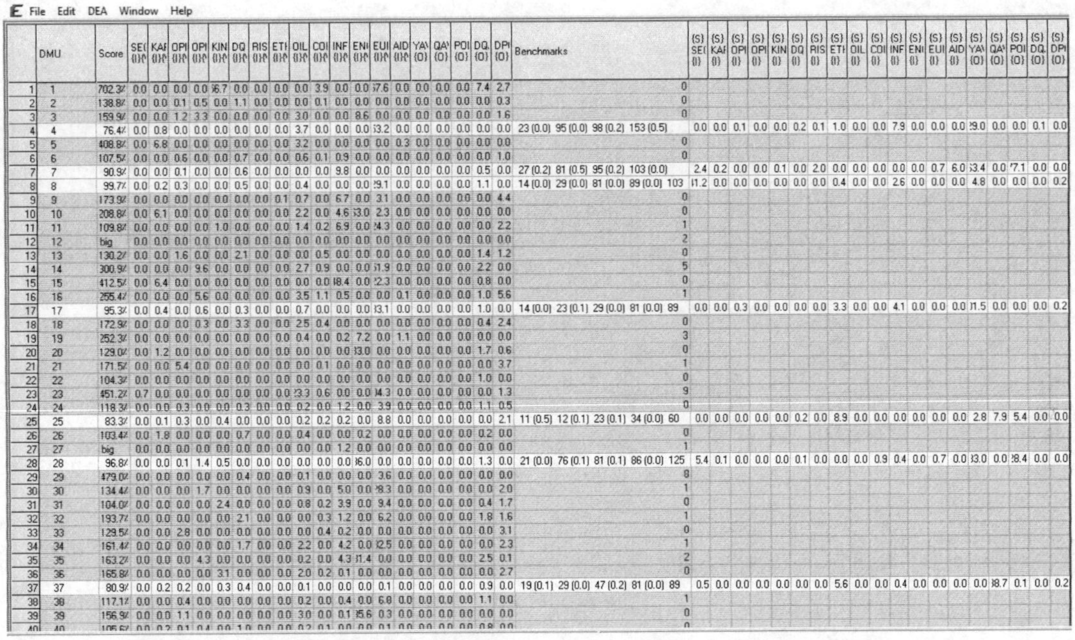

Figure 1. Efficiency calculation for input oriented CRS.

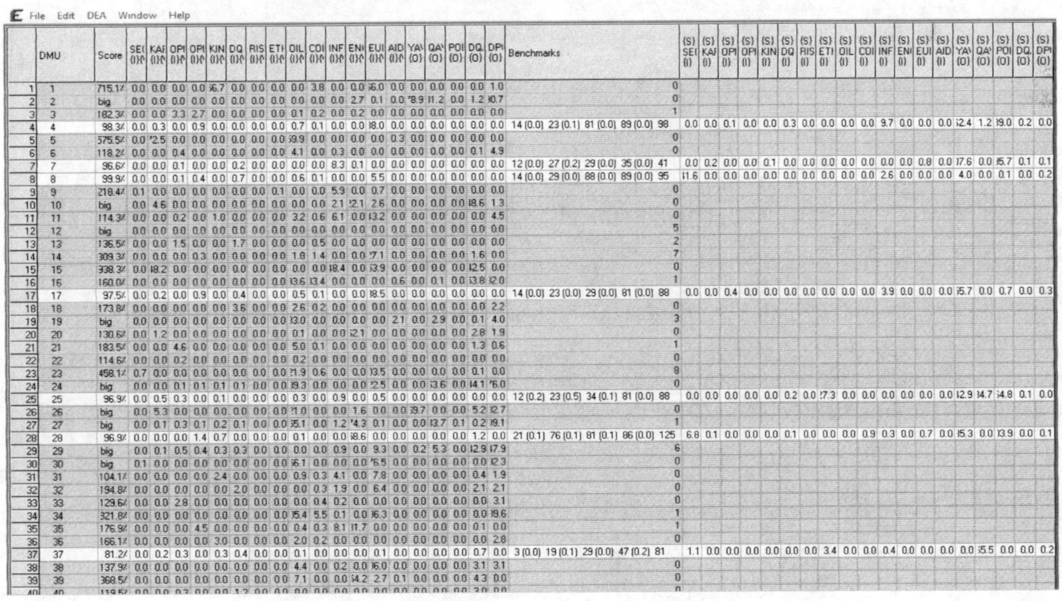

Figure 2. Efficiency calculation for input oriented VRS.

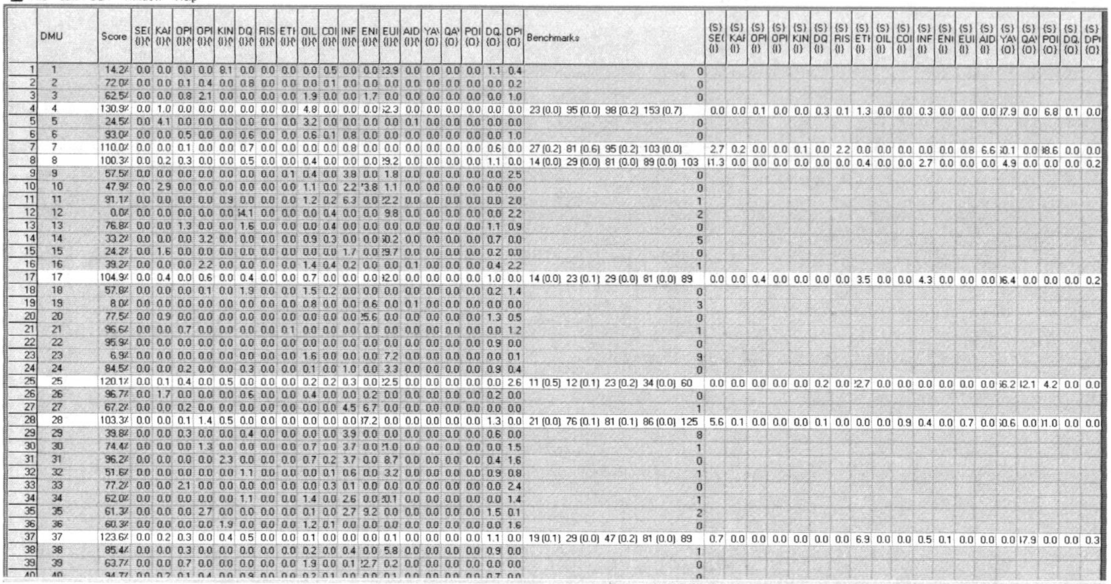

Figure 3. Efficiency calculation for output oriented CRS.

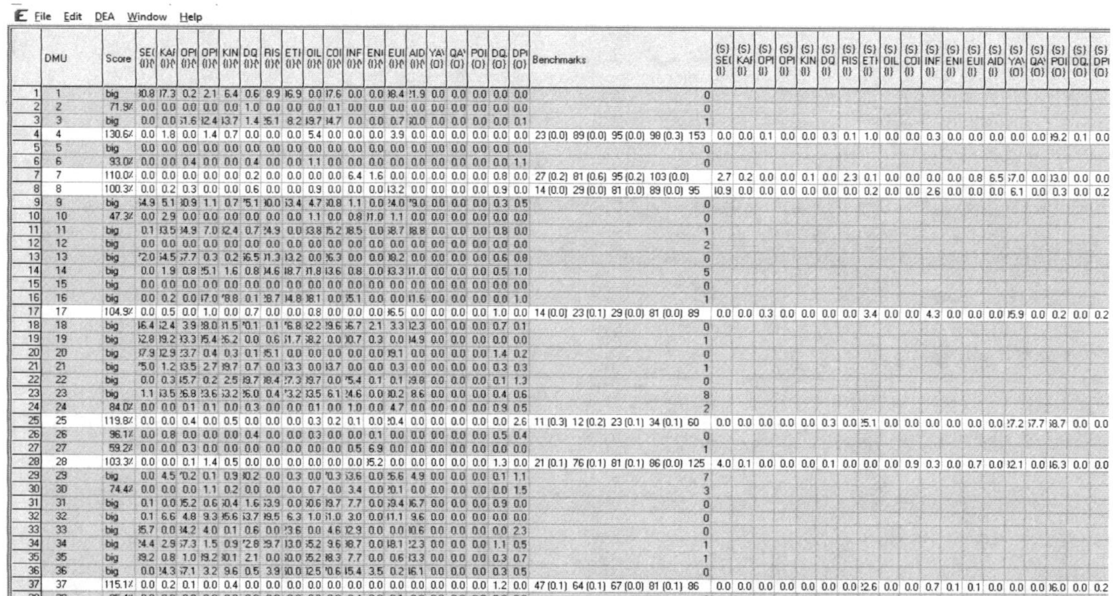

Figure 4. Efficiency calculation for output oriented VRS.

The figures show the number of efficient and inefficient DMUs as follows:

- The calculation for input oriented.
  a. Number of efficient DUMs are 153.
  b. Number of inefficient DUMs are 23.
  c. For both CRS, VRS.
- The calculation for output oriented.
  a. Number of efficient DUMs are 152,154.
  b. Number of inefficient DUMs are 24, 22.
  c. For CRS and VRS.

C. *Classification:* After calculating the efficiency for the DMUs, the approach classifies the DMUs using the DMUs efficiency as a class using the classifiers: Decision tree (J48), K nearest neighbor (KNN), Support vector machine (SVM) and Naïvebayse using both cross-validation and percentage-split for the dataset. The classification result is shown in table 1.

Table 1. The classifiers along with the accuracy using both Cross-validation and percentage-split

	Classifier	Accuracy Cross Validation	Accuracy Percentage Split
1	J48	85.2273 %	75%
2	Naive Bayes	74.4318 %	78.3333 %
3	SVM	86.9318 %	75%
4	KNN-9-1	86.9318 %	75%

As shown in the table, the best classifiers are the SVM and KNN with K=9 which achieved an accuracy of 86.9318 %. The decision tree was the next and achieved an accuracy of 85.2273 %. Since the class is imbalanced in the dataset with 153 as efficient country and 23 as inefficient country, the SMOTE filter is used in Weka to balance data. Table 2 shows the results after balancing the data.

Table 2. The classifiers along with the accuracy using both Cross-validation and percentage-split for each classifier after balancing the data for each class.

	Classifier	Accuracy Cross Validation	Accuracy Percentage Split
1	J48	87.2131 %	89.4231 %
2	Naive Bayes	74.4262 %	75 %
3	SVM	79.6721 %	75 %
4	KNN-4	83.2787 %	79.8077 %

The table shows that the best classifier is the decision tree with accuracy of 89.4231 % followed by KNN with K=4 and accuracy = 83.2787 %.

D. *Clustering:* After the classification process, the K-mean cluster is used for the clustering process. The results are shown in Table 3, where the clustering accuracy is 54.7541%.

Table 3. The clustering accuracy.

	Clusterer	Accuracy
1	K-mean	Incorrectly clustered instances: 45.2459%
		Correctly clustered instances : 54.7541%

4. Conclusion

This paper has discussed the envelopment analysis for a number of cities using a number of classifiers. The paper has shown that both input and output oriented DEA achieves almost the same result. The best classifiers were the support vector machine and k nearest neighbor in the unsmoothed data with accuracy of 86.9318 %. In the smoothed data, the best classifier was the decision tree with accuracy of 89.4231 %. However, the clustering shows a low accuracy of 54.7541% using the K-means.

References

[1]. Adler N, Lea Friedman L and Sinuany-Stern Z 2002 Review of ranking methods in the data envelopment analysis context *European Journal of Operational RESEARCH*

[2]. Kaoru Tone K and Miki Tsutsui M An epsilon-based measure of efficiency in DEA revisited-A third pole of technical efficiency Tokyo, Japan.

[3]. Berg S. 2010. Water Utility Benchmarking: Measurement, Methodology, and Performance Incentives. *International Water Association.*

[4]. Vera A.; Kuntz L. 2007. Process-based organization design and hospital efficiency. *Health Care Management Review*. 32 (1), pp. 55–65.

[5]. Thanassoulis E 1995. Assessing police forces in England and Wales using data envelopment analysis. *European Journal of Operational Research*. 87 (3), pp. 641–657.

[6]. Jacobs I and Bean C 2004 Data Envelopment Analysis, Theory Techniques for Economic and Operation Research, *Cambridge University*.

[7]. Fang, C., et al., 2013. Input output efficiency of urban agglomerations in China: an application of data envelopment analysis (DEA). *Urban Study*, 50 (13), pp. 2766– 2790.

[8]. Bian, Y., Yang, F., 2010. Resource and environment efficiency analysis of provinces in China: a DEA approach based on Shannon's entropy. *Energy Policy*, 38, pp.1909–1917

[9]. Kuosmanen, T., Kortelainen, M., 2005. Measuring eco-efficiency of production with data envelopment analysis. *J. Ind. Ecol.* 9 (4), pp.59–72

[10]. Loikkanen, H., Susiluoto, I., 2002. An evaluation of economic efficiency of Finnish regions by DEA and TOBIT models. *In: 42nd Congress of the European Regional Science Association, Dortmund.*

[11]. Raab, R., Lichty, R., 1997. An efficiency analysis of Minnesota counties: a data envelopment analysis using 1993 IMPLAN input–output analysis. *J. Reg. Anal. Policy,* 27 (1), pp. 75–93.

[12]. Bal, H., Örkcü, H.H., 2011. A new approach to cross efficiency in data envelopment analysis and performance evaluation of Turkey cities. *Gazi Univ. J. Sci.,* 25 (1), pp. 107–117

[13]. Suzuki, S., Nijkamp, P., Rietveld, P., 2011. Regional efficiency improvement by means of data envelopment analysis through Euclidian distance minimization including fixed input factors: an application to tourist regions in Italy. *Reg. Sci.*, 90 (1), pp. 67–89.

[14]. Easterly W and kraay A 2000. Small States, Small Problems? Income, Growth, and Volatility in Small States, *The World Bank.*

IOP Conf. Series: Materials Science and Engineering **459** (2019) 012003 doi:10.1088/1757-899X/459/1/012003

Speeding DBLP querying using hadoop and spark

M Alsmirat, Y Jararweh and M Al-Ayyoub

Jordan University of Science and Technology, Computer and Information Technology, Computer Science Department, Jordan

E-mail: masmirat@just.edu.jo

Abstract. Big data is becoming bigger every day. Even for simple applications such as the Digital Bibliography & Library Project (DBLP) database, the data is becoming unmanageable using the conventional databases because of its size. Applying big data processing methods such as Hadoop and Spark is becoming more popular because of that. In this work, we investigate the use of Hadoop and Spark in the querying process of big data and we compare the performance of them in terms of their execution time. We use the DBLP database as a case study. Results show that Hadoop and Spark enhances the query execution time significantly when compared with conventional database management systems. We also found that Spark enhances the execution time over Hadoop.

1. Introduction

Big data is a term for data sets that are so large or complex for traditional data processing application software to deal with them [1]. Relational database management systems and desktop statistics and visualization packages often have difficulty handling big data. This processing may require parallel software running on maybe thousands of servers. Big data solutions can be classified according to the approach used in storing and processing these data. The first approach is the traditional enterprise approach where an enterprise have a mainframe computer to store and process big data. The second approach is the MapReduce algorithm [2], which was proposed by Google. MapReduce divides the task into small parts and assigns them to many computers, and collects the results from them which when integrated, form the result dataset. Apache implemented the MapReduce approach into Apache Hadoop [3] and Apache Spark [4]. DBLP is a computer science bibliography website which index all important journals and conferences on computer science. In this paper, we loaded the DBLP dataset [5] in the aforementioned two approaches. More specifically, we used Oracle 12c to implement the first approach and we used Apache Hadoop with Apache Hive [6] and Apache SparkeSql [7] to implement the second approach. We compare the execution time of querying the DBLP dataset using the three approaches.

2. Background Information

2.1. Apache Hadoop with Apache Hive

Hadoop is an open source framework by Apache that is implemented in java. This framework allows decentralized processing of large datasets over distributed computers. The Hadoop framework also provides distributed storage to be able to facilitate its distributed computation. Furthermore, Hadoop is designed to be scalable in a way that it can be run on a single machine or thousands of machines which offer its own computation power and storage. Hadoop framework consists of the following two major layers.

(i) Processing/Computation layer (MapReduce)
(ii) Storage layer (Hadoop Distributed File System).

Hadoop does not have internal data warehouse. Usually, use Apache Hive is used as Hadoop data warehouse. Hive is a tool that is usually used to be able to process structured data in Hadoop. Hive can be seen as a layer above Hadoop facilitate querying and analyzing Hadoop data easy. Initially, Hive was developed in Facebook. From there, Apache adapted it and developed it as an open source with the name Apache Hive. Currently, Hive is being used by different companies such as Amazon in Amazon Elastic MapReduce. Hive provides an SQL like language, called HiveQL or HQL, for querying data. Figure 1 shows the Hive internal architecture [1].

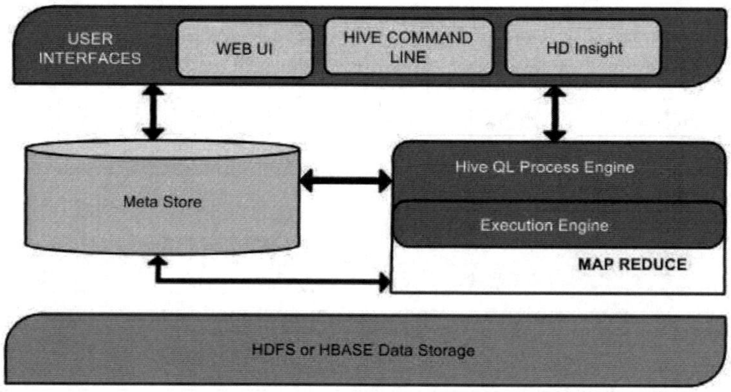

Figure 1. Hive Architecture

2.2. Apache Spark

Apache Spark came to speed the processing time of Hadoop and to efficiently use more types of operations. These operation may include interactive queries and stream processing. In these operations, the main concern is to maintain high speed in processing large datasets to make the waiting time between queries and the waiting time to run a program very short. As Spark has its own computation distribution manager, it can be used without Hadoop. With that said, Spark still can be implemented using Hadoop. Spark may use Hadoop for storage or for processing. Spark main components are shown in Figure 2 [2]. As Spark has its own internal SQL components, it does not need any warehouse to be install with.

3. Materials and Methods

We installed Hadoop and Hive on a laptop that has 5th generation quad core i7, 16 GB or RAM, and an NVIDIA GTX 960M. We then downloaded the DBLP database as an XML file. The file contains around 5 million records.

To use Hadoop with Hive on a single machine, it should be first installed on the machine and then Hive is installed above it. After that, Hive should be run using the command hive on a terminal. After that, to convert any XML data you want to use to Hive required format, an

[1] https://www.tutorialspoint.com/hive/hive_introduction.htm

[2] https://intellipaat.com/tutorial/spark-tutorial/apache-spark-components/

IOP Conf. Series: Materials Science and Engineering **459** (2019) 012003 doi:10.1088/1757-899X/459/1/012003

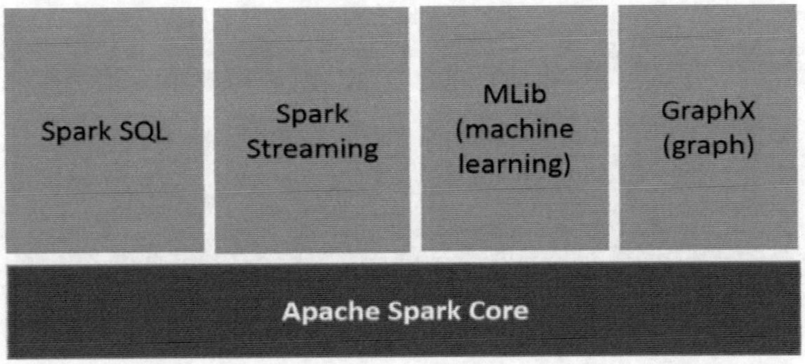

Figure 2. Spark Components

external tool called XmlSerDe [3] that is implemented as a jar file can be used. In our case, the DBLP dataset was converted, loaded, and processed using the following code.

```
CREATE EXTERNAL TABLE PUBLICATIONS
(author array<STRING>, title STRING, journal STRING)
ROW FORMAT SERDE 'com.ibm.spss.hive.serde2.xml.XmlSerDe'
WITH SERDEPROPERTIES
("column.xpath.author"="/article/author/text()",
"column.xpath.title"="/article/title/text()",
"column.xpath.journal"="/article/journal/text()")
STORED AS INPUTFORMAT 'com.ibm.spss.hive.serde2.xml.XmlInputFormat'
OUTPUTFORMAT 'org.apache.hadoop.hive.ql.io.IgnoreKeyTextOutputFormat'
LOCATION '/test/book6'
TBLPROPERTIES ("xmlinput.start"="<article","xmlinput.end"= "</article>");
```

After running the previous code, SQL queries can be used to query the data. In this paper, we used the queries in Figures 3, 4, and 5.

To use We use eclipse IDE with its APIs to run our code on it. A Maven project is created and pom.xml (the dependencies XML file) is edited to configure Spark and SparkSQL.

```
<dependencies>
    <dependency>
        <groupId>org.apache.spark</groupId>
        <artifactId>spark-core_2.10</artifactId>
        <version>2.0.2</version>
    </dependency>
    <dependency>
        <groupId>org.apache.spark</groupId>
        <artifactId>spark-mllib_2.10</artifactId>
        <version>2.0.2</version>
    </dependency>
    <dependency>
        <groupId>org.apache.spark</groupId>
        <artifactId>spark-sql_2.10</artifactId>
        <version>2.0.2</version>
    </dependency>
```

[3] https://github.com/dvasilen/Hive-XML-SerDe

IOP Conf. Series: Materials Science and Engineering **459** (2019) 012003 doi:10.1088/1757-899X/459/1/012003

```
    <dependency>
        <groupId>com.databricks</groupId>
        <artifactId>spark-xml_2.10</artifactId>
        <version>0.4.0</version>
     </dependency>
     <dependency>
        <groupId>se.uu.farmbio</groupId>
        <artifactId>cp</artifactId>
        <version>0.0.1-SNAPSHOT</version>
     </dependency>
</dependencies>
```

The following Scala language script is written to load the dataset and be able to run the same queries in Figures 3, 4, and 5 on it.

```
package com.spark_example
import org.apache.spark._
import org.apache.spark.{SparkConf , SparkContext}
import org.apache.spark.sql.{DataFrame , SQLContext}
import com.databricks.spark.xml
import javafx.scene.input.DataFormat
import org.codehaus.jackson.format.DataFormatDetector
import org.apache.spark.sql.DataFrameReader
import com.sun.xml.internal.ws.wsdl.writer.document.Import
object main{
    def main(args: Array[String]) {
        val conf = new SparkConf().setAppName("PUBLICATIONS").setMaster("local")
        val sc = new SparkContext(conf)
        val sqlContext :   SQLContext = new org.apache.spark.sql.SQLContext(sc)
        loadBookData(sqlContext)
    }
    def loadBookData(sqlContext : SQLContext){
        var df:DataFrame = null
        var newDF:DataFrame = null
        import sqlContext.implicits._
        df = sqlContext.read.format("xml").option("rowTag", "article").load("dblp.xml")
        df.registerTempTable("dblp")
        sqlContext.sql("SELECT author FROM dblp where journal
                            = 'Acta Inf.' and year = '2000'").show()
    }
}
```

```
SELECT author
      FROM PUBLICATIONS
      WHERE journal = 'Acta Inf.' and year = '2000';
```

Figure 3. The SQL Query 1.

IOP Conf. Series: Materials Science and Engineering **459** (2019) 012003 doi:10.1088/1757-899X/459/1/012003

```
SELECT author
    FROM PUBLICATIONS
    WHERE journal = 'IEEE Trans. Knowl. Data Eng.'
```

Figure 4. The SQL Query 2.

```
SELECT author
    FROM PUBLICATIONS
    WHERE title = 'Ranking Z-numbers with an improved ranking
    method for generalized fuzzy numbers.';
```

Figure 5. The SQL Query 3.

Table 1. Execution time of different implementations for Query 1

Implementation	Average Execution time (Seconds)
Oracle	10821.03133
Hadoop	792.4905
Spark	363.2

Table 2. Execution time of different implementations for Query 2

Implementation	Average Execution time (Seconds)
Oracle	16336.938
Hadoop	1014.6898
Spark	503.3

Table 3. Execution time of different implementations for Query 3

Implementation	Average Execution time (Seconds)
Oracle	10825.19267
Hadoop	568.6398
Spark	503.3

4. Experiments and Results

On the three implementations, we run the queries shown in Figures 3, 4, and 5 each for ten times and we measure the execution time of each run. We then calculate the average of the 10 runs of each query on each implementation. The results are shown in Tables 1, 2, and 3. As can be seen clearly from the results, the Spark implementation outperformed all other implementations significantly. It reduces the execution time by almost 2X when compared with Hadoop and by almost 32X when compared with Oracle.

References

[1] Hilbert M and Lopez P 2011 *science* 1200970
[2] Dean J and Ghemawat S 2008 *Commun. ACM* **51** 107–113 ISSN 0001-0782 URL `http://doi.acm.org/10.1145/1327452.1327492`
[3] Apache Hadoop URL `http://hadoop.apache.org/`
[4] Apache Spark URL `https://spark.apache.org/`
[5] DBLP dataset URL `https://dblp.uni-trier.de/faq/How+can+I+download+the+whole+dblp+dataset`
[6] Apache Hive URL `https://hive.apache.org/`
[7] Apache Spark sql URL `https://spark.apache.org/sql/`

A Mutual Debt Cut Algorithm for a Group of Countries

S Fatouros[1], P Papadopoulos[2], N L Matiadou[2] and S Vassiliadis[2]

[1]Department of Informatics and Computer Engineering, University of West Attica, Greece
[2]Department of Electrical and Electronics Engineering, University of West Attica, Greece

Abstract. In the recent years the financial debate has focused on the national debt of the countries. These debts consist of different debts to banks, institutions and other countries. In the current work we investigate a method for a mutual national debt cut amongst a group of countries using as a deposit capital equal to the minimum of the debts within the group. The only condition is that all countries own and possess debt within the group. The deposit capital is circulated to repay the debts. At the end remains intact, it works in fact as a catalyst. The algorithm introduced in this paper is based on a simple transfer table. The advantage of the algorithm is that it is not actually based on financial resources. On the other hand depends on countries will for cooperation.

Keywords: Mutual Debt Cut, transfer table

1. An Example of Mutual Debt Cut

We consider a group of countries as in table 1 where we denote the debt of one country to another. In the (i,j) cell we denote by d_{ij} the debt of the i-Country to the j-Country

Table 1

	C1	C2	C3	C4	C5	
C1	0	31	25	16	8	80
C2	11	0	23	10	7	51
C3	4	22	0	10	0	36
C4	13	25	22	0	11	71
C5	22	90	112	43	0	267
	50	168	182	79	26	505

On the column to the right of the table we denote the Total Debt $D_i = \sum_{j=1}^{n} d_{ij}$ of every country to the rest of the group and below the table we denote the amount $P_j = \sum_{i=1}^{n} d_{ij}$ that every country processes. We also denote as M the minimum amongst the total Depts and Possess of every country i.e. $M = \min_{\substack{i=1...n\\ j=1...n}} \{P_i, D_j\}$. among the countries.

As we see in our example $M = 26$

We pick the *"poorest" country* Cj that currently has the maximum total debt and we grand it with a deposit equal to M under the restriction that the whole amount should be used to reduce the country's debt starting with the country's minor creditor. In our example C_1 county is C_5 's *minor creditor* with 22 units. Using the deposit C_5 buys that debt sending the corresponding amount of deposit to C_1 as seen on tables 2a and 2b.

Table 2a

	C1	C2	C3	C4	C5	Current Debt	Deposit	Used Credit	Rem. Credit
C1	0	31	25	16	8	80			26
C2	11	0	23	10	7	51			26
C3	4	22	0	10	0	36			26
C4	13	25	22	0	11	71			26
C5	22	90	112	43	0	267	26		26

Table 2b

	C1	C2	C3	C4	C5	Current Debt	Deposit	Used Credit	Rem. Credit
C1	0	31	25	16	8	80	22		26
C2	11	0	23	10	7	51			26
C3	4	22	0	10	0	36			26
C4	13	25	22	0	11	71			26
C5	0	90	112	43	0	245	4	22	4

We repeat the procedure of buying debt from the poorest country (amongst the countries that have Deposit) to its major creditor. If a country completes the higher available credit it is excluded from the procedure. This is shown in Table 3 of our example where C_5 country have reached the maximum Debt Cut

Table 3

	C1	C2	C3	C4	C5	Current Debt	Deposit	Used Credit	Rem. Credit
C1	0	31	25	16	8	80	22		26
C2	11	0	23	10	7	51			26
C3	4	22	0	10	0	36			26
C4	13	25	22	0	11	67	4		26
C5	0	90	112	39	0	245	0	26	0

We should also take into account that the sum of the available deposit of a country and the cut to its debt should not exceed the maximum allowance (26 in our example)

The procedure continues with the remaining countries up to the point where according to the restrictions there are no moves available.

In table 4 transfer from C_4 to C_1 is not allowed as for the latter the sum of its deposit and of its used credit has reach to M

Table 4

	C1	C2	C3	C4	C5	Current Debt	Deposit	Used Credit	Rem. Credit
C1	0	31	25	0	8	64	10	16	10
C2	11	0	23	10	7	51			26
C3	4	22	0	10	0	36			26
C4	9	25	22	0	11	67	16		26
C5	0	90	112	39	0	241	0	26	0

Eventually we will come to a point where all the countries but one will have used the maximum deposit in or where there are no moves available as shown in table 5. A new decision on how this deposit will be distributed has to be taken. Thus we evaluate the differences between the initial and the current debt for each of the countries. This difference should balance the countries debt cut.

Table 5

	C1	C2	C3	C4	C5	Current Debt	Deposit	Used Credit	Rem. Credit
C1	0	31	15	0	8	54	0	26	0
C2	11	0	23	10	7	51	6	6	20
C3	4	22	0	4	0	30	20	6	20
C4	9	19	6	0	11	45	0	26	0
C5	0	90	112	39	0	241	0	26	0
Cur. Poss	24	162	156	53	26				
In.Poss.	50	168	182	79	26				
Difference	-26	-6	-26	-26	0				

We observe that countries have used less of the deposit while the country has not cut any of its possession of debts. The aim of this phase is to balance each country's cut of debt and cut on possession of debt. Thus we transfer deposit from the countries that have not used it (C_2) to the countries that have cut more of their own debt that of the debt of the others (C_5). After some iterations of the procedure we reach table 6 where new decisions have to be taken as this phase also comes to an end. In table 6 we don't have direct moves that will deliver the balance. thus we do indirect transfers from C_2 to C_5 via available indirect paths ($C_2 \rightarrow C_4 \rightarrow C_5$) in order to eliminate the possession of C_5.

Table 6

	C1	C2	C3	C4	C5	Current Debt	Deposit	Used Credit	Rem. Credit
C1	0	31	15	0	8	54	0	26	0
C2	11	0	23	10	0	44	19	6	20
C3	4	2	0	4	0	10	0	26	0
C4	9	19	6	0	11	45	0	26	0
C5	0	90	112	39	0	241	7	26	0
Cur. Poss	24	142	156	53	19				
In.Poss.	50	168	182	79	26				
Difference	-26	-26	-26	-26	-7				

Finally we come to a balanced solution where every country has cut an amount of its own debt equal to the cut on its possession of other countries debt as shown in table 7

Table 7

	C1	C2	C3	C4	C5	Current Debt	Used Credit
C1	0	31	15	0	0	46	34
C2	3	0	22	0	0	25	26
C3	4	2	0	3	0	9	27
C4	9	19	6	0	0	34	37
C5	0	90	112	39	0	241	26
Cur. Poss	16	142	155	42	0		
In.Poss.	50	168	182	79	26		
Difference	-34	-26	-27	-37	-26		

It is obvious that last line's differences match each county's debt cut.

2. Summary and Conclusions

The procedure that is introduced is based on the mutual will of a group of countries to exchange debts. The whole procedure consists of the following phases:

a. We grant the "poorest country" with a deposit capital M the minimum amongst the total Debt and Debt Possess of every country.
b. Transfer of deposit capital from the country with the current maximum total debt to its minor creditor under the condition that every country's deposit and used deposit should not exceed M. We repeat until there are no moves available.
c. We transfer deposit directly or via paths until we balance every countries cut on own debt and possessed debt. At this phase some countries cut exceed M.

The end of the procedure finds all countries with a reduced debt. We have used simple transfer tables and the simple idea of debt transfer and exchange in the sense that it is used by merchants and providers for centuries. Alternative steps can be introduced by different criteria that might lead to bigger cuts or fewer steps. What probably the countries lack in order to proceed to mutually beneficial agreements is the trust and the political will.

References

[1] Kavaljit Singh, Global Research, "Who owns Greece's Debt", https://www.globalresearch.ca/who-owns-greeces-debt/5460265
[2] International Monetary Fund, "World Economic Outlook Database" July 2017

Speech recognition in flight simulator

S Rustamov[1,3], E Gasimov[1], R Hasanov[1], S Jahangirli[2], E Mustafayev[3], D Usikov[2]

[1]ADA University, School of Information Technologies and Engineering, 11 Ahmadbey Aghaoglu St, Baku, Azerbaijan

[2]Azerbaijan National Aviation Academy, Azerbaijan

[3]Institute of Control Systems of ANAS, Azerbaijan

srustamov@ada.edu.az, ADA University

Abstract. The article is devoted to the investigation of speech recognition in Flight Simulator cockpit. We have done research and developed software for speech recognition in Flight Simulator with limited vocabulary in C# from scratch. The speech recognition system works in real time and offline mode in Windows platform. We used Mel Frequency Cepstral Coefficients and Linear Predictive Coefficients feature extraction algorithms and trained the system by multilayer artificial neural networks. The User Interface of platform is highly functional and allows users to update system parameters through the interface. Our Speech Recognition system results are compared to the results of Microsoft Speech SDK and received satisfactory achievement.

1. Introduction

Virtual Reality (VR) creates an opportunity for users to interact with three-dimensional digital representations of products through natural human means. Through combining speech interface with a feasibility to interact directly with objects in the virtual environment leads to a multimodal interface wherein users are able to interact with the Virtual Reality surroundings by the means of physical or speech commands [1].

Two approaches are available for applying speech recognition to a virtual environment: speech of fully interactive and speech of "command and control". Notwithstanding the fact that fully interactive speech makes possible to recognize numerous words and phrases, it requires a speaker-dependent system to be applied. The speech of "command and control" needs just a minor set of predefined commands and allows using speaker-independent methods [9]. As the illustrated approach demands only a limited vocabulary, we have chosen the command and control approach combined with speaker-independence. The main purpose of the article is to realize the system with a menu of speaker-independent, command and control speech recognition for Flight Simulator which one of the VR applications. As experiment we recognize 41 Flight checklist commands. The major function of the flight deck checklist is to ensure that the crew will properly configure the airplane for any given segment of flight.

2. Related works

There are many researches have been investigated for speech recognition.

D Doye et. al. suggested non-linear time alignment model for the Marathi language. As feature extraction algorithms they used Mel Frequency Cepstral Coefficients (MFCC), Linear Frequency Cepstral Coefficients (LFCC) and Linear Prediction Coefficient (LPC) [12].

In [13] authors used DTW approach based on MFCC features for the Speech Recognition of Tamil database.

Elyes et. al. suggested hybrid approach of SVM/HMM for Arabic ASR based on triphones modelling. They showed SVM/HMM hybrid model gives better results than HMMs and MLP/HMM models [14].

J. Padmanabhan et.al used Gaussian mixture model and HMM for automatic speech recognition [15].

In [16] authors proposed sparse auto-encoder using Convolution Neural Network.

In [17] authors suggested hybrid HMM/DTW approach by using kernel adaptive filters for speech analysis and recognition.

Aida-zade et al. applied multi-structured neural networks to improve SR for Azerbaijani language [20][22].

The authors from [19] developed End-to-End Speech Recognition with Recurrent Neural Networks. They trained SR from raw data without intermediate phonetic representation. The suggested system is based on a combination of the deep bidirectional LSTM and the Connectionist Temporal Classification objective function. They achieved 27.3% WER on the Wall Street Journal corpus with no prior linguistic information, 21.9% with only a lexicon of allowed words, and 8.2% with a trigram language model. Combining the network with a baseline system further reduces the error rate to 6.7%.

Xiang-Lilan et. al. introduced merged-weight dynamic time wrapping algorithm and showed this approach gives better result than DTW [20].

3. Feature extraction

The following tasks have been carried out for calculation of characteristics of speech.

Cleaning noise from speech. As speech signal varies in low frequency, for cleaning high frequency noise of it, speech is filtered in first order high frequency filter.

Identification of the end points of speech. Identifying the start and end points of pronounced speech is one of the main problems in Speech Recognition. VAD (Voice Activation Detection) is one of the methods used to identify the end points of speech signal. Our system provides user with three parameters to identify the end points of speech accurately.

1. The number of blocks that speech is divided to identify the end points of speech. In our system, for silence 100 partitions have been spared. Having high value helps to get the pauses between speech's phonemes.

2. The number of blocks in which environmental sounds occur. In our system, 5 blocks have been spared for silence (30msecs).

3. Speech separation coefficient. Changing this coefficient according to speech recording sensitivity, we can test the correctness of identification of end points of speech signal.

Dividing words which establish speech into classes, and finding their average length: As the recognition system that we create is realized with neural networks model, the input of neural networks should be collection of data taken in equal length. That's why speech parts with different length are resulted in matching length. Lagrange's interpolation method is used for this problem.

Speech framing. According to the practices, speech signal keeps its stationarity in some interval, for approximately 20 ms. Using speech signal's quasi stationary characteristics, to analyse speech, it is divided into frames. In the next steps, before applying discrete Fourier transformation, speech signal is multiplied by Hamming window. After Hamming window, in some part of speech signal information decreases considerably. That's why, to avoid the risk of losing useful information of speech signal, frames are approximately overlapped.

Feature extraction from speech frames. In our system, two types of features extraction algorithms are used in parallel: Mel Frequency Cepstral Coefficients (MFCC) and Linear Predictive Coding (LPC) features.

Cleaning speech signal from channel effect. While recording speech signal some noise occur, which is called channel effect. During system training and afterward usage of it different channel effects decreases the system recognition accuracy. To avoid this, cepstral mean subtraction method is used.

4. Recognition module

Depending on the pronunciation of speech, system divides speech samples into classes according to their length: short and long. Artificial neural networks is used to train the system. Separate network is created for each class. The outputs of neural networks are number of words in class, and input of neural networks are number of features of speech signal. The neural network that we use is fully connected graph and trained by conjugate gradient method.

As mentioned above, in the speech pre-processing block, its MFCC and LPCC features are calculated. Training and recognition are done based on both two features. Results are compared based on recognition according to both two features.

Recognition process is done in two steps:

1. Parallel recognition processes are done in MFCC-based and LPC-based subsystems.
2. Calculated recognition results are compared in MFCC-based and LPC-based subsystems, and speech recognition system accept the result of two subsystem.

5. Experiment results and Conclusion

We have done research and developed software for speech recognition in Flight Simulator with limited vocabulary in C# from scratch. As experiment we recognize 41 Flight checklist commands. The major function of the flight deck checklist is to ensure that the crew will properly configure the airplane for any given segment of flight. The list of checklist commands given in the table 1. We processed approximately 33000 speech samples from 80 people and applied artificial neural networks to train collected data. We divide commands into two classes (long and short) and trained neural networks for each class. Some words included in both classes. At first, the system checks uttered command length, then call the appropriate neural network for recognition.

Table 1. The list of checklist commands

Preflight inspection	Cockpit
Towbar	Cockpit checklist completed
Weight and balance	One
AC documents	Two
Circuit Breakers	three
Seats & Belts	Four
Cabin doors	Five
Fuel selector	Six
All switches	seven
Completed	eight
On	nine
Off	zero
Closed	decimal
Checked	removed
Fuel Shutoff Valve	aboard
Shut-off cabin heat	adjusted & locked

Alternate air door	locked
Battery+Main bus	open
Fuel Quantity	In
Fuel Temperature	Sufficient
Flight controls	

The speech recognition system works in real time and offline mode in Windows platform. We used Mel Frequency Cepstral Coefficients and Linear Predictive Coefficients feature extraction algorithms and trained the system by multilayer artificial neural networks. The User Interface of platform is highly functional and allows users to update system parameters through the interface. Our Speech Recognition system results are given in table 2 and 3. We tested Microsoft Speech SDK for our data set and received 89.6% correct recognition, 2% error recognition and 8.4% rejection rate. It is not fair to compare directly our results with Microsoft Speech SDK due to data collection conditions. Our collected data and testing data were recorded in same environments and by the same microphone.

Table 2. Recognition result of the training samples

	Accuracy (%)	Rejection (%)	Error (%)
Short words with MFCC	99.86	0.14	0.00
Short words with LPC	98.85	1.12	0.03
Long words with LPC	99.97	0.03	0.00
Long words with MFCC	99.97	0.03	0.00

Table 3. Recognition result of the test samples

	Accuracy (%)	Rejection (%)	Error (%)
Short words with MFCC	94.85	4.36	0.79
Short words with LPC	93.86	4.56	1.58
Long words with LPC	97.56	1.72	0.72
Long words with MFCC	97.56	1.72	0.72

Acknowledgement
This work has been carried out in Center for Data Analytics Research at ADA University.

References

[1] Scott Mc and Tomas A 2007 A Speech Interface to Virtual Environments. http://citeseerx.ist.psu.edu/viewdoc/download?doi=10.1.1.39.9070&rep=rep1&type=pdf
[2] Rami M, Svitlana M, Vyacheslav V L, Nataliya V B 2017 Speech Recognition Systems: A Comparative Review *IOSR Journal of Computer Engineering (IOSR-JCE)* **19** 5 pp 71-79
[3] Aman A, Sonu K M, Rinaz Sh, Chandraketu K G, Prakhar M, Soudamini P 2016 A Survey Paper on Acoustic Speech Recognition Techniques *Int. Journal of Recent Advances in Engineering & Technology* **4** 7
[4] Dinesh K D, Yogesh R 2015 A Survey Paper on Automatic Speech Recognition by Machine *Int. Journal of Computer Science and Information Technologies* **6** 3 pp 2918-2922
[5] Suma S and Ramakrishnan K 2013 An Efficient Speech Recognition System *Computer Science & Engineering Int. Journal* **3** no 4
[6] José M C, Manuel D, José M P 2012 Automated Speech Recognition in ATC Environment. *ATACCS'2012*

[7] Neha Ch, Gangwar R, Rajeev B 2015 Current Challenges and Application of Speech Recognition Process using Natural Language Processing: A Survey. *Int. Journal of Computer Applications* **131** no 11

[8] Shane S, Jonathan R, Takaaki H, Shinji W, John R H 2018 End-to-End Multi-Speaker Speech Recognition *IEEE Int. Conf. on Acoustics, Speech, and Signal Proc.*

[9] Denis V D, Judy M V 2002 Implementing Speech Recognition in Virtual Reality *Mechanical Engineering Conference Presentations, Papers, and Proceedings*

[10] Ayushi Y V, Krina A S, Yesha A K, Nirali P 2017 Survey paper on Different Speech Recognition Algorithm: Challenges and Techniques *Int. Journal of Computer Applications* **175** no 1

[11] Kelvin V, Kevin Ch, and Endra J 2017 Virtual Reality Flight Simulator *Internetworkıng Indonesıa Journal* **9** no 4

[12] D D Doye, T R Sontakke, Smita N 2015 The Nonlinear Time Alignment Model for Speech Recognition System *IETE Journal of Research, Taylor & Francis* pp 271-275

[13] Kavitha R, Nachammai N, Ranjani R, Shifali J 2014 Speech Based Voice Recognition System for Natural Language Processing *Int. Journal of Computer Science and Information Technologies* **5** 4 pp 5301-5305

[14] Elyes Z, Yassine A, Faiez G 2014 Hybrid continuous speech recognition systems by HMM, MLP and SVM: a comparative study *Int. Journal of Speech Technology* **17** 3 pp 223-233

[15] Jayashree P and Melvin J P K 2015 Machine Learning in Automatic Speech Recognition: A Survey *IETE Technical Review, Taylor & Francis* pp-1-13.

[16] Qirong M, Ming D, Zhengwei H, and Yongzhao Z 2014 Salient Features for Speech Emotion Recognition Using Convolutional Neural Networks *IEEE transactions on multimedia* **16** no 8.

[17] Siva Prasad Nandyala and T. Kishore Kumar 2014 Hybrid HMM/DTW based Speech Recognition with Kernel Adaptive Filtering Method *Int. Journal on Computational Sciences & Applications* **4** no 1

[18] Aida-Zade K, Ardil C, Rustamov S 2007 Investigation of combined use of MFCC and LPC Features in Speech Recognition Systems *World Academy of Science, Engineering and Technology Int. Journal of Computer and Information Engineering* **1** pp 2647-2653

[19] Alex G, Navdeep J 2014 Towards End-to-End Speech Recognition with Recurrent Neural Networks *Proc. of the 31 st Int. Conference on Machine Learning, Beijing, China. JMLR: W&CP* **32**

[20] Xiang-Lilan, Zhang, Zhi-Gang, Luo, Ming Li 2014 Merge-Weighted Dynamic Time Warping for Speech Recognition *Int. Journal of Computer Science and* Technology **29** 6 pp 1072-1082

[21] Aida-Zade K, Rustamov S 2008 The Principles of Construction of the Azerbaijan Speech Recognition System *The 2nd Int. Conf. "Problems of Cybernetics and Informatics"* **1** pp 183-186

[22] Aida-Zade K, Rustamov S, Mustafayev E 2009 Principles of Construction of Speech Recognition System by the Example of Azerbaijan Language *Int. Symposium on Innovations in Intelligent Systems and Applications* pp 378-382

[23] Ayda-zade K, Rustamov S 2005 Research of Cepstral Coefficients for Azerbaijan speech recognition system *Transactions of Azerbaijan National Academy of sciences."Informatics and control problems* **25** no 3 pp 89-94.

Automated News Categorization using Machine Learning methods

U Suleymanov[1] and S Rustamov[2,3]

[1] State Agency for Public Service and Social Innovations under the President of the Republic of Azerbaijan, Baku, Azerbaijan
[2] School of Information Technologies and Engineering, ADA University, Baku, Azerbaijan
[3] Institute of Control Systems of ANAS, Baku, Azerbaijan

Email: srustamov@ada.edu.az, ADA University

Abstract. Being one of the most linguistically rich languages, Azerbaijani has been researched less in the context of natural language processing area. The text corpus created from Azerbaijani news articles is designed to apply supervised machine learning approaches for the case of automatic news labeling. Chi-squared test and LASSO methods have been implemented for feature selection and pre-processing. The application of supervised machine learning approaches to the text corpus allowed us to compare the performance results of well-established supervised machine learning approaches in the domain of Azerbaijani language.

1. Introduction

In text classification, different statistical and machine learning methods are applied in order to automatically assign one of the predefined labels to a given element of the unlabeled document space. More formally, if some d is a document within the entire set of documents D and C = {c1, c2, c3, …, cn} is the set of all the categories, the text classification assigns one category ci to the given document d. Classifier γ trained on a training set D of labeled documents, determines the label of the previously unseen document according to the learned parameters. The name of supervised learning is obvious from the above definition, namely, machine learning models learn best parameters for predication under the supervision of training data set [1].

Text classification is not a linear process rather it requires iterative approach. Namely, for the predictive model to be able to learn optimal parameters and to predict accurate results, the statistical trends and patterns observed during predictive model training phases should be applied to data preprocessing and vice versa. Iterative approach contributed a lot to the development of news corpora and to the determination of optimal categories under which news articles have been grouped.

Feature extraction is the first and one of the integral parts of the text classification. Traditional machine learning models cannot work with textual data, and in order for them to be able to learn parameters and make predictions, numerical features of the document should be extracted. Bag of words is a general approach taken for feature extraction [1]. The feature space generated by bag of words approach, especially for bigrams and n-grams are very huge and sparse. Therefore, for enabling machine learning models to work with this type of high dimensional features and for optimization of model performance, the selection of essential features becomes priority [2].

Text classification problem can be solved by a number of machine learning approaches such as support vector machines, artificial neural networks, decision tress and etc. Text classification problems are distinguished by their high dimensional feature space from other machine learning problems. Therefore, algorithms which can work with thousands of features efficiently performs best in the text classification.

2. Related work

Text classification begins with feature extraction. Application of different future extraction methods yields different performance results even if we apply the same classification algorithm in the next step. Traditional approach to feature extraction is bag of words where each word is represented as a feature [1]. Bag of words approach does not take the order of words into account and represents the count of each word in the document as feature [1]. Term-frequency inverse-document-frequency approach of bag of words weights documents not only by their counts but also by their frequency in the whole training dataset. This allows to weight some words such as 'and' appropriately as these types of words are common in whole dataset but carries little semantic information about the actual category of the document.

Feature selection is next step in text classification which helps to distinguish and extract relevant features from the huge feature space. Researchers have developed several feature extraction methods in order to feed classifiers with necessary feature set and to decrease error rate. In filter based feature selection schemas features are evaluated and assigned weights which at the last step are taken as selection criteria [3]. Least absolute shrinkage and selection operator (LASSO) is another approach implemented widely for feature selection as it penalizes the coefficients of features in the cost function [4]. This penalization makes coefficients of features approach zero and for features which are weakly related to document category the coefficients become zero if penalization parameter is chosen adequately [4].

Text classification has been approached by a number of classification methods by researchers. Naive Bayes classifier is based on Bayes theorem and tries to solve the classification problem by probabilistic approach. [5]. Naive Bayes approach assumes that there is no correlation between features in other words they are independent. In spite of the fact that words in the given document are semantically connected, Naïve Bayes approach still performs reasonably well for text classification [5].

Another approach generally applied to text classification is Support Vector Machines [6]. Joseph (2015) analyzes the application of Support Vector Machines with word2vec and tf-idf approaches.

Artificial neural networks are researcher's choice of classifiers in recent years. Lai (2015) presents the application of recurrent convolutional neural networks and its comparison with other well established text classification methods. Recurrent convolutional neural networks can learn contextual information from text and outperforms other classifiers in several datasets [1]. Several novel approaches have been implemented for artificial neural networks as well. As seen in [7], [8] and [9] neural fuzzy systems and their hybrids have been implemented successfully for sentiment analysis.

3. Data preparation and feature extraction

As classification algorithms work with numerical data, it is necessary to handle textual data properly and extract relevant features from it. Moreover, having deep understanding of dataset enables applying more suitable feature extraction and selection techniques to classification problem. We will discuss different feature extraction and selection methods. In the classifiers section we will provide the analysis of applying different feature extraction methods for naive Bayes, support vector machines (SVM) and artificial neural networks classifiers.

3.1. Dataset

In order to be able to apply feature extraction as well as machine learning approaches 130000 news articles have been gathered along with their assigned categories. The documents are grouped under 8 mutually exclusive categories. This allows us to train and test different text classifiers and build a predictive model. Moreover, the dataset has been splitted randomly into train and test sets so that the accuracy of classifiers can be determined.

3.2. Count vectorization

For the machine learning methods to be able to learn, the features have to be extracted from documents. Bag of words approach considers each unique word as feature and expresses each document as the set of features. By this way feature space has as many dimensions as the number of unique words in the whole dataset. Count vectorization method assigns the count of the word in the document to its corresponding feature. Therefore, count vectorization is not able to represent word orderings or specific word combinations. As most of the words in the whole dataset dictionary is not present in the specific document, most of the feature values are zero for given document.

3.3. TF-IDF vectorization

Term frequency inverse document frequency approach not only considers the frequency of the given word but also weights it inversely by the word's frequency in the whole dataset. This allows to reduce the weights of the common words which is not influential for the category of the document [10]. Term frequency inverse document frequency method is able to weight common words by its IDF part:

$$idf(t) = \lg\frac{1+n_d}{1+df(d,t)}+1 \quad (1)$$

In equation (1), t is the term in the given document d, while n_d and $df(d, t)$ expresses the whole document count and the number of documents containing t respectively. Thus, if a word is frequent in most of the documents, the denominator and numerator gets close to each other and IDF score approaches zero. Thus words which is not discriminative enough get close to zero weights [10]. We have applied both types of feature extraction methods and trained two classifiers naive Bayes and SVM. Comparative analysis of these methods for the SVM classifier shows that SVM classifier has 2.5 % more accuracy while working with features generated by tf-idf vectorization method rather that count vectorization.

3.4. Stop word removal

Classifier assigns different weights to each word while defining the category of the document. For instance, the weight assigned to the word "aktyor" (actor) for economics category should be small, while its weight for art category should be high. However, some words in the document are so common that they do not play any role in determining document category. The words "üçün" (for) and "hər" (each) are examples of such words. These types of words are commonly referred as stop words and removing them generally benefits the classification accuracy as this reduces the feature space. We have applied the stop word removal technique and get 0.4% improvement in accuracy for support vector machines classifier. The same technique for naïve Bayes classifier gave 0.2% improvement.

4. Feature selection

4.1. Chi-squared

The feature selection techniques reduce the burden of the classifiers in terms of choosing relevant features for classification. Namely, classifiers have been shown to perform better when they are provided with less and more important features. Chi-squared is a well-known statistical approach to measure the relation between each feature and target class. By taking the chi-squared test values of each feature and ordering them by descending order, we can select n features with highest values. Extracting relevant features, we got 88.6% accuracy with SVM which is 0.7% more than the accuracy result of SVM with tf-idf vectorization.

IOP Conf. Series: Materials Science and Engineering **459** (2019) 012006 doi:10.1088/1757-899X/459/1/012006

4.2. LASSO

Least absolute shrinkage and selection operator is a well-known approach to feature selection. It applies the following l1 penalty to ordinary minimized sum of squares method [11]:

$$\lambda \sum_{i=1}^{N} |\beta_i| \qquad (2)$$

N denotes the number of features and λ denotes the regularization parameter. Choosing high values of λ makes the weights assigned to features go to zero. If we choose the parameter appropriately, only the features relevant for the classification have coefficients different from zero. In our case, extracting features with LASSO increased the accuracy of naïve Bayes classifier from 71.5% to 76.9% while it increased the SVM classifiers accuracy from 87.9% to 88.5%.

5. Classifiers

5.1. Naïve Bayes

Firstly, we trained the naive Bayes classifier with count vectorization method. Bayes rule is at the core of naïve Bayes Classifier. In order this rule to work, it is assumed that features are independent of each other. As words are connected semantically in news articles, this assumption does not hold true. Regardless of that, for text classification problems naïve Bayes classifiers gives reasonable accuracy results. For instance, on famous 20 Newsgroups and WebKB datasets, it shows 85% and 86% prediction accuracy respectively [12].

By applying naive Bayes, the highest accuracy we got was 80.4%. This accuracy is achieved by applying count vectorization as feature extraction. The least accuracy observed was by using tf-idf approach with Naïve Bayes as seen in Fig.1. We can also observe from figure 1 that the most successful feature selection approach was chi-squared test with accuracy result of 79.6%.

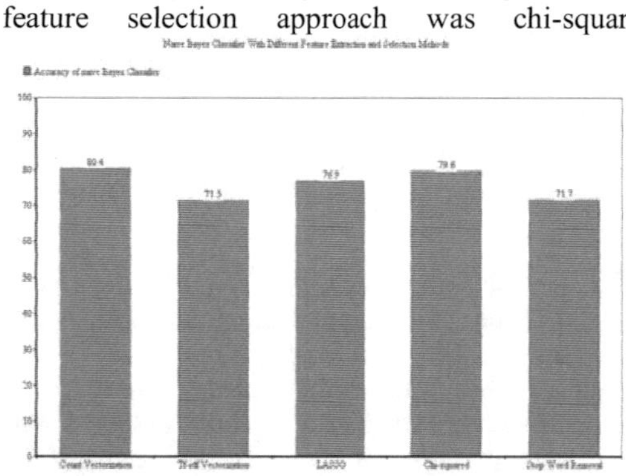

Figure 1. Accuracy of naive Bayes classifier with different feature extraction and selection methods.

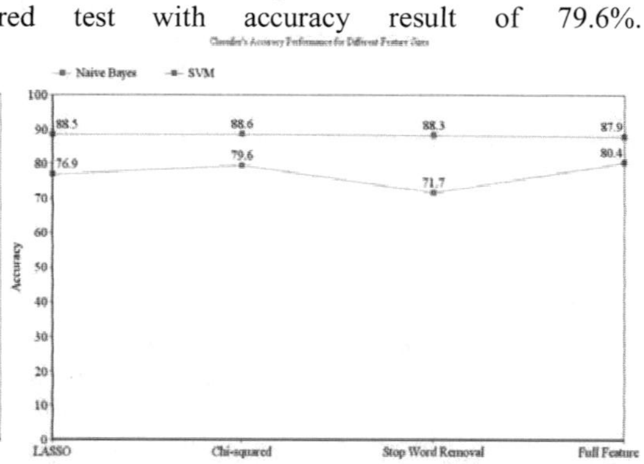

Figure 2. Accuracy of naive Bayes and SVM classifier with different feature sizes.

Applying different feature selection methods produce different feature set as well as feature sets with different sizes. In our case minimum feature space size is generated by LASSO while maximum was generated by applying just feature extraction without using feature selection. We decided to measure the effect of feature space size on classification accuracy. Figure 2 demonstrates naïve Bayes and SVM classifiers' accuracy measures with different feature space sizes.

It should be noted that in case of applying no feature selection, we took the maximum accuracy results of count and tf-idf vectorizer as they have the same feature space size. Upon analysing figure 2 we can come to the conclusion that for our case chi-squared performed best amongst feature selection methods.

5.2. Support vector machines

Applying support vector machines (SVM) was the next goal for us. The prediction accuracy of SVM for other datasets is shown to be very high as it can handle classification problems with high feature space sizes. For our case also, SVM showed high performance results. Figure 3 describes SVM classifier's accuracy in different data pre-processing settings.

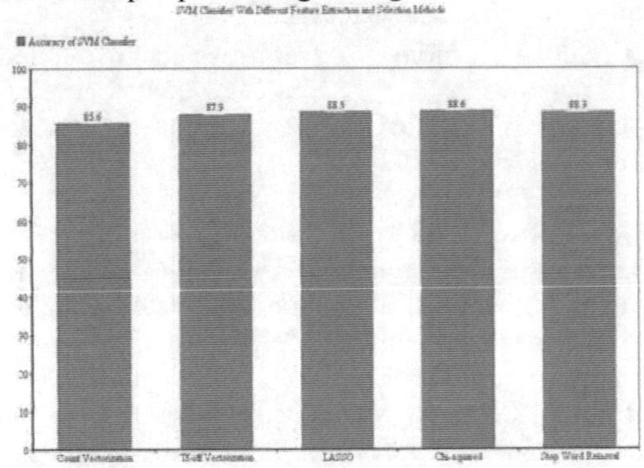

Figure 3. Accuracy of SVM classifier with different feature extraction and selection methods.

As figure 3 suggests SVM classifier performs reasonably well in all feature extraction and selection settings. However, we saw the greatest increase in accuracy when we changed the feature extraction approach from count to tf-idf.

5.3. Artificial neural network

Artificial Neural networks have been implemented in various settings and for solving a variety of problems. Text classification is amongst them and artificial neural networks is proven to perform well in this domain [13]. Artificial neural networks can be arbitrarily complex in terms of architecture. Therefore, getting highly accurate results can be computationally expensive. In order to make the calculations less expensive and to make the architecture more compact, the feature selection is a reasonable approach in text classification as it enables to deploy less and more influential features to the artificial neural network. Artificial neural network model gave 86.3% accuracy result on Azerbaijani news article dataset. Application of feature selection namely, Chi squared test increased the accuracy of artificial neural networks by 2.8%.

6. Conclusion

Because of the dataset being new, comparing the results obtained by implementing various pre-processing methods with other benchmarks is difficult for us. Therefore, we tried to compare the accuracy results of classifiers with the accuracy results obtained on famous datasets. Moreover, the dataset can be made available to other researchers if they request it by contacting us via email.

As shown in the classification accuracy of naïve Bayes and SVM, feature extraction and selection is an important step for text classification. Moreover, it is impossible to say which pre-processing approach is best, because of two factors. Firstly, choosing pre-processing approach depends heavily on dataset properties such as size, the length distribution of documents within dataset and etc. Secondly, even if we keep dataset constant, the different classifiers performs differently with the features provided by various pre-processing approaches. For example, if we compare text vectorization approaches for naïve Bayes and SVM, it becomes apparent that using count vectorization yields best accuracy result for naive Bayes, while for SVM count vectorization yields lowest accuracy.

Acknowledgement
This work has been carried out in Center for Data Analytics Research at ADA University.

References

[1] Siwei L, Liheng X, Kang L and Jun Z 2015 Recurrent Convolutional Neural Networks for Text Classification *AAAI'15 Proc. of the Twenty-Ninth AAAI Conf. on Artificial Intelligence* (Austin: AAAI Press) pp 2267–73

[2] Girish C and Ferat S 2014 A survey on feature selection methods *Computers & Electrical Engineering* **40** 16–28

[3] Alper K U 2016 An improved global feature selection scheme for text classification *Expert Systems with Applications* **43** 82–92

[4] Fonti V and Belitser E 2017 Paper in Business Analytics Feature Selection using LASSO

[5] Jain A and Mandowara J 2016 Text Classification by Combining Text Classifiers to Improve the Efficiency of Classification *International Journal of Computer Application* **6** 1797–2250

[6] Lilleberg J, Zhu Y and Zhang Y 2015 Support vector machines and word2vec for text classification with semantic features *IEEE 14th Int. Conf. on Cognitive Informatics & Cognitive Computing (Beijing)* pp 136–140.

[7] Aida-zade K, Rustamov S, Clements M and Mustafayev E 2018 *Recent Developments and the New Direction in Soft-Computing Foundations and Applications* vol **361** pp 63–70

[8] Rustamov S 2018 A Hybrid System for Subjectivity Analysis *Advances in Fuzzy Systems* vol **2018**

[9] Rustamov S, Mustafayev E and Clements M A 2018 Context Analysis of Customer Requests using a Hybrid Adaptive Neuro Fuzzy Inference System and Hidden Markov Models in the Natural Language Call Routing Problem *Open Engineering* vol **8** pp 61–68

[10] Neto J L, Santos A D, Kaestner C A A, Alexandre N, Santos D, Celso A A, Kaestner A , Freitas A A and Parana C 2000 Document Clustering and Text Summarization

[11] Urda D, Franco L and Jerez J M 2017 Classification of high dimensional data using LASSO ensembles *IEEE Symposium Series on Computational Intelligence (Honolulu)* pp 1–7

[12] Aida-zade K and Rustamov S Learning 2016 User Intentions in Natural Language Call Routing Systems *Recent Developments and New Direction in Soft-Computing Foundations and Applications* pp 37–46

[13] Rustamov S, Mustafayev E and Clements M A 2013 Sentiment Analysis using Neuro-Fuzzy and Hidden Markov Models of Text 2013 *Proc. of IEEE Southeastcon(Florida)* pp 1–6

IOP Conf. Series: Materials Science and Engineering **459** (2019) 012007 doi:10.1088/1757-899X/459/1/012007

Performance enhancement of a DFIG wind energy conversion system using phase advanced network

M Abozekry[1], D S M Osheba[2], H Z Azazi[2], and S M Osheba*[2]

[1]Site Engineer, Zafarana wind farm, Renewable Energy Authority, Egypt.
[2]Electrical Engineering Dept., Faculty of Engineering, Menoufia University, Shebin El-kom*(32511), Egypt.*
*** Corresponding author's e-mail address Oshebas@Yahoo.com**

Abstract. This paper presents the development of a phase advance network to enhance the performance of a variable-speed doubly-fed induction generator (DFIG) driven by a wind turbine under various operating conditions at Zafarana wind farm, Egypt. Firstly a fairly detailed simulation of a real DFIG-based wind energy conversion system is built using Matlab/SimPower environment. The simulation output is verified by comparing the simulation output with similar practical results and also with that published by Gamesa G5x standard test results. A phase advance network is designed and implemented to enhance the DFIG-based Gamesa G5x wind energy conversion unit. The FRT characteristics is examined when the system is subjected to a three phase short circuit test and the results obtained using the various enhancement techniques have presented in a comparative form with similar results using the implementation of the phase advance network. The results illustrate significant improvement in overall system performance and superior operation of the wind conversion unit using the phase network.

1. Introduction

With the continuous increasing in the demand of electrical energy and the depletion of conventional energy sources besides the emissions of carbon dioxide, all of that have been attracted dire needs to find another source of energy. Renewable energy sources offer an alternative solution because they are continuous, environmentally clean and available worldwide. Among renewable energy sources, wind energy attracted a lot of researchers due to its competitive cost. Various techniques have been implemented in wind energy conversion systems (WECS) and it is expected that wind power generation capacity will reach 2,000 GW by 2030 and increased to 25-30% of worldwide electricity supply by 2050 [1]. Recently variable speed doubly fed induction generators DFIGs wind energy conversion systems have become widely implemented due to its flexible control, low cost, high efficiency and many others salient features[2]. However, many challenges face wind energy such as supplying power at point of common connection (PCC) with fast response through a wide range of wind speeds, fluctuation in generated power that produce undesirable effects on voltage and frequency levels and consequently performance enhancement during fault conditions as well as changes in wind speed is of prime importance in the operation of WECS.

Various techniques have been developed and implemented for the enhancement of the performance of wind energy conversion systems [3-9]. These include the use of proportional plus integral (PI) control, conventional flywheel, super capacitor and superconducting magnetic energy storage (SMES). Probably the most widely used is the PI control as it offers the fast recovery response due to the proportional gain and elimination of the steady state error due to the integral action. However, this technique does not effectively cope during fault conditions due to difficulties in tuning the PI gains. Super capacitor and superconducting magnetic energy storage (SMES) are also used as a short term storage systems which enhance wind turbines transient characteristics. However, super capacitors suffer from low energy density while the use of SMES is not well established due to its high implementation cost. Regular flywheel acts as the energy storage system and achieves a smooth power output during wind speeds variation but very poor performance during short circuit characteristics. Moreover, wind turbine generators is required to feed reactive power during disturbances to support the voltage at the point of common coupling (PCC) and resume active power generation after fault clearing. Grid codes in countries with high penetration of wind power generation require that all new installed wind turbines should have the capability of low voltage ride through (LVRT) [3]. This capability requires an increase in the stability margin and low voltage conditions must be kept within certain limits.

This paper presents the design and implementation of a phase advance network to enhance the performance of a DFIG-based wind energy conversion system. A MatLab simulation is built for a DFIG-based wind energy conversion system and then verified by comparing the simulation output with that of Gamessa Turbine standard test and also with that obtained experimentally at Zafarana site. Various performance enhancement techniques have been discussed, implemented and their response is compared with that obtained using the proposed phase advance network when the system is subjected to a step change in wind speed and a three phase short circuit. The results illustrated superior enhanced of the performance of the WECS using the phase advance in comparison over previous techniques.

2. System Description

The largest wind farm in the Middle East is located at Zafarana, Egypt with total power 545MW . The most important wind turbine in Zafarana wind farm is Gamesa G5X with 850KW total power, it also the newest wind turbine in Zafarana wind farm and the only one that contains the advance technology equipped with a double fed Induction Generator (DFIG). A layout of the WECS is shown in Fig. 1. The DFIGs rotor and stator are both connected to electrical sources. The rotor is connected to grid through AC/DC/AC converter that consists of two components: the rotor side converter RSC (C rotor) and Grid side converter RSG (C grid)[10,11] Fig.1.

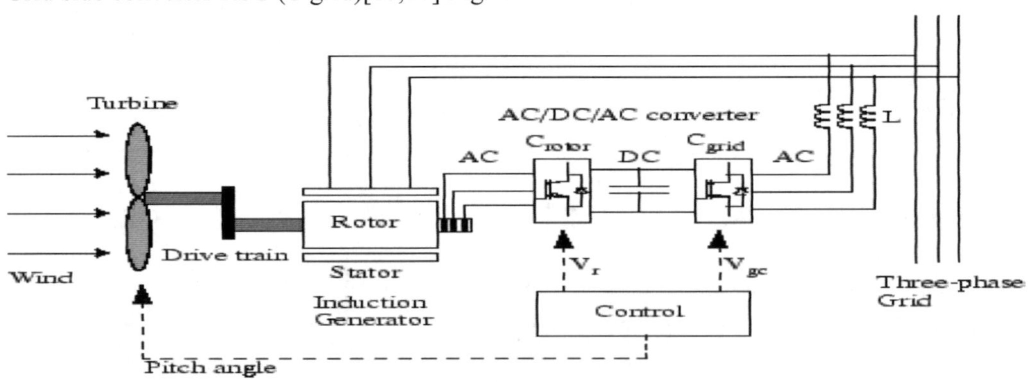

Figure 1. Typical DFIG Wind Energy Conversion System

A coupling inductor L is used to connect the grid converter to the grid. The three phase rotor winding is connected to the Crotor by slip rings and brushes. The basic principle used in DFIG is to control a frequency converter amongst the changeable frequency induction generator (it is mainly for injecting the current in the rotor circuit towards frequency compensation) as well as permanent frequency grid.

Converter works equally in both super and sub synchronous speed ranges. The slip power flow in both directions to the rotor from supply and from supply to the rotor side. DFIG is able to deal with wide variation of wind speeds with two modes of operation (sub and super synchronous speed). In the sub synchronous mode, the slip power flows from grid to the rotor, when rotor speeds less than the synchronous speed. However, when the rotor speed exceeds the synchronous speed, (super synchronous mode), the slip power flows from rotor to the grid.

3. Modelling of Zafarana Wind ECS

3.1. Modeling of wind turbine

The power in the moving air (P) can be written as follows [12]:

$$P = \frac{1}{2} \rho \ A \ V^3$$

Where: A = the area swept by the rotor blades.
ρ = the air density= 1.225 kg/m³ is the air density
ν = the wind speed in (m/sec).

The actual Power mechanical (P_w) extracted by blade can be written as follows:-

$$P_w = \frac{1}{2} \rho \ A \ V^3 \ C_p$$

C_p is the fraction of wind power captured by the rotor blades.

The power coefficient varies according to the turbines manufacture. Also, it varies according to the Tip Speed Ratio (TSR), of the wind turbine which can be expressed as:

Where R is the blade radius of the wind turbine, ω_T is rotational speed of the rotor in rad/s and V is the wind speed in (m/sec). The power coefficient can be defined as a function of the tip-speed ratio and the blade pitch angle as follows

$$C_p(\lambda, \beta) = C_1\left\{\left(C_2 \frac{1}{\lambda_i}\right) - (C_3\, \beta) - C_4\right\} e^{-C_5 \frac{1}{\lambda_i}} + C_6\, \lambda$$

Where the coefficients C1 to C6 for the Gamesa G5x Turbine are: C1= 0.5176, C2= 116, C3= 0.4, C4= 5, C5= 21, C6=0.0068, β is the pitch angle of the blade in degrees and λ is the tip speed ratio.

3.2. Modeling of doubly fed induction generator

Figure 2 shows the DFIG representation in a synchronously rotating [4], where the machine equations can be written in stator and rotor reference frame as follows:

$$V_{qs} = R_s \ I_{qs} + \frac{d}{dx}\varphi_{qs} + \omega_s \ \varphi_{ds}$$

$$V_{ds} = R_s \ I_{ds} + \frac{d}{dx}\varphi_{ds} + \omega_s \ \varphi_{qs}$$

$$V_{qr} = R_r \ I_{qr} + \frac{d}{dx}\varphi_{qr} + \omega_r \ \varphi_{dr}$$

$$V_{dr} = R_r \ I_{dr} + \frac{d}{dx}\varphi_{dr} + \omega_r \ \varphi_{qr}$$

The flux linkage expressions in terms of current can be written as follows:

$$\varphi_{qs} = L_s \; I_{qs} + M \; I_{qr}$$

$$\varphi_{ds} = L_s \; I_{ds} + M \; I_{dr}$$

$$\varphi_{qr} = L_r \; I_q + M \; I_{qs}$$

$$\varphi_{dr} = L_r \; I_{dr} + M \; I_{ds}$$

The mechanical equation is given by:

$$J \frac{d\Omega}{dt} + f \; \Omega = T_e - T_d$$

$$T_e = N_p \frac{M}{L_s} (\varphi_{qs} \; I_{dr} - \varphi_{ds} \; I_{qr})$$

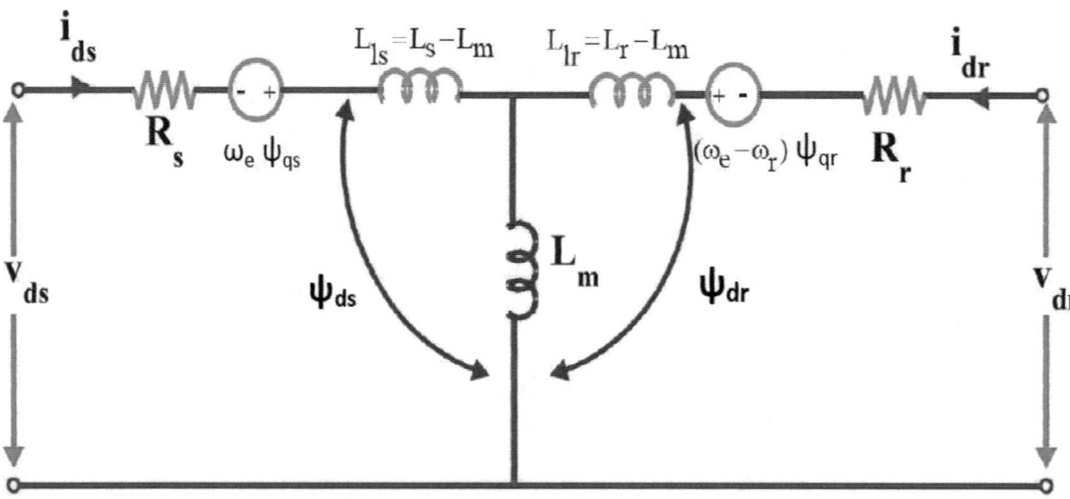

Figure 2. D-Q equivalent circuit of DFIG

Where:

V_{sdq}, I_{sdq}, φ_{sdq} refer to stator voltages, currents and flux in direct and quadrature axis; V_{rdq}, I_{rdq}, φ_{rdq} are referring respectively the rotor voltages, currents and flux in direct and in quadrature; Rs, Rr Refer respectively to the resistance of stator and rotor windings; Ls, Lr, M Refer respectively to the inductance of stator and rotor windings and mutual inductance ; ω_s ω_r Refer respectively to the stator electrical speed and rotor electrical angular velocity; Te, Td Refer respectively to electromagnetic torque and driving torque; J refers to total inertia; Ω refers to rotor angular velocity; Np is pole pair number and f is the Friction coefficient.

4. Simulation of Gamesa G5x based WEC system

Nominal power of the WEC units at at Zafarana wind farm in Egypt is Gamesa G5x 850KW. A comprehensive simulation is built using Matlab includes the Gamesa G5x turbine interior system, DFIG circuit, crowbar and control system as shown in Fig.3. The unit output voltage is 690V which stepped up to a 23KV bus and is then connected to the 220KV point of common connection (PCC) at the Egyptian grid via a transfers power via a 10 km underground cable as shown in Figure 4.

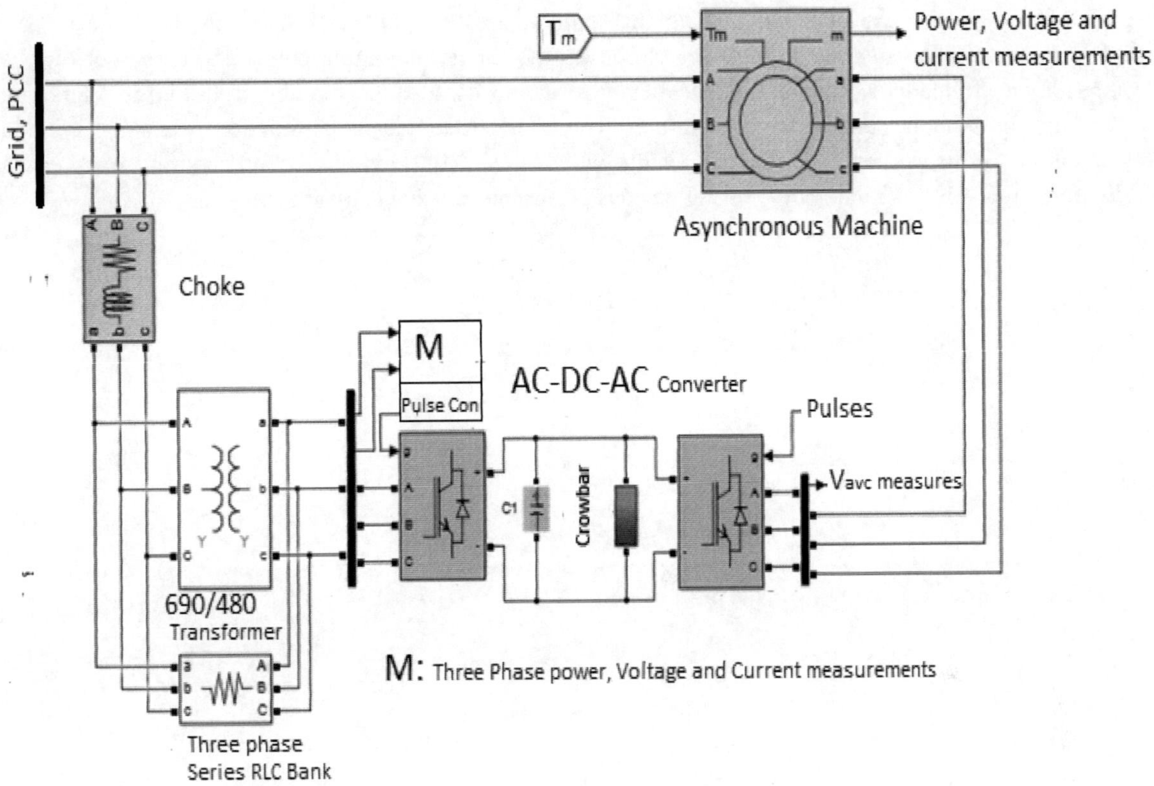

Figure 3. Simulation of the Gamesa G5x based WECS

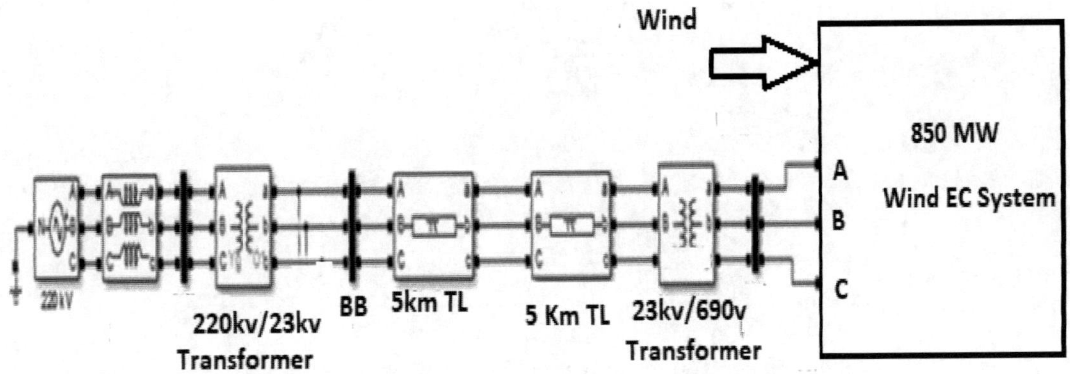

Figure 4. Connections with the Unified Egyptian Grid

5. Simulation Verification

Before dealing with detailed performance analysis, the simulation is assessed by comparing the simulation output with similar results obtained experimentally. Initially, the simulation is verified by comparing the simulation results of the wind turbine output obtained at air density equals 1.225 kg/m^3 with that obtained experimentally using the standard test conditions as listed in the Gamesa G5x Turbine data sheet as shown in Figure5. Moreover, the simulation output power and current at various speeds are compared with similar experimental results using field tests as shown in Figure 6 ignoring changes in wind air density. Examining these results illustrate that the simulation results are in good agreement with similar experimental results which gives confidence with other simulation results which intended to evaluate output using various performance enhancement techniques

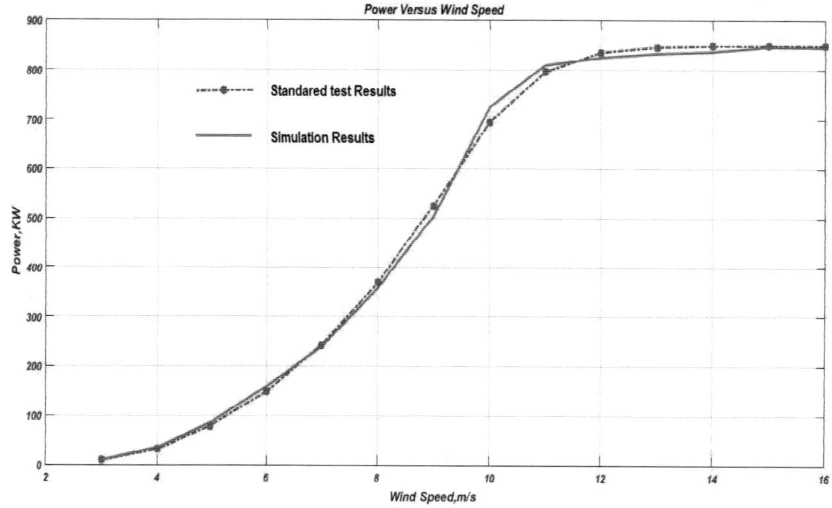

Figure 5. Simulated Output Power versus Output Power obtained at standard Test (Data Sheet)

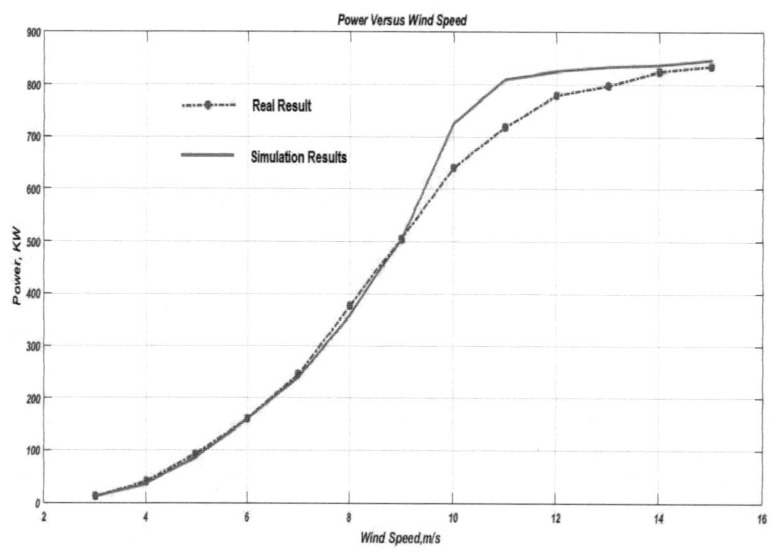

Figure 6. Simulated and experimental Output Power at Zafarana Site

6. Performance Enhancement Methods

6.1. Crowbar

Crowbar is an electronic circuit consists of thyristors, diodes, and resistors in order to protect rotor circuit in the case of overvoltage, which is 1325V in Gamesa G5x. Simply when the wind turbine sense an overvoltage in the rotor circuit, it activates crowbar which provides an alternative path with low resistor to absorb high current

6.2. Proportional Integral control

Proportional plus Integral(PI) controller used in WECS is a feedback control loop mechanism which consists of a proportional gain connected in parallel with an integrator gain, where the proportional gain improves the response and stability of the system, the integral gain when added to proportional gain speed up the movement towards set point and reduces the steady state error.

6.3. Superconducting magnetic energy storage (SMES)

Superconducting magnetic energy storage (SMES) is a superconducting coil stores electric energy into magnetic field with the basic configuration shown in Figure 7. However, this technique suffers from the high cost of SMES initial investment, and maintenance required, required low cooling system for the superconducting coil[9]

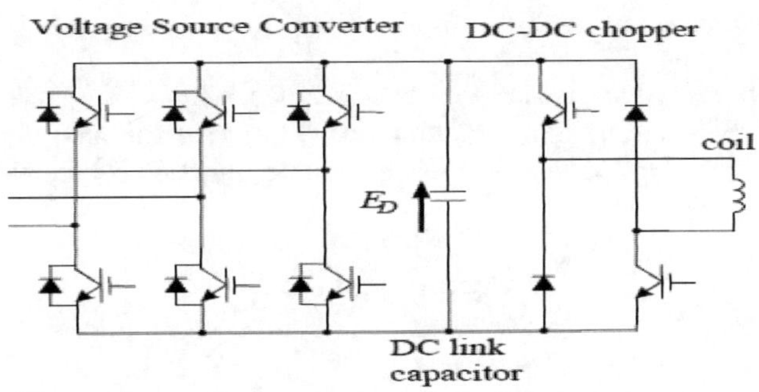

Figure 7. Superconducting magnetic energy storage (SMES)

6.4. Flywheel

Flywheel is a rotating mass that can be used to store and restore energy and consequently exchange the power with wind turbine during variable wind speeds. Regular FESS [10] enhances dynamic performance of wind turbine by minimizing the dynamic stress on wind turbine and smoothing the output power. However, FESS technique results in a steady state error in the DC-link and hunting oscillations following short circuit conditions. It minimizes the dynamic stress on wind turbine, besides the long live and the low overall cost of flywheel. Despite of the advantages the FESS it suffers from short discharge time, high mechanical stress, selecting the tensile strength of flywheel material, possible hazard situations and bad short circuit performance.

6.5. Super capacitor

Super capacitor is a double layer capacitor with very high capacitance [8]. It does not have a conventional dielectric and instead it has plates that are filled with two layers of an identical substance, that give the chance to super capacitor plates to pack with a larger surface area, thus resulting in very high capacitance. Super capacitors are used in wind turbine as a storage energy devices as enhance wind turbines performance by reducing the fluctuation of wind speed and also can be used to support DC-link voltage during transient operation which enhances wind turbines low voltage ride through (LVRT) capability, as shown in Figure 8.

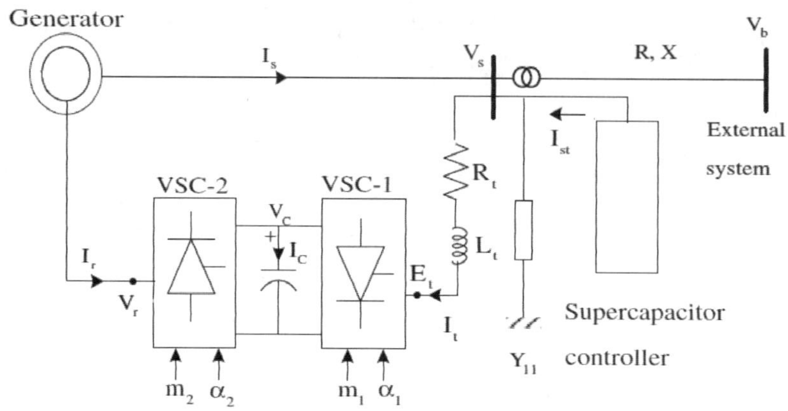

Figure 8. DFIG with a Super capacitor

6.6. Proposed performance enhancement technique

Alternative, the paper suggested the phase advance network as technique to improve the performance of the WECS during both variations in wind speed and also during short circuit faults. The phase advance is a lead lag compensator whose transfer function can be written as follows:

$$G(s) = K\frac{Ts + 1}{Ts + 1}$$

This advance network is used to replace the DC-Link PI controller as shown in Fig 10.

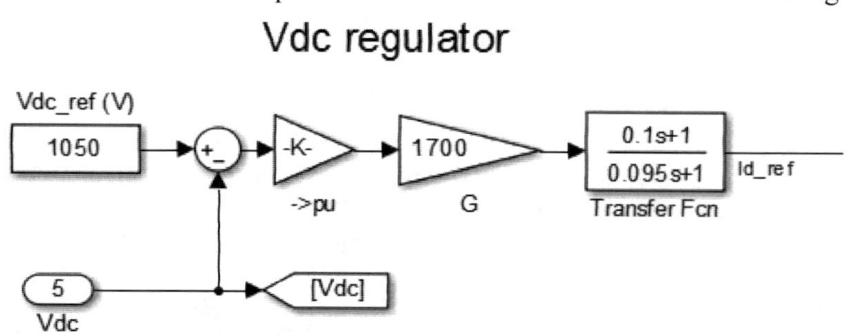

Figure 9. Phase advance Network inserted in the DC-Link

7. Simulation Results

The simulation model of grid connected Gamesa G5x wind turbine with DFIG that is shown in Figures 5 and 6. The different enhancement techniques are implemented by computer simulation to evaluate their performance in comparison with the proposed method. The system response with various enhancement techniques is described subsequently in comparison with the proposed technique under step changes in wind speed and also during three phase short circuit.

7.1. Operation Under step change in wind speed

The system is subjected to about 20% increase in wind speed. The system response with various enhancement techniques are shown in comparison with the system response with the proposed technique is shown in Figure 10. Examining these results illustrate excellent performance enhancement of the system performance with the proposed technique in comparison with all previous methods. This is clear with the very small fluctuation in rotor speed, grid voltage, DC-link voltage, active power and reactive power. Moreover, it's enhance wind turbine tracking to wind speeds change, that means the wind turbine will reach to the maximum power of wind turbine at currently wind speed faster and that lead to produce more power during same wind speed variations. Also, Phase advance enhance DC-Link performance by increasing the stability and response of DC-Link voltage in the start of wind turbine and also during wind speed changes.

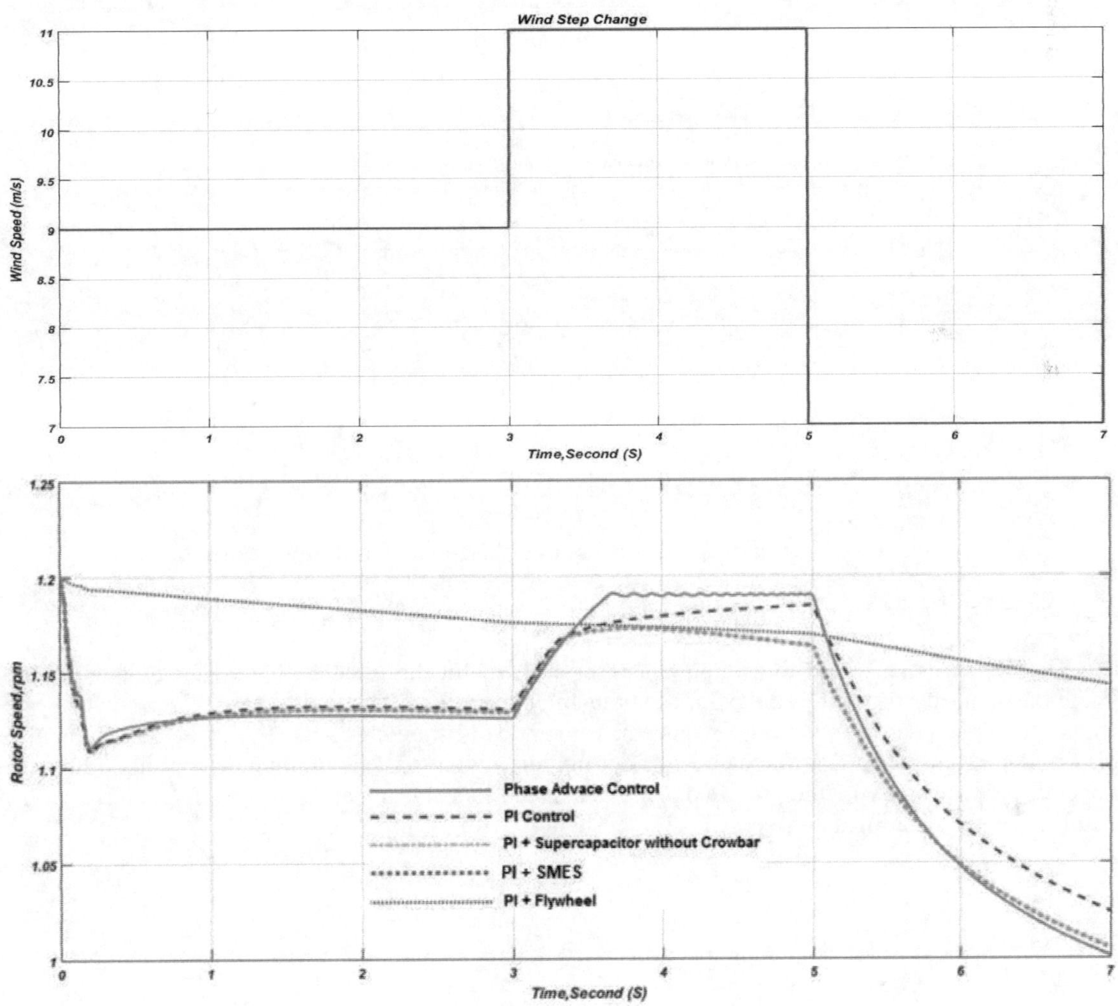

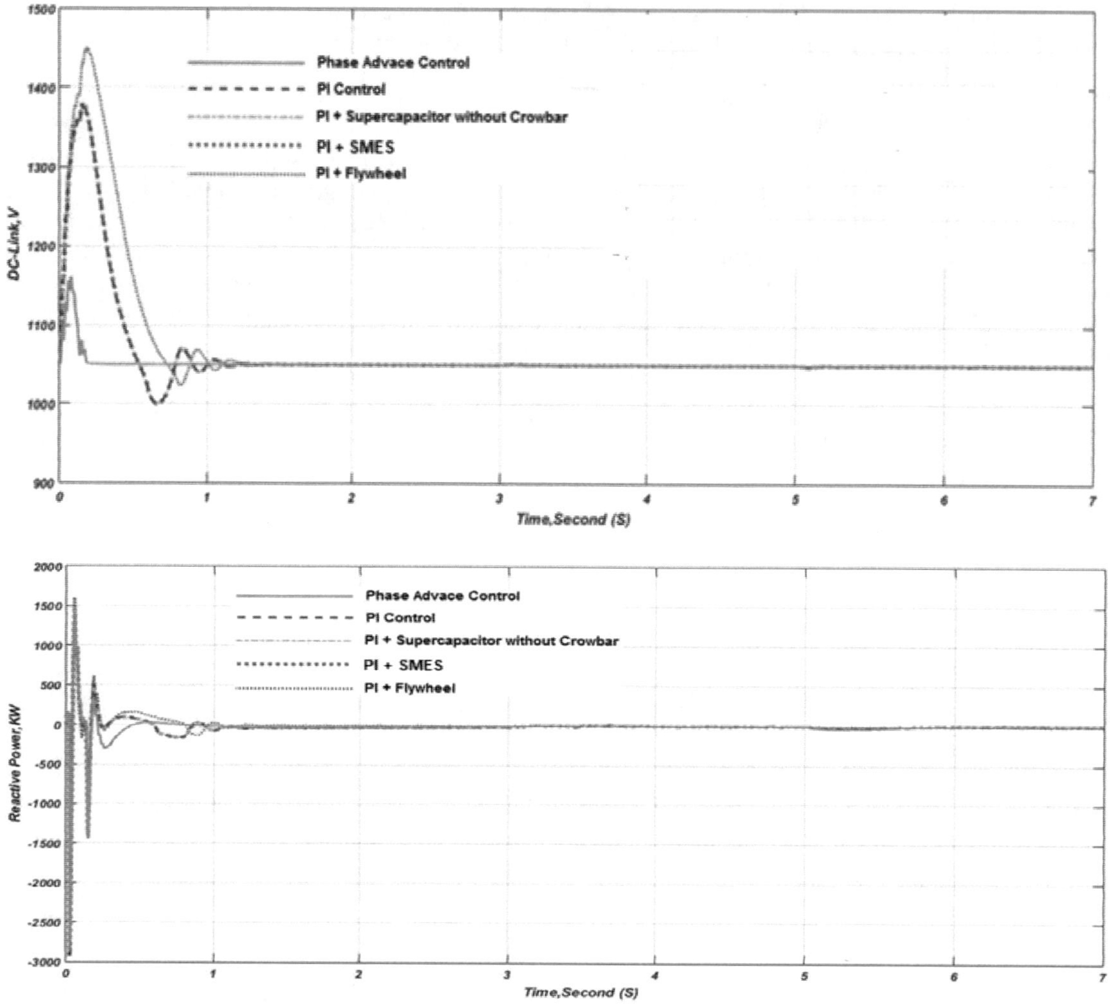

Figure 10. System responses to about 20% step increase in wind speed.

7.2. Operation Under Transient Fault Condition

The system is subjected to a three phase short circuit in the middle of the 10 km underground cable that connects the unit to the point of common connect(PCC The fault sustained for 5 cycles and the system response with various enhancement methods is shown in Figure 11. These results also show the performance at starting as well as during fault conditions. Examining these results illustrate superior performance with the proposed technique under both starting and during fault conditions. The results also show significant increase in system stability with the proposed technique.

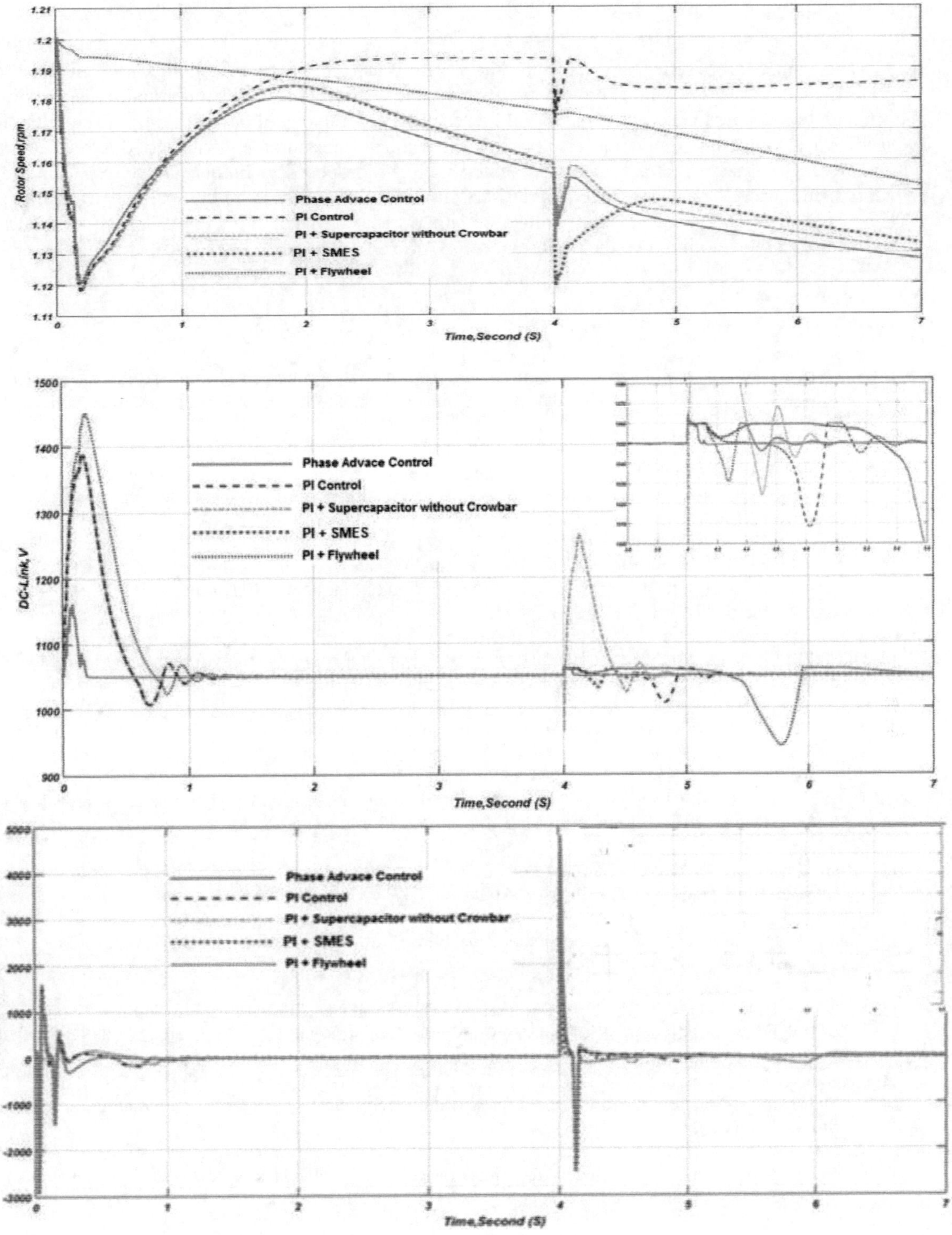

Figure 11. System response during fault

8. Conclusion

The paper proposed phase advance network as an alternative technique to enhance the performance of a DFIG wind energy conversion system. A fairly detailed simulation model is built to simulate the Gamesa G5x DFIG based wind energy at Zafarana wind farm. Egypt. The simulation is verified by comparing simulation results with experimental measurements and good agreements are obtained. The system performance is presented with the proposed technique under step increase in wind speed as well as under three phase short circuit fault conditions in comparison with previous enhancement techniques. The results illustrate superior performance with the proposed method and form a useful guide to improve the performance of WECS under wide range of disturbances.

9. Reference

[1] Yingzhen L, Ronghai Q, Jin Wang, and Hong Chen 2015 Influences of Generator Parameters on Fault Current and Torque in a Large-Scale Superconducting Wind Generator Ieee Transactions On Applied Superconductivity, 25 no 6

[2] Jafar Mohammadi, Sadegh Vaez-Zadeh, Esmaeil Ebrahimzadeh and Frede Blaabjerg 2018 Combined control method for grid-side converter of doubly fed induction generator-based wind energy conversion systems *IET Renewable Power Generation* 12 no pp 943-952

[3] Kasem A H, El-Saadany E.F, El-Tamaly H H and Wahab M A 2008 An improved fault ride-through strategy for doubly fed induction generator-based wind turbines IET IET Renewable Power Generation 2 no 4 pp 201–214

[4] Huchel L, El Moursi M, Zeineldin HH 2015 A parallel capacitor control strategy forenhanced FRT capability of DFIG IEEE Trans Sustainable Energy 6 no 2 pp 303–315

[5] Lin Y, Ang K H, and Chong G C Y 2006 Patents, software, and hardware for PID control," IEEE Control System Mag. 26 no 1 pp 42–54

[6] Lin Y, Ang K H, and Chong G C Y 2006 PID control system analysis and design, problems, remedies, and future directions IEEE Control Syst. Mag. 26 no 1 pp 32–41

[7] Liyan Q and Wei Qiao 2011 Constant Power Control of DFIG Wind Turbines with Supercapacitor Energy Storage IEEE Transactions on Industry Application 47 no 1

[8] Chad A and Géza Joos 2007 Supercapacitor Energy Storage for Wind Energy Applications IEEE Transactions on Industry Application 43 no 3

[9] Mohamed H A and Dougal R A 2010 Comparison of SMES and SFCL for Transient Stability Enhancement of Wind Generator System IEEE Energy Conversion Congress and Exposition pp 3382-3387

[10] Ayman S, and . Massoud A M 2013 Fault Current Contribution of Medium Voltage Inverter and Doubly FedInduction-Machine-Based Flywheel Energy Storage System IEEE Transactions on Sustainable Energy 4 no 1 pp 58 - 67

[11]R. Pe˜na J and AsherG 1996 Doubly-fed induction generators usingback-to-back PWM converters and its applications to variable-speed windenergy generation Proc. Inst. Electr. Eng. 153 pp 231–241 .

[12] Tapia A, Tapia G, Ostolaza J X , and Saens R J 2003 Modeling and control of a wind turbine driven doubly fed induction generator IEEE Transactions on Energy Conversion 118 no 2 pp 194-204

Speed control of 3-phase induction motor using rotor impedance

A S Zein El Din and A E El-Sabbe
Menoufia University, Faculty oj Engineering, Department of Electrical Engineering, Gamal Abdel-Naser Street, P.O.32511, Shebin El Kom, Egypt

ashrafzeineldin2002@yahoo.com

Abstract. This paper introduces a method of rotor impedance control by using controlled capacitor/inductor connected in series in the rotor circuit. The proposed system can be operate as a speed control system or a soft starting system to limit the starting current of the motor. An approach to a three MOSFETs-based a rotor circuit of induction motor is presented. The experimental realization of the drive posed some serious problems, notably with regard to the successful driving pulses of MOSFETs over a wide speed range. Analysis of the secondary MOSFETs-controlled motor raised many interesting challenges. It was decided to use an equivalent circuit approach to the analysis because the primary voltage and current were both largely sinusoidal. The equivalent circuit used is based on the single-phase equivalent of a balanced sinusoidal three-phase system. The effective value of the external rotor inductance or capacitance is calculated. Moreover, the motor performance specially speed, motor starting current, and torque during the control of duty cycle of MOSFETs are studied. Also, MOSFETs performance has been studied. Closed loop speed control of the motor speed have been investigated using hysteresis control based on controlling MOSFETs duty cycle. Experimental results have been carried out in the Laboratory to verifying the analysis. The simulation results are given and proved to yield good agreement when compared with the experimental results.

Keywords: Microprocessor, Mosfets, Induction Motor, and Rotor Impedance Control

1. Introduction

The induction machine is a rugged, reliable and less expensive ac machine. It has been used for high performance drive applications. Different control methods of varying degrees of complexity have been proposed and used for the control of induction machines [I, 10, 11, 12, 13, 14].

Stator voltage control of three-phase induction motor is a popular scheme used in industry for torque and speed control. A common method of controlling the stator voltage is by use phase angle control of thyristor circuit inserted in series between the supply and motor. Other methods use AC chopper with pulse width modulation PWM technique to control the applied voltage of the motor [2-4]. Method of the motor control based on the rotor circuit is used, such as using a thyristor-controlled resistive network in each rotor circuit [5-6]. This method will increase the rotor losses (heating). Operating the motor close to its rotor resonance by using reactive circuits in the rotor [7] gives high starting and breaking torque. This technique is limited as the rotor resonance is related to the varying motor speed. This paper introduces a method of rotor impedance control. By using controlled capacitor/inductor connected in series in the rotor circuit. The proposed system can be operated as a speed control system or a soft starting system to limit the starting current of the motor.

2. System description

A schematic diagram of the proposed system is shown in Figure 1. It is composed of a power electronic converter of three switching legs and the controller. Each converter leg consists of one MOSFET transistor, and a single-phase diode bridge. A three-phase stator windings of the slip-ring induction motor is connected directly to a three-phase sinusoidal a-c supply. On the other hand, the rotor windings are connected to three-phase R or R-L or RC load in parallel with the three switching legs. The equivalent circuit of the induction motor (rotor side) with the series element is shown in Figure 2. The speed controller receives the error signal between the preset speed (reference speed) and the actual measured speed of the motor shaft, then transfers it to the microprocessor through an Anaiog/Digital (ND) converter. The controller is interfaced to the power converter through a three-channel gate drive circuit. Each channel consists, as shown in Figure 3, of an optocoupler isolator and open buffer gates to deliver the control signals to the MOSFET. The input signals to the gate drive circuit are obtained from the microprocessor output port to give the speed control action according to the input error signal. An assembly program is designed and written for the microprocessor to apply the hysteresis method and the ON/OFF control. The microprocessor controls the switching process according to the following conditions:

If error signal $>$ zero, the MOSFETs are ON
If error signal $\leq$ zero, the MOSFETs are OFF (1)

3. Motor mathematical model

The following main assumptions are taken into consideration:

The. space harmonic, saturation effects, and iron losses are neglected. Also, the rn.m.f. is sinusoidal distributed. The voltage balance equations for a three-phase motor are [8]:

$$\begin{aligned}
e_{sa} &= \rho\, \psi_{sa} + r_s\, i_{sa} \\
e_{sb} &= \rho\, \psi_{sb} + r_s\, i_{sb} \\
e_{sc} &= \rho\, \psi_{sc} + r_s\, i_{sc} \\
e_{ra} &= \rho\, \psi_{ra} + r_r\, i_{ra} \\
e_{rb} &= \rho\, \psi_{rb} + r_r\, i_{rb} \\
e_{rc} &= \rho\, \psi_{rc} + r_r\, i_{rc}
\end{aligned} \tag{2}$$

In matrix notation, the relation between the phase flux linkages and the phase currents may be expressed as:

$$\psi = [L]\, I \tag{3}$$

Where

$$\psi = [\psi_{sa} \quad \psi_{sb} \quad \psi_{sc} \quad \psi_{ra} \quad \psi_{rb} \quad \psi_{rc}]^T \tag{4}$$

and $\mathbf{i}=[i_{sa} \quad i_{sb} \quad i_{sc} \quad i_{ra} \quad i_{rb} \quad i_{rc}]^T$ (5)

and $[\mathbf{L}] = \begin{bmatrix} L & L \\ L & L \end{bmatrix}$ (6)

Where

$$L11=\begin{bmatrix} L & Ls & Ls \\ Ls & L & Ls \\ Ls & Ls & L \end{bmatrix}$$

$$L12=\begin{bmatrix} M\ .c\ \theta & M\ .c\ (\theta+\frac{2}{3}) & M\ .c\ (\theta-\frac{2}{3}) \\ M\ .c\ (\theta-\frac{2}{3}) & M\ .c\ \theta & M\ .c\ (\theta+\frac{2}{3}) \\ M\ .c\ (\theta+\frac{2}{3}) & M\ .c\ (\theta-\frac{2}{3}) & M\ .c\ \theta \end{bmatrix}$$

$$L21=\begin{bmatrix} M\ .c\ \theta & M\ .c\ (\theta-\frac{2}{3}) & M\ .c\ (\theta+\frac{2}{3}) \\ M\ .c\ (\theta+\frac{2}{3}) & M\ .c\ \theta & M\ .c\ (\theta-\frac{2}{3}) \\ M\ .c\ (\theta-\frac{2}{3}) & M\ .c\ (\theta+\frac{2}{3}) & M\ .c\ \theta \end{bmatrix}$$

$$L22=\begin{bmatrix} L & Lr & L \\ Lr & L & Lr \\ Lr & Ls & L \end{bmatrix}$$

Instantaneous torque and speed equations are given by

$$Te=[(i_{sa}.i_{ra} + i_{sb}.i_{rb} + i_{sc}.i_{rc}) \sin\theta + (i_{sa}.i_{rb} + i_{sb}.i_{rc} + i_{sc}.i_{ra}) \sin(\theta + 2\pi/3) + (i_{sa}.i_{rc} + i_{sb}.i_{ra} + i_{sc}.i_{rb}) \sin(\theta - 2\pi/3)]\ M_{sr} \tag{7}$$

and the mechanical torque equation of the motor and load is given by:

$$J\frac{d^2\theta_m}{d\ ^2} + K\frac{d\theta_m}{d} + T_m - T_e \tag{8}$$

$$\theta = \frac{P}{2}\theta_m \tag{9}$$

In solving the above differential equations, the Runge-Kutta method has been employed.

4. Microprocessor-based controller

A microprocessor-based controller [9] is a microprocessor based controller (MCU). Its detailed circuit schematics is shown in Figure 4.

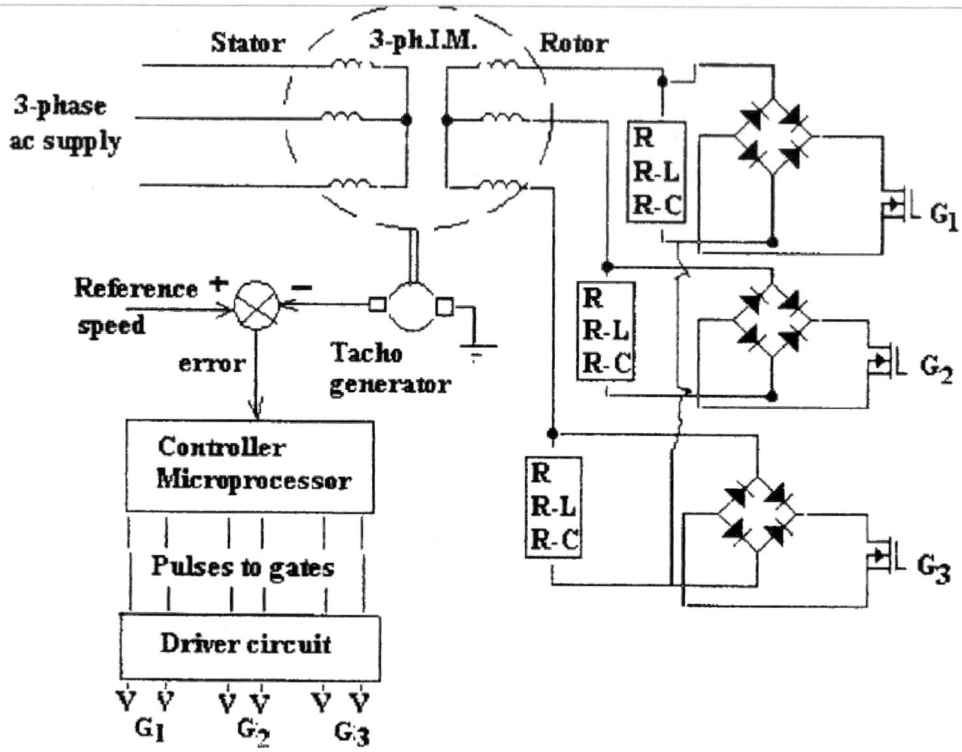

Figure 1. A proposed system

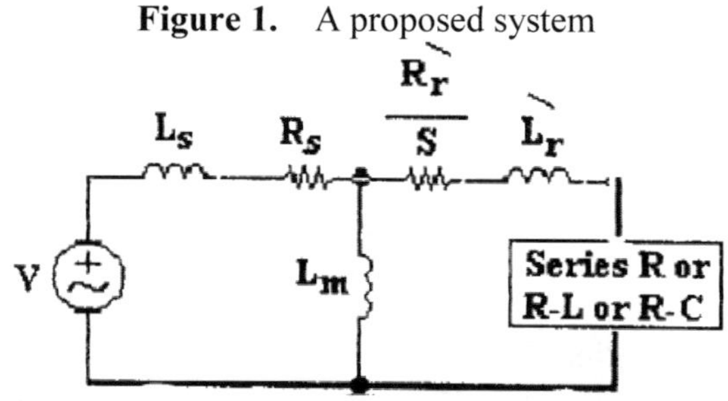

Figure 2. Induction Motor (rotor side) equivalent

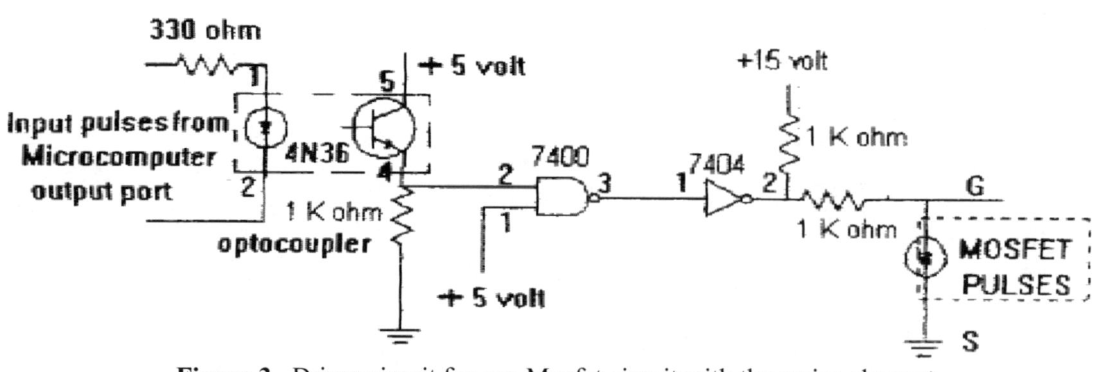

Figure 3. Driver circuit for one Mosfet circuit with the series element

It consists of central processing unit (CPU), which includes the 80386 microprocessor, 80387

IOP Conf. Series: Materials Science and Engineering **459** (2019) 012008 doi:10.1088/1757-899X/459/1/012008

numerical data processor (co-processor), 256K dual-port RAM, 256K EPROM, 8254 programmable interval timer iPIT), and 8259 programmable interrupt controller (PIC). The other is named input/output (I/O) unit which may include programmable digital I/O controller, programmable duty-cycle controller, *AID,* D A, etc. In the proposed system, the analog error signal between the reference motor speed and actual motor speed is transferred to digital value through AD channel I, and store the value in a memory location of RAM. The comparator is used to compare the error signal with the zero according to hysteresis control technique, then generates three pulses from output port to be deliverd to the gates of MOSFETs through driving circuits. A system controller, which resides on a 32-bit microprocessor, provides task coordination, external interfaces, motion profile control, parameter tuning, and signal recording, and can also be used to monitor the operational status of individual processors.

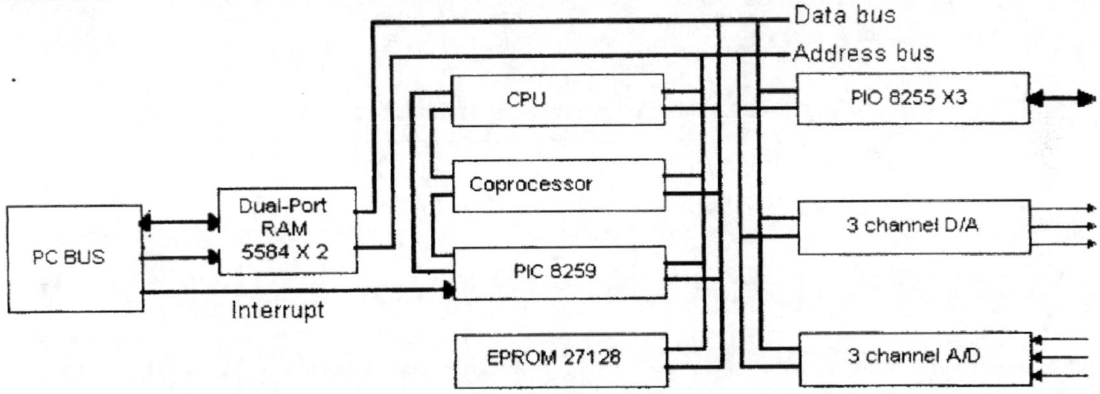

Figure 4. A Microprocessor based controller module

5. System modelling and modes of operation

There are two modes of operation, one of them occurs when Mosfets are turned-on and the other mode occurs at Mosfets are turned-off as follows:

5.1. Mosfets are turned-on

When series R or R-L or R-C is disconnect to rotor impedance. The solution of equations from I to 9 is obtained numerically using Rung-Kutta fourth-order numerical integration method.

5.2. Mosfets are turned-off

Series R or R-L or R-C is connect to rotor impedance and hence some equations 1-9 are modified in order to study motor performance at this mode of operation.
At R-series only is connected to each rotor phase, Eqn.2 is modified by replacing R+ rr instead of rr. But if R-L series is connected to each rotor phase, motor Eqns. are modified as following:

Lrr + Land R + rr are substitute instead of Lrr and rr respectively. When R-C is connected in series with each rotor phase, $R+r_r$ instead of rr, and added the following Eqns.

$dVca /dt = ira / C$
$dVcb /dt = irb / C$
$dVcc /dt = i_{rc} / C$
and modify eqs. era, erb, erc to become as following:
$er_a = \rho\psi\ r_a + (R+r_r)\ Ira + Vca$
$er_b = \rho\psi\ r_b + (R+r_r)\ Irb + Vcb$
$er_c = \rho\psi\ r_c + (R+r_r)\ Irc + Vcc$

IOP Conf. Series: Materials Science and Engineering **459** (2019) 012008 doi:10.1088/1757-899X/459/1/012008

6. Simulation results

The proposed system is designed and built in the laboratory to verify the analysis. It consists of a slip-ring induction motor. The motor parameters and data are listed in tables 1, 2

Table 1. Data of induction motor

Name plate data	Value
Input Voltage	220 Volt.
Frequency	50 Hz.
Motor Connected	3-phase star
No. of Poles	4 poles
Output Power	0.75 HP
Rated speed	1450 r.p.m.
Rated Current	1.2 Amp.

Table 2. Nominal parameters of induction motor

Nominal parameters	Value
Lss	0.4671 H
Lrr	0.4671 H
Lsm	0.23155 H
Lrm	0.23155 H
Msr	0.42887 H
R1	19 ohm
R2	10 ohm
K	0.0001 Kg. m^2/s
J	0.004 Kg.m^2

Series R, L and C connected to rotor phase are 42ohm, 0.2H and 50 pf respectively. The hardware setup consists of a microcomputer based on microprocessor 80386, 12-bits *A/D* card, 12-bits *D/A* card, three MOSFETs with type IRFP740, three uncontrolled single-phase bridges, and three R-C circuits. The microprocessor-based control system designed for the proposed system is shown in Fig.l , the control algorithm is implemented on an EPROM interfaced to the CPU of a microcomputer. The proposed control scheme was executed every 0.5 m.sec The error signal between command speed and actual motor speed is transferred through 12-bit *A/D* converter, the generated pulses to MOSFETs are processed from output port depending upon the error signal and hysteresis control. To check the validity of the model described above; a set of simulation tests have been carried out to analyse the system under steady state and transient conditions using C Program designed by the authors to solve the given non-linear differential Equations. Experiment is carried out to determine the characteristics of the proposed system. First, Open loop speed control of three-phase induction using controlled series impedance connected with the rotor circuit has been studied. Switch on and off of the three MOSFETs in rotor circuits can control the motor speed. The different duty ratios ton /T are selected, Where ton is turn on of the Mosfet and T is the total tum on and tum off of the Mosfet. As a practical example, 20 period (T) have been selected every one cycle of input supply frequency (50 Hz.), so each period of T equals I m.sec. Figure (5) shows the computed and experimental three-phase steady-state rotor currents at duty ratio = 20 and 50 of controlled only resistance respectively. It should be noted that fluctuating of rotor currents is appeared due to control of adding and shorting series resistance in rotor circuit. Figure 6 shows the computed and experimental three-phase steady-state rotor currents at duty ratio = 20 and 50 of controlled resistance and inductance (R-L) respectively. It can be noted that the distortion of the rotor currents, this because of connecting and shorting the series R-L with rotor impedance in very short time 0.2 m.sec and 0.5 m.sec. Control of resistance in series with the rotor will increase the rotor losses. Controlled R-L in rotor circuit causes the starting torque of the motor decreased whatever it gives nearly sinusoidal supply and rotor currents, for these reasons the controlled capacitance in the rotor circuit is used. So, We have replaced R-L by resistance and capacitor (R-C) series til rotor circuit, rotor currents are obtained in Figure.7 computational and experimental. It can be concluded that the computed results are in good correlation with the results obtained experimentally. Figure 8 shows Mosfet terminal voltage and current during open loop when add or short R-L and R-C to rotor circuit with different duty ratios. Figure 9 shows the computed and experimental steady-state results of three-phase stator voltage and current at different duty cycle control of series R-C connected to rotor circuit. Spectrums of both supply voltage and current waveforms confirm the computational results. Closed loop speed control of motor has been studied experimental and computational, by connecting ON or OFF R-C series with rotor circuit depending upon the error between Reference speed and feed back motor speed taken from a Tacho-generator. As shown In Figures l0-11, Simulation and Experimental of motor performance (speed, stator current and rotor current), pulses to mosfet during step change in load torque by 20 of its rated. It can be cleared that the controller used is effective and faster. Figures 12-13 show another study of motor performance (speed, stator current and rotor current), pulses of Mosfet at a certain operating case when a reference speed is changed from 1000 r.p.m. to 800 r.pm. suddenly and set at this value for 200 m.sec. then return back to its original steady state speed value. Controller used here is fast and accurate for (nearly 160 m.sec for reach required reference speed). Figure 14 shows simulation study of transient rotor speed at different duty cycle of R-C series controlled in rotor circuit. On the other hand, an experimental start-up of rotor speed at duty cycle = 20 of R-C series in rotor circuit is shown in Figure 15. It can be concluded that, increasing duty ratio of R-C series in rotor circuit non-linear proportionally with rotor speed and raising time is the best when using duty ratio $\leq$ 100% .

7. Conclusion

Series controlled conservative element in the rotor circuit of the induction motor will control the rotor equivalent circuit impedance. The proposed control technique used to limit the motor starting current. In steady state operation the proposed system can be operate as speed control system. Hysteresis control (ON/OFF) algorithm is used to control operation of the switches to control the effective rotor impedance value. The computed results, which are obtained from the simulation results, are proved to yield good agreement with the relevant experimental results.

8. Appendix

a.b,c: First, second, and third phases of three-phase system
C: Series Capacitance in rotor circuit per phase.
Dl,G I,SI: Drain, gate and source of MOSFET 1 D2,G2,S2: Drain, gate and source of MOSFET 2
D3,G3,S3: Drain, gate and source of MOSFET 3
K: viscus friction constant, seconds.
i: Instantaneous current
J: Inertia of motor and connected load
L: Series inductance in rotor circuit per phase.
Lss, L_{rr}: Self-inductances of three-phase stator and rotor circuits, respectively.
Lsm: Mutual inductance between stator phases Lrm: Mutual inductance between rotor phases
M_{sr}: Mutual inductance between three-phase stator and rotor circuits
p: Differential operator, d/dt
P: Number of poles
R: Series resistance in rotor circuit per phase.
s, r: Suffixed denoting stator and rotor, respectively
r: Resistance
s: Instantaneous flux linkage
0: Electrical angle denoting instantaneous rotor position, radians
0_m: Mechanical angle denoting instantaneous rotor position, radians
T e: Electrically developed torque
T, Tm: Superscript-denoting transpose of matrix and External load torque respectively

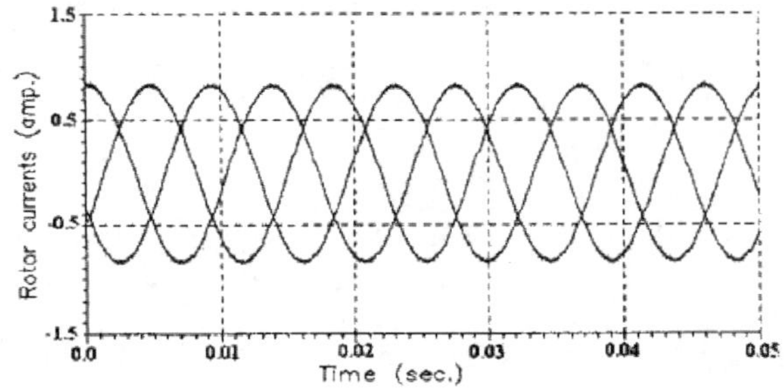

(a) Rotor currents at duty ratio = 20%

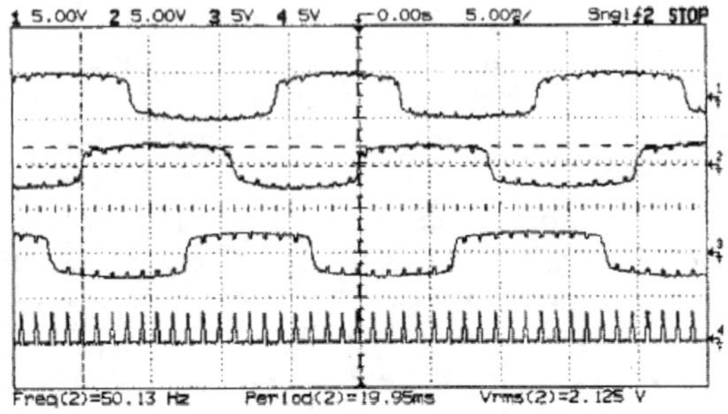

(c) Three-phase rotor currents and MOSFET pulses of one phase at duty ratio = 20%

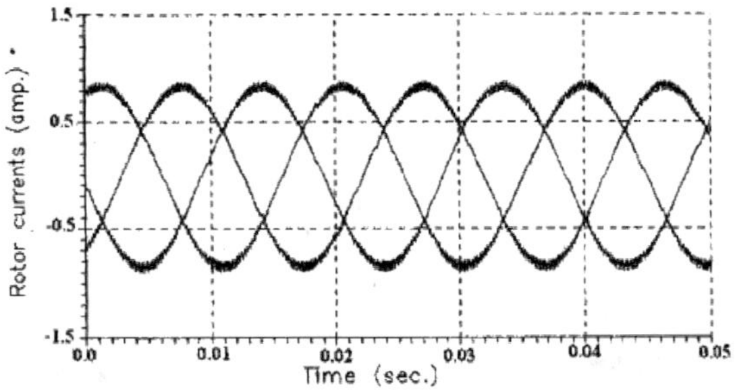

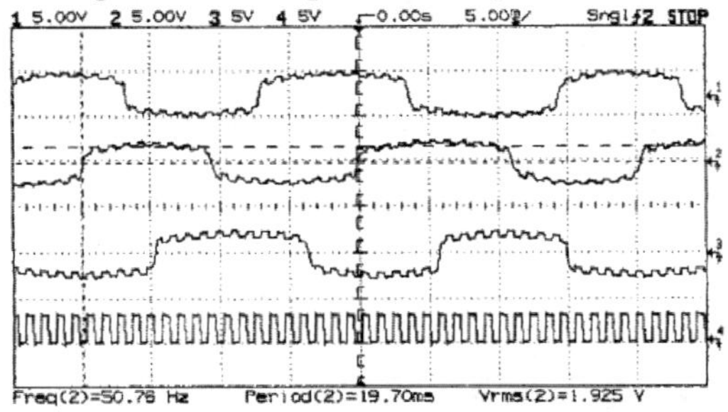

(b) Rotor currents at duty ratio = 50%

(d) Three-phase rotor currents and MOSFET pulses of one phase at duty ratio = 50%

Simulation results Experimental results

Figure 5. R-control of rotor resistance with different duty cycle ratios

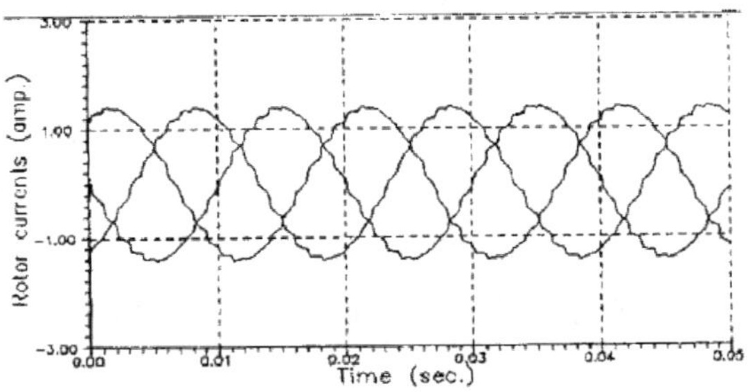

(a) Rotor currents at duty ratio = 20%

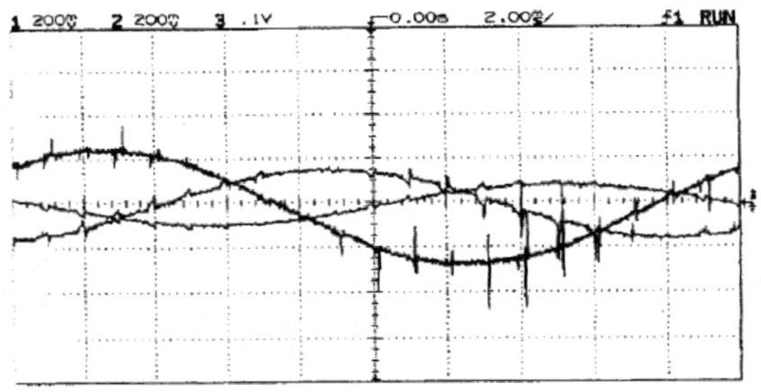

(c) Three-phase rotor currents and MOSFET pulses of one phase at duty ratio = 20%

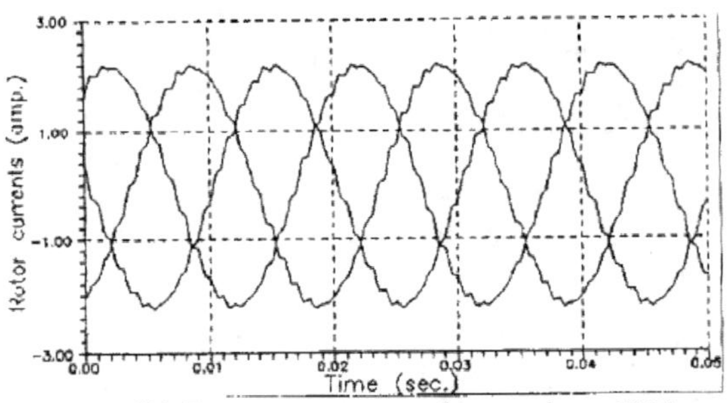

(b) Rotor currents at duty ratio = 50%

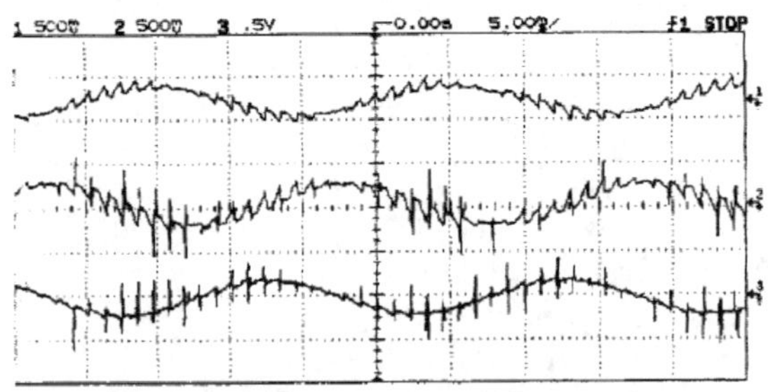

(d) Three-phase rotor currents and MOSFET pulses of one phase at duty ratio = 50%

Simulation results Experimental results

Figure 6. R-L control of rotor resistance with different duty cycle ratios

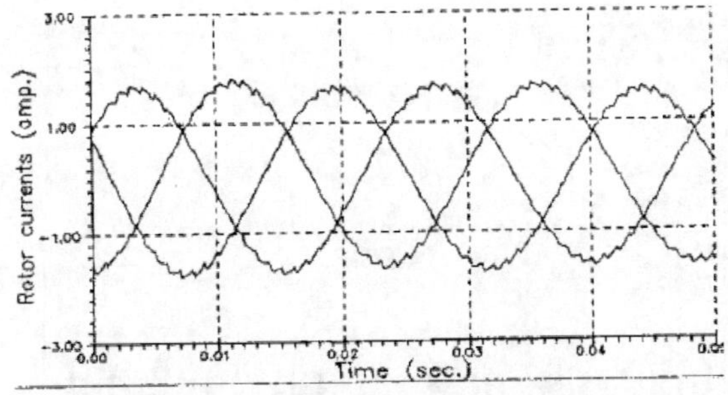

(a) Rotor currents at duty ratio = 20%

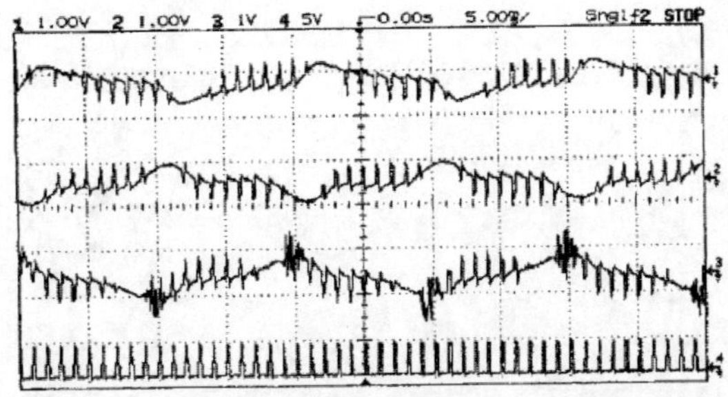

(c) Three-phase rotor currents and MOSFET pulses of one phase at duty ratio = 20%

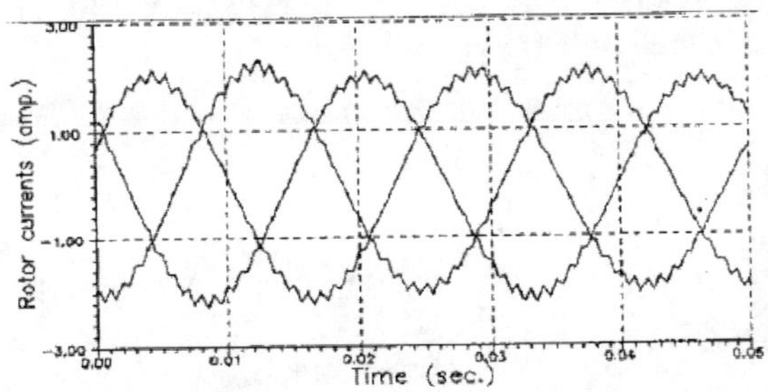

(b) Rotor currents at duty ratio = 50%

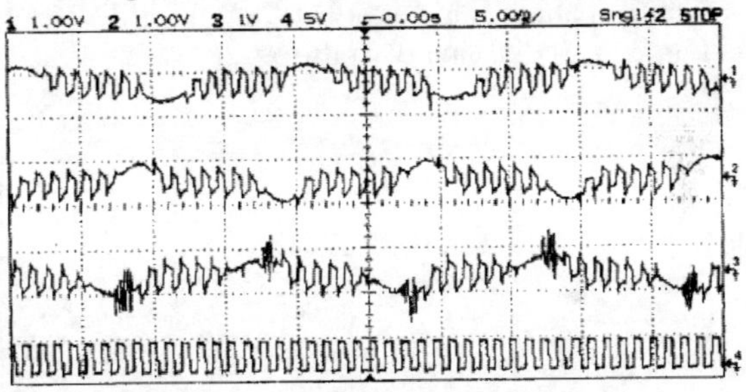

(d) Three-phase rotor currents and MOSFET pulses of one phase at duty ratio = 50%

Simulation results Experimental results

Figure 7. R-C control of rotor resistance with different duty cycle ratios

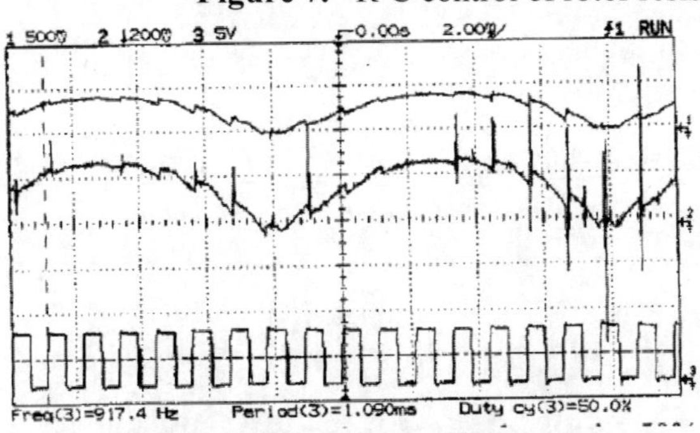

(a) MOSFET voltage and current at duty ratio=20% (R-L Control)

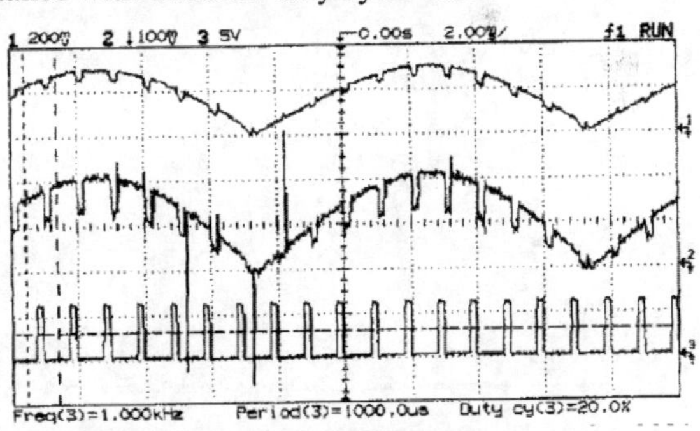

(b) MOSFET voltage and current at duty ratio=50% (R-L Control)

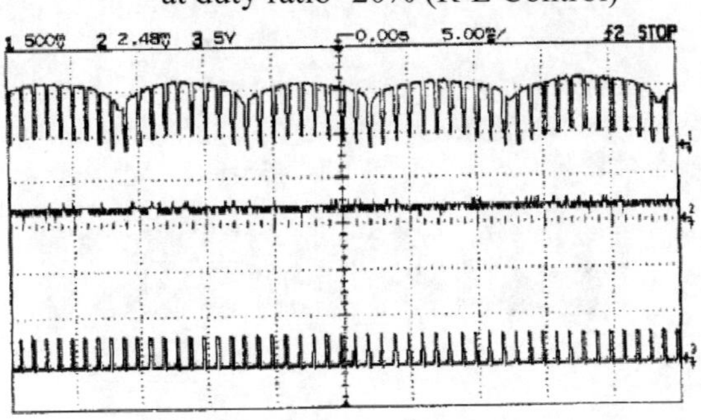

(c) MOSFET voltage and current at duty ratio=20% (R-C Control)

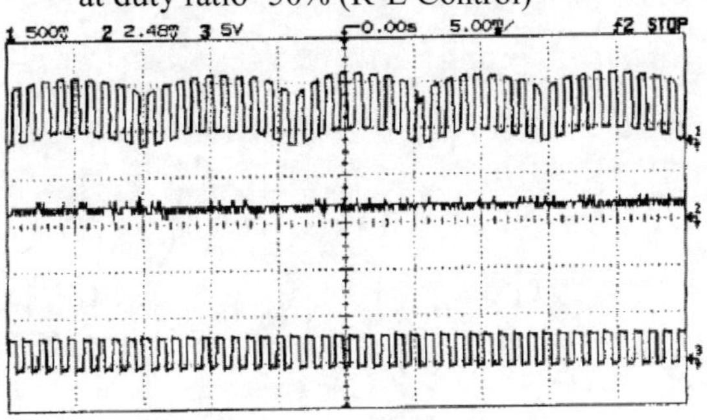

(d) MOSFET voltage and current at duty ratio=50% (R-C Control)

Figure 8 MOSFET performance at R-L and R-C control of rotor impedance

IOP Conf. Series: Materials Science and Engineering **459** (2019) 012008 doi:10.1088/1757-899X/459/1/012008

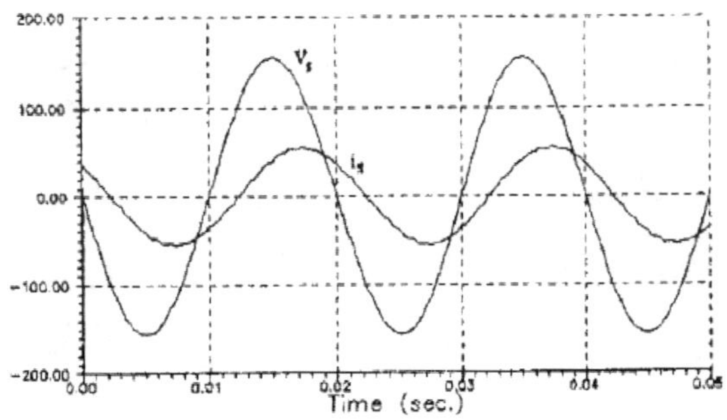

(a) Supply voltage and supply current (is, I amp. =50 div.) at duty ratio = 20%

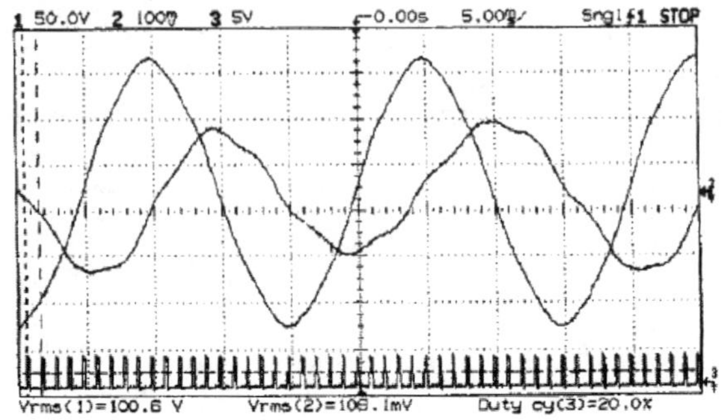

(b Supply voltage, stator current and MOSFET pulses at duty ratio = 20%

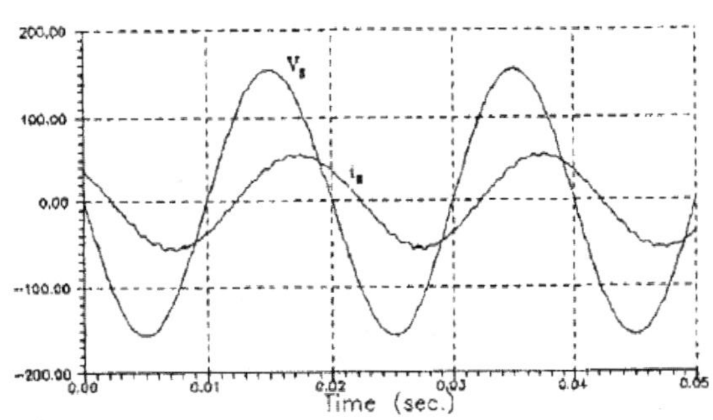

(c) Supply voltage and supply current (is, I amp. =50 div.) at duty ratio = 50%

Simulation Results

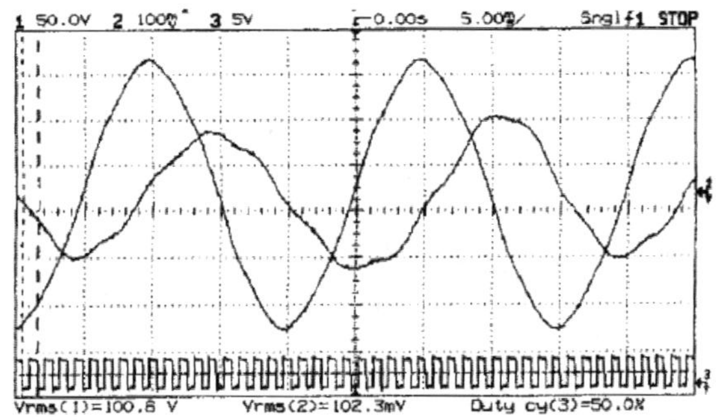

(d) Supply voltage, stator current and MOSFET pulses at duty ratio = 50%

Experimental Results

Figure 9. Supply voltage and stator current at different duty ratio R-C control of rotor impedance

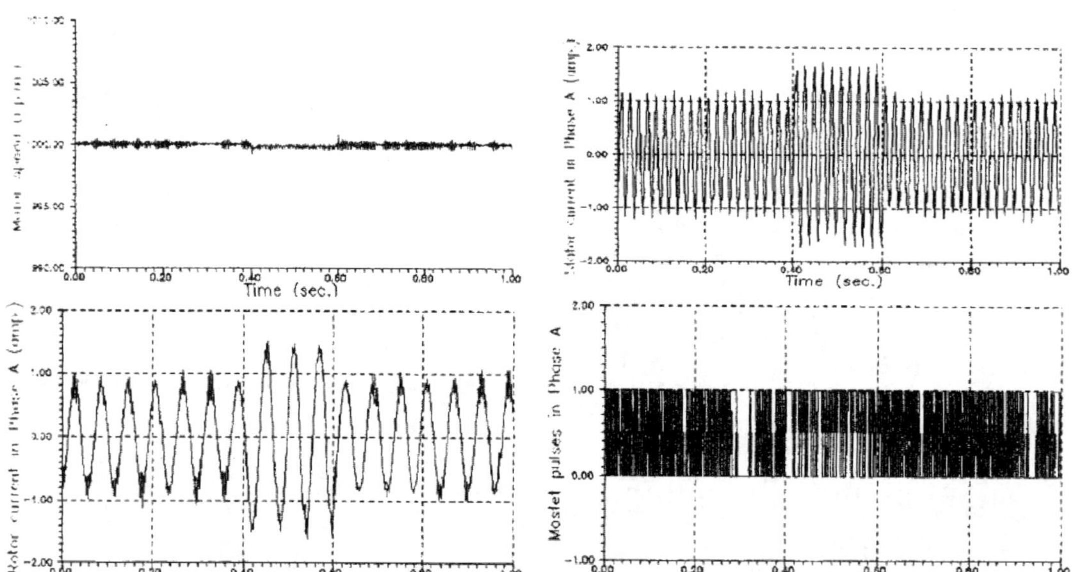

Figure 10. Motor performance and MOSFET pulses during step change in load torque by 20% (Simulation results)

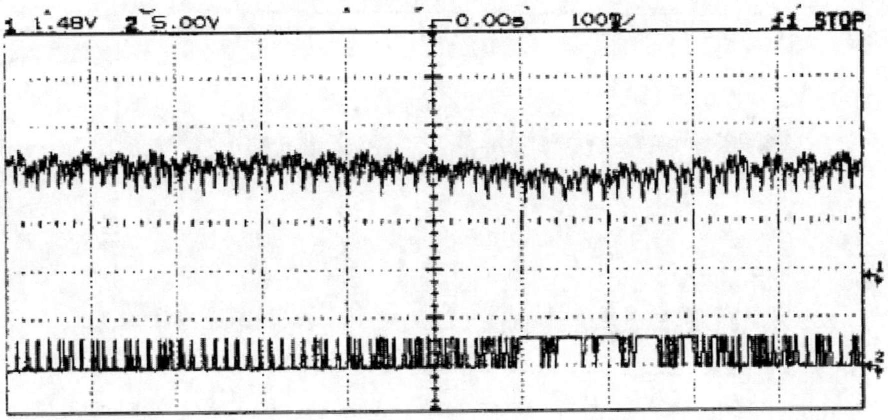

Figure 11. Motor speed and MOSFET pulses during step change in load torque by 20% (Experimental results)

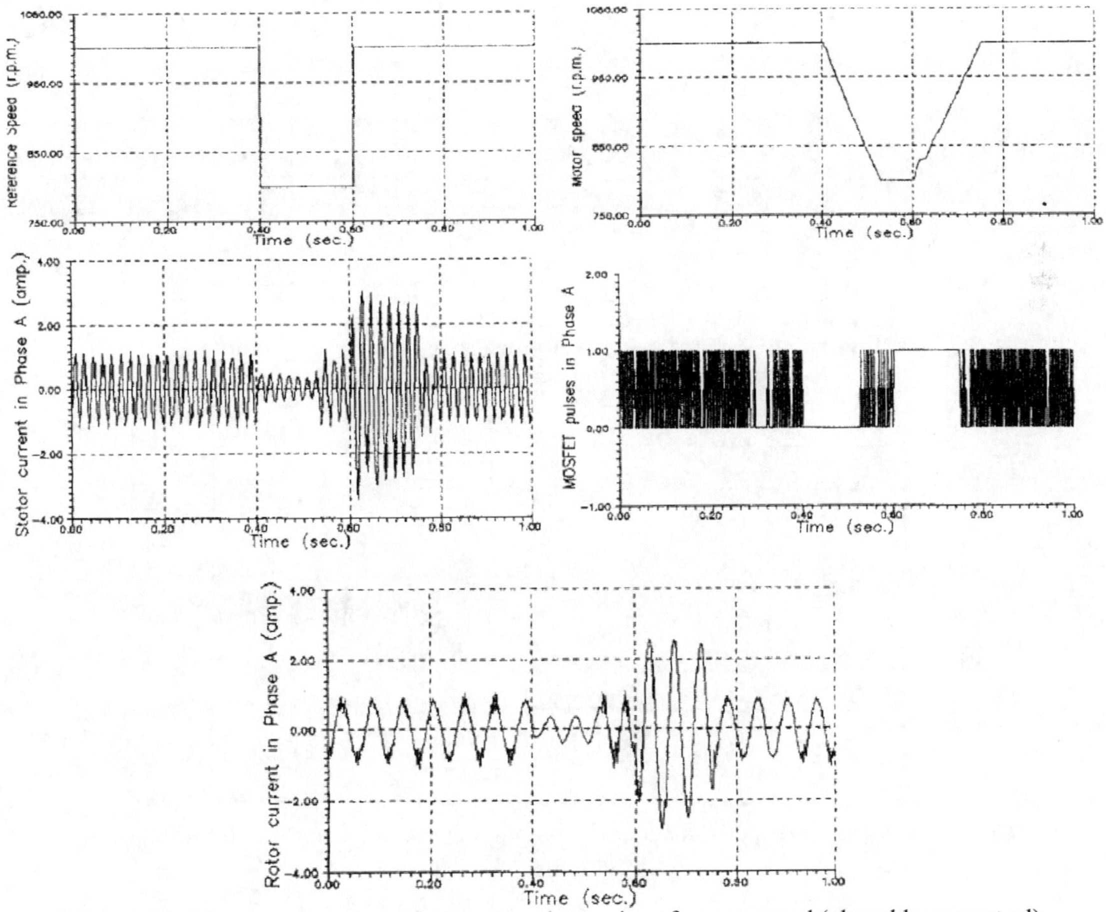

Figure 12. Motor performance during step change in reference speed (closed loop control) from 1000 to 800 r.p.m, Then return back to 1000 r.p.m,

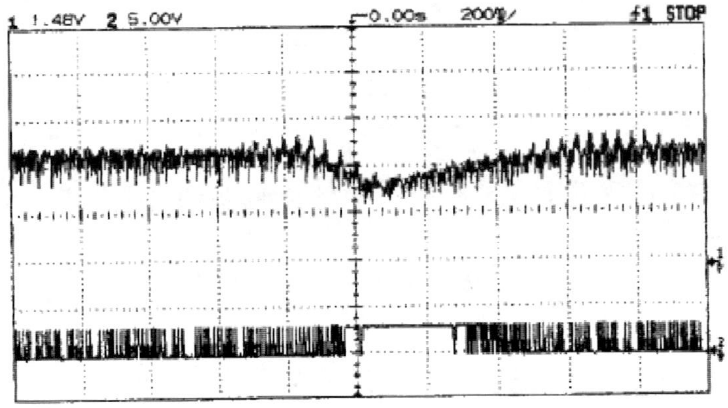

Figure 13. Motor speed and MOSFET pulses during step change in reference speed (Experimental results)

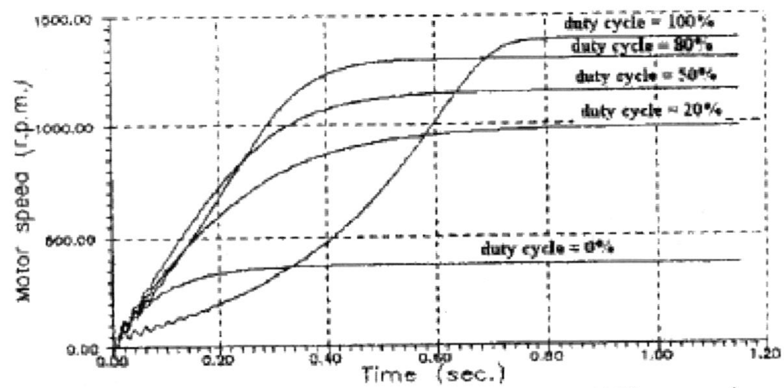

Figure 14. Motor start-up performance at different duty ratio R-C control of rotor impedance (Simulation results)

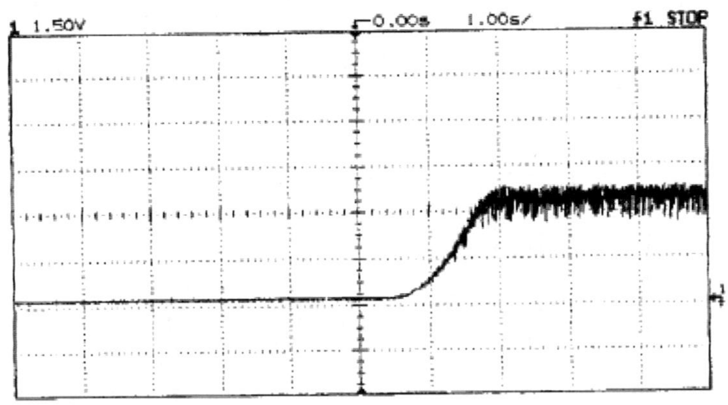

Figure 15. Motor start-up performance at duty ratio= 20 of R-C control of rotor impedance (Experimental results)

References

[1] Deghedie S, Ahmed M. and Abdelhamid T.H., 2001 "Performance of variable speed induction motor controlled by hysteresis current controller", MEPCON'2001, Egypt, pp.85-89,.

[2] Tsai Minga-Fa, and Tzou Ying-Yu, 1997 "A Transputer-Based Adaptive Speed Controller for AC Induction Motor Drives with Load Torque Estimation", *IEEE Trans. On Industry Applications*, **Vo1.33, No.2**.

[3] Ramamoorty M., and Arunachalam M., 1979 "Dynamic performance of a closed-loop induction motor speed control system with phase-controlled SCR's in the rotor", *IEEE Trans. On Industry Applications*, **VoL. lA-15, No.5**. .

[4] Hess Herbert L., Divan Deepakraj M., and Xue Yanhong, 1994 "Modulation strategies for a new SCR-based Induction motor drive system with a wide speed range", *IEEE Trans. On Industry Applications*, **Vo1.30, No.6**.

[5] Eren Halit, 1998, "Computer control of thyristorized induction motor and on-line performance - predictions", Electric Machines and Power Systems, pp.873-884

[6] Lesan Saeed Samiai , Mohamed Salah, and Shephers William 1996, ,"Control of wound rotor induction motor using thyristors in the secondary circuits", *IEEE Trans.On Industry Applications*, **VoI.32**,No.2.

[7] Reinert Juregea, and Parasely George Michael James, 1995 "Controlling the speed of an induction motor by resonating the rotor circuit", *IEEE Trans. On Industry Applications*, **Vol. 31, No.4**

[8] Asish K. De Sarkar and Gunnar J. Berg, 1970 "Digital Simulation of Three-Phase Induction Motors", *IEEE Transactions on Power Apparatus and Systems*, **Vol. PAS-89, No.6**, pp.1031-1037.

[9] Brey Barry B, 2006 "The intel microprocessor 8086/8088, 80186, 80286, 80386, and 80486", Maxwell Macmillan International Edition.

[10] Kailaswar Shilpa V. and Keswani R.A. , 2013 "Speed control of three phase induction motor by v/f method for batching motion system", *International Journal of Engineering Research and Applications (IJERA) ISSN: 2248-9622 www.ijera.com*, **Vol. 3,** Issue 2, pp.1732-1736.

[11] Devi Kamini, Gautam Shailendra and Deepak Nagaria,2014 " Speed control of 3-phase induction motor using self-tuning fuzzy PID controller and conventional PID controller", *International Journal of Information & Computation Technology. ISSN 0974-2239*, **Volume 4, Number 12**, pp. 1185-1193.

[12] Jasim Ali Abdulrazzak, Sultan Ahmed Jasim, and Kadhim Kadhim Karrar Hameed, 2017 " Speed control of three phase induction motor by v/f method with arduino Uno.", *Int. Journal of Engineering Research and Application www.ijera.com, ISSN : 2248-9622*, **Vol. 7, Issue 12, Part -5**, pp.62-66.

[13] Kannan Kaliappan K.,Anusiya S., Kavipriya K., Logeswari R., and Suganya V., February 2017 " DsPIC based speed control of three phase induction motor using v/f method", (*International Journal of Advanced Research in Electrical, Electronics and Instrumentation Engineering,* **Website:** *www.ijareeie.com*, **Vol. 6, Issue 2**

[14] Chande Sharda and Khanke Pranali, 2017 " Fuzzy logic controller based speed control of three phase induction motor", *IJARIIE-ISSN(O)-2395-4396, www.ijariie.com,* **Vol-3 Issue-4**, pp.2463-2472

IOP Conf. Series: Materials Science and Engineering **459** (2019) 012009 doi:10.1088/1757-899X/459/1/012009

EMI protection elements on cadmium telluride thin films

M V Kirichenko, G S Krypunov, M G Khrypunov, R V Zaitsev and A N Drozdov

National Technical University «Kharkiv Polytechnic Institute»

kirichenko.mv@gmail.com

Abstract. The amplitude-time characteristics of cadmium telluride thin films switching were investigated at the influence of single impulses duration 1 μs. It has been founded that with an increase of cadmium telluride layer thickness from 3 μm up to 8 μm, an increase of the operating threshold from 70 V to 105 V is es-tablished. The maximum residual sample voltage was change in the range from 12 V to 40 V, the minimum - from 5 V to 20 V. Samples switching time was no more than 2 nanoseconds; the samples interelectrode capacity does not exceed 2 pF. All test samples were operated without failure up to 20 times. Based on the results of cadmium telluride films structural studies by X-ray diffractometry and scanning electron microscopy we proposed a mechanism of cadmium telluride films with columnar structure monostable switching based on the formation of melted high-conductivity channels in cadmium tel-luride grains oriented in the [111] direction.

1. Introduction

The problem of ensuring radio-electronic equipment (REE) electromagnetic stability is because under the influence of electromagnetic impulses (EMI), overvoltage impulses are induced in the circuits, which can have a serious damaging effect on the elements of the REE [1]. Semiconductor devices are particularly susceptible to damaging effects of EMI thankfully to properties of the p-n junction and to the specific heat conductivity of semiconductor materials. With a decrease of semiconductor device structures size, the level of their damage is reduced and for integrated circuits is from 10^{-3} J to 10^{-7} J. To ensure the protection of electrical circuits, practically the REE protection elements from EMI are used. The most important property of them, is their ability to reduce resistance R_e from 10^4 - 10^{10} Ω for a short time τ_{sw} (switching time or response time) to a value significantly lower than input resistance of the protected REE, when the voltage U_i in the circuit exceeds the threshold voltage U_t, called the switching threshold [2]. The most widely used restrictive silicon diodes, since they have τ_s at a level of 1 nanosecond. However, they can shunt a limited amount of energy and have an interelectrode capacitance at a level of 20 pF, which limits their application for the microwave REE protection. Therefore, in order to create protection elements for microwave REE should be perspective, to carry out research aimed to creation EMI protected elements on the base of cadmium telluride thin films.

2. Experimental technique

For deposition base layers of cadmium telluride on 10cm × 10cm substrates from electrolytic polished molybdenum foil, we used industrial vacuum unit (Fig.1a) and method of thermal vacuum deposition (Fig.1b) from 99.999% purity powder with a particle size of 10 mm. The initial vacuum was $6{\cdot}10^{-6}$ mm Hg, the working pressure in the vacuum chamber during the deposition was maintained at $1{\cdot}10^{-5}$

mm Hg. For thermal evaporation of cadmium telluride we use a graphite evaporator with indirect heating from two electrically insulated heaters (Fig. 1b).

a) b)

Figure 1. Overall view of vacuum chamber (a) and graphite evaporator of cadmium telluride (b).

The evaporation temperature was controlled by a thermocouple installed in the volume of the heater, the time of evaporator heating to working tempera-ture (700 ± 750) ° C was 260 to 275 sec. The accuracy of evaporator working temperature control was not more than 5 °C. The thickness of the deposited layers of cadmium telluride, which was set by the deposition time, was 3-10 μm. To control the cadmium telluride films thickness we used a quartz thickness meter KKT-40.

The crystalline structure of cadmium telluride layers was investigated by X-ray diffractometry by the θ-2θ method with a step of 0.01 degrees in the radiation of the copper anode.

The control of cadmium telluride film layers initial electrical resistance was carried out at a tempera-ture 20 °C using a digital ampere-voltmeter MS8040. Along with the initial electric resistance of the samples to a direct current, at a frequency of 107 Hz and temperature of 20 °C, their electrical capaci-tance was measured by the L2-28 capacitance meter device.

The switching characteristics of the samples were studied on a test bench with subsequent stress im-pulses parameters: amplitude Ui from 100 V to 1400 V with a duty ratio of 2·107, rise time of the pulse to the amplitude value was 2.5 nanoseconds, impulse voltage decreased exponentially to 0.5 in a time of 100 nanoseconds. The amplitude-time characteristics of samples switching process were determined from experimental oscillograms obtained on the digital oscilloscopes.

To connect to a coaxial line with a wave impedance of 50 Ohm, film CdTe samples were placed in modified bodies of microwave diode D403B type (Fig.2) with interelectrode capacitance at 10^7 Hz is only 0.2 pF.

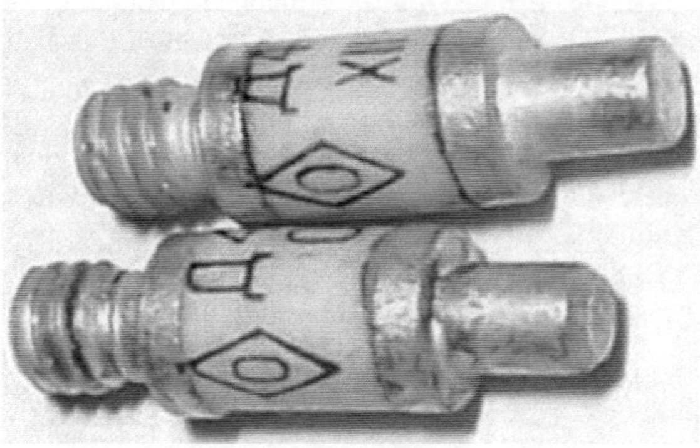

Figure 2. The microwave diode bodies prepared for installing cadmium telluride films.

The specificity of the electrodes design was that it should gently pressurize to the cadmium telluride film surface and should be able to freely move along the perpendicular direction of layer to track changes in film thickness caused by its heating and cooling during the forward and backward switching stages. A schematic illustration of the assembled case with a thin-film layer is shown on Fig. 2.

3. Results and their discussion

Using X-ray diffractometer it was found that when deposition temperatures less than 300 °C, a two-phase cadmium telluride film was formed, which contains, along with a stable cubic modification, a metastable hexagonal phase. At substrate temperature above 350 °C, the CdTe layer growth rate sharply decreases. Therefore, the substrate temperature during depositing cadmium telluride films for switched layers was (320-330) °C. Film layers were made which had a thickness (d) from 3 μm up to 8 μm. Investigations of the crystal structure shown that all the layers have a stable cubic modification, as is unambiguously confirmed by the presence of reflections from the (111), (200) (311), (400), (331), and (422) planes (Fig. 3a). The intensity ratio of the detected peaks differs from the theoretical ones and indicates the advantage of [111] direction of the films orientation. The morphology of cadmium telluride films surface, studied with the Philips CM30 raster microscope, indicates that the grain size in cadmium telluride layer of is about 1 μm (Fig. 3b).

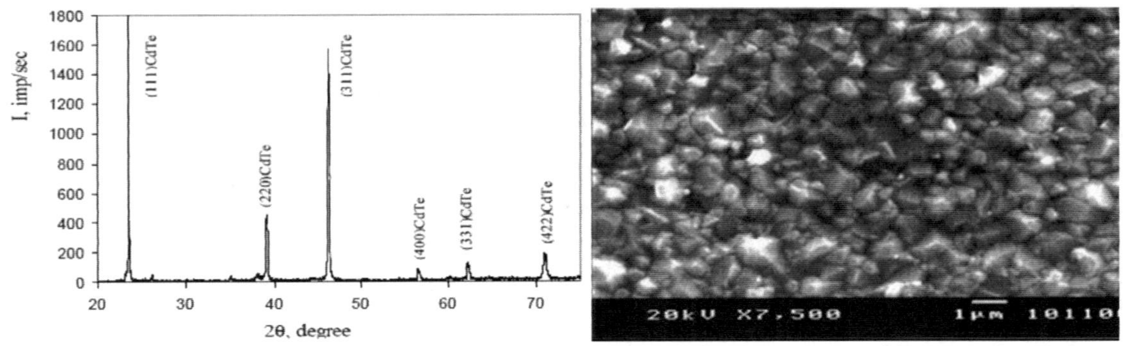

a) b)

Figure 3. X-ray diffractograms (a) and surface (b) of 4 μm thick cadmium telluride layer

According to the literature data [5, 6], cadmium telluride films oriented in the [111] direction have a columnar structure. In the initial state, the electrical resistance R_c of the samples exceeds 10^9 Ohms and did not depend on the thickness of cadmium telluride layer. For the initial electrical capacitance, the value of which was from $0.4 \cdot 10^{-12}$ F to $1.2 \cdot 10^{-12}$ F, there was also no dependence from the layer thickness. The absence of electrical resistance and capacitance dependence from the cadmium telluride layer thickness on our opinion is due to the variation of the electrode contact area with the cadmium telluride switching layer.

Experimental oscillograms were used to study the amplitude-time characteristics (ATC) of the switching process in the resulting thin films of cadmium telluride. The qualitative form of a typical oscillogram of the experimental stress diagram on film samples is shown in Fig. 4a (curve 1). This figure also gives a qualitative view of a of the voltage impulses acting on the samples typical oscillogram (curve 2). On the Fig.4b represents the typical electromagnetic pulse shapes which are acting on real semiconductor devices and caused by nature or human influence.

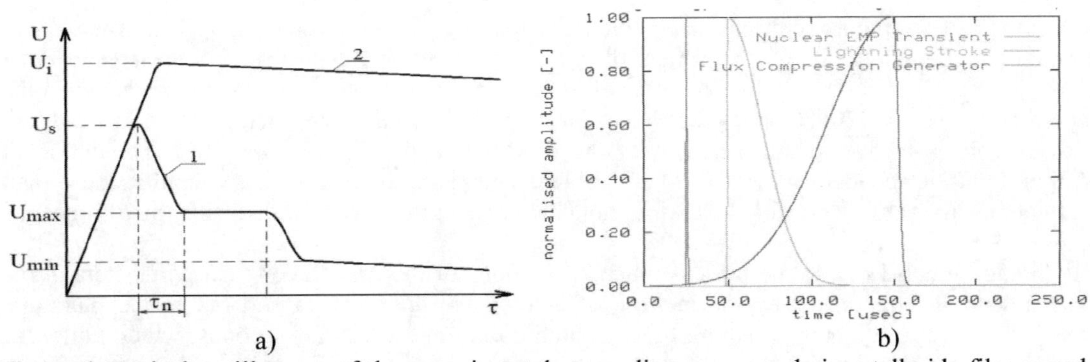

Figure 4. Typical oscillogram of the experimental stress diagram on cadmium telluride film samples (a) and typical electromagnetic pulse shapes (b).

The determination of the ATC for switching processes was preceded by the determination of the pickup threshold, which was carried out by feeding the forming impulse to the samples. The shaping impulse had the amplitude, which was minimum necessary for the first operation. Then several measuring impulses were fed with the minimum amplitude necessary for the next trips, which was identified with the value of the threshold voltage U_t. The results of ATC research are given in Table 1. The table shows the threshold voltage U_t, the maximum voltage on the samples U_s, the maximum residual voltage on the samples U_{max}, the minimum residual voltage on the samples U_{min}, the time of switching to the low-resistance state τ_s.

Also in Table 1 gives the electrical resistivity of the samples to the direct current R_e after a 20-fold impulse action with amplitude U_i. It was found that the values of ATC do not depend from the polarity of the current impulses acting on the film samples.

Table 1. – Main switching parameters of thin-film cadmium telluride samples and their electrical resistivity to a direct current after the 20 impulses influence

d (μm)	U_t (V)	U_s (V)	U_{max} (V)	U_{min} (V)	τ_s (sec)	U_i (V)	R_e (Ohm)
3	75	42	20	10	<2	285	$1.7 \cdot 10^5$
3	75	38	20	10	<2	271	$2.0 \cdot 10^5$
4	70	40	13	5	<2	214	$1.3 \cdot 10^6$
4	70	50	12	5	2	211	$7.7 \cdot 10^6$
6	70	52	15	7	2	216	$5.6 \cdot 10^6$
6	75	51	15	7	2	225	$5.6 \cdot 10^6$
7	80	55	20	10	<2	240	$1.5 \cdot 10^6$
7	70	53	20	10	2	219	$1.7 \cdot 10^6$
8	105	120	40	20	2	316	$2.3 \cdot 10^6$

Analysis of Table 1 shows that with an increase in thickness from 3 μm to 8 μm, an increase in the operation threshold from 70 V to 105 V is observed. The maximum residual voltage on the sample varies from 12 V to 40 V, the minimum - from 5 V to 20 V. The switching time of the samples was no more than 2 nanoseconds. All the test samples were operated without failure 20 times.

We thinks that in the investigated cadmium telluride films, the effect of a monostable switching from a low-conducting state to a high-conducting state is observed, which is realized due to the appearance of a reversible instability in the ionic subsystem. The instability in the ionic subsystem occurs when the semiconductor layer melts. The source of heat for monostable switching is the Joule heating of cadmium telluride layer by high-density current, which occurs in the initially high-resistance material. According to the literature data, when heating to 800 °C [7], the specific electrical conductivity of cadmium telluride films increases exponentially. According to [8], melting of cadmium telluride due to the rearrangement of the crystal structure leads to an abrupt increase in the electrical conductivity by more than an order of magnitude from its electrical conductivity. When the temperature exceeds 120 °C, the coefficient of electrical conductivity temperature dependence increases noticeably due to the inclusion of the dissociation process [9].Since the dissociation of cadmium telluride occurs congruently, when both elements pass to the gas phase simultaneously and the stoichiometry of the remaining layer does not change, the samples showed stability at 20 times the impulse action.

It should be noted that at the present time, calculations of the ATC theoretical parameters in the monostable switching caused by the melting of layers have not been carried out in the literature. Nevertheless, the fixed short switching times, which do not exceed 2 nanoseconds, which limits the amount of Joule heat released, suggest that the switching does not occur simultaneously over the entire layer, but due to the appearance of liquid phase channels in the polycrystalline film layer.

Moreover, taking into account the columnar structure of the cadmium telluride layer, it can be assumed that the dimensions of the channels correspond to the grain sizes, which according to the results of structural studies are about 1 μm. According to the results of X-ray diffraction studies, the obtained cadmium telluride films are predominantly oriented in the [111] direction and in [10] has been noted that biphasic, twinning and high concentration of packing defects are characteristic for cadmium telluride films predominantly oriented in the [111] direction (see, for example, in [6]), which is caused by errors in the stacking sequence of close-packed planes to which the (111) plane refers to. It is quite obvious that the defective grains of cadmium telluride films oriented in the [111] direction will melt first of all, forming highly conducting channels of molten cadmium telluride.

4. Conclusion

We have found that CdTe layers with a thickness from 3 to 7 μm placed in the body of microwave diodes under the action of electric impulses duration up to 1 μs had a switching time no more than 2 nanoseconds and capacity up to 2 pF. As a result such films can be used as a base layers in order to creation EMI protection elements for ultrahigh-frequency radioelectronic equipment.

Investigations of the crystal structure have made it possible to propose a mechanism for monostable switching in CdTe films by the formation of molten high-conductivity channels in grains oriented in the [111] direction, which have a columnar structure under the action of an electromagnetic impulse.

The fixed failure-free operation of protection elements on the CdTe films basis for 20 cycles of impulse action is due to the congruent melting of CdTe films, which ensures the preservation of switching layer stoichiometry after the action of a high-frequency electric impulse.

References

[1] Gurevich V 2011 *Elect Electr Eng* **5** pp 23-28
[2] Solodov A, Pirogov Y, Klyuchnik A 2012 *Stability of IC in electromagnetic fields of pulsed radiation* (Lambert Academic Publishing)
[3] Pergament A, Stefanovich G, Malinenko V and Velichko A 2015 *Advances in Condensed Matter Physics* **2015** 26.
[4] Skornyakov S, Pavlov V, Rakhmanov A, 2008 *Components and Technologies* **3** pp 70-74
[5] Terheggen M, Heinrich H, Kostorz G, Romeo A, Tiwari A, 2001 *Proceeding of the 17th European Photovoltaic Solar Energy Conference* (Munich, Germany) pp 1188-1191
[6] Romeo A, Batrner D, Zogg H, Tiwari A, 2001 *Proceeding of 16th European Photovoltaic Solar Energy Conference* (Glasgow, Scotlan:2001) pp 843-846
[7] Ching-Hua S 2015 *AIP Advances* **5** p 057118
[8] Belas E, Grill R, Franc J, Turjanska L, Turkevych I, Moravec P and Hoschl P 2003 *J Electron Mater* **32 7** pp 752-755
[9] Glazov V, Chizhevskaya S, Glagoleva H, 1967 *Liquid semiconductors, (*Moscow: Science)
[10] Panchekha P, Alaverdova O, Gnidash V, 2000 *Ukraine Journal of Physics* **45** 1 pp75-80

Thermoelectric heat patch for clinical and self-management of post melanoma excision wound care

I Logothetis[1], S Vassiliadis[2], E Siores[3] and E Pirogova[1]
[1]RMIT University, School of Engineering, Discipline of Electrical and Biomedical Engineering, Plenty Road, Bundoora, 3083, Australia
[2]Piraeus University of Applied Sciences, Engineering, Electronics Engineering, Thivon 250, Egaleo 12241, Greece
[3]Bolton University, Institute of Materials Research and Innovation, Deane Rd, Bolton BL35AB, UK

irini.logothetis@student.rmit.edu.au

Abstract. Thermodynamics as a physical and healing therapy is considered to have potential beneficial effects when applied to soft tissue. Normothermia is the state of a body being at a normal temperature and is essential to healthy cell functioning and wound healing. This article presents the application of thermoelectric elements on a conductive textile; focusing on the heat transfer of five elements over varied areas. By setting the temperature of the thermoelectric elements relative to skin temperature, a thermal camera captured the heat transfer. An increase in heat transfer was created by lining the conductive textile with commercial polyester wadding forming a quilted patch. Thus, by adapting this patch, the maximum distance between the elements relative to a minimum heat loss was observed at 3 cm. Any distance less than 3 cm covers a small area, whilst an increase in heat loss occurs above 3 cm. Further research can be conducted by implementing a pulse width modulating (PWM) system to maintain a constant temperature between 32 and 34 °C. With the addition of a temperature sensing device, a preventative system can be developed for chronic wounds essential for diabetic patients.

1. Introduction

Thermoregulation is the process in which the body works to maintain its core body and skin temperature within the required boundaries [1]. The human body achieves thermal balance from heat exchange that occurs between the body and its surrounding thermal environment. Autonomic thermoregulation relates to the autonomic nervous system composed of the hypothalamus that unconsciously controls the body thermal balance.

Perioperative hypothermia caused from anaesthesia has become a major concern in the medical field. In fact by administrating anaesthesia, thermoregulation is impaired; in addition to exposure to a cold environment, administration of intravenous fluids and evaporation from incisions, hypothermia is induced. Associated with major surgical procedures, such as cardiac and spinal surgery [2-5], it is vital to foster normothermia. Normothermia is the state of a body being at a normal temperature essential to healthy cell functioning [6] and wound healing.

Research into normothermia for regional anaesthetised surgical wounds such as melanoma excisions has yet to be studied. Melanoma excisions induce a local wound environment influenced by the anaesthesia for 4 - 6 hours post administration. Post-surgery, the wound is stitched as the site undergoes the healing process. Wound healing consists of four overlapping phases: haemostasis, inflammation, proliferation and remodeling. Haemostasis initiates the healing process preparing the phases that follow. During this phase, the inflammatory phase also begins. Once haemostasis has been completed, regeneration occurs through proliferation and remodeling [7-9]. Skin temperature is vital in monitoring wound healing; it can be applied to prevention of complications such as infections. Despite unimpaired epidermis layer, in the case of diabetic patients measuring skin temperature in the insensate foot has been addressed resulting in a relationship between an increase in localized temperature and localized pressure [10].

A skin temperature differential of 1 - 3 °C of the wound bed indicates a healthy healing state. Wounds that do not heal within three months are considered chronic; generally caused by a complication in one or more of the healing phases. In this case, skin temperature of the wound bed can increase up to 4 - 5 °C accounting for an infected wound. Postinfection temperature increases range only 0.8 - 1.1°C [11].

Skin acts as a thermal sensor for the body giving rise to the thermoregulation process; moreover, a variation of core body and skin temperature exists around the body [12]. Thus for normothermic wound therapy, temperatures approaching the skin temperature when the body is thermally balanced are ideal. Zhan et al. [13] identifies 33.4 °C as the ideal skin temperature (T_{sk}) where the human body is unaware of the temperature change with a 1.5 - 3 °C allowance. Discomfort is felt by the human body when the temperature increases or decreases more than 4.5 °C. Thomas et al. [14] reported the importance of duplicating normal physiological conditions in order to provide an optimum wound environment to improve wound healing. Current approaches into normothermic wound management apply heat to the wound inducing a moist environment whilst preventing hypothermia. However, the applied temperature is relative to the body core temperature and not the skin temperature [14-17]. Moreover, the materials and methods adapted are inconvenient requiring regular wound cover changes.

This study aims to provide a basis for development of an efficient wearable normothermic wound therapy patch by applying thermoelectric elements to electrical and thermal conductive textiles. This paper identifies the optimum positioning of thermoelectric elements for an even temperature distribution on a conductive patch. By applying this patch onto a wound, the temperature is distributed to the wound bed and periwound area maintaining thermophysiological skin temperature post anaesthetic administration thus preventing hypothermia. Subject to the results of this project, the proposed outcome is to effectively construct a single patch that can be applied to wounds of various sizes. Analysis of the heat distribution across the patch relative to the patch size is performed by identifying the optimum distance between the thermoelectric elements and patch size. The textile patch offers adaptability, providing a customised or patient-specific solution for improved wound management practice. This research has taken into consideration extensive wound sizes by positioning the thermoelectric elements in a matrix that is repeatable; thus, enabling versatile tailored textile patches for wounds of various sizes that can be place on any part of the body. Moreover, it has the potential for use in clinical or home settings.

2. Materials and Methods

To perform the experiment with precision, the thermophysiological skin temperature of about 33 °C was adapted as a reference model; thus, analysis of heat transfer on conductive textiles in the form of a patch was studied by setting the thermoelectric elements at 30, 32, 34 and 36 °C. To follow, a description of the materials selected is presented including the methodology of the experiment.

2.1. Materials

2.1.1. Thermoelectric Elements

Thermoelectric elements (product number QC-17-1.0-3.0) were purchased from Quick-Ohm Küpper & Co. GmbH of dimensions (l×w×h) 12×12×4 mm. To ensure optimum performance of the thermoelectric modules, the maximum current and voltage should not surpass 70% of the specifications. The specifications for the module are as follows; I_{max} = 3.3 A, V_{max} = 2 V, Q_{max} = 3.9 W with a ΔT_{max} of 71 °C.

IOP Conf. Series: Materials Science and Engineering **459** (2019) 012010 doi:10.1088/1757-899X/459/1/012010

2.1.2. Textiles and Materials

For efficient thermal diffusion, conductive metalized nylon textile was purchased from SHIELDEX. Its rip stop weave is beneficial for commercial use due to its ideal properties, such as strength, durability and resistance to wear and tear; having an abrasion resistance of 500,000 cycles, one hundred times better than most conventional fabrics. Consisting of three fully plated metalized layers of 99% pure silver and coated with copper and tin, it is lightweight weighing 77 gsm with a thickness of 0.1 mm; in addition, it has a low surface resistance measuring at less than 0.02 Ω and an operating temperature ranging between -30 to 90 °C. It possesses electromagnetic and radio frequency-interference (EMI/RFI) shielding properties whilst eliminating passive static and minimising infrared radiation (IR) in addition to its antimicrobial properties; thus, it is well suited to wearable biomedical garments, where interference is common in medical centres. To improve the thermal diffusion, a thermal insulating textile was used as to line the conductive textile. This textile, a traditional polyester non-woven fabric purchased commercially, was selected for its high thermal diffusion value [18]. Aluminium tape was selected to fix the thermoelectric elements onto the nylon textile. It was adapted for this study due to its thermal properties and its common use as a heat sink.

2.2. Method

A thermal pocket consisting of five layers was prepared. Top and bottom layers were formed by folding the conductive metalized nylon fabric; thus creating the first and fifth layer respectively, to house the thermoelectric elements. Adding the polyester non-woven textile to the first layer produced the second layer. By applying the hot side of the thermal elements onto the polyester non-woven fabric, this third layer formed the basis of the experiment. The fourth layer of aluminium tape provided a heat sink for the system in addition to keeping the thermoelectric modules fixed.

Five thermoelectric elements were selected and positioned on a conductive textile as presented in figure 1 (a) and (b). Four thermoelectric elements were positioned in a square array with a fifth in the middle. Varying the distance *(d)* equally between the four outer elements deduced a set of six samples: sample 1, $d = 1$ cm; sample 2, $d = 2$ cm; sample 3, $d = 3$ cm; sample 4, $d = 4$ cm; Sample 5, $d = 5$ cm and sample 6, $d = 10$ cm. This pattern permits repeatability relative to any textile surface area whilst a middle node assists in an evenness of heat transfer through the textile. Figure 1 (c) presents the repeatability horizontaly. Thermal imgages were captured using the Keysight Technologies U5855A Thermal Imaging Camera and its software TrueIR Analysis and Reporting Tool was used to analyse the heat transfer of line and segent and produce histograms.

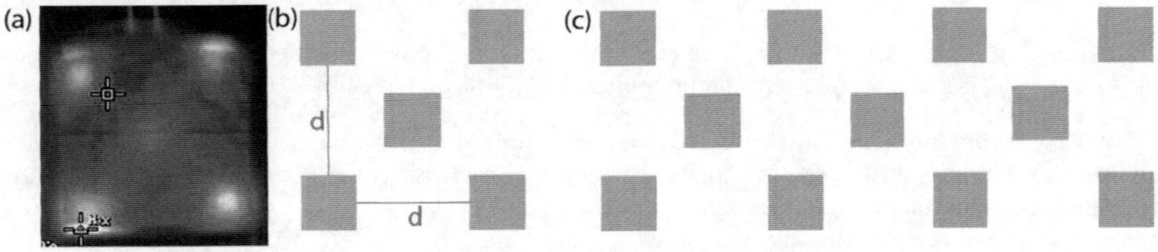

Figure 1. Positioning of thermoelectric elements: (a) thermal photograph showing the working elements, (b) the matrix setup where the distance *d* varies and (c) horizontally repeated pattern.

3. Results and Discussion

To determine the position of the thermoelectric modules for optimum heat diffusion within the conductive and insulating textiles, samples 1 and 6 were initially made with the thermoelectric elements electronically connected in parallel. By using the boxed analysis feature for samples 1 and 6 at the temperature of 34 °C, the sample range was determined. Boxed analysis is the method by forming a box over the area under study in which the temperature distribution across this selected area is represented using a histogram. By evaluating the histograms for these samples (refer to supplementary document), it was concluded that for sample 6 the thermoelectric peltiers were not efficiently utilised, with the temperature of the conductive patch ranging from 29 to 34 °C and a slightly left skewed distribution. By contrast sample 1 represented by a trimodal distribution, possesses a wider distribution approaching the desired temperature through the patch and a restricted temperature range of approximately 2.5 °C, relatively half of sample 6. From this, samples 2 to 5 were built for further study using line segment analysis at the predetermined temperatures 30, 32, 34 and 36 °C. A current of 3A was supplied for samples 1 to 5 for temperatures of 30 and 32 °C whilst 5 A was supplied for 34 and 36 °C.

3.1. Line Segments

By varying the distance between the thermoelectric elements, ΔT_{max} was calculated across the vertical and horizontal line segments measured from the middle of two elements and diagonally, L1, L2 and L3 respectively, refer to figure 2 (a). Measurements were also taken and the ΔT_{max} was recorded across line segments between the outer and inner elements, L1 and L2, refer to figure 2 (b).

Line segmental analysis was identical for L1 and L2 in figure 2 (a) and in figure 2 (b); thus, a single ΔT_{max} was recorded for each set. These results along with ΔT_{max} recorded for L3 in figure 2 (a) were compared and presented as a percentage difference from the smallest ΔT_{max} to the largest ΔT_{max} in order to verify the optimum distance between thermoelectric modules, refer to figure 2 (c).

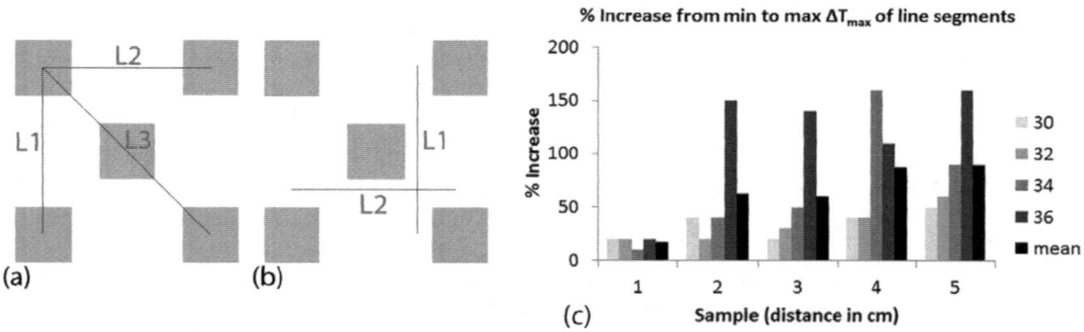

Figure 2. Line segment analysis: (a) connecting elements, (b) between elements and (c) graph of % increase from min to max ΔT_{max} of line segments.

From the results presented in figure 2 (c), it can be deduced that sample 3 with a 3 cm distance between the thermoelectric elements appears to be ideal relative to the distance. For example, the percentage difference across the line segments is comparable to samples 1 and 2, contrast to samples 4 and 5. Taking into account that the heat spreading across a larger surface area is desired, samples 1 and 2 are relatively small. For samples 4 and 5 ΔT_{max} varies greatly with an increase in temperature. By taking the average of the percentage increases, sample 3 performs better than samples 2, 4 and 5. Sample 1 proved to be the most efficient. Thus, further analysing the system, the power per square meter was calculated for each distance. Taking sample 1 as a reference, a comparison of the samples is shown in figure 3; presenting the power in $W.m^{-2}$ and the percentage difference of the samples in comparison to sample 1. Comparing the samples to sample 1, sample 3 appears to be more efficient relative to the surface area.

IOP Conf. Series: Materials Science and Engineering **459** (2019) 012010 doi:10.1088/1757-899X/459/1/012010

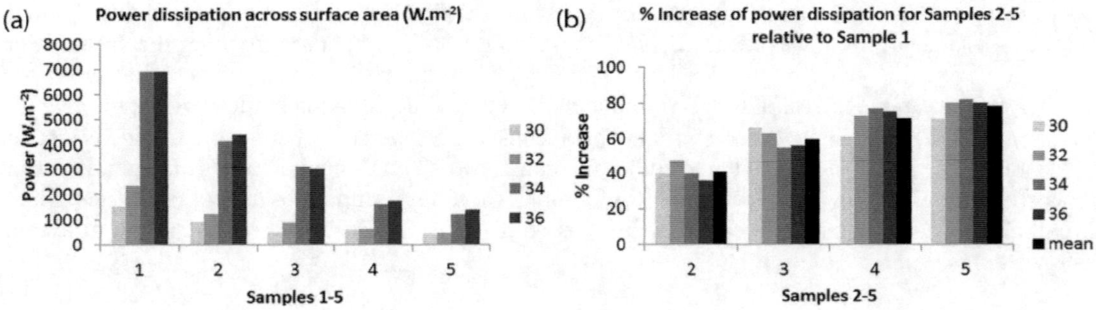

Figure 3. Power dissipation (a) measured across surface area (W.m^{-2}) and (b) % increase of samples 2-5 relative to sample 1.

3.2. Histograms

Analysing sample 3, histograms were produced for all studied temperatures. It can be seen from figure 4 that for the distance of 3 cm, a wide normal distribution occurs implying a constant heat spread. Although at 34 and 36 °C the heat spread is slightly skewed to the left and right respectively, the distribution is still wide. Comparing sample 3 to the other samples (refer to supplementary document), the distribution shows more symmetry and is closer to a uniform distribution.

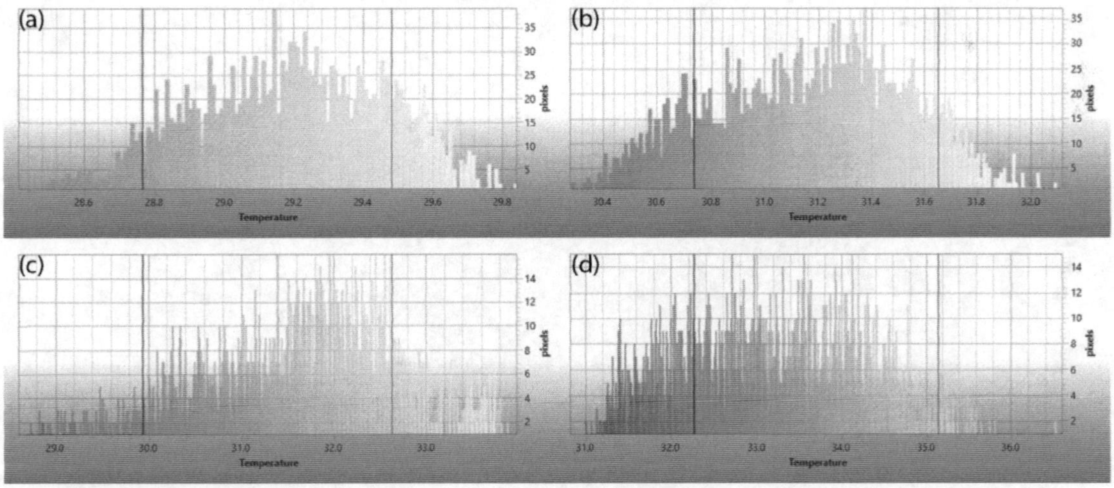

Figure 4. Histograms for sample 3: (a) at 30 °C, (b) 32 °C, (c) 34 °C, and (d) 36 °C.

4. Conclusion

Thermoelectric elements, when applied to attire, can provide thermophysiological comfort to humans maintaining normothermia around locally anaesthesised wounds. Equations governing thermophysiological comfort depend on skin temperature and not core temperature; thus, for this study the temperature setting for the thermoelectric elements are ideal to restore the skin temperature to its physiological state. By building a thermal pocket using five thermoelectric elements, the optimum distance between the elements was analysed. This was completed via recording the temperature distribution through a conductive textile, in the form of line segment analysis and power dissipation per square meter. It can be concluded that by applying thermoelectric elements to provide thermal comfort, a distance of 3 cm was ideal to cover a patch 55 × 55 mm, whilst providing a constant heat spread. However, this setup required a large amount of current giving rise to ergonomic issues of powering this wearable patch. Further research will require the implementation of a pulse width modulator (PWM) system in order to maintain a constant temperature between 32 and 34 °C to provide a portable and cost effective normothermic patch. Moreover, a temperature sensing device can be incorporated as a preventative feature for chronic wounds essential for diabetic patients. Heat distribution is vital to many applications;thus, this study can be adapted by researches from a vast variety of fields.

Supplementary Document

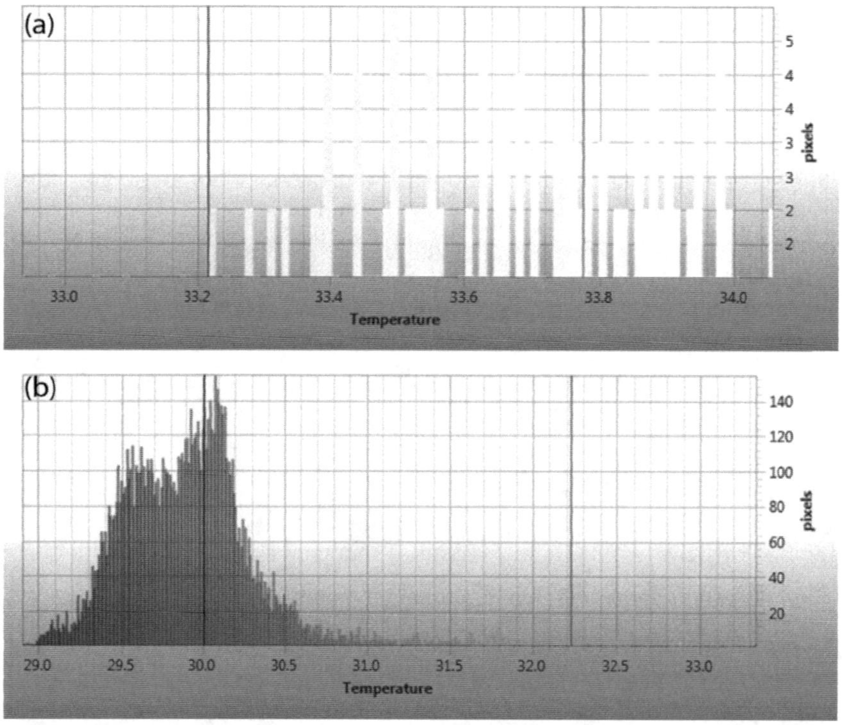

Figure 5. Histograms at 34 °C for samples (a) 1, $d = 1$ cm and (b) 6, $d = 10$ cm.

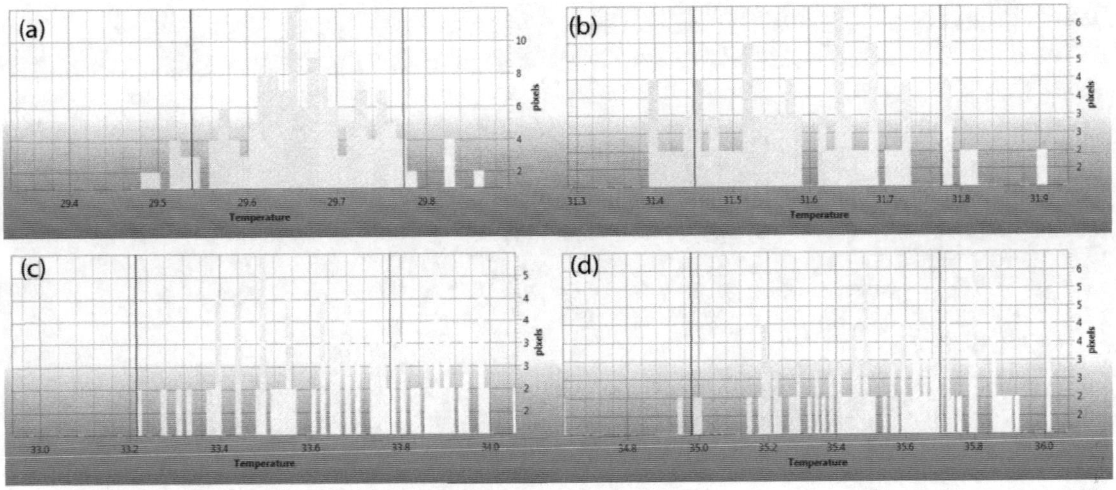

Figure 6. Histograms for sample 1: (a) at 30 °C, (b) 32 °C, (c) 34 °C, and (d) 36 °C.

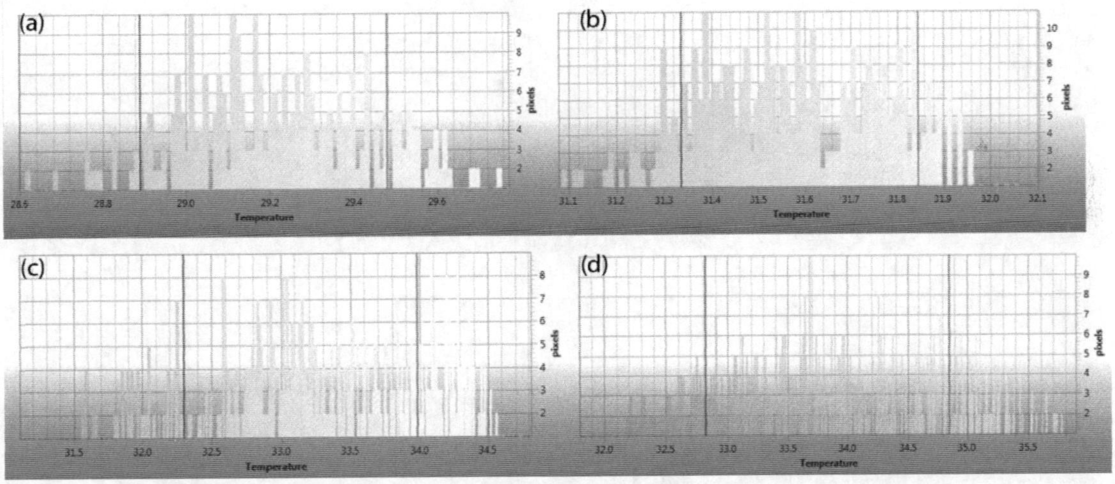

Figure 7. Histograms for sample 2: (a) at 30 °C, (b) 32 °C, (c) 34 °C, and (d) 36 °C.

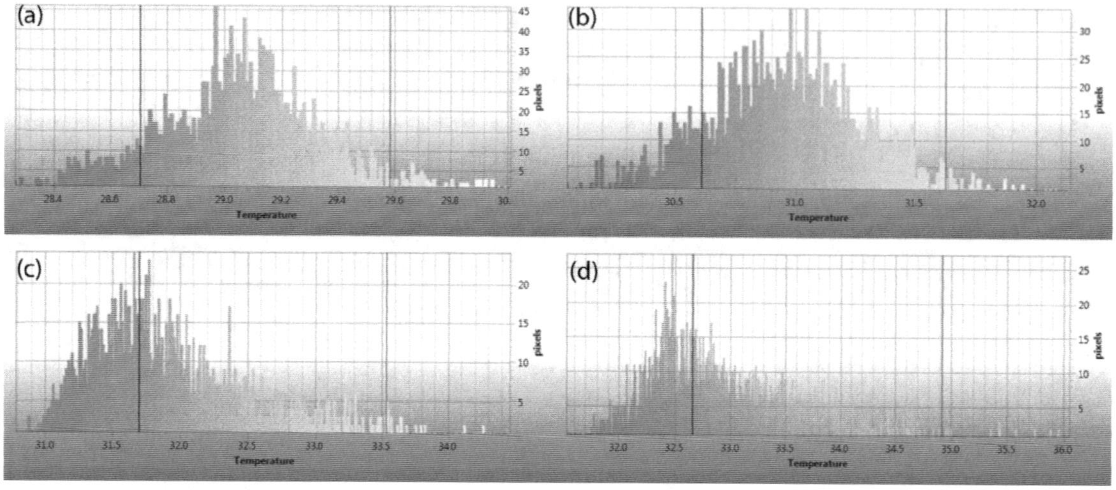

Figure 8. Histograms for sample 4: (a) at 30 °C, (b) 32 °C, (c) 34 °C, and (d) 36 °C.

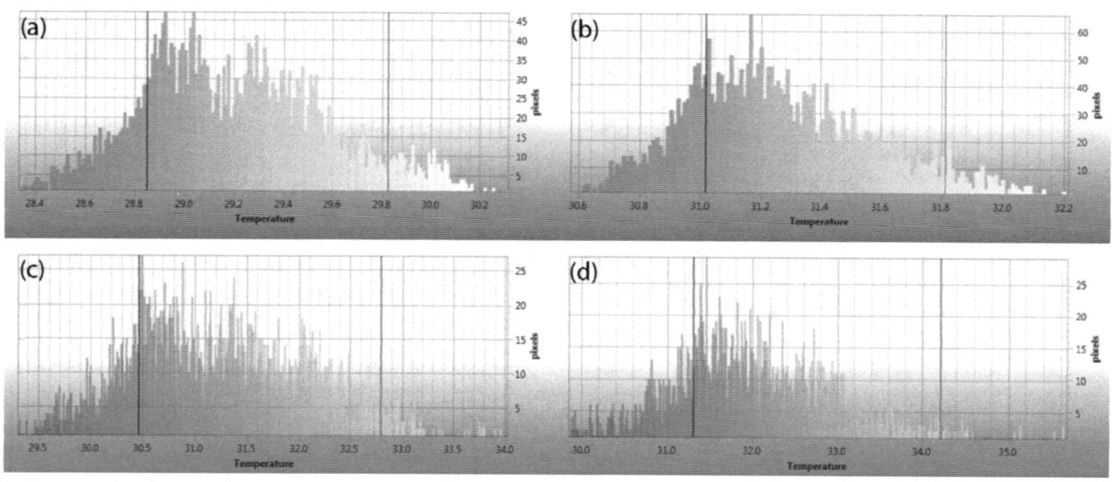

Figure 9. Histograms for sample 5: (a) at 30 °C, (b) 32 °C, (c) 34 °C, and (d) 36 °C.

References

[1] Zhongping L and Shiming D 2008 *Build Environ* **43** 70-81.
[2] Kurz A 2008 *Best Pract Res Clin Anaesthesiol* **22(1)** 39-62.
[3] Kurz A, Sessler DI and Lenhardt R 1996 *N Engl J Med* **334(19)** 1209-15.
[4] Teodorczyk JE, Heijmans JH, van Mook W, Bergmans D and Roekaerts P 2012 *Open J Anesthesiol* **2(3)** 65-9.
[5] Allen TK and Habib AS 2018 *Anesth Analg.* **126(1)** 7-9.
[6] Buggy DJ and Crossley AWA 2000 *Br J Anaesth* **84(5)** 615-28.
[7] Ud-Din S and Bayat A 2016 *Exp Dermatol* **25** 579-85.

[8] Gurtner GC, Werner S, Barrandon Y and Longaker MT 2008 *Nature* **453** 314-321.
[9] Rivera AE and Spencer JM 2007 *Clin Dermatol* **25(1)** 39-48.
[10] Houghton VJ, Bower VM and Chant DC 2013 *J Foot Ankle Res* **6(31)**
[11] Chanmugam A, Langemo D, Thomason K, Haan J, Altenburger EA, Tippett A, Henderson L and Zortman TA 2017 *Adv Skin Wound Care* **30(9)** 406-414
[12] Romanovsky AA 2014 *Acta Physiol* **210** 498–507
[13] Zhang X 2001 *Smart Fibres, Fabrics and Clothing* vol 1, ed Tao X (Elsevier) 34-57
[14] Thomas RT, Diebold MR and Eggemeyer LM 2005 *J Am Med Dir Assoc* **6(1)** 46-9.
[15] Alvarez OM, Rogers RS, Booker JG and Patel M 2003 *J Foot Ankle Surg* **42(1)** 30-5.
[16] Whitney JD, Salvadalena G, Higa L and Mich M 2001 *J Wound Ostomy Continence Nurs* **28(5)** 244-52
[17] Altura D 2003 *US Patent No. US 6,613,953 B1.*
[18] Matusiak M 2006 *Fibers Text East Eur* **14(5)** 98-102

Contribution of Azerbaijan to the energy security of the European Union

S I Azakov

Baku Higher Oil School, Petroleum Engineering Department, Khojali Avenue 30, AZ 1025, Baku, Azerbaijan,
E-mail address: azakov_s@hotmail.com

Abstract. Despite all progress connected with use of renewable energy sources it is widely known that fossil fuels will remain dominating in short - and medium-term prospects. Presently fossil fuels are made more, than three quarters of a global power mix, and in the next decade the level of their use will probably be the same. Among them there are crude oil and natural gas as widely used and globally sold energy resources which reserves unevenly distributed, are responsible to a large extent for energy security in the world. Therefore the relations between the European Union, one of the largest oil and gas consumers, and Azerbaijan, the supplier of energy whose location isn't really far from the EU attract special attention of many politicians and creators of public opinion in Europe. The Southern Gas Corridor which ceremony of start of the first stage was held on May 29, 1918 in Baku is one of priority projects for the EU, and provides transportation of 10 billion cubic meters of the Azerbaijani gas from the Caspian region through Georgia and Turkey to Europe by 2020 will be the first real diversification of gas coming in from the East. At an initial stage, the gas extracted within the second phase of development of the Azerbaijani gas-condensate field "Shah Deniz" is considered as the main source for the corridor. At later stage other sources can be attracted to the project. Gas within the second stage of Shah Deniz field will be exported to Turkey and to the European markets by expansion of the South Caucasian gas pipeline and construction of Trans-Anatolian (TANAP) and Trans-Adriatic (TAP) pipelines. The project involves two EU member states, Italy and Greece, candidate countries, Turkey and Albania, and Azerbaijan, the EU strategic partner in energy. Four different types of countries were involved: member states, candidate countries, associated partners and future strategic partners. The EU supported Southern Gas Corridor is directed to curtail Russian energy leverage over Eastern and Southeastern European countries which are heavily reliant on Russian gas deliveries. It opens up the direct access for the land locked Caspian states, primarily Turkmenistan and Kazakhstan, to the European gas markets. This article describes development of the energy relations of the EU and the Republic of Azerbaijan. The energy needs and potential of these two partners are investigated and their main interests are defined. The main problems, opportunities and obstacles for power cooperation of the EU and Azerbaijan are discussed, and views of development of their principles of energy policy in the near future are considered.

1. Introduction

Unsecured by a sufficient amount of energy resources, the European Union (EU) countries are forced to look for sources of their imports and establish relations with supplier countries. This is manifested not only in the framework of economic cooperation, but also becomes the basis for a political dialogue with partners.

The issue of energy security in Europe became especially acute after the first gas crisis in relations between Russia and Ukraine in 2006. Although before that, European politicians believed that in the event of a conflict between states and supply disruptions, the EU countries might suffer, but after these events the issue of energy security has acquired political significance, becoming "special." The EU identified the problem of dependence on Russia as a challenge, seeking to diversify energy sources and liberalize the gas market.

The main problem of energy security in European countries, as the European Commission (EC) stresses, is the EU's dependence on energy imports. More specifically, the EU's dependence on energy imports from Russia, while the EC views Russia as a country that tends to use its energy resources as a political tool.

After the gas crises of 2006 and 2009, much has been done to strengthen energy security, but the EU still feels vulnerable. In 2014, the EU adopted the Energy Security Strategy [1], aimed, in particular, at reducing dependence on "certain types of fuel, suppliers and routes." The main tasks to be fulfilled in order to achieve the goal set in the document are to increase domestic production capacities, enhance energy efficiency, develop renewable energy sources (RES), and diversify suppliers and infrastructure.

The development of renewable energy is particularly relevant in the context of the dominance of the trend to reduce the use of nuclear power after the accident at the Japanese nuclear power plant Fukushima-1, and also because of the impossibility of using large volumes of coal, as this contradicts the decarbonization policy that became actual after the entry into force of the Paris Agreement on climate in 2016.

The share of renewable energy in the energy mix of Europe is indeed growing and the EU's success in this is undeniable, but they still cannot completely replace traditional energy resources, which is the main problem of non-traditional energy.

Another way to reduce dependence on Russia, the EU sees in diversification of gas suppliers, both on new pipelines and liquefied gas.

As for such gas pipelines, the most attractive and economically profitable for Europe would be a gas pipeline from the Caspian region (Azerbaijan, Turkmenistan, Kazakhstan, Uzbekistan and Iran) rich in hydrocarbon resources.

As for liquefied natural gas (LNG), Europe primarily relies on LNG imports from the US. The EU actually received the first LNG shipments in 2016, but gas volumes are still insignificant.

In addition, Europe expected that Qatar will increase exports to the EU. By 2020, Qatar plans to increase the volume of produced gas by 20 billion cubic meters per year. However, gas consumption is increasing in Pakistan, the UAE, and also in Qatar itself. Therefore, even if production volumes really increase, this does not mean that they will all go to the European market.

In addition to the United States and Qatar, the EU viewed the countries of North Africa as a tool to reduce dependence on Russia. It is obvious that from Libya, where the war goes for years, gas will not be supplied in significant volumes. The lack of investment in the gas industry of Algeria also in the short term will not allow a serious increase in the extraction and export of "blue fuel". In addition, the growing demand for gas within the country itself also plays a role. According to forecasts, it will exceed the increase in production up to 2040. Hopes for gas supplies from Egypt also did not materialize, because domestic consumption increased in the country.

The EU makes maximum efforts and takes various measures to ensure energy security, but it faces a number of problems, primarily the lack of a coordinated energy policy within the EU. The replacement of Russian gas with alternative energy is also not yet realized. American LNG due to high prices in comparison with Russian gas and still insufficient volumes is not able to displace Russian gas from the European market. Algeria, Qatar, Iran, Libya and Egypt for a number of reasons cannot become reliable suppliers of gas for the EU.

2. What is energy security and why it is so important for any country?

The term "energy security" emerged in the 1970s when a number of oil-producing countries of the Middle East imposed an embargo on the supply of oil to Europe during the escalation of the Arab-Israeli conflict, and as a result, its price increased threefold.

In modern terminology, energy security in the broad sense means a balance between the demand and the supply of energy raw materials. It is the belief that energy will be available in the quantity and quality that are required under given economic conditions.

At the same time, the adequacy of the supply of fuel and energy resources and moderation of their demand are taken into account. And all this happens with the depletion of relatively cheap fossil fuels in most regions of the world, the need to develop their more expensive reserves, on the one hand, and a significant increase in the world economy's demand for primary energy resources (faster in developing countries), on the other.

The complexity, scale and depth of the energy problems of the modern world put energy security in the list of the most important components of national security.

In order to ensure energy security, a specific policy of the state and the implementation of special measures are required: joint use of all regulatory mechanisms is necessary. Energy security is the security of supply and the reliability of transit, control over pipelines, the rejection of energy blackmail, the avoidance of sharp jumps in prices, and so on.

Energy security is associated with the energy independence of the state, its subject or its region; it is in the first place, a political problem. The sustainable development and stability of energy producing states (Azerbaijan among them) is of great importance for ensuring economic energy security. Such stability may be under threat in connection with the need to conduct exploration and production of energy resources in increasingly difficult conditions, which is possible only with the use of ultra-modern technologies, causing damage to the environment and entailing difficulties in transportation.

Energy security can be defined as:

- the energy independence of the state;
- a characteristic of the country's heat and power complex;
- a state of security of civilians, society and the state against the threat of deficit of energy and fuel and resources;
- a state of society and economy, which allows to maintain the necessary level of energy consumption;
- a set of conditions for which there is no energy deficit;
- a means of economic and political influence, etc.

Today, the problem of energy security acquires a special relevance, which is due, on the one hand, to the depletion of proven reserves of natural energy resources, and on the other, to a constant increase in fuel consumption and various types of energy.

The EU which includes the most developed economies of the world is one of the leading consumers of energy resources. Its lucrative market remains attractive to many energy exporters, especially those who depend on energy- born revenues. In 2015 the EU consumed 1 629.5mln t of oil equivalent, which makes up 11.9% of the global energy mix [2].

For the EU, the issue of energy security is one of the most important ones because it has very limited access to oil and gas reserves, has to secure large energy supplies from different sources at reasonable and stable prices in order to sustain economic performance and growth.

The issue of energy security is also quite acute for Azerbaijan. This is due to the resource orientation of the country's economy, which naturally requires minimizing various kinds of political, economic, military and technological risks in the sale, transit and delivery of oil and gas to the consumer.

For Azerbaijan, intensification of energy contacts with the EU gives an opportunity not only to get hard currency, but also modernize its economy.

3. Why natural gas is best as a fuel and a source of energy used today (although there are other energy sources used: coal, oil, hydro, renewable energy sources)?

Currently, natural gas is considered as one of the main types of fuel, after oil products. Its main "plus" is that this substance can be called almost environmentally friendly. Natural gas is up to 50 percent cleaner than other fossil fuels [3]. Compared to average emissions from coal-fired power plants, natural gas produced less than a third as much nitrogen oxides and 1 percent as much sulphur oxide [4].

Compared to production of energy or heat from coal or oil, natural gas does not only reduce greenhouse gas emissions – it hugely improves air quality, reducing smog and dangerous particulate matter with very real benefits for human health [5].

Nowadays it is very important, especially when the problem of ecology and pollution of the environment is very actual. For heating and cooking, gas is used much more often than solid fuels and all because its combustion does not form smoke and does not leave ash behind.

Basically, the fossil gas consists of gaseous methane. It also includes such gases as propane, ethane and butane. The concentration and amount of all these substances in the formula of natural gas depend on the place of its extraction.

Today the market of gas fuel is very wide. A lot of countries in the world appreciated the advantages of using "blue" fuel. It is used not only as fuel; it is used in household and chemical industry. In addition, during combustion, gas produces virtually no harmful substances, so its advantages, compared to other types of fuel, are significant. Gas is much easier to ignite and it is more convenient to control the combustion process, for example, with coal and other solid fuels, this is more difficult to do.

Modern life of a person is very difficult to imagine without the use of natural gas. Due to its environmentally friendly composition, as a source of energy, gas is used much more often than other substances. Rich deposits allow us to use these minerals for a long time.

While the advanced countries of the world are developing all new ways of obtaining "clean energy", most of the world still consumes such energy sources as firewood, coal and oil. Even in developed countries, coal still retains its positions.

Therefore, on a global scale, the transition to gas fuel is considered a major step forward due to its energy efficiency and environmental purity.

The International Energy Agency (IEA) predicts that by 2040 the demand for energy resources in the world will grow by 30% [6].

Gas is one of the few types of energy that can be used in all sectors of the economy, be it transport, housing or manufacturing. The world's gas reserves are huge and located in many regions.

The IEA estimates that technically recoverable natural gas reserves in the world will last at least 220 years at the current production level [7].

Natural gas will continue to increase its share in the global energy mix, growing at 2 percent a year until 2020 [8].

It has an important role to play in decarbonisation of Europe's energy systems -countries such as Canada have successfully reduced overall greenhouse gas emissions [9] while increasing energy use, through inclusion of more natural gas in their energy mix.

4. How much natural gas does the European Union produce and consume?

The EU possesses less than one per cent of global crude oil and natural gas proven reserves. The discovery of the hydrocarbon offshore fields under the North Sea in the 1970s has led to the development of gas markets in the United Kingdom, Netherlands and Denmark. These countries are still the most important EU producers; however, their reserves are expected to be finished soon. Except for the recent discoveries of offshore fields in Cyprus, the other EU member states have a very narrow gas production base.

Own gas production in the EU is steadily falling by an average of 3% per year and in 2017 amounted to 128 bcm [10].

Reduction of production in the Dutch Groningen by 10% and the closure of the largest gas storage Rough in the UK put serious pressure on the market and lead to price volatility. There is depletion of gas deposits in the North Sea.

And, despite the fact that Norway is currently at the peak of production, its reserves of reserves are only 1900 bcm. Obviously, with the level of today's production, which is about 120 bcm per year, Norwegian resources are rapidly depleted, or production will begin to decline and, therefore, exports.

5. How much gas does Europe import, in what form and from what sources? What is the forecast for imports in the future until 2035?

The import of natural gas by the EU countries in 2017 amounted to about 360 bcm. Russia retains leadership in supplying gas to Europe. So, in 2017, Russia delivered to the European market 194.4 bcm of gas, and the US - less than 2 million tons of LNG (about 3 bcm). Now the market share of Gazprom in Europe is about 35%, the share of American LNG is less than 1%.

Last several years the EU imported about 350-400 bcm per year. Russia remains the EU main supplier of natural gas, providing 30-40 % of all Union imports (in 2012: 25%). According to Brussels, the EU's overall demand projected for 2030 will remain more or less the same as today, and will amount to 380-450 bcm per year [11].

To ensure reliable gas supply, Gazprom, Wintershall, E.ON, Gasunie and GDF Suez (now "Engie") have built the Nord Stream pipeline, with a capacity of up to 55 bcm per year.

Norway with 33% share is in the second place among the exporters of gas to the EU. This year, the Norwegian Petroleum Directorate forecasts production in the country at 114.5 bcm, which is 2% less than in 2016.

Deliveries of pipeline gas from Algeria declined in the second quarter of 2017 by 27% after two years of growth. The import of Algerian gas to Italy fell by about 32%.

Along with the supply of natural gas through pipelines, from 2015 LNG shipments by tankers also play an important role. The main suppliers of LNG to Europe are Qatar, Nigeria and Algeria. However, the US also began to supply LNG to Europe.

Currently, LNG supplies to the EU countries are declining due to higher prices for natural gas in the markets of Asia or South America.

In the future, first of all, Great Britain, France, Spain and Italy will increase the import of LNG.

6. Gas Potential of Azerbaijan

The main volumes of gaseous fuel of Azerbaijan are concentrated in the giant gas-gas condensate field "Shah Deniz". Located in the southern part of the Caspian Sea, 70 km to the south-east of Baku (in water depths ranging from 50 to 500 m), the hydrocarbon structure of Shah Deniz was discovered by Azerbaijani geologists in 1954.

After Azerbaijan gained independence, it was decided to attract foreign investors to its development. The contract for the Shah Deniz field was signed in Baku on June 4, 1996 and ratified by the Milli Mejlis (Azerbaijani Parliament) on October 17, 1996.

The contract for its development was the second after the contract for the development of the Azeri-Chirag-Guneshli (ACG) oil field, which is large in terms of financing and politically important PSA signed by the government of the country with foreign enterprises (valid from 1996 to 2031).

In 1997, three-dimensional geophysical studies were carried out on the structure. Then for 18 months and 145 million US dollars, paid by the project participants, the floating semi-submersible drilling rig "Shelf-5" was upgraded, later renamed into "Istiglal", which started drilling on the contract area.

In 1999, the first well SDX-1 was drilled in the contract area. As a result, the largest gas condensate field with carbon stocks estimated at a first approximation of 1 trillion tons of oil equivalent was found on Shah Deniz (1.7 trillion cubic meters of natural gas and 400 million cubic meters of gas condensate).

In April 2000, the 2nd SDX-2 well was drilled, which clarified the hydrocarbon potential of Shah Deniz. In 2001 SDX-3 was drilled, which turned out to be empty, but allowed to clarify the contours of the field.

At present, the structure of the group's capital developing the Shah Deniz field is as follows: the British BP (28.8% operator), AzSD (10.0%), SGC Upstream (6.7%) (subsidiaries of SOCAR), the Malaysian Petronas (15.5%), the Russian LUCOIL (10.0%), the Iranian NICO (10.0%), the Turkish TPAO (19.0%).

Development of the Shah Deniz field reserves is carried out in three stages. At the first stage (Shah Deniz-1), a sea platform was built, with which it was planned to drill 15 wells in the course of the work, with a deviation from the vertical. Industrial production of hydrocarbons began in 2006. Within the first stage of development, i.e., until 2031, it is planned to extract 178 billion cubic meters (bcm) of gas and 34 million tons of gas condensate [12].

In 2010 - 2016 years gas production in Azerbaijan increased by more than 50% on average over the year and was characterized by the following indicators (in bcm): 2010 - 6.9, 2011 - 6.67, 2012 - 7, 73, 2013 - 9.8, 2014 - 9.9, 2015 - 9.9, 2016 - 10.7. In the future, before the implementation of the second stage of development of the field (Shah Deniz-2), the concern intends to keep production at the level of at least 9 bcm per year to ensure production and export plans.

In 2017, about 10.2 bcm of gas were produced from the Shah Deniz field. The maximum capacity of the mining system reached 29.5 million cubic meters (mcm) per day. As of the end of January 2018 about 88.5 bcm of gas and about 22 million tons of condensate were produced from the Shah Deniz field.

Under the project, the Sangachal terminal, which is located about 45 km south of Baku, forms the central point for the collection, processing and export of all gas produced at the Shah Deniz field, in addition to the oil produced at the ACG. In the period of maximum production, the total throughput of the terminal is expected to be about 1 million barrels of oil and more than 28.3 mcm of gas per day.

To transport gas from Shah Deniz to Turkey in 2005-2006, the South Caucasus Pipeline (SCP) was laid with a length of 690 km, between the Sangachal terminal and the Turkish Erzurum. The project cost of all works on the Baku-Tbilisi-Erzurum route was about one billion dollars.

The maximum capacity of the pipeline is 20 bcm gas per year. After carrying out all the works in December 2006, the first gas from the Shah Deniz field was obtained. In July 2007, gas reached the territory of Turkey.

With an optimistic scenario, the maintenance of production at the level of 9 billion cubic meters per year is possible in the period until 2021-2022.

In the second stage (Shah Deniz-2, cost $ 28 billion), it is planned to implement a set of measures for the construction of two offshore platforms that support the drilling of 26 subsea wells, the expansion of the offshore pipeline network and the capacity of the Sangachal terminal.

It is assumed that within the framework of Shah Deniz -2, by 2020, gas production will increase to 16 bcm a year, of which 10 bcm will be exported to the EU, 6 bcm to Turkey. The specified volume of production is planned to be kept for about 10 years - until 2028-2030 [13].

In order to preserve the competitiveness of the national gas sector in the long term, the government of Azerbaijan and the participants of the relevant concern, headed by BP, are considering the possibility of developing and implementing the third stage of development of this structure (Shah Deniz-3). By 2020, the decision will be made. Currently, the results of seismic survey conducted using the 3D technique are being studied. Based on the results of this analysis, it will be decided whether to conduct exploratory drilling.

Thus, in the next eight years in Azerbaijan, with an optimistic development of events, average annual gas production can increase to 34-35 billion cubic meters. In the second half of the next decade, this indicator is seen to decrease by 5-10% per year.

Almost all of the gas that will be produced in the second stage of development of Shah Deniz is contracted for deliveries to the countries of the EU.

In the country, gas production is also carried out on the block ACG and some small structures. At the ACG, part of the extracted gas is pumped back into the reservoir to maintain working pressure, the rest of the fuel is sent to the Azerbaijani side on a no-cost basis. Until 2030, it is possible to organize the production of gaseous fuels in the deep water areas of ACG. According to some information, gas deposits there may amount to 200-300 billion cubic meters.

The production of gaseous fuels is associated with the implementation of costly measures to permanently stabilize recovery volumes due to the geological features of the Caspian offshore fields. An important task of the government of Azerbaijan and the consortium Azerbaijan International Operating Company (AIOC) is to keep average annual production at the level of 8.5-9 bcm of gas in the period until 2019-2020, i.e. before the commencement of commercial fuel production in the second phase of works (Shah Deniz-2) in order to ensure the export and domestic needs of the country.

In addition to the main Shah Deniz gas-gas-condensate field, Azerbaijan also has other gas deposits in the Caspian.

One recent discovery, announced in 2010, is the Umid gas field in the south Caspian Sea, estimated to contain 200 bcm of gas and 40 million tons of condensate; the country's state energy company, SOCAR, has been producing gas there since 2012, and has recently finished drilling the third well. Also of great promise is the adjacent Babak field, yet to be developed, with potential gas reserves of 400 bcm, plus 80 million tons of condensate. Further off the coast, the Shafag-Asiman field holds an additional 300 bcm of gas, which Azerbaijan plans to extract in an equal-share partnership with BP. (This field may contain oil too, according to a seismic survey.) And another offshore deposit, Nakhichevan, is being jointly explored with the German company RWE.

The difference-maker for Azerbaijan's ambitions as a gas exporter is the Absheron field, discovered in 2011. This field's development, currently in its first phase, could greatly boost potential supplies (its reserves are estimated at 350 bcm of gas and 45 million tons of condensate).

Certain gas deposits are located on the southern shelf sites located in disputed areas, which are very likely to pass to Azerbaijan after the delimitation of the marine area.

All in all, according to SOCAR's strategy document, these steps could expand Azerbaijan's gas export volumes to a massive 40 bcm per year.

7. The Southern Gas Corridor project

In order to ensure its energy security the EU adapted a diversification strategy which was laid down in the 2008 Second Strategic Energy Review of the EC [14]. It envisions opening up a Southern Gas Corridor (SGC); with the Shah Deniz gas field is in the forefront.

The history of SGC started, it can be said, in 2006 with the beginning of the development of the Shah Deniz field by Azerbaijan.

In 2004, the EC launched the Baku Initiative, a policy discussion with Caspian states. In 2006, an EC energy policy document foresaw Black Sea and Caspian countries working with the EU to "boost new [energy] supplies from central Asia to the EU and further projects from the Caspian basin" [15]. In 2011, a detailed communiqué on external energy policy described the SGC as a "key infrastructure priority" and "a supply route for roughly 10-20 per cent of the EU estimated gas demand by 2020" [16].

In 2013, a consortium of companies developing the Shah Deniz -2 field finally approved the route of the SGC. The project eventually included the already existing South Caucasus pipeline (Baku-Tbilisi-Erzurum) and the new ones – Trans-Anatolian (TANAP) and Trans Adriatic (TAP). In the same year, the EC for 25 years withdrew the TAP from the action of the Third Energy Package, which required providing the capacity of the pipeline to third suppliers.

The foundation of the gas pipeline was laid in 2015 by the presidents of Turkey, Azerbaijan and Georgia, and the construction work itself began in 2016. The gas corridor will stretch almost 3.5 thousand kilometers from Baku to the south of Italy through Georgia, Turkey, Greece and Albania and will be used for transportation of "blue fuel" from Azerbaijan to the Balkans and the Apennines. In this regard, the project plans to install interconnectors (two-way highways) that connect to South Caucasus, which will in particular allow gas to be supplied to Bulgaria.

According to the SGC 16 billion cubic meters of gas per year should be transported. Of these, 6 billion are due to Turkey, the remaining 10 billion will go further, to Southern Europe. According to preliminary calculations, 8 billion cubic meters per year will be exported to Italy; 1 billion will be received by Greece and Bulgaria.

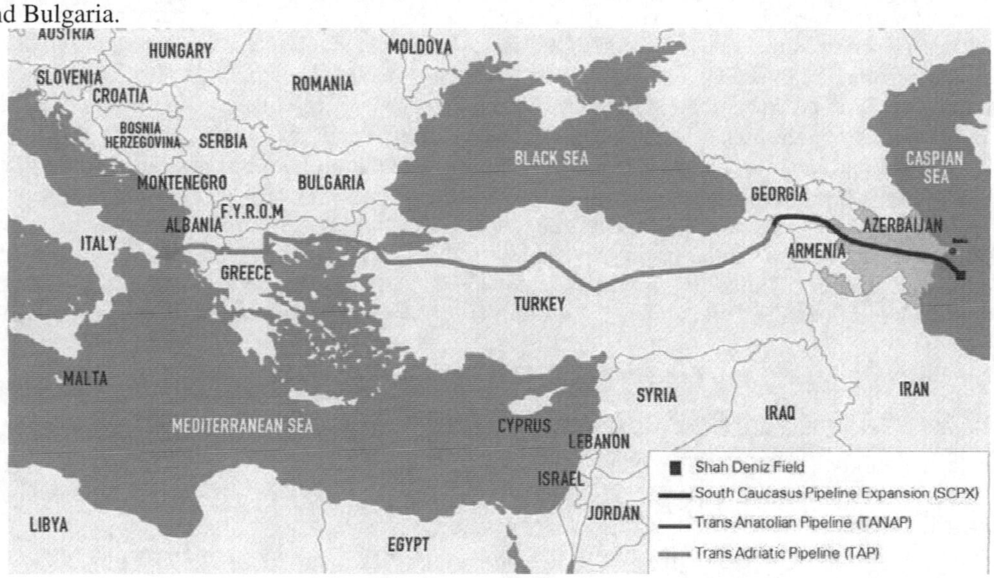

Figure 1.Southen Gas Corridor

IOP Conf. Series: Materials Science and Engineering **459** (2019) 012011 doi:10.1088/1757-899X/459/1/012011

The costs for the implementation of the SGC project are periodically reassessed. To date (August 2018), the SGC project is estimated at about $ 45 billion.

Azerbaijan's investment contribution to the SGC is estimated by experts at about $ 10.5 billion. This amount was calculated on the basis of the addition of shares of SOCAR in various constituent parts of the SGC.

SOCAR owns 16.7% of Shah Deniz-2 gas field, the raw material base of the future pipeline, 58% in TANAP and 20% in TAP gas pipelines. The total amount of investments in the development of the second stage of Shah Deniz (together with the expansion of the South Caucasus pipeline Baku-Tbilisi-Erzurum is $ 23 billion, the cost of TANAP is estimated at $ 9.3 billion, TAP -$ 6 billion

Not everything, of course, went smoothly, but the Azerbaijani project was necessary for Europe and after all the changes in November 2017, the EC approved an updated list of "Projects of Common Interest" in which the SGC project was left.

On June 12, 2018, the opening of the TANAP gas pipeline was held in the Turkish city of Eskisehir. Presidents of Azerbaijan Ilham Aliyev, Turkey Recep Tayyip Erdogan, Ukraine Petro Poroshenko, Serbia Alexander Vučić, Turkish Republic of Northern Cyprus Mustafa Akynchi, as well as the Prime Minister of Bulgaria Boyko Borisov, BP General Manager Robert Dudley, the energy ministers of a number of countries, heads of energy companies took part in the ceremony.

The foundation of this project was laid in Kars on March 17, 2015 with the participation of the presidents of Azerbaijan, Turkey and Georgia.

Prior to the opening ceremony of TANAP on May 29 of this year, the Sangachal terminal officially opened the SGC.

TANAP, which is one of the branches of the SGC, runs through Turkey. On the Turkish-Greek border, it will connect to the TAP, which provides for the transportation of Azerbaijani gas to Europe.

The transit of natural gas through Turkey via TANAP from 2019 on will be underpinned by an intergovernmental agreement (IGA) between Azerbaijan and Turkey [17].The pipeline will be governed by the provisions of the Energy Charter Treaty of 1991, which does not include a third party access regime and unbundling. This is why it is possible for SOCAR to hold 51% of TANAP and an additional 7% through its Turkish unit SOCAR Turkey Energy and to also operate the pipeline. The IGA between Azerbaijan and Turkey even states that “The States agree that the participating interest of state entities owned by the Republic of Azerbaijan in the TANAP Project Entity shall not be less than 51% (fifty one per cent) of the total participating interest"[18]. Under the EU regulation it would not be possible that a production company such as SOCAR operates pipeline infrastructure.

With its majority share and due to the fact that it operates TANAP, SOCAR thus controls the gas transit through Turkey. Therefore, Azerbaijan is in a position to allow or refuse the feed-in of additional gas quantities from other sources. This is of importance as the EU in particular speculates on the feed-in of gas from additional sources such as Turkmenistan, Iraq and Iran. Moreover, Azerbaijan may set the transit fees.

In Turkey, TANAP will have two exit points - in Eskishehir and Trakia provinces. The sale of gas in Turkey will be carried out by BOTAS company. The total length of the gas pipeline is 1850 km.

The pipe diameter of the gas pipeline section from the Georgian-Turkish border to Eskisehir (Phase-0) is 56 inches, the length is 1340 km. Hence the gas will be supplied to Turkey. The diameter of the Eskishehir-Edirne-Europe section (Phase-1) is 48 inches, the length is 476 km. Phase-1 will be ready to receive gas for Europe in 2019.

This branch is the longest and largest in terms of the volume of pipelines built up in Turkey so far.

By realizing this project, Azerbaijan has once again proved its commitment to security, ensuring mutually beneficial economic cooperation throughout the world. TANAP is a step against hegemony and monopoly in the energy market. For many years the West has been carrying out activities to diversify energy supplies. Implementations of this idea are the Baku-Tbilisi-Ceyhan and TANAP.

TANAP demonstrates to the world the importance of the tripartite partnership between Azerbaijan, Georgia and Turkey. This unity reaffirms that regional cooperation can bring huge benefits to security and economic relations in a broad format. The regional partnership can help to peacefully resolve conflicts in the region.

Commercial gas supplies from Phase 2 of the Shah Deniz field development to Turkey under TANAP began on June 30, 2018.

The cost of the project is estimated at $ 7.99 billion. In the future, the capacity of the pipe will be brought to 24 billion cubic meters of gas per year, and then to 31 billion cubic meters.

The shareholders of the TANAP project are SOCAR (51%), SOCAR Turkiye Enerji (7%), BOTAS (30%), BP (12%).

Until 2026, it is possible to increase the capacity of the gas pipeline to 31 billion cubic meters of gas per year. At subsequent stages, the issue of delivering gas to Turkmenistan from the EU, Turkmenistan, Iran and Iraq and the Eastern Mediterranean via the pipeline may be considered.

TAP approximately 878 km long pipeline ships the gas coming through TANAP from the Turkish-Greek border at Kipoi cross Greece and Albania and the Adriatic Sea, before coming ashore in Southern Italy. TAP's shareholding is comprised of BP (20%), SOCAR (20%), Snam (20%), Fluxys (19%), Enagás (16%) and Axpo (5%).

TAP is based on a 2013 intergovernmental agreement between Albania, Italy and Greece [19]. As Italy and Greece are EU member states and Albania is Energy Community member, the EU's internal market regulation applies to TAP. In 2013, TAP secured a third party access exemption for half of the initial capacity of 20 billion cubic meters annually for gas volumes from Azerbaijan supplied under the relevant Shah Deniz gas sales agreements over a period of 25 years. The third party access exemption was granted under Article 36 of the Third Gas Directive.

TAP will be connected to the Italian natural gas grid, operated by Snam Rete Gas, allowing for gas to reach other European countries through Italian "exit points." Gas flowing through TAP can also reach Austria and other Central European countries through the gas hub in Baumgarten, Austria through the Trans Austria Gas (TAG) pipeline.

For the transport of natural gas northwards the Ionian-Adriatic Pipeline (IAP) is supposed to be built. It would run from Fier in Albania through Montenegro, and Bosnia and Herzegovina, to Split in Croatia. In Fier, IAP would be connected with the planned TAP. IAP is based on a Ministerial Declaration between Ministers from Albania, Croatia and Montenegro [20] and was struck within the framework of the Energy Community. The stipulations of the EU's internal market regulations would fully apply.

Both TAP and IAP have received funding or special status from the EU.

Commissioning of the TAP pipeline is expected in 2020. As of the end of June 2018, 74 per cent of works were completed for the TAP project, including engineering, procurement and construction work [21].

SGC has become a reality; its implementation cannot be stopped. Azerbaijan has succeeded in explaining to Europe the advantages and benefits that this grand pipeline gives to it, competitive and at the same time not competing with anyone.

For Azerbaijan and Turkey, TANAP, as well as its continuation - TAP, is a purely commercial project, the task of which is to deliver gas from the Caspian fields to the European consumer. In Azerbaijan they do not try to "displace" anyone from the market and in general use pipelines as a means of pressure or blackmail.

In August 2018 Bulgaria joined the project. ICGB AD, which is the operator of the Greece-Bulgaria interconnect project (IGB), and the consortium for the construction of TAP have reached an agreement on the interconnection of pipelines. IGB is a gas pipeline that will allow Bulgaria to receive Azerbaijani gas, in particular, extracted from the second stage of the Shah Deniz gas condensate field. The IGB will be connected to the TAP [22].

8. US interests

Of particular interest is the interest shown to this Azerbaijani project by the United States and personally by President Donald Trump. As S & P Global Platts informed in August 7, 2018 with reference to Trump's appropriate order, new US sanctions against Iran will not affect the SGC project. The phrase "exclusion of the natural gas project", which means the SGC, is mentioned at the disposal [23].

As it is known, new US sanctions on the energy sector, which will come into force on November 4, 2018, will affect companies that continue to conduct business with Iran. In this regard, BP wanted to get permission for the nonproliferation of sanctions for the Shah Deniz field development project, whose gas will go through the SGC. The matter is that (as noted earlier), the Iranian NICO has a 10% stake in the project of the second stage of development of the Shah Deniz field.

With this unprecedented decision, the US administration stressed the importance of the Azerbaijani project and the importance of Azerbaijan as a guarantor of Europe's energy security. The US President demonstrated that this project and cooperation with Azerbaijan is much more important for him than anti-Iranian sanctions.

The fact that anti-Iranian sanctions were not extended to the SGC project is very important from the point of view of a correct assessment of the existing geostrategic reality and taking into account the interests of Europe in energy security.

The US president supports the Azerbaijani project in any way, which in fact, does not seem to affect the US interests and, on the contrary, even to some extent contradicting the American plans for deliveries of overseas LNG to Europe. Nevertheless, a global view of things makes it clear that Azerbaijan, with its capabilities, is becoming a very serious player in the energy market. Moreover, it is very profitable to conduct business with our country, since its projects are not directed against anyone's interests, but, on the contrary, integrate the interests of the whole region and are always open to expanding this space.

Accepting the Greek Prime Minister Alexis Tsipras in Washington in October 2017, Trump thanked Athens for "contributing to the European energy security through the support of the Trans Adriatic Pipeline," hinting at the US interest in this project. And in August 2018 at a press conference with Italian Prime Minister Giuseppe Conte the Head of White House made a very remarkable statement: "Mr. Prime Minister, I hope that you will be able to build this competitive pipeline."

It should be noted that certain circles of Italy at times are at the peak of the project, declaring its alleged threats to the environment, tourism and so on. These vain fears were always refuted by specialists with figures and facts. Despite the unexpected announcement by the new government of Italy of the intention to "reconsider" its participation in the TAP construction project, during the visit to Baku, the President of this country, Sergio Mattarella, expressed full support for the TAP project and noted the importance of the Caspian region and Caspian gas for Europe [24].

In early August 2018, Premier Giuseppe Conte met with the mayor of Melendugno, through which the pipeline will pass. During the meeting, Conte reminded the mayor that the diversification of energy supplies to Italy is a strategic goal of the country [25].

It seems that after the start of the construction of the Italian site, fears of ecologists and other opponents of the TAP will disperse. And the unequivocal interest shown to the project by the United States will once again convince those doubting the global nature of this enterprise. It is no coincidence that in 2018 the European Investment Bank issued one of the largest loans in its history - 1.5 billion Euros for the implementation of the TAP [26].

9. Conclusion

The EU, which includes the most developed economies of the world and has very limited access to oil and gas reserves, has to secure large energy supplies from different sources at reasonable and stable prices in order to sustain economic performance and growth. Also, for Azerbaijan, intensification of energy contacts with the EU gives an opportunity not only to get hard currency, but also modernize its economy.

As of the beginning of July 2018, the partners have invested more than $ 30 billion in the SGC project. Of these, 17.7 billion account for the Shah Deniz-2 project, the total cost of which is $ 22.7 billion, $ 4.2 billion for the expansion of the South Caucasus pipeline, worth $ 4.6 billion, $ 4 billion for the TANAP project, and 2.9 billion for the TAP project [27].

The SGC is being built with the support of Azerbaijan and consortium companies including SOCAR. If the EU needs additional volumes of gas for its market in future and still wants to diversify supply sources using new routs it will need an expensive pipeline track. Basically, as long as gas pipelines exist the EU and Turkey are provided with an infrastructure for long term gas transportation from neighboring countries through the Caspian region for many years to come. There would be no pipelines like that without Azerbaijan, and the EU itself would have to invest billions in the construction of a new pipeline system due to the growing demand for imported gas. In other words, the SGC has freed the EU and European companies from multibillion investments in the creation of strategically important pipeline infrastructure [28].

10. References

[1] European Commission, Energy Security Strategy COM (2014) 330, 28 May, 2014, https://eur-lex.europa.eu/legal-content/EN/TXT/PDF/?uri=CELEX:52014DC0330&from=EN (last accessed: 12 August 2018).
[2] http://ec.europa.eu/eurostat/statistics-explained/index.php?title=File:Gross_inland_consumption,_2015_world18.png (last accessed: 17 August 2018).
[3] http://www.igu.org/the-case-of-natural-gas (last accessed: 12 August 2018).
[4] http://www.epa.gov/cleanenergy/energy-and-you/affect/natural-gas.html#footnotes (last accessed: 12 August 2018).
[5] https://www.aga.org/environmental-benefits-natural-gas (last accessed: 12 August 2018).
[6] https://www.iea.org/weo2017/ (last accessed: 14 August 2018).
[7] International Energy Agency, Outlook of Natural Gas, Excerpt from World Energy Outlook 2017, http://www.iea.org/publications/freepublications/publication/WEO2017Excerpt_Outlook_for_Natural_Gas.pdf (last accessed: 12 August 2018).
[8] http://www.iea.org/topics/naturalgas/ (last accessed: 12 August 2018).
[9] http://www.cga.ca/wp-content/uploads/2016/10/ICF-Report-Executive-Summary-Reducing-GHG-Emissions-Find-2016.pdf (last accessed: 12 August 2018).
[10] BP Statistical Review of World Energy 2018 (last accessed: 15 August 2018).
[11] Anouk Honore, Natural gas demand in Europe in 2017 and short-term expectations (Oxford: Oxford Energy Insight 35, April 2018), https://www.oxfordenergy.org/wpcms/wp-content/uploads/2018/04/Natural-gas-demand-in-Europe-in-2017-and-short-term-expectations-Insight-35.pdf (last accessed: 10 August 2018).
[12] https://www.bp.com/en_az/caspian/operationsprojects/Shahdeniz/SDstage1.html (last accessed: 1 August 2018).
[13] https://www.bp.com/en_az/caspian/operationsprojects/Shahdeniz/SDstage2.html (last accessed:

1 August 2018).
[14] European Commission, Second Strategic Energy Review , COM(2008) 781, 13 November 2008, https://eur-lex.europa.eu/LexUriServ/LexUriServ.do?uri=COM:2008:0781:FIN:EN:PDF , (last accessed 15 August 2018).
[15] European Commission, External energy Relations –from principles to action, COM (2006) 590, 12 October 2006, http://aei.pitt.edu/39622/1/COM_(2006)-590.pdf, (last accessed 15 August 2018).
[16] European Commission, On security of energy supply and international cooperation – The EU Energy Policy: Engaging with partners beyond our borders, COM (2011) 539, 7 September 2011, https://eur-lex.europa.eu/legal-content/EN/TXT/PDF/?uri=CELEX:52011DC0539&from=EN, (last accessed 15 August 2018).
[17] TANAP, "Intergovernmental Agreement between the Government of the Republic of Turkey and the Government of the Republic of Azerbaijan concerning the Trans Anatolian Natural Gas Pipeline System", 26.06.2012, (last accessed 12 August 2018).
[18] TANAP, ibid, p 7.
[19] TAP AG, www.tap-ag.com, "Albanian Parliament ratifies Intergovernmental Agreement on the Trans Adriatic Pipeline", 26.03.2013 (last accessed 15 August 2018).
[20] Energy Community, "Ministerial Declaration on the Ionian-Adriatic Gas Pipeline Project", 25.09.2007 (last accessed 15 August 2018).
[21] https://ru.sputnik.az/economy/20180630/416003935/TANAP-postavka-gaza-turkey-nachal (last accessed 15 August 2018)
[22] https://www.azernews.az/oil_and_gas/133448.html (last accessed 15August 2018).
[23] https://www.spglobal.com/platts/en/market-insights/latest-news/natural-gas/080718-us-confirms-iran-sanctions-waiver-for-azerbaijan-natural-gas-pipeline-to-turkey-europe (last accessed 15August 2018).

[24]https://azertag.az/en/xeber/Italian_media_highlight_President_Sergio_Mattarellas_visit_to_Azerbaijan-1181603 (last accessed 14 August 2018)
[25] https://uk.reuters.com/article/uk-italy-tap/italy-pm-non-committal-after-discussing-tap-pipeline-dispute-with-mayor-idUKKBN1KN1VD (last accessed 11 August 2018).
[26] https://www.reuters.com/article/eu-energy/update-1-eib-approves-1-5-bln-euro-loan-for-tap-gas-pipeline-idUSL8N1PW6EY (last accessed 11 August 2018).
[27] https://ru.sputnik.az/economy/20180630/416003935/TANAP-postavka-gaza-turkey-nachalo.html(last accessed 8 August 2018).

[28] Gulmira Rzayeva: "Azerbaijan will ensure Europe's energy security for many years to come" http://vestnikkavkaza.net/interviews/Gulmira-Rzayeva-Azerbaijan-will-ensure-Europe-s-energy-security-for-many-years-to-come.html (last accessed: 5 August 2018).

IOP Conf. Series: Materials Science and Engineering **459** (2019) 012012 doi:10.1088/1757-899X/459/1/012012

Seebeck coefficient of thermopile made of nickel-coated carbon fiber

H Hardianto[1,2], B Malengier[1], G De Mey[3], C Hertleer[2] and L Van Langenhove[2]

[1] Polytechnic of Textile Technology (Politeknik STTT Bandung), Jalan Jakarta 31, Bandung 40272, Indonesia
[2] Ghent University, Department of Materials, Textiles and Chemical Engineering, Technologiepark 907, 9052 Zwijnaarde, Belgium
[3] Ghent University, Department of Electronics and Information Systems, Technologiepark 15, 9052 Zwijnaarde, Belgium

E-mail: hardiant.hardianto@ugent.be

Abstract. A textile-based thermopile has been fabricated by etching the nickel layer of a nickel coated carbon fiber (NiCF) selectively to form a series of CF-NiCF junctions along the NiCF. The pristine NiCF was inserted into the polyester woven fabric manually. Each half part of the float yarns was covered with Lurapret® D579 dropwise to form a 36-pair CF-NiCF thermopile. After drying in the oven, the sample was etched in the etching solution, then rinsed with water and air dried. The Seebeck coefficient resulting from 36-pair CF-NiCF thermopile is 93.04 μV/K. This proofs that creating a flexible thermoelectric generator from a conductive textile yarn is possible.

1. Introduction

Thermoelectric has attracted a lot of attention because of its considerable potential to be utilized as a renewable and sustainable energy source. Since 2001, research on harvesting and converting heat into electrical energy in a wearable thermoelectric generator has been conducted [1]. Recently, a number of studies on the thermoelectric generator (TEG) by way of harvesting body heat have been reported [2–6] but the TEG they used were made of inflexible materials and non-textile materials. Some papers reported the energy harvesting of the human body using different structures of textile-based TEG [7–9], however, none of these studies used the thermopile principle made from a single yarn.

According to our previous finding, Nickel-coated Carbon Fiber (NiCF) and Carbon Fiber (CF) are good candidates for fabricating a textile-based thermoelectric generator because they can be inserted into a fabric structure [10]. In this paper, we try to utilize the NiCF as a starting material and convert it into a textile-based thermopile. The objective of this work is to fabricate a textile-based TEG from a single yarn i.e. NiCF in a textile fabric by means of the etching process to form a series of CF-NiCF junctions in order to obtain a thermopile and study its Seebeck coefficient. In this paper, the term NiCF corresponds to Nickel-coated Carbon Fiber in the form of multifilament yarn.

2. Experimental

In this experiment, NiCF yarn from Toho Tenax Germany GmbH (Tenax®-J HTS40 A23 12K 1420tex MC) was used. The thickness of the nickel is around 0.25 μm [11]. The textile-based TEG was prepared by manually inserting the NiCF yarn into a polyester fabric by means of a needle to form a pattern as shown in Figure 1. Each row was circa 8 cm in length and each float of the yarns was around 1 cm. The distance between the rows was 1 cm.

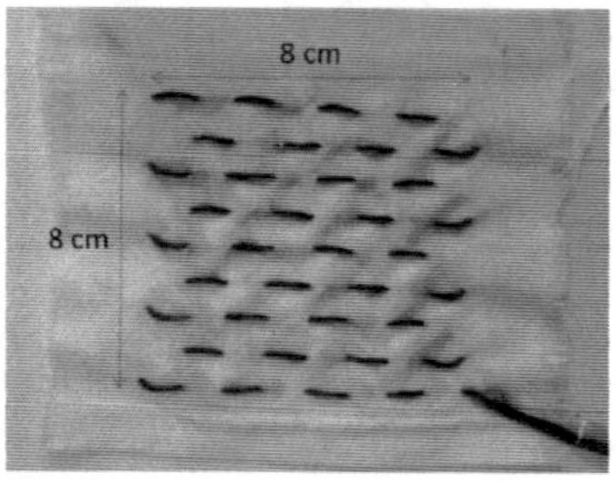

Figure 1. A textile thermopile sample having 36 pairs of CF-NiCF junction.

Then, this NiCF was covered with a polymer called Lurapret® D579 dropwise on the half part of the float yarns on top and bottom sides of the fabric as shown in Figure 2. Next, the sample was dried in the oven at 130°C for 15 minutes. Then, the etching process was performed according to the literature [10] where the sample was dipped in 10% H_2O_2 + 37% HCl (1:1) for 30 minutes without heating, rinsed with running water and air dried for 24 hours (minimum) before use. This process allows the etching chemical to remove the nickel selectively from the NiCF where the nickel was not covered with Lurapret® D579. In this way, a textile-based TEG from a single yarn containing CF-NiCF junctions was obtained.

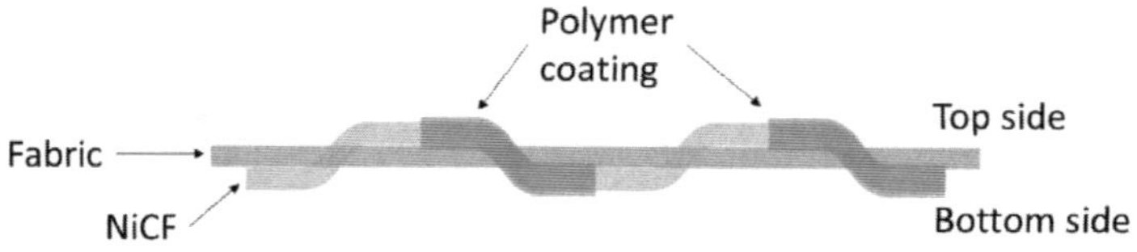

Figure 2. Illustration of the covered and uncovered area of NiCF with polymer before etching.

Microscope images were taken with an Olympus SZX10 stereomicroscope equipped with ToupView camera control software. Voltage measurement was performed by placing the sample on a hot plate and placing a cooling fan on the sample. The output voltage of the sample was measured with a nanovoltmeter NV 724 at different ΔT of hot and cold sides. Measurement of hot and cold temperatures was done with a Fluke 54 II B digital thermometer. From the gradient of the voltage and temperature difference graph, the Seebeck coefficient was obtained.

3. Results and discussion

In this paper, we aim at fabricating a textile-based thermopile from NiCF filaments by removing the nickel layer of NiCF at specific zones by an etching technique. Some zones were covered with a polymer called Lurapret® D579 before the etching process. As a result, only nickel in the area where there is no polymer on its surface would have a contact with the etching chemical forming an intermittent pattern of CF-NiCF along the yarn.

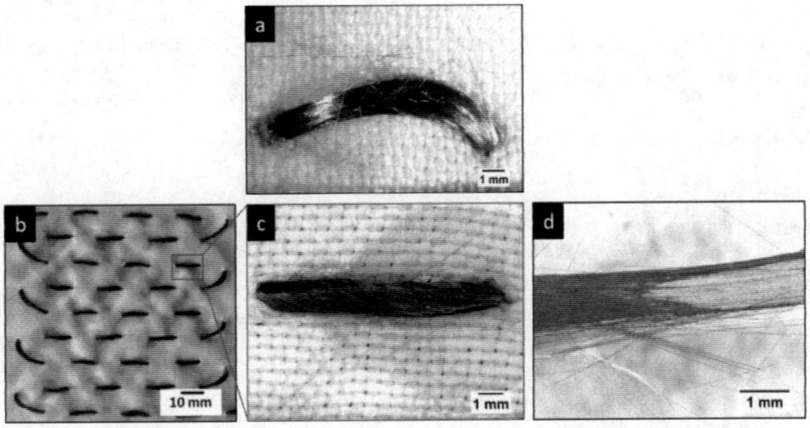

Figure 3. Microscope images of NiCF yarn samples in the fabric: (a) before and (b)-(d) after etching. Image (c) is the magnification image of the selected area from image (b).

Figure 3 shows microscope images of NiCF yarn samples in a polyester fabric before and after the etching process. It is clear that the NiCF yarn colour is brown before etching (Figure 3a) and it becomes black after etching process (Figure 3b-3d). Since the original colour of carbon fiber is black, it is observed that after the etching process the yarn changes to black indicating the nickel has been removed from the surface of carbon fiber. In Figure 3b and 3c, the polymers Lurapret D579 added dropwise on the sample are also observed; each in an oval shape and is seen darker than the normal colour of the polyester fabric. Theoretically, the nickel on the surface of the carbon fiber beneath the Lurapret D579 polymer will remain brown if it is not removed by the etching chemical. Unfortunately, from the image (b) and (c), we cannot observe clearly the colour difference between the covered and uncovered part of the yarn after the etching process. Due to that issue, a sample of the floated yarn was cut so that the inner part of the filaments bundle was revealed as shown in image (d). From this image, it is obvious that there are two different colours i.e. black and pale brown which confirm that the nickel in the area covered with the polymer still remains there.

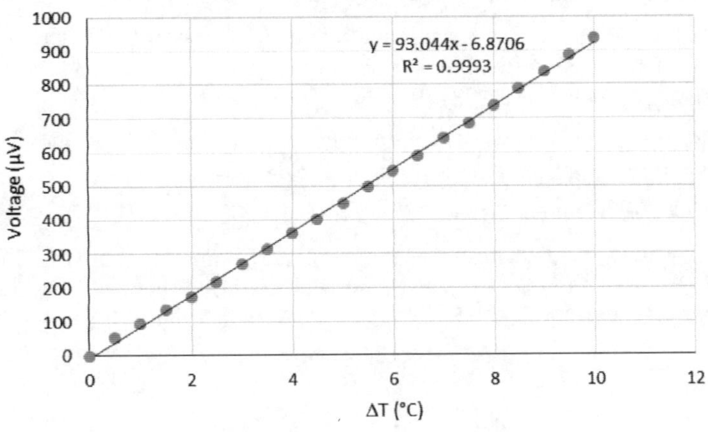

Figure 4. Graph of voltage versus temperature difference from 36-pair CF-NiCF junction after the etching process.

The Seebeck coefficient resulting from 36-pair CF-NiCF thermopile is 93.04 µV/K as presented in Figure 4. The result is lower than the theoretical calculation i.e. 648 µV/K as the single thermocouple from CF-NiCF was found to be around 18 µV/K [12]. This indicates the junctions are not at full potential when constructed as given.

4. Conclusion

In this work, a 36-pair CF-NiCF thermopile from single NiCF yarn has been fabricated using the etching technique and can generate a Seebeck coefficient of 93.04 μV/K. This proofs that it is possible to create a flexible thermoelectric generator from a conductive textile yarn. However, the results were still far from the theoretical maximum.

5. Acknowledgements

Mr. Hardianto gratefully acknowledges the Indonesia Endowment Fund for Education (LPDP) for making this work possible by financially supporting the research activities.

References

[1] Siddique A R M, Mahmud S and Heyst B Van 2017 A review of the state of the science on wearable thermoelectric power generators (TEGs) and their existing challenges *Renew. Sustain. Energy Rev.* **73** 730–44

[2] Myers A, Hodges R and Jur J S 2017 Human and environmental analysis of wearable thermal energy harvesting *Energy Convers. Manag.* **143** 218–26

[3] Myers A C and Jur J S 2017 Effects of thermal energy harvesting on the human - Clothing - environment microsystem *IOP Conf. Ser. Mater. Sci. Eng.* **254** 072015

[4] Leonov V, Vullers R J M and Hoof C Van 2012 Thermoelectric generator hidden in a shirt with a fabric radiator *AIP Conference Proceedings* vol 1449 pp 556–9

[5] Hyland M, Hunter H, Liu J, Veety E and Vashaee D 2016 Wearable thermoelectric generators for human body heat harvesting *Appl. Energy* **182** 518–24

[6] Leonov V 2013 Thermoelectric energy harvesting of human body heat for wearable sensors *IEEE Sens. J.* **13** 2284–91

[7] Kim M K, Kim M S, Lee S, Kim C and Kim Y J 2014 Wearable thermoelectric generator for harvesting human body heat energy *Smart Mater. Struct.* **23** 105002

[8] Lu Z, Zhang H, Mao C and Li C M 2016 Silk fabric-based wearable thermoelectric generator for energy harvesting from the human body *Appl. Energy* **164** 57–63

[9] Raihan A, Siddique M, Rabari R, Mahmud S and Heyst B Van 2016 Thermal energy harvesting from the human body using flexible thermoelectric generator (FTEG) fabricated by a dispenser printing technique *Energy* **115** 1081–91

[10] Hardianto A, Hertleer C, Mey G De and Langenhove L Van 2017 Removing nickel from nickel-coated carbon fibers *IOP Conf. Ser. Mater. Sci. Eng.* **254** 072010

[11] Tenax T 2011 *Tenax®-J HTS40 A23 12K 1420tex MC*

[12] Hardianto H, De Mey G, Wróbel I C, Hertleer C and Van Langenhove L 2018 Seebeck Coefficient of Thermocouples from Nickel-coated Carbon Fibers: Theory and Experiment *Materials (Basel).* **11** 922

Hybrid photoenergy installation development

R V Zaitsev, M V Kirichenko, G S Krypunov and L V Zaitseva

Materials for Electronics and Solar Cells Department, National Technical University "Kharkiv Polytechnic Institute", 2 Kirpicheva Street, Kharkiv, 61002, Ukraine

zaitsev.poman@gmail.com

Abstract. The current problem is the autonomous supply thermal energy and hot water for civilian and military consumers. The most commonly used for this purpose are installations based on solar thermal collectors, which at the same time require an external power source up to 100 W for electronic control components and a circulating pump, which does not allow them to be fully considered as autonomous. One of the ways to solve this problem is to combine in one installation solar panels and heat collectors. It has been proposed the constructive-technological solution of the hybrid photoenergy system with flexible solar cells based on cadmium telluride, which allows getting the total efficiency of such a system up to 73% for the transformation of solar energy into heat and electric energy.

1. Introduction

The current problem is the autonomous supply thermal energy and hot water for civilian and military consumers. The most commonly used for this purpose are installations based on solar thermal collectors, which at the same time require an external power source up to 100 W for electronic control components and a circulating pump, which does not allow them to be fully considered as autonomous.

One of the ways to solve this problem is to combine in one installation solar panels and heat collectors. Such hybrid thermo-photoenergy installations are usually made from solar cells (SC) based on monocrystalline silicon and their electrical characteristics are much more prevalent [1]. In addition, the silicon based SC, due to the material properties, has an intense absorption of the near-infrared solar radiation component with a wavelength more than 1000 nm. Such absorption is photovoltaically ineffective and leads to parasitic SC heating and decrease efficiency of thermal part such hybrid installation. This necessitates the organization of efficient heat removal from SC to the heat exchanger through intermediate thermo-interfaces, which leads to complication of the installation design. At the same time, the maximum absorption of flexible thin-film SC based on the CdS/CdTe system is concentrated in the range 400-800 nm, and for near-infrared component, they are practically transparent.

Therefore, using such SC in hybrid photovoltaic installations will provide free access to the heat exchanger of the longwave solar radiation component, and efficient heat energy generation [2]. Using flexible thin-film CdS/CdTe based SC on the polyimide substrate in the electrical part of the hybrid installation will allow to mount such SC by gluing on a protective glass directly on the standard solar thermal collectors photoreceiving surface. Therefore, the development of an autonomous hybrid photoenergy installation equipped with flexible CdS/CdTe based SC, adapted for the mounting directly on the protective glass of standard solar thermal collectors, is perspective.

IOP Conf. Series: Materials Science and Engineering **459** (2019) 012013 doi:10.1088/1757-899X/459/1/012013

2. Results and their discussion

The main optimization problem for the development of such a system is the determination the level of the collector working temperature influence on the rate of thin film SC efficiency reduction. In order to determine the temperature effect on thin film SC parameters, it has been carried out a study of industrial samples of SC based on CdS/CdTe. For samples of series 1-3 at temperatures from 0 °C to 60 °C, illuminated current-voltage characteristics (CVC) were measured [3]. As a result of further analytical processing such experimental illuminated CVC, the output parameters and light diode characteristics of the SC (Table 1) were determined.

Table 1. Output parameters and light diode characteristics of thin film SC based on CdS/CdTe at temperature 25°C

Output and light diode parameters	Samples serie		
	1	2	3
J_{SC}, mA/cm^2	19.3	19.4	19.6
U_{OC}, mV	733	756	762
FF, rel. un.	0.59	0.51	0.58
Efficiency, %	8.5	7.2	8.7
J_{PH}, mA/cm^2	19.6	19.9	19.8
R_S, Ohm·cm^2	5.8	8.2	8.7
R_{SH},Ohm·cm^2	2300	1520	1890
A_i, rel. un.	2.2	1.8	1.9
J_0, A/cm^2	$1.1 \cdot 10^{-9}$	$5.9 \cdot 10^{-9}$	$2.3 \cdot 10^{-9}$

The analysis shows that with temperature increase for all investigated samples, similarly to silicon samples, there is practically a linear decrease of the efficiency (Fig. 1, a). At the same time, the efficiency reduction coefficient, which describes the relative efficiency change when temperature change on one degree, is only -0.14 rel%/$_o$C, which is much lower than the corresponding parameter for silicon samples [4]. For open circuit voltage and short-circuit current density, there is also a decrease in their values when the working temperature is increased (Fig. 1, b, c). Experimentally it has established that the CVC fill factor practically does not change with temperature increase (fig. 1, g). The analysis of light diode parameters shown that the observed efficiency decrease is accompanied by an increase of diode saturation current density and by decrease of shunt resistance.

The analysis of the obtained temperature dependences for thin film solar cells shows the traditional decrease of parameters with increasing temperature, which correlates with data of the leading world producers (Calyxo GmbH, Germany). In such solar cells, in comparison with silicon, there are absence of additional mechanisms that can reduce the efficiency with increasing temperature. For the investigated samples of thin film SC based on CdS/CdTe structure, there is no significant decrease in efficiency with increasing temperature up to 60 °C.

Optical studies have shown that the solar energy absorption coefficient for flexible SC based on cadmium telluride in the visible range (400-800 nm) is up to 94-96%, and the reflection coefficient in the infrared part of spectrum does not exceed 7-8% (Fig. 2), which allows excluding the selective coating from collector design.

The conception of a photoenergy system based on a thin film SC is in the development a design solution for the placement of flexible thin film SC based on CdS/CdTe system which adapted for the mounting directly to the plate of solar collector. Such autonomous combined photoenergy installation for military and civilian applications should be based on standard solar thermal collectors. The main part of the installation is the photoenergy module, which is a standard solar thermal collector on the photovoltaic surface of which would be mounted a flexible solar cell based on the CdS/CdTe system.

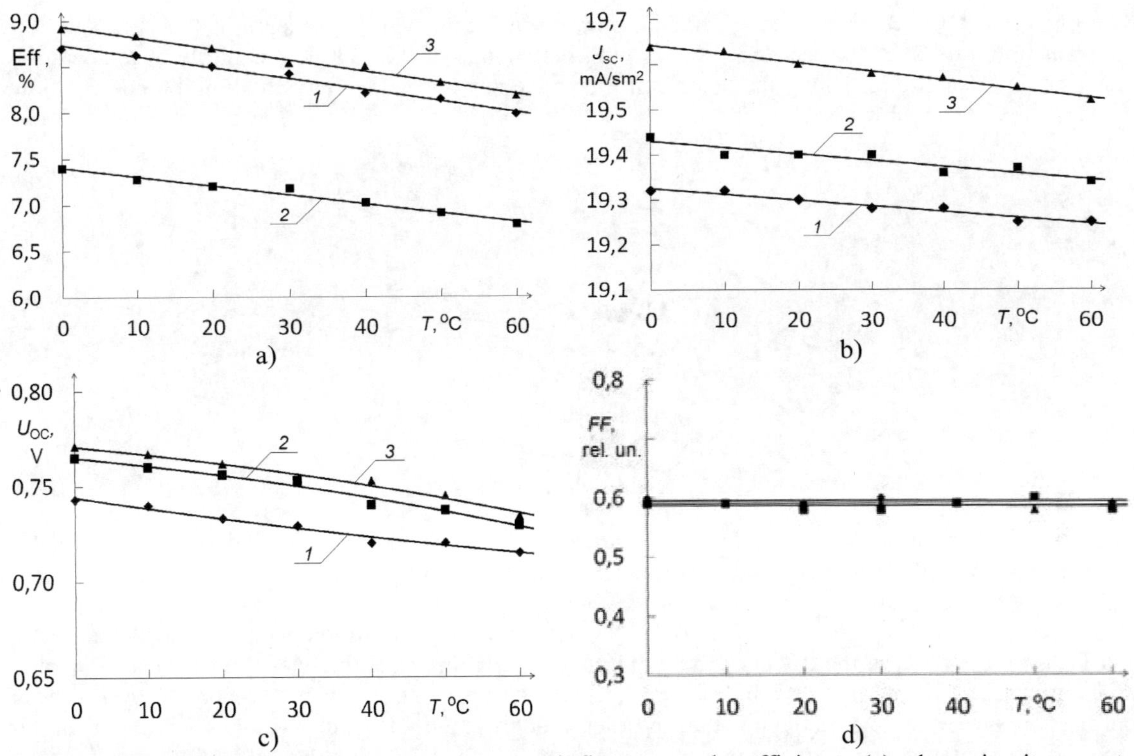

Figure 1. The character of working temperature influence on the efficiency (a), short-circuit current density (b), open circuit voltage (d), and fill factor of CVC (d) for the investigated thin film SC based on CdS/CdTe, series 1-3.

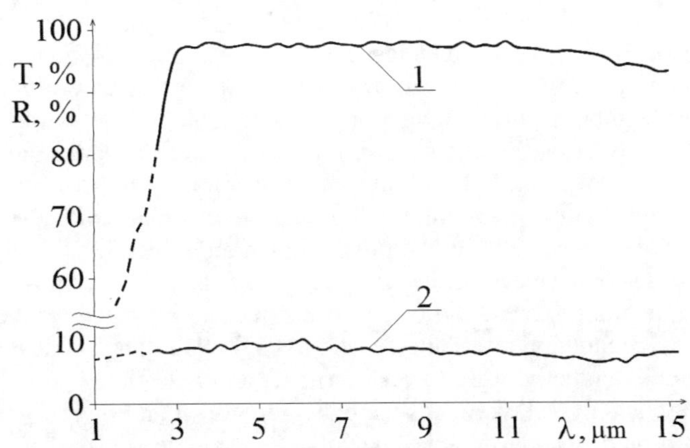

Figure 2. Spectral dependences of the transmission (1) and reflection (2) coefficients for SC based on the CdS/CdTe structure.

As a cooling system, it has been proposed the using of standard solar collector plates [5]. To determine the possibility and effectiveness of their using the simulation of thermal processes during the working of such photoenergy system was carried out.

A series of model experiments with different parameters, namely the plate thickness and the coolant flow velocity in the system, were carried out to detect the optimal heat exchange regime between the SC, the plate and the pipe. The aim of this work was the finding out the optimal balance between these parameters, the uniformity of heat transfer and the cost of such system. The thickness of the plate was changed in the range from 0.5 mm up to 5 mm with increments of 0.5 mm, and at a coolant flow rate from 0.2 m/s up to 1 m/s with increments of 0.2 m/s. An example of temperature distribution visualization along the solar collector plate surface with 1 mm thick and a coolant flow velocity 0.4 m/s is shown on Fig. 3a.

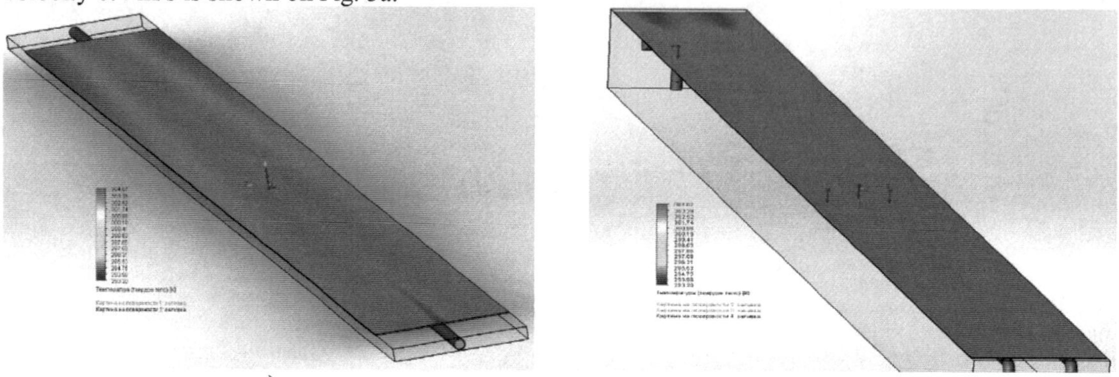

a) b)

Figure 3. Distribution of temperatures on the surface of a single-tube (a) and two-tube (b) solar collector plates with 1 mm thick and a coolant flow velocity 0.4 m / s

Based on the received visualizations of the temperature distribution along the collector plate surface, we can observe the tendency that with increasing of the plate thickness and coolant flow rate, the temperature distribution over the surface becomes uniform. This is due to the basic physical laws of heat conductivity. The problem is that in order to achieve a coolant flow velocity of more than 0.6 m/s, it is necessary to use a powerful circulation pump, which will result in to unnecessary consumption of electric energy and the impossibility of autonomous installation creating. In addition, if the collector plate thickness is increased up to 2 mm, use of such plate will be inappropriate, because it will have a very large mass and cost.

Therefore, based on this results it has been proposed to improve the system of solar collector plate in such by means the organizing heat exchange between the plate and the system through the two pipes instead one pipe, as it was made in industrial solutions. In this way, a uniform distribution of heat should be achieved without using high velocities of the coolant flow and large thicknesses of the collector plate. This will lead to some increase in the complexity of plant manufacturing due to the failure of the standard plate collector, but will allow to develop the system with maximum efficiency. An example of temperature distribution along the surface of solar collector two-tube plate with thickness 1 mm and a coolant flow velocity 0.4 m/s is shown on Figure 3, b.

Based on the carried out research, a test solar cell based on the ITO/CdS/CdTe/Cu/Au heterostructure was made and mounted through heat-conducting interface on the two-pipe plate of a solar collector. As a heat conduction interface we using a Stars 922, this is a heat conductive adhesive, which has a high thermal conductivity up to 1.2 W/(m·K) and a sufficient strength about 15 kg/cm^2. Schematic representation and overall view of such plate has shown on Figure 4.

As a result of carried out research of such type collector plate, it was shown that the technology of its manufacture is significantly simple. The proposed design and technological solution of such photoenergy module with flexible SC is optimal and allows getting the total efficiency of such type installation up to 73% for thermal and electric energy [6].

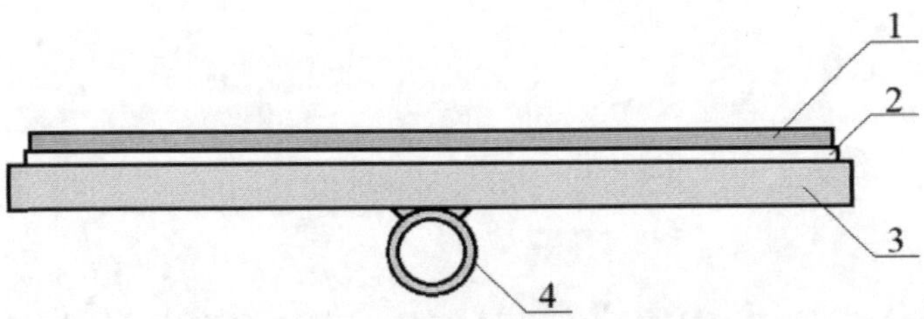

1 – SC with ITO/CdS/CdTe/Cu/Au heterostructure;
2 – heat conductive dielectric glue;
3 – copper collector plate;
4 – copper tube with coolant.

a)

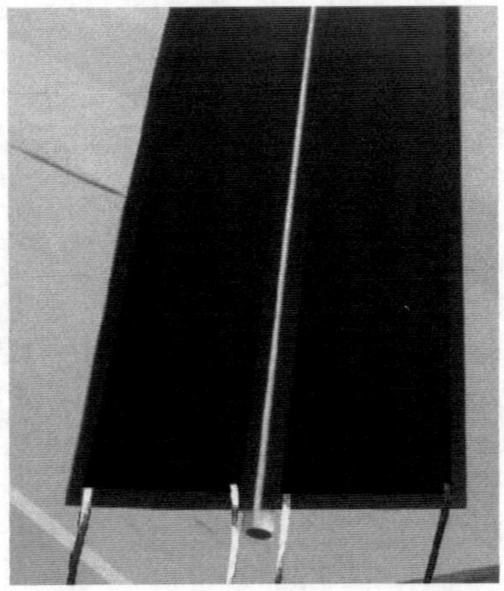

b)

Figure 4. Schematic representation (a) and overall view (b) of solar collector plate with mounted SC based on ITO/CdS/CdTe/Cu/Au heterostructure.

3. Conclusions

By the comparison of efficiency temperature dependences for thin film SC based on CdTe compounds produced in laboratory conditions it has been shown that when the temperature of such devices is changed on 50 °C, the efficiency decreases only on 1%, and the relative rate of efficiency decrease is -0.14 rel %/C. This is significantly less than the similar indicator for other types of solar cells.

After the analytical processing and analysis of light diode characteristics influence on the efficiency of cadmium telluride based SC, it has been established that the temperature stability of their efficiency is ensured by the diode saturation current density. When the temperature rises from 20 °C to 50 °C, the diode saturation current density increases only on 50 % from $1.9 \cdot 10^{-9}$ A to $2.7 \cdot 10^{-9}$ A, which is much lower than for silicon SC, for which the diode saturation current increases on 300 %.

The construction-technological solution of the photoenergy system with flexible cadmium telluride based solar cells has been proposed, which allows getting the total efficiency of such type system up to 73% for thermal and electric energy. The installation of a solar battery is carried out by using a heat conductive adhesive, which has a high thermal conductivity up to 1.2 W/(m·K) and a sufficient strength about 15 kg/cm^2.

References

[1] Colangelo G 2016 *Renewable and Sustainable Energy Reviews* (Amsterdam: Elsevier) vol. **57**, p. 1141-1159.

[2] Zaitsev R V 2017 *First Ukraine Conference on Electrical and Computer Engineering* (Ukraine, Kiyv: NTUU "KPI") pp. 360-364.

[3] Bashahu M, Habyarimana A 1995 Review and test of methods for determination of the solar cell series resistance. *Renewable Energy*, **6**(2), p. 129-138.

[4] Raushenbach H S 1980 *Solar Cells Array Design* (New York: Litton Uducation Publishing).

[5] Azad E 2018 *Renewable and Sustainable Energy Reviews* (Amsterdam: Elsevier) vol. **82**. part 3. p. 4320-4325.

[6] Zaitsev R V 2017 *IEEE International Young Scientists Forum on Applied Physics and Engeneering* (Ukraine, Lviv: LNU) p. 112-115.

Numerical investigation of swirl flow effect on heat exchanger efficiency according to different inlet position and Reynolds number

A B Atalar[1] and M Cakan[2]

[1] M.Sc., Department of Mechanical Engineering, Istanbul Technical University, Istanbul, 34437, Turkey
[2] Ph.D., Department of Mechanical Engineering, Istanbul Technical University, Istanbul, 34437, Turkey

atalara@itu.edu.tr

Abstract. The aim of this investigation is to determine the configuration that will enable a micro heat exchanger system to function more efficiently. Heat exchangers are devices that can be used for exchanging energy between flows, which can have a wide variety of uses and different configurations; therefore, studies are carried out in order to increase the efficiency of heat exchangers. In this study, heat transfer capacity and system efficiency of the swirl flow were investigated using CFD as compared to the parallel flow heat exchanger, which is another configuration. The air drawn by the fan is sucked into the holes and enters the system, thereby air entering the system impinges on the heat exchanger tube. The flow character has a major role on the Nusselt number and heat transfer coefficient. The main flow variables are Reynolds number and the orientation of the inlet holes. The boundary conditions were determined according to the fan capabilities and power of resistance. The positions and types of the holes were adjusted to allow for axial and tangential impact onto the heat exchanger tube, and the holes were also located in order to obtain the best impingement angle. CFD simulations were carried out with ANSYS Fluent and k-epsilon turbulent model was used with coupled algorithm. The effect of laminar flow on the heat transfer were observed at different Reynolds numbers. Also the effect of swirl flow characteristics on system efficiency is clearly observed as compared to parallel flow.

1. Introduction

Energy efficiency is one of the significant factors for developing electronic devices. Especially heat exchangers are used commonly worldwide in many electronic devices, whose sizes are becoming smaller day by day.

Swirl flows in pipes; can be defined as a flow type in which the fluid particles make both a translational movement along the pipe axis and a rotation about the pipe axis. For this reason, cylindrical coordinates are used in the analysis of rotational flows. The effect of the radial velocity component is neglected in most studies, while the influence of the axial and angular velocity components is primarily investigated in the cylindrical coordinates (Scott 1973). Rotational flows enable to increase the heat transfer and swirl flows are used in heat and mass exchangers, cooling of turbine blades, combustion chambers, rockets, centrifugal separators and pneumatic conveying technology (Zaherzadeh 1975). Generation of swirl flows in heat exchanger configurations can be achieved in two ways. The first way is to form helical grooves to induce continuously swirl motion. This configuration creates a non-decaying swirl flow motion. The other method, which creates a decaying swirl motion is obtained through tangential flow inlets (Neto et al. 1998).

The induced rotational flows can be examined in two different subgroups. According to the geometry of the inlet type: there are tangential inlet types that may be rectangular or circular. Or according to the flow regime: laminar or turbulent (Hay et al. 1975). In addition to this, heat transfer and impingement effect were investigated according to various number of tangential inlets configurations and Reynolds number (Rao et al. 2017).

The present study investigates the role of decaying laminar swirl flow on heat transfer enhancement from the inner surface of a coaxial heat exchanger as compared to laminar flow that flows parallel to the pipe axis. The Reynolds number based on flow passage hydraulic diameter is kept below 2300 in order to ensure laminar flow regime.

2. Methodology

The study uses two different types of flow entrances to induce laminar flow in between the inner boundaries of the co-axial tube arrangement.

2.1. *Assembly of Experimental Setup*

The heat exchanger system consists of an axial fan, different types of inlet holes, and co-axially arranged inner and outer pipes. The inner aluminium pipe which is heated internally simulates the "warm" heat exchanger fluid, while the unheated outer Plexiglass pipe bounds the heat exchanger fluid. Axial fan aspirates the air which enters from inlet openings and the flow can be adjusted for several Reynolds numbers by means of axial fan frequency converter.

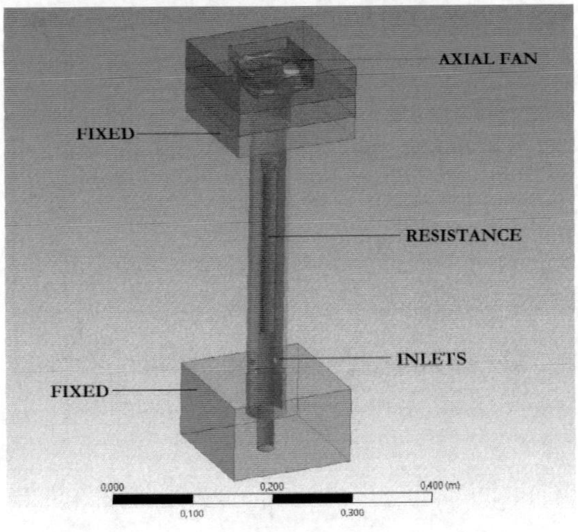

Figure 1. *Assembly of heat exchanger*

Parts are shown in Figure 1 for heat exchanger assembly. The fan is located above the resistance and air is aspirated along the resistance. The resistance starts from the fixed plate and continues until the fan section.

2.2. Tested Configurations

There are radial and tangential impingement possibilities on the outer surface of the inner (heated) pipe. This makes 12 different configurations because of two different types of inlet holes and various Reynolds numbers that are chosen to be investigated.

Radial and tangential entrance hole arrangements are shown in Figure 2 and it was defined respective to the axis of the heated tube. Table 1 summarizes the configurations that are examined in this investigation.

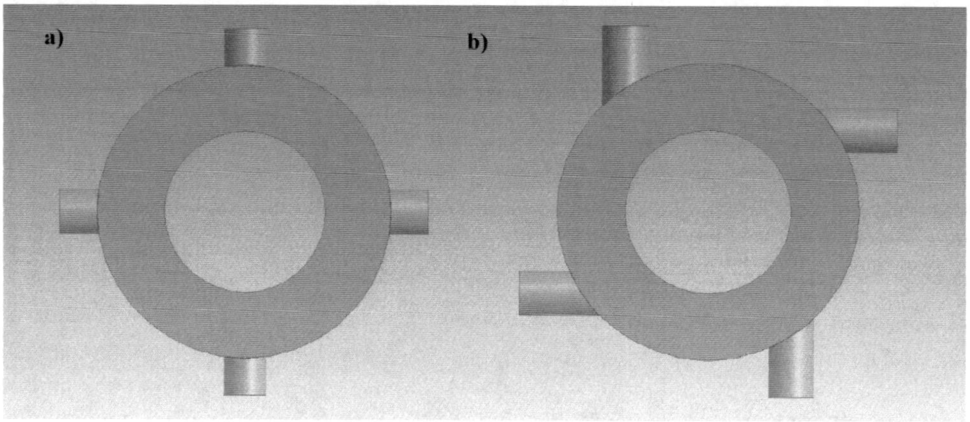

Figure 2. *Top view (z-axis) of impingement configuration for (a) Radial (b) Tangential*

Table 1. Numerically tested configurations

Reynolds no.	Hole orientation	Number of holes
	Radial	4
775	Radial	2
	Tangential	4
	Tangential	2
	Radial	4
1160	Radial	2
	Tangential	4
	Tangential	2
	Radial	4
1550	Radial	2
	Tangential	4
	Tangential	2

Mass flow rate, which is adjusted by the frequency control of the axial fan, is kept constant for 4-sided and double sided inlet configuration.

3. Computational Model

Simulations were carried out using ANSYS Fluent software with computer which have 8 core-3.6GHz processor and 16 GB of RAM.

3.1. Computational Geometry and Grid

The actual geometry (Fig. 3) consists of two co-axially arranged pipes and the gap between these pipes is set as the computational domain. The aspirated air enters through the entrance holes and flow in between the pipes. The radii of inner pipe (heated) and outer pipe are respectively 11 mm and 20 mm whereas the length of both pipes is 600 mm. The diameter of inlets is 6 mm and it extends along 5 mm because of the thickness of the Plexiglass.

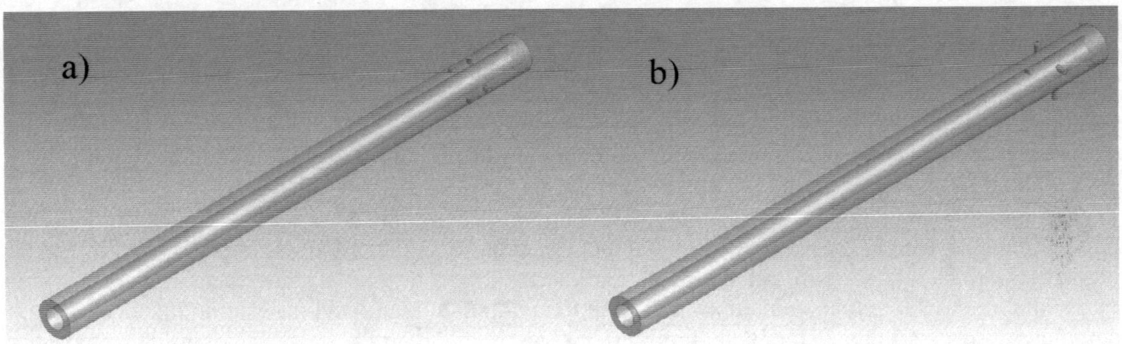

Figure 3. *Isometric view of 4-sided configuration of actual geometry for (a) Radial (b) Tangential entrances*

The isometric view of the computational domain is shown in Figure 4. Computational grid is unstructured and it consists of average 1,885,000 elements. There are about 15,000 elements difference between the radial and tangential grid models; however, this is not a significant value for the accuracy of the calculations.

Figure 4 shows grid model for radial configuration and there are 6 inflation layers, which have 0,15 mm thickness with 1,05 growth rate, in order to obtain the required y plus value.

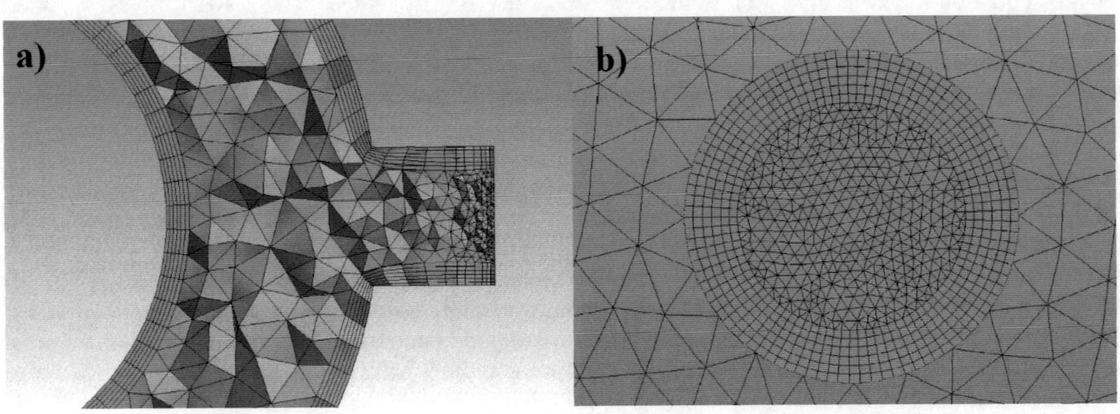

Figure 4. *Computational grid for (a) Horizontal cross section (b) Inlet section*

3.2. Boundary Conditions

The heat flux of resistance is held constant for all simulations; so that the inlet hole type and Reynolds numbers stay as main parameters. All boundary conditions which is used in CFD simulations are listed in Table 2. Reynolds numbers are respectively 775, 1160 and 1550 and the heat flux which is applied on resistance is 150 W/m^2.

Table 2. Boundary conditions for different configurations

Reynolds Number	Heat Flux of Resistance	Inlet Configuration Type
775	150 W/m^2	4-Sided & Double Sided
1160	150 W/m^2	4-Sided & Double Sided
1550	150 W/m^2	4-Sided & Double Sided

Simulations were carried out with 4-sided and double sided, radial and tangential inlet types and all configurations were investigated for each Reynolds numbers.

In order to apply the fan aspiration as the boundary condition, outlet is defined as a mass flow outlet. The characteristic length (D) is defined as the difference in radius between the two pipes.

Reynolds number depends on characteristic length, velocity and kinematic viscosity for internal flow. D represents characteristic length in [m] and U denotes the velocity of air in [m/s], υ is kinematic viscosity in [m^2/s]. If Re is under the 2300 for internal flow, flow is designated as laminar. In this study, the air, which is aspirated from inlet, is 20°C. Also the boundary condition type for inlets is defined as pressure inlet. With all these assumptions, the boundary conditions for heat exchanger's experimental setup are most closely matched for CFD simulations.

3.3. Solver Settings

Numerical computations were carried out by solving the 3D model, incompressible flow and steady state. There are also laminar and turbulent flow, so turbulent form of the Reynolds-Averaged Navier–Stokes (RANS) equations were performed. The flow is turbulent at the inlet section and impingement surface while the flow is laminar at the exit of the heat exchanger. Navier–Stokes equations including energy terms were performed by finite volume discretization method using ANSYS Fluent 18.0 software package.

Two equation RNG k-epsilon turbulence model is used for computations. The most significant difference between standard and RNG k-epsilon turbulent models is the additional strain rate term in epsilon equation. There is an option to modify turbulent viscosity to account for swirl. Pressure based coupled algorithm is used for pressure-velocity coupling. For the discretization of momentum, k and epsilon equations, a second order upwind scheme is used. Moreover, enhanced wall treatment is used and it is recommended to keep y+ values between 1 and 11 for enhanced RNG k-epsilon turbulence model.

4. Results and Discussions

In order to obtain the best heat transfer enhancement performance, different configurations were simulated using CFD software. Distributions of Nusselt number and surface temperature on the outer surface of the inner pipe is the main parameter, which indicates cooling efficiency.

The method was tested by investigating y+ value especially on the inner pipe surface. The highest Reynolds number configuration (Re=1550) was investigated because of the fact that higher velocity enables to higher y+ values. Therefore, if y+ is in the standard interval, the lower Reynolds numbers will not exceed interval of ideal y+.

Figure 5 shows that the maximum y+ occurs around the inlet section and the y+ value decreases as the air velocity decreases. In general, y+ value is less than 1 along the heated wall; however, in the impingement zone this value stays between 2 and 3.

Figure 5. y+ *distributions for (a) 4-Sided Radial (b) 4-Sided Tangential*

Distribution of Nusselt number on the heated wall for different Reynolds numbers and inlet configurations are shown respectively in Figure 6, Figure 7 and Figure 8. Nusselt number is scaled between 0 and 40 in the figures and it is decreasing from inlet to outlet.

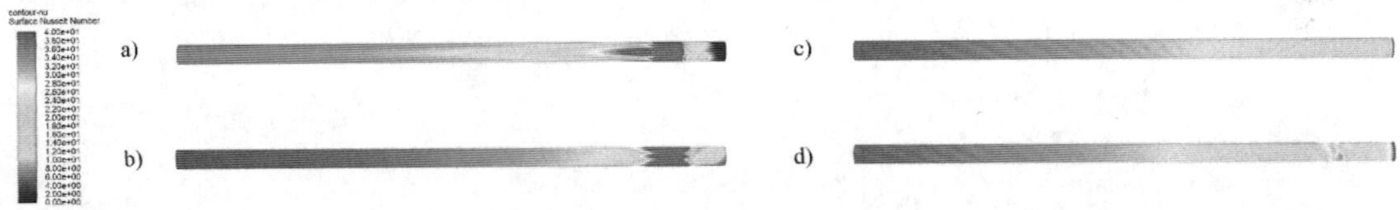

Figure 6. *Distributions of Nusselt number on the heated wall at Re=775 for (a) 4-Sided radial (b)Double sided radial (c) 4-Sided tangential (d) Double sided tangential configuration.*

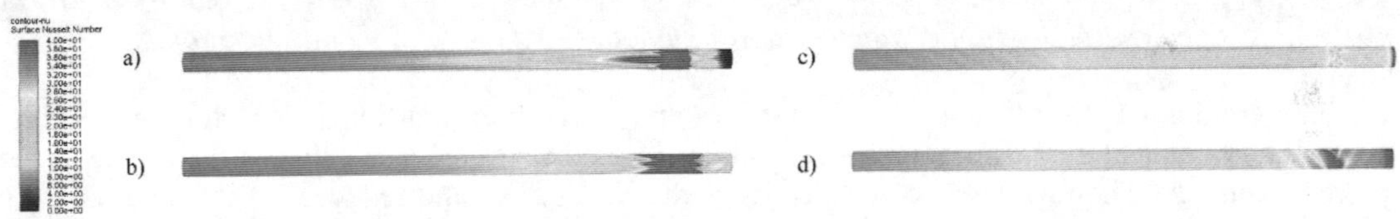

Figure 7. *Distributions of Nusselt number on the resistance at Re=1160 for (a) 4-Sided radial (b)Double sided radial (c) 4-Sided tangential (d) Double sided tangential configuration.*

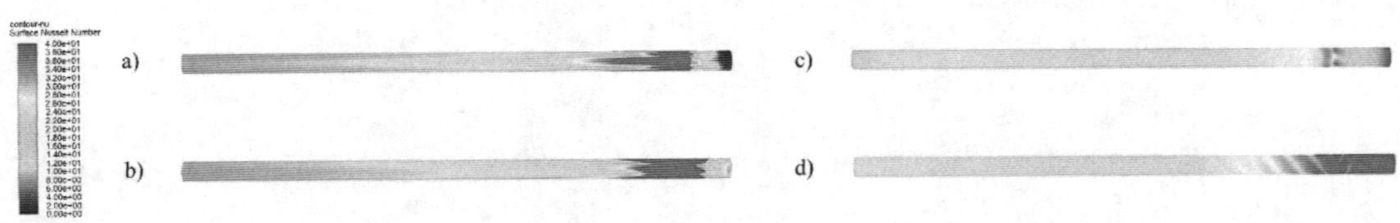

Figure 8. *Distributions of Nusselt number on the resistance at Re=1550 for (a) 4-Sided radial (b)Double sided radial (c) 4-Sided tangential (d) Double sided tangential configuration.*

While the Nusselt number exhibits values up to respectively 120 for tangential and 230 for radial at the impingement surface, it goes down to 10 towards the end of the heat exchanger. Also, these distributions are more homogenous in tangential configurations in which swirl flow dominates.

Distribution of temperature on the heated wall for different Reynolds numbers and inlet configurations are shown respectively in Figure 9, Figure 10 and Figure 11. Temperature is scaled between 293 [K] and 310 [K] in the figures and it increases as the flow picks heat and warms up from inlet to outlet.

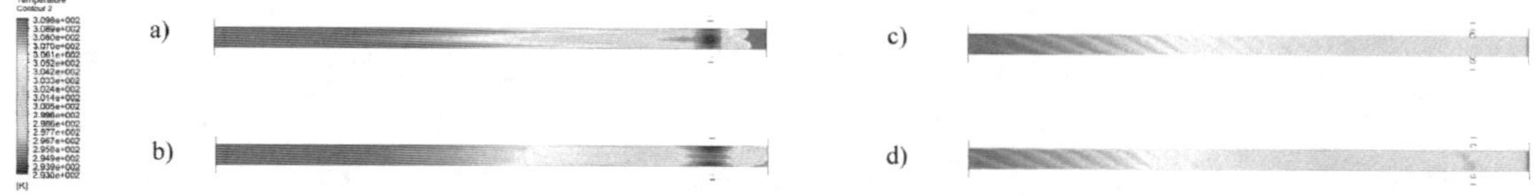

Figure 9. *Distributions of temperature on the resistance at Re=775 for (a) 4-Sided radial (b) Double sided radial (c) 4-Sided tangential (d) Double sided tangential configuration.*

Figure 10. *Distributions of temperature on the resistance at Re=1160 for (a) 4-Sided radial (b)Double sided radial (c) 4-Sided tangential (d) Double sided tangential configuration.*

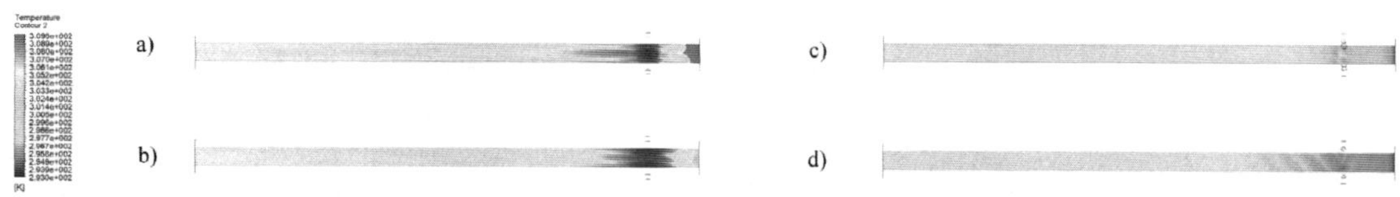

Figure 11. *Distributions of temperature on the resistance at Re=1550 for (a) 4-Sided radial (b)Double sided radial (c) 4-Sided tangential (d) Double sided tangential configuration.*

Temperature and Nusselt number distributions are inversely proportional to each other. Where the Nusselt number is high, the heat convection is higher and the surface of the inner heated pipe is cooled more effectively. The relatively colder air strikes the hot surface and reduces the temperature by heat transfer through the heat exchanger.

Particularly when the temperature distributions of Figure 9 are examined, the temperature distributions occur along the streamlines. For all inlet configurations, streamlines are shown in Figure 12 and temperature distributions on resistance is related to the formation of streamlines.

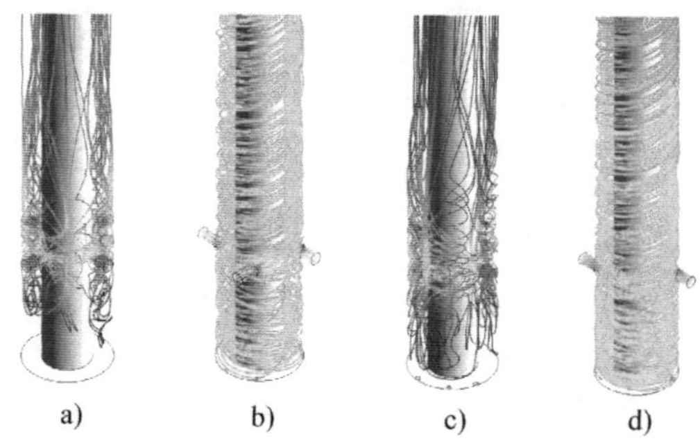

Figure 12. *Streamlines for (a) 4-Sided radial (b) 4-Sided tangential (c) Double sided radial (d)Double sided tangential configurations.*

Radial and tangential air inlet types exhibit different cooling performances from each other. Moreover, 4-Sided inlet hole configurations show worse performance than double-sided in all cases. The average heated wall temperature drop difference is 2°C between tangential and radial

IOP Conf. Series: Materials Science and Engineering **459** (2019) 012014 doi:10.1088/1757-899X/459/1/012014

impingement configurations. However, there is also 0.5-1°C difference between double and 4-sided model in the same locations.

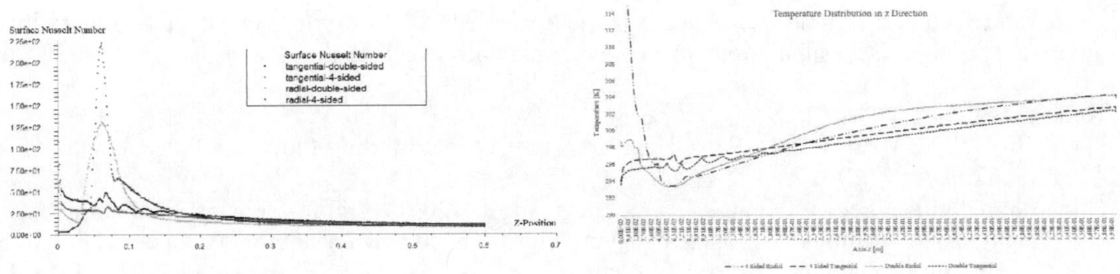

Figure 13. *Distributions of Nusselt number(left) and temperature(right) along z-direction on the heated wall surfaces at Re=1550 for hole section.*

Figure 13 was plotted on the heated wall surface and along the length of the pipe. The temperature on the resistance goes up to 31°C for radial configurations; and up to 29 °C for tangential configurations. In addition, Nusselt numbers were determined as a 9 at the outlet for radial type and 13 for tangential inlet types. All results clearly show that the tangential configurations are more effective than the radial ones on average.

5. Conclusion

Energy efficiency is one of the most significant factors for electronic devices and optimization studies are becoming indispensable to obtain an effective design. As heat exchangers are reduced in size and efficiency is increased, areas of use will become even more widespread.

In this study, the best configuration was obtained according to cooling performance. There are two main design options: one of them is radial air entrance and the other is tangential entrance type. These definitions are made according to the axis of the heated pipe. The variable speed fan which is located at the top of the heat exchanger aspirates air from inlet holes. A total of 12 different simulations were carried out for 3 different Reynolds number and 4 different types of designs. Reynolds numbers are respectively 775, 1160 and 15550. Moreover, there are 4-sided and double sided inlet types for tangentially and radial impingement cases. Heat flux of resistance is constant as 150 W/m^2.

Two equations RNG k-epsilon turbulence model was used in the computations. Moreover, pressure based coupled algorithm and second order upwind schemes were used in the solver. Distributions of Nusselt number and temperature were obtained for all configurations. The most effective cooling is double sided tangential type because flow streamlines are spreading independently and in a more homogenous manner in double sided configurations than 4-sided ones. When radial and tangential inlet types are compared, there are also some significant differences.

First of all, there is high Nusselt number on the impingement surface for radial inlet type because of the fact that the aspirated air which enters from inlet hits to heated surface directly. However, tangential inlet configurations enable a swirl flow and air is passing on the heated surface in a well-distributed fashion. Also the swirl flow effect can be seen clearly at the temperature distributions for tangential types.

Maximum Nusselt number reaches up to 230 at the impingement surface for radial configurations whereas the Nusselt number stays around 120 for tangential configurations. As the flow proceeds towards the outlet, Nusselt number decreases gradually and drops to 9 for radial type and 13 for the tangential type. In terms of temperature, radial inlet type configurations enable surface temperatures that vary from 293.2 [K] to 304.1 [K]. However, the heated wall temperatures vary from 295.8 [K] to302.3 [K] in tangential configurations.

A tangential model should be preferred for a well distributed cooling, but a radial inlet type model can be used if a sudden cooling around the entrance hole region is required. Also closed-bottom side of the heat exchanger has higher temperatures in radial inlet configurations and it has negative effect on general for averaged cooling. The best cooling performance was obtained with double sided tangential configuration at the highest Reynolds number.

In future studies, optimization studies will be carried out by trying new inlet configurations to attain ideal design. CFD simulations will be experimentally verified after the test setup is established.

References

[1] Hay, N., & West, P. D. (1975). Heat Transfer in Free Swirling Flow in a Pipe. *Journal of Heat Transfer,97(3)*

[2] Neto, S. R., Legentilhomme, P., & Legrand, J. (2001). Finite element simulation of mass transfer in laminar swirling decaying flow induced by means of a tangential inlet in an annulus. *Computer Methods in Applied Mechanics and Engineering*,190(35-36), 4713-4731. doi:10.1016/s0045-7825(00)00341-8

[3] Scott, C. J. (1973). *Turbulent annular swirl flows*. Heat Transfer Lab., Univ. of Minnesota.

[4] Rao, Y., Biegger, C., & Weigand, B. (2017). Heat transfer and pressure loss in swirl tubes with one and multiple tangential jets pertinent to gas turbine internal cooling. *International Journal of Heat and Mass Transfer*, 106, 1356-1367. doi:10.1016/j.ijheatmasstransfer.2016.10.119

[5] Zaherzadeh, N., & Jagadish, B. (1975). Heat transfer in decaying swirl flows. *International Journal of Heat and Mass Transfer,18*(7-8), 941-944. doi:10.1016/0017-9310(75)90187-8

IOP Conf. Series: Materials Science and Engineering **459** (2019) 012015 doi:10.1088/1757-899X/459/1/012015

Negative pressure waves in fluid systems – natural and technological phenomena

F H Veliyev

Professor of Baku Higher Oil School, Department of "Oil and Gas Engineering", AZ1025, Khojalu ave.30, Azerbaijan

fuadveliev@hotmail.com

Abstract. The possibility of creation of direct wave of negative pressure in real heterogeneous liquid systems was confirmed experimentally under the certain kinetic and hydraulic conditions. The negative pressure can be considered as a factor of both useful and destroying energies. The new approach to generation of the negative pressure waves in impure, unclean fluids has allowed the creation of principally new energy saving technologies and installations to increase the effectiveness and efficiency of different production processes. It was proved that the negative pressure is one of the main factors causing hard troubles in some technological and natural processes. Received results emphasize the necessity to take into account the role of the negative pressure as an energy factor in evaluation of many transient thermohydrodynamic processes in the Nature and production systems.

1. Introduction

There are numerous results of the experimental work of static and dynamic character, where negative pressure has appeared in one or another degree. In laboratory conditions, negative pressure apparently was first revealed in the experiences made by F. M. Donny (1843), who used degassed sulfuric acid and obtained negative pressure only –0.012 MPa. Among the further attempts of receiving bigger negative pressure, it is worth mentioning the experiences made by O.Reynolds, M.Bertelot and J.Meyer. Basing upon a centrifugal method and using mercury, L.J.Briggs has obtained the record value of negative pressure (- 42.5 MPa) [1].

However, in fact, beginning from the first experiences by F. M. Donny, the main conditions in the investigations for the appearance of negative pressure has been the homogeneous character of the liquid and high degree of the purity the liquid–vessel system. Significant negative pressure has been obtained under those conditions, but the results have not been practically implemented, as real liquids in the Nature and technological processes are mainly heterogeneous and compound systems.

2. Experimental results and their application

The goal of our investigations was to obtain negative pressure in real impure, unclean, heterogeneous liquid systems. The main idea here was reaching the negative pressure due to the sudden character of extending efforts.On this base long-term experimental work has been done to create impulsive negative pressure in real heterogeneous compound liquid systems, such water, oils, solutions and to use the phenomenon of negative pressure for raising the effectiveness of various technological processes.

Figure 1. represents the typical variation of pressure with time in crude oil (ρ=934kg/m^3) stream in two mentioned test points under the initial values of pressure and temperature P_0=0.7MPa, T_0=298K.As one can see, the generated negative pressure appears for a short time, reaching the maximum value of –1.5 MPa during of ~0.04 sec. Nevertheless, it leads to the significant overheating of the liquid and to spontaneous steam generation and gas emission. The process is followed by quick considerable fall in the temperature of the stream (~7°-10°) after which it is slowly restored apparently due to heat transfer process accompanied by processes of cavitation, reverse condensation of steam, dissolving of the extracted gas (Fig.2).

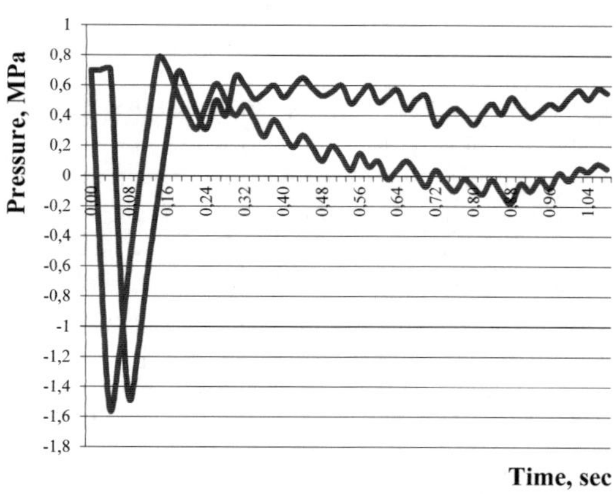

Figure 1. Variation of pressure in two test points of a stream

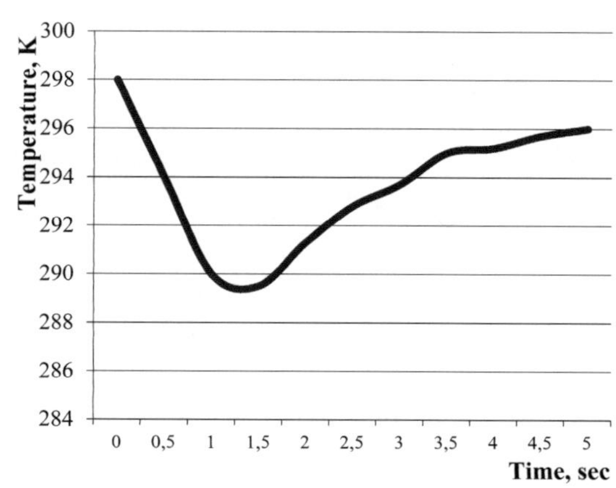

Figure 2. Variation of a stream's temperature

The analogous results are obtained for a tap water, clay solutions.

The important result of the investigation was the discovery of possibility of generation of negative pressure waves in real liquid systems. As one can see from the Figure 1, the starting phase shift between the pulses of negative pressure in two tested points is about 0.03 sec, i.e., the negative pressure as a single wave spreads with an acoustic sound velocity. The wave of negative pressure is like a turned soliton wave with one hump that is negative. It is a rather conservative wave, which mainly keeps its shape reaching the second sensor almost without loss.

On the basis of the results obtained, a method was developed for the artificial generation of negative pressure waves in liquids.The essence of the method is that negative pressure waves can be generated by means of discharge in hydraulic systems when the drop of the pressure takes place during the characteristic time much less than that of pressure relaxation in the system. The greater is the volume of hydraulic system and the higher is the depression of the pressure, the more intensively the negative pressure wave may manifest itself.

IOP Conf. Series: Materials Science and Engineering **459** (2019) 012015 doi:10.1088/1757-899X/459/1/012015

The new approach to generation of the negative pressure waves in impure, unclean fluids has allowed the creation of principally new energy saving technologies and installations to increase the effectiveness and efficiency of different production processes. It was proved that the negative pressure is one of the main factors causing hard troubles in some technological and natural processes.

3. Conclusion

Received results emphasize the necessity to take into account the role of the negative pressure as an energy factor in evaluation of many transient thermohydrodynamic processes in the Nature and production systems and the great profit that can be received from practical adoption of this remarkable phenomenon.

Reference

[1] Hayward, A.T.J., *Negative Pressure in Liquids: Can it be Harnessed Serve Man. American Scientist* **59**, 1971.

IOP Conf. Series: Materials Science and Engineering **459** (2019) 012016 doi:10.1088/1757-899X/459/1/012016

Evaluation of aloe vera extract loaded polyvinyl alcohol nanofiber webs obtained via needleless electrospinning

S Jegina[1], S Kukle[1] and J Gravitis[2]

[1] Riga Technical University, Faculty of Materials Science and Applied Chemistry, Institute of Design Technologies, Kipsalas street 6, Riga, Latvia
[2] Latvian State Institute of Wood Chemistry, Dzerbenes street 27, Riga, Latvia

E-mail: Sandra.Jegina@rtu.lv

Abstract. In this study aloe vera liquid extract was added to PVA to obtain nanofiber mats with integrated bioactive compounds. Nanofibers are effective in drug release process and the most cost and time effective way to produce them is through roller electrospinning method. Several concentrations of PVA and aloe vera extract have been prepared, as well as pure 10 wt% polyvinyl alcohol solution to compare their properties. Viscosity and electroconductivity of solutions were measured, atomic force microscopy was used for evaluation of nanofiber mats morphology and diameter measurements, samples were analysed and compared by Fourier transform infrared spectroscopy and mechanical properties of the nanofiber mats were tested.

1. Introduction

Aloe vera is widely known for its health benefits. Since ancient times people used it topically to treat skin problems and to heal digestive disorders. Nowadays there are many consumer products available in different forms- gels, creams, drinks etc. [1]. Although, there are concerns regarding possible toxicity of orally administrated aloe vera [2], the topical applications are considered as safe unless a person has allergic reaction to the plant itself or the herbal is not applied correctly and cause adverse reactions [3].

To improve topical effectiveness of the biologically active ingredients, researchers worldwide test the possibility of integrating aloe vera extract into nanofibers [4,5].

Electrospinning equipment with a syringe is the most common device mentioned in the published papers, therefore in this work researchers used syringeless (needleless) electrospinning method to evaluate differences in morphology, mechanical and bioactive possessions of the nanofiber mats obtained by added distinct aloe percentage in spinning solution. Needleless electrospinning compared to the conventional syringe-based system has such advantages as high scalability, productivity, web and fibres diameters uniformity, wide choice of polymers and substrates, as well as economical operation and easy maintenance [6].

2. Materials and methods

Polyvinyl alcohol (PVA; 81365 Aldrich Mowiol® 18-88 Mw ~130,000) and liquid aloe extract (AV) obtained from local shop "BB Factory Cosmetics" were used to prepare spinning solutions in different concentrations (Table 1.).

Table 1. Samples of prepared polyvinyl alcohol (PVA) and aloe vera liquid extract (AV) spinning solutions

Composition	AV content wt%
PVA (10 wt%)	0
PVA (10 wt%) and AV	3; 5; 10; 15

PVA and AV were dissolved in distilled water using Biosan magnetic stirrer with hot plate MSH-300. PVA was stirred (1000 rpm) for 2 hours at 120-135 °C and then cooled down to 20°C. AV was stirred (1000 rpm) for 30 minutes without heating in order not to impact bioactive compounds.

Electroconductivity and viscosity were measured before electrospinning as the both values show the possibility of obtaining nanofibers before the production. It is known that for the solution to be spinnable, viscosity value must be 100-2000 millipascal-seconds (mPa-s) [6]. Electroconductivity was measured using universal conductivity meter Greisinger GLF 100.

Nanofiber mats were obtained using Nanospider™ LAB 200 (Elmarco, Czech Republic) with following settings- distance between electrodes- 17 cm, roller speed- 3 rpm, applied voltage- 65 kV, temperature- 20 C°. It is important to set appropriate electrode rotation speed to ensure maximum production rate [7] and distance between electrodes for the solvent to evaporate [8].

Morphology of the nanofiber mats was evaluated and fibers diameters measured on atomic force microscope Dimension Edge Veeco (Bruker, USA) with silicone cantilever OTESPA-R3, f0=300kHz, k=26N/m in tapping mode.

Fourier Transform Infrared (FTIR) spectroscopy analysis was performed on Thermo Scientific Nicolet 6700 spectrometer to characterize electrospun mats.

Tensile strength and elongation at break were tested on Zwick Roell Z2020 universal testing machine.

3. Results and discussion

3.1. Viscosity and conductivity.

Prior to preparing compositions, PVA 10wt% and undiluted AV were measured. Viscosity (mPa-s) of PVA 10 wt% was 411, AV- 0,7, conductivity (μS/cm) of PVA 10 wt% was 390, AV 1084. As shown in Figure 1., viscosity decreases (by 22 %) and conductivity 1.8 times rises when AV concentration in the solution is increased from 0 to 10 wt%.

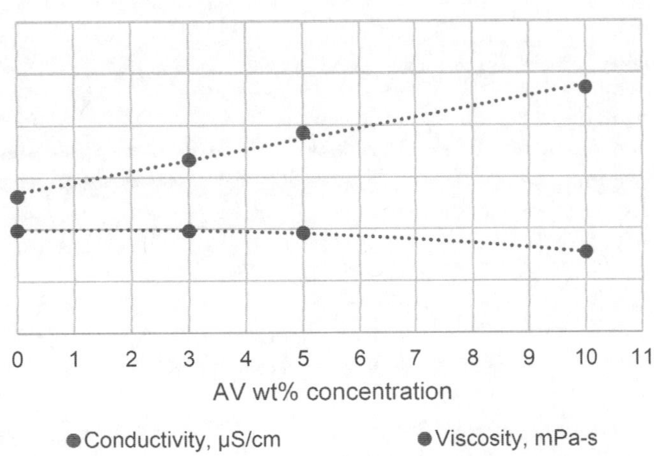

Figure 1. PVA and AV spinning solution properties

Authors concluded that the ratio of conductivity 942 µS/cm and viscosity 395 mPas·s does not maintain the stability of production process as AV 15 wt% and PVA solution did not form quality web.

Growth of conductivity in AV range under investigation could be described by linear equation (1) and increase of viscosity by nonlinear equation (2).

$$Y_C = 41.40\ x + 540.22 \qquad (1)$$
$$Y_V = -1.22\ x^2 + 3.49\ x + 394.46 \qquad (2)$$

3.2. Morphology of nanofiber mats
Morphology of 10 wt% PVA and 3 wt%, 5 wt% and 10 wt% AV is shown in Figure 2. As AFM micrographs show, fibers are well defined. As AV concentration increases, fibers show a slight tendency to become finer.

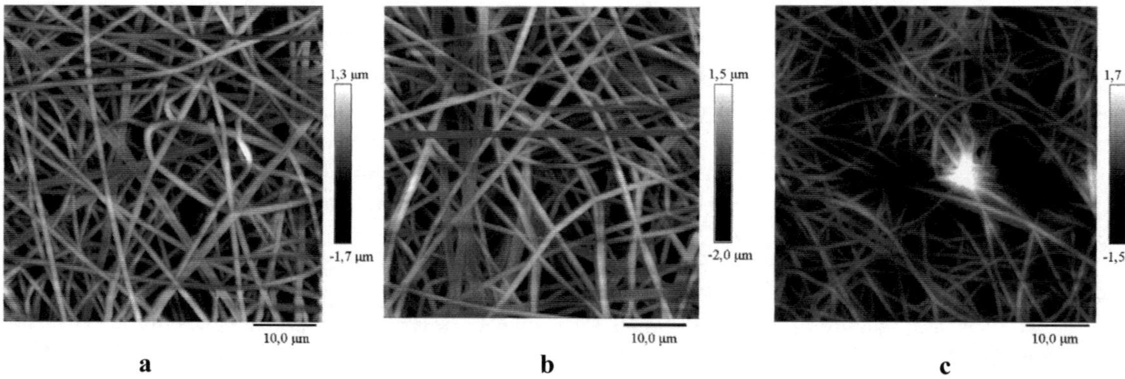

Figure 2. Morphology of nanofiber mats- **a** (PVA+3% AV), **b** (PVA+5% AV), **c** (PVA+10% AV)

3.3. Diameters of nanofibers
Fiber diameters were measured from AFM images.100 diameter measurements of each sample were taken to ensure data reliability, relative error do not exceed 3.5%. In medical applications fibres diameters are important parameters in drug release process [9].

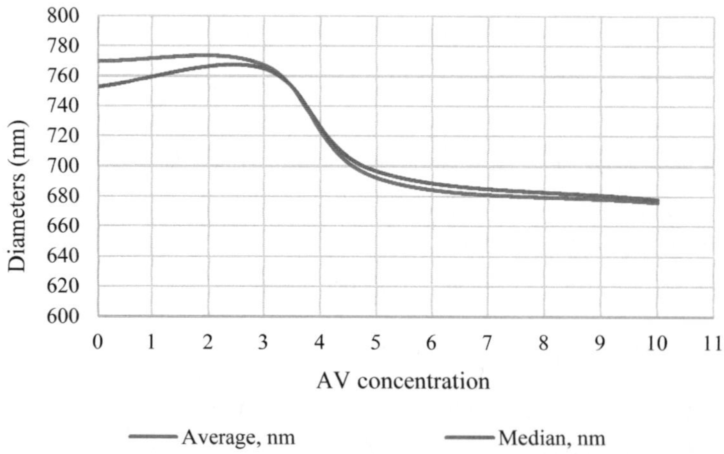

Figure 3. Average and median diameters of nanofibers

As AV concentration rises, diameter values decrease (Figure 3), average and median values are very similar which means that diameters distributions are symmetrical. It is seen from graphs that AV 3wt% is too small concentration to influence fibres diameters. The average diameters drop down by 10 % when AV concentration increase to 5 wt %. Decrease of diameters are moderate in a range of AV concentration increase from 5 to 10 wt%.

While average diameter size of syringe electrospun nanofibers were 123 nm (10% PVA and 5% AV powder) [10], 10% PVA and 5 wt% AV composite nanofibers obtained by needleless electrospinning have average value of 692 nm.

3.4. Fourier transform infrared (FTIR) spectroscopy

FTIR spectra of aloe vera values are in wave length range of 330.69 cm^{-1} to 3398.3 cm^{-1} and the chosen preparation method can either sustain or damage bioactive compounds of the plant [11].

In this study FTIR measurements show (Figure 4.) that there are three new peaks (921, 922.12 and 923.67) in PVA and AV composites compared to pure PVA 10 wt% sample. The results confirm that AV is successfully integrated into nanofiber mats.

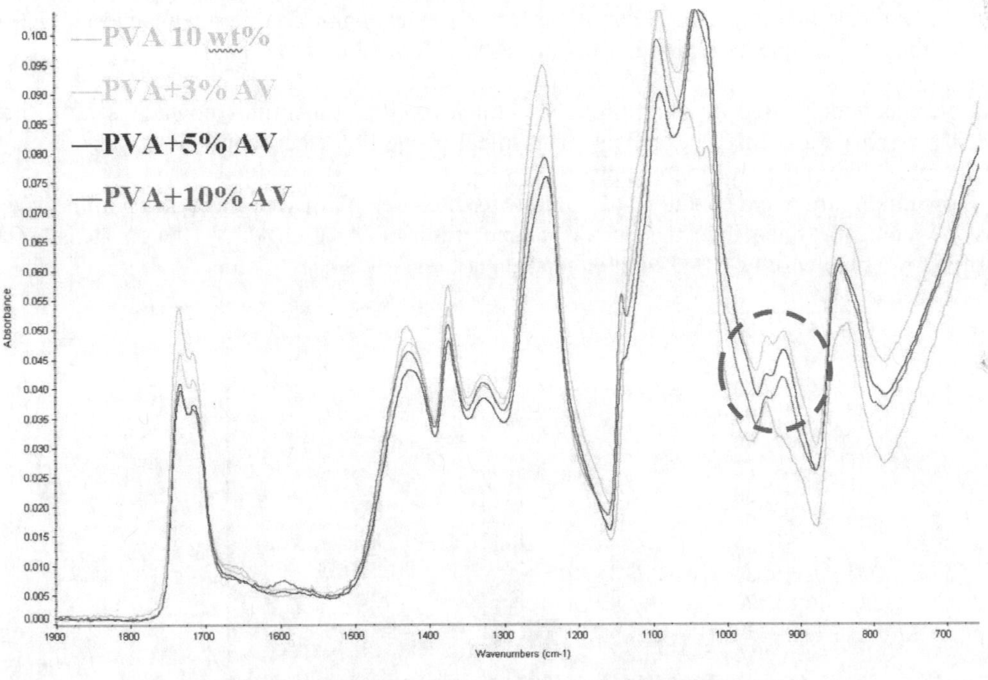

Figure 4. Fourier Transform Infrared (FTIR) spectroscopy

3.5. Mechanical properties of nanofiber mats

Tensile strength and elongation at break are techniques to characterise mechanical properties of nanofibers and nanofiber mats. It is suggested to place fabricated samples on a frame or paper (Figure 5.) to avoid the breakage or slippage of the fibers at the grips and misalignment of the sample as mechanical grippers of the tested device are not suitable for nanofibers [12].

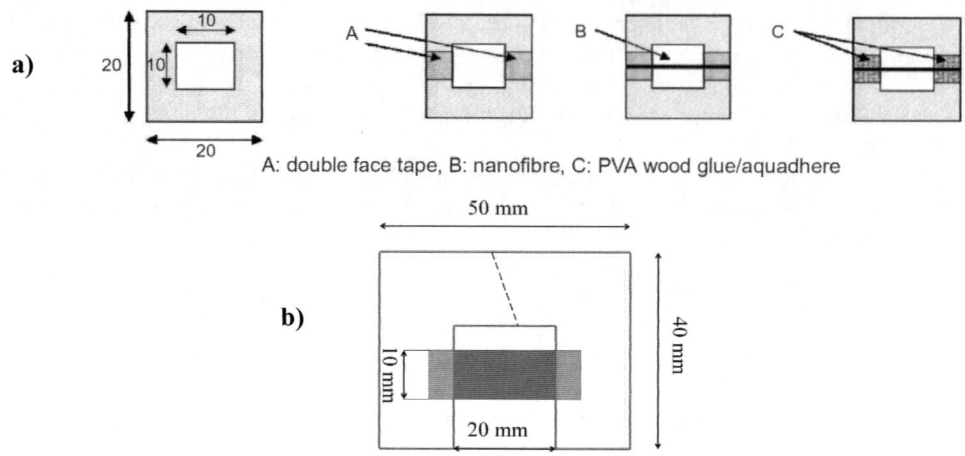

Figure 5. Preparation of **a)** a single nanofiber for tensile testing in Chatillon tensile tester [12] and **b)** nanofiber mat sample in Instron Universal Tester (Model 2519-107) [13]

Aloe vera is widely used in nanofibrous structures for skin substitutes nowadays [14], therefore produced materials should have excellent mechanical properties when applied on skin to guarantee comfort while delivering active ingredients to cells.

As shown in Figure 6.and equations (3) and (4) tensile strength of web increases nonlinear with the increase AV content. At max investigated AV concentration (PVA 10wt %/AV 10 wt %) tensile strength of nanofiber mats increase by 210 % and elongation at break by 80%.

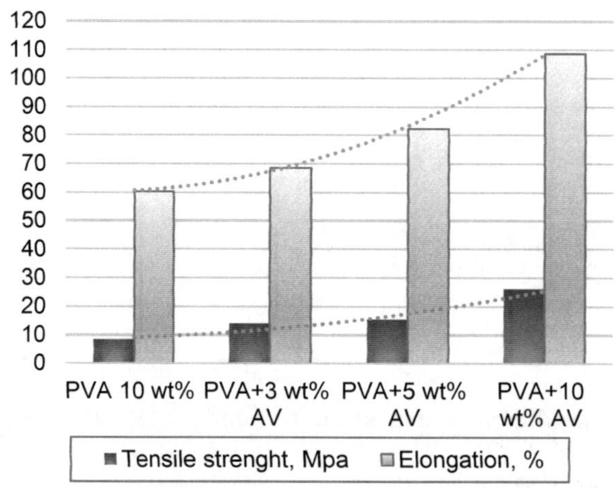

Figure 6. Mechanical properties of nanofiber mats

$$Y_S = 1.25\,x^2 - 0.82\,x + 22 \quad (3)$$
$$Y_E = 4.57\,x^2 - 6.94\,x + 63.04 \quad (4)$$

Where: Y_S and Y_E tensile strength and elongation at break, x - AV concentration, %.

4. Conclusion

The results of this study show that AV in concentrations 3, 5 and 10 wt% added to 10 % PVA solution can be electrospun by roller electrospinning method, but 15 wt% AV does not form quality nanofiber mat.

AV influences viscosity and electrical conductivity of the spinning solution, diameters of nanofibers and morphology and mechanical properties of nanofiber mats.

AV decreases viscosity and increases conductivity of the spinning solution.

Nanofiber diameters tend to decrease when AV concentration is increased.

FTIR spectroscopy analysis confirmed that AV was successfully integrated into PVA nanofiber mat.

Mechanical test results show that AV enhances tensile strength and elongation at break values of PVA nanofiber mat.

Acknowledgments

This research was partially supported by Latvian State Institute of Wood Chemistry (Bioeconomy grant TeKPoD Nr. 03-18).

References

[1] Maharjan H Radha and Nampoothiri P Laxmipriya, 2015, Evaluation of biological properties and clinical effectiveness of Aloe vera: A systematic review *Journal of Traditional and Complementary Medicine* **vol 5** issue 1, pp 21-26.

[2] A Xiaoqing Guo and Nan Mei, 2016, Aloe vera: A review of toxicity and adverse clinical effects *Journal of Environmental Science and Health* Part C **34:2**, pp 77-96.

[3] Ekor M, 2014, The growing use of herbal medicines: issues relating to adverse reactions and challenges in monitoring safety. Frontiers in Pharmacology, 4:177.

[4] Rahman S, Princeton C and Narayan B, 2017, Aloe Vera for Tissue Engineering Applications *Journal of Functional Biomaterials* **vol 8 (1)** 6, pp 1-17.

[5] Balaji A, Vellayappan M V, John A A et al., 2015, Biomaterials based nano-applications of Aloe vera and its perspective: A review. RSC Advances, **Vol.5**, pp. 86199-86213, doi: 10.1039/c5ra13282g

[6] Jegina S, Sutka A, Kukle S and Zelca Z, 2016, The effect of sodium chlorophyllin on polyvinyl alcohol electrospun nanofiber diameters (CYSENI 2016) (Kaunas: Lithuanian Energy Institute), pp 234 – 240.

[7] Yener F., 2014, New methods in the study of roller electrospinning mechanism. Summary of the thesis. Liberec: Technical University of Liberec.

[8] Sutka A, Kukle S, Gravitis J et al., 2013, Nanofibre Electrospinning Poly (vinyl alcohol) and Cellulose Composite Mats Obtained by Use of a Cylindrical Electrode. *Advances in Materials Science and Engineering,* **vol. 2013**, Article ID 932636, 6 pages.

[9] Hrib J, Sirc J, Hobzova R et al., 2015, Nanofibers for drug delivery – incorporation and release of model molecules, influence of molecular weight and polymer structure. Beilstein *Journal of Nanotechnology,* **vol. 6**, pp 1939-1945.

[10] Abdullah N A @Shukry, Ahmad Sekak K, Ahmad M R and Bustami Effendi T J, 2014, Characteristics of electrospun PVA-aloe vera nanofibres produced via electrospinning (ICTEFAD 2014),(Springer Science+Business Media Singapore) pp 7-11.

[11] Narsih A, 2016, Evaluation of bioactive compounds of aloe vera extract using subcritical water method, *BioTechnology An Indian Journal*, Volume **12** Issue 3, pp. 113-120. ISSN: 0974-7435.

[12] Woodhead Publishing Series in Textiles, Volume Number: 186. Electrospun Nanofibers. Edited by Mehdi Afshari, 2017, published by Woodhead Publishing in association with The Textile Institute, pp. 30-31, ISBN: 978-0-08-100907-9 (print)

[13] Sutka A, 2015, Technologies, Structure and Properties of Lignocellulosic Nanofibre Bicomponent Filaments. PhD Thesis. Rīga: [RTU], pp 80.

[14] Baghersad S, Hajir Bahrami S, Mohammadi M R et al., 2018, Development of biodegradable electrospun gelatin/aloe-vera/poly(ε-caprolactone) hybrid nanofibrous scaffold for application as skin substitutes, *Mater. Sci. Eng. C* 2018, **93**, 367–379, doi:10.1016/J.MSEC.2018.08.020.

IOP Conf. Series: Materials Science and Engineering **459** (2019) 012017 doi:10.1088/1757-899X/459/1/012017

Electrical connection issues on wearable electronics

D Matsouka[1], S Vassiliadis[2] X Tao[3], V Koncar[3], S K Bahadir[4], F Kalaoğlu[4] and S Jevšnik[5]

[1]Thetametrisis SA, Polidefkous 14, Athina 122 43, Greece
[2] Piraeus University of Applied Sciences, Department of Electronics Engineering, Thivon 250, Egaleo 122 41, Greece
[3] ENSAIT, 2 Allée Louise et Victor Champier, 59056 Roubaix, France
[4] İTÜ Tekstil Teknolojileri ve Tas. Fakültesi, İnönü cd. No.65 Gümüşsuyu-Taksim 34437 İstanbul Turkey
[5] INLAS, Grajska ulica 3, Slovenske Konjice, Slovenia

ptosky@gmail.com

Abstract. The drive for integration of electronic components into textile substrates has been a major aspect of research into production of smart textiles. Electronic components as such, are rigid, inflexible, do not easily follow body movements and can cause hypersensitivity of the skin upon prolonged contact due either to the presence of irritants or the rigid nature of the materials. Another important disadvantage of electronic components, in wearable applications, is that while textiles and textile like materials can withstand the usual cleaning/ care treatments, those same treatments can have detrimental effects on the transmission lines and on any exposed electronic parts. This paper investigates a method for the protection of transmission lines made of conductive yarn seams on specimens containing both electronic and textile elements integrated on a textile substrate as well as the protection of the electronic parts. The reliability and durability of the transmission line on the combined specimen is determined by the measurements of the value of an electrical resistance located on a rigid PCB, after repeated wash cycles.

1. Introduction

Innovative smart textile structures [1-4] that endeavour to incorporate electronic components usually face a host of difficulties that stem from the fact that textile structures and textile materials in general are characterised by a quiet different set of characteristics, that can seem incompatible with the characteristics of electronic components. Apart from the requirements for high flexibility, conformity to body shape, in wearable applications, while textiles and textile like materials are designed to withstand the usual cleaning/ care treatments, those same treatments can have detrimental effects on the transmission lines and on any exposed electronic parts [5-7].

The research described in this paper, was carried out within the framework of Project Welding of E-Textiles for Interactive Clothing-E-TexWeld. The aims of the project are i) the design of transmission lines of e-textile structures, ii) integration of electronic elements (different sensors, actuators, microprocessors, data transmission and power supply systems) to textile structures, and ii) the design and development of whole e-textile system for protective clothing applications including interactive protective garments and shoes. A significant application of personal protective equipment is the bunker gear or turnout gear used by firefighters when they need to handle an incident. This turnout gear mainly consists of a multilayered-jacket and multilayered-trousers (as well as boots). The layers of the uniform usually consist of fabrics made off a mixture or pure aramid fibres. One of the most well-known companies producing fibres and fabrics specifically for firefighter PPE (personal protective equipment) is DuPont with the trademarked Nomex flame-resistant meta-aramid material.

Protection of the electronic components of the combined specimens was carried out using two strategies; the rigid PCB containing the electrical resistance was encapsulated using silicon, and the conductive seams, comprising the transmission lines were covered on both sides of the textile substrate (single jersey knitted fabric), using TPU (thermoplastic polyurethane) thermoplastic film.

2. Materials - Methods

2.1. Materials

The combined specimens (Figure 1) consist of textile and electronic elements integrated on a single jersey knitted fabric (100% Nomex). The electronic elements comprise of a rigid printed circuit board (PCB) that includes a 10 Ohm resistance. The PCB is soldered onto electrodes made of flexible PCB (Kapton). Then the PCB is encapsulated in silicon. The textile elements are textile-based EKG electrodes (provided by TTRI Taiwan Textile Research Institute), which are connected by conductive thread seams, to the flexible PCB electrodes. The conductive thread (Shieldtex 234/34-2 ply HC) is silver coated polyamide. The thread is two-ply with a finesse of 234dTex and 32 filaments per thread ply (total of 68 filaments); its linear resistance is less than 100Ω/m. All conductive thread seams are covered by a thermoplastic polyurethane (TPU) film (thickness 1mm) on both sides of the seam. The TPU film is thermoset using a press at 180 ^{0}C for 20sec.

Figure 1. Test specimen before encapsulation of the rigid PCB

2.2. Methods

Washing test. In order to test the durability of the combined specimen, successive washing cycles were carried out. Washing was realized according to ISO 6330 [8] which describes standard washing and drying procedures for textiles. The washing temperature was 40^{0}C and the duration was 30 mins using a Datacolor Ahiba IR laboratory dying machine. Between washing cycles, but not necessarily between every cycle, specimens were drip dried and their functionality was determined. Fifty wash cycles were carried out in total.

Measurement of electrical resistance. The value of the resistance on the rigid PCB was measured on the EKG electrodes after each wash, up to and including the tenth wash and every five washes after that up until the fiftieth wash cycle. The nominal value of the resistance on the dummy PCB is 10 Ohm however when measured on the ECG electrodes the value is 12.83 Ohm. This resistance is the sum of resistance on the rigid PCB, the resistance of the flexible PCB, the resistance of the conductive yarn seams, the resistance of the ECG electrode as well as the sum of the contact resistances between the conductive thread the flexible PCB and the ECG electrodes.

3. Results and Discussion

The evolution of the value of the mean resistance measured on the EKG electrodes during the full set of the 50 wash cycles can be seen on Figure 2. The values of the minimum and maximum resistances measured are also plotted on Figure 2. All measurements have been normalised by dividing them with the value of the resistance measured in the pristine condition. The values of the minimum, average and maximum values measured on the EKG electrodes don't appear to vary significantly as can be seen in Figure 2. This can be attributed to the very high conductivity of the EKG electrodes.

After the fiftieth wash cycle the mean resistance was approximately four times the value of the resistance of the pristine specimen, which is considered an acceptable value for EKG applications. The behaviour of the resistance of the combined sample is in concordance with the behaviour observed in the research by Tao et al. [9] who also carried out washing tests on samples containing conductive threads and flexible PCB electrodes. This increase in the value of the resistance is considered to be connected with an increase of the contact resistance between the fibres of the conductive thread and the EKG electrodes. From the macroscopic examination of the specimens no defects where detected either on the areas covered by the TPU tape (peeling, cracking), or on the silicon coating of the rigid PCB.

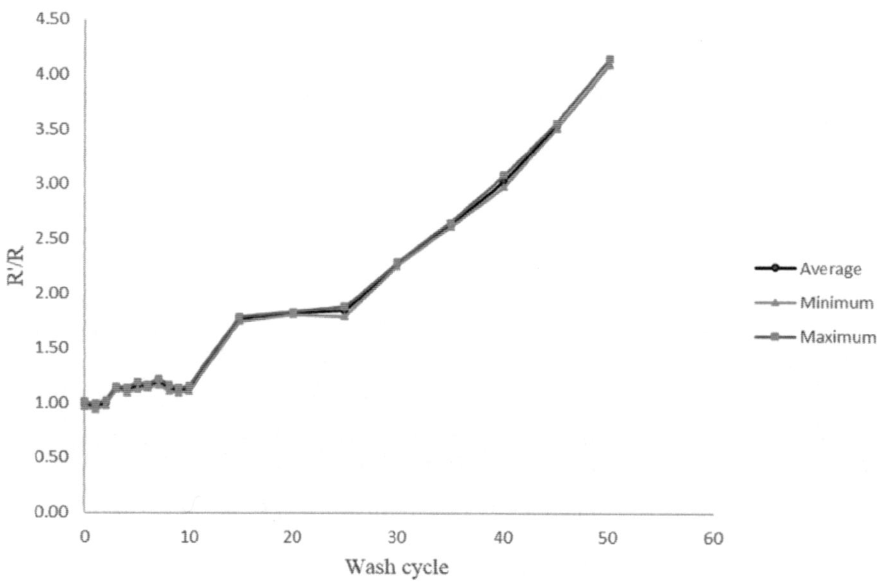

Figure 2. Resistance measurements vs number of wash cycles

4. Conclusions

This study was carried out within the framework of ETexWeld project. One of the goals of the project is the integration of electronic elements (various sensors, actuators, microprocessors, data transmission and power supply systems) to textile structures. Integration usually fails due to the weakness of electronic elements to the usual cleaning/ care procedures commonly used for garments (most often washing in a detergent solution with added agitation). Based on the results of this study, we can conclude that use of silicon to encapsulate electronics vulnerable to contact with water and mechanical action appears to ensure their functionality even after 50 wash cycles. Moreover, the thermoplastic polymer film offers satisfactory protection for the conductive silver coated thread used for the transmission lines.

Acknowledgments
The research, described in this paper, relates to the project, which has received funding from the European Union's Horizon 2020 research and innovation program under the Marie Sklodowska-Curie grant agreement no. 644268, project title: Welding of E-Textiles for Interactive Clothing-E-TexWeld.

References

[1] Balint R, Cassidy N J, Cartmell S H 2014 Conductive polymers: towards a smart biomaterial for tissue engineering *Acta Biomater.*, **10**, 2341-53
[2] Yang C M., Yang T L, Wu C C, Hung, S H, Liao M H, Su M J, Hsieh H C 2014 Textile-based capacitive sensor for a wireless wearable breath monitoring system *Proc. 2014 IEEE Int. Conf. on Consumer Electronics (ICCE)* (Las Vegas, NV) (Piscataway, IEEE), 232-3
[3] Takamatsu S, Kobayashi T, Shibayama N, Miyake K, Itoh T 2012 Fabric pressure sensor array fabricated with die-coating and weaving techniques *Sens. Actuators Phys.*, **184**, 57-63
[4] Vassiliadis S, Jevšnik S, Potirakis S, Stathopoulos N 2016 Textile based discreet electrical components *Proc. 16th World Textile Conf. Autex* (Ljubljana, Slovenia), (Ljubljaba, Univerza v Ljubljani, Naravoslovnotehniška fakulteta)
[5] Linz T, et al. 2005 Embroidering electrical interconnects with conductive yarn for the integration of flexible electronic modules into fabric *Proc. Ninth IEEE Int. Symp. on Wearable Computers (ISWC'05)* (Osaka, Japan) (Los Alamitos, California, Washington, Tokyo, IEEE) 86
[6] Stoppa M, and Chiolerio A 2014 Wearable Electronics and Smart Textiles: A Critical Review *Sensors*, **14**(7), 11957.
[7] Kallmayer C, et al.2003 New Assembly Technologies for Textile Transponder Systems *Proc. 53rd Electronic Components and Technology Conf.* (New Orleans, Louisiana, USA) (Piscataway, IEEE) 1123
[8] ISO, ISO 6330:2012 Textiles -- Domestic washing and drying procedures for textile testing,2012
[9] Tao X, Koncar V, Huang T, Shen C, Ko Y and Jou G 2017 How to Make Reliable, Washable, and Wearable Textronic Devices *Sensors*, **17**(4), 673.

IOP Conf. Series: Materials Science and Engineering **459** (2019) 012018 doi:10.1088/1757-899X/459/1/012018

Numerical investigation drag coefficient of micro-encapsulated roughened PCM particles in laminar flow

E P Papazoglou[1], D Bouris[2] and K S P Nikas[1]
[1]Heat Transfer Laboratory, Mechanical Engineering Department, University of West Attica, Thivon & P. Ralli Ave. 250, Aigaleo, Greece
[2]Laboratory for Innovative Environmental Technologies, School of Mechanical Engineering, National Technical University of Athens, 157 80, Zografou, Greece

ksnikas@puas.gr

Abstract. Phase Change Materials (PCM) have received considerable attention in recent years in many thermal energy storage applications, due to the isothermal phase change process and large storage capacities. This study deals with the hypothetical structure of micro encapsulated phase change materials of matrix type and the particle's surface roughness influence on its laminar drag coefficient. The investigation has been performed by numerical simulation using commercially available software. Results show that the drag coefficient decreases as particles' surface roughness increases, even at low Reynolds numbers.

1. Introduction

Phase Change Materials (PCM) are substances, which have the ability to absorb and release large amounts of heat energy, by melting and solidification, under almost constant temperature. This feature makes them suitable forapplications of thermal comfort in buildings, thermal protection, cooling, air conditioning, solar heating systems, etc,gathering the interest of several research teams worldwide, especially in the last decade.

One of the productiontechniques used to avoid loss of material and to reduce any interaction (beyond heat) with the environment, whilealso controllingthe volume change during the phase change process, is that of encapsulation. PCM particles are surrounded by a continuous membrane of inorganic or polymeric material, resulting in a capsulein the size range of 10^{-6}-10^{-3}m. These capsules may have spherical, tubular, oval or irregular shape and their internal morphology may be polynuclear, mononuclear or matrix. The encapsulated type is used either in the form of a powder or dispersed in fluids. In the latter case, problems may occur, such as the increase of the fluid viscosity (with impact on pumping), the reduced resistance to deformation, with consequent breakage and subcooling, etc. The recent production of nano-capsules promises to resolve some of these problems [1, 2].

Although the phase change materials' melting and solidification processes have been studied macroscopically both experimentally and computationally, thermofluidic phenomena have not been studied, until now, at the level of micro (nano) capsules [3].

Encapsulated PCM production may be performed by various physical, chemical and physico-chemical methods. In the case of spray dryers, PCM capsules are produced in matrix and polynuclear type. The nature and possible periodicity of the microstructure of the outer surface is unknown, [4].

For the computational investigation of the drag coefficient ofa nano capsule PCM, a realistic and precise geometry of the outer surface of the particle needs to be simulated. Due to the very small size of the capsule, the surface stresses are expected to play a crucial role in the formation of the stereo structure.

2. Problem Description

The scope of this work is to determine the drag coefficient of particles of different surface roughness at low Reynolds numbers and to investigate any possible relationship between these quantities.

The simulation is based on PCM particles produced by the substances [5], paraffin wax (as phase change material) and polyethylene (as matrix material), duringevaporation of organic solvent heptane, in a spray dryer.The ratio of PCM to matrix is varied from 2:1 up to 1:1 by weight, converted to volume ratio using the density of the relative substances.

Since the repetitive spatial geometry of the produced encapsulated PCM particles in a spray dryer is unknown, a hypothesis is made, that these may be in the form of (a) the Weaire-Phelan structure and (b) a Cartesian one, as illustrated in Figure 1.

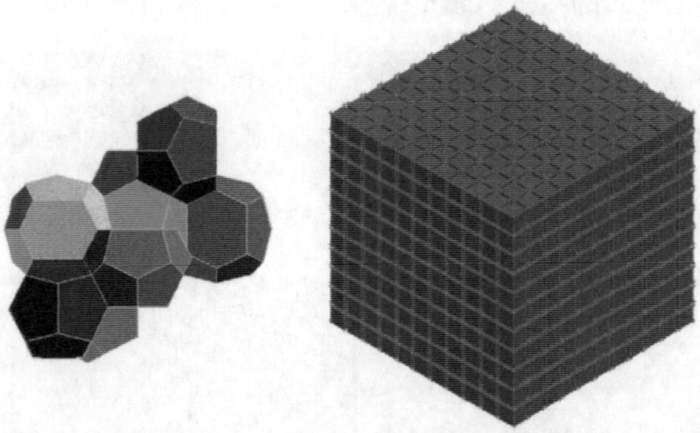

(a) (b)

Figure 1: (a) Weaire – Phelan structure (b) Cartesian structure

Both geometries have been generated in the commercial software Gambit by Ansys, spatially repeated and then intersected with spheres of various diameters, in order to obtain particles with different roughness. The resulting outer surface cavities represent empty spacei.e. without phase change material, framed by rigid polyethylene,duringproduction in the spray dryer. The final particles generated are illustrated in Figure 2.

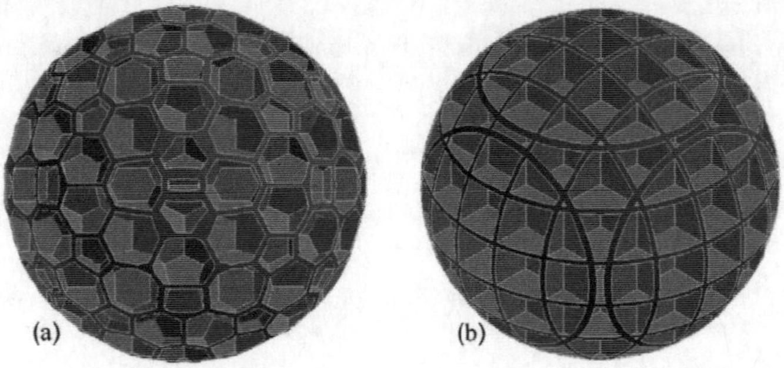

Figure 2: (a) Weaire – Phelan rough particle (b) Cartesian rough particle

The mean diameter, $D_m=2r_m$, of each generated particle has been definedas the one that splits the total volume (V_{rough}) of peaks and voids, of the outer particle surface, intotwo equal parts. The roughness, ε, is then calculated using the following formula, as defined in ISO 25178-2 standard [6]:

$$\varepsilon = \frac{V_{rough}}{4\pi r_m^2} \tag{1}$$

The relative roughness of each particle is determined as ε/D_m.

3. Numerical Details

A 3D flow field of adequate dimensions 20D x 20D x 20D has been introduced, so as to avoid blockage effects. The particle center is at 5D from the inletin the X direction.

The flow field has been meshed using mostly hexahedral cells of various sizes, whilst the cube surrounding the spherical particle has been meshed using unstructured tetrahedral cells. The overall mesh varied from 2.1M up to 12M grid cells for the particles under investigation. In order to improve solution accuracy,grid refinement was applied tothe area around the particle, where the velocity and pressure gradients are expected to be more pronounced. Correspondingly, larger grid cells were located near the sides of the flow field. Figure 3 illustrates the mesh generated for the particles.

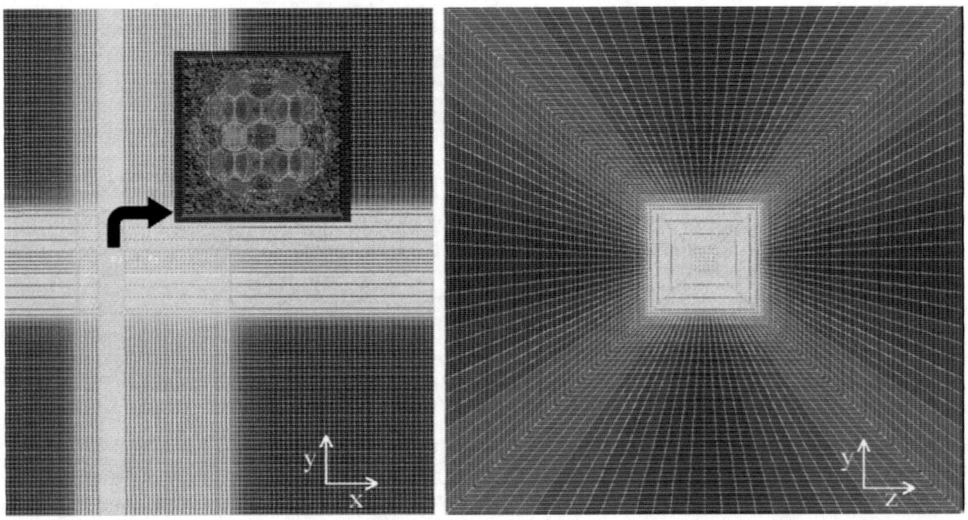

Figure 3: Computational grid in YX and YZ planes. Flow is always in the +X direction

Boundary conditions for the computational investigation were a prescribed uniform velocity (U_o) at the inlet corresponding to $Re=U_oD_m/\nu=4$, a constant gradient outlet and symmetry conditions on allother sides of the computational domain. All the computations were performed assuming steady state conditions and incompressible air flow, whilst a second-order discretisation scheme was used.

To check whether the solution was dependent on the grid created, the mesh was refined in all planes resulting in ahigher number of cells and the flow field for this grid was calculated as well. The difference between the two solutions was negligible. Indicatively, a 55% increase on the overall grid size, resulted in 0.012% change of the drag coefficient.

Convergence of the numerical solution was monitored by means of continuity residuals, the net imbalance between incoming and outgoing flow rates, as well as the drag coefficient stabilisation.

4. Numerical Results

The predicted drag coefficient, c_d,at Re=4, is presented in Table 1 together with the calculated mean diameter of each particle, its relativeroughness ε/D_m, and ratio of total particle surface (A_t) per smooth sphere surface, A_{sphere} (based on mean diameter for all particles).

Table 1: Results of Drag Coefficient

D_m (×1e-06)	ε/D_m	A_{total}	A_{sphere}	A_t/A_s	c_d
3.044	0.132	14698.5647	2914.04061	5.04	7.14
3.681	0.092	14134.5194	4256.2279	3.32	7.56
3.28	0.073	9383.65012	3378.86189	2.78	7.50
4.337	0.064	16137.3082	5909.63639	2.73	7.66
2.917	0.055	43488.0416	10695.7379	4.07	7.69
2.726	0.052	23073.4912	9334.94573	2.47	7.87
3.139	0.042	32522.9769	12380.4706	2.63	7.92
3.8	0	3378.861888	3378.861888	1	8.67

The drag coefficient results, at Re=4, are plotted versus roughness as illustrated in Figure 4.

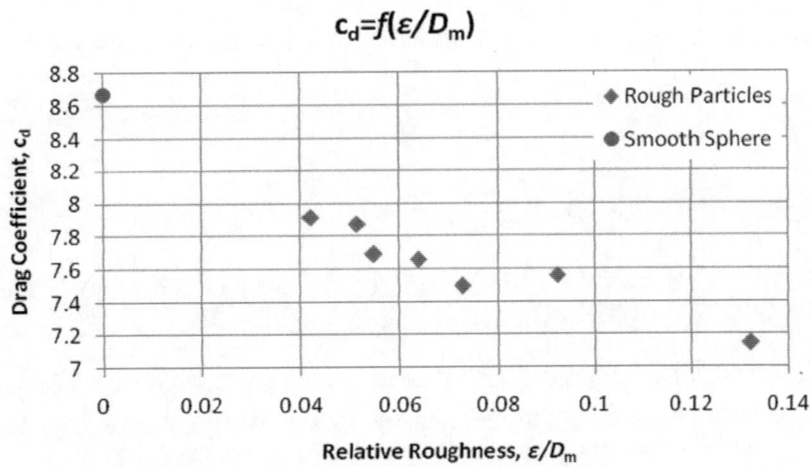

Figure 4: Drag coefficient computation results plot

The results show that the drag coefficient of rough particles is lower than the one for the smooth sphere (c_d=8.67).Moreover, the drag coefficient decreases as the particle's roughness is increased.

5. Conclusions

Although, it is considered that roughness does not influence the drag coefficient in laminar flow [7], according to the computational results, the particles' roughness affects the drag coefficient, even at low Reynolds numbers.

Further investigation can be made at lower values of roughness, as well as at different Reynolds numbers in order to verify the range of the observed behavior. A further parametric analysis could be made by investigating the effects of different PCM to matrix ratios. A future work will include the thermal response of the particles as a function of time.

References

[1] Jamekhorshid, A., Sadrameli, S.M., Farid, M. (2014). A review of microencapsulation methods of phase change materials (PCMs) as a thermal energy storage (TES) medium.*Renewable and Sustainable Energy Reviews*, **31**, pp. 531–542

[2] Jessica Giro-Paloma, MònicaMartínez, Luisa F.Cabeza, A. Inés Fernándeza, Types, methods, techniques, and applications for microencapsulated phase change materials (MPCM): A review, Renewable and Sustainable Energy Reviews **53** (2016) 1059–1075

[3] E. Assis, L. Katsman, G. Ziskind, R. Letan, Numerical and experimental study of melting in a spherical shell, International Journal of Heat and Mass Transfer **50** (2007) 1790–1804

[4] Particle Sciences, Technical Brief 2010 Volume **7**, [online] available from: http://www.particlesciences.com/docs/technical_briefs/TB_2010_7.pdf [accessed 28/02/2017]

[5] Gravalos M., Calvo, H., Mieres R., Cubillo C., Borreguer S., Carmona F., Rodriguez R., Valverde P. (2009). Procedure for microencapsulation of phase change materials by spray-drying. *EUROPEAN PATENT APPLICATION,* EP 2 119 498 A1

[6] International Organization for Standardization (2012). Geometrical product specifications (GPS). Surface texture. Areal. Terms, definitions and surface texture parameters (ISO 25178-2:2012). Retrieved from https://www.iso.org/standard/42785.html

[7] Munson, Young, Okiishi's, (2016) *Fundamentals of Fluid Mechanics*. Edition 8: Wiley.

The impact of wear on mechanical properties of roving

M Pechočiaková[1] **and M Novotná**[2]
[1]Technical University of Liberec, Textile Faculty, Department of textile material engineering, Studenstká 2, Liberec, Czech Republic
[2]LUKOV Plast spol. s r.o., Zámecká 191, Český Dub, Czech Republic

miroslava.pechociakova@tul.cz

Abstract. The paper deals with the influence of mechanical wear out on the resulting mechanical properties of roving. The paper deals with the scraping of roving from inorganic fibres. Glass, carbon and basalt roving are used for this experiment. Rovings are mainly used for the production of composites. The preparation of the roving for processing into composites includes their coiling of the supplied large coils on the coils used in the manufacture. With this rewinding, mechanical damage to roving can occur, which could result in loss of strength. The aim of this work is to determine whether the values of the mechanical properties change depending on the wound generated during winding.
For this purpose, a special Cycle Abrasion Apparatus has been developed to simulate wiping during winding. The work is first aimed at finding the optimal measurement procedure on the Abrasion Apparatus. Further, the results of weight loss, loss of strength, particle removal and measurements on the Zweigle G522 instrument are processed.

1. Introduction

1.1 Description of used material

It was focused on carbon-fibre, basalt and glass (GF-glass fibres) in the form of roving, which are mainly used in the production of composites. The mechanical properties of several types of fibres from different manufacturers were selected and tested. These are in particular carbon fibres from four different manufacturers, glass fibres and basalt fibres. Description of used fibres you can see on table 1.

Table 1. Sample of rovings

Material	Type	Number of fibril	Fineness [tex]
Glass 907	Glass-E PR 440	16 K	2400
Glass 908	Glass-E PR 440	16 K	1200
Carbon	E HTS40 F13	24 K	1600
Carbon	J STS40 F13	24 K	1600

1.2 Measurement description Apparatus

For measurement of roving we have specially built an abrasive equipment. It consists of a pulley and a large wheel that are powered by a motorcycle. It also has two pulleys that are firmly attached. The fibres are guided over the wheel and between the pulleys. After abrasion, roving was captured in frames and subjected to a tearing test.

Figure 1. Apparatus for simulation of wear of roving

1.3 Measurement description Zweigle G522

The planes were scraped off by roller-coated sandpaper, which rotated when moving, see figure 2. The roller travelled in a certain section under roving back and forth, that is, in the cycle. Due to the wicking of the separated tear-off fibres on the roller, the abrasion to the entire roving layout was not possible and only cycling was selected.

Figure 2. Zweigle G 522 with carbon roving

2. Experiment

2.1 Results of measurement after simulation on Apparatus

Simulation was done in 50 cycles. Abrasions samples was use for testing of mechanical properties. Each sample were fixed with glue to paper frame for simple manipulation. Measurement was make for 10 sample of each type of testing sample.

Table 2. Weight decrease [‰] of glass and carbon samples after using apparatus for wears simulation

Speed of simulation [m.s^{-1}]	Glass 907	Glass 908	Carbon E	Carbon J
0,7	0,07	0,34	0,19	0,19
1,4	0,18	0,54	0,35	0,28
2,1	0,22	0,21	0,28	0,58

Table 3. Average length/ minimal length [mm] of glass and carbon samples after using apparatus for wears simulation

Speed of simulation [m.s^{-1}]	Glass 907	Glass 908	Carbon E	Carbon J
0,7	8,4/1	3,6/2	14,6/1	17,8/1
1,4	5,5/1	6,3/1	14,4/1	11,6/1
2,1	4,7/2	11,3/2	11,5/1	14,7/1

Table 4. Strenght [N] of glass and carbon samples after using apparatus for wears simulation

Speed of simulation [m.s^{-1}]	Glass 907	Glass 908	Carbon E	Carbon J
without simulation ("raw")	1476,4	735,8	1900,9	1596,7
0,7	1425,3	669,1	1999,3	1733,1
1,4	1431,6	716,6	2026,5	1784,9
2,1	1392,1	718,0	2005,6	1746,0

Table 5. Elongation [%] of glass and carbon samples after using apparatus for wears simulation

Speed of simulation [m.s^{-1}]	Glass 907	Glass 908	Carbon E	Carbon J
without simulation ("raw")	3,0	2,8	1,8	1,8
0,7	3,3	2,9	2,1	1,9
1,4	3,3	3,0	2,0	1,9
2,1	3,3	3,1	2,0	1,9

Table 6. Module of elasticity [GPa] of glass and carbon samples after using apparatus for wears simulation

Speed of simulation [m.s^{-1}]	Glass 907	Glass 908	Carbon E	Carbon J
without simulation ("raw")	51,0	55,5	117,6	100,4
0,7	45,8	48,4	105,6	101,7
1,4	45,6	49,9	111,4	102,3
2,1	43,8	48,5	110,5	103,1

2.2. Results of measurement after simulation on Zweigle G522

Table 7: Strenght [N] of glass and carbon samples after abrasion on Zweigle G522

Number of abrasion	Glass 907	Glass 908	Carbon E	Carbon J
0	1476,4	735,8	1900,9	1596,7
50	-13,4 %	-20,0 %	-3,4 %	-3,3 %
100	-13,8 %	-20,5 %	-5,2 %	-7,4 %

You can see different behaviour due to tensile testing of glass and carbon rovings on figures 3 and 4.

Figure 3. Behaviour of Glass roving in process tensile testing

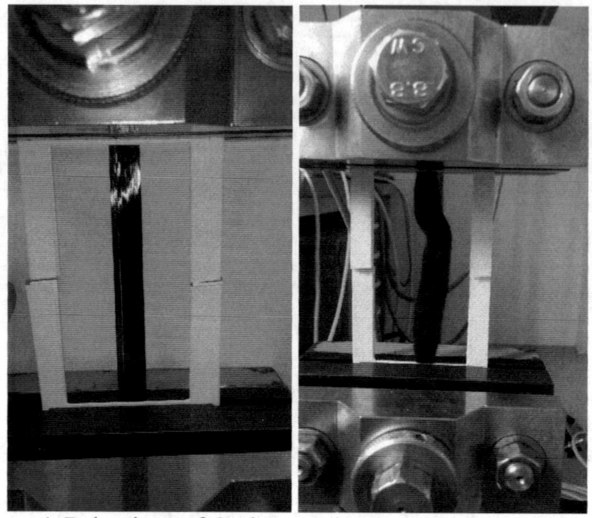

Figure 4. Behaviour of Carbon roving in process tensile testing

3. Conclusion

As a result of the wicking of individual fracture fibres on the pulleys, there was a greater serious drop in some roving. Part of the measurement was to determine the lengths of the contaminated particles during device measurements. Particle collection was performed in the three most undesired places under the Aparatarium. For the wear of roving, the Apparatus for abrasion proved to be the most suitable way of rubbing.

As results of measurement tensile properties of Glass and Carbon roving shows change which is not so significant. We can suppose: Ordinary damage of rovings, during processing, are not significant for properties of final products.

References

[1] Chul-Jin M, In-Hoon K, Bae-Hyeon Ch, Jin-Hwe K, Jin-Ho Ch., *Buckling of filament-wound composite cylinders subjected to hydrostatic pressure for underwater vehicles applications.*, (Composite Structures **92**- Republic of Korea) (2010) 2241–2251

[2] Martynkova S G, *Nové technické materiály.* 1. vydání. Brno: Akademické nakladatelství CERM, s.r.o., 2010. 99 s. ISBN 978-80-7204-714-7

[3] Gay D., Gambelin, J, *Dimensionnement des Structures, une Introdution,* Hermes Science Publishing Ltd., London, 1999. p. 680, EAN13:9782746200494

[4] Grégr J, Slavík M, Šilcová M, *Povrch uhlíkových vláken* (Technical University Liberec) 2007

[5] Kratochvil, B., Švočík V, Vojtěch D, *Úvod do studia materiálů.* Praha: Vysoká škola chemicko-technologická, 2005.

[6] Ptacek L, *Nauka o materiálu. II. 2.* opr. a rozš. vyd. Brno: CERM, 2002. 392 s. ISBN 80-7204-248-3

[7] Chung, Deborah D L, *Composite materials; functional materials for modern technologies.* London: Springer, 2003. 289 s. ISBN 185233665X

[8] Neckar B, *Morfologie a strukturní mechanika obecných vlákenných útvarů.* Fakulta textilní TUL, 1998.

[9] Hashin Z, Rosen B W, *The elastic moduli of fiber reinforced materials*, Journal of Applied Mechanics, Trans ASME, Vol.**31**, 1964, pp.223-232.

[10] Ernestova M, Syrovatkova M, *The simulation of fiber wear on a revolving pulley* (8th International Conference Texsci 2013) ISBN 978-80-7372-989-9

[11] Novotna M, *Comparison properties of the selected inorganic fibers* (Proceedings of the 8th Aachen-Dresden, International Textile Conference) 2014. ISSN 1867-6405

Environmental durability of externally natural fabrics reinforced polymer concrete

G Priniotakis[1], H Bouguessir[2], E Harkati[2], M Rokbi[3] and S Vassilliadis[4]

[1] Department of Textile Engineering, University of West Attica, Thivon 250, Egaleo 122 41, Greece.
[2] Department of Civil Engineering, Faculty of Sciences and Technology, Larbi Tebessi TEBESSA University 12002 Constantine's Road, TEBESSA, Algeria
[3] Mechanical Engineering Department, Faculty of Sciences and Technology, Mohamed Boudiaf M'SILA University. 28000, M'SILA, Algeria
[4] Department of Electronic Engineering, University of West Attica, Thivon 250, Egaleo 122 41, Greece.

gprin@uniwa.gr

Abstract. The quick emergence of composites since their invention due to their high specific properties do not gainsay the health problems associated with the use of synthetic fibers as reinforcing agent of almost all commercially available polymer composites. To avoid those hazardous effects, the current global tendency is the incorporation of lignocellulosic fibers instead of these mineral ones owing to their biodegradability, low cost and low density, renewability and aspect ratio in addition to their high mechanical properties. A new material's concept currently applied in Algeria, based on an externally GFRP polymer concrete, proved its efficiency in piping systems. The design merging composite materials with polymer concrete in laminated form is used to get the most out of these two materials such as: more economic final material with high mechanical properties in tension also in bending in addition to a better durability properties. The aim of the present work is the substitution of glass fabrics and short fibers reinforcing the polymer materials in the current material concept by natural fibers and fabrics made up of jute plant. To assess the possibility of using this new concept reinforced with jute fabrics in civil engineering applications, we carried out tensile and interlaminar shear strength tests. Then, in the second part of this research work, mechanical properties durability of this new material was followed up through its immersion in a harsh environment simulating theoretically 56 years of aggressive exposure. The obtained results evidence high tensile and interlaminar shear properties. On the other hand, the residual tensile and ILS strengths allow the consideration of this new material's concept reinforced with natural jute fabrics as suitable for civil engineering applications for non-bearing load elements.
Key words: natural fibers, BPREX, eco-friendly laminate, SFR polymer concrete, environmental durability.

1. Introduction

Developed from a laboratory curiosity in the time since the Second World War, fiber reinforced composites became the cornerstone of a colossal industry delivering millions of tons every year worldwide [1]. Composite materials find their use in more and more applications, from the manufacture of aircrafts, cars, and bicycles to superconductor components. They have been proven to be stronger yet more flexible than steel and other widely used materials [2]. Nowadays, Fiber reinforced Polymers (FRPs) invade almost all of our daily life applications thanks to their specific characteristics. FRPs have a versatility concerning their bonding to different complex shapes, the highest strength-to-weight ratios, good mechanical characteristics especially in tention. In addition, they are lightweight and are resistant to corrosion, ultraviolet rays and the phenomenon of fatigue. Furthermore, they also offer the reduction of mandatory work and downtime to their implementation; their free shapes and adaptable design capability and chemical inertia exceeding that of all conventional materials [3], [4], [5], [6]. The lightweight of composite materials made them attractive for applications in any region of the world especially where construction cost penalties increase tremendously with remoteness of the locations and heaviness of the construction materials and equipment [7].

However, the quick emergence of composites since their invention due to their high specific properties does not gainsay the health problems associated with the use of synthetic fibers as a reinforcing agent of almost all commercially available polymer composites. To avoid those hazardous effects, the current global tendency is the incorporation of lignocellulosic fibers instead of these mineral ones owing to their biodegradability, low cost and density, renewability and aspect ratio in addition to their high mechanical properties. Recently, there has been an increase interest in the use of bio-composites due to environmentally aware consumers to preserve the world [8]. Natural fibers are low-priced and sustainable natural resources. With increasing environmental protection consciousness, natural fibers as a relatively new group of environmental friendly materials are in considerable demand in recent years by unifying technological, economic and ecological aspects [9].

A new material's concept currently applied in Algeria and in many other countries, based on an externally GFRP polymer concrete, proved its efficiency in piping systems. We have already presented the concept of BPREX combining the polymer concrete with Glass fiber reinforced laminates. The design merging composite materials with polymer concrete in laminated form is used to get the most out of these two materials in the same time such as: more economic final material with high mechanical properties in tension also in bending in addition to a better durability properties. The applicability of this new concept in civil engineering in addition to its durability tracking in harsh environment was previously studied [10].

The aim of the present work is the substitution of glass fabrics and short fibers reinforcing the polymer materials in the current material concept by natural fibers and fabrics made up of jute plant. Keeping the same concept would ensure high level strengths and manufacture easiness in addition to the most important aim which is getting a more eco-friendly material and so, to have a compromise between sufficient performances and environmental protective aspect. The material constitution is shown on the figure below:

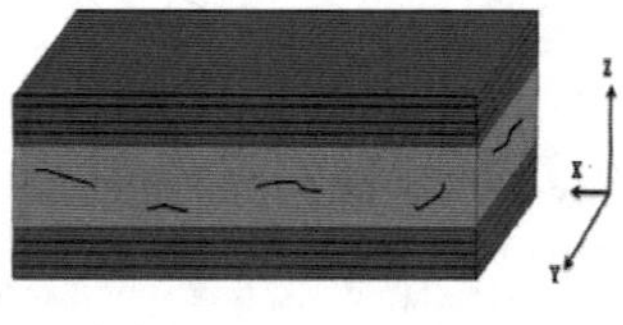

Figure 1. The concept of BPREX developed during our current project; two skins made of JFRP and an SFR polymer concrete core.

Where: JFRP: Jute fabric reinforced polyester resin

SFR polymer concrete: Short fibres reinforced polymer concrete

Although composite materials exhibit superior performances compared to conventional materials because they possess the potential to revolutionize engineering materials and technologies [11], previous researchers have proven that harsh environmental factors adversely affect the material properties of these polymer based materials [12]. The vulnerability of FRPs to such environmental factors has created a critical need for cost-effective and durable materials [13]. In order to use efficiently this concept reinforced with natural fibers in civil engineering, a detailed mechanical characterization accompanied with a durability assessment is imperative.

To assess the possibility of using this new concept reinforced with jute fabrics in civil engineering applications, we carried out tensile and interlaminar shear strength tests upon specimens cut from its laminates. Then, in the second part of this research work, because durability properties of materials are crucial for its consideration as sustainable construction material, mechanical properties durability of this new material was followed up through its immersion in a harsh environment simulating theoretically 56 years of aggressive exposure as recommended by Francesco Micelli et al [14]. The obtained results evidence high tensile and ILSS properties. On the other hand, the residual tensile and ILS strengths allow the consideration of this new material's concept reinforced with natural jute fabrics as suitable for civil engineering applications for non-bearing load elements.

2. Experimental procedure

2.1. Materials

2.1.1. Jute woven fabric. A bidirectional Commercial Jute yarns fabric was used for the reinforcement of our BPREX in the current research work. According to the specifications provided by the supplier, the mass per unit area of the fabric is 460 g/m2 and it was produced by 100% Jute yarns in both warp and weft direction with similar yarns. Before its use as a reinforcing agent, we carried out a full characterization of it in our previous work [15]

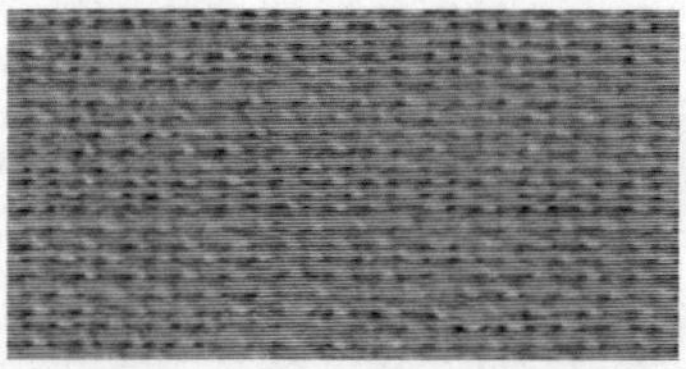

Figure 2. Jute bidirectional fabric used as reinforcing agent

2.1.2. Polymer Binder. The thermosetting Isophtalic unsaturated polyester UPE resin has a density of 1200 kg/m3, viscosity at 25° equals 426 mPas and a gel time at 25° equals 14'30''. The UPE are the most frequently used thermosetting matrices owing to their low cost and adaptability to be transformed into large composite structures. The Catalyst and accelerator used for this investigation are Methyl Ethyl Ketone Peroxide (MEKP) and Cobalt in weight ratios 1:0.03:0.2 respectively.

2.1.3. Silica sand. The mineral fillers consist of standardized silica sand 200-500 μm. It is specially treated for the manufacture of the polymer mortar. This sand is produced by the Algerian company Adwan Chemical Industries Co. Ltd.

Figure 3. BPREX specimens developed and characterized during the current research work

2.2. Methods

All the quasi-static and durability tests were done at an ambient temperature on a computerized Universal Testing machine UTM YLE 25 capable to loading to 250 KN with displacement control.
The long-term durability of the elaborated BPREX material was characterized by following the detrimental effects of its immersion in an alkaline liquid previously used by Francesco Micelli et al. [14] for which the formulation will be presented below.

3. Results and discussion

Experimental results of both virgin and immersed specimens are presented and discussed hereafter. It is well known that the performances of natural fibers reinforced composite materials are governed by the properties of lignocellulosic material and those of the polymer binder [16].

Figure 3 presents average virgin BPREX tensile curves. It is easily observable that BPREX specimens exhibit a brittle behavior till reaching their maximum bearing load before encountering an abrupt failure i.e. a linear proportionality between stresses applied and their corresponding strains followed by a fall of strength without showing any plastic deformations. This behavior is typical for almost all FRP materials and it is governed by the brittleness of its basic constituents; fibers and polymer binder.

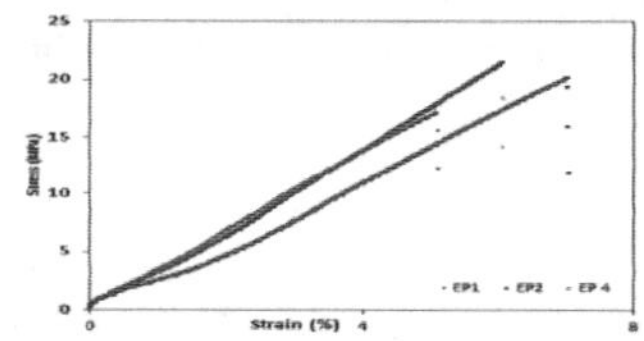

Figure 4. Tensile stress-strain curves of virgin BPREX specimens

Due to the relatively weaker interfaces among the layers of polymer composite laminates compared to the interfaces in their plane, the interlaminar shear strength test (ILSS) is typically an important design factor [17]. Thereupon, it is of great importance to understand comprehensively the ILSS behavior of our developed BPREX material. Typical ILSS curves of BPREX test samples are shown on the figure 4. The brittle behavior is reproduced during ILSS tests for all BPREX characterized test samples.

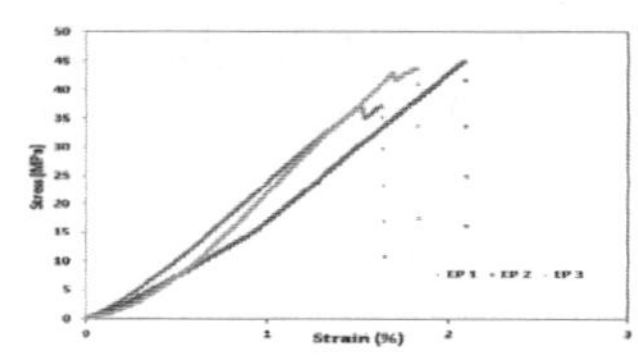

Figure 5. Interlaminar Shear Strength curves of virgin BPREX specimens

The small non-linear regions in tensile and ILSS curves of laminates made from fragile materials, similar to those of our work, occurs from the development of damage from pre-existent defects in the matrix (polyester resin) and in the fiber / matrix interface. These microscopic defects become, with the loading increments, macroscopic order cracks causing the total failure of specimens subjected to the tests [18].

The second step in the presentation of our elaborated BPREX material reinforced with natural jute fabrics as a construction material for civil engineering applications is its long term behavior. For this purpose, the BPREX laminates were immersed in a harsh liquid environment for 84 days which theoretically equals 56 years in natural conditions.

The alkaline liquid has a pH = 13 and it is obtained following the equation below:

$$\boldsymbol{Distilled\ water} + \mathbf{0.16}\,\%\ \boldsymbol{Ca}\,(\boldsymbol{OH})_2 + \mathbf{1}\%\ \boldsymbol{Na}(\boldsymbol{OH}) + \mathbf{1.4}\%\ \boldsymbol{K}(\boldsymbol{OH}) \qquad (1)$$

After the immersion period, BPREX laminates should be carefully washed with distilled water to remove any chemical traces. Afterward, the composite laminates were cut into standardized dimensions prescribed in ASTM D3039 and ASTM D2344 for tensile and ILSS tests respectively.

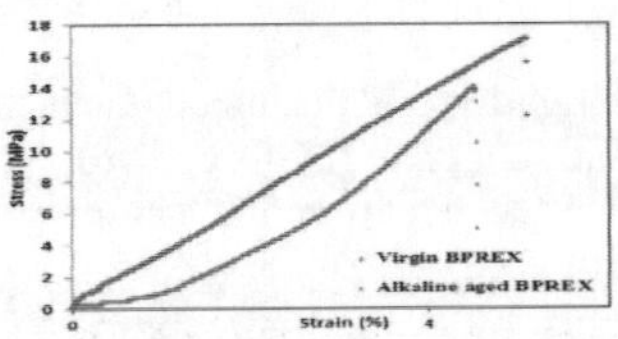

Figure 6. Comparative tensile curves for virgin and aged BPREX samples

A comparison between virgin and aged BPREX tensile behaviors is drawn on the figure 6. A substantial change in tensile behavior occurred with reference to the effects of alkaline immersion period; the elastic linear line shape is substituted by a totally nonlinear curve. The enhanced nonlinearity noticed in aged test samples is attributed to the viscoelastic behavior of the polyester resin and, furthermore, to its plasticization and its reinforcement debonding [19] as a consequence of the alkaline liquid absorption.

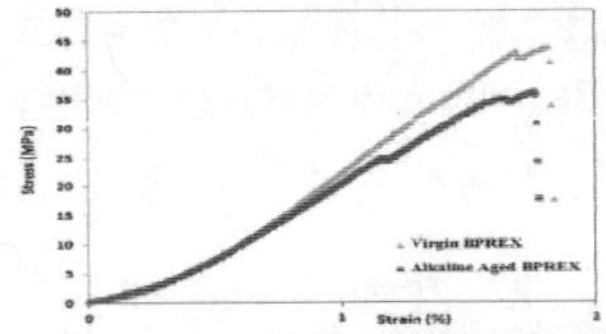

Figure 7. Comparative ILSS curves of virgin and aged BPREX samples

For both virgin and aged BPREX specimens, similar curve patterns are obtained further to interlaminar shear strength tests. On the figure 7 the reduction in the slope of the stress-strain curve for aged samples evidences a loss in the ILSS. This drop in interlaminar shear strength is of the order of 17% with respect to reference values reached with virgin BPREX samples.
It is clear that the alkaline liquid environment has deteriorating effects on the mechanical performances of the developed BPREX. Nevertheless, the tested specimens have significant residual stresses as shown on the following histogram:

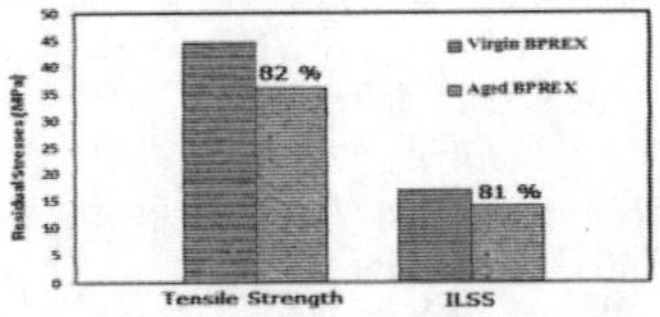

Figure 8 Residual tensile and ILS stresses of BPREX specimens

Even though the fact that the environmental tolerance and overall durability of the composites are dominated by the polymer matrix while the reinforcing fibers are responsible for carrying the structural loads [20], for the particular case of bio-composites, the major sources of degradation amidst ageing process come from the degradation of the natural fibers and the weakening of their interface with the polymer binder [21]. Natural fibers degrade by decomposition into thin fibrils and detached layers. Thwe et al. [22] presumed that the drop in tensile strength is attributed to the breaking of the Si-O-Si bonds on the fibers surface as a result of their attacking by alkaline solution molecules. Taking into account the aforementioned assumptions, we could assume that the tensile curve layout change in addition to the loss of the ILSS after immersion period could be hypothesized to be associated mainly with the degradation of the jute fabrics. This natural reinforcement necessitates its chemical treatment prior to their insertion in polymer composites in order to improve the waterproofness without altering its surface wettability by polymer matrices [23]. Moreover, the inclusion of a compatibilizer in the polymer matrix improves its interfacial bonding with the fibrous phase and, hence, enhances significantly the mechanical properties [24].

4. Conclusion

The present paper sets out an investigation on the mechanical properties along with the durability performances of an elaborated polymer concrete reinforced with natural jute fabrics to which we have given the name BPREX. Based on the experimental campaign conducted during this study, the following conclusion could be made:

- The substitution of glass fibers by natural jute fabrics for BPREX reinforcement allowed attaining good mechanical properties in terms of tensile and interlaminar shear strengths which have values equal to 21±0.9 MPa and 44±0.6 MPa respectively.
- Ageing in alkaline solution induced a drop in BPREX performances for both tensile and interlaminar shear strengths.
- After immersion period simulating theoretically 56 years in natural environmental conditions, the BPREX laminates have important residual stresses percentages of 82% for tensile strength and 81% for ILSS.

The good mechanical properties achieved with the developed polymer laminates in addition to their ecofriendly aspect thanks to their reinforcement with natural jute fibers make the natural fibers reinforced BPREX a low energy and low cost construction material useful in applications requiring sufficient mechanical strengths and good durability concomitantly.

Acknowledgments

Authors would like to express their sincere acknowledgments to Mr Baali Brahim & Mr Riadh Bensaadia and all the staff of Maghreb Pipe Industry Company for their technical support and materials supply.

References

[1] Summerscales, John. *Non-Destructive Testing of Fibre-Reinforced Plastics Composites*. London: Elsevier Applied Science, 1990.

[2] Isaac M. Daniel, Ori Ishai. *Engineering mechanics of composite materials*. 2nd edition. ISBN 978-0-19-515097-1.

[3] Mohammad Reza Aram, Christoph Czaderski, Masoud Motavalli. *Debonding failure modes of flexural FRP-strengthened RC beams.* Composites: Part B **39** (2008) 826–841.

[4] J.F. Chen, J.G. Teng. *Shear capacity of FRP-strengthened RC beams: FRP debonding.* Construction and Building Materials **17** (2003) 27–41.

[5] M. Maalej, K.S. Leong. *Effect of beam size and FRP thickness on interfacial shear stress concentration and failure mode of FRP-strengthened beams.* Composites Science and Technology **65** (2005) 1148–1158.

[6] Lelli Van Den Einde, Lei Zhao , Frieder Seible. *Use of FRP composites in civil structural applications.* Construction and Building Materials **17** (2003) 389–403.

[7] Piyush K. Dutta. *Low temperature and freeze-thaw durability of thick composites*. Composites: Part B **27B** (1996) 371-379.

[8] A. Atiqah, M.A. Maleque, M. Jawaid, M. Iqbal. *Development of kenaf-glass reinforced unsaturated polyester hybrid composite for structural applications.* Composites: Part B **56** (2014) 68–73.

[9] Andrzej K. Bledzki, WenyangZhang, Andris Chate. *Natural-fibre-reinforced polyurethane microfoams. Composites Science and Technology* **61** (2001) 2405–2411.

[10] H Bouguessir et al 2017 *IOP Conf. Ser.: Mater. Sci. Eng.*254 022002.

[11] Houssam A. Toutanji. *Durability characteristics of concrete columns confined with advanced composite materials.* Composite Structures **44** (1999) 155-161.

[12] José P. Gomez, Casto B. *Freeze-Thaw durability of composite materials*. Proceddings of the 1st international conference on composites in infrastructures, ICCI, Tucson, Aris. 947-955.

[13] V.M.Karbhari, M.Engineer, D.A.Eckel. *On the durability of composite rehabilitation schemes for concrete: use of a peel test.* Journal of materials science 32 (1997) 147Ð156.

[14] Francesco Micelli, Antonio Nanni. *Durability of FRP rods for concrete structures.Construction and building Materials* 18 (2004) 491-503.

[15] Bouguessir Hamza, Harkati ElHadi, Rokbi Mansour, Georgios Priniotakis, Savvas Vassilliadis, Sozon Vasilakos, Hadjer Boughanem and Lahcene Fellah. *Physico-chemical and mechanical characterization of Jute fabrics for civil engineering applications.* Journal of Computational Methods in Sciences and Engineering -1 (2018) 1–19.

[16] Ryszard Kozlowski, Maria Wladyka-Przybylak. *Flammability and fire resistance of composites reinforced by natural fibers: Review.* Polymers for advanced technologies 2008; **19**: 446–453.

[17] Xin Wang, Xing Zhao, Zhishen Wu, Zhongguo Zhu and Zihao Wang. *Interlaminar shear behavior of basalt FRP and hybrid FRP laminates.* Journal of Composite Materials 0(0) 1–12.

[18] R.G. Cuntze, A. Freund. *The predictive capability of failure mode concept-based strength criteria for multidirectional laminates.* Composites Science and Technology **64** (2004) 343–377.

[19] Paul Böer, Lisa Holliday, Thomas H.-K. Kang. *Independent environmental effects on durability of fiber-reinforced polymer wraps in civil applications: A review.* Construction and Building Materials **48** (2013) 360–370.

[20] Omar Faruk, Andrzej K. Bledzki, Hans-Peter Fink, Mohini Sain. *Biocomposites reinforced with natural fibers : 2000–2010.* ProgressinPolymerScience **37** (2012) 1552–1596.

[21] Moe Moe Thwe, Kin Liao. *Effects of environmental aging on the mechanical properties of bamboo-glass fiber reinforced polymer matrix hybrid composites.* Composites: Part A **33** (2002) 43-52.

[22] Moe Moe Thwe, Kin Liao. *Durability of bamboo-glass fiber reinforced polymer matrix hybrid composites.* Composites Science and Technology **63** (2003) 375–387.

[23] Yanjun Xie, Callum A.S. Hill, Zefang Xiao, Holger Militz, Carsten Mai. *Silane coupling agents used for natural fiber/polymer composites: A review.* Composites: Part A **41** (2010) 806–819.

[24] A. K. Rana, A. Mandal, B. C. Mitra, R. Jacobson, R. Rowell. A. N. Banerjee. *Short Jute Fiber-Reinforced Polypropylene Composites: Effect of Compatibilizer.* Journal of Applied Polymer Science, Vol. **69**, 329-338 (1998).

IOP Conf. Series: Materials Science and Engineering **459** (2019) 012021 doi:10.1088/1757-899X/459/1/012021

Potential roles of fluorine-containing sol-gel coatings against adhesion to control microbial biofilm

G Rosace[1], P Cardiano[2], C Urzì[2], F De Leo[2], M Galletta[2], I Ielo[3] and MR Plutino[3,*]

[1]University of Bergamo, Department of Engineering and Applied Sciences, Viale Marconi 5, 24044 Dalmine (Bg), Italy
[2]Department of ChiBioFarAm, University of Messina, Viale F. Stagno d'Alcontres 31, Villa S. Agata, 98166 Messina, Italy.
[3]Institute for the Study of Nanostructured Materials, ISMN – CNR, O.U. Palermo, c/o Department of ChiBioFarAm, University of Messina, Viale F. Stagno d'Alcontres 31, Villa S. Agata, 98166 Messina, Italy.

rosaria.plutino@ismn.cnr.it

Abstract. Treatment of materials by sol-gel technique can be an excellent tool to convey new properties to surfaces, therefore the hybrid organic-inorganic materials show the properties of both phases, contributing for example to the obtainment of an anti-fouling coating. In this research, the explored procedure includes the co-condensation of silane coupling agents with epoxide and amine tail-groups, (*3*-Glycidyloxypropyl)-trimethoxysilane (GPTMS) and (*3*-Aminopropyl)-triethoxysilane (APTES), respectively, in combination with two perfluorosilane precursors, namely glycidyl-2,2,3,3,4,4,5,5,6,6,7,7,8,8,9,9-hexadecafluorononylether and trimethoxy-(*3,3,3*-trifluoropropyl)-silane, either individually or together. This synthetic approach allows collecting stable hydrophobic, non-toxic, anti-fouling coatings that were investigated to study their morphology and chemical structure by different physical-chemical techniques. The anti-fouling properties were evaluated through test on treated glass slides in different microbial suspension in seawater-based medium per 24 h at room temperature. During tests, each suspension was maintained in continuous agitation to simulate the natural movement of seawater, and the attachment of cells on bare degreased glass slides is compared with that occurring on the treated slides. Results show that the fluorinated coatings have good antimicrobial activities and low adhesive properties, no biocidal effects towards the studied microorganisms were observed.

1. Introduction

The undesired colonization of submerged surfaces by marine micro- and macro-organisms such as bacteria, diatoms, algae, barnacles, and seaweeds is called biofouling [1]. It has detrimental effects among which on aquaculture systems and oceanographic sensors. Moreover, this natural phenomenon on shipping and leisure vessels causes severe problems for marine industries due to corrosion and hydrodynamic drag, which leads to elevated fuel consumption and higher maintenance costs [2-4]. In order to reduce both economic and environmental penalties, the primary strategy for inhibit marine fouling is to use biocide-containing paints [5-8]. At the same time, environmental business and legislation are driving science and technology towards non-biocidal solutions based only on physical-chemical and materials properties of coatings [9-14]. From this point of view, advances in nanotechnology and recent knowledge of marine chemistry and biology are giving a significant contribution to the development of a new generation of bioinspired surface designs by the nature [15-17]. An approach to the development of novel coatings is to create a 'deterrent' surface that inhibits the initial stage of microorganism settlements. Currently, deepen research has been realized on newer anti-fouling (AF) technologies using fluoropolymers [18-19]. Fluoropolymers can be used to form anti adhesion surfaces with low critical surface energy, thanks to the presence of the exposed CF_2 and CF_3 moieties at the interface, which reduce the attachment of fouling (i.e. FR or fouling release coatings) depending also on the fluorine atoms limited mobility resulting around the backbone bonds [20-21]. Moreover, amphiphilic coatings, which incorporate some of the benefits of both hydrophobic and hydrophilic functionalities, have been developed in order to create an engineered surface with local variations in surface chemistry, topography and mechanical properties. Sol-gel technique for the treatment of materials can be an attractive tool to impart new and interesting properties to their surfaces, particularly if organic components are incorporated into the formulation.

2. Results and Discussion

2.1. *Sol-gel synthesis and characterization*

The newly synthesized sol–gel coating was made from silicon trialkoxide [R'Si(OR)$_n$] (**Figure 1**) undergoing simultaneously hydrolysis–condensation reactions in several steps. Associated condensation reactions are directly related to the removal of water and corresponding alcohol from the polymer matrix resulting in the formation of more rigid silica networks [7-8]. Four separate solutions, namely G/A, G/A_F3, G/A_F16, G/A_F3_F16, all containing GPTMS and APTES in 2/1 molar ratio, and F3 and/or F16 at a total 0.5 wt% were prepared.

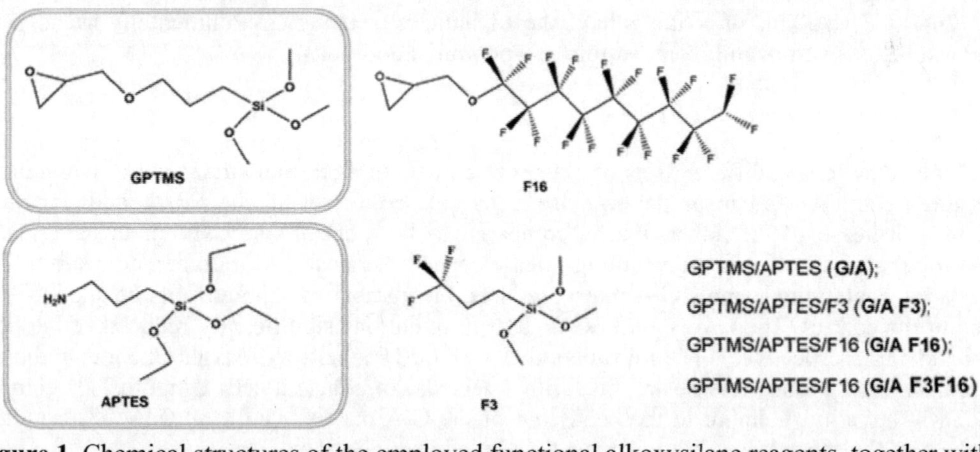

Figure 1. Chemical structures of the employed functional alkoxysilane reagents, together with the adopted acronyms, employed for the tested anti-fouling coatings.

The four G/A, G/A_F3, G/A_F16 and G/A_F3_F16 sols were fully characterized by FT-IR spectroscopy and Nuclear magnetic resonance (NMR). Coated glass was analyzed through FT-IR spectroscopy. ^{1}H and ^{13}C NMR provide a useful tool to detect the epoxy consumption and the hydrolysis reaction. The first step of the reactions, namely the opening of the oxirane ring of GPTMS, has been followed by ^{1}H and ^{13}C NMR spectra on freshly prepared samples from the bulk reaction mixtures that have been taken and dissolved in $CDCl_3$ [22-25]. The formation of the R'N(H)CH_2-fragment is associated to the progressive lowering of the signals due to the epoxy rings. The spectra also reveal a concomitant but partial hydrolysis of the methoxy and ethoxy groups of GPTMS and APTES which occurs at a lower rate [24]. Before application of coatings, the glass slides were cleaned with a concentrate sulfuric acid/ potassium permanganate solution, then washed several times with ultrapure water and dried in an oven at 80°C for 24 hours.

2.2 Application on glass slides
In order to test the efficacy of the products and the lack of toxicity, plate tests have been set up by testing treated and non-controlled slides immersed in microbial suspensions. In particular, the toxicity was evaluated by comparing the number of cells/mL in the suspension both before and after the test, while the antifouling capacity comparing the number of microbial cells adhered on the control slides compared to those treated with the various products. Preparation of the slides was carried out starting from a thorough cleaning of the latter: the glass slides were cleaned with a concentrate sulfuric acid/potassium permanganate solution, then washed with ultrapure water and stored at 80°C for 24 hours. Dip coatings were performed with a NIMA Instruments dip coater at a 20 mm/min speed for a total of 20 immersion/withdrawal cycles for each. After each coating the glass slides have been thermally cured at 80°C for 10 minutes, then at 160 °C for 3 minutes. To explore the nature of the coatings and to confirm their successful deposition, ATR FT-IR spectra of silane xerogels applied and annealed on glass slides were registered and investigated. The nature of the hybrid structure of silica precursors (**G, A, F3, F16)** is strongly influenced by the epoxy ring opening that can succeed different reaction pathways: (a) hydrolysis with diol formation; (b) alcoholysis with ethyl ether terminal groups; (c) consecutive polymerization steps to give a hybrid poly(ethylene)oxide 3D-network; (d) reaction with primary amino group, thus transformed in secondary one. Actually, during the final thermal curing step, each silanol group can react with each other to form stable siloxane bonds (Si-O-Si). At the same time, glycidoxy groups can react both with themselves and with hydroxyl groups of hydrolyzed precursors. Therefore, the presence of Si-O-C bonds can increase the flexibility of the structure, favoring a great homogeneity between the organic and inorganic components of the 3D network [26-28]. These kind of coatings have the advantages of being environmentally friendly, non-toxic and can be obtained by mild temperature and pressure conditions.

2.3 Evaluation of anti-fouling properties

Results obtained by testing different sets of glasses in contact to microbial strains were shown in table 1 and Figures 2 and 3. As can be deduced from the data collected, all the FR formulations tested resulted in a decrease of the adhered cells compared to the control (only slide), in particular the G/A_F3 formulation showed the best fouling-release capacity against the Gram negative strain (Table 1). The coating containing a single G/A matrix reduced the number of adhered cells on slide to 76.5% compared to the control. The G/A_F3 and G/A_F16 formulations significantly reduced cell adhesion to 29.5% and 53%, respectively. The formulation G/A_F3_F16 albeits to a lesser extent than the other formulated, showing fouling release activity, with a decrease of adherent cells equal to 71% compared to the control, but almost similar to the behaviour of the G/A matrix. Thus, the G/A_F3 formulation showed the best fouling-release ability against the Gram negative strain.

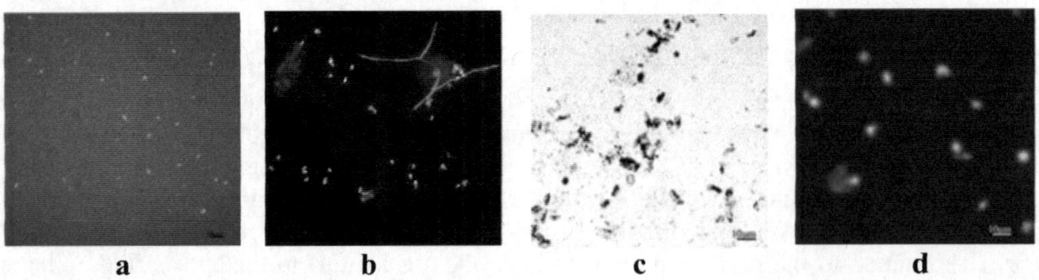

Figure 2. Images of the glass slide treated with **G/A_F3** showing diffuse cells of the Gram- strain *S. maltophilia BC 658* observed under EM 40x (**a**). Glass slide treated with **G/A_F16** observed under EM with the Gram+ strain *B. aquimaris BC 660*. Note the formation of spores and chain of cells 40x (**b**). Images of the treated slides

IOP Conf. Series: Materials Science and Engineering **459** (2019) 012021 doi:10.1088/1757-899X/459/1/012021

with **G/A_F3** in contact with the cells of *Navicula sp.* obtained with the LM **(c)** and EM microscopes **(d)** in which is evident cells sufference. Magnification 20X.

Table 1. Percentage of the adhered cells onto the untreated and treated glass slides after incubation with the strains *S. maltophilia (BC 658)* and *B. aquimaris (BC 660)*, and *Navicula sp.*

	S. maltophilia (BC 658)	*B. aquimaris* (BC 660)	*Navicula* sp.
Samples	% of adhering cells*		
CTRL	100	100	100
G/A	76.5	4.3	149
G/A F3	29.5	0	10.5
G/A F16	53	2.7	25.5
G/A F3F16	71	0	181

*As determined by direct counts under LM and/or EM microscopy. Values above 100% mean a higher number of cells adhering to the treated slides than to the glass slide controls.

Regarding the toxicity, none of the products seems to show a detrimental effect on the growth of bacteria and diatoms as clearly shown in Figure 3.

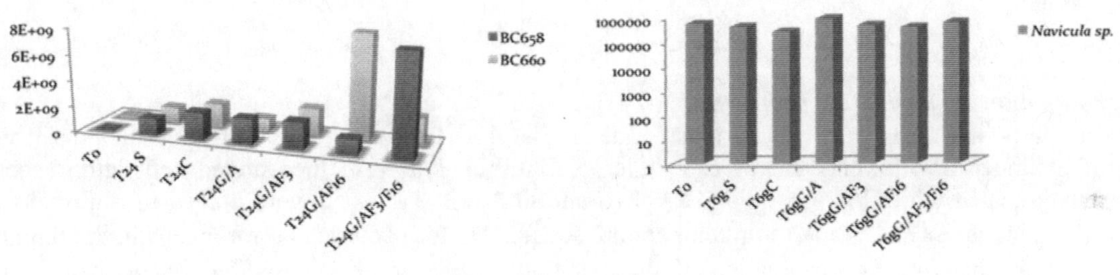

Figure 3. Number of viable cells of the strains *S. maltophilia (BC 658)* and *B. aquimaris (BC 660)* (**left**) and *Navicula sp.* (**right**), before and after incubation with the treated and untreated glass slides.

3. Conclusions

The aim of the present work was to investigate the most significant factors to obtain efficient and stable hybrid sol-gel GPTMS-based FR coatings [22-23]. The FR effects of the silica containing coatings depend on the possibility to realize chemical interactions between the hydroxyl group and/or epoxy group of the precursor with functional groups belonging to the polymers [24-25]. Room temperature reaction of the GPTMS with the bifunctional APTES leads to the formation of a hybrid silica-epoxy polymers. In whose network the amine acts as a network former while the epoxy compounds mainly acts as a network modifier leading to a different development of the polymeric backbone. The structural differences reflect the thermal properties and hydrophobic properties of material particularly influenced by the addition of perfluorinated species. The nature of the surface of the materials obtained from GPTMS/APTES. 2:1 following the addition of the long-chained F16 and F3, changes to more hydrophobic state, whereas GPTMS/APTES 5:1 exhibits an opposite trend. The different hydrorepellent activity is explained with a different assembly and alignment of the fluorinated chains at the polymer-air interphase. These formulations have been properly designed for fouling release purposes. These sol-gels (G/A, G/A_F3, G/A_F16, G/A_F3_F16) were applied on slides using a dip coater and anti-fouling and toxicity tests were performed by using pure microbial cultures. Cell adhesion experiments using microorganisms isolated from marine environment such as *Bacillus aquimaris*, *Stenotrophomonas maltophilia,* and *Navicula* indicate the considerable anti-fouling capacity against both Gram-positive and Gram-negative bacteria as well as against diatoms

[29]. These properties were tested on treated glass slides in different microbial suspension in seawater-based medium for 24 h at room temperature. The fluorinated coatings have good antimicrobial activities and low adhesive properties towards the studied bacteria, furthermore the eventual biocide effect due to the product release in the liquid medium evaluated by counting the microbial cells before and after the period of incubation showed no biocidal effects [30].

4. Experimental Details

(3-aminopropyl)-triethoxysilane (APTES, labelled **A**), (3-glycidyloxypropyl)-methyldiethoxysilane (GPTMS, **G**), 3,3,3-trifluoropropyl-trimethoxysilane (**F3**), glycidyl-2,2,3,3,4,4,5,5,6,6,7,7,8,8,9,9-hexadeca fluorononyl ether (**F16**) (**Figure 1**), absolute Ethanol (HPLC grade) were purchased from Sigma Aldrich and used without any other purification. In a typical procedure of G/A_F3, 2.013 g of GPTMS were mixed with 0.943g of APTES in 37.48 g of ethanol under stirring. Then 0.2 g of F3 was added to the clear ethanol solution and left at room temperature under stirring for 24 h. The other functional alcoxysilane sol gels, G/A, G/A_F16, G/A_F3_F16, were synthetized in the same way.
FT-IR spectra were acquired by means of a Thermo Avatar 370, equipped with Attenuated Total Reflection (ATR) accessory. A diamond crystal was used as internal reflectance element on the ATR accessory. Spectra were recorded, at rt, in the range from 4000 to 650 cm^{-1}, with 32 scans and a resolution of 4 cm^{-1}. ^{1}H and $^{13}C\{^{1}H\}$ NMR spectra were performed in methanol-d_4 at 298.2 (±0.1) K with a Varian 500 spectrometer, equipped with a 5 mm OneNMR (TM) probe operating at 500.1 and 125.7 MHz.

4.1. Evaluation of anti-fouling properties

Different sets of glasses were placed in separate sterile glass Petri dishes (∅ 120 mm) and sterilized overnight under UV. Twenty ml of each microbial suspension were then added and Petri dishes, previously coated with G/A, G/A_F3, G/A_F16 and G/A_F3_F16 sols, were placed in a horizontal shaker and incubated for 24h at room temperature set at 25°C for bacteria and for 6 days in the light at 30°C for the diatoms. The evaluation of the anti-fouling properties was carried out with a Gram negative strain *Stenotrophomonas maltophilia* BC658, a Gram positive spore forming strain *Bacillus aquimaris* BC 660, a diatom strain *Navicula* sp, in sea-based culture media; all isolated from marine habitat. Antifouling properties of the coatings was determined comparing cells adhesion to the untreated and treated glasses observed under Light microscopy, (LM) and Epifluorescence microscopy (EM) a. The toxicity of the coatings was evaluated by the measurement of the density and/or by counting the cells before and after the incubations in the liquid medium.
After incubation time, glasses were rinsed with sterile PBS and prepared for three different treatments: one set was air dried by placing the glass directly to the Bunsen flame for 30 seconds, observation was carried out in LM or under epifluorescence microscope (EM) after staining with Acridine Orange (AO).
The second set was prepared for epifluorescent microscopy and fixed in a solution of a fixation buffer (4% paraformaldehyde in PBS, pH 7.2) with PBS (3:1 molar ratio) and incubated for 90 min at 4 °C. Bacterial cells were dyed with Acridine Orange (AO)(0.1mg/ml) diluted in sterile distilled water 1:2 (v/v) for 3-4 min, [11] and the last set of glasses was prepared for SEM microscopy by fixing in glutaraldehyde 2,5% in Phosphate buffer (PB) 0.1M pH 7.4 per 6h at 4°C, and then air dried.

Acknowledgments
Authors wish to thank: MIUR and CNR for financial support.

References

[1] La Russa MF, Ruffolo SA, Ricci S, Davidde B, Barca D, Ricca M, Caprisco V, *Period Miner.*, 82, **2013**, 101-111

[2] Dobretsov S, Thomason JC, *The Journal of Bioadhesion and Biofilm Research*, 27:8, **2011**, 869-880.
[3] Langer R, *Nature,* 392, **1998**, 5–10.
[4] La Cono V, Urzì C, *Journal of Microbiological Methods,* 55, **2003**, 65-71.
[5] Plutino MR, Colleoni C, Donelli I, Freddi G, Guido E, Maschi O, Mezzi A, Rosace G, *Journal of Colloid and Interface Science*, 506, **2017**, 504–517.
[6] Guido E, Colleoni C, De Clerck K, Plutino MR and Rosace G, *Sensors and Actuators B Chemical, 203*, **2014**, 213–222.
[7] Rosace G, Trovato V, Colleoni C, Caldara M, Re V, Brucale M, Piperopoulos E, Mastronardo E, Milone C, De Luca G, Plutino MR, *Sensors and Actuators B,* 252, **2017**, 428-439.
[8] Trovato V, Colleoni C, Castellano A, Plutino MR, *Journal of Sol-Gel Science and Technology* 87, **2018**, 27–40.
[9] Detty MR, Ciriminna R, Bright FV, Pagliaro M, *ChemNanoMat,* 1, **2015**, 148 – 154.
[10] Brouwers JRBJ, *Pharm. World Sci*., 18, **1996**, 153–162.
[11] Uhrich KE, Cannizzaro SM, Langer RS, Shakesheff KM, *Chem. Rev*. 99, **1999**, 3181–3198.
[12] Brannon-Peppas L, *ACS Symposium Series*, 520, **1993,** 42–52.
[13] Levy, R.; Nichols, M.; Miller, T W *ACS Symposium Series*, **1993**, 202–212.
[14] Pistone A, Visco AM., Galtieri G, Iannazzo D, Espro C, Marino Merlo F, Urzì C, De Leo F. *International Journal of Polymer Analysis and Characterization*, 21: 4, **2016**, 327-336.
[15] Callow JA, Callow ME, *Nature Communications*, 2, **2011**, 244.
[16] Nierstrasz VA, *AUTEX Research Journal*, 4, **2002**, 2.
[17] Radu CD, Popa M, Parteni O, Salariu M, Lupuşoru EC, Ghiciuc C, Foia L, Chiriac A, Lupusoru R, Oproiu L, Ulea E *The Open Conference Proceedings Journal*, 5, **2014**, 1-8
[18] Finnie A, Williams DN, In: *Biofouling* (eds Durr S. & Thomason J.C.) (Wiley-Blackwell, **2010**)
[19] Chung C, Lee M, Choe EK, *Carbohydr. Polym*. 58, **2004**, 417–420.
[20] Nurioglu AG, Esteves ACC, de With G, *J. Mater. Chem. B*, 3, **2015**, 6547-6570.
[21] Eduok U, Suleiman R, Gittens J, Khaled M, Smith TJ, Akid R, El Alia B and Khalile A, *Royal Society of Chemistry Adv.*, **2015**, 5, 93818–93830.
[22] Cardiano P ,Lo Schiavo S,Piraino P, *Journal of Non-Crystalline Solids* 356, **2010**, 917–926.
[23] Cardiano P, Ponterio RC, Sergi S, Lo Schiavo S, Piraino P, *Polymer* 46, **2005**, 1857-1864.
[24] Cardiano P, Mineo P, Sergi S, Ponterio RC, Triscari M, Piraino P, *Polymer* 44, **2003**, 4435–4441.
[25] Cardiano P, *Journal of Applied Polymer Science*, 108, **2008**, 3380–3387.
[26] Munch AS, Wolk M, Malanin M, Eichhorn KJ, Uhlmann FS, Uhlmann P, *Journal of materials chemistry,* 6, **2018**, 830.
[27] Caldara M, Colleoni C, Guido E, Re V, Rosace G, *Sensors Actuators, B Chem*. **2016**, 222, 213–220.
[28] Lim CS, Lee SSC, Leong W, NG YX, Teo SLM *Indian Journal of Geo-Marine Sciences*, 43(11), **2014**, 2067-2074.
[29] Magin CM, Cooper SP, Brennan AB, *Mater. Today*, 13, **2010**, 36-44.
[30] Lejars M, Margaillan A, Bressy C. *Chem. Rev*. 112, **2012**, 4347–4390.

Study of Properties of Arselon Spun Yarn

D B Ryklin[1] and S S Medvetski[2]

[1] Head of Department of Textile Technology, Vitebsk State Technological University, Vitebsk, Belarus
[2] Assistant Professor of Department of Textile Technology, Vitebsk State Technological University, Vitebsk, Belarus

E-mail: Ryklin-db@mail.ru

Abstract. The paper is devoted to development of the range of ring-spun yarns of Arselon fibers. Arselon is a polyoxadiazole heat-resistant fiber produced at JSC "SvetlogorskKhimvolokno" (Belarus). The investigated yarn count range was from 16 to 29 tex (Ne37 – Ne20). It was determined the influence of linear density of the Arselon yarn on its mechanical properties and evenness. In addition, it was shown that winding and assembling lead to significant reducing the number of frequently-occurred faults of 2-ply yarns. The developed range of Arselon yarns is the base for creation of new textile products such as woven and knitted fabrics and heat resistant goods which can be made of them.

1. Introduction

Arselon is a polyoxadiazole heat-resistant fiber produced at JSC "SvetlogorskKhimvolokno" (Belarus). As the production process was designed alternatively to meta-aramid fibers, thus by its properties Arselon is very similar to various meta-aramid fibers. The staple fiber Arselon is a basis for spun yarns and felts which further can be processed into many kinds of materials for thermal application.

The manufacturer indicates the following main features of Arselon:

- Oxygen index (LOI) is 30%;
- No melting;
- Temperature of use is 250 °C;
- Heat shock temperature is 400° C;
- Flexible at high low temperatures;
- Low thermal expansion coefficient;
- Friction coefficient (metal surface) is 0,2;
- Good adhesion to rubber;
- Moisture level is 10-12% (similar to cotton);
- High electrical insulation properties.

Manufacturer indicates the following applications of Arselon:

- Heat resistant spun yarns and fabrics (protective apparel, gloves, home and office textile);
- Woven and nonwoven filtering materials, bag filters (for metallurgy, cement, asphalt);
- Composites (sealing, sliding bearings, packing).

Despite of the mentioned features of these fibres before 2017 Belarusian mills produced only one kind of Arselon spun yarn. It was carded yarn 29,5 tex×2 (Ne 20/2). Moreover, the number of publications devoted to investigation of Arselon fibers and yarns processing is very limited. There are papers describing twisting of mentioned type Arselon spun yarns [1, 2], also Arselon filament yarns texturizing [3] and properties of textiles and composites produced on the base of these fibers [4, 5].

The aim of presented research was evaluating of Arselon yarns count on their properties. Development of wide range of Arselon yarns will create the base of new textile products such as woven and knitted fabrics and goods from them.

2. Materials and methods

Properties of used Arselon fibers are presented in Table 1. Cased on these data it is possible to make the conclusion about possibility to use machinery of short-staple fibers processing for manufacturing of Arselon yarn.

Table 1. Properties of Aselon fibers

Property	Value
Length, mm	36
Linear density, mtex	162
Breaking tenacity, cN/tex	32,7
Elongation, %	38
Crimps per 1 cm	3,2

Samples of ring-spun carded yarns were produced using the following machinery:

- Blowroom (Rieter);
- Card C60 (Rieter);
- Drawframe SB-D45 (Rieter);
- Drawframe RSB-D45 (Rieter);
- Roving frame 668 (Zinser);
- Ring spinning machine G35 (Rieter).

For 2-ply Arselon yarns manufacturing the following machinery were used:

- Automatic winder Polar L (Savio);
- Assembly winder TW-2D (SSM);
- Twisting machine Geminis S261 (Savio).

To obtain yarns with the highest achievable properties during our research in each stage of the technological process experiments were carried out.

Strength tester RM-3 and evenness tester Uster Tester 5 were applied for yarns quality assessment.

3. Properties of ring-spun Arselon yarn

The yarn count range was from 16 to 29 tex (Ne37 – Ne20). Mechanical properties of ring-spun yarns and results of their testing using Uster Tester 5 are presented in Table 2.

Experimental data presented on the Table 3 show that breaking tenacity and elongation of Arselon ring-spun yarn increase with raise of its linear density.

Table 2. Properties of ring-spun Aselon yarns

Property	Value				
Nominal linear density, tex	16	20	22,2	24	29,5
Actual linear density, tex	15,8	19,4	21,6	24,1	29,5
Breaking force, cN	297,0	364,7	438,5	503,7	601,8
Breaking tenacity, cN/tex	18,8	18,8	20,3	20,9	20,4
Coefficient of variation of breaking force, %	10,8	12,1	10,7	10,9	9,1
Elongation, %	12,8	14,0	13,9	15,6	15,1
Coefficient of variation CVm, %	16,7	14,7	14,3	13,2	12,2
Thin places (-50 %) per 1 km	114,2	33,2	101,4	4	10,8
Thick places (+50 %) per 1 km	174	64,2	53,2	29,8	13
Neps per (+200 %) 1 km	259,2	168,2	144,4	49,2	26,2

For evenness and frequently occurring yarn faults the following regression equations were obtained:

— Coefficient of variation CVm (%):

$$CVm = \frac{66,3}{\sqrt{T}}, \tag{1}$$

— Thick places (+50 %) per 1 km:

$$Thick_{(+50\%)} = 4950 \cdot e^{-0,21 \cdot T}, \tag{2}$$

— Neps (+200 %) per 1 km:

$$Neps_{(+200\%)} = 2469 \cdot e^{-0,14 \cdot T}, \tag{3}$$

where T is actual linear density of Arselon ring-spun yarn (tex).

Obtained equation (1) for evenness CVm calculation looks like Martindale's formula for the limiting irregularity calculation.

A similarity of these formulas indicates the stability of the technological process of Arselon yarns manufacturing. Using equation (1) and Martindale formula we found that for whole range of Arselon yarns count index of irregularity is about 1,65.

On the figure 1 the spectrogram of yarn 22,2 tex. For the rest samples almost similar spectrograms were obtained. On presented spectrogram there is no high peaks corresponding to disturbing periodic yarn faults. Dip on the wavelength at 36 mm (actual fibers length) indicates the stability of the drawing process on machinery of spinning production.

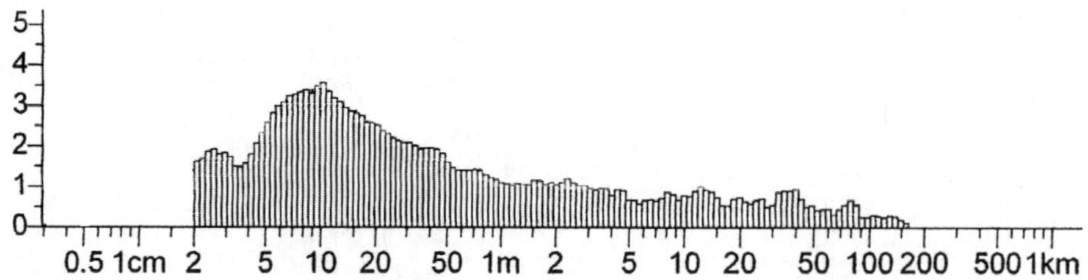

Figure 1. Spectrogram of Arselon ring-spun yarn

Additional task solved within research is determining minimal linear density of yarn, which satisfies the following criteria:

- Breaking tenacity must be not less than 16 cN/tex;
- Coefficient of variation CVm must appropriate Uster® Statistics Percentile not higher than 50 %.

All investigated samples of yarns satisfied the requirements for yarns strength.

For evaluation of yarns evenness we used information from Uster® Statistics 2013 (US 2013) about cotton carded yarn for weaving. It is due to the lack of information about heat-resistant fibers. On the other hand, data about properties of polyester yarn were not used as the base for evaluation because presented data related to yarn manufactured of fibers 133 tex with length 38 – 40 mm. As we used Arselon fibers 162 tex the yarn evenness must be significantly higher. So, we cannot recommend use-developed technology for producing Arselon yarn with linear density less than 20 tex.

On the figure results of CVm evaluation are presented. The diagram shows that only for yarn 16 tex coefficient of variation CVm exceeds the limit: USP 50 %. So, we cannot recommend producing Arselon yarn with linear density less than 20 tex using for developed technology.

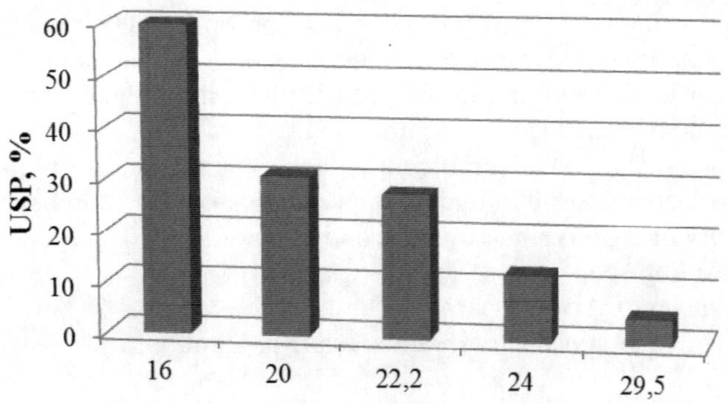

Figure 2. Uster® Statistics Percentiles for CVm of Arselon yarns of different linear density

4. Properties of 2-ply Arselon yarn

Properties of 2-ply yarns obtained after their twist optimisation are presented in Table 3.

Table 3. Properties of 2-ply Aselon yarns

Property	Value		
Nominal linear density, tex	16×2	20×2	22,2×2
Actual linear density, tex	31,7	39,7	43,3
Breaking force, cN	713,5	909,6	930,0
Breaking tenacity, cN/tex	22,51	22,91	21,48
Coefficient of variation of breaking force, %	4,97	8,17	8,03
Elongation, %	16,74	17,10	18,12
Coefficient of variation CVm, %	13,14	12,30	11,32
Thin places (-50 %) per 1 km	0,0	0,0	2,0
Thick places (+50 %) per 1 km	54,0	20,0	11,0
Neps per (+200 %) 1 km	57,0	35,0	8,0

Analysing presented data the following conclusions were drawn:

- The breaking force of 2-ply yarn increases almost directly in proportion to its linear density, while the breaking tenacity in the investigated range changes insignificantly;
- Rise of yarn linear density leads to its elongation increasing, to reducing irregularity and the frequently-occurred faults. Decreasing of coefficient of variation CVm is due to increasing of fibers number on the yarn cross-section;
- Comparison of data presented in the tables 2 and 3 shows that breaking tenacity of 2-ply yarns 16 tex×2 and 20 tex×2 by 20 % higher than breaking tenacity of corresponding single ring-spun yarns;
- Yarns winding and assembling lead to significant reducing the number of the frequently-occurred faults. Thin places are almost eliminated, thick places and neps number reduced by more than 3 times.

5. Conclusions

1. Developed technology allows manufacturing Arselon ring-spun yarns 20 – 29,5 tex of acceptable quality. Coefficient of variation CVm of these yarns appropriate USP not higher than 50 %. It is proven that index irregularity of the yarns for whole range is 1,65 that can be used for Cvm prediction for yarns of other linear densities. For yarn of linear density less than 20 tex it was recommended to correct technology on the further stage of research.

2. Winding and assembling lead to significant reducing the number of frequently-occurred faults of 2-ply yarns. Thin places are almost eliminated, thick places and neps number reduced by more than 3 times. Breaking tenacity of 2-ply yarns 16 tex×2 and 20 tex×2 by 20 % higher than breaking tenacity of corresponding single ring-spun yarns.

3. The developed range of Arselon yarns is the base for creation of new textile products such as woven and knitted fabrics and goods which can be made of them.

References

[1] Mazurik V D, Rodionov V A 2004 Optimization of Processing of Twisted Arselon Yarn *Fibre Chemistry* **№ 2 (vol 36)** (Springer New York Consultants Bureau) pp 104–105.

[2] Mihalko M N, Sadovski V V, Dokuchaev V N 2010 Optimization of twisting of heat-resistant yarns produced on the base of fibers Arselon-S Vestnik of Vitebsk State Technological University **№ 18** (Vitebsk State Technological University) pp 58–63.

[3] Medvetski S S, Ryklin D B, Kogan A G 2004 Air Texturing of Arselon Thread *Fibre Chemistry* **№ 2 (vol 36)** (Springer New York Consultants Bureau) pp 101–103.

[4] Erofeev O O, Voloshchik T E, Kozinda 2010 Study of Properties of Needle-Punch Nonwovens Made of Heat-Resistant Fibers *Technology of Textile Industry* **№ 6** (Ivanovo) pp 46–49.

[5] Adashkevich S V, Bakayev A G, Gordienko A I, Markevich M I, Stelmakh V F, Chaplanov A M and Shcherbakova E N 2015 Magnetic Resonance Diagnostics Radio-Absorbing Composite Materials *Polymer Materials and Technologies* **№ 1 (vol 1)** (Gomel: Belyi Metal Polymer Research Institute of National Academy of Sciences of Belarus) pp 71–75.

IOP Conf. Series: Materials Science and Engineering **459** (2019) 012023 doi:10.1088/1757-899X/459/1/012023

N-Thiocarbazole-based gold nanoparticles: synthesis, characterization and anti-proliferative activity evaluation

C Saturnino[a], M S Sinicropi[b*], D Iacopetta[b], J Ceramella[b], A Caruso[b], N Muià[b], P Longo[c], G Rosace[d], M Galletta[e], I Ielo[f,*], and M R Plutino[f,*]

[a]University of Basilicata, Department of Science, Via Nazario Sauro, 85, 85100 Potenza Italy.
[b]University of Calabria, Department of Pharmacy and Nutrition and Health Science, via Pietro Bucci, 87036, Arcavacata di Rende (Cs) Italy.
[c]University of Salerno, Department of Pharmacy, Via Giovanni Paolo II, 132, 84084 Fisciano (SA) Italy.
[d]University of Bergamo, Department of Engineering and Applied Sciences, Viale Marconi 5, 24044 Dalmine (Bg) Italy.
[e]Institute for Polymers, Composites and Biomaterials, IPCB-CNR, P.le Fermi 1, Portici, 80125 Napoli Italy.
[f]Institute for the Study of Structured Nanomaterials, ISMN – CNR, c/o Department of chemical, biological, pharmaceutical and environmental science. (UniME), Viale F. D'Alcontres n.31, 98166-Vill.S. Agata (ME), Italy.

*s.sinicropi@unical.it; ielo@pa.ismn.cnr.it; rosaria.plutino@ismn.cnr.it

Abstract. Carbazoles are aromatic heterocyclic compounds derived from the fusion of a benzene ring with an indole nucleus in 2,3 position. Today, many carbazole derivatives are widely studied as anticancer, anti-fungal, antioxidants, photoconductor, anti-bacterial, anti-malarials, anti-Alzheimer, anti-tuberculosis, anti-HIV agents and for the treatment of obesity. Some of them, differently substituted on carbazole nucleus, have been synthetized and their biological activity have been evaluated. The purpose of this work is the study of new species obtained binding, a series of carbazole derivatives in which the carbazolic nitrogen has been functionalized with different alkyl-thiol chains, with gold nanoparticles, synthesized in organic solvent. The formation of self-assembled monolayers of these ligands is obtained on the spherical surface of gold nanoparticles, with a stabilizing effect against aggregation. These systems might present a double innovative function, matching an antineoplastic activity typical of carbazoles, with the photothermal effects of gold nanoparticles.

1. Introduction

Pharmaceutical research is focused on the design of new antineoplastic drugs [1] with higher selectivity on tumoral cells, which are able to overcome any resistance of the diseased cells and that cause bland side effects [2]. Carbazoles were largely studied for all their properties [2-4], [27-30], which can implement by changing functionalized groups on the nitrogen atom or inserting suitable substituents on carbazole core, with the purpose of obtaining new and unique properties as antioxidant or antimicrobial [5-7]. Carbazoles derivatives has become important for their efficient topoisomerase, tubulin, telomerase, kinase and integrase inhibition activity [8, 27, 31]. These compounds induce antiproliferative activity and significant apoptotic response approaching cancer cells selectively when compared to the normal ones [9-11]. A new way of administering therapeutic agents, as carbazoles, involves the use of nanocarriers [12]. Recently, a lot of delivery carrier based on nanometric size, as nanotubes [13], nanorods [14], dendrimers [15] and lysosomes [16] have been designed; among these methods gold nanoparticles (Au-NPs) are good carrier of different payloads into target cells [17]. Gold nanoparticles (Au-NPs) take advantage of their unique chemical and physical properties to carry and release drugs, they are able to carry different size of drug molecules from little one to large biomolecules as DNA, RNA or proteins. A promising aspect in the treatment of cancer disease is the photothermal damage to cells caused by the presence of gold nanoparticles [20]. Au-NPs photothermal therapy is a technique for selective damage of cancer cells that is based on the use of gold nanospheres irradiated with 20 ns laser pulses (λ= 532 nm) to generate local heating [21-22]. A condition for an effective therapy provides for an efficient release of the drug, which can be released thanks to external stimuli such as light [18], or from the inside through variations in the pH levels. This paper describes a method to synthetize Au-NPs functionalized with thio-carbazole derivatives, as potential antiproliferative agents against breast and uterine cancer cell lines-without affecting non-tumoral cell lines viability.

2. Results and Discussion

2.1. Synthesis and Biological evaluation of N-thioalkylcarbazole derivatives (1a-1c, 2a, 3a).

The N-thioalkylcarbazole derivatives were synthesized by the reaction of 9-(bromo-alkyl)-9H-carbazoles (P1a-P1c, P1a-P3a) with thiourea using i-PrOH as solvent under reflux for 12 hours, in a nitrogen atmosphere (Figure 1).

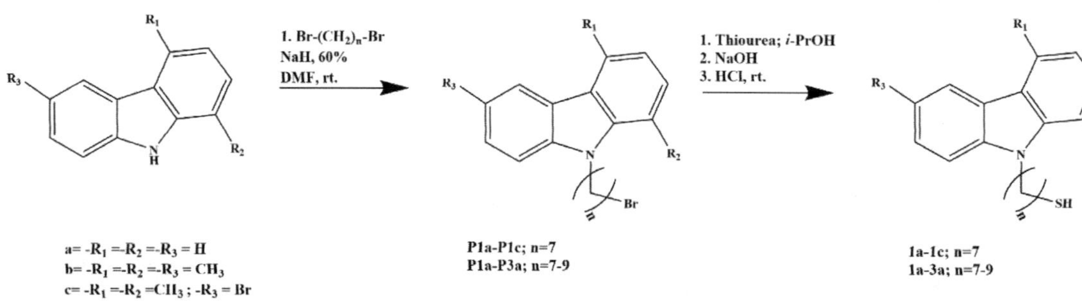

Figure 1. General procedure of synthesis of N-thioalkylcarbazole derivatives (1a-1c, 2a, 3a).

After removal of the i-PrOH, the residue was treated with 6N NaOH under reflux for 5 h, and then HCl 3N has been added at rt. The obtained product was purified by PTLC in a mixture n-hexane/ethyl acetate (97/3=v/v) to collect the pure compounds. The intermediates 9-(bromo-alkyl)-9H-carbazoles (P1a-P1c, P1a-P3a) were prepared according to the general synthetic methods [23]. In fact, the proper carbazole derivatives (a–c) were stirred at rt with dry DMF until became clear. Then, NaH 60% oil dispersion and, successively, the appropriate terminal dibromoalkane [$(-CH_2)_n$- n=7-9] were added at 0 °C. The mixture was stirred for 5 hours at rt, then water was added and the resulting mixture was extracted with EtOAc. The obtained residue was purified by silica gel column chromatography (Et_2O/hexane, 2/3 as eluent) to give the pure compounds.

The new synthetized carbazole derivatives (1a-1c, 2a, 3a) have been examined for their antiproliferative activity against two human breast cancer cell lines, namely MCF-7 and MDA-MB-231 and two human uterine cancer cell lines, cervix epithelium HeLa and endometrial Ishikawa. Only 1c, demonstrated a significant anti-proliferative activity on uterine cancer cells and, to a lesser extent, on breast cancer cell lines. Moreover, the assayed compounds did not show effects on the proliferation of non-tumoral MCF-10A and 3T3-L1 cells (Figure 2). Next, the molecular mechanism of compound

1c has been studied, proving that the observed anti-tumoral effects on HeLa cells depend on the induction of apoptosis [8].

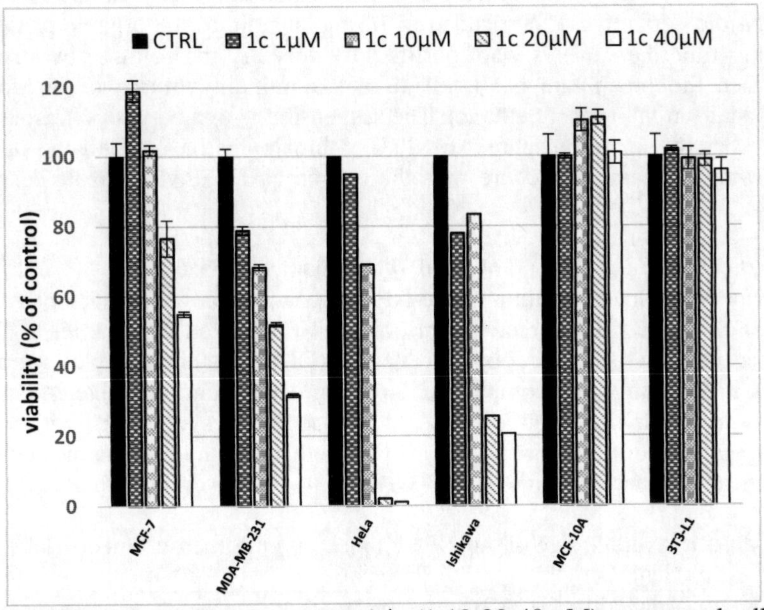

Figure 2. Antiproliferative activity of compound 1c (1-10-20-40 μM) on tumoral cell lines (MCF-7, MDA-MB-231, HeLa, Ishikawa) and non tumoral cells (MCF-10A, 3T3-L1).

Compound 1c induces apoptosis through the intrinsic pathway, also known as mitochondrial pathway, producing a clear increase of caspases 3/7 activity, which are cleaved by the initiator caspase-9 and that, in turn may activate other pro-apoptotic proteins and cleaves Parp-1, a protein involved in repairing DNA damages.

2.2 Synthesis of Au-NPs functionalized with N-thioalkylcarbazole derivatives.

The Au-NPs are synthesized by modifying the procedure described by Brust [24] (Figure 3), using the appropriate amount of binders.

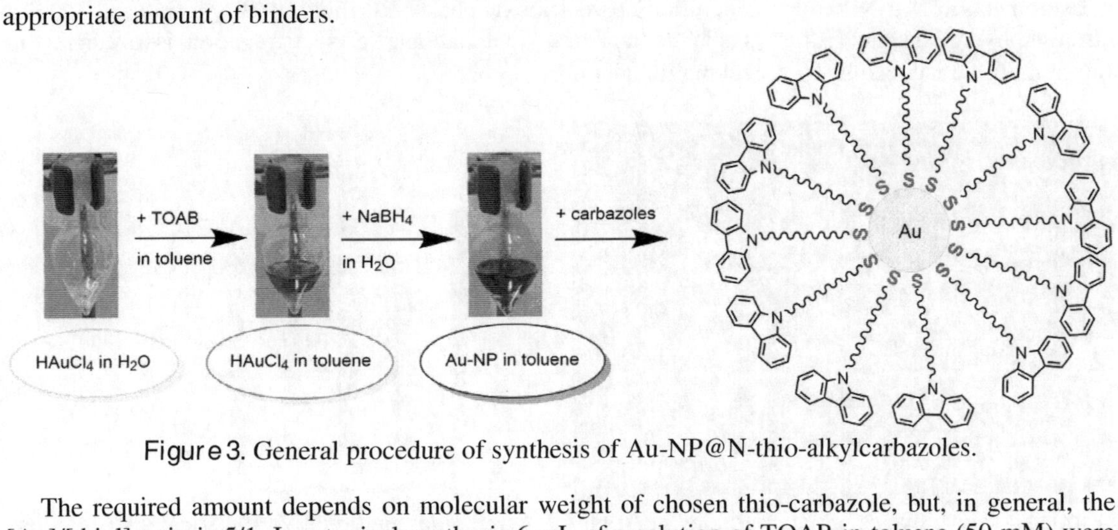

Figure 3. General procedure of synthesis of Au-NP@N-thio-alkylcarbazoles.

The required amount depends on molecular weight of chosen thio-carbazole, but, in general, the [Au]/[thiol] ratio is 5/1. In a typical synthesis 6 mL of a solution of TOAB in toluene (50 mM) were added under stirring to 2.3 mL of an aqueous solution of 30 mM $HAuCl_4$ to extract Au(III) from the aqueous phase; then 1.91 mL of an aqueous solution of $NaBH_4$ 0.5 M was added to the heterogeneous mixture, under vigorous stirring to reduce the Au(III) to Au (0). In a short time, a dark red organic phase is obtained, which is kept under stirring at room temperature for ~ 1h. After this period of time the thio-alkylcarbazole derivative 1a are added in 1/5 molar ratio to the organic phase and stirred for 24 hours. After this time the Au-NPs were purified, by very dry the solution by a rotary evaporator under vacuum; then the precipitate is dissolved in a minimum volume of dichloromethane and reprecipitated by addition of 10mL of ethanol. The suspension is treated with ultrasound and then it is centrifugated in order to isolate the pure Au-NPs@N-thioalkylcarbazole. All the other Au-NPs are synthesized following the same procedure with the chosen N-thioalkylcarbazole derivatives (1a- 1c, 2a, 3a).

2.3 Chemical-Physical Characterization of AuNP@N-thioalkylcarbazole.

The UV-Vis spectra of a toluene solution of Au-NP@carbazoles show a surface plasmonic resonance (SPR) band at 536 nm, which is characteristic of about 10 nm-sized nanoparticles. The particle size was also measured using a Dynamic Light Scattering (DLS) and the samples were prepared with TOAB or without it, to control and compare the growing of nanoparticles size. As shown in table 1 the difference between the hydrodynamic radius, the nanoparticles synthesized with the use of TOAB present a smaller aggregation therefore form particles with lower hydrodynamic radius compared to the same nanoparticles synthesized without the TOAB.

Table 1. Average size of Au-NP obtained by DLS measurements (left).

Name Sample	R_{hyd}/nm	% polydispersity
AuNP@1a (TOAB)	60	25%
AuNP@1a	100	10%
AuNP@2a (TOAB)	70	15%
AuNP@2a	140	20%
AuNP@3a (TOAB)	35	15%
AuNP@3a	35	10%

Freshly prepared colloids have been then casted onto stubs for SEM imaging, and particle analysis indicates that the diameter of all Au-NPs can be described approximately as Gaussian distribution (Figure 4). SEM images show different size of nanoparticles after toluene evaporation and DLS measurements of Au-NPs toluene solutions reveal a two peaks distribution (R_{hyd}/nm≅50 and 150), at different hydrodynamic radius, which demonstrates that nanoparticles aggregation is occurring in solution maybe induced by π–π stacking interaction.

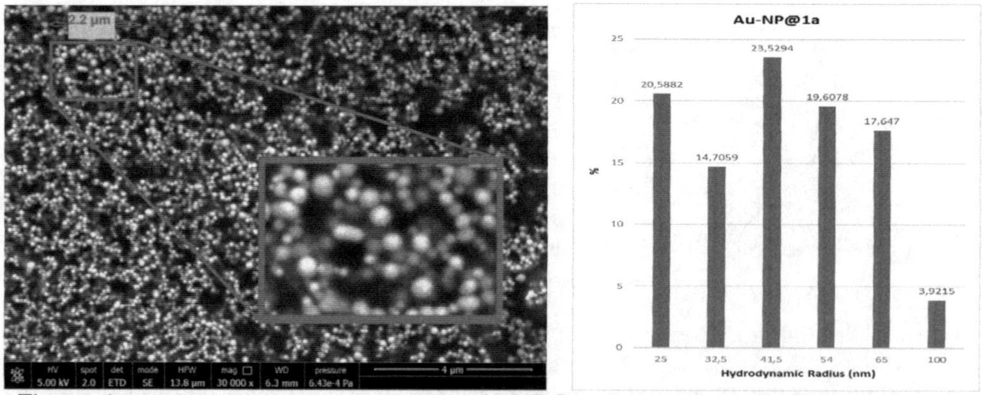

Figure 4. SEM images of nanoparticles of AuNP@1a (left side), and relative statistical distributions of the Au-NPs diameters size in the selected area (right side).

3. Conclusions

Cancer is one of the principal deathly disease in developed countries, probably due to lifestyle choices, such as smoking, lack of physical activity and unbalanced diets [25]. Compound **1c** induces apoptosis through the mitochondrial pathway, producing a clear increase of caspases which is involved in repairing DNA damages. New thiol-functionalized carbazoles, in which the carbazole nitrogen has been functionalized with alkyl-thiol chains, have been prepared [22], and used to stably functionalize Au-NPs. They are soluble in organic solvents, and are stable over weeks on the air. Au-NP@carbazoles show typical band in UV-Vis spectra. Morphological images of these nanoparticles were carried out in order to determine the gold dimension, while DLS studies were run directly in solutions. The cytotoxicity evaluation of the carbazole-capped Au-NP is actually in progress. These systems might present a double innovative function, combining an antitumor activity common of carbazoles, with the photothermal effects of gold nanoparticles.

4. Experimental Details

All reagents, excluding carbazole binders, and all solvents has been bought from Sigma-Aldrich and they are used without any other purification. Carbazole derivatives has been synthesized as already reported [25, 31, 32, 33]. Au-NPs are prepared by changing the reported literature method of Brust and co-workers [24].

The UV-vis absorption spectra were recorded using an HP spectrophotometer mod. 8451, capable of operating between 190 and 1100 nm. The set-up of the Dynamic Light Scattering (DLS) experiments is constituted of a He-Ne laser source (λ=632.8 nm and power of 35 mW), linearly polarized, focused onto the sample, and of a single photon counting detector (avalanche photodiode) for the scattered light. The detection technique is based on a self-beating detection mode, the scattering angle is fixed at 90° and temperature at T=24°C. A Malvern 4700 correlator is used to obtain the scattered intensity autocorrelation function. The hydrodynamic radius of particles RH is obtained from the normalized scattered electric field autocorrelation function by using a second-order cumulant expansion. The size distribution of the particles in solution was obtained through the Laplace inversion method. SEM morphology and SEM-EDS of the investigated samples were obtained using a FEI Quanta FEG 450 microscope.

Cell cultures, cell viability assay, tunel assay, western blot analysis, immunocytochemistry and caspase assay were performed as already described in Iacopetta et al. and Tundis et al.

Acknowledgments

Authors wish to thank: MIUR and CNR for financial support, Prof. C. Milone and Dr. E. Piperopoulos (Dept. of Engineering UniME, Italy) for SEM characterizations, Dr. N. Micali (CNR-IPCF Messina, Italy) for DLS measurements.

References

[1] Saturnino C, Caruso A, Longo P, Capasso A, Pingitore A, Caroleo MC, Cione E, Perri M, Nicolò, F, Nardo VM, Scolaro LM, Sinicropi MS, Plutino MR, El-Kashef H; Current Topics in Medicinal Chemistry 2015, 15 (11), 973-979.

[2] Caruso A, Iacopetta D, Puoci F, Cappello AR, Saturnino C, Sinicropi MS; Mini Rev Med Chem 2016;16:630–43.

[3] Bashir M, Bano A, Ijaz AS, Chaudhary BA. a review. Molecules 2015;20:13496–517.

[4] Gluszynska A. Eur J Med Chem 2015;94:405–26.

[5] Carocci A, Catalano A, Sinicropi MS. Clin Pharmacol 2014;6:127–37.

[6] Iacopetta D, Grande F, Caruso A, Mordocco RA, Plutino MR, Scrivano L, Ceramella J, Muià N, Saturnino C, Puoci F, Rosano C, Sinicropi MS , Future medicinal chemistry, 2017, 9, 17

[7] Pandey S, Agarwal P, Srivastava K ,RajaKumar S , Puri S K, Verma P , Saxena JK , Sharma A , Lal J , Chauhan PMS, Eur J Med Chem 2013;68:81–8.

[8] Sinicropi MS, Iacopetta D, Rosano C, Randino R, Caruso A, Saturnino C, Muià N, Ceramella J,

Puoci F, Rodriquez M, Longo P, Plutino MR, Journal of Enzyme Inhibition and Medicinal Chemistry 2018, 33 (1), 434–444.

[9] Iacopetta D, Rosano C, Puoci F, Sinicropi MS, Randino R, Caruso A, Saturnino C, Muià N, Ceramella J, Puoci F, Rodriquez M, Longo P, Plutino MR, Eur J Pharmaceut Sci 2017;96:263–72.

[10] Vann KR, Ergün Y, Zencir S, Oncuoglu S, Osheroff N, Topcu Z. Bioorganic & Medicinal Chemistry Letters. 2016 26: 1809-12.

[11] Maji B, Kumar K, Muniyappa K, Bhattacharya S. Org BiomolChem 2015;13:8335–48.

[12] Andrews WJ, Panova T, Normand C, Gadal O, Tikhonova IG, Panov KI. J Biol Chem 2013;288:4567–82.

[13] Ferrari M, Nat. Rev., Cancer 2005; 5 161–171.

[14] Wu W, Wieckowski S, Pastorin G, Benincasa M, Klumpp C, Briand JP, Gennaro R, Prato M, Bianco A. Chem., Int. Ed. 44; 2005; 6358–6362.

[15] Salem AK, Searson PC, Leong KW, Nat. Mater. 2; 2003; 668–671.

[16] Paleos CM, Tsiourvas D, Sideratou Z, Tziveleka L, Biomacromolecules, 5; 2004; 524–529.

[17] Paciotti GF, Myer L, Weinreich D, Goia D, Pavel N, McLaughlin RE, Tamarkin L, Drug Development Research, 11; 2004; 169–183.

[18] Paciotti GF, Kingston DGI, Tamarkin L, Drug Dev. Res. 67; 2006; 47–54.

[19] Han G, You CC, Kim BJ, Turingan RS, Forbes NS, Martin CT, Rotello VM, Angew. Chem., Int. Ed. 45 (2006) 3165–3169.

[20] Kennedy LC, Bickford LR, Lewinski NA, Coughlin AJ, Hu Y, Day ES, West JL, Drezek RA, Small, 2011, 7, 169–183.

[21] Huang X, Jain PK, El-Sayed IH, El-Sayed MA, Lasers Med. Sci, 2008, 23, 217–228.

[22] Torre LA, Bray F, Siegel RL, Ferlay J, Lortet-Tieulent J, Jemal A. Global cancer statistics, CA Cancer J Clin 2012. 2015;65:87–108.

[23] Chimento A, Saturnino C, Iacopetta D, Mazzotta R, Caruso A, Plutino MR, Mariconda A, Ramunno A, Sinicropi MS, Pezzi V, Longo P, Bioorganic & Medicinal Chemistry,, 2015, 23, 7302-7312.

[24] Brust M, Walker M, Bethell D, Schiffrin DJ, Whyman R J. Chem. Soc., Chem. Commun., 1994, 801-802.

[25] Pitsillides CM, Joe EK, Wei X, Anderson RR, Lin CP, Biophys. J., 2003, 84, 4023–4032.

[26] Saturnino C, Iacopetta D,. Sinicropi MS, Rosano C, Caruso A, Caporale A, Marra N, Marengo B, Pronzato MA, Parisi OI, Longo P, Ricciarelli R. Molecules, 2014,19, 9307-9317

[27] Panno A, Sinicropi MS, Caruso A, El-Kashef H, Lancelot J-C, Aubert G, Lesnard A, Cresteil T, Rault S. Journal of Heterocyclic Chemistry, 2014, 51, E294-E302

[28] Sinicropi MS, Lappano R, Caruso A, Santolla MF, Pisano A, Rosano C, Capasso A, Panno A, Lancelot J-C, Rault S, Saturnino C, Maggiolini M. Current Topics in Medicinal Chemistry, 2015, 15, 11, 1035-1042

[29] Iacopetta D, Rosano C , Puoci F, Parisi OI, Saturnino C, Caruso A, Longo P, Ceramella J, Malzert-Fréon A., Dallemagne P,Rault S, Sinicropi MS. European Journal of Pharmaceutical Sciences, 2017, 96, 263–272

[30] Saturnino C, Palladino C, Napoli M, Sinicropi MS, Botta A, Sala M, Carcereri de Prati A, Novellino E, Suzuki H. European J. Med. Chem. 2013, 60, 112-119

[31] Parisi O I, Morelli C, Puoci F, Saturnino C, Caruso A, Sisci D, Elvi Trombino G, Picci N,Sinicropi MS. Journal of Materials Chemistry B, 2014, 6619-6625

[32] Caruso A, Lancelot J C, El-Kashef H, Sinicropi M S, Legay R, Lesnard A, Rault S. Tetrahedron, 2009, 65, 10400-10405.

[33] Iacopetta D, Mariconda A, Saturnino C, Caruso A, Palma G, Ceramella J, Muià N, Perri M, Sinicropi MS, Caroleo MC and Longo P. ChemMedChem, 2017, 12, 2054-2065

IOP Conf. Series: Materials Science and Engineering **459** (2019) 012024 doi:10.1088/1757-899X/459/1/012024

Limits of carbon micro/nano particles utilization to improve properties of polymer matrices in fibre reinforced composites

B Tomková, J Novotná, M Pechočiaková

Dept. of Material Engineering, Technical University, Liberec (Czech Rep.)

blanka.tomkova@tul.cz (Author)

Abstract. This work focuses on the changes in properties of epoxy resin filled with five carbon forms in several volumes (graphene nanopellets, carbon particles from acrylic waste, and 3 different types of recycled carbon fibres). Main purpose is to enhance knowledge in influence of various carbon forms and their volume on final matrix structure and properties, and their applicability in fibre reinforced composites. Current measurements show that all types of carbon particles influence matrix properties to a certain extent, as their presence activates overall structural changes in epoxy resins curing.

1. Introduction

The development of new composite materials is one of the fastest growing sectors in industrial research. As the textile reinforcements remain more or less the same except for development of "green" composites, many recent studies are focused on improvement of mechanical, thermal, electrical or wear properties using different combinations of nanofillers in polymer matrices [1-9]. The selection of the filler type depends, in the first place, on the property we aim to improve.

It is known that mechanical and thermal properties mainly depend on level of matrix cross-linking, and interface quality (optimized contact) between the matrix and applied nanoparticles [10-11], while transmission of electrical charges in polymer composites depends not only on quality of their structure but also on electrical properties of individual components [12].

For instance, if we mean to increase thermal/mechanical endurance along with electrical conductivity, we must consider the fact that most polymeric materials belong into category of insulators [13] with electrical conductivity lower than 10^{-14} S/m. To increase both, electrical conductivity, and thermal/mechanical properties, it is necessary to fill polymer matrix with electro-conductive particles and/or nanoparticles, that concurrently support arrangement of matrix structure. From this viewpoint, electro-conductive carbon/graphite particles of various shapes and sizes seem like very promising material [14]. Their market offer is very wide [15], from chopped or milled carbon fibers, over graphen nanopellets, carbon black, multiwalled or singlewalled carbon nanotubes, to natural fullerens, and they can be variously combined. As such they are promising filling material for variety of technical applications. This study focuses on properties of epoxy resin filled with five carbon forms (graphene nanopellets, carbon particles from acrylic waste, and 3 different types of recycled carbon fibres). Aim was to study the influence of particles structure and volume on epoxy resin properties, and analyse its applicability for improvement of fibre reinforced plastic composites (FRPC) in perspective of appropriate balance between the cost of the production and required properties of final composites.

2. Experimental

2.1. Material

Epoxy specimens were made from Bisphenol A-based low viscosity epoxy resin MGS LR 285 [16], and cyklo-alifatic polyamine curing agent H 508 (mixing ratio 100:40 by weight). As fillers were used following carbon materials: graphene nanopellets (GNP, Figure 1, a) [17], milled carbon particles from acrylic waste (mCPAN, Figure 1, b) [18], commercially available Carbiso Milled Carbon Fibres (CMF, Figure 1, c) [19], milled Carbiso CF CMF (mCMF, Figure 1, d) [20], and milled recycled carbon fibre from epoxy resin composites (mRCF, Figure 1, e) [21].

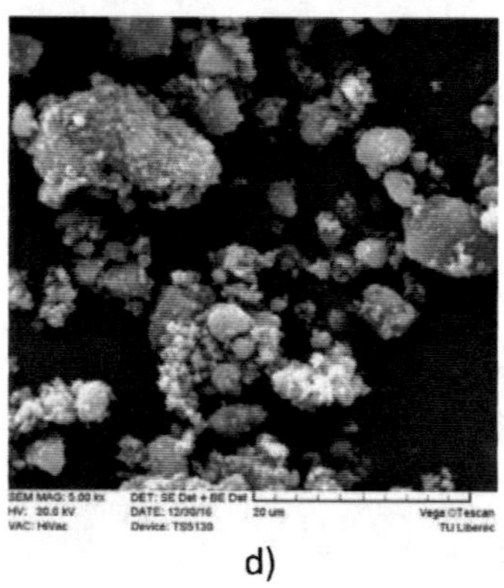

Figure 1. a) Graphene nanopellets (GNP), b) Carbon particled from acrylic waste (mCPAN), c) Carbiso Fibres (CMF), d) Milled Carbiso (mCMF), e) Milled recycled CF (mRCF).

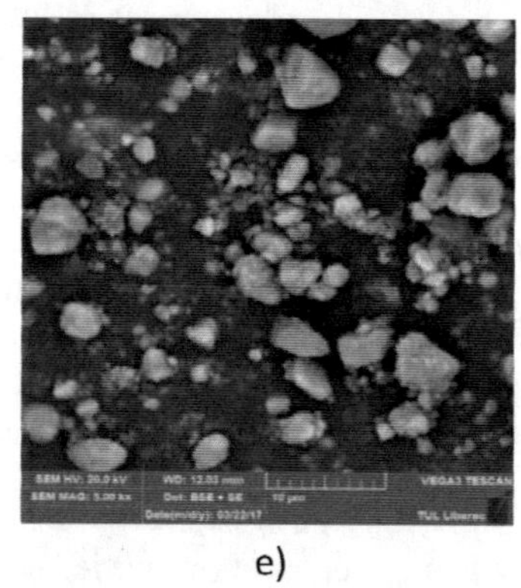

One specimen was neat epoxy, the others were filled with carbon particles in following concentrations: 0.5, 1.0, 1.5, 2.0, 2.5 and 3.0 wt% respectively. Higher concentration was not used due to striking increase of resin viscosity in 3wt%, which means it is useless as a fiber bonding in FRPC.

2.2. Milling

Nanoparticles from Carbiso (mCMF), recycled CF (mRCF) and carbonized acrylic fibrous waste (mCPAN) were made by the dry pulverization using high energy planetary ball milling of Fritsch pulverisette 7, Germany. Carbon material was placed in 80 ml sintered corundum container and milled by zirconium balls (∅10 mm) for 30 minutes in dry state. Selected balls/material ratio was 10:1, rotation speed was set at 850 rpm. Samples of Carbiso Milled CF (CMF), and graphene nanopellets (GNP) were not milled.

2.3. Size of fillers

Parameters of Carbiso milled carbon fibres are specified by producer [19], who states average diameter about 7 μm and average length about 100 μm. Size of other samples was determined from structural micrographs using image analysing software NIS Elements [22]. Micrographs were scanned on SEM microscopes VEGA 3 TESCAN and VEGA TS 5130.

Table 1. Structural geometry of used carbon micro/nano fillers.

Properties	GNP	mCPAN	CMF	mCMF	mRCF
Equivalent particle diameter [μm]	0.36	0.62	7.0	3.2	1.51

2.4. Methods of measurement

Alternating current (AC) conductivity

AC conductivity σAC [Sm^{-1}] of carbon filled specimen was measured using AGILENT 4294. The measurements were made in the frequency range 100 Hz–3MHz according to ASTM D150-98:2015. A precision analyzer was used to measure the sample capacitance C [F] and the loss tangent tan δ directly. The total conductivity σ_{AC} [Sm^{-1}] is calculated using following equation:

$$\sigma_{AC} = 2\pi f \varepsilon_0 \, \varepsilon''$$

where, f [Hz] is the measurement frequency and ε_0 [Fm^{-1}] is free space permittivity (the electric constant, approx. 8.85×10^{-12} Fm^{-1}). Imaginary part of permitivity ε'' [Fm^{-1}] is calculated using following relation:

$$\varepsilon'' = \varepsilon' \, tg\delta$$

where the real part of permittivity ε' is determined from the relation:

$$\varepsilon' = \frac{C.d}{\varepsilon_0 S}$$

d [m] is the sample thickness and S [m^2] the sample cross-sectional area.

Charpy Impact test

Impact strength and energy were tested to analyse the influence of carbon nanofillers on epoxy matrix impact resistence. The test was run according to ISO 179-1:2010, where specimen is laid into horizontal position on Charpy's hammer struts. When the ram is released, the hammer freely falls into the center of the specimen between the supports, and the amount of energy, consumed for specimen breaking, is set in Joules. Impact strength is then determined from following relation:

$$a_{CN} = \frac{E_C}{h * b} * 1000$$

where, a_{CN} [$kJ.m^{-2}$] is the Impact strength, E_C [J] is Impact energy, h [mm] is specimen thickness, and b [mm] is specimen width.

Analysis of thermo-mechanical and thermal properties

Differential Scanning Calorimeter DSC6 (Perkin Elmer) was used to analyse glass transition of analysed set of filled epoxy specimen. Samples of 10 mg weight were placed in aluminum pans and sealed. The specimens were heated in an inert nitrogen atmosphere from room temperature (25°C) up to presumed softening temperature of used epoxy resin [16] (120°C) at a heating rate of 10°C/min.

Using dynamic mechanical analysis DMA DX04T (RMI) in 3PB mode the changes of elastic/plastic properties were scanned. These changes indicate structural transformation of matrix due to carbon filling. The specimen has been loaded as free-fixed beam. The tests have been realized only up to 60°C (working temperature of used resin) at the heat rate of 3°C/min under the 1 Hz load frequency.

Heat transfer of epoxy specimens was tested with Isomet 2114 - Thermal Properties Analyser to estimate the progress of thermal conductivity due to carbon filling in epoxy. The amount of transferred heat may be distorted due to weak insulation of the device, and non-homogenity of material, so the final value of thermal conductivity serves only for monitoring, as it is not confirmed yet.

3. Results

The measurements of epoxy resin filled with various carbon forms proved, that increase of particles' volume changes the properties of epoxy resin, as shown in Table 2 (data for mCMF were selected).

Table 2. Properties of neat and mCMF filled epoxy.

Properties of mCMF	Neat Epoxy	0.5wt%	1wt%	1.5wt	2wt%	2.5wt%
AC conductivity x 10^{-8} [$S.m^{-1}$] at 10 kHz	6.94	7.32	7.66	7.81	8.12	8.34
Flexural Modulus [GPa]	3.28	3.32	3.38	3.42	3.51	3.52
Impact strength [$kJ.m^{-2}$]	38.27	38.62	38.95	39.28	39.41	39.43
Thermal conductivity [$W.m^{-1}.K^{-1}$]	0.101	0.159	0.183	0.199	0.201	0.208

As other carbon fillings showed more or less similar behavior, and with respect to processing requirements on low viscosity of resin used as a composite bonding, we suggested 2.5wt% carbon filling as an optimum combination for both property improvement and matrix processability, and further experiments were done only for this filling volume. A comprehensive overview of the results for 2.5wt% carbon filled epoxy resins is presented in Table 3 (in Discussion part).

Results for mechanical properties (see Figure 2) show remarkable effect only for CMF filling, which is standard short fiber carbon reinforcement (even in low concentration). Other particles filling have little or rather weakening effect, except for GNP in impact measurements. AC conductivity of specimens are in detail shown in Figure 3. It is obvious that only Carbiso filling remarkably shifted the AC conductivity of the resin. Other carbon particles show only slight improvement of epoxy no matter, what type of particle was used. DSC analysis showed, that all carbon fillers have plasticizing effect on used epoxy resin (see Figure 4). Same behavior also showed up on DMA analysis in 60°C. While neat epoxy is still stiff, filled specimen already show viscous behavior. As for thermal conductivity, all fillings increased the conductivity approx. twice.

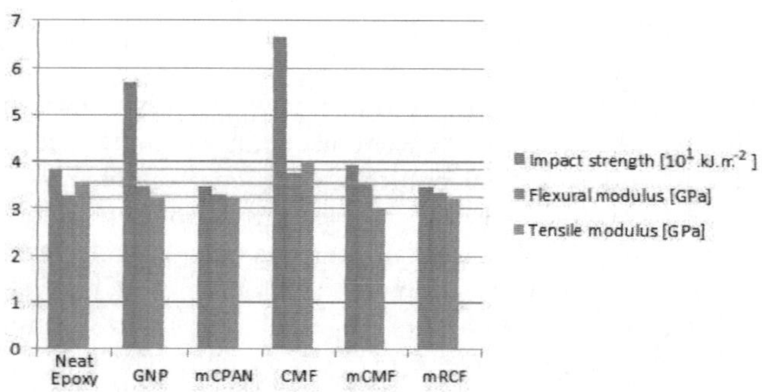

Figure 2. Selected mechanical properties of specimen (Impact strength, flexural, and tensile moduli).

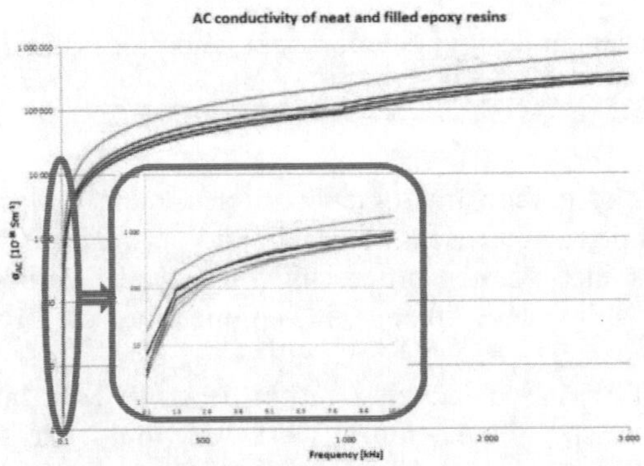

Figure 3. AC coductivity progression in frequency range from 100 to 3000 kHz

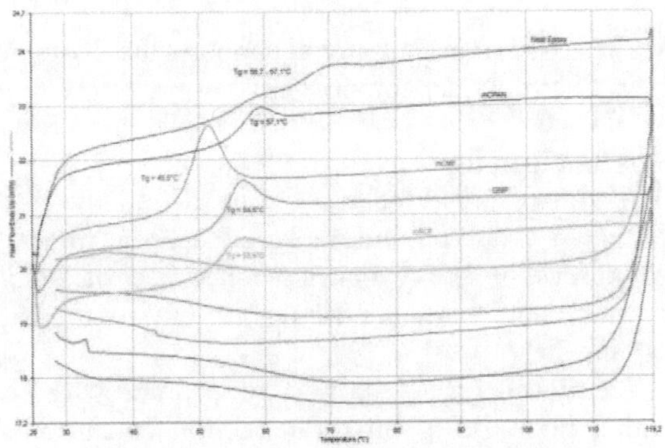

Figure 4. DSC analysis of neat, and filled epoxy specimen

4. Discussion

Experimental measurements showed that use of carbon micro/nanoparticles in the same volume (2.5wt%) influences matrix properties to a certain extent depending on the carbon particle morphology. Remarkable effect showed mainly CMF filling, which is standard short fiber carbon reinforcement (even in low concentration). Other particles' filling exhibit little influence on electrical properties, in case of thermal/mechanical properties it sometimes exhibits even weakening effect (e.g. mCPAN, or mRCF). Comprehensive overview of the results is presented in Table 3.

Table 3. Selected properties of neat and carbon particles filled epoxy resin.

Properties	Neat Epoxy	GNP	mCPAN	CMF	mCMF	mRCF
AC conductivity x 10^{-8} [S.m^{-1}] at 10 kHz	6.94	9.12	7.55	18.2	8.34	8.51
AC conductivity x 10^{-7} [S.m^{-1}] at 100 kHz	8.32	11.2	9.10	22.1	9.51	10.0
AC conductivity x 10^{-5} [S.m^{-1}] at 1000 kHz	0.89	1.24	0.99	2.35	1.06	1.07
Impact strength [kJ.m^{-2}]	38.27	57.02	34.6	66.41	39.43	34.50
Flexural modulus [GPa]	3.28	3.45	3.31	3.78	3.52	3.33
Tensile modulus [GPa] calculated from 3PB	3.55	3.21	3.22	4.01	3.04	3.23
Tg [°C]	58.75	54.56	57.12	57.46	49.54	53.63
Thermal conductivity λ [W.m^{-1}.K^{-1}]	0.101	0.215	0.195	0.203	0.208	0.192

Achieved results show, that limits of carbon micro/nanoparticles utilization as matrix filling in FRPC are derived from processing parameters based on resin viscosity, and from morphology of carbon particles. The fibrous particles (2D) exhibits better influence comparing to bulk particles (3D). Significant is also market price of selected fillings, which points to milled recycled fibers (CMF) as not only cost-effective but also environmentally friendly filling prepared from recycled carbon fibre wastes. Use of such material supports the ecological aspects in design, development, and processing of new fiber/polymer composites.

5. Conclusions

The aim of presented work was to find the use limits of carbon micro/nanoparticles, mainly those prepared from ground recycled carbon fibers, to improve electrical, thermal and/or mechanical properties of epoxy matrices. The use of such fillers incorporates processing of recycled carbon wastes, and may contribute to environmentally friendly development in engineering of FRP composites.

Achieved results show a good processing potential of milled recycled carbon fibers (CMF) but rather poor use of further reduction of carbon particles size. In the future work we would like to analyse this behavior more deeply, and we will also test selected fillers' influence on different types of resin systems, and incorporate achieved results in future design and development of novel fiber reinforced polymer composites. Finally we would like to transfer our findings into practical industrial applications. This is the main challenge for future work.

Acknowledgements

Authors are grateful for the financial support by the Ministry of Education, Youth and Sports of the Czech Republic and the European Union - European Structural and Investment Funds in the frames of Operational Programme Research, Development and Education - project Hybrid Materials for Hierarchical Structures (HyHi, Reg. No. CZ.02.1.01/0.0/0.0/16_019/0000843).

References

[1] Kotsilkova R, Fragiadakis D and Pissis P 2005 *J. of Polymer Science* B **43(5)** 522
[2] Singh S, Srivastava VK and Prakash R 2015 *Applied Nanoscience* **5(3)** 305
[3] Zhou T, Wang X, et al. 2013 *Polymer Letters* **7** 585
[4] Safdari M, Al-Haik MS 2013 *Carbon* **64** 111
[5] Zhang K, Zhang Y, Wang S 2013 *Carbon* **65** 105
[6] Guo M, et al. 2014 *Comp. Sci. Technol.* **97** 27
[7] Hussan SI 2012 *Al-Mustansiriya J. of Science* **23(3)** 133
[8] Du FP, Tang H, Huang DY 2013 *Int. J. of Polymer Sci.* **2013** 1
[9] Seiler J, Kindersberger J 2013 *Proc. of 18. International Symposium on High Voltage Engineering* (Seul, Korea) p 1297
[10] Ellis B 1993 *Chemistry and Technology of Epoxy Resins* (Dordrecht: Springer Netherlands)
[11] Fiedler B, et al. 2006 *Composite Sci. Technol.* **66** 3115
[12] Aneli JN, Khanasvili LM, Zaikov GE 1998 *Structuring and Conductivity of Polymer Composites* (NY, USA: Nova Science Publishers)
[13] Peliskova M, Saha P 2012 *Chem. Letters* **106** 1104
[14] Burchell TD 1999 *Carbon Materials for Advanced Technologies* (Oxford, UK: Elsevier)
[15] Delgado JL, et al. 2008 *J. of Materials Chemistry* **18** 1417
[16] EPIKOTETM 2017 www.swiss- composite.ch/pdf/t-Epoxyd-Harz-L-285-LF-e.pdf
[17] KOH-I-NOR GRAFIT 2017 *http://www.grafitnetolice.cz/grafit_en.php*
[18] Baheti V, et al. 2015 *Fibers and Polymers* **16(10)** 2193
[19] Carbiso 2017 *http://www.easycomposites.co.uk /#!/resin-gel-silicone-adhesive/filler-powders-and-additives/general-fillers-and-additives/milled-carbon-fibre-powder.html*
[20] Novotna J, et al. 2017 *Proc. of 9th Central European Conference* (Liberec, Czech Rep.)
[21] Tomkova B, Militky J 2010 *World J. of Engineering* **7(2)** 232
[22] Tomkova B, Grabmullerova J, Novakova, J 2013 *Proc. of 5th International Conference on Experiments/Process/System Modeling/ Simulation/ Optimization* (Athens, Greece, on CD)

Textile waste for chemical and textile industries feedstock (presentation of EU project)

B Voncina[1], J Valh Volmajer[1], S Vajnhandl[1], A Majcen Le Marechal[1], A P Aneja[1] and A Lobnik[1,2]

[1]University of Maribor, Institute of Engineering Materials and Design, Laboratory for Chemistry and Environmental Protection,
[2]Institute for Environmental Protection and Sensors, Maribor, Slovenia

bojana.voncina@um.si

Abstract The RESYNTEX project aims at designing, developing and demonstrating a new industrial symbiosis between textile waste and the chemical industry. The new original symbiosis is based on the chemical/enzymatic transformation of textile waste in a form that facilitates the easy take up as feedstock by the chemical industry in order to produce high added value chemicals. The parallel production of various high added value products ensures competitive production costs for the chemical market. As a result, economic advantages can be provided besides prevention of industrial environmental problems. The project will consider and demonstrate the whole value chain starting from the citizen behaviour change and the textile collection of unwearable textiles, improving and automatizing the industrial sorting, demonstrating the production of the transformed textile components and the symbiosis with the obtained chemical products and finally analysing the best economic models and policy actions for a successful introduction in EU markets.

Keywords: textile waste, recycling, chemical depolymerisation, enzymatic transformation, circular economy.

1. Introduction

According to the statistic [1] we can estimate that textile waste annual volumes will reach 17 Mt in 2020. Targeting a significant increase in collection rate (50 % in 2020) and considering that 23 % of textile waste is constituted of re-wearable clothes and remaining waste could be recycled at a rate of 95 %, we can estimate a potential of 6.2 Mt of fibers per year that could be converted into new feedstock. In the long term (post 2020), reaching 100 % collection would give access to more than 12.4 Mt of fibers. As a consequence, there is a strong need to develop a new industrial symbiosis based on this textile waste stream associated with adapted processes in order to achieve an efficient recycling especially for non-wearable textile waste and move away from landfilling and energy recovery.

The textile recycling process could therefore be considered as one of the effective methods in reducing the production of virgin materials and minimising the environmental impact, i.e. use of water, energy and chemicals in the textile production chain, as well as in reducing the amount of textile waste.

In general, the post-consumer textile waste can be recycled by 4 main methods:

- Defibrilation/shredding/pulling/carding into fibres
- Energetic recycling
- Thermo-mechanical recycling
- Chemical recycling

Defibrillation/carding is carried out by breakdown of fabric to fiber through cutting, shredding, carding, and other mechanical processes. The fiber is then re-engineered into value-added products. These value-added products include stuffing, automotive components, carpet underlays, building materials such as insulation and roofing felt, and low-end blankets. The majority of this category consists of unusable garments - garments that are stained, torn, or otherwise unusable. Several methods are available for recovery of energy from waste such as gasification, pyrolysis, carbonisation etc. [3].

Thermo-mechanical recycling is re-melting the sorted waste of thermoplastic. This method is mainly used for bottle–to–fiber technologies, where sorted and cleaned PET bottles are re-extruded into products for non-food application, to textile fibers.

Chemical recycling (also called feedstock or tertiary recycling) allows the recovery of more value products from plastic wastes than incineration. By more restrictively definition, the chemical recycling may be defined as 'the production of chemical products of value from waste polymeric materials by economically feasible processes'. This definition, which requires the recovery of products of value, excludes from chemical recycling both biodegradation and combustion, and limits chemical recycling to those processes that are also economically feasible [4]. The products of chemical recycling are easily reintroduced into the production cycle, without any problems of market saturation; another benefit is that the crude products resulting from chemical breakdown can be used without further purification.

1.1 RESYNTEX Consortium

RESYNTEX has 20 project partners from across 10 different EU member states (Figure 1). Partners include industrial associations, businesses, SMEs and research institutes. Working together, the group creates an effective model for the whole value chain.

Figure 1. RESYNTEX Consortium

2. Experiments

The main objective of WP4, where University of Maribor is currently mostly involved, is optimisation of a physicochemical recovery/transformation process via depolymerisation of textile waste.

Scope and goal: Total chemical depolymerisation of PET or PA material after sequential protein and cellulose removal.

This eco-friendly process comprises of:

- polyamide recovery with the production of PAs oligomers and
- polyester recovery and the production of PET monomers

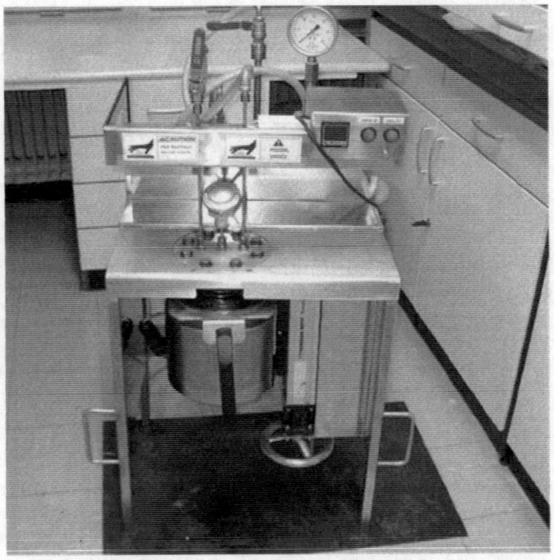

Figure 2. Laboratory scale high pressure, high temperature reactor.

Relevant issues:

- influence of impurities on PAs and PET depolymerisation, yield, efficiency and purity; impact of other organic and inorganic waste on chemical transformation.
- optimization of production of added value feedstock for chemical industry in a pilot plant.
-

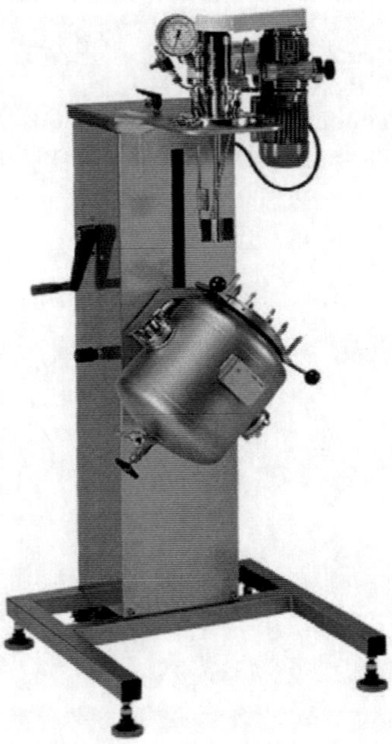

Figure 3. Pilot high pressure and high temperature plant.

3. Results and discussion

3.1 Polyamide recovery with the production of PAs oligomers

Chemical depolymerisation of PA6 was performed – promising results were obtained (liquid fractions which precipitate after few hours were obtained) by high temperature and high pressure hydrolysis with high excess of water. By ^{1}H NMR it was proved that mainly PA6 dimers were produced. With less harsh conditions, several water non-soluble oligomers were produced.

3.2 Polyester recovery and production of PET monomers

Testing of PET depolymerisation efficiency according to experimental conditions P, T and t and PET type (high and low viscosity type), with and without the present of catalyst was carried out. The degree of chemical depolymerisation of virgin PET polymers was preliminary determined gravimetrically, by carboxylic group number, FTIR spectroscopy and DSC.

After depolymerisation of PET textile materials 95 % of monomer (terephthalic acid) was obtained (proved by ^{1}H NMR).

Acknowledge

RESYNTEX project received funding from the European Union's Horizon 2020 research and innovation programme under grant agreement 641942.

References

[1] JCR Scientific and Technical Report, 2010
[2] Grand Agreemnet Number 641942-Resynex
[3] Nunes,L.J.R., et al. J. Clean. Production 171 (2018), 1353-1360.
[4] Wang, Y., 2006. Recycling in Textiles - 1st Edition. Woodhead Publishing.

IOP Conf. Series: Materials Science and Engineering **459** (2019) 012026 doi:10.1088/1757-899X/459/1/012026

Comparison of the photochromic behaviour of dyes in solution and on polyester fabric applied by supercritical carbon dioxide

M T Abate[1,3,4], S Seipel[1], M Viková[2], M Vik[2], A Ferri[3], G Jinping[4], G Chen[4] and V A Nierstrasz[1]

[1]Textile Materials Technology, Department of Textile Technology; Faculty of Textiles, Engineering, and Business; University of Borås; 50190 Borås, Sweden
[2]Department of Material Engineering; Faculty of Textile Engineering; Technical University of Liberec; 461 17 Liberec, Czech Republic
[3]Department of Applied Science and Technology, Politecnico di Torino, Corso Duca degli Abruzzi 24, 10129 Torino, Italy
[4]College of Textile and Clothing Engineering, Soochow University, 215006 Suzhou, Jiangsu, China

Email: molla.abate@polito.it, molla.tadesse_abate@hb.se

Abstract. Photochromic textiles are of considerable interest for smart and functional textile applications due to their remarkable dynamic colour changing effect when irradiated with light of a certain wavelength. The use of resource efficient processes, such as digital inkjet printing and supercritical carbon dioxide ($scCO_2$) dyeing techniques enables an economic production of those high-end functional products with high material costs. In this study, photochromic polyester fabric has been prepared by applying two commercially important photochromic dyes from spirooxazine (SO) and naphthopyran (NP) dye classes using $scCO_2$-dyeing technique. The properties of $scCO_2$ dyed photochromic fabrics were compared with the properties of the same dyes in a non-polar solvent, hexane. UV-Vis spectroscopy and a specially designed online colour measurement system capable of simultaneous UV irradiation and colour measurement were used to evaluate the photochromic colour behaviour. Both photochromic dye types embedded in textile as well as in solution showed significant reversible colour changing properties when exposed to UV light and revert to their original non-coloured form when the UV light is removed. The $scCO_2$ dyed polyester fabrics exhibited similar trends of colour build-up as in solution, while contrasting behaviour was observed in terms of colour changing rates compared to their behaviour in solution.

Keywords: supercritical CO_2 ($scCO_2$), dyeing, polyester, photochromic, spirooxazine, naphthopyran

1. Introduction

Photochromism is a phenomenon, in which the absorption spectrum of a molecule or a crystal changes reversibly when the sample is irradiated by light of a certain wavelength. Photochromic materials undergo a colour change when irradiated with UV light, most commonly from colourless to coloured, which can be reversed either by the influence of temperature or wavelengths of visible light [1]. The ability of photochromic compounds to switch between different molecular states makes them interesting and feasible for many commercial applications as ophthalmic lenses, optical data recording and storage, memories, light control filters, sensors and displays. In recent years, there has been a growing interest in the development of photochromic textiles for smart and functional textile applications such as brand protection, responsive camouflage and as UV sensors [2, 3]. Several methods have been used to apply photochromic dyes to various textile substrates such as exhaust dyeing [4-6], screen and digital inkjet printing [7-11], sol-gel coating [12], pad-dry method [13, 14], electro spinning [15, 16] and mass colouration [17, 18]. However, due to technical and performance limitations, it appears that the search for an appropriate production technique that meets the necessary commercial requirements is still underway.

In this investigation, we studied the potential application of photochromic dyes to polyester fabric using $scCO_2$ dyeing technique and compared the photochromic behaviour in solution and on polyester fabric. The use of a resource efficient technique as $scCO_2$ dyeing to develop photochromic fabrics is an attractive alternative to other methods, which use both water and organic solvents, as $scCO_2$ dyeing offers significant environmental and economic benefits. CO_2 is inexpensive, non-toxic, non-flammable, and inert with easily accessible critical conditions (Tc=31 ˚C and Pc=7.38 MPa). The process completely avoids the use of water and the associated wastewater pollution, simpler dye formulation without auxiliary chemicals, shorter dyeing times, and drying is not required with 50 % lower cost [19].

The kinetics of photo-colouration and bleaching were analysed by using UV-Vis spectroscopy for photochromic dyes in solution and a specially designed online colour measurement system, which is capable of simultaneous UV irradiation and colour measurement for photochromic fabrics. Both photochromic dye types in textile and in solution showed significant reversible colour changing properties when exposed to UV light and revert to their original non-coloured form when the UV light is removed. The $scCO_2$ dyed polyester fabrics exhibited similar trends of colour build-up as in solution, while contrasting behaviour was observed in terms of colour changing rates compared to their behaviour in solution. Ruby red (RR) from naphthopyran dye class displayed fast colouration and decolouration in solution but its speed of colour conversion become slower when incorporated to polyester fabric. For Sea green (SG) from the spirooxazine family, opposite trends with slower kinetics in solution and faster photochromic conversion on the fabric were observed. This study provides important information for the application of photochromic materials in solid state embedded within a crystalline polymer matrix.

2. Experimental

2.1. Dyeing process in $scCO_2$

Two commercial photochromic dyes namely ruby red (RR) and sea green (SG) from Vivimed labs, UK were used to dye a plain-woven polyester fabric of 147 gsm (FOV fabrics, Sweden) in $scCO_2$ medium. The dyeing was performed using a batch type $scCO_2$ dyeing apparatus equipped with a high-temperature oil (glycerine) bath, a rotary wheel where the vessels are mounted, a motor, temperature and time controller, a heater, and cooling element. The polyester fabric (21×30 cm, ca. 9 g) was wrapped around a Teflon mesh, suspended inside the dyeing vessel (internal volume of 290 mL) and a small amount of photochromic dye (0.027 % owf ~ 2.56 mg) was placed below the fabric at the bottom of the vessel. Dyeing was carried out at 120 ˚C temperature and 25 MPa pressure for one hour, which is the optimum dyeing condition selected based on our preliminary study. After dyeing, the vessels were removed from the oil bath and the CO_2 was vented slowly until the pressure reached atmospheric pressure and the samples were taken out from the vessels for further analysis without reduction cleaning procedure.

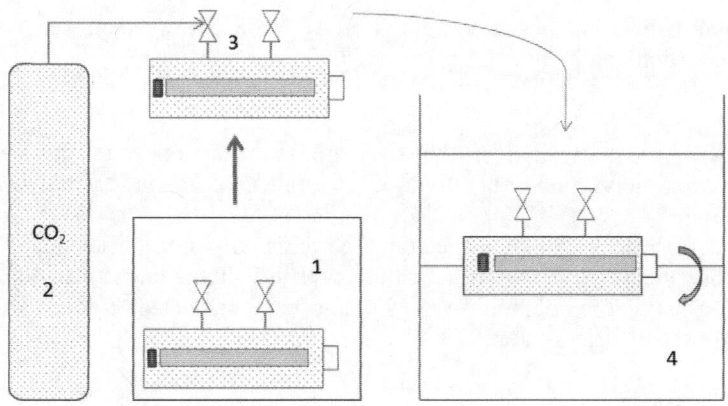

Figure 1. Schematic $scCO_2$ lab dyeing procedure, (1) freezer, (2) CO_2 tank (3) dyeing vessel, (4) high-temperature oil bath

2.2. Colour measurement

UV-Vis spectra of the photochromic dyes in solution were measured at their specific dominant wavelength (absorbance maximum), defined by a wavelength scan between 400 and 800 nm after UV activation using an Evolution 200 (Thermo scientific) UV-visible spectrophotometer at 21°C. The performance of each dye to undergo reversible photo-isomerization under the influence of UV and visible light was studied at their absorbance maximum wavelength using six different exposure times (30, 60, 120, 180, 210, and 300 s). The dye solutions (0.05mg/ml) were irradiated using a UVA light torch with maximum emission wavelengths at 395-400 nm. Upon UV-irradiation the torch was placed on top of the PMMA cuvettes containing the dye solutions in the UV-Vis measurement cell with open lid. After exposure time, the lid of the measurement cell was closed and the measurement started with a short delay (ca. < 2 s) between exposure and measurement. The absorbance was recorded after each exposure time and the fading performance was estimated qualitatively by observing the time required to revert into their original uncoloured state and comparing each other. The colouration and decolouration properties of the photochromic textiles were measured using an LCAM Photochrome 3 spectrophotometer (Technical University of Liberec, Czech Republic), which is capable of measuring the colour development and fading performance continuously over various cycles of UV exposure and relaxation. The measured reflectance values (R) are recorded in 2-second intervals of continuous UV irradiation and the K/S values are calculated from the reflectance data using the Kubelka-Munk function using equation (1)

$$\frac{K}{S} = \frac{(1-R)^2}{2R} \qquad (1)$$

where: R is the reflectance percentage at the maximum absorption wavelength, K is the absorption coefficient, and S is the scattering coefficient of the dyed samples.

Table 1. Activation and deactivation intervals (in seconds) of UV light during continuous colour measurement

Measurement settings	Samples	
	Sea green (SG)	Ruby red (RR)
Activation	300	500
Deactivation	400	1000

The colouration and fading kinetics, determined from the K/S-time data sets, were fitted to an extension of the first-order kinetic model, equation (2).

$$\frac{K}{S} = \left(\frac{K}{S_0} - \frac{K}{S_\infty}\right) \times e^{-kt} + \frac{K}{S_\infty} \quad (2)$$

where: K/S_0 and K/S_∞ are the initial and final (equilibrium) colour intensity values respectively attained upon exposure or relaxation of UV light, k is the rate constant and t is the time of exposure.

The photochromic performance of $scCO_2$ dyed photochromic polyester fabrics were evaluated in terms of the average colour yield ($\Delta K/S$) upon activation with UV-light, rate constants of colour growth ($k_{colouration}$) to achieve maximum colouration (K/S_∞), and rate constant of colour decay ($k_{decolouration}$) to revert to the ring-closed colourless state (K/S_0).

3. Results and discussion

The absorbance maxima of the dyes were defined by a wavelength scan between 400 and 800 nm after UV activation of the dye solutions. The obtained absorbance maxima were 472 and 600 nm for SG

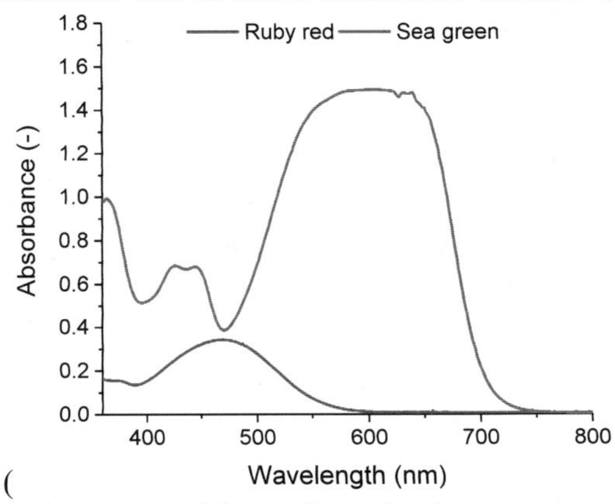

and RR dyes, respectively (
Figure **2**) and these wavelengths were used in our investigations.

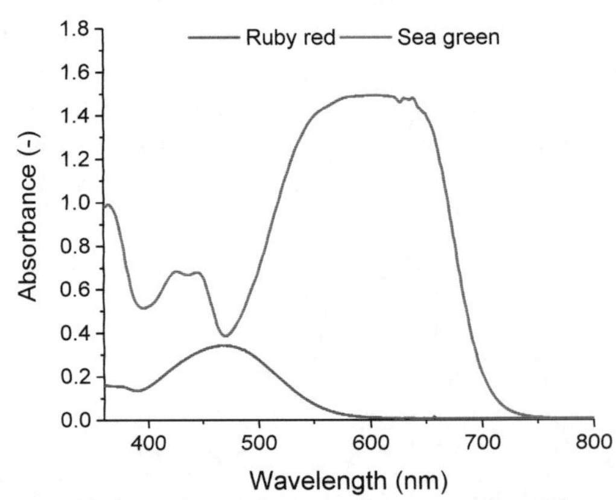

Figure 2. UV-Vis spectra of sea green and ruby red in hexane (0.05mg/ml)

The photochromic behaviour of the dyes in solution and on polyester fabric has been compared. A non-polar solvent, hexane, was selected as a solvent due to its similar polarity to CO_2 [20] to minimize the solvent effect for comparison purposes. The photochromic compounds in both media, in solution as well as in textile, showed reversible colour changing properties when exposed to UV light, and revert to their original colourless state when UV light is removed. SG exhibited higher colour intensity than RR in both mediums with $\Delta K/S$ of 0.096 in textile and an absorbance of 1.89 in solution while

RR has ΔK/S of 0.088 in textile and lower absorbance of 0.27 in solution. A direct and quantitative comparison of ΔK/S and absorbance values is not possible also due to different UV-sources and powers for sample activation. However, the data gives sufficient information on trends of maximum colour yields. In terms of rate kinetics, RR showed faster photo-colouration and fading rates in solution and slower colour conversion rates in textile and the opposite is true for SG (Figure 3 and Figure 4). In solution, the plateau of maximum absorbance of RR was already reached after 30 s of UV activation (linear fit), SG reached saturation of maximum absorbance after 60 s (exponential fit) (figure 3 A). Whereas, when applied in textile, SG achieved maximum colouration at a rate of 0.141 s^{-1}, while fabrics dyed with RR are remarkably slower with $k_{colouration}$ of 0.020 s^{-1}.

Figure 3 B shows the reversion to uncoloured state of the dyes in a nonpolar solvent, hexane. As illustrated in the picture, all dye solutions before UV irradiation were colourless and upon exposure to UV at their wavelength maxima, SG becomes deep blue and RR becomes light orange. It can be clearly seen that, all cuvettes with SG dye solution still show a colour (top) while for RR dye solution (bottom), only the cuvette used for the last activation was coloured, which confirms slow decolouration for SG and fast fading rate for RR in solution. However, opposite fading rates were observed when incorporated into the fabric (Figure 4 b) compared with their decolouration speed in solution. For SG dyed samples it took less than 20 s to revert to its uncoloured sate while for RR dyed samples it needs more than 200 s to fade back to their original colour, which is significantly slower.

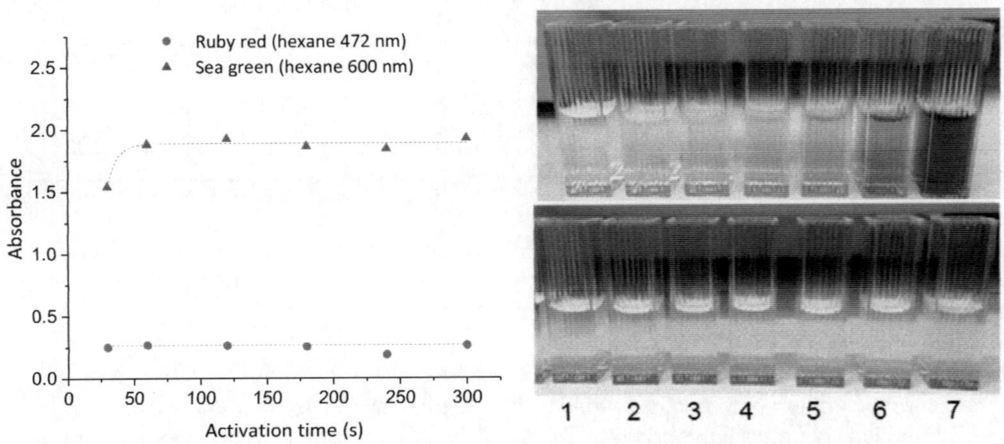

(A) (B)

Figure 3. (A) Absorption spectra of Ruby red and Sea green in hexane at different UV activation times, (B) Colour reversion of the dye solutions directly after removal of the 300 s activation; Top is Sea green and the bottom is Ruby red, both in hexane. Cuvettes from left: (1) reference cuvette with solvent only, (2) 30 s, (3) 60 s, (4) 120 s, (5) 180 s, (6) 240 s and (7) 300 s activation.

According to previous studies, dyed photochromic fabrics generally tend to give slow colour switching speed, because the dye molecules incorporated into rigid polymer matrices are physically restricted, which slows down the photochromic transformations [2, 21]. Napthopyran dyes printed on textiles generally are more durable and slower switching compared to spirooxazines, which are fast-switching, but more temperature-sensitive [22]. This agrees with our observation of photochromic dyed polyester fabrics, but opposite behaviour was observed when the dyes are dissolved in hexane as presented above. In solution, one explanation for the slow switching speed for Sea green may be related to the structural differences between the two dye classes. The spirooxazines are inherently rigid in their structure because of the indoline ring, whereas the naphthopyrans are relatively flexible [22, 23]. Thus, spirooxazines are more restricted during the photochromic conversion due to their rigid nature, which makes the conversion speed relatively slower. While in textile, SG from the spirooxazine family, are believed to be located close to the fabric surface due to the non-planarity and bulky molecular characteristics of some spirooxazine derivatives, which could limit efficient penetration of the dye molecules into the tightly packed crystalline structure of polyester fiber [6]. This provides relatively more free space for SG dye molecules, which facilitated faster photochromic conversion. However, for NP dyed samples, due to their flexible structure, the dye molecules have a higher chance to penetrate inside the fiber structure, which limits the configuration change of photochromic molecules resulting in a slower speed of photo-isomerization. Generally, differences in kinetic behaviour of photochromic dyes in solution compared to in solid state and applied in matrices are reported. Whereas photo-colouration and thermal reversion of spirooxazines and napthopyrans in solution follow mono-exponential kinetics, more complex, multi-exponential behaviour of the dyes can be expected in polymer matrices[24]. At this point, however, conclusions on the specific effect of the polyester matrix on the multiplicity of the rate constants involved in the photo-colouration process of SG and RR dyes are impossible.

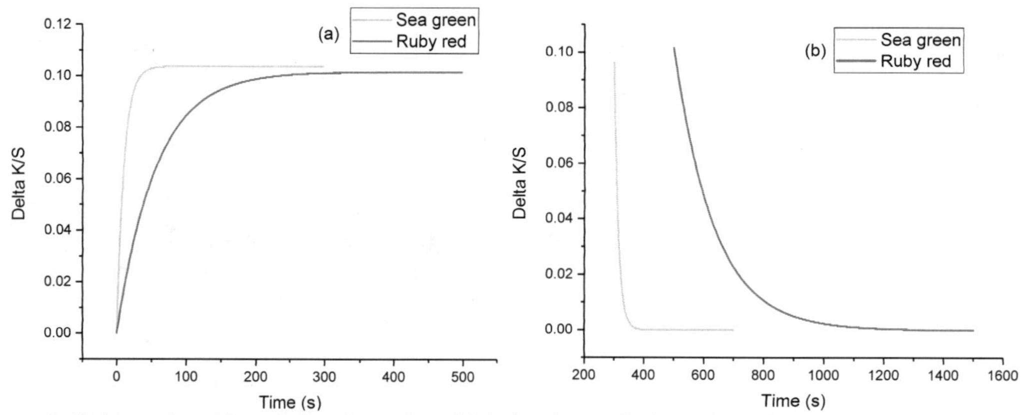

Figure 4. Colouration (a) and decolouration (b) behaviour of photochromic polyester fabrics dyed in $scCO_2$. Experimental data is fitted to first order kinetic model (Equation (2)) and normalized to have the same initial value.

4. Conclusions

Two photochromic dyes belonging to the spirooxazine and naphthopyran dye classes were successfully applied to polyester fabric using $scCO_2$ dyeing technique. Both photochromic dyes in textile and in solution showed significant reversible colour changing behaviour. The dyed polyester fabric exhibited similar level of colour strength to that of the photochromic dyes in solution while opposite photochromic behaviour was observed in terms of rate kinetics. The rate of colour conversion for both dyes to undergo photo-isomerization depends on the media (in solution or textile) as well as their inherent structural differences. The effect of the Polyester matrix on the multiplicity of the rate constants needs further investigations which will be dealt in our future study.

Acknowledgment

The project is funded by the European Union under the framework of Erasmus Mundus program; Sustainable Management and Design for Textile (SMDTex) grant number 2016-1353/001-001-EMJD. All the Universities collaborating on the program are gratefully acknowledged.

References

[1] Crano JC and Guglielmetti RJ 1999 *Organic Photochromic and Thermochromic Compounds: Physicochemical Studies, Biological Applications, and Thermochromism* Vol. **2** (New York:Springer Science & Business Media)

[2] Wang PY and Wu CJ 1997 Photochromic behaviour of some phenoxyanthraquinone dyes in solution and on polyester substrate *Dyes Pigments* **35** pp 279-88.

[3] Chowdhury MA, Joshi M and Butola BS 2014 Photochromic and Thermochromic Colorants in Textile Applications *J Eng Fiber Fabr.* **9** pp 107-23.

[4] Shah PH 1985 Azodisperse dyes with photochromic mercury (ii)-dithizonate moiety for dyeing polyester, nylon and cellulose triacetate fibres *Indian J Fibre Text* **10** pp 179-82.

[5] Lee S-J, Son Y-A, Suh H-J, Lee D-N and Kim S-H 2006 Preliminary exhaustion studies of spiroxazine dyes on polyamide fibers and their photochromic properties *Dyes Pigments* **69** pp 18-21.

[6] Billah SMR, Christie RM and Shamey R 2008 Direct coloration of textiles with photochromic dyes. Part 1: Application of spiroindolinonaphthoxazines as disperse dyes to polyester, nylon and acrylic fabrics *Color. Technol.* **124** pp 223-8.

[7] Little AF and Christie RM 2010 Textile applications of photochromic dyes. Part 2: Factors affecting the photocoloration of textiles screen-printed with commercial photochromic dyes *Color. Technol.* **126** pp 164-70.

[8] Kozicki M and Sąsiadek E 2012 UV-assisted screen-printing of flat textiles *Color. Technol.* **128** pp 251-60.

[9] Feczkó T, Samu K, Wenzel K, Neral B and Voncina B 2013 Textiles screen-printed with photochromic ethyl cellulose-spirooxazine composite nanoparticles *Color. Technol.* **129** pp 18-23.

[10] Aldib M 2015 Photochromic ink formulation for digital inkjet printing and colour measurement of printed polyester fabrics *Color. Technol.* **131** pp 172-82.

[11] Seipel S, Yu J, Periyasamy AP, Viková M, Vik M and Nierstrasz VA *Resource-Efficient Production of a Smart Textile UV Sensor Using Photochromic Dyes: Characterization and Optimization* 2018 ed Kyosev Y, Mahltig B, Schwarz-Pfeiffer A In: Narrow and Smart Textiles (Springer) p 251-7.

[12] Cheng T, Lin T, Brady R and Wang X 2008 Fast response photochromic textiles from hybrid silica surface coating *Fiber Polym* **9** pp 301-6.

[13] Cheng T, Lin T, Brady R and Wang X 2008 Photochromic fabrics with improved durability and photochromic performance *Fiber Polym* **9** pp 521-6.

[14] Aldib M and Christie RM 2013 Textile applications of photochromic dyes. Part 5: application of commercial photochromic dyes to polyester fabric by a solvent-based dyeing method *Color. Technol.* **129** pp 131-43.

[15] Lee EM, Gwon SY, Son YA and Kim SH 2012 Modulation of a fluorescence switch of nanofiber mats containing photochromic spironaphthoxazine and D-π-A charge transfer dye *J. Lumin.* **132** pp 1427-31.

[16] Lee E-M, Gwon S-Y, Ji B-C, Wang S and Kim S-H 2012 Photoswitching electrospun nanofiber based on a spironaphthoxazine–isophorone-based fluorescent dye system *Dyes Pigments* **92** pp 542-7.

[17] Viková M, Periyasamy AP, Vik M and Ujhelyiová A 2017 Effect of drawing ratio on difference in optical density and mechanical properties of mass colored photochromic polypropylene filaments *J Text I* **108** pp 1365-70.

[18] Little AF and Christie RM 2016 Textile applications of commercial photochromic dyes. Part 6: photochromic polypropylene fibres *Color. Technol.* **132** pp 304-9.

[19] Van der Kraan M. Process and equipment development for textile dyeing in supercritical carbon dioxide [Ph.D. dissertation]: Technische Universiteit Delft; 2005.

[20] Mourier P, Sassiat P, Caude M and Rosset R 1986 Retention and selectivity in carbon dioxide supercritical fluid chromatography with various stationary phases *J. Chromatogr. A* **353** pp 61-75.

[21] Bamfield P 2010 *Chromic phenomena: technological applications of colour chemistry* (Cambridge:Royal Society of Chemistry)

[22] Little AF and Christie RM 2011 Textile applications of photochromic dyes. Part 3: factors affecting the technical performance of textiles screen-printed with commercial photochromic dyes *Color. Technol.* **127** pp 275-81.

[23] Periyasamy AP, Vikova M and Vik M 2017 A review of photochromism in textiles and its measurement *Textile Progress* **49** pp 53-136.

[24] di Nunzio MR, Gentili PL, Romani A and Favaro G 2010 Photochromism and Thermochromism of some Spirooxazines and Naphthopyrans in the Solid State and in Polymeric Film *J. Phys. Chem. C* **114** pp 6123-31.

Shape memory clay flaps assisted body cooling fabrics

S B Y Abeywardena[1,2], S Perera[3], P V T Weerasinghe[4], K M Nalin de Silva[2,5], S Walpalage[1], M C W Somaratne[3], N M Hettiarachchi[2] and S Jeyasakthy[2]

[1]Department of Chemical and Process Engineering, Faculty of Engineering, University of Moratuwa, Sri Lanka.
[2]Sri Lanka Institute of Nanotechnology (Pvt) Ltd, Nanotechnology and Science Park, Sri Lanka.
[3]Division of Polymer, Textile and Chemical Engineering Technology, Institute of Technology, University of Moratuwa, Sri Lanka.
[4]Department of Textile and Clothing Technology, Faculty of Engineering, University of Moratuwa, Sri Lanka.
[5]Department of Chemistry, Faculty of Science, University of Colombo, Sri Lanka.

E-mail: sujani.abeywardana@yahoo.com

Abstract. Mechanical responses of inorganic clay to external stimuli have been rarely implemented into devices that interact with the human body. We demonstrate that the hygroscopic and free swelling behaviours of Na-bentonite were engineered to design wearables, which give multifunctional responsiveness to human sweat. By printing bentonite nanoclay (BNC) composite on flaps of fabrics, they can reversibly change shape within a few seconds in response to environmental humidity gradients. Body's need for cooling is facilitated through controlled ventilation by opening and closing of these clay printed flaps. This study provides a background for developing new shape memory clay composite for many applications in garment industry.

Keywords: Bentonite nanoclay; Mechanical response; Moisture management; Polyester fabric; Clay swelling

1. Introduction

High swelling property of Na-bentonite clay is rarely used in applications, which lead to mechanical responses in relation to hydration. The moisture gradient is a stimulating factor, which can trigger shape transformation in clay due to mechanical amplification of moisture-induced strains in swellable clay. Similar observation can be found in nature in pine cone scales and wheat awns [1,2]. Moreover, it has been reported that shape changing polymers could be triggered by the moisture gradient in the environment [3,4]. However, a powder compound such as clay is an unsupervised material when it comes to shape changing material. Different types of clays are ubiquitous in nature. From those, bentonite nanoclay (BNC) exhibits high cation exchange capacity and high swelling capacity in relation to hydration [5]. The structure of bentonite is composed of a three-layer platelet with an octahedral aluminum hydroxyl sheet sandwiched between two layers of silicon-oxygen tetrahedral. The nano size space between adjacent platelets of BNC comprises of exchangeable cations such as Na^+, which draw water, and form a rigid network made up of water layers [6]. Hence, this nanospace is vital for BNC's incomparable hydrophilicity. In addition, Na-bentonite consists of negatively charged aluminosilicate layers, which are kept together by cations. Hence, their ability to absorb water between the layers, results in strong repulsive forces and clay expansion [7]. Further, the hydroxyl groups on edges of clay provide excellent sites for water adsorption [8]. Not only excellent hydrophilicity of bentonite clay, but also it inherents excellent health benefits in terms of skin related problems [9,10].

Here in, we report a moisture responsive Na-bentonite composite, which changes its shape and functions simultaneously with humidity change. Na-bentonite makes a perfect composite with silicone rubber, which assists to expand and contract when composite absorbs and removes water, respectively. Hence, this moisture responsive material can be used to improve moisture management property of sportswear. Moisture management is one of the key performance criteria in today's apparel industry [11]. However, highly demanded synthetic fabrics are lack in moisture management property due to its hydrophobic nature [12,13]. Na-bentonite composite is localised on ventilating flaps that can automatically adjust the moisture transfer of the fabric through change of skin exposure percentage by controlling the area of exposed skin via shape transformation. This will open up new avenues in using hygroscopic transforming materials to tackle problems related to moisture, such as in sports wearables and indoor humidity control.

2. Methodology

A dispersion of bentonite nanoclay was obtained by dissolving 4 g of nanoclay in 100 ml of deionized water. Particles of bentonite nanoclay, 100 nm in size, were obtained by ball milling (FRITSCH PULVERISETTE 7 premium line grinder). 4 g of bentonite clay (hydrophilic bentonite clay-Aldrich) in 100 ml of deionized water was activated with 2 g of NaCl (Aldrich). Then, reaction was kept stirring and boiling for one hour. After one hour, the dispersion was diluted with water. Dilution with water was repeated to make sure that all impurities soluble in water were removed. The dispersion was allowed to settle for 48 hours. The settled clay was centrifuged and allowed to dry in the oven at 105 °C. Then, powdered sample and unmodified clay were dispersed in silicon rubber solution, followed by adding curing agent to the reaction. Obtained Na-bentonite and unmodified clay composites were localised on one side of woven polyester fabrics and kept in the oven at 60 °C for 3 hours. The samples were characterized using Scanning electron microscopy (SEM) and Energy dispersive X-ray spectroscopy (EDX-HITACHI SU6600 SEM coupled with EDX detector). X-ray powder diffraction (XRD) profiles were recorded on a Bruker PW 1050 diffractometer using Cu-Ka radiation (k = 0.154 nm), over a 2θ range of 3°–70°, step size of 0.01, and a step time of 2 s. The bending angles of rectangular samples (40 mm in length, 5 mm in width, and 0.2 mm in thickness) were measured to understand the mechanical performances of the samples at different time intervals. Bending angles were then calculated using a theoretical formula (Eq. 1) based on the classical Stoney formula [14]. The theoretical formula relates the bending angle, θ, to the expansion strain of the composite due to wetting (supplementary Fig. 1.).

$$\theta = 360\,\frac{L}{2\pi R} \qquad (1)$$

where L is the length of the strip; R is the bending radius. The weight swelling ratio of the Na-betonite composite samples were calculated by measuring their weight before and after immersion in deionized water at different time intervals. The weight swelling ratio of absorbed water during immersion relative to the dry mass is given by;

$$\text{Weight swelling ratio} = \frac{\text{Mass (wet)} - \text{Mass (dry)}}{\text{Mass (dry)}}\ 100\% \qquad (2)$$

where mass (wet) refers to the weight of the sample submersed in water at different time intervals; mass (dry) refers to the weight of the sample dried at 60 °C under vacuum for 36 hours. Clay composite was localised on prototype polyester fabric using a stencil (supplementary Fig. 2.). The fabric sample was kept in the oven at 60 °C for 3 hours. Stencil was removed from the fabric, and rectangular shape Na-bentonite composite was partially separated from the fabric to act as ventilating flaps. Finally, shape transformation of these ventilating flaps were observed.

3. Results and discussion

The SEM images in Fig. 1. (A) and (B) proved that Na-bentonite is uniformly distributed in silicon rubber composite, and intercalation of Na^+ is proved from EDX spectra in Fig. 1. (C) and (D). After intercalation, Na Wt% has risen from 3.4% to 5.3%. This results in enhancing water absorption in bentonite clay, and porous nature of clay promotes water travelling both inside the structure and outer surface. Consequently, Na-bentonite swelled rapidly by absorbing water. XRD spectrum (supplementary Fig. 3. A) shows the typical peak of (001) plane, which corresponds to the basal spacing of bentonite clay which exhibits a peak at $2\theta = 7.8°$, indicating the basal spacing of 1.4 nm. XRD spectrum of Na-bentonite (supplementary Fig. 3. B) shows a significant shift in corresponding peak of (001) plane, which has increased to 1.8 nm. Hence, above results confirm the successful intercalation of Na^+ in bentonite clay.

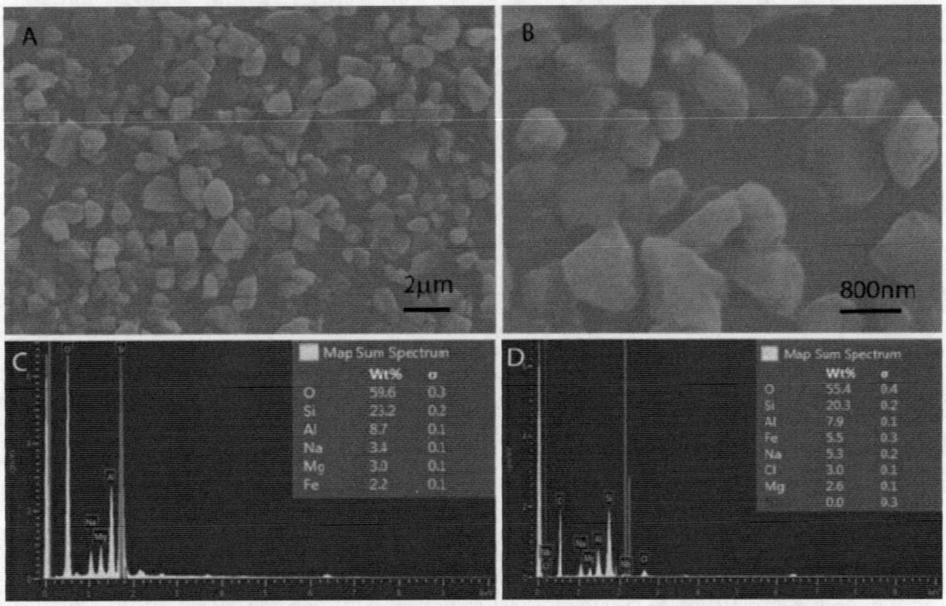

Figure 1. (A) and (B) SEM images of Na-bentonite composite. SEM-EDX spectra of unmodified bentonite clay (C) and Na-bentonite clay (D).

Fig. 2. shows the shape change of Na-bentonite composites upon exposing to water vapour. The original dimensions of the sample were 40 mm in length, 5 mm in width, and 0.2 mm in thickness. The non-stretching woven polyester fabric on top of the composite is restricted in expanding and as a result, the straight (original shape) specimens are bent upon absorbing moisture by Na-bentonite composite. The bent specimen fully regained its original shape quickly after being removed from the moisture environment. The weight swelling ratio of the samples were measured after immersed in water for 2 min, as a percentage with respect to the initial weight. Unmodified clay composite showed 10% increase in weight swelling ratio, while 23% increase in Na-bentonite composite. This clearly reveals that there is an obvious increase in weight, when the Na-bentonite composite is exposed to water compared to unmodified clay.

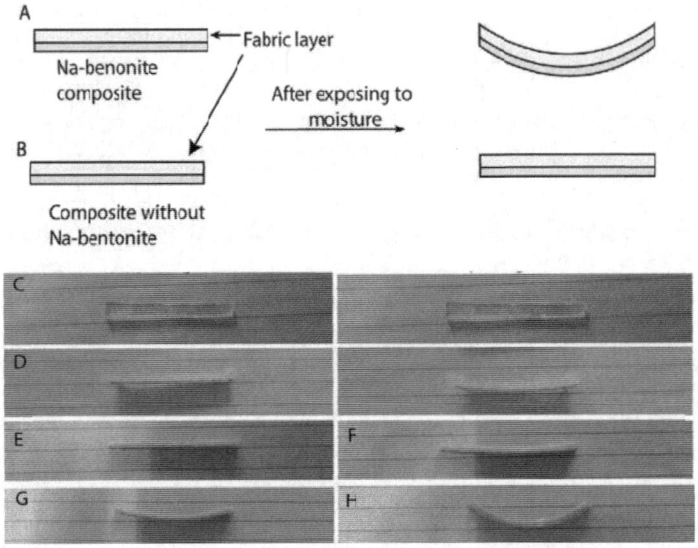

Figure 2. (A) Schematic diagram of shape transformation of Na-bentonite composite printed fabric when exposed to moisture. (B) Schematic diagram of inability in shape transformation of silicon rubber printed fabric when exposed to moisture. (C) and (D) Images of shape transformation of silicon rubber and Na-bentonite printed fabrics when exposed to moisture for 2 min, respectively. (E), (F), (G) and (H) Images of shape transformation of Na-bentonite composite printed fabrics after exposing to moisture for 0 s, 30 s, 1 min and 2 min, respectively.

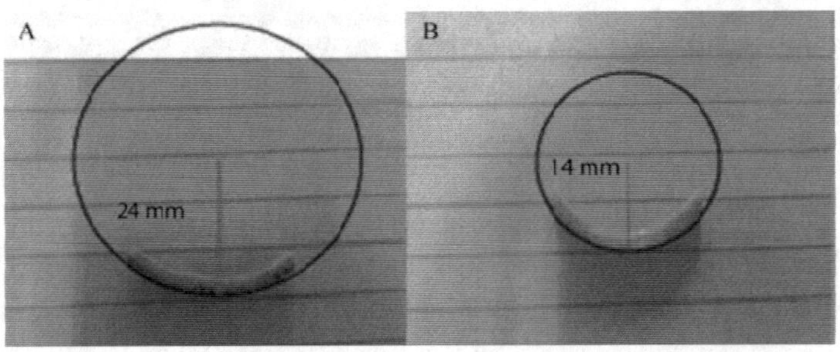

Figure 3. Images of radii of curvatures of (A) unmodified bentonite and (B) Na-bentonite composites.

The radius of curvature is the reciprocal of the curvature, and for a curve, it equals the radius of the circular arc which best approximates the curve at that point (Fig. 3.). Subsequently, completing the imaginary circle of the curve, the mean values of radii of the curvatures were measured as 24 mm for unmodified bentonite and 14 mm for Na-bentonite composite (Table 1.). Bending angles of unmodified and Na intercalated clay are 95.5° and 163.6°, respectively. This implies that Na-bentonite composite is more prone to absorb water compared to unmodified bentonite, as the lowest radius of curvature and the highest bending angle belongs to Na-bentonite composite. Considering this hydrophilic nature, Na-bentonite composite was used for localisation on prototype polyester fabric sample.

Table 1. Weight swelling ratios, bending radii and bending angles of unmodified bentonite and Na-bentonite composite at different time intervals.

Time (s)	Bentonite			Na Bentonite		
	Weight swelling ratio (%)	Bending radius (mm)	Bending angle (degrees)	Weight swelling ratio (%)	Bending radius (mm)	Bending angle (degrees)
30.0	5.0	60.0	38.2	8.5	48.0	47.7
60.0	7.5	40.0	57.3	15.5	38.0	60.3
90.0	9.0	28.0	81.8	20.0	24.0	95.5
120.0	10.0	24.0	95.5	23.0	14.0	163.6

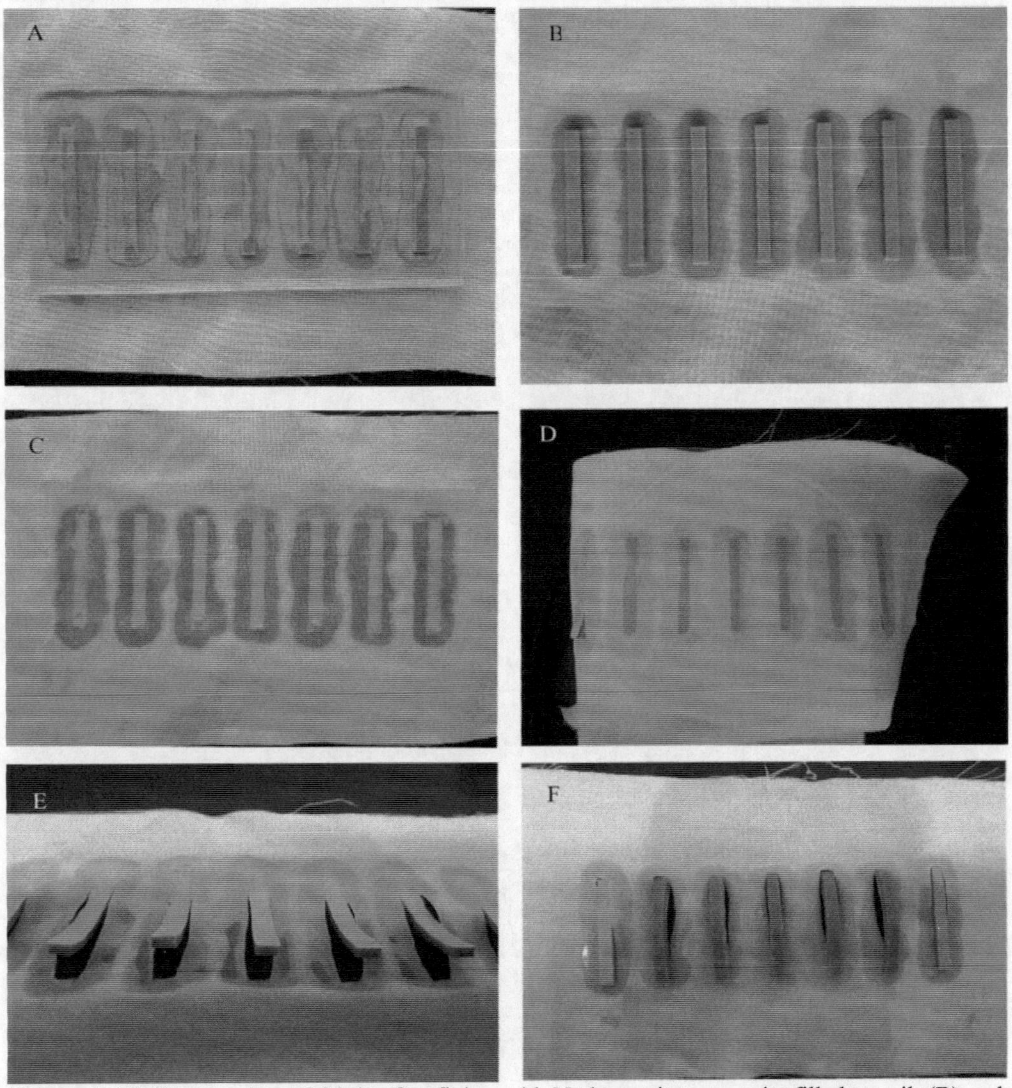

Figure 4. (A) Bottom view of fabric after fixing with Na-bentonite composite filled stencil. (B) and (C) Bottom and top views after removing the stencil, respectively. (D) Image when clay flaps were exposed to water vapour. (E) and (F) Top and bottom views of open clay flaps after exposed to water vapour, respectively.

Making sweat-responsive wearables modified with Na-bentonite composite was replicated in laboratory, by printing the clay composite on prototype fabric sample using a stencil. Fig. 4. clearly illustrates different stages from the printing of Na-bentonite composite on fabric to hydroscopic mechanical responsiveness of it. Shape transformation occurs in a flat clay composite layer when exposed to moisture. Fig. 4. (D) illustrates the flat ventilation flaps of prototype fabric sample before exposing to moisture, and Fig. 4. (E) and (F) illustrate the curved ventilation flaps after exposing to moisture. Owing to combination of non-stretching polyester and expansion of the composite, the ventilation flaps open on wetting. Na-bentonite composite printed flaps opened in 50 s when exposed to water vapour, then regained their original shape by closing the flaps in 65 s when removed from the moisture environment. This resembles the opening of flaps in the garment with increased human sweat, and closing of flaps with cooling of the human body.

Future works will focus on finding a more feasible printing method, which prints thin layers of Na-bentonite composite on fabrics in a range of designs. We predict that this scalable and adaptable technology can be used in industries beyond fabricating wearables, where moisture gradient is a key factor, such as hydration enhancement in skin care products, humidity control in smart homes, and moisture removal in biomedical textiles.

4. Conclusions

The designed novel Na-bentonite composite mechanically responds to moisture. SEM proved uniform distribution of Na-bentonite in the composite. Intercalation of Na^+ into bentonite was proved by EDX and XRD spectra. Weight swelling ratio and bending angle proved its expanding ability over unmodified bentonite composite. Na-bentonite printed flaps opened when absorbing the moisture and closed when it is dried. We harnessed the swelling behaviour of Na-bentonite clay, which provides us with a new perspective in using clay minerals for making moisture-responsive wearables. Our work will make it feasible to utilize Na-bentonite composite in many fields, with shape-changing interfaces in moisture changing environment.

Acknowledgement

This work was financially supported by the grant 16-101, received from National Research Council, Sri Lanka. We would also like to thank Sri Lanka Institute of Nanotechnology (Pvt) Ltd (SLINTEC) for providing facilities for the research.

Supplementary Material

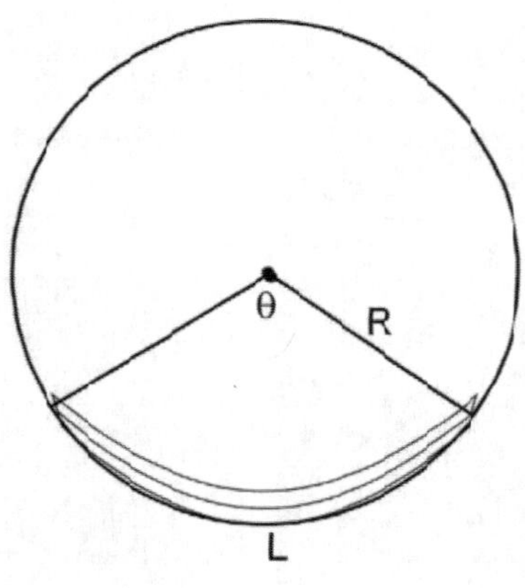

Fig. 1. Sketch of bending angle of Na-bentonite composite printed fabric sample.

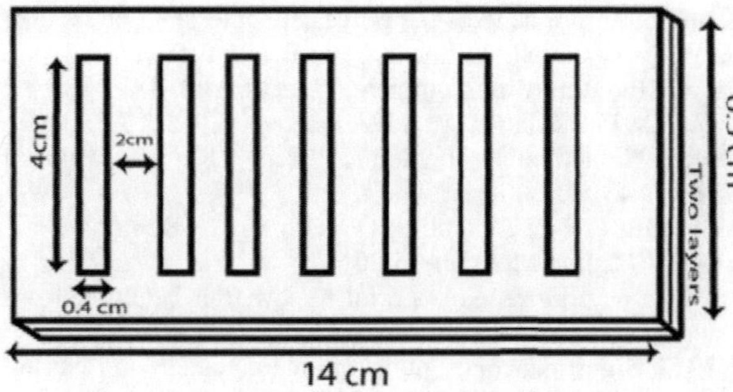

Fig. 2. Sketch of the stencil used in printing the clay composite on prototype fabric sample.

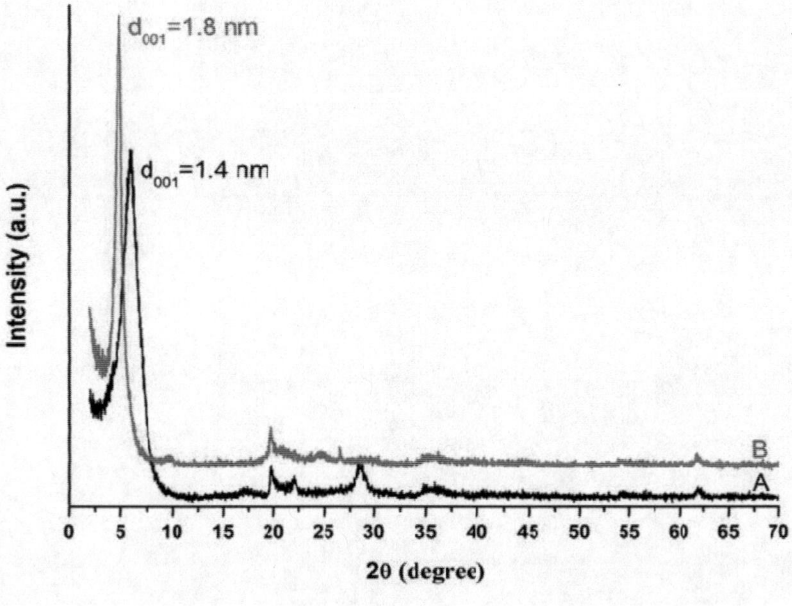

Fig. 3. XRD diffraction patterns of (A) unmodified bentonite clay and (B) Na-bentonite clay.

References

[1] Dawson C, Vincent J F V and Rocca A M 1997 How pine cones open Nature 390 668.
[2] Elbaum R, Zaltzman L, Burgert I and Fratzl P 2007 The role of wheat awns in the seed dispersal unit Science 316 884.
[3] Hu J, Meng H, Li G and Ibekwe S I 2012 A review of stimuli-responsive polymers for smart textile applications Smart Mater. Struct. 21 053001.
[4] Le X, Lu W, Zheng J, Tong D, Zhao N, Ma C, Xiao H, Zhang J, Huanga Y and Chen T 2016 Stretchable supramolecular hydrogels with triple shape memory effect Chem. Sci. 7 6715.

[5] Abeywardena S B Y, Perera S, de Silva K M N, Walpalage S and Somaratne M C W 2017 Mimicking elephant mud bathing to produce wettable polyester Materials Letters 205 90.
[6] Hensen E J M and Smit B 2002 Why clay swells J. Phys. Chem. B 106 12664.
[7] Uddin F 2008 Clays, nanoclays, and montmorillonite minerals Metall. Mat. Trans. 39A 2804.
[8] Shah L A, Khattak N S, Valenzuela M G S, Manan A and Valenzuela F R 2013 Preparation and characterization of purified Na-activated bentonite from Karak (Pakistan) for pharmaceutical use Clay Minerals 48 595.
[9] Fowler J F 2001 A skin moisturizing cream containing Quaternium-18-Bentonite effectively improves chronic hand dermatitis J Cutan Med Surg 5 201.
[10] Maryam M 2017 Bentonite clay as a natural remedy: a brief review Iran J Public Health 46(9) 1176.
[11] Chinta S K and Gujar P G 2013 Significance of moisture management in textiles Int. J. Innovative Research in Science, Eng. and Technology 2 2104.
[12] Giwa A, Isa M T and Idris U A 2013 Analysis of some physical and mechanical properties of selected fabrics International Journal of Textile Science 2(3) 49.
[13] Sparavigna A Plasma treatment advantages for textiles.
[14] Stoney G G 1909 The tension of metallic films deposited by electrolysis Proc. R. Soc. A Math. Phys. Eng. Sci. 82 172.

Fabric physical properties and clothing comfort

M Abu-Rous[1], I Dabolina[2] and E Lapkovska[2]

[1] Lenzing Aktiengesellschaft, Werkstraße 2, 4860 Lenzing, Austria
[2] Riga Technical University, Faculty of Material Science and Applied Chemistry, Institute of Design Technologies, Kipsalas 6-220, Riga, LV-1048, Latvia

m.aburous@ lenzing.com

Abstract. The role of fiber content regarding the textile performance and physiological and sensorial comfort attributes of single jersey fabrics, typically used in skin-near applications such as underwear, was investigated. Fabrics made of 100% lyocell and 50:50 lyocell/polyester blends were compared to 100% polyester fabrics, typically used for institutional underwear. Blending polyester to the cellulosic fiber improved fabric tenacity and dimensional stability, while the 100% polyester fabric remained unmatched. While all tested fabrics showed low vapor transfer resistance and hence high breathability at a hotplate test, lyocell-containing fabrics proved superior in terms of vapor uptake as well as in skin sensorial aspects. A synergy effect was observed in the blended fabrics in terms of more efficient liquid transportation and faster fabric drying.

1. Introduction

Clothing comfort is one of the most important attributes of textile materials. It implies understanding the interrelationship between fiber material, yarn structure, fabric structure, transmission characteristics (air, heat and moisture) and tactile aspects of textile materials on thermo-physiologicall and neurophysiological processes [1]. Fabric performance in terms of better physiological and sensorial comfort ought to be an essential requirement of materials, which are used in contact with skin (i.e. lingerie, sportswear and bed clothing). Wear comfort has been listed as the most important property of clothing demanded by users and consumers according to recent studies [2].

An attribute essential to physiological comfort is the ability of the textile material to remove moisture generated by perspiration away from the skin surface, thus supporting the body cooling mechanism. Vapor uptake is decisive for the performance of textile under "comfort" conditions before the body comes into sweating, while liquid uptake, transportation and evaporation are the mechanism required for performance textiles [3].

Earlier works have shown the consequence of the enhanced moisture uptake of the wood-based cellulosic fibers lyocell and modal for the comfort properties in textile applications [4]. Moisture absorption has also an effect on the sensorial perception of the surface. Hygroscopic materials are known to provide a cool touch [5].

Thermo-physiological properties of clothing are often assessed by the hotplate test, where the "breathability index" is the ratio between evaporative resistance and thermal insulation. However, it is known, that a standard hotplate measurement, which is performed under equilibrium conditions, does not yield all information, especially when hygroscopic fibers are involved [6,7]. Under transient conditions, studies recommended considering moisture buffering function as an additional comfort characteristic [7,8].

A further important aspect is the skin-sensorial comfort. Various physical methods to assess fabric handfeel by mimicking hand actions have been developed in the last 50 years such as Kawabata, Phabrometer, drape coefficient etc. [9]. A new approach to assess fabric handfeel is offered by the Tissue Softness Analyzer (TSA) by Emtec Electronic GmbH. The method is based on measuring the sonic waves generated by applying a friction on the fabric. [10]. Originally established for the quality control in the hygiene tissue sector, this method is beginning to claim its place in nonwovens and textile sectors recently [11,12]. A rotating part of the TSA, while moving over the fabric surface, generates noise, which is recorded and analyzed into its amplitude signals. In the resulting sonic spectrum, the signal peak at 750Hz and is a measure for the fabric vibration under the rotating part, while the peak at 6500Hz occurs through the vibration on the rotating part moving above the fabric surface itself. Both are correlated with fabric surface parameters. The lower the generated noise, the smoother resp. softer is the fabric. A handfeel value (HF) can be calculated based on the TS values and the fabric weight and thickness.

The aim of this research is to investigate the role of fiber materials in fulfilling textile performance and physiological and sensorial comfort requirements in skin-near workwear clothing.

2. Materials and methods

Knitted fabrics of different construction as in table 1 were investigated.

Table 1. Single jersey fabrics used in this study.

	Blend	Weight [g/m²]	Thickness [mm]	Air permeability (50 Pa) [mm/s]	Drape [%]	R_{et} [m² Pa / W]	Fi [gVapor/ m²]
PES1	Polyester filament	132	0,48	617,9	20,8	1,46	0,1
PES2	Polyester filament	144	0,65	>835	24,3	1,57	0,7
CLY/ PES	Lyocell: Polyester 50:50	145	0,55	>835	22,7	2,13	4,5
CLY	Lyocell:Elastane 94:04	178	0,68	369	30,6	3,08	15,5

Vapor transfer resistance (R_{et}) was measured by a hotplate test according to ISO 110921 [13]. Additionally, moisture buffering capacity (Fi) was measured by the weight difference due to the vapor uptake of the fabrics before and during R_{et} measurement. To measure the drying rate, the fabric was placed on a scale connected to PC for digital weight registration. 500mg deionized water were dropped on the fabric surface and the weight loss by evaporation was measured every minute. Skin-sensorial aspects were investigated objectively using Tissue Softness Analyzer (TSA) as well subjectively by a handfeel panel of 7 female and 3 male experienced assessors. The fabrics were rated on a scale from 1 to 10 based on 3 descriptor pairs (rough/smooth; stiff/drapy; warm/cool) as well as on the over-all "softness" rating. Elasticity indicators of all fabrics were obtained according to EN 14704-1:2005 [14] for testing the elongation rate (%) which is a ratio of the extension of the specimen to its initial length, and the un-recovered elongation rate – the ratio of un-recovered extension after cycling [14]. Testing the changes in the linear dimensions of fabrics after laundering (30° C; natural drying) was done according to the standard ISO 6330 [15].

3. Results & discussion

As shown in table 2 and figure 1, the indicators of elongation for all fabrics are estimated as sufficient for single jersey fabrics. The recovering of a material to its initial size and shape without permanent deformation is essential for product development – taking into account the potential deformations of fabrics and products after extension. Polyester fabrics show the permanent deformation only in weft direction, in turn lyocell blends show more significant deformation in both directions.

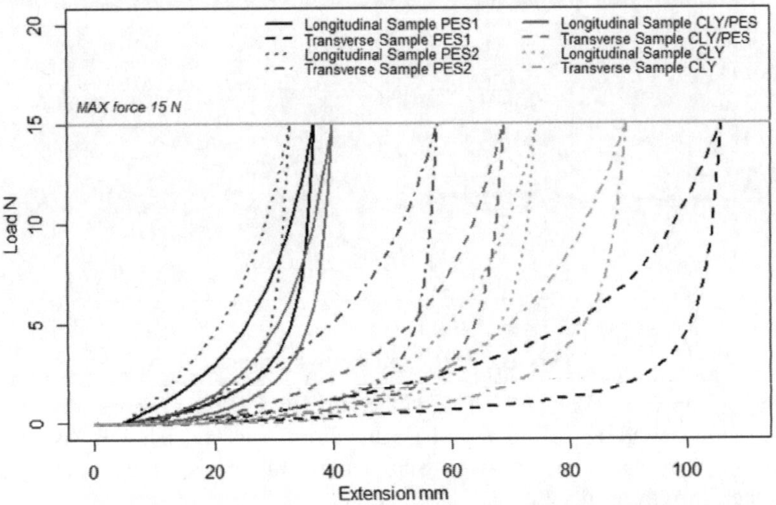

Figure 1. Extension at maximum force on the 5th cycle average 5 specimens.

Polyester materials show no significant shrinkage in either directions (warp and weft). The lyocell and elastane blend indicates 4,5 % of surface shrinkage (table 2). Significant shrinkage may cause the changes of products initial size and form.

Addition tests of fabric propensity to surface pilling were carried out according to ISO 12945 [16]. After certain visual assessment, both polyester fabrics reached 5 points - the highest rating: at all stages of the test no visual changes occurred. Lyocell blend specimens reached 4 points due to the slight surface fluffs (table 2) indicating a possible loss of visual appearance and fabric durability compared to polyester materials.

Table 2. Fabric testing results.

	Elongation [%]		Permanent deformation [%]		Shrinkage [%]	Pilling [7000 cycles]
	↕	↔	↕	↔		
CLY/PES	40%	69%	3%	6%	3,50%	4
CLY	75%	90%	6%	12%	4,50%	4
PES1	37%	106%	0%	6%	0,00%	5
PES2	33%	58%	0%	2%	0,00%	5

R_{et} measurements showed that all fabrics lay within the range of "very breathable" (R_{et}<6) as defined by Hohenstein Institute [17]. Fabrics containing absorbing fibers show slightly higher vapor transfer resistance, as the test typically doesn't consider the amount of vapor absorbed inside the fabric. On the other hand, Fi values show the dependence of vapor absorption (buffering capacity) on the amount of the cellulosic lyocell fiber in the fabric.

If considered isolated from other properties, a polyester fabric would be misleadingly expected to provide more "breathability" than a cellulosic (see 'Figure 2'). Therefore, moisture buffering capacity (Fi) was also considered as a relevant criterion expressing the ability of the fabric material to absorb vapor off skin surface and hence supporting cooling mechanism.

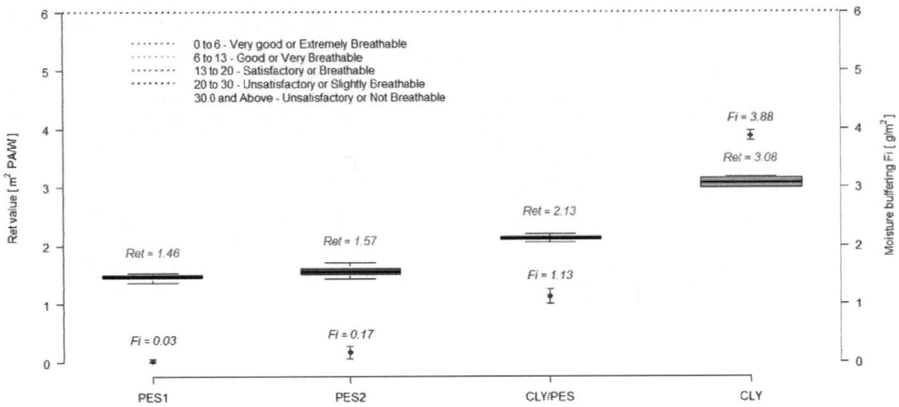

Figure 2. R_{et} values of samples.

The drying rate is shown in 'Figure 3'. This value is related to the evaporation surface provided by the transportation and spreading of the liquid in the fabric surface. Though hydrophilic and water absorbing, cellulosic material tends to swell and narrow the transpiration channels. Non-swelling polyester shows therefore faster drying.

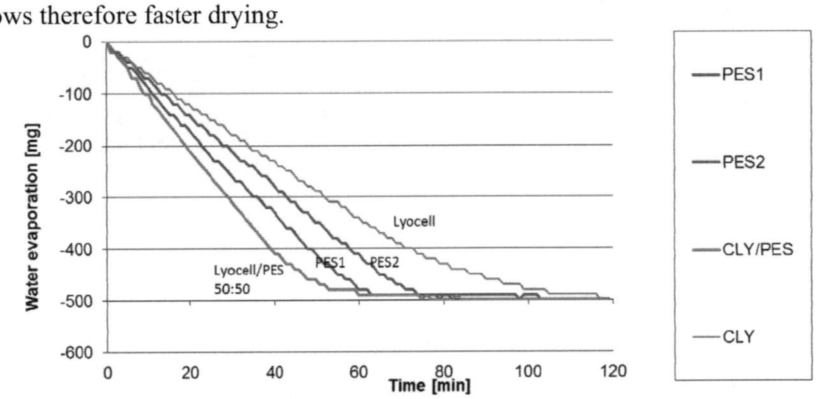

Figure 3. Drying rate.

Interestingly, the yarn blend of lyocell and polyester shows elevated liquid transportation and hence faster drying. This can be attributed to synergy effect of the absorbing cellulose and the polyester fiber acting as spacer between the swelling fibers and keeping transportation channels open. An explanation of this phenomenon is provided in 'Figure 4', where an environmental scanning electron microscopy (ESEM) imaging of the swelling shows the absorption inside the cellulose as a driving force and the spacer-role of the polyester to keep the capillaries open [18].

Figure 4. Electron microscopy (ESEM) imaging of a cross-section of a lyocell (round) and polyester (star form) blended yarn in in dry (left) and wet (right) condition [source: Lenzing AG].

'Figure 5' shows the handfeel objective assessment as received by TSA. Based on the acoustic signals and fabric weight and thickness a handfeel value (HF) is calculated. For this assessment, the basic Algorithm Q1 (light tissues) as provided by the device was used. Lyocell knit showed lower acoustic signal intensity and hence the "softest", the highest HF values, followed by the lyocell/polyester blend. Polyester showed lower HF values.

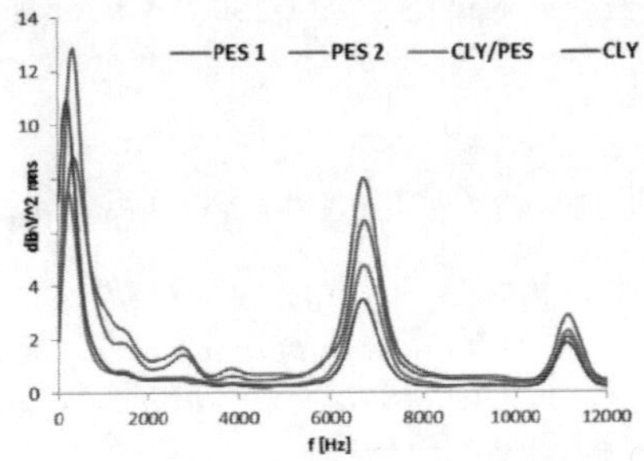

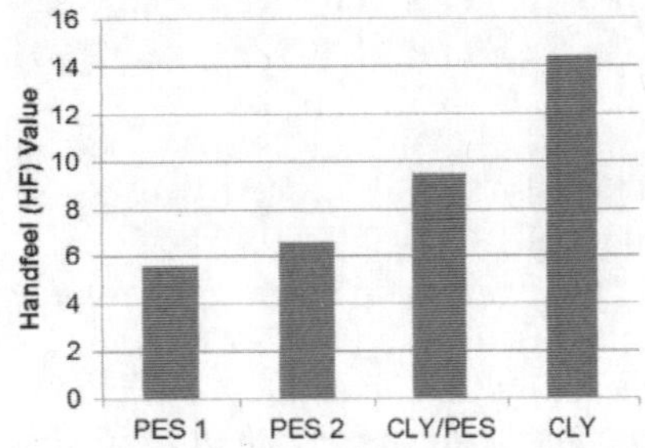

Figure 5. TSA skin sensoric assessment: Sonic spectrum of the fabric surface (left) and the resulting calculated handfeel values HF (right).

Subjective handfeel assessment ('Figure 6') results are more complex. Though abstract smoothness and drape evaluations show no significant differences between the cellulosic and the polyester fabrics, human hand seems to recognize synthetic as such and assessors provide higher all-over ratings for cellulosic materials. This recognition and preference seem to be mostly influenced by the warm/cool perception of the fabric. Moisture absorbing surfaces remove vapor from the hand surface and provide a cool impression at first contact. The polyester non-absorbent surface feels warm.

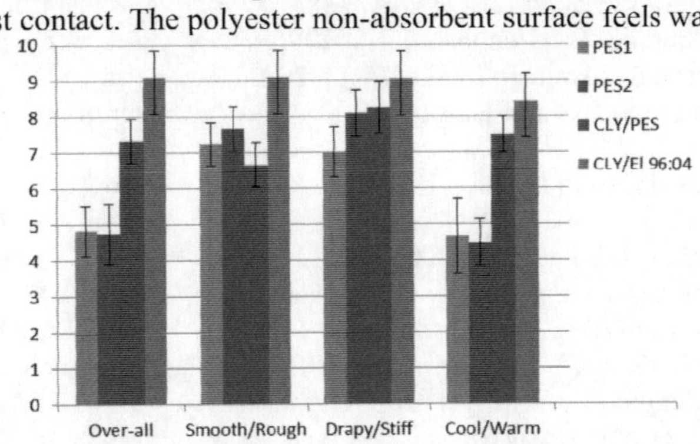

Figure 6. Subjective handfeel assessment by expert handfeel panel.

4. Conclusions

Workwear requirements foresee high textile performance properties of the used fabrics, while comfort and skin sensorial aspects receive less consideration. The requirements for underwear also do not take into account the period of the effective wearing down (months of usage, frequency of washing).

Taking vapor transfer resistance as measured by the hotplate as the sole criterion for "comfort" can be misleading, as it doesn't consider dynamic conditions, where the immediate reaction of the textile on vapor development is relevant. While polyester fabrics are superior in terms of dimensional stability and mechanical resistance and hard to be matched by cellulosics, the lyocell fabrics are superior in terms of moisture buffering and sensorial comfort. An optimum of mechanical performance and physiological and sensorial comfort can be reached in yarn blends. In such blends, an additional benefit is the received synergy effect observed on liquid transportation and drying speed.

To determine the level of comfort, objective as well as subjective methods must be used to determine the correct psychological, physiological and physical factors. It is not possible to obtain a complete quantitative assessment of comfort using isolated test methods. The effects of all five senses, the environment and textiles can have a significant effect on individual perception of clothing comfort.

Acknowledgments
The authors wish to acknowledge Carina Wenger and Gabriele Palenik for performing the TSA and the hotplate measurements and Herta Pixner and her team at Lenzing AG for the handfeel panel evaluation.

References

[1] Apurba D and Alagirusamy R 2010 *Science in Clothing Comfort* (Woodhead publishing India in textiles) ISBN 1845697898 p 175
[2] Song G 2011 *Improving comfort in clothing* (Woodhead Publ. Ltd.) ISBN 1845695399 p 496
[3] Ha M, Takura H, Yanai Y, Moriyama T, Tuchiya N 1999 *Combined effects of fabric air permeability and moisture absorption on clothing microclimate and subjective sensation during intermittent exercise at 27°C* (Ergonomics **42**(7)), p 964-979
[4] Schuster K C, Suchomel F, Männer J, Abu-Rous M, Firgo H 2006 *Functional and Comfort Properties of Textiles from TENCEL® Fibres Resulting from the Fibres' Water-Absorbing Nanostructure: A Review, Macromol.* Symp. **244**, p 149-165
[5] Li Y, Holcombe V 1996 *Enhancement of coolness to the touch by hygroscopic fibers* (Textile Research Journal **66**(9)), p 587-594
[6] Kim J O, Spivak S M 1994 *Dynamic Moisture Vapor Transfer Through Textiles (*Textile Research Journal **64**(2)) p 112-121
[7] Hong K, Hollies N R S, Spivak M 1988 *Dynamic Moisture Vapor Transfer Through Textiles, Part I: Clothing Hygrometry and the Influence of Fiber Type* (Textile Res. J. **59**) p 697-706
[8] Plante A M, Halcombe B, Stephans L G 1995 *Fiber Hgyroscopicity and perceptions of dampness* (Textile Research Journal **65**(5)) p 293-316
[9] El Mogahzy Y E, Kilinc F S and Hasan M 2005 *Developments in measurement and evaluation of fabric hand Effect of Mechanical and Physical Properties on Fabric Hand ed M Behery* (Cambridge:Woodhead Publishing Limited) chapter 3 p 45–65
[10] Grüner G and Grüner A 2011 *Method and device for determining the softness of sanitary papers and textiles* (US Patent US8082791-B2)
[11] Abu-Rous M Liftinger E, Innerlohinger J, Malengier B and Vasile S 2017 *A new physical method to assess handle properties of fabrics made from wood-based fibers* (World Textile Conference AUTEX 2017, Korfu, Greece. IOP Conf. Ser.: Mater. Sci. Eng. 254 14200)
[12] Abu-Rous M, Malengier B, Liftinger E and Innerlohinger J 2017 *Handfeel of Single Jersey Fabrics as Assessed by a New Physical Method* (J Fashion Technol Textile Eng) S4
[13] ISO 11092:2014 (standard) *Textiles - Physiological effects - Measurement of thermal and water-vapour resistance under steady-state conditions (sweating guarded-hotplate test)*
[14] EN 14704-1:2005 (standard) *Determination of the elasticity of fabrics – part 1: Strip tests*
[15] ISO 6330:2012 (standard) *Textiles - Domestic washing and drying procedures for textile testing*
[16] ISO 12945-2:2000 (standard) Textiles - *Determination of fabric propensity to surface fuzzing and to pilling - Part 2: Modified Martindale method*
[17] Williams J T 2009 *Textiles for Cold Weather Apparel* (Woodhead Publishing in Textiles) p 159
[18] Varga K 2009 *Man-made cellulosics, Thermodynamic of water soprption, structure visualisation and modification by conductive polymers* PhD Thesis, Leopold Franzenz University, Inssbruck, Austria

The comparative study of nursing pads by electrospun cellulose acetate, polyethylene oxide and thermoplastic polyurethane nanofibers

C Akduman[1], E P Akcakoca Kumbasar[2] and I Ozguney[3]

[1]Pamukkale University, Denizli Vocational School of Technical Sciences, Department of Textile Technology, 20100, Denizli, Turkey
[2]Ege University, Faculty of Engineering, Department of Textile Engineering, 35100, Izmir, Turkey
[3]Ege University, Faculty of Pharmacy, Department of Pharmaceutical Technology, 35100, Izmir, Turkey
cakduman@pau.edu.tr

Abstract. This study summarizes the general information about nursing pads and novel electrospun nanofiber mats as potential component for nursing pads. It also compares electrospun thermoplastic polyurethane (TPU), cellulose acetate (CA) and polyethylene oxide (PEO) nanofibers with a polypropylene conventional disposable nursing pad (NP) in terms of hydrophilicity, breathability, air permeability and swelling properties. Nanofiber mats prepared by the electrospinning method have unique properties such as smooth surface, high specific surface area and high porosity with fine pores which will lead to improved wicking properties. These properties make nanofibers potential component for disposable nursing pads. Mean diameters of produced nanofibers were 284.39, 609.70 and 219.30 nm for CA, TPU and PEO, respectively. Water contact angle measurement revealed that these nanofibers show good wettability properties better than commercial nonwoven nursing mat and air permeability results revealed that these nanofibrous mats have considerably adequate permeability. Besides, water vapor permeability results showed these nanofibers still show good breathability despite their compact structure.

1. Introduction

Nursing or breast pads are commonly used by nursing women to prevent milk, which may leak from the woman's breasts, from staining clothing and bed linens when the woman is not nursing [1]. These nursing pads generally divided into two subcategories; reusable pads and disposable pads [2]. Typically, a disposable breast pad contains a multilayered design, including a non-permeable layer for preventing transfer of breast milk from the liner to clothing, an absorbent layer for holding the milk within the liner, and a wicking layer to draw the liquid away from the breast and into the absorbing layer. A non-permeable layer may also be disposed about the outer periphery of the liner so that a reservoir is formed between the outer surface non-permeable layer. An adhesive may be applied to the outer portion of the liner to hold the liner in place in a bra cup [3]. Wicking top layer and the absorbing layer are the most critical part of these disposable pads. Top layer should deliver the absorbed milk immediately to the inner absorbing layer to prevent the discomfort of the mother. Wet nipples also may increase the sensitiveness of the skin.

In that point, nanofiber mats prepared by the electrospinning method have unique properties such as smooth surface, high specific surface area and high porosity with fine pores which will lead to improved wicking properties. These properties make nanofibers potential component for disposable nursing pads. Electrospinning method can be applied to synthetic and natural polymers, as well as polymer blends. Polymer solutions could also be loaded with functional nanoparticles, or active agents [4]. A large number of drugs [5-8], vitamins [9,10] could also be immobilized in polymer fibers [11] which make them functional other than casual nonwovens. More than fifty different polymers have been successfully electrospun into ultrafine fibers using this technique [12]. In this study, thermoplastic polyurethane (TPU), cellulose acetate (CA) and polyethylene oxide (PEO) nanofibers were selected and compared with a commercial nursing pad in means of hydrophilicity, breathability, air permeability and swelling properties.

2. Experimental

2.1. Materials

The TPU used in this study was the commercial Pellethane 2103-80AE, which is based on 4,4-methylene bisphenylene isocyanate, polytetramethyleneoxide and 1,4 butanediol, and was provided from Velox (Lubrizol Advanced Materials) and PEO (Polyox WSR N750) with approximate molecular weight of 300,000 was provided from Abalıoğlu Teknoloji as a gift. CA with number average molecular weight (Mn) of ~30,000 g/mol and acetylation degree of 39.8 %wt, acetone, dimethylacetamide (DMAc), N,Ndimethylformamide (DMF) were purchased from Sigma Aldrich Chemical Company. A commercial nursing pad was purchased and top layer of polypropylene mat was removed for comparisons. It is coded as NP through the study.

2.2. The preparation of electrospinning solutions

TPU solution, coded as TPU was prepared by dissolving TPU granules in the concentrations of 9% w/w in DMF at room temperature. Solution was stirred for at least 12h. CA solution, coded as CA was prepared by dissolving CA powder in the concentrations of 16% w/v in acetone/DMAc in the volume ratio of 2:1 at 50 ^{o}C. Solution was stirred for 3h. PEO solution, coded as PEO, was prepared by dissolving PEO powder in the concentrations of 8% w/w in distilled water. Solution was stirred for 2h at 50°C than cooled down to the room temperature and stirred further 2h. Polymer concentration and solvents are given at Table 1.

Table 1. Polymer concentration, solvents and solvent ratios

	Polymer concentration (%)	Solvents	Solvent ratio
TPU (w/w)	9	DMF	1
CA (w/v)	16	Aceton/DMAC	2:1
PEO (w/w)	8	Distilled water	1

2.3. Electrospinning

Electrospinning of the polymer solutions was carried out by a set-up consisting of a syringe (10 mL) with a stainless steel needle (22 gauges, and flat tip), a ground electrode and a high voltage supply (Simco, MP Series CM5 30 P, Charging Generator Output 30 kV DC). All solutions were electrospun at a voltage of 15 kV and a tip-to-collector distance of 15 cm with a feeding rate of 0.4-0.5 ml/h. A grounded rotating metal drum collector covered by a 90 mesh of plain weave monofilament %100 polyester screen mesh fabric was used as deposition material. Each polymer solution was electrospun for 2 hours.

2.4. The WCA measurements

For determination of hydrophilicity of the nanofiber membranes, the WCA were measured by using CAM 200 contact angle meter (KSV Instruments, Helsinki, Finland). The contact angles with distilled water were measured on the upper surface of the electrospun nanofiber membranes. The measurements were performed by the sessile drop method at ambient temperature. A real-time camera captured the image of the droplet.

2.5. SEM analysis

The morphology of TPU, CA, PEOX nanofibers and nursing pad top layer fibers were observed by a scanning electron microscope (SEM; Phenom G2pro scanning electron microscope). The samples were sputtered by Quorum Q150R S ion sputtering device with a thin layer of gold prior to SEM observation. The mean diameter of the nanofibers was calculated from measurements on SEM images of 5000× magnification by using Image J program. Approximately 50 measurements were carried out from the different parts of each sample.

2.6. Swelling of nanofibrous mats

Swelling of the TPU, CA, PEOX nanofiber and nursing pad top layer mats were measured upon exposure in distilled water for 24 h. The samples were removed from the water and carefully blotted with tissue paper to remove the excess after from the surface. The degree of swelling due to the water uptake was determined and weight loss was calculated at the end after drying for 4 h at 70 °C in an oven. The percentage of degree of swelling was calculated as follows [13]:

$$\text{Degree of swelling}(\%) = \frac{M-M_d}{M_d} \times 100 \tag{1}$$

where M was the weight of each nanofiber mat sample upon exposure in distilled water for 24 h, M_d was the weight of the samples after swelling and subsequent drying.

2.7. Characterisation of air permeability and water vapor permeability

Air and water vapor permeability of electrospun nanofibers and NP were tested by Textest FX3300 in accordance with EN ISO 9237 (5 cm^2 surface area and 200 Pa pressure drop) and by SDL-Atlas M261 in accordance with BS 7209, respectively.

3. Results and Discussion

Water contact angle measurement (WCA) is used to determine the hydrophility of the produced nanofiber mats. The contact angles between water droplets and the TPU, CA, PEOX nanofiber mats and NP were measured and pictures were taken after 0.128 s. The average WCA values of the right, middle and left are reported in Table 2. It also shows the droplet images after the water droplets were placed on the surface of TPU, CA and PEO nanofibrous mats and NP. It can be seen that the water contact angle of TPU and NP are the highest which are 119.8° and 109.50° respectively. These WCAs reveal the hydrophobic characters of TPU and NP. Since CA has an acetylation degree of 39.8%, it is more hydrophobic than PEOX with WCA of 86.21°. PEOX is the most hydrophilic mat with WCA of 29.53. When we compare the new developed nanofiber mats with NP, TPU nanofibers show similar, CA nanofibers show better, PEO nanofibers show very high wetting tendency. However, the reason for high wetting tendency of PEO is that, it is soluble in aqueous environment and in case of water contact it loses its integrity which restricts its usage.

Table 2. Water contact angle (°) and droplet images of TPU, CA, PEO nanofibers and NP mat

Mat Code	WCA (°)	Droplet image	Mat Code	WCA (°)	Droplet image
TPU	119.80	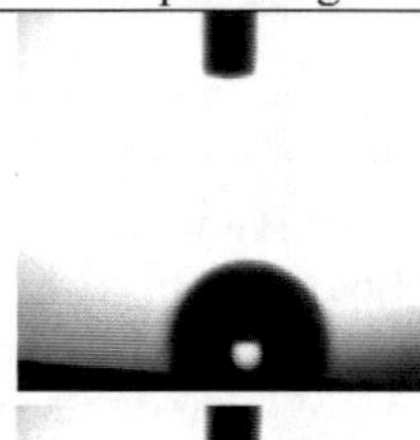	**CA**	86.21	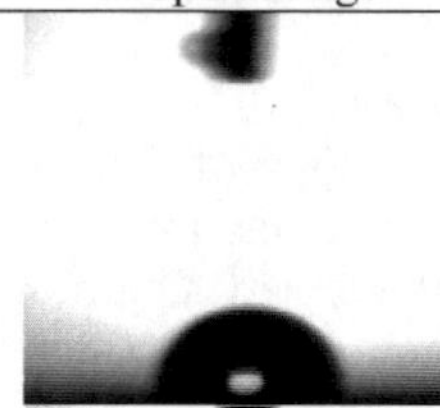
PEO	29.53	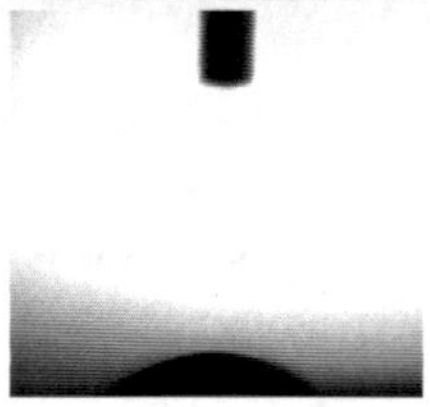	**NP**	109.50	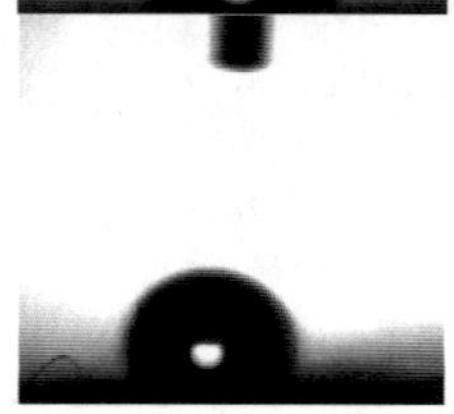

The surface morphology of electrospun TPU, CA and PEO nanofibers and NP were investigated by SEM imaging. In Figure 1 two different magnifications of SEM images are given for NP. Spot weldings of binding agent could be seen approximately with 1 mm distance from each other. From the measurements of 26 fibers, mean diameter of the NP was calculated as 14.65μm. SEM images of TPU, CA and PEO nanofibers are given in Figure 2. According to these images bead free, smooth nanofibers were produced and mean diameters were calculated as 284.39, 609.70 and 219.30 nm for CA, TPU and PEO, respectively. Finest nanofibers were obtained with 8% of PEO solutions.

a) b)

Figure 1. SEM images NP with a magnification of a) 78x b) 420x

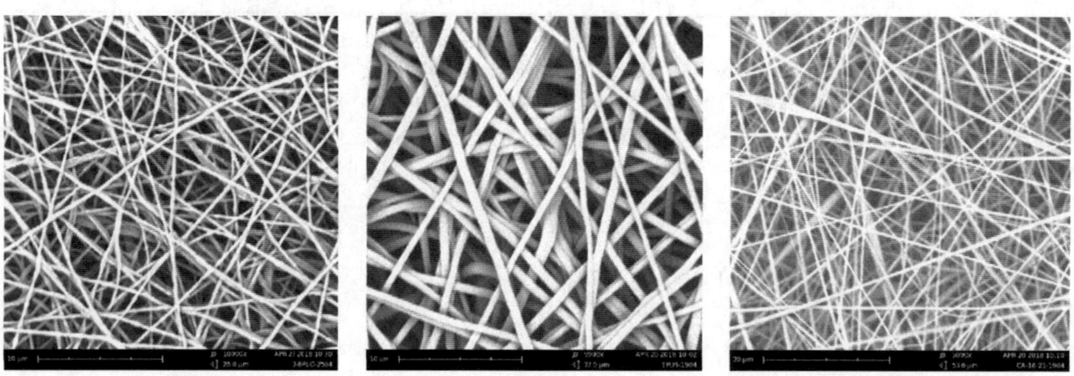

a) b) c)

Figure 2. SEM images a) CA, b) TPU and c) PEO

IOP Conf. Series: Materials Science and Engineering **459** (2019) 012029 doi:10.1088/1757-899X/459/1/012029

Table 3 lists the diameters of TPU, CA, PEO nanofibers and NP fibers, results of their air permeability and water vapor permeability results.

Table 3. The Diameters, Air Permeability Results of Water Vapor Permeability and Thickness of TPU, CA, PEO nanofibers, NP fibers and blank 90 mesh PES fabric

	Nanofiber Diameter *(nm)*	Air Permeability *(L/m^2s)*	Water Vapor Permeability *(g/m^2day)*
TPU	609.70± 140.03	42.78	655.85
CA	284.39 ± 143.01	46.44	637.88
PEO	219.30± 42.82	26.16	637.88
NP	14.65± 1.44	4006	691.78
90 mesh		4740.00	664.83

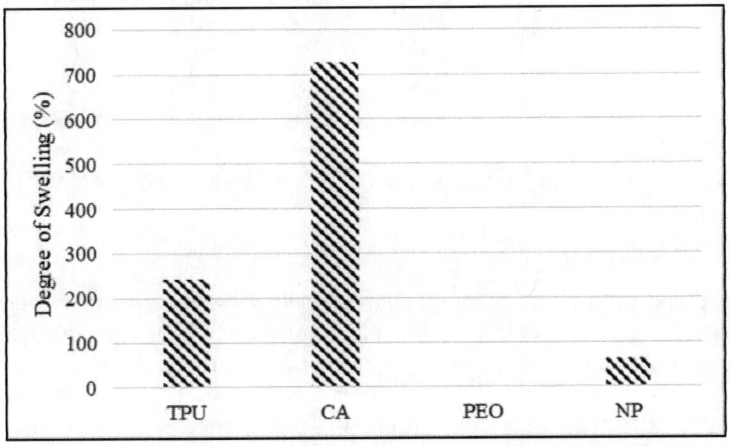

Figure 3. Swelling of TPU, CA, PEO nanofibers and NP mat

Swelling percentages of TPU, CA, PEO nanofibers and NP mat in distilled water for 24h are given in Figure 3. Due their hydrophobic characters TPU nanofibers and NP mat show lower swelling percentages as ~244% and ~63%, respectively. CA nanofibers showed considerable high swelling percentages of ~726% because of its hydrophilic structure and showed no weight loss while PEO nanofiber mats totally dissolved immediately after putting the mat into the water. The results showed that electrospun CA nanofiber membranes could be preferred when better hydrophilicity and swelling are required. Whenever more elastic and rigid mats are required TPU nanofiber mats could also be used with tolerable swelling.

Differences in air permeability was analyzed by one-way analysis of variance (ANOVA) followed by a Duncan test for pairwise comparison. There was a statistically significant difference between the air permeability of TPU, CA, PEO nanofibers and NP mat as determined by a one-way ANOVA ($p < .001$). Nanofibrous mats exhibited very low air permeability when compared to the NP mat and base fabric (4740 $/m^2/s$). The pores of micrometer dimensions for TPU, CA and PEO nanofiber mats, which are quite smaller than those in NP and the base fabric, would be too small to allow air flow at a prescribed pressure differential; thus result in very low air permeability [14]. Duncan post hoc test revealed that the air permeability values of TPU, CA and PEO nanofibers were found to be in same subset despite the differences in fiber diameter and statistically significantly lower than NP mats. Nano diameters of TPU, CA and PEO nanofibers decreased the air permeability values which would provide a barrier effect for the skin.

In case of water vapor permeability, polymer type or fiber diameters did not have a significant effect, all TPU, CA and PEO nanofiber mats showed similar results and they were not statistically different from each other ($p < .001$). Nanofibrous mats gave very high water vapor transmission like NP mat (691.78 g/m^2day) and base fabric (664.83 g/m^2day). Fine pores allowed water vapour transmission in all cases (Figure 2). According to this result (Figure 4), electrospun TPU, CA and PEO nanofibrous mats can be considered to have high water vapor transmission values despite their different hydrophilic characters. Water vapor transport properties of electrospun nanofibrous mats are excellent and indicate that layers based on electrospun fiber technology will be thin, lightweight, and very ''breathable'' with respect to allowing cooling to take place through a barrier material [15].

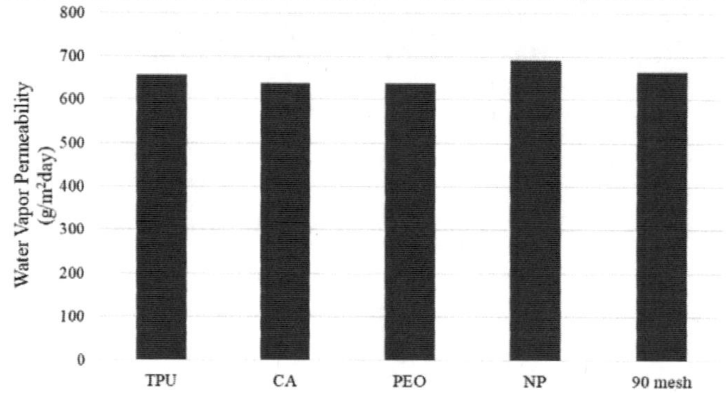

Figure 4. The effect of collection period (h) on water vapor permeability (g/m^2day)

4. Conclusion

Nanofiber mats prepared by the electrospinning method have smooth surface, high specific surface area and high porosity with fine pores which could improve wicking properties of a top layer of a disposable nursing pad. Electrospinning method can be applied to synthetic and natural polymers, as well as polymer blends with various active ingredients. In this study, CA, TPU and PEO nanofibers were selected and compared with a commercial nursing pad in means of hydrophilicity, breathability, air permeability and swelling properties. It was seen from the WCA measurement that when compared with NP, the new developed TPU nanofibers have similar, CA nanofibers have better, PEO nanofibers have very high wetting tendency. However, high solubility of PEO restricts its usage. Nano diameters of TPU, CA and PEO nanofibers decreased the air permeability values which would provide a barrier effect for the skin. In case of water vapor permeability, polymer type or fiber diameters did not have a significant effect, all TPU, CA and PEO nanofiber mats showed pretty high water vapor permeability results. Fine pores allowed water vapour transmission in all cases which indicates that layers based on electrospun fiber technology will be thin, lightweight, and very ''breathable'' and could be a potential candidate for disposable nursing pads.

Acknowledgments

The authors also would like to thank TUBITAK (Grant No. 217M133) for financial support. C.A. Author would like to thank to HYFIBER-Abalıoğlu Teknoloji for the support given.

5. References

[1] Spiezio C L, Spiezio M, 2004, Disposable Nursing Breast Pad with Medication, US 2004/0058619 Al
[2] Hebert C A, 2000, Nursing Pad Bra Liner, US6074272A
[3] Foley R M, Nanna K A, Thomas S, 2006, Medicated Breast Pad, US 2006/0052033 Al
[4] Greiner A, and Wendorff J H 2007 Angewandte Chemie International Edition. 46(30) 5670-5703.
[5] Zeng J, Yang L, Liang Q, Zhang X, Guan H, Xu X and Jing X 2005 J. Control. Release 105(1)

43-51.
[6] Taepaiboon P, Rungsardthong U and Supaphol P 2006, Nanotechnology 17 2317-2329.
[7] Akduman C, Özgüney I and Akçakoca Kumbasar E P 2016 Materials Science and Engineering: C 64 383-390.
[8] Kenawy E R, Bowlin G L, Mansfield K, Layman J, Simpson D G, Sanders E H and Wnek G E 2002 J. Control. Release 81 57–64.
[9] Taepaiboon P, Rungsardthong U and Supaphol P 2007 Eur. J. Pharm. Biopharm. 67(2).387-397.
[10] Sheng X, Fan L, He C, Zhang, K, Mo X and Wang H 2013 International journal of biological macromolecules 56 49-56.
[11] Sen R, Zhao B, Perea D, Itkis M E, Hu H, Love J and Haddon R C 2004 Nano letters. 4(3) 459-464.
[12] Huang Z M, Zhang Y Z, Kotaki M and Ramakrishna S 2003 Composites science and technology 63(15) 2223-2253.
[13] Cay A, Kumbasar EPA, Keskin Z, Akduman Ç and Urkmez A Ş 2017 *J Matter Sci* **52(20)** 12098-12108
[14] Lee S, and Obendorf S K 2007 *Fibers and polymers* **8(5)** 501-506.
[15] Akduman C, Oglakcıoğlu N, Kumbasar E P A and Sarı B 2018 *J Fashion Technol Textile Eng* **S4** 015

IOP Conf. Series: Materials Science and Engineering **459** (2019) 012030 doi:10.1088/1757-899X/459/1/012030

Use of DMDHEU with novel FR chemical for CO/PET blends

R Atakan[1], E Celebi[2], G Ozcan[1], N Soydan[2]
[1]Istanbul Technical University, Faculty of Textile Technologies and Design, Department of Textile Engineering, Gumussuyu, 34437, İstanbul, Turkey
[2]Eksoy Kimya Sanayi ve Ticaret A.S., Tekirdag, Turkey
ratakan@itu.edu.tr

Abstract. In present study, DMDHEU (dimethylol dihydroxyethyleneurea), which has been used for wrinkle-resistant finishing in textile, was used in FR finishing of CO/PET blends as a crosslinker with novel FR chemical (Fire-off). Flame retardancy performance of CO/PET fabrics treated with Fire-off and Fire-off/DMDHEU were investigated and compared. In addition, wrinkle resistant test was performed for untreated and Fire-off/DMDHEU treated fabrics to investigate the effect of the treatment on wrinkle recovery performance of fabrics. Results showed that Fire-off/DMDHEU system is a more effective nitrogen provider than Fire-off alone to enhance the flame retarding performance of the treated CO/PET fabrics through phosphorus–nitrogen synergism, therefore the presence of DMDHEU in the flame retardant finishing system increases the flame resistance of the treated CO/PET fabric. In addition, DMDHEU does not significantly change the wrinkle recovery of fabrics when used with Fire-off.

1. Introduction

Reducing the flammability of textile fibers and fabrics has been one of the major challenges facing the scientific and industrial researchers. Flame resistance is a desirable property that can be imparted to fabrics by means of chemical finishing. Due to the fact that conventional organo halogen-based flame retardants are getting banned more and more [1-4], several halogen-free substitutes have been developed, for instance, polyphosphates, organic phosphates, or nitrogen compounds [5-6]. However, the reactive organophosphorus flame retarding system contains significant levels of formaldehyde, a known carcinogen. Precondensate/ammonia process (known as ''Proban'' technology) requires the use of an ammoniation chamber, which is not compatible with most of the finishing operations in the industry. Those finishing treatments may also cause excessive fabric strength loss and adverse effects on fabric handle properties [7]. In this context nitrogen and phosphorus containing chemicals are especially in importance, because of their P–N synergistic effect in flame retardant applications [8]. However, their low stability with regard to washing and mechanical abrasion is limiting their applicability [9-10]. Using crosslinkers in FR systems is an option to lead better bonding properties of FR agents on textile fibers. N-methylol reagents, such as DMDHEU (dimethylol dihydroxyethyleneurea) have long been used in the textile industry as the crosslinking agents for cotton to produce wrinkle-resistant cotton fabrics and garments [11]. Few studies have indicated that it was also suitable as a crosslinker for flame retardant agents on cotton fabrics [12-13].

In our previous study [14], a nonformaldehyde flame retarding system, PVP (PR)-P-DCDA was synthesized by polyvinyl alcohol, hydrophilic polyester resin, phosphoric acid, and dicyandiamide. 100% polyester, 100% cotton, and 50/50% cotton/polyester fabrics were treated with PVP (PR)-P-DCDA by impregnation method. Flammability characteristics, thermal decomposition, surface morphology, and chemical structure of treated and untreated fabrics were investigated by vertical flammability test, limiting oxygen index, differential scanning calorimetry, thermogravimetric analysis, scanning electron microscopy, and Fourier-transform infrared, respectively. The results showed that PVP (PR)-P-DCDA are able to protect all three types of fabrics against their thermal degradation and favoring the formation of a stable char. The results suggested that the addition of FRs onto the cotton, polyester, and CO/PET might have reduced the flammability via PVP (PR)-P-DCDA dehydration into char. In addition to being an eco-friendly flame retardant system, a good char-forming flame retardant agent, PVP (PR)-P-DCDA has also superior ease of application for cotton, polyester, and cotton/polyester blends unlike commercially available "Proban" technology. At larger scale, Eksoy Chemical Company commercialized PVP (PR)-P-DCDA under the name Fire-off EBR.

In this study, a new FR finishing system based on Fire-off and DMDHEU were developed in order to improve durability for cotton part of the blend. CO/PET fabric samples (292 and 300 g/m^2) were treated Fire-off and Fire-off/DMDHEU FR systems. Flammability characteristics, thermal stability and chemical structure of treated fabrics were investigated by LOI, Micro Cone Calorimeter, DSC and FTIR, respectively. The wrinkle resistance properties of untreated and Fire-off/DMDHEU treated fabrics were also evaluated.

2. Experimental Study

2.1. Materials

The experiments were carried out using scoured 35/65% CO/PET (300 g/m^2, twill) and 50/50% CO/PET blend fabrics (292 g/m^2, twill) supplied from Ata Textile in Turkey. Experimental fabrics selected for this study represent those used in apparel and home textiles. The hydroxy-functional FR agent with the commercial name of "Fire-off" (supplied by Eksoy Chemical-Turkey) and DMDHEU with the commercial name of Reaknitt TW (supplied by CHT Chemical in Turkey) were used.

2.2. Fabrics Finishing Procedure

The fabrics were first immersed in a finish solution containing 350 g/L Fire-off and DMDHEU (10%) (Including a catalyst) (in two solutions: pH:3 and pH:6), then passed through a laboratory padder with two dips and two nips, dried at 100°C for 3 min, and finally cured in an oven at 165°C for 10 min.

In order to compare FR results of Fire-off with Fire-off/DMDHEU systems, the fabrics were also treated with Fire-off only (pH:6) via pad-dry-cure process (350 g/L Fire-off, 100°C for 3 min drying and 180°C for 3 min curing).

2.3. Characterization techniques

The chemical structure of the untreated and treated fabrics was characterized by Fourier-transform infrared spectroscopy (FTIR) using Perkin Elmer spectrum One FTIR-ATR System.

The thermal behaviours of untreated and treated CO/PET fabric samples were investigated by differential scanning calorimeter (DSC). DSC thermograms for CO/PET samples were obtained using a Perkin- Elmer model differential scanning calorimeter, from 30ºC to 275ºC with a heating and cooling rate of 10ºC/min. Samples of about 4-5 mg were sealed in aluminum pans. Nitrogen was used as the purge gas to provide an inert atmosphere and to prevent sample degradation. All DSC thermograms were analyzed using Pyris software for quantitative evaluation of melting, crystallization temperatures, and their associated heat capacities and enthalpies.

2.4. Fabric Performance Evaluation

Limiting oxygen index (LOI) test was carried out according to BS 4589-2 standard for flammability characteristics of fabrics. In addition, Micro Cone Calorimeter (MCC) was performed to investigate heat release properties of treated fabrics in accordance with ASTM D7309.

IOP Conf. Series: Materials Science and Engineering **459** (2019) 012030 doi:10.1088/1757-899X/459/1/012030

The conditioned wrinkle recovery angle (WRA) was measured according to ISO 2313 "Textile Fabrics: Determination of the recovery from recovery creasing of a horizontally folded specimen by measuring the angle of recovery" using SDL Atlas Recovery Tester.

3. Results and Discussion

3.1. FTIR analysis of untreated and treated fabrics

FTIR spectra of untreated and treated CO/PET fabrics are given in Figure 1.

50/50% CO/PET (292 g/m²)

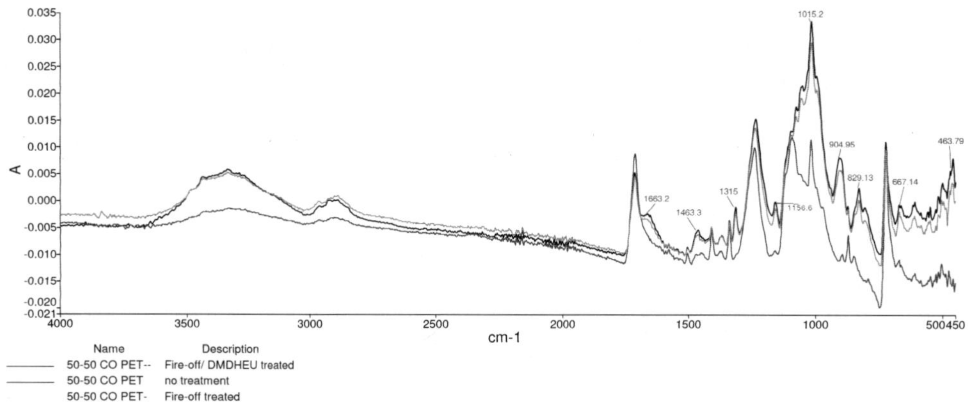

35/65% CO/PET (300 g/m²)

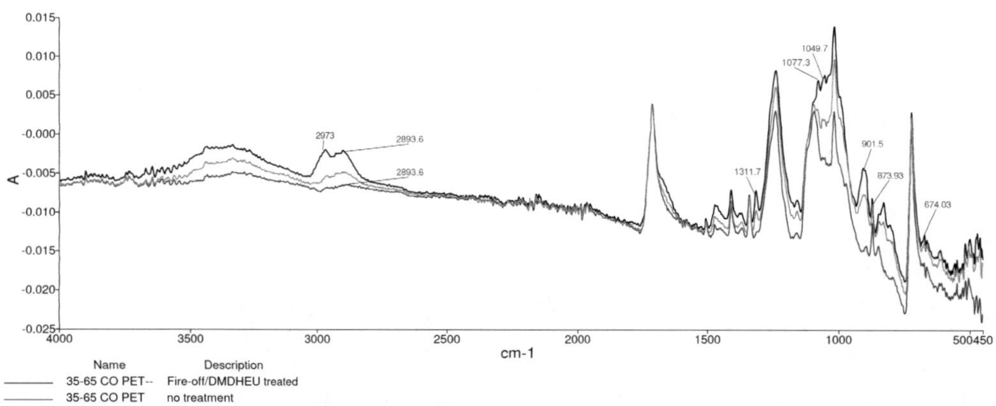

Figure 1. FTIR spectra of untreated and treated CO/PET fabrics (cm^{-1}: data is wavenumber, A: data is absorbance %).

As seen in Figure 1, FTIR spectra of Fire-off and Fire-off/DMDHEU treated fabrics are almost overlapped. However, FTIR spectrum of Fire-off/DMDHEU treated fabrics have sharper absorption peaks at 2973 (N), 901 cm^{-1} (P) and extra peaks at 1077, 1049 cm^{-1} (C–O–P), which indicate that Fire-off/DMDHEU treatment provides more phosphor and nitrogen than the Fire-off treatment into the fabric construction.

3.2. Differential scanning calorimeter results of treated and untreated fabrics

Table 1 shows the collected data of DSC.

IOP Conf. Series: Materials Science and Engineering **459** (2019) 012030 doi:10.1088/1757-899X/459/1/012030

Table 1. Collected data of DSC.

Fabric Type	Treatment	T_m (°C) / ΔH_m (J/g)	T_c (°C) / ΔH_c (J/g)
50/50 % CO/PET (292 g/m^2)	-	255.12/38.44	208.91/27.45
35/65 % CO/PET (300 g/m^2)	-	253.44/40.17	208.59/29.30
50/50 % CO/PET (292 g/m^2)	Fire-off	250.71/36.33	211.92/33.48
35/65 % CO/PET (300 g/m^2)	Fire-off	251.16/44.95	212.27/40.38
50/50 % CO/PET (292 g/m^2)	Fire-off / DMDHEU	248.63/37.45	209.08/33.69
35/65 % CO/PET (300 g/m^2)	Fire-off / DMDHEU	249.63/43.07	209.92/40.15

It is seen from the Table 1 that both Fire-off and Fire-off /DMDHEU treatments lead a decrease in melting temperatures, an increase in crystallinity temperatures. There are also an increase in enthalpy values with treatments during crystallinity. These results indicated that thermal stability of CO/PET fabrics has changed with Fire-off and Fire-off /DMDHEU treatments. It is also obvious that decrease in melting temperatures for Fire-off /DMDHEU treated samples are higher than the Fire-off treated ones.

3.3. Evaluation of the flame retardancy of treated fabrics

Table 2 shows LOI test results of untreated and treated cotton and CO/PET fabric samples, which were also associated with related add-ons (%).

Table 2. LOI values of untreated and treated CO/PET fabrics.

No	Fabric Types	Treatment	Add-on (%)	LOI (%)
1	50/50 % CO/PET (292 g/m^2)	-	-	18.8
2	35/65 % CO/PET (300 g/m^2)	-	-	18.6
3	50/50 % CO/PET (292 g/m^2)	Fire-off pH:6	19.3	27.0
4	35/65 % CO/PET (300 g/m^2)	Fire-off pH:6	17.2	26.0
5	50/50 % CO/PET (292 g/m^2)	Fire-off / DMDHEU pH: 3	16.5	26.7
6	35/65 % CO/PET (300 g/m^2)	Fire-off / DMDHEU pH: 3	15.2	26.2
7	50/50 % CO/PET (292 g/m^2)	Fire-off / DMDHEU pH: 6	23	27.2
8	35/65 % CO/PET (300 g/m^2)	Fire-off / DMDHEU pH: 6	23.8	27.2

It is seen from the Table 2, in general, LOI values of treated fabrics increase with the increase in add-ons. Fire-off treatment leads an increase in LOI values of CO/PET fabrics from 18.6 to 26-27, whereas Fire-off/DMDHEU system increase LOI values of CO/PET samples up to 27.2. Fire-off / DMDHEU treatment imparts flame retardancy to CO/PET fabrics at around 15 add-on values (at 15.2 add-on LOI: 26.2). However, add-on of Fire-off treatment for LOI of 26 should be at least 17.2 for the same CO/PET fabric. Thus, addition of DMDHEU in finishing bath has a positive effect on flame retardancy properties of fabrics. These results are in line with the FTIR and DSC results. In addition, pH of the finishing solution affects the ease of application of treatment.

Test results of MCC test are shown in Table 3. As seen from Table 3, there is a decrease in Heat Release Capacity (HRC), peak heat release rate (PHRR), time to peak heat release rate (T_{max}), total heat release rate (THR) and mass loss values for treated samples, which indicates that flammability and fire risk of fabrics decrease. Decrease in mass loss shows that FR treatments leads char formation. When compared Fire-off and Fire-off/DMDHEU treatments, it is seen that addition of DMDHEU in finishing bath leads more decrease in HRC, pHRR and mass loss of CO/PET fabrics, which indicate this FR system imparts better flame retardancy on CO/PET blends. These results are also in line with LOI test results.

Table 3. MCC Test Results of untreated and treated fabrics.

Fabric Type	Treatment	HR Capacity (J/k K)	PHRR (W/g)	T max (°C)	THR (kJ/g)	Mass loss (%)
50/50% CO/PET-292 g/m²	-	196.33	181.27	379.13	16.27	90.98
50/50% CO/PET-292 g/m²	Fire-off	136	123.53	437.77	13.53	83.63
50/50% CO/PET-292 g/m²	Fire-off /DMDHEU	141.67	129.9	438.63	13.87	81.94
35/65% CO/PET-300 g/m²	-	224.33	206.77	452.2	17.1	90.32
35/65% CO/PET-300 g/m²	Fire-off	185.33	170.9	440.27	14.83	84.27
35/65% CO/PET-300 g/m²	Fire-off /DMDHEU	178	163.83	442.9	14.5	83.34

3.4. Evaluation of wrinkle recovery of untreated and treated fabrics

The results of WRA of CO/PET fabrics in warp (W) and filling (F) directions are shown in Table 4.

Table 4. Wrinkle recovery angles of CO/PET fabrics.

50/50% CO/PET (292 g/m²)

Untreated				Fire-off/ DMDHEU treated			
face to face		back to back		face to face		back to back	
W1: 75°	F1:120°	W1: 105°	F1:95°	W1: 70°	F1:118°	W1: 105°	F1:99°
W2: 70°	F2:120°	W2: 110°	F2:95°	W2: 70°	F2:120°	W2: 105°	F2:104°
W3: 80°	F3:125°	W3: 105°	F3:95°	W3: 62°	F3:118°	W3: 105°	F3:107°
W4: 78°	F4:127°	W4: 100°	F4:95°	W4: 62°	F4:117°	W4: 105°	F4:94°
W5: 78°	F5:125°	W5: 100°	F5:95°	W5: 60°	F5:127°	W5: 105°	F5:102°
avg: 76.2°	**avg: 123.4°**	**avg: 104°**	**avg: 95°**	**avg: 64.8°**	**avg: 120°**	**avg: 105°**	**avg: 101.2°**

35/65% CO/PET (300 g/m²)

Untreated				Fire-off/ DMDHEU treated			
face to face		back to back		face to face		back to back	
W1: 100°	F1:135°	W1: 128°	F1:132°	W1: 85°	F1:130°	W1: 113°	F1:105°
W2: 82°	F2:136°	W2: 125°	F2:110°	W2: 85°	F2:130°	W2: 113°	F2:100°
W3: 98°	F3:130°	W3: 120°	F3:110°	W3: 85°	F3:130°	W3: 111°	F3:108°
W4: 110°	F4:136°	W4: 109°	F4:130°	W4: 85°	F4:129°	W4: 111°	F4:117°
W5: 88°	F5:155°	W5: 113°	F5:115°	W5: 82°	F5:128°	W5: 115°	F5:110°
avg: 96°	**avg: 134.4°**	**avg: 119°**	**avg: 119.4°**	**avg: 84.4°**	**avg: 129.4°**	**avg: 112.6°**	**avg: 108°**

It is seen that, Fire-off/ DMDHEU treatment lead a slight increase in WRA values of CO/PET fabrics. However, this change could be negligible. Wrinkle resistance of fabrics are not deteriorated dramatically with FR/DMDHEU treatment.

4. Conclusion

In this research, both heat release and flammability properties of CO/PET fabrics, which can be used in apparel and home textiles treated with a novel flame-retardant agent,-"Fire-off"- and Fire-off/DMDHEU were investigated. Results showed that DMDHEU functions as nitrogen provider for the flame retardant system, thus enhancing its flame retarding performance. Flammability test results also demonstrated that Fire-off/DMDHEU treated CO/PET blend fabrics generated less heat of combustion and obtained better flame retardancy than Fire-off treated ones, which can be also confirmed by the increase of LOI values at lower add-ons. Both Fire-off and Fire-off/DMDHEU treatments induced a significant decrease in PHRR and THR, and increase in char formation for CO/PET blend fabrics. Consequently, DMDHEU could be used for cellulosic blends with Fire-off in finishing formulation and Fire-off/DMDHEU treatment does not significantly deteriorate wrinkle recovery of fabrics.

Acknowledgments
The authors gratefully acknowledge the funding by I.T.U BAP, under grants of 39775 project. We wish to thank Ata Textile for fabric supply. We extend our appreciation to Prof. Dr. Nevin Çiğdem Gürsoy for laboratory facilities to perform chemical application.

References

[1] Zhao X 2010 Synthesis and application of a durable phosphorus/silicon flame-retardant for cotton. *The Journal of the Textile Institute* **101**: 538-546.
[2] Neisius M, Stelzig T, Liang S and Gaan S 2015 Flame retardant finishes for textiles. In: Roshan P (eds) *Functional Finishes for Textiles.* Cambridge-England: Woodhead Publishing Ltd, pp.429-452.
[3] Van der Veen I and De Boer J 2012 Phosphorus flame retardants: Properties, production, environmental occurrence, toxicity and analysis. *Chemosphere* **88**:1119–1153.
[4] Cordner A, Mulcahy M and Brown P 2013 Chemical regulation on fire: rapid policy. *Advances on Flame Retardants. Environ. Sci. Technol.* **47**: 7067−7076.
[5] Gaan S, Mauclaire L, Rupper P, Salimova V, Tran T-T and Heuberger M 2011 Thermal degradation of cellulose acetate in presence of bis-phosphoramidates. *Journal of Analytical and Applied Pyrolysis* **90**:33–41.
[6] Morgan A B and Wilkie C A 2014 The non-halogenated flame retardant handbook. *Wiley* Chichester, UK.
[7] Horrocks A R (1986). Flame-retardant Finishing of Textiles, Rev. Prog. Coloration, 16(3): 62–66.
[8] Bakoš D, Košík, M, Antoš K, Karolyová M and Vyskočil, I. 1982 The role of nitrogen in nitrogen−phosphorus synergism. *Fire Mater*. **6**:10−12.
[9] Horrocks A R 2003 In textile finishing; Heywood, D., Ed.; Society of Dyers and Colourists: Bradford, UK, pp 214−250.
[10] Schindler W D, Hauser P J, In Chemical Finishing of Textiles, Schindler, W. D., Hauser, P. J., Eds. Woodhead Publishing: Cambridge, UK, 2003; Chapter 8, pp 98−116.
[11] Weil E D 1973 US Patent 3,746,572, July 17.
[12] Wu W and Yang C Q 2004 Comparison of DMDHEU and Melamine-formaldehyde as the Binding Agents for a Hydroxy-functional Organophosphorus Flame Retarding Agent on Cotton. *Journal of Fire Sciences* **22**.
[13] Wu W 2004 Flame retardant finishing for cotton using a hydroxy-functional organophosphorus oligomer. PhD thesis, the University of Georgia.
[14] Atakan R, Bical A, Celebi E, Ozcan G, Soydan N and Sarac A S, *Development of a flame retardant chemical for finishing of cotton, polyester, and CO/PET blends.* Journal of Industrial Textiles, 2018: p. 1528083718772303.

IOP Conf. Series: Materials Science and Engineering **459** (2019) 012031 doi:10.1088/1757-899X/459/1/012031

Development of special fabrics protecting from electromagnetic radiation

B H Baymuratov[1], S Sh Tashpulatov[1], R D Akbarov[1], M Ilhamova[1], G A Yusuphodjaeva[1], U T Uzakov[1], N A Yusuphodjaeva[1]

[1]Tashkent Institute of Textile and Light Industry, Faculty of Technology of Textile Materials, Weaving Technology Department, 5 Shohjahon, Yakkasaroy dist., Tashkent, Uzbekistan

E-mail: bakhodir_b@mail.ru

Abstract. The influence of electromagnetic radiation (EMR) on a person is connected with the existence of electromagnetic field - a special form of matter which conducts the effect of electrically charged particles. The causes of electromagnetic field formation are defined by the fact that the occurrence of electric field changing in time is the reason of magnetic field formation and magnetic field occurrence is the cause of vortex electric fields; both fields change in time and excite each other. Electromagnetic waves formed during these processes exist irrespective of a source of their initiation. Continuous impact of electromagnetic radiation on a person negatively affects various organs of human body. EMR leads to headaches, fatigue of the organism, disorders of the body. The maximum influence of EMR occurs on "blood and eyes, the incidence of cancer and cataract increases, the number of people suffering from skin diseases increases". An important area today is the creation of fabrics protecting from electromagnetic radiation.

1. Introduction

At present, with development and improvement of technology for the production of polymeric materials, in particular chemical fibers, intensive research is being carried out to impart special properties to these materials: heat resistance, high strength, electrical conductivity, etc. One of the representatives of such materials is electrically conductive fibers. They have unique properties: high electrical conductivity characteristic for metal, lightness, elasticity and other valuable characteristics inherent in textile materials.

Electroconductive fibers and materials obtained on their basis are used in various branches of the national economy. One of the widest areas of application of electrically conductive fibers is the production of antistatic non-electrifying materials. Fabrics with low electrical resistance are also used to create special clothing that protects from the effects of an electric field and shields electromagnetic waves. On the basis of electrically conductive fibers, lightweight and flexible woven electric heaters for various purposes are produced [1].

Electroconductive fibers are mainly produced and widely used in highly developed countries such as the USA, England, Japan, Germany, France, etc. In 1990, our Republic also developed the technology of production and created a unique in the CIS pilot production line for the production of electrically conductive fiber nitron (ECFN). ECFN was used mainly to create special radio engineering materials and products for the protection of special equipment.

An important area today is the creation of protective fabrics from electromagnetic radiation. It is necessary to face many problems, since there are no established methods for designing such fabrics; their properties have not been studied sufficiently yet, the requirements not clearly defined.

IOP Conf. Series: Materials Science and Engineering **459** (2019) 012031 doi:10.1088/1757-899X/459/1/012031

Today, protective (shielding) properties are achieved either by applying special preparations to the fabric (at the stage of finishing) or by applying metal coatings to the surface of fabric; as a result, protective properties of the product have no permanence to severe conditions of operation-maintenance.

2. Statement of the problem, aims and objectives

The purpose of this work is the development and introduction into production of new special fabrics for protecting a person from electromagnetic radiation, greatly weakening electric and electromagnetic fields, and to offer a technology for their production.

The choice of cotton fiber for the production of mixed electro-conductive yarns is due, firstly, to the availability of rich source of raw material, and secondly to the high hygienic properties of cotton fiber, given that the materials created will be used to fabricate special clothes shielding electrical fields. The mixture of certain compositions has been prepared from the above fibers; they have been processed into yarns on an express spinning installation. Electrical resistance of the yarn samples was measured at a direct current in accordance with the procedure described in [2], and in the microwave range at a wavelength of λ = 0.03 m (at a frequency of 10 GHz), according to the method developed by the Radio Engineering Research Institute (Moscow), by measuring the coefficient of radio wave passing. Physical and mechanical properties of yarns have been determined in accordance with the State standards - 1119-80. The coefficient of electrical resistance variation has been calculated basing on results of 100 measurements [3].

To determine the dependence of the properties of mixed yarns on its composition, yarn samples with a linear density of 50±2 tex with different content of components have been obtained and investigated. Twist number is 500 twist/m. The dependence of physical and mechanical characteristics of mixed yarn on its composition is shown in Table 1.

Table 1 shows that an increase in amount of electroconductive fiber in yarn composition causes a slight decrease in its tensile strength and an increase in elongation.

Below are the results of strength calculation of mixed yarn, depending on its composition. In calculations, it is assumed that the yarn rupture occurs only due to the rupture of elementary fibers of the electroconductive nitron fibers (ECFN) and cotton fibers entering the yarn.

In calculations, the following formula for calculating the cross-sectional area of fiber is used [3].

$$F = \frac{1000T}{\rho}, \text{ mkm}^2$$

where: F – is the fiber cross-sectional area, mkm^2;

ρ - is the density of fiber, g/cm^3;

T – is the linear density of fiber, tex.

Table 1. Dependence of physical-mechanical characteristics of mixed yarn on its composition.

Ratio of components in yarn, ECFN /cotton, %	Breaking load, cN	Specific breaking load, cN/Tex	Breaking elongation, %	Uneven breaking load, %	Uneven tensile elongation, %
10/90	625	12,5	10,8	7,2	16,8
20/80	560	11,2	11,5	8,1	16,0
30/70	525	10,5	12,1	8,4	14,8
40/60	450	9,0	14,0	9,0	14,2
50/50	430	8,6	15,5	9,2	13,8
60/40	422	8,4	15,6	10,2	13.1
80/20	405	8,1	16,2	11,8	9,2

In calculations the following data are used:
The linear density of yarn - 50 tex;
Breaking load of ECFN -9,2 cN;
Density of ECFN -1,73 g/cm^3;
ECFN cross-sectional area – 312 mkm^2;
Breaking load of cotton fiber - 4,0 cN;
Density of cotton fiber -1,52 g/cm^3;
Cotton fiber cross-sectional area - 105,3 mkm^2

The data presented in table 1 show that breaking load of yarn decreases at increase in ECFN amount. This is explained by the fact that an increase in breaking load due to increase in ECFN amount cannot compensate its decrease due to reduction of cotton fibers in yarn.
These data, though being approximate (displacements of fiber relative to each other at rupture are not considered), give the chance to judge regularities of strength change in mixed yarn depending on its composition [4].
Study of the effect of yarn composition on its electric properties is of essential interest (table 2).

Table 2. Electrical-physical characteristics of yarn depending on its composition.

Ratio of components in yarn, ECFN /cotton, %	Electrical resistance per unit length, Rp		Variation coefficient of Rp at direct current, K_v, %
	at direct current lgRp (Rp, kOhm/m)	at λ = 3 cm (Rp, kOhm/m)	
10/90	2.2	120	30.4
20/80	1.6	48	26.5
30/70	1.3	41	20.2
40/60	1.2	35	18.2
50/80	1.2	35	15.3
60/40	1.3	33	14.5
80/20	1.2	35	13.0

Data presented in table 2 indicate that the introduction of ECFN to yarn composition causes sharp decrease in its electric resistance, especially at ECFN amount up to 20-25%.

3. Experimental researches

3.1. Study of physical-mechanical properties of electroconductive fabrics

To obtain experimental samples of electroconductive fabrics a conductive yarn has been used. Its characteristics are:

Composition - 60% of cotton, 40% of electroconductive fiber (ECFN);
Linear density - 50±2 tex;
Specific tensile strength – 9 g/tex
Breaking elongation - 14%;
Electrical resistance per unit length measured at λ=- 0,03 m - 35 kOhm/m
Number of twists - 500 twist/m.

A cotton yarn with linear density of 29.4 tex is used as the second component of woven product. Five experimental samples of cotton fabrics of a plain weave with different amounts of electroconductive yarn have been prepared. Electroconductive yarn was introduced into the composition of fabric on a warp and weft at the set distance from each other. Interweaving density is:
on a warp - 180±3, on a weft - 150±3 yarn/dm.

Samples No. 1 and No. 5 consist completely of electroconductive yarn and cotton yarn, respectively. In other cases the distances between electroconductive yarns are the following: sample No. 2 - 2,5 mm; sample No. 3 - 5,0 mm; sample No. 4 - 10,0 mm.

Results of properties research of fabric samples are given in figures 1-3. As seen from figures 1-3 the breaking load, breaking elongation and fabric rigidity increase with an increase in content of electroconductive yarn. The minimum values of these indicators are observed at pure cotton fabric (sample 5), and the maximum values - at fabric of electroconductive yarns only (sample1). The same regularity could be observed on indicators of surface density and air permeability of fabrics (table 3).

These facts can be explained by the effect of specific properties of the electroconductive Nitron fiber which is a component of conducting yarn and has a greater strength, elongation and rigidity than cotton fiber [5-7].

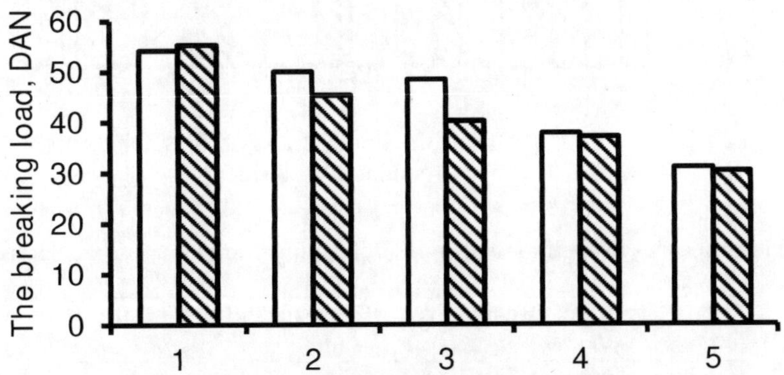

Figure 1. The values of the breaking loads of the samples of conductive fabric

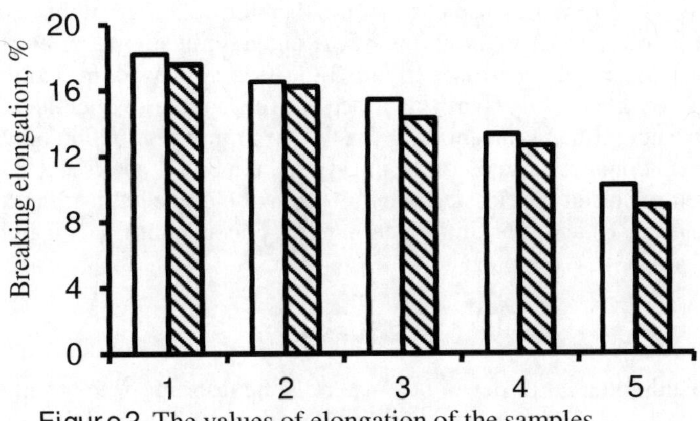

Figure 2. The values of elongation of the samples of conductive fabric

□ - warp; ▧ - weft

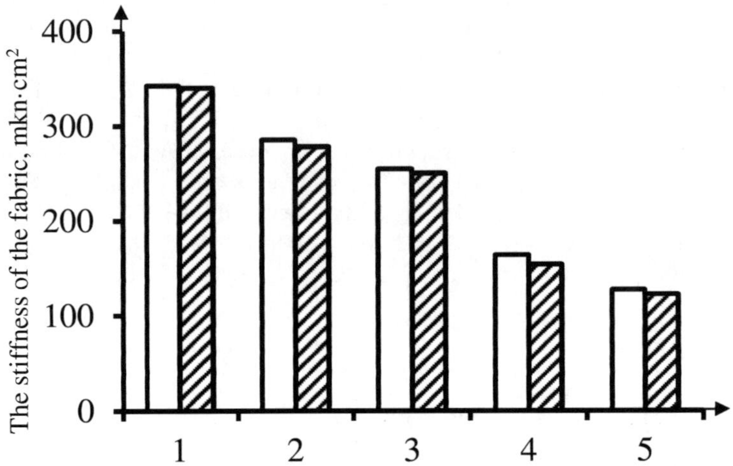

Figure 3. The values of stiffness of the samples of conductive fabric

Table 3. Change of properties of the fabric samples depending on the amount of electroconductive yarn.

Surface density g/m^2	Air permeability dm^3/m^{2}s
180±5	605
148±5	226
140±5	185
135±5	173
125±5	150

Electroconductive yarn is more loose in structure, than the cotton yarn; that explains the reduction in air permeability of fabric samples in the series 1-5. At increase in amount of electroconductive yarn the durability of fabric considerably decreases (figure 3). It is obviously connected with the attrition of thin metal layer applied on elementary synthetic fibers in the course of metallization and formation of microparticles which act further as an abrasive, leading to a rapid destruction of material. The efficiency of materials with semiconducting and conducting yarns, aimed to shield electric fields, is checked by their placement in strong electric fields and determination of electric field strength in the set point before introduction of material (E) and after introduction (Eэ). The shielding factor in dB is determined by a formula:

$$K\text{э} = 20\lg\frac{E}{E\text{э}}$$

To study the shielding effect of developed materials the test unit of the Center of high tensions and big currents of the Institute of Power Engineering (Energy) and Automatic Equipment of the Academy of Sciences of the Republic of Uzbekistan has been used.

En increase in amount of electroconductive yarn leads to a sharp decrease in surface electric resistance of fabric and an increase in its shielding ability (table 4). Surface electric resistance of fabric samples is measured at the wavelength λ=0.03 m. The shielding factor of fabrics in dB is determined at a tension of 80 kV.

Table 4. Electric properties of fabrics.

Surface electrical resistance, Ohm	The shielding factor of electric field, Ke, dB , U=80 kV
34	58,1
152	50,0
296	46,0
630	28,1
$>10^8$	-

It should be noted that according to existing norms, the protective clothes shielding electric fields have to possess the shielding factor no less than 50. Protective clothes or material with shielding factor equal to 50 weaken the electric field strength by 320 times. It means that even at tension of 400 kV/m, field tension inside the clothes won't exceed 1,3 kV/m; that allows the person to stay in electric field without threat to health for rather long period. Naturally at tension decrease the duration of person's stay in electric field considerably increases. Data in the table 5 show that the samples of fabrics No. 2 and No. 3 with electroconductive yarn on a warp and weft located at distance of 2.5 mm and 5 mm, respectively, from each other meet the requirements imposed to materials used for creation of protective clothes shielding electric fields of high tension. Increase in amount of electroconductive yarn in fabric leads to increase in the shielding factor up to 58 and higher (sample1), at the same time the cost of fabric and its weight increases, which is naturally undesirable.

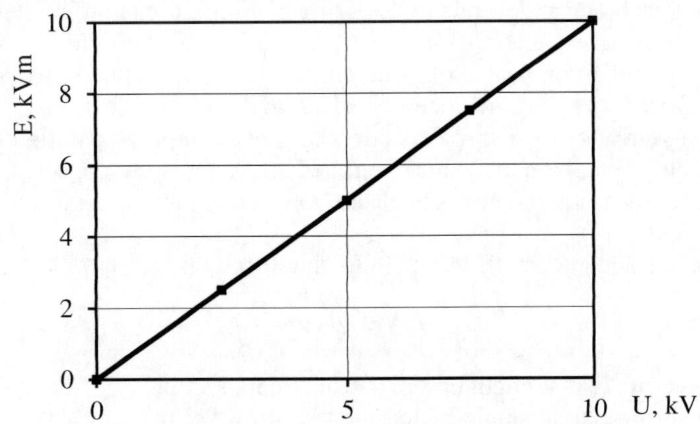

Figure 4. Dependence of electric field on the voltage

Dependence of the shielding ability of materials on tension is of essential interest [8-10]. Dependence of electric field strength on the value of high voltage attached to electrodes without shielding material is given in figure 4. As seen from the figure this dependence has a linear character and at voltage of 10 kV on electrodes the indication of the measuring device reaches the maximum value of -10 kV/m. Quite another picture we observe when shielding the measuring device by electroconductive materials developed in the current study. Results of measurements of electric field strength are given in figure 5 in the same system of electrodes that has been described at data acquisition in figure 4, when shielding the measuring device by electroconductive material.

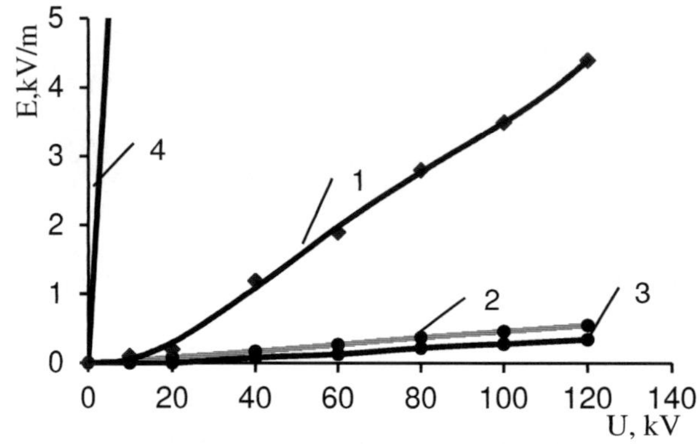

Figure 5. Dependence of electric field on the

In case of material No1 with conductive threads 10 mm apart from each other, both on a warp and weft, (dependence 1), it is established that at tension between electrodes 10 kV/m, the device screened by this material shows the tension 0,1 kV/m only, i.e. even at such sparse arrangement of conductive yarns in fabric as in this option, the shielding factor of electric fields is about 40. It means that this electroconductive fabric reduces the influence of electric fields of alternating voltage of industrial frequency almost by 100 times. However, at further increase in tension electric field inside the screen gradually increases, reaching the value of 2,4 kV/m at tension in interelectrode space of 75 kV/m.
Essential decrease in electric field strength behind the screen is observed when screen material No 2 with more dense arrangement of conductive threads is taken, for example, when these threads in fabric are 5 mm apart from each other (dependence 2, figure 5) So if at tension in interelectrode space of 75 kV/m in a case of conductive threads 10 mm apart the tension of the field is 2,4 kV/m, then reducing this distance to 5 mm, at the same value of field tension in interelectrode space, it equals to 0.03 kV/m behind the screen. Greater decrease in electric field strength happens in case of material No 3, at compaction of conducting threads to 2,5 mm (dependence 3). For comparison in figure 5, the dependence of measured electric field strength on the value of applied tension without the screen is given (dependence 4). Shielding factors of the studied samples of electroconductive fabrics are specified in table 5 at various tensions.
Shielding factor Ke is calculated in dB by the following formula:

$$K\text{э} = 20\lg\frac{E}{E\text{э}}$$

where: E – is the electric field strength before the shielding, kV/m.
Ee - is the electric field strength after the shielding, kV/m.

As seen from table 5 electroconductive materials samples No 2 and No 3 possess rather high factors of shielding even at relatively high values of tension. They can be used to create special clothes shielding electric fields of high tension.

Table 5. Shielding factors of various samples of materials, Ke, dB.

Voltage, kV	Without shielding		Sample 1		Sample 2		Sample 3	
	E, kV/m	Kэ	E, kV/m	Kэ	E, kV/m	Kэ	E, kV/m	Kэ
10	10	0	0,1	40,0	0,03	50,4	0	-
20	20	0	0,2	40,0	0,08	48,0	0	-
40	40	0	1,2	30,5	0,17	47,4	0,08	53,4
60	60	0	1,9	30,0	0,27	46,9	0,13	53,3
80	80	0	2,8	29,1	0,37	46,8	0,22	51,2
100	100	0	3,5	29,1	0,46	46,8	0,28	51,1
120	120	0	4,4	29,0	0,55	46,7	0,34	51,0

The studies carried out on the properties of electroconductive fabrics of various compositions have allowed to establish the following:
- an introduction of electroconductive yarn into the composition of cotton fabric leads to an increase in breaking load, breaking elongation, rigidity, air permeability and surface density of fabric. At the same time some decrease in durability of fabric on attrition is observed. The most rational from the point of view of electric properties is the fabric in which electroconductive yarns are located on a weft and warp 2,5÷5 mm apart from each other. Such fabric can provide the shielding factor of electric field more than 50; that corresponds to the existing standards.

References

[1] Gordeev V A, Dudko D A and Mokeev M N 1989 Woven Assembling. Moscow: Legprombitizdat, p 192

[2] Kuhn H H and Child A D 1998 Electrically Conducting Textiles. - *Handbook of Conducting Polymers*. Edited by Skotheim, Reynolds. p 168

[3] Pakshver A B and Meos A I 1966 Technological Calculations in Production of Chemical Fiber. *Chemistry*. p 324

[4] Akbarov D, Baymuratov B, Akbarov R, Kiekens P, Westbroek Ph and De Clerck 1998 Development of Electroconductive Polyacrylonitrile Fibers through Chemical Metallization and Galvanisation *J of Applied Electrochemistry*, UK, pp 411-418

[5] Baymuratov B, Akbarov D, Akbarov R D et al 1998 Study of Property of Fabrics Possessing Enhanceable Conductivity: *Proc. of the Republican scientific conference*. Tashkent, pp 58 - 59.

[6] Akbarov D and Baymuratov B 2000 Receipt and Research of Properties of Fabrics on a Warp of Electro-Conducting Fibers 2000 *Process - 2000 : Proc. of the Republican scientific conference on May, 26-27*. 2000. Bukhara, Uzbekistan.

[7] Baymuratov R, Akbarov D and Akbarov R 2000 Research of Properties of Electro-Conductive Fabric. *Actual problems of flax processing in modern conditions*: *Proc. of the International scientific conference*. Kostroma. pp. 125-128

[8] Akbarov R.D, Baymuratov B H and Ilhamova M 2017 Investigation of Electrical Characteristics of Electrically Conducting Yarns and Fabrics *17th World Textile Conference AUTEX* Textiles - Shaping the Future. IOP Conf. Series: Materials Science and Engineering. Greece. 2017.

[9] Akbarov N D, Baymuratov B H, Akbarov R D and Gafurov J K 2015 Patent IAP 05107. Shielding clothing. Uzbekistan.

[10] Akbarov R D 2004 Development and Study of Properties of Special Fabrics with Desired Electrophysical Characteristics. Diss. PhD. Tashkent. Uzbekistan. p 121

The effects of core material parameters on the mechanical properties of double core and single core spun yarns

T Bedez Ute[1] and H Kadoglu[1]
[1]Ege University, Department of Textile Engineering, İzmir, Türkiye

tuba.bedez@ege.edu.tr

Abstract
Comfort and aesthetic appearance are the basic features that consumers expect from their clothes. For improving body movement comfort in apparel, elastic core-spun yarns can be used to produce elasticity of the woven and knitted fabrics. In this study, the effects of elastane linear density on the mechanical properties of double core and single core yarns, spun with different core materials, were investigated. According to statistical evaluations, significant differences were found between yarn properties.

1. Introduction
Consumers want to feel comfortable as well as they want their clothes to look good. While improving the aesthetic appearance of the garments is of interest to fashion designers, it is the focus of textile engineers to enhance comfort. For improving body movement comfort in apparel, elastic core-spun yarns can be used to produce elasticity of the woven and knitted fabrics. In these yarns, elastane as the core is covered with staple fibres. On the other hand, bagging which is a three dimensional fabric deformation, is an undesirable appearance of fabrics [1]. The lack of dimensional stability or recovery after prolonged or repeated pressure on the fabrics causes bagging deformation [2]. In recent years, double core yarns have begun to be produced, especially for denim jeans to overcome the mentioned problem.

Core-spun yarns consist of at least two different components; a staple sheath and a filament core [3]. While, double-core (dual-core) yarns made of three components; an elastic polyurethane filament (such as Lycra®, Creora® or Inviya® I-300) and a multifilament (such as Lycra® T400®) are used in the core, covered by a staple sheath. Double-core yarns are used for high quality denim fabrics [4].
Many researchers have been focused on core-spun and elastic yarns' properties [5-7]. Sarıoğlu and Babaarslan investigated the fatigue behavior of rigid core-spun yarn structure containing PET textured filament yarn (PET DTY) with respect to different filament fineness and yarn count [8]. Celik and Kaynak examined the effects of elastane draw ratio on air permeability of denim woven fabrics [9]. Ertaş et al [10] analyzed the effect of the density changes in the use of the double-core threads used in denim fabrics. Telli et al, focused on the usage of core and double-core yarns containing tungsten for electromagnetic shielding [11]. Qadir et al. investigated the effects of elastane linear density and draft ratio on the physical and mechanical properties of single core-spun yarns. According to literature survey, elastic material parameters of the double core spun yarns have not be investigated previously.

2. Experimental
In this study, the effects of elastane linear density and elastane type on the properties of double core and single core yarns, spun with different core materials, were investigated. The performance characteristics of the two most widely used elastane types in the market, expressed as elastan A and

IOP Conf. Series: Materials Science and Engineering **459** (2019) 012032 doi:10.1088/1757-899X/459/1/012032

elastane B, have also been examined. For this purpose 44 dtex, 78dtex and 117 dtex elastane filaments were used in the production of elastic core spun yarns. Elastic core spun yarns containing single core will be referred to as "soft core" in this article. Besides, 55 dtex polyester (PES) or 55 dtex elastomultiester (EME) filaments were used for accompanying elastane in double core spun yarn production. Additionally, PES or EME containing core spun yarns produced without using elastane will be referred to as "rigid core". Sheath material was cotton for all yarn types.

Yarns were spun in the same yarn count and twist coefficient (Ne 18, α_e 4,2) on a ring spinning frame. Yarn samples were conditioned under standard atmospheric conditions, thereafter basic yarn characteristics such as evenness, imperfections, hairiness, breaking strength and breaking elongation, were tested. Yarn uniformity, the IPI values and yarn hairiness were measured on Uster Tester 5, tensile properties were measured with Uster Tensorapid 4.

Table 1. Experimental plan and some test results.

		Rigit Core material	**Elastik Core material**	**Elastan dtex**	**Yarn unevenness CVm (%)**	**Number of thin places (-50%)**	**Number of thick places (+50%)**	**Number of nep (+200%)**	**Yarn hairiness (H)**
Single core	Rigid core yarn	PES	-		10,91	0,00	20,00	17,50	5,69
		EME	-		10,20	0,00	8,75	12,50	5,82
	Soft core yarn	-	Elastan A	44	10,68	0,00	13,75	16,25	6,22
		-	Elastan A	78	10,51	0,00	9,38	10,63	5,97
		-	Elastan A	117	10,56	0,50	10,50	5,50	6,13
		-	Elastan B	44	10,62	0,00	6,25	10,63	5,79
		-	Elastan B	78	10,15	0,00	8,75	9,38	6,23
		-	Elastan B	117	10,68	0,00	18,75	12,50	6,45
Double core		EME	Elastan A	44	9,97	0,00	6,25	11,88	5,54
		EME	Elastan A	78	9,24	0,00	2,50	8,75	5,04
		EME	Elastan A	117	8,85	0,00	4,38	6,88	5,41
		EME	Elastan B	44	9,47	0,00	7,00	11,00	5,64
		EME	Elastan B	78	9,90	0,00	8,75	5,63	5,94
		EME	Elastan B	117	9,59	0,00	6,88	6,88	5,81
		PES	Elastan A	44	10,24	0,00	15,00	15,00	5,88
		PES	Elastan A	78	10,22	0,00	13,13	13,75	5,86
		PES	Elastan A	117	9,68	0,00	6,25	9,38	5,67
		PES	Elastan B	44	10,72	0,00	17,50	16,88	5,72
		PES	Elastan B	78	10,40	0,00	16,25	14,38	5,65
		PES	Elastan B	117	9,81	0,00	9,38	10,63	5,82

Test results, given in Table 1, were evaluated statistically with SPSS software. With the aim of determining the statistical importance of the differences between the yarn characteristics, ANOVA (Analysis of variance) was performed. F values were obtained from analysis of variance and their statistically significance were evaluated ($p>0,05$).

3. Results

The effects of elastane type, elastane linear density and different core materials on the properties of core spun yarns on the yarn evenness, yarn hairiness, breaking strength and elongation results are given in Figure 1.

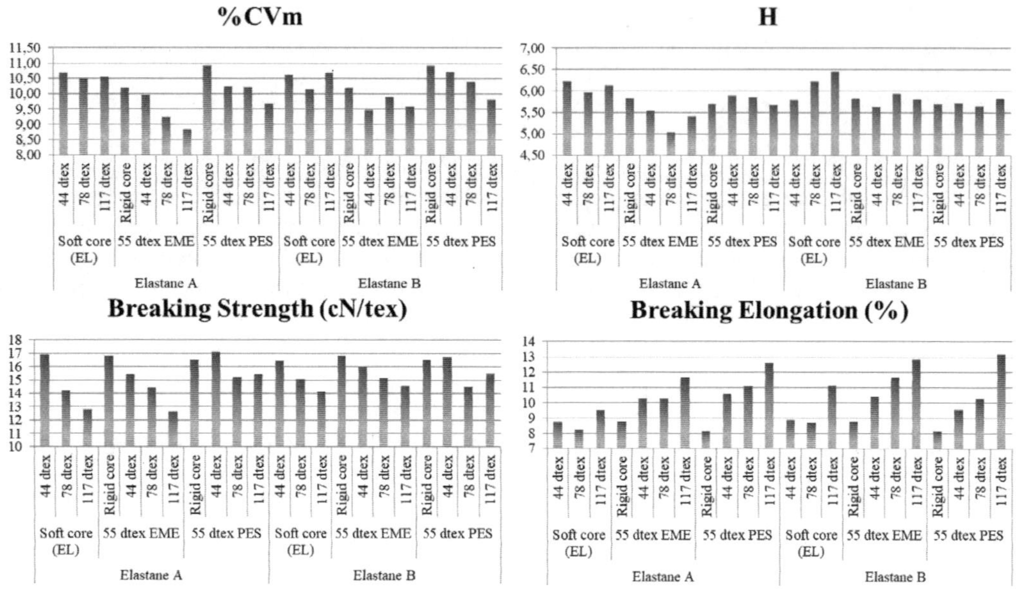

Figure 1. Yarn unevenness, yarn hairiness, breaking strength and elongation values of the single and double core spun yarns.

According to test results and statistical evaluations, it is found that yarn evenness decreases with the increase of elastane thickness. Dual core spun yarns have lower yarn unevenness values than single core spun yarns, including both soft core and rigid core spun yarns. These are thought to be caused by the decrease in the amount of sheath cotton fibre. Dual core yarns produced with 117 dtex elastane A and B have the lowest unevenness values whereas dual core spun yarns produced with 44 dtex elastane or rigid core spun yarns had the highest unevenness values. Yarns containing elastane A showed lower unevenness values than the yarns containing elastan B, in general. Soft core yarns produced with only elastane have the highest Uster CV% values whereas, dual core yarns produced with EME and elastane yarns showed the lowest unevenness values compared to others.

In terms of yarn hairiness, as a result of higher fibre movement, soft core yarns have higher hairiness values than rigid core and double core spun yarns. EME containing double core yarns showed lower hairiness values than PES containing yarns. It is found that the effect of elastane thickness on yarn hairiness was not statistically significant in general.

It is found that the breaking strength and yarn unevenness of the core spun yarns were decreased while breaking elongation values increase with the increase of the elastane thickness. For 44 dtex and 78 dtex fineness, elastane A caused higher elongation but lower breaking strength compared to elastane B. On the other hand, 117 dtex elastane B showed statistically the highest elongation values for both PES and EME containing double core spun yarns (Table 2) and also single cored elastic yarns.

Table 2. Student-Newman-Keuls (SNK) post-hoc results for yarn elongation at break (effect of elastane)

	2nd core material-PES						2nd core material-EME			
1st core material	**1**	**2**	**3**	**4**	**5**	**6**	**1st core material**	**1**	**2**	**3**
- (rigid core)	8,12						**- (rigid core)**	8,78		
Elastane B-44dtex		9,52					**Elastane A-44dtex**	10,25	10,25	
Elastane B-78dtex			10,23				**Elastane A-78dtex**	10,27	10,27	
Elastane A-44dtex			10,50				**Elastane B-44dtex**	10,41	10,41	
Elastane A-78dtex				11,08			**Elastane A-117dtex**		11,64	11,65
Elastane A-117dtex					12,58		**Elastane B-78dtex**		11,65	11,65
Elastane B-117dtex						13,20	**Elastane B-117dtex**			12,81
Sig.	1,0	1,0	0,32	1,00	1,00	1,00	**Sig.**	,058	,174	,151

Double core yarns have been observed to have higher elongation values, independent of the elastane brand and thickness, than single core yarns. In terms of hard core materials, PES containing double core yarns have higher elongation and strength values than EME containing double core yarns, in general.

Table 3. Student-Newman-Keuls **(**SNK) post-hoc results for yarn elongation at break (effect of rigid core material)

		44 dtex			**78 dtex**			**117 dtex**		
Elastane A		**1**	**2**		**1**	**2**		**1**	**2**	**3**
	Soft core	8,74		**Soft core**	8,24		**Soft core**	9,50		
	EME		10,25	**EME**	10,27	10,27	**EME**		11,64	
	PES		10,50	**PES**		11,08	**PES**			12,58
	Sig.	1,00	0,31	**Sig.**	1,000	,312	**Sig.**	1,00	1,00	1,00

		44 dtex				**78 dtex**				**117 dtex**	
Elastane B		**1**	**2**	**3**		**1**	**2**	**3**		**1**	**2**
	Soft core	8,84			**Soft core**	8,69			**Soft core**	11,12	
	PES		9,52		**PES**		10,23		**EME**		12,81
	EME			10,41	**EME**			11,65	**PES**		13,20
	Sig.	1,00	1,00	1,00	**Sig.**	1,00	1,00	1,00	**Sig.**	1,00	0,59

According to Table 3, soft core yarns produced with only elastane had statistically significantly lower breaking elongation values than elastic double core spun for both elastane A and B for all elastane thicknesses. On the other hand PES containing double core yarns have the highest elongation values when elastane A is used whereas EME containing double core yarns have the highest elongation values when elastane B is used.

4. Conclusion

Nowadays, consumers are increasingly demanding products with improved elasticity as well as reduced bagging problems. For this reason, double core yarns, especially for denim jeans, are of interest. In this study, the effects of elastane linear density and elastane type on the properties of double core and single core yarns, spun with different core materials, were investigated. It is found that the type core materials and the thickness of the materials have significant effects on yarn properties.

Acknowledgments

We would like to present our thanks to the Scientific and Technological Research Council of Turkey-TÜBİTAK, Project number: 116M687 for its financial support. A special thanks to Mesut Çelik from Orta Anadolu Tekstil for yarn spinning.

References

[1] Şengöz N G 2004 Bagging in textiles *Textile Progress* **36** 1 pp 1-64
[2] Uçar N Realff M L Radhakrishnaiah and P Ucar M 2002 Objective and subjective analysis of knitted fabric bagging *Textile Research Journal* **72** 11 pp 977-982
[3] Technical Document 2006 Producing Core-Spun Yarns Containing Lycra® Elastane Fiber *Invista* February
[4] http://www.tekstilteknik.com.tr/fashionable-dual-core-yarnsperfectly-spliced-wound/ (Available on 25.06.2017)
[5] Su C and Yang H 2004 Structure and Elasticity of Fine Elastomeric Yarns *Textile Research Journal* **74** 12 pp 1041-1044
[6] Su C I Maa M C and Yang H Y 2004 Structure and performance of elastic core-spun yarn *Textile Research Journal* **74** 7 pp 607-610
[7] Babaarslan O 2001 Method of Producing a Polyester/Viscose Core-Spun Yarn Containing Spandex Using a Modified Ring Spinning Frame *Textile Research Journal* **71** 4 pp 367-371

[8] Sarıoğlu E and Babaarslan O 2017 Fatigue behaviour of core-spun yarns containing filament by means of cyclic dynamic loading *Autex 2017* Greece

[9] Çelik H İ and Kaynak H K 2017 An investigation on effect of elastane draw ratio on air permeability of denim fabrics *Autex 2017* Greece

[10] Ertaş O G Ünal B Z and Çelik N 2016 Analyzing the effect of the elastane-containing dual-core weft yarn density on the denim fabric performance properties *The Journal of The Textile Institute* **107**, 1 pp 116–126

[11] Telli A Daşan Y Babaarslan O and Karaduman S 2017 Usage of Core and Dual-Core Yarns Containing Tungsten for Electromagnetic Shielding *Advance Research in Textile Engineering* **2** 1 p1013

[12] Qadir M B Hussain T Malik M Ahmad F and Jeong S H 2014 Effect of elastane linear density and draft ratio on the physical and mechanical properties of core-spun cotton yarns *The Journal of The Textile Institute* **105** 7 pp 753-759

IOP Conf. Series: Materials Science and Engineering **459** (2019) 012033 doi:10.1088/1757-899X/459/1/012033

Evaluation of the knitted fabrics stiffness through dynamic testing

M Blaga[1] R Harpa[1] N-E Seghedin[2] A Marmarali[3] G Ertekin[3]

[1] "Gheorghe Asachi" Technical University of Iasi, Romania, Faculty of Textiles, Leather and Industrial Management, [2]Faculty of Machine Manufacturing and Industrial Management, 56 D. Mangeron Street, 700050, Iasi, Romania, [3] Ege University, Department of Textile Engineering, Bornova, Izmir, Turkey

mirela_blaga@yahoo.com

Abstract. The paper aims at introducing a qualitative method for the evaluation of warp knitted spacer fabrics stiffness, through the measurement of their natural frequencies, considering the direct relationship between natural frequencies of the textile fabrics and their stiffness. Additionally, the goal of this work was to comparatively evaluate the results obtained by two testing methods that utilise different measurement principles. The experimental design included the free vibration method, employed to measure the natural frequencies of the spacer mesh structures produced on K. Mayer RD 6 N double needle bar machine, manufactured in two variants of thickness and widths.The natural materials frequencies have been recorded coursewise, walewise and perpendicular on the fabric surface. The testing results confirmed the influence of the fabric thickness, the samples having lower thickness being characterized by the highest level of natural frequencies and stiffness. In parallel, the analysis of the knitted spacer fabrics' flexibility was performed, allowing a comparison of fabrics stiffness and validating the dynamic test results.

1. Introduction

Warp-knitted spacer fabrics are three-dimensional textile structures formed by two parallel-knitted fabrics connected by a yarn layer called as spacer yarn. Spacer yarn plays an important role on the fabric's properties such as high compressibility, breathability, impact resistance and elastic recovery characteristic. In the production of warp knitted spacer fabrics, especially polyester multifilament and monofilament yarns are used as surface layers and spacer yarn, respectively. These fabrics are taken from the knitting machine in the form of closed structures. After knitting process, the warp knitted spacer fabrics are normally subjected to a heat-setting treatment in order to increase the structural stability and to achieve a determined open form [1].

The stiffness of the textile fabrics can be in some cases a basic feature, determining their suitability for a specific end use. Generally, rigidity is defined as the capacity of one material to resist under the influence of different forces, such as: uniaxial tension, compression, bending, shearing and vibration. Most methods described and used in textile laboratories are based on the principle of indirect stiffness assessment by determining the dependence of strain and the force that induces this strain. Some methods use tensile testing machines for precise measurement of the dependence of the force and strain of textile fabrics [2].

The researchers have focused their interest on the knitted fabrics characterization through their natural frequencies, which were determined by employing the free vibrations method [3-5]. The results confirmed that knitted fabrics can be engineered and exploited as structures with vibration absorption capabilities. The weft spacer fabrics containing monofilaments as spacer yarns, have been subjected to the dynamic testing and the results indicate that the differences of the materials natural frequencies are related to the spacer monofilament yarns diameter, patterning of the spacer yarns, threading of monofilaments per needle and machine gauge [3]. In case of warp spacer knitted fabrics, it has been found that number of spacer yarn connections in the unit area are contributing to a higher rigidity of the fabrics on perpendicular testing direction, while the natural frequencies measured wale and course wise are comparable. The finishing process, by its chemical agent's action proved a certain influence on the higher natural frequencies values of the tested fabrics. The consolidation process of yarns contacts is taking place, determining a lower yarns friction and thus an increased rigidity of the structure [4]. The level of natural frequency of each material is giving indications about its rigidity, the higher is the natural frequency the higher rigidity is expected [5].

This relationship was used in the present study and warp knitted spacer fabrics were investigated through the free vibration method, for measuring their natural frequencies. In parallel, the flexibility analysis was carried out for the same fabrics, to establish the stiffness materials hierarchy. The aim of this work was to comparatively evaluate the stiffness of the warp spacer knitted fabrics, determined by two testing methods using different measurement principles, in order to validate the dynamic testing results.

2. Material and method

2.1. Materials used

Spacer fabrics were knitted on a Karl Mayer RD 6 double-needle bar Raschel machine equipped with six yarn guide bars (GB). While a 150D/48x4 polyester PTY and 300D/72x3 polyester PTY multifilaments were used to create the top outer and bottom outer layers, a polyester monofilament of 0.243 mm in diameter was used as spacer yarn to connect two outer layers. The chain notation of the guide bars are displayed in Table 1.

Table 1. The chain notation of the guide bars

GB1	2-2 / 2-2 / 0-0 / 0-0 / 2-2 / 2-2 / 0-0 / 0-0 / 2-2 / 2-2 / 1-1 / 1-1 / 3-3 / 3-3 / 1-1 / 1-1 / 3-3 / 3-3 / 1-1 / 1-1 //
GB2	0-1 / 1-1 / 1-0 / 0-0 //
GB3	0-1 / 0-1 / 1-0 / 1-0 //
GB4	0-1 / 0-1 / 1-0 / 1-0 //
GB5	0-0 / 0-1 / 1-1 / 1-0 //
GB6	1-1 / 2-2 / 2-2 / 0-0 / 0-0 / 2-2 / 2-2 / 0-0 / 0-0 / 2-2 / 2-2 / 1-1 / 1-1 / 3-3 / 3-3 / 1-1 / 1-1 / 3-3 / 3-3 / 1-1 //

Brückner (Vn-Sfp-24/ 6-Q99) heat-setting machine was used to increase the structural stability and to achieve a determined final width (stretching). The samples were horizontally (course-wise) pinned on the sliding aluminium frame pins and heat set at the temperatures of 150°C and 180°C for time durations of 1.8 and 3 min. The tension applied to the samples during heat-setting was arranged so that the final width of the samples was 110 and 160 cm while the width of unset fabrics which refers to the fabric width after knitting process was 60 cm [1]. After heat-setting, the fabric samples were allowed to cool down at room temperature for 24 hours. The samples obtained with these parameters are represented in Figure 1.

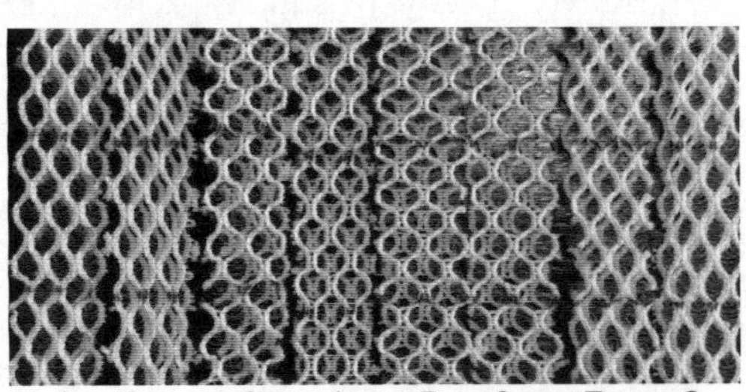

Figure 1. Variants of spacer fabrics

The mass per unit area values are determined according to the TS EN12127 standard. Fabric density was calculated by dividing fabric weight to fabric thickness. The porosity characteristic of the fabrics is calculated by using Equation (1):

$$\varepsilon = \frac{\rho}{\rho} \ (\%) \qquad (1)$$

Where: ε is the fabric porosity (%), ρ_a is the fabric density (g/cm^3) and ρ_b is the fibre density (g/cm^3).

The details of production process and the characteristics of the samples are presented in Table 2.

Table 2. Fabric details and characteristics

Fabric code	Thickness (mm)	Temperature of thermosetting (^{0}C)	Stretching width (cm)	Mass per unit area (g/m^2)	Fabric density (g/cm^3)	Porosity (%)
F1.	13	150	110	737.952	0.0533	96.14
F2.	10	150	110	721.888	0.0697	94.95
F3.	13	150	160	588.832	0.0435	96.85
F4.	10	150	160	570.912	0.0531	96.15
F5.	13	180	160	577.824	0.0424	96.93
F6.	10	180	160	515.264	0.0523	96.21
F7.	13	180	110	716.736	0.0564	95.91
F8.	10	180	110	711.854	0.0744	94.61

The stretching parameter affected the mass per unit area and the fabric density values significantly. The higher the stretching (samples F3-6), the lower the mass and fabric density values of the fabrics. For the same fabrics, their porosity increases with the stretching parameter increasing. The thicknesses of all fabrics are smaller than the distance set (12.5 and 15 mm) between the two needle bars of the machine. This is an expected situation because after the knitting process, the spacer yarns tended to relax into bended forms, conducting to the decreasing of the fabric thickness, respectively to 10 and 13 mm [6].

2.2. Testing method and equipment

2.2.1. Fabrics natural frequencies measurement

The vibration tests of warp knitted spacer fabrics were conducted by using the free vibration method [3-5]. According to the system theory, an elastic system out of the position of stable equilibrium then released, produces free vibrations. In the presence of friction forces, mechanical energy is dissipated and the vibration is damped by a certain number of cycles. Free vibration frequencies depend on the mass, stiffness and damping of the system. They are independent of the initial conditions of the motion and of the system external forces. Therefore, their frequencies are called natural frequency of vibrations. For a given system, they have constant values well defined [7].

In this study, the natural frequencies of the spacer knitted fabrics have been measured by testing the dynamic behaviour of one metallic piece, 30x70x30 mm dimensions, fixed through an adhesive, directly on the fabric's surface. The textile material is also fixed on the heavy plate, the relative movements between the piece-knitted fabric-plate being avoided. The exciter used was an impact hammer Piezotronics and frequencies were measured with an accelerometer PCB B52 Piezotronics type. The signal was processed with one data acquisition card 6023 National Instruments. In order to determine the natural frequencies of the system, the Fast Fourier Transformation-FFT has been applied and the Spectrum Analyzer application from the LabView 8.2 software has been employed.

2.2.2. Fabrics flexibility measurement

Fabric stiffness tester provides a quick and accurate method of determining the fabric flexibility according to internationally recognized test standard ASTM D1388-2007. Employing the principle of cantilever bending, a rectangular specimen (200mm x 25 mm) is supported on a smooth low-friction horizontal platform. A weighted slide is placed over the specimen and advanced at a constant rate, allowing a narrow strip of fabric (10mm) to bend under its own weight. As the leading edge of the specimen projects from the platform, it bends under its own weight. For each bending length of 1cm, the material flexes and touches the bending angle that allows an easy read of the corresponding angle fabric bends. The higher the bending angle, the more flexible is the fabric.

$$H = \frac{A}{A_{abs}} \cdot 100 = \frac{1}{9}\left(\varphi_1 + \varphi_2 + ... + \varphi_9 + \frac{\varphi_{10}}{2}\right)\% \qquad (2)$$

Where: H is fabric flexibility (%), A represents the area of the "bending angle variation diagram" for the testes sample, Abs is the corresponding are of the "bending angle variation diagram" area for the absolutely flexible specimen. In this case, the bending angle is 90^0 and corresponds to any free length of the specimen from 10 mm to 100 mm.

3. Results and discussions

Three readings of the frequencies have been done and the average has been considered. The vibrations were produced and measured on three directions: wale wise, course wise and perpendicular on the fabric surface. The frequency at the highest peak represents the natural frequency of the fabric, as shown in Figure 2.

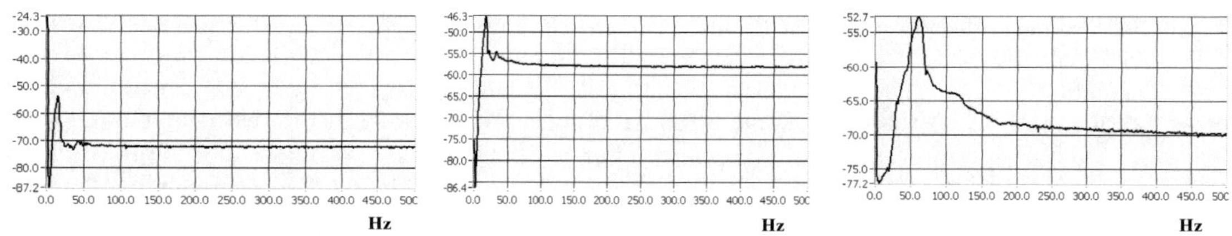

a) Course wise b) Wale wise c) Perpendicular

Figure 2. Natural frequencies of the spacer warp knitted fabric [Hz]

The values of the natural frequencies measured for the warp spacer fabrics are illustrated in Figure 3. The results indicate that perpendicular on the fabrics surface, the measured values are higher compared to other two directions. This can be justified by the position of the spacer yarns, which play an important role on the fabric's properties, such as: compressibility, breathability, impact resistance and elastic recovery characteristic. In this case, the monofilaments positioning inside the fabric layers influence the assembly behaviour to the dynamic testing, by increasing the fabric rigidity and level of natural frequencies recorded on this direction. With these special properties, spacer fabrics have great potential use against vibration protection applications.

On perpendicular and wale testing direction, it can be observed that the fabric's thickness is influencing the level of natural frequencies, the thinner fabrics have higher values of the frequencies and are consequently higher rigidity. This is also confirmed by the flexibility measurements for these fabrics on wale direction, which has lower values for the same fabrics.

With the increase of the final width of the samples, the size of holes on fabric surface of the samples increases also. The fabrics with 160 cm width have lower values of the frequencies and higher values of flexibility, against the samples with 110 cm. Among them, those with the thickness of 10 mm have higher values of the frequencies compared to the 13 mm thickness ones.

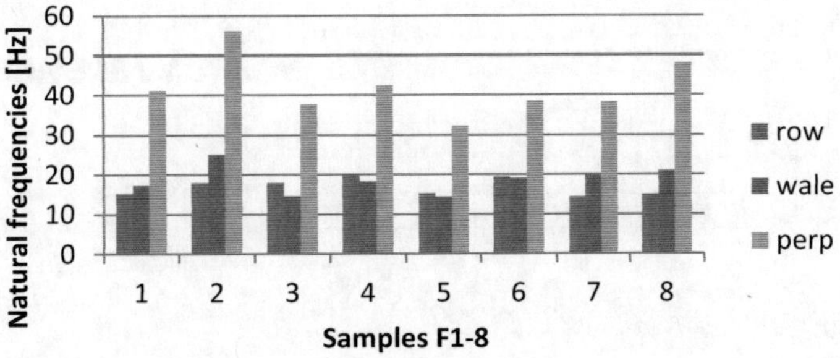

Figure 3. Natural frequencies of the warp spacer fabrics (Hz)

Heat-setting under drawing of spacer fabrics increases the porosity and air permeability of sample while decreases the thickness and mass per unit area of sample [8]. In this study, the analysed samples with higher heat-setting temperature were manufactured with 2 values of thickness and 2 different mass per unit. On wale direction, the fabrics F7-F8, which are heavier than F5-6, have a higher stiffness and a lower flexibility, as shown by the measurements displayed in Figure 4. Wale wise, the fabrics F4 and F6 with 10 mm thickness, show a higher rigidity and lower flexibility in comparison with the pair F5 and F7, with 13 mm value of thickness. The highest values of the natural frequencies have the fabrics F2 and F8, manufactured with the lower thickness and lower stretching width, which are also characterized by a higher fabric density and bigger mass per unit.

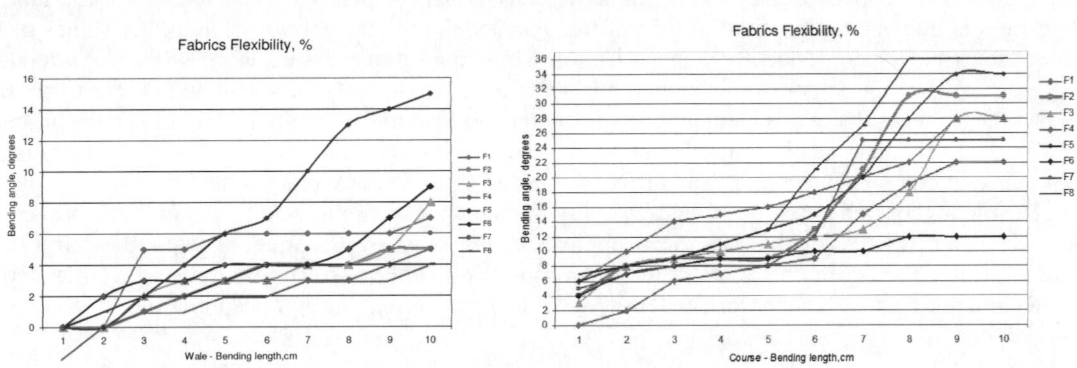

Figure 4. Fabrics flexibility (%)

Figure 4 reveals a different flexural behaviour of the fabrics, walewise being more stable, due to the particular warp knitting principle, with the parallel yarns manufactured on vertical direction. On wale direction, the hierarchy of the samples starts with those having 13 mm thickness and 160 cm width after stretching and ends up with the variants of 10 mm thickness and 110 cm width. On course direction, the materials having 10 mm thickness and 110 width are more flexible, compared to ones produced with 13 mm distance between the layers. It can be concluded that thickness and stretching width are adjustable influence parameters for controlling the knitwear flexibility.

Figure 5 presents the values of the fabrics stiffness determined by the two described methods in the paper. The shape of the two graphs demonstrates the indirect relationship between the measurements, when the fabric has a high value of the natural frequency and consequently high stiffness, the measured flexibility is low.

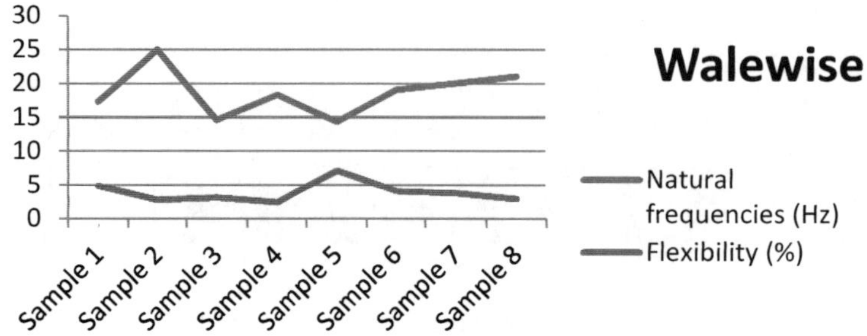

Figure 5. Warp knitted spacer fabrics stiffness

This fact validates the assumption and results of the stiffness evaluation of the fabrics, strengthening the sampled hierarchy and recommending the method of free vibration for the qualitative evaluation of the warp spacer knitted fabrics stiffness.

4. Conclusions

This study reports an investigation on the warp spacer knitted fabrics stiffness determined through the method of free vibrations. Within this method, each recorded curve consists of a certain number of waves, the highest peak of the first one in horizontal direction, being the natural frequency of the fabric, measured in [Hz]. The level of natural frequency for each material is giving indications about its rigidity, the higher is the natural frequency the higher rigidity is expected. It has been found that the fabric's thickness is influencing the level of natural frequencies, the thinner fabrics have higher values of the frequencies and are consequently higher stiffness. More, the samples with a lower stretching width and bigger mass per unit demonstrated higher level of frequencies and higher stiffness. In order to confirm and validate the results of this method, a second method known in the textile field, based on the principle of cantilever bending, has been used to determine the fabrics flexibility. The measurements confirmed the fabrics hierarchy concerning their stiffness obtained through the vibrations testing.

The method of free vibrations found its utility in mechanical engineering field and to the best of our knowledge, in textile field it has been applied so far only by the authors, with the purpose of characterizing the textile materials in their selection for vibration isolation purposes, such as people protection during a particular work or sensitive devices protection during transportation.

References

[1] Ertekin G and Marmarali A 2017 Impact resistance behaviour of silicone coated warp knitted spacer fabrics used for protective clothing *The Journal of The Textile Institute* 108:12 2123-2131

[2] Goetzendorf-Grabowska B Karaszewska A Vlasenko V and Arabuli A T 2014 Bending Stiffness of Knitted Fabrics – Comparison of Test Methods *Fibres & Textiles in Eastern Europe* 22 1(103): 43-50

[3] Blaga M et al 2014 Weft knitted spacer fabrics response to vibrations 14th Autex World Textile Conference (Bursa: Turkey)

[4] Blaga M et al 2013 Dynamic testing of the warp knitted spacer fabrics 13th Autex World Textile Conference, (Dresden: Germany)

[5] Blaga M and Seghedin N E 2017 Knitted Spacer Fabrics Behaviour at Vibrations *J Textile Eng Fashion Technol* 3(2): 00092

[6] Ertekin G and Marmarali 2016 The Effect of Heat-Setting Conditions on the Performance Characteristics of Warp Knitted Spacer Fabrics *Journal of Engineered Fibers and Fabric* 11 3 64-71

[7] Goldman S 1999 Vibration Spectrum Analysis, A Practical Approach, Industrial Press. Inc.

[8] Borhani S Seirafianpour S Ravandi SAH Sheikhzadeh M and Mokhtari R 2010 Computational an experimental investigation of moisture transport of spacer fabrics *Journal of Engineered Fibers and Fabrics* 5(3) 42-48.

IOP Conf. Series: Materials Science and Engineering **459** (2019) 012034 doi:10.1088/1757-899X/459/1/012034

The importance of a comparative thermophysiological study in female jeans

I Braga[1, 2], M J Abreu[2] and M Oliveira[3]
[1]Federal University of Piauí, Curse de Bach .in Fashion, Design and Stylish , Campus Universitário Ministro Petrônio Portella, Bairro Ininga, Teresina /PI, Brazil
[2]2C2T-Center for Textile Science and Technology, University of Minho, Campus de Azurém, Guimarães, Portugal
[3]University of Minho, Campus Gualtar, Braga, Portugal

iarabraga@yahoo.com.br

Abstract. The comfort of clothing is essential as to performance relative to the wearer and is considered a quality factor in the choice of a particular garment. That is why the science of comfort is used to evaluate jeans, a symbol of popular fashion and the consumption of the poor classes of Brazil [1]. Therefore, the present paper consists of the presentation of the experimental phase of a PhD research on the study of the comfort of Brazilian female jeans, specifically referring to thermal comfort tests through the use of equipment such as Permetest and Thermal Manikin for measuring thermophysiological comfort properties.

1. Introduction

The popular jeans in question refers to jeans created, produced and consumed by the popular poor class of the Northeast region of Brazil, with the main focus being the items sold at fairs and popular markets in Fortaleza, Ceará state, where fashion popular jeans have stood out due to economic incomes, job creation and income for the poor populations of the peripheral and rural urban areas of the cities of Ceará [2].

The aim of this research is to compare five different models of jeans used by consumers in order to carry out an analysis of the products offered in popular markets, taking into account thermo-physiological aspects, such as air permeability, thermal resistance and moisture behavior, in order to establish an understanding of the comfort characteristics that the female target population using this specific product choose.

The relevance of this research lies in the analysis of this type of jeans that are being offered in the popular markets, determining the most demanded jeans regarding comfort conditions in the scope of combine the study of science of comfort and fashion concepts in the Brazilian popular fashion market and even worldwide.

2. Materials and methods

The study is based on the methodology of ISO 11092: 2014 [3]. The methodology of evaluation of the physiological evaluation of clothing is organized in five levels. For this stage of research, we applied the techniques level 1, the physical-mechanical analysis of the materials (using testing equipment) and level 2, the biophysical analysis of the garment (using thermal manikin). Five models of female's jeans were purchased in the popular shopping centers of Fortaleza, Buraco da Gia and Beco da Poeira in Forteleza/Brazil.

In order to analyse the thermal properties of these fabrics, the relative water vapour permeability (RWVP) and the absolute water vapour permeability (AWVP) were measured using PERMETEST[4], and the air permeability was measured on a Textest FX 3300 instrument [5] and the measurement of thermal insulation with the use of Thermal Manikin[6] Maria of the University of Minho PT-Teknik made in Denmark, in a thermal chamber with controlled environment with temperature and humidity between 28ºC with a variation of ± 2ºC and 65%, with variation of ± 2%. The thermal resistance tests were performed at 37°C corresponding to the internal temperature of the human body.

The data were analysed statistically using Excel and SPSS software version 23 (Statistical Package for the Social Sciences), using the SPSS in ANOVA analysis of variance, applying the later tests, Tukey tests, with significance level of 0.05.

3. Results

The Popular jeans were initially analysed for fabric composition, structure, count threats and industrial washing types are presented in table 1.

Table 1. Jeans description

Jeans Models	Jeans 1	Jeans 2	Jeans 3	Jeans 4	Jeans 5
Image	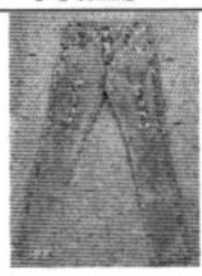			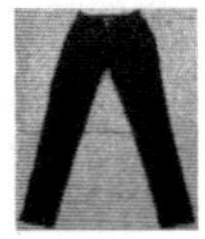	
Composition	98%contton 2%spandex	77%contton 21%polyester 2% spandex	98%contton 2% spandex	96,5%contton 3,5% spandex	77%contton 21%polyester 2% spandex
Industrial wash	Acid wash and destroyed - hand scraping	Stone wash and acid wash (permanganate)	Dark water jet wash and overall crinkle (resin)	Dark sulfur dyeing	Rinse wash and softener
Structure	Twill (2/1 1)	Twill (2/1 1)	Twill (3/1 1)	Twill (3/1 1)	Twill (3/1 1)
Count threats (yarns/cm)	29 x 21	37 x 23	39 x 22	40 x 26	38 x 23

After that, the results of the ANOVA variance analysis were calculated using the models of jeans and thus comparing the values of the characterization measurements materials and of the data of the jeans thermophysiological tests. With the test of homogeneity of variances, the Levene test has as hypothesis H0 that there is homogeneity of variances between the different treatments, when analysing the data, it was observed the existence of both homogeneity and heterogeneity between the averages of the treatments performed. To better emphasize the relationship between the different jeans as the physical characteristics the analysis of multiple comparisons of the means was done through the Tukey test. The comparison of the means in homogeneous subsets, with harmonic mean = 10,000.

The analysis of the comparison of the measures means of the mass per unit area are expressed in table 2 and the comparison of the homogeneous groups are shown in figure 1.

Table 2. Mass/unit area (g/m^2)

Tukey HSD[a]	**Mass/unit area (g/m^2)**		
Models	Subset for alpha= 0.05		
	1	2	3
Jeans 1	281,43		
Jeans 4	289,08		
Jeans 3		321,93	
Jeans 5		323,52	
Jeans 2			332,08
Sig.	0,052	0,976	1,000

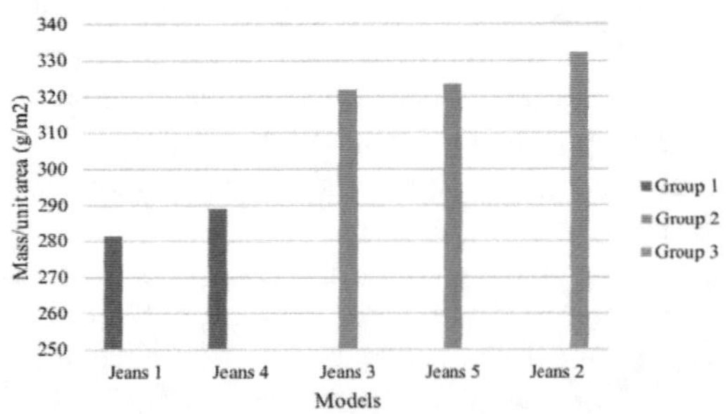

Figure 1. mass/unit area(g/m^2) homogeneous groups

IOP Conf. Series: Materials Science and Engineering **459** (2019) 012034 doi:10.1088/1757-899X/459/1/012034

Evaluating the mass/unit area values of the 5 jeans models, group 1- jeans 1 and jeans 4 had lower mass values, although jeans 4 presented 7.65 (g/m^2) higher than jeans 1, they are considered statistically homogeneous the value of significance = 0.052. Group 2 - jeans 3 and jeans 5 are also evaluated as homogeneous with significance = 0.976, when purchased with the other jeans showed values of mass greater than group 1 and smaller than group 3. Group 3 - jeans 2 presents higher mass/unit area (g/m^2).

The analysis of the comparison of the averages as the thickness measures are expressed in table 3, the data shown in figure 2.

Table 3. Thickness(mm)

Tukey HSD[a]	Thickness (mm)			
	Subset for alpha= 0.05			
Models	1	2	3	4
Jeans 1	,6970			
Jeans 4		,7170		
Jeans 3			,7490	
Jeans 5			,7530	
Jeans 2				,8340
Sig.	1,000	1,000	,973	1,000

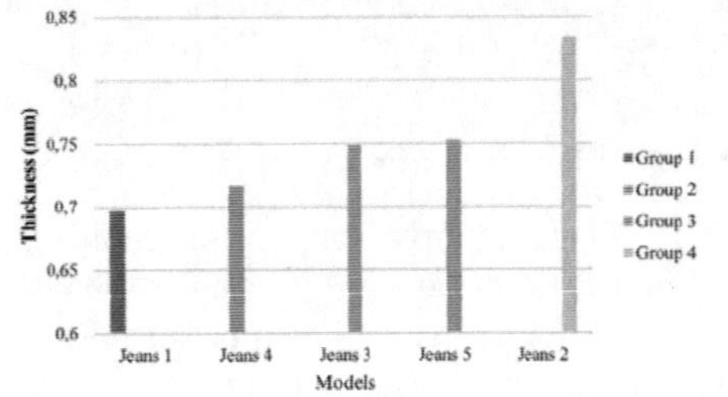

Figure 2. Thickness(mm) homogeneous groups

Evaluating the comparison of the fabric thickness values, the group 1 - jeans 1 0.6970 mm, smaller value; group 2 - jeans 4, 0.7170 mm; group 3 - jeans 3 and 5 are also evaluated as homogeneous with significance = 0.973; and group 4 - jeans 2 .0,8340 mm, with bigger value.

The comparison of the means data of the tests thermophysiological properties analysis as air permeability, relative water vapour permeability (RWVP) and absolute water vapour permeability (AWVP) are evaluated in homogeneous subsets, with harmonic mean = 10,000.

The analysis of the means measurements air permeability are shown in table 4, with the data shown in figure 3.

Table 4 - Air Pemeability (l/m^2/s)

Tukey HSD[a]	Air Pemeability (l/m^2/s)	
	Subset for alpha = 0.05	
Models	1	2
Jeans 5	37,19	
Jeans 1	38,23	
Jeans 2		48,69
Jeans 4		49,62
Jeans 3		49,89
Sig.	,897	,838

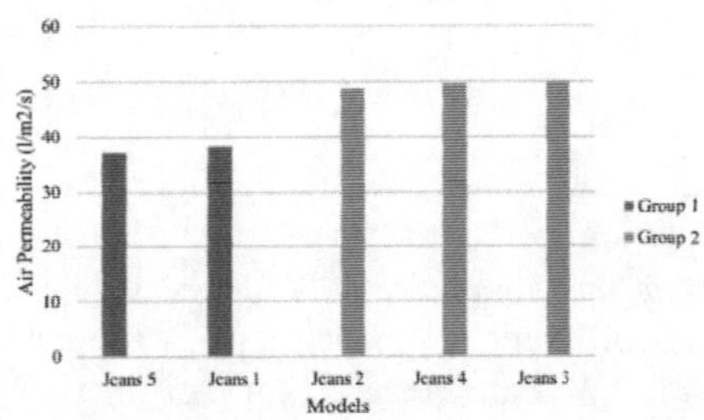

Figure 3. Air Pemeability (l/m^2/s) homogeneous groups

Comparing the air permeability values, we observed homogeneity of the values divided into two groups: group 1, with lower values - jeans 5 (37,19 l/m^2/s) and jeans 1 (38,23 l/m^2/s) and group 2: jeans 2 (48.69 l/m^2/s), jeans 4 (49.62 l/m^2/s) and jeans 3 (49.89 l/m^2/s), the latter showing higher air permeability value.

The analysis of the relative water vapor permeability values are shown in table 5, with the data shown in figure 4.

IOP Conf. Series: Materials Science and Engineering **459** (2019) 012034 doi:10.1088/1757-899X/459/1/012034

Table 5. Relative Water Vapour Permeability (RWVP) (%)

Tukey HSD[a]	**Relative Water Vapour Permeability (RWVP) (%)** Subset for alpha= 0.05			
Models	1	2	3	4
Jeans 2	56,28			
Jeans 1		60		
Jeans 5		61,66	61,66	
Jeans 3			63,76	
Jeans 4				68,98
Sig.	1,000	,434	,209	1,000

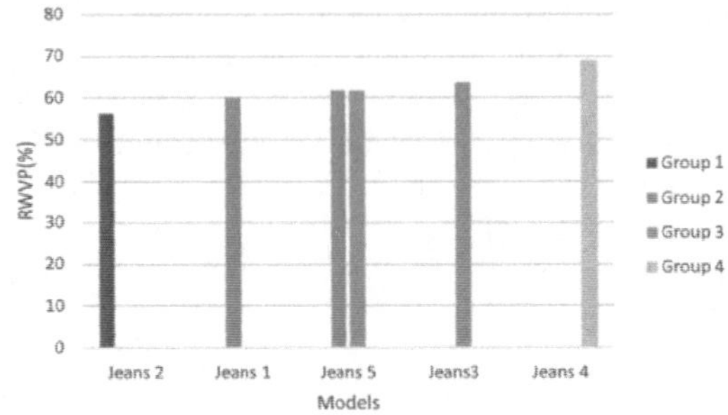

Figure 4. Relative Water Vapour Permeability (RWVP) (%) homogeneous groups

Evaluating the comparison of the relative water vapor permeability values, homogeneity was observed in 4 different groups. Group 1 - jeans 2 (56.28%) with lower value; group 2 - jeans 1 (60%) and jeans 5 (61.66%), jeans 5 also belonging to group 3 due to being statistically very close to jeans 3 (63.76%) and group 4 - jeans 4 (68.98%) with higher relative water vapor permeability value.

The analysis of the relative water vapor permeability values are shown in table 6, with the data shown in figure 4.

Table 6. Absolute Water Vapour Permeability (RWVP) ($Pa/m^2/W^{-1}$)

Tukey HSD[a]	**Absolute Water Vapour Permeability (RWVP)** ($Pa/m^2/W^{-1}$) Subset for alpha= 0.05		
Models	1	2	3
Jeans 4	3,3		
Jeans 3		4,0	
Jeans 5		4,4	
Jeans 1			5,5
Jeans 2			5,5
Sig.	1,000	,271	1,000

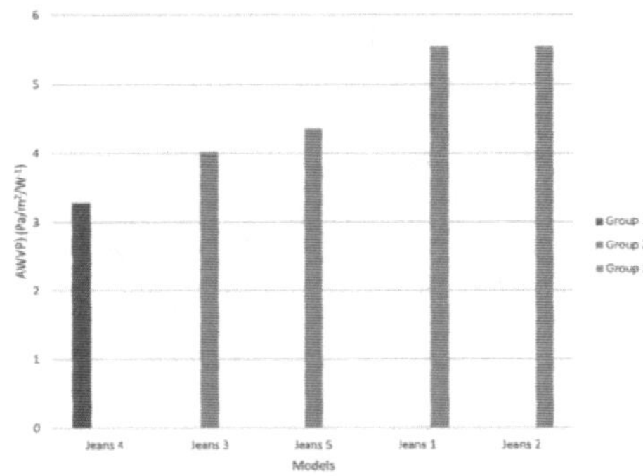

Figure 5. Absolute Water Vapour Permeability (RWVP) ($Pa/m^2/W^{-1}$) homogeneous groups

Evaluating the comparison of the absolute water vapour permeability values, homogeneity is observed in 3 different groups. Group 1 - jeans 4 (3,3 Pa / m^2 / W^{-1}) with lower value; group 2 - jeans 3 (4,0 Pa / m^2 / W^{-1}) and jeans 5 (4,4 Pa / m^2 / W^{-1}) the group 3 - jeans 1 and jeans 2 equal values (5,5 Pa / m^2 / W^{-1})) with bigger value of absolute water vapor permeability.

The analysis of the comparison of the heat loss values referring to the different parts of the body are expressed in figure 4.

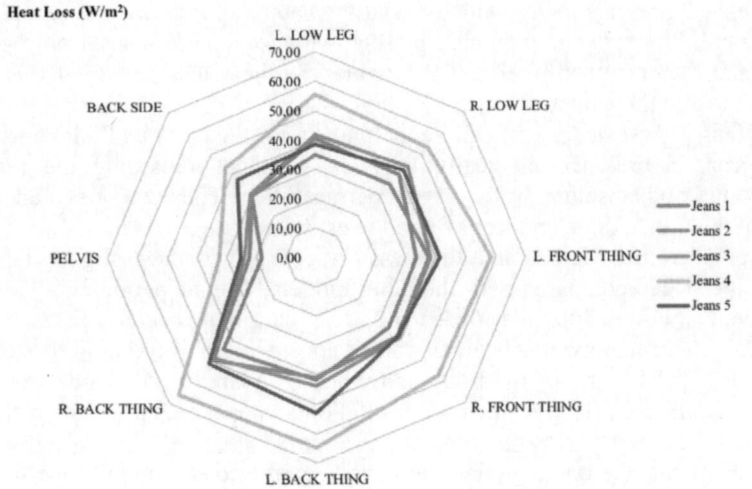

Figure 6. Heat loss from body section of manikin dressed with five jeans models

Evaluating the comparison of the heat loss values, we can observe values homogeneity, with lower values of heat loss - jeans 1, jeans 2 and jeans 3, being the order from the lowest to the highest, then jeans 5 and finally the jeans 4 standing out with greater value of loss of heat in all parts of the body.

The analysis of the comparison of the total *(It)* and effective *(Icle)* thermal insulation values of each jeans model are expressed in figure 5.

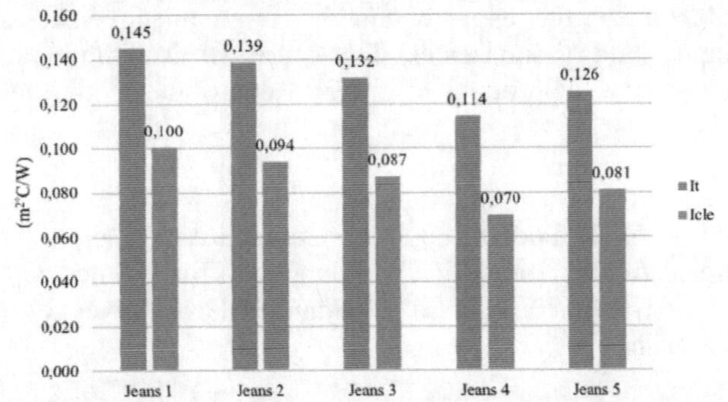

Figure 7. Total thermal insulation and effective clothing insulation of five jeans models

In the comparison of the total and effective insulation values, it is observed that jeans 1 *It* (0,145 m^2.°C/W) and *Icle* (0,100 m^2.°C/W), jeans 2 *It* (0,139 m^2.°C/W) and *Icle* (0,094 m^2.°C/W), and jeans 3 *It* (0,132 m^2.°C/W) and *Icle* (0.087 m^2.°C/W) this tree models presented the highest values for total thermal insulation *It*, and also effective thermal insulation *Icle*. Following in descending order, jeans 5 with *It* (0,126) and *Icle* (0,081 m^2.°C/W) and jeans 4 *It* (0,114 m^2.°C/W) and *Icle* (0,070 m^2.°C/W) are presented, showing the lowest total and effective thermal insulation values.

4. Conclusions

Through the evaluations of the parameters described above it is possible to highlight the following conclusions:

• The jeans 1, although presenting lower values of mass and thickness and low capacity of air permeability. The jeans 1 presents the second lowest value of relative water vapor permeability and the highest absolute permeability value to water vapor and higher values of total thermal insulation *It* and effective *Icle*.

• Jeans 2 presented higher values of mass and thickness, the air permeability presented higher value than jeans 1. Jeans 2 presents lower value of relative water vapor permeability and as for the value of absolute permeability water vapour jeans 2 presents values equal to jeans 1 with greater permeability to absolute water vapor. When analysed the values for heat loss and total thermal insulation and presented data close to the values of jeans 1.

• Jeans 3 and jeans 5 presented statistically equal values for the mass and thickness of the fabrics. The jeans 3 and jeans 5 present statistically very close values regarding the relative water vapor permeability value and absolute water vapor permeability. The heat loss and total and effective thermal insulation. higher value and jeans 5 the lowest value.

• The jeans 4 for the values of mass and thickness presented low values, slightly larger than the jeans 1 and smaller values when compared with the other models. The air permeability values, presented the second lower value, with data similar to jeans 2. The jeans 4 presented highest relative water vapor permeability value and the lowest absolute water vapor permeability value. It stood out with higher values of heat loss in all parts of the body and when compared to all models of jeans. It presented lower value of total and effective thermal insulation, when comparing the values with the other models.

The results show that the raw material, the structures, the construction of the fabrics, as well as the type of industrial washing influence the thermophysiological properties of the different tested models of jeans. It was possible to conclude that the objective evaluation of the comfort of popular clothing and specifically of jeans designed, produced and sold in Brazilian popular markets is unprecedented and that there is much to research, study and to learn from the popular fashion universe of the Brazilian Northeast.

Future perspectives

In further studies we will compare the results obtained through objective evaluation with data obtained through evaluation using a inquiry and scales. This study is also important for the producers and traders of Fortaleza of this type of product in order to contribute to the development of this huge market.

Acknowledgments

This work is financed by FEDER funds through the Competitivity Factors Operational Programme - COMPETE and by national funds through FCT – Foundation for Science and Technology within the scope of the project POCI-01-0145-FEDER-007136" and Federal University of Piauí, Curse de Bach. in Fashion, Design and Stylish.

References

[1] Braga I, Abreu MJ. Social aspects in denim production. 2017.

[2] Braga I, Abreu MJ. The jeans in the popular Brazilian panorama. In: TIWC 2016. Poznan: 2016. page 534–40.

[3] ISO. ISO 11092:2014 - Textiles – Physiological effects – Measurement of thermal and water-vapour resistance under steady-state conditions (sweating guarded-hotplate test). 2014.

[4] Bogusławska-Bączek M, Hes L. Thermophysiological Properties of Dry and Wet Functional Sportswear Made of Synthetic Fibres. [Internet]. In: Tekstilec, 2017. 2017 [cited 2018 Jul 5]. page 331–8.Available from: http://www.tekstilec.si/wp-content/uploads/2017/12/331-338.pdf

[5] Mangat A, Bajzik V, Lubos H, Buyuk F. Impact of surface profile of polyester knitted rib structure on its thermal properties. Ind. Textila [Internet] 2016 [cited 2017 Jun 20];67:103–8. Available from: http://media.proquest.com/media/pq/classic/doc/4060570821/fmt/pi/rep/NONE?cit%3Aauth=Mangat%2C+Asif%3BHes%2C+Lubos%3BBajzik%2C+Vladimir%3BBuyuk%2C+Funda&cit%3Atitle=Impact+of+surface+profile+of+polyester+knitted+rib+structure+on+its+...&cit%3Apub=Industri

[6] Abreu I, Ribeiro P, Abreu M. The issue of thermal confort of medical clothing in the operating room. DYNA 2014;81:1–2.

Flexible piezoresistive pressure sensors for smart textiles

H Carvalho[1], D Tama[1,2], P Gomes[1], M J Abreu[1], Yu Yao[3] , A P Souto[1]

[1] Univ. of Minho, Center of Science and Textile Technology, Guimarães, Portugal

[2]EgeUniversity, Department of Textile Engineering, 35100 Bornova, İzmir, Turkey

[3]Univ. of Minho, Dep. Industrial Electronics, Guimarães, Portugal

helder@det.uminho.pt

Abstract. The development of smart textiles relies in many applications on of textile-integrated sensors. Flexible piezoresistive pressure sensors have many potential applications, for instance in sports science. In this study, flexible pressure sensors are built using piezoresistive polymer film and conductive fabric. Tests using a universal testing machine show that the sensors are functional, accurate, although showing some hysteresis. However, methods for joining the electrode and piezoresistive layers are necessary to assure mechanical stability of the sensor, without affecting electrical contact between layers. Several methods were tested and results are reported here. The use of thermofusible bonding nets and webs has been found to be an interesting solution.

1. Introduction and state-of-the-art

This paper presents the development of polymer-based flexible piezoresistive sensors appropriate for integration in e-textiles. The transducing element used is a volume-conductive carbon-loaded polyethylene film with commercial name Linqstat from Caplinq.

The sensor construction comprises the transducing element as well as two electrodes for electrical connection, as shown in Figure 1. The electrode elements should also be flexible. In this work, silver-plated conductive fabric was used. Other materials have been tested in [1]. The sensors present a variation of electrical resistance when subjected to pressure, namely decreasing electrical resistance when pressure increases.

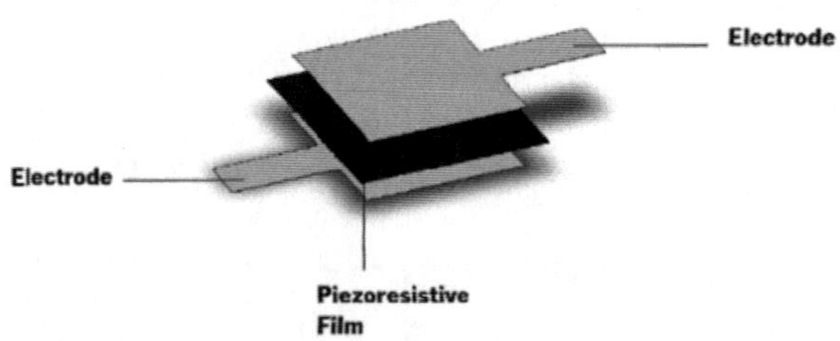

Figure 1. Construction of the flexible pressure sensors

Pressure sensors based on this principle have been presented by other researchers. Mueller *et al* [2] developed a pressure sensor array for use in robotics, based on Velostat, a similar piezoresistive material. Electrodes were set up with stripes of copper-coated fabric, with the upper layer oriented perpendicularly to the lower layer, thus forming a matrix.

In [3], a model for piezoresistive sensors of this type was developed and compared with a sensor fabricated using copper plates as electrodes. The authors found that contact resistance is not very important at high resistance (low force) but becomes significant at low resistance. This means that the sensor construction and electrode preparation may have an important influence on the performance of the sensor.

A construction for integration of the sensors in textiles was presented in [4]. The piezoresistive film was inserted into a pocket in a knitted structure using the intarsia technique. This allowed to insert an area of conductive yarn at each side of the film, thus forming the two electrodes.

The use of different electrode materials was compared in [1], where a conductive woven fabric, a conductive knitted fabric and copper tape were used as electrodes. It was shown that the behaviour of the copper tape is very similar to that of the woven fabric, with some spread and hysteresis. Using a specifically designed circuit for conditioning of the sensor, it was possible to obtain an almost linear relationship between Force and output voltage. Using the knitted fabric, however, the relation turned non-linear, although with less spread and hysteresis.

In the previous experiments, the layers of the sensor were simply overlaid, without any binding between them. This results in instability of the zero point, because contact between the layers varies when pressure is removed. Moreover, to provide mechanical stability of the assembly, which assures robustness for practical use of the sensor, a method of joining the layers should be found. Several solutions have been tried, such as fusing the fabric with Linqstat using a hot press, preparing the electrodes with conductive silicone or with conductive ink, all of them presenting significant drawbacks [5][6]. In this work, thermoplastic bonding nets and webs were used to achieve bonding between layers.

2. Materials and methods

The sensors were constructed according to the following table:

Table 1. Materials and conditions.

	Material	Conditions
Electrodes	Sn/Cu/Ag plated polyamide fabric: Zell by Statex	Cut smaller than piezoresistive layer, see Figure 2
Piezoresistive material	Linqstat (Caplinq)	Dimensions square 3x3cm
Bonding material 1	Thermoplastic web based on polyolefin, manufacturer 1	Press, 5 bar, 10 sec, 110 ºC Oven, 10 minutes, 110 ºC Glass plate on sensor assembly
Bonding material 2	Thermoplastic net based on polyolefin, manufacturer 2	Press, 5 bar, 10 sec, 85 ºC Press, 5 bar, 10 sec, 110 ºC Oven, 10 minutes, 110 ºC Glass plate on sensor assembly

Figure 2 shows the sensor layers overlaid before the bonding operation:

Figure 2. Overlay of the sensor components: Electrode fabric, piezoresistive layer, bonding net (left). Setup for compression test (right)

Figure 3 shows photographs of the two bonding materials used

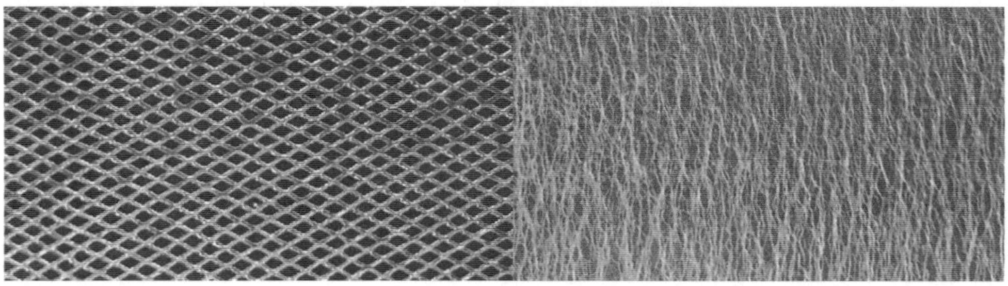

Figure 3. Bonding materials: Thermoplastic net (left) and thermoplastic web (right)

To provide mechanical protection and amplitude for the compression test, a layer of 3mm EPDM (ethylene propylene diene monomer rubber) foam was placed on each side of the sensor. The sensors were then connected to the sensor signal conditioning equipment and subjected to cyclic compression in a Houndsfield universal testing machine, at 5 and 50 mm/sec, 0 to 100 N. During the test, the output voltage was acquired by a data acquisition board and stored. Finally, the data of force and voltage during the tests are brought together in a spreadsheet program for presentation. Two samples for each sensor were tested.

3. Results and discussion

3.1. Bonding Material 1 – Thermoplastic web.

The bonding material 1 is specified with a melting temperature of 105-110 ºC. The layers were gathered in the press during 10 s with a pressure of 5 bar and a temperature of 110 ºC. Given that Linqstat has a melting temperature of 120 ºC, the temperature was selected to avoid its fusion. Adhesion between layers was good, but the sensor was not functional, i.e. its resistance did not vary with applied pressure (sample 1), or the behavior was very unstable (sample2).

The same assembly was tried with the oven at 110 ºC, during 10 min, with a minimum of pressure conveyed by a glass plate that was overlaid. In this case, the results were good, with the sensor responding evenly to the applied pressure at both speeds of 5 and 50 mm/min, as can be seen in figures 4 and 5.

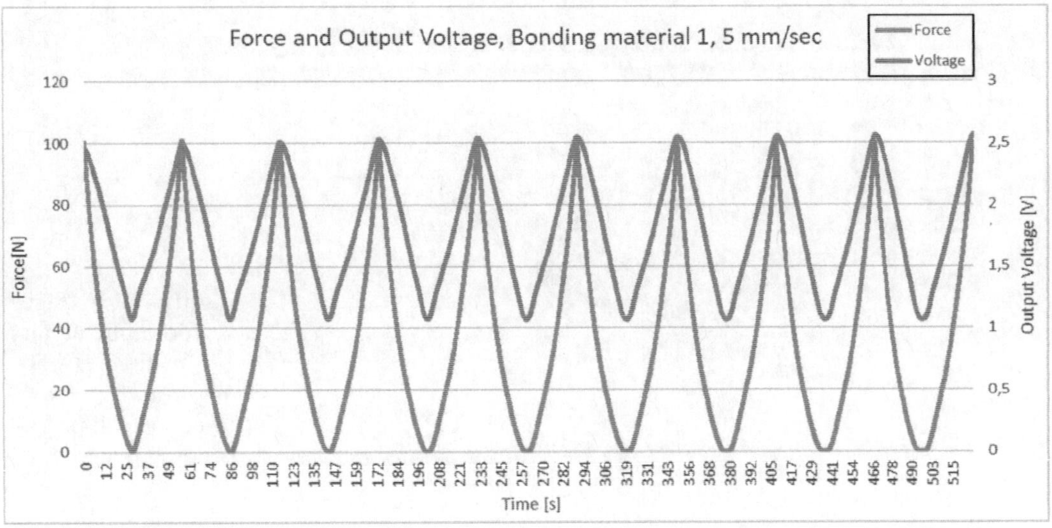

Figure 4. Force and voltage output for 10 cycles at 5 mm/min, max. force of 100 N

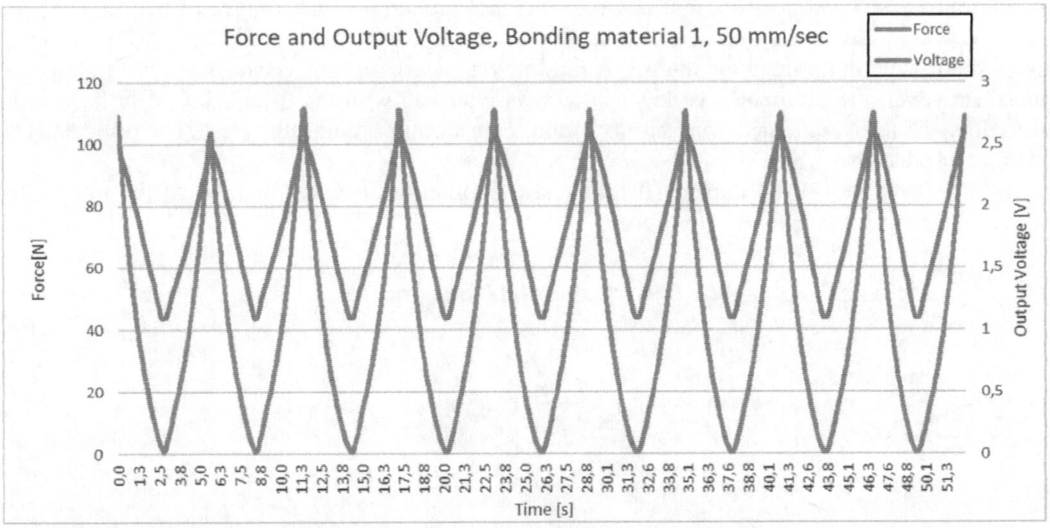

Figure 5. Force and voltage output for 10 cycles at 50 mm/min, max. force of 100 N

Although the result seems regular, it is still possible to observe, in both cases, that the voltage peak as well as the valley are rising at every cycle; also, the shape of the voltage output is not exactly the same at the rising as at the falling segment of the signal, meaning that there is some hysteresis present. This can be quickly confirmed with the voltage versus force graph depicted in Figure 6:

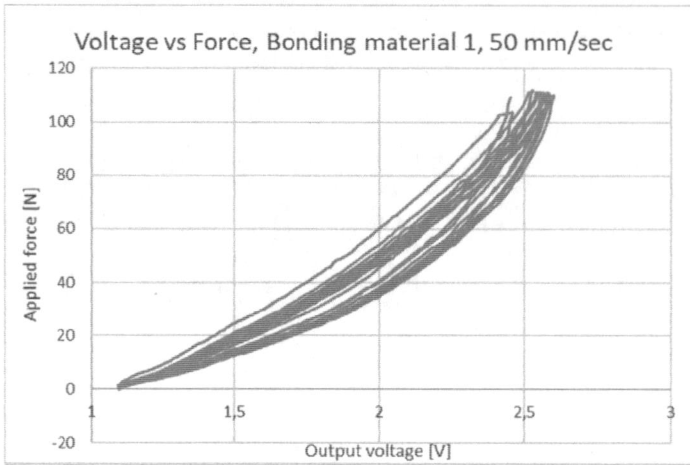

Figure 6. Voltage vs force output for 10 cycles at 50 mm/min, max. force of 100 N, bonding material 1

3.2. Bonding Material 2 – Thermoplastic web

The bonding material 2 has a melting point of 78-88 ºC and was applied to the sensor layers using the press and the oven methods. With the press method, temperatures of 85 ºC and 110 ºC were tried. The temperature of 85 ºC was used since it is within the melting temperature ranges of the material, while the 110 ºC were used to test the adhesion behaviour and to obtain comparative values with the other samples made with the same temperature. Adhesion was good, and the sensor reacted to the pressure applied. However, the piezoresistive layer and/or its interface with the electrodes seem to have been modified by the pressure/temperature combination, considering the unstable response, especially with the sample produced at 110 ºC.

Using the oven at 110 ºC during 10 minutes, a similar result was obtained as the one with the bonding material 1, as can be seen in Figure 7:

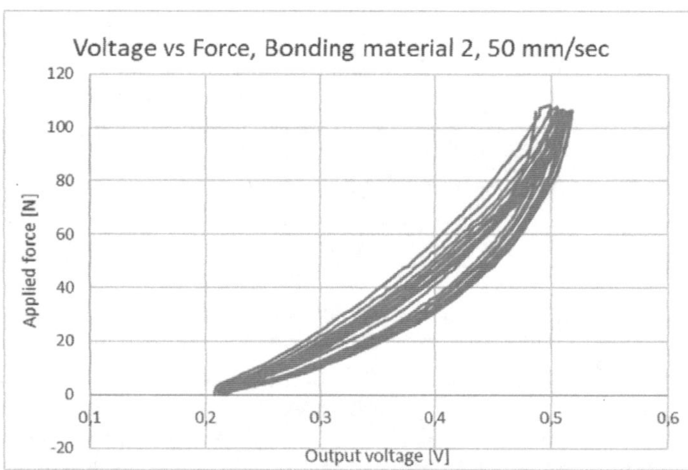

Figure 7. Voltage vs force output for 10 cycles at 50 mm/min, max. force of 100 N, bonding material 2

Further experimentation has to be carried out to determine if the hysteretic behavior and increase of the response is due to the properties of the sensor itself, or if it is related to the viscoelastic properties of the EPDM foam used to encapsulate it. Moreover, it should be determined if this behavior stabilizes with use or if it floats over time.

3.3. Comparison with unbonded sensor
The results obtained using the bonding materials are qualitatively very similar to the ones obtained for sensors with simply overlaid layers [1]. The same type of hysteretic behaviour and increase of maximum output voltage over time has been observed. A major difference is the resistance values measured. As expected, with the bonding layer partly isolating the electrical contact between the piezoresistive and the electrode layers, the resistance values are higher for the bonded sensors. Using bonding material 1, resistance values at 100 N were 125 and 101 Ω for sample 1 and 2, respectively. With bonding material 2, resistance values at 100N were 206 and 142 Ω. Sensors without bonding material presented values of about 50 Ω.

4. Conclusions and future work
In this work, a method for joining the layers of a flexible, polymer-based piezoresistive sensors has been found. Previous efforts had not been successful [5][6]. This is an important achievement for the practical use of such a sensor, allowing it to be placed in any position, avoiding the displacement of the layers and assuring a stable zero point.

Adhesion between the layers seems to be sufficient in all cases, but has to be further evaluated and quantified. Plasma treatment of the piezoresistive layer can help improve adhesion and will be used in future experiments. Future work will also focus on the observed effect of decreasing electrical resistance when cyclic loading is applied.

Acknowledgements
This work is financed by Project "Deus ex Machina", NORTE-01-0145-FEDER-000026, funded by CCDRN, through Sistema de Apoio à Investigação Cientifica e Tecnológica (Projetos Estruturados I&D&I) of Programa Operacional Regional do Norte, from Portugal 2020 and by FEDER funds through the Competitivity Factors Operational Programme - COMPETE and by national funds through FCT – Foundation for Science and Technology within the scope of the project POCI-01-0145-FEDER-007136. Derya Tama thanks FCT for fellowship 2C2T-BPD-08-2017.

References
[1] Carvalho H, Yao Y and Gonçalves M 2017, Flexible force sensors for e-textiles, *17th World Textile Conference AUTEX 2017- Textiles - Shaping the Future, Conf. Series: Materials Science and Engineering* **254** (2017)
[2] Mueller S, Schroeter C, Gross HM 2015 Smart Fur Tactile Sensor for a Socially Assistive Mobile Robot, Intelligent Robotics and Applications, Volume **9245** of the series *Lecture Notes in Computer Science* pp 49-60, Springer
[3] Kalantari M, Dargahi J, Kovecses J, Mardasi M G, and Nouri S 2012 A New Approach for Modeling Piezoresistive Force Sensors Based on Semiconductive Polymer Composites, IEEE/ASME Transactions on Mechatronics, Vol. **17**, No. 3
[4] Lin X, Seet B C 2015A Linear Wide-Range Textile Pressure Sensor Integrally Embedded in Regular Fabric, IEEE Sensors Journal, Vol. **15**, No. 10
[5] Tama, D, Gomes, P, Yao, Y, Abreu, M.J, Souto, A P, Carvalho, H 2017 Developing Smart Body Protector For Sports Karate, *Proc. of the XIVth International Izmir Textile and Apparel Symp.*, Izmir, Turkey
[6] Gomes, P, Tama, D, Yao, Y Abreu, M J, Souto, A P, Carvalho, H 2018 Development of pressure sensors for smart textiles, *Proc. of the AUTEX 2018 World Textile Conference*, Istambul, Turkey

IOP Conf. Series: Materials Science and Engineering **459** (2019) 012036 doi:10.1088/1757-899X/459/1/012036

Assessment of acrylic based textile wastes as energy source

A Çay[1], J Yanık[2] and A Hanoğlu[2]

[1] Ege University, Department of Textile Engineering, İzmir, Turkey
[2] Ege University, Faculty of Science, Department of Chemistry, İzmir, Turkey

E-mail: ahmet.cay@ege.edu.tr

Abstract. The properties of biochars obtained by the torrefaction of acrylic based textile wastes were investigated. Waste fibres were torrefied at 300-400 °C and the results indicated that acrylic based textile wastes could be converted into biochar with a high mass yield. Resultant biochars had similar properties to bituminous coal.

1. Introduction

Textile fibre consumption continuously increases in parallel with the increase in population, standards of living and fast fashion effects [1-3]. Therefore, millions of tonnes of textile waste are generated annually. Since the re-using and recycling ratio of textile wastes is very low, large amounts of wastes are disposed by landfilling. Moreover, when recycled textile materials are considered, after completion of their primary post-recycling lifetime, a secondary recycling is not generally possible due to the substantial fibre damage. For this reason, recycled textile materials also become waste for disposal. Therefore, rather than disposal, the use of textile wastes for energy conversion will be advantageous in terms of natural resource consumption.

Emissions of greenhouse gases resulting from the use of fossil fuels are causing environmental problems such as global warming, acid rain and air pollution. For a sustainable production, increasing the share of the use of renewable energy sources is of great importance. An important class of renewable energy sources is biomass. Hence, textile wastes have the potential to be utilized as biomass and can be used as energy source. On the other hand, due to being heterogeneous, bulky and porous, energy density of textile wastes are low and it is difficult and problematic to process textile wastes at existing power generation facilities. Therefore, they should be converted into homogeneous, dense and carbon-rich structure to be used as energy source, called biochar. Biochar is a renewable feedstock with reduced CO_2 emissions and is used either for energy output or as carbon source [4]. This study investigates the fuel characteristics of biochars obtained from acrylic based textile wastes through low temperature pyrolysis (torrefaction).

2. Experimental

Textile wastes composed of 100% acrylic fibres (PAC) and their blends with polyester (PAC/PES, 50/50%) and viscose (PAC/CV, 50/50%) were collected. The torrefaction experiments were carried out in a laboratory scale fixed bed reactor in nitrogen stream. Torrefaction temperatures were selected to be 300, 350 and 400 °C. Textile wastes and resultant biochars were characterised through biochar mass yield, elemental analysis, heating value and energy efficiency determination.

Biochar mass yield (MY, %) was calculated as,

$$M \;\; = (M_b/M_t) \times 100 \tag{1}$$

where Mb is the mass of biochar and Mt is the mass of textile waste.

Elemental compositions of textile wastes and resultant biochars were tested by using an elemental analyser (Leco CHNS 932). Using elemental analysis results, higher heating value (HHV, MJ/kg) of textile wastes and resultant biochars were calculated as:

$$HHV = 0{,}3491C + 1{,}1783H + 0{,}1005S - 0{,}10340O - 0{,}0151N - 0{,}0221A \tag{2}$$

where C, H, S, O, N and A denotes the mass ratio of carbon, hydrogen, sulphur, oxygen, nitrogen and ash, respectively [5].

Energy density (ED) and energy yield (EY, %) of resultant biochars were determined by using Equations (3) and (4), respectively,

$$E \;\; = H \;\; _b/H \;\; _t \tag{4}$$

$$E \;\; = M \;\; \times E \tag{5}$$

where HHV_b is the higher heating value of biochar and HHV_t is the higher heating value of textile waste.

3. Results and Discussion

Elementel analysis, HHV, mass yield, ED and EY values of torrefied acrylic based textile wastes were given in Table 1. Mass yields of acrylic based biochars were obtained in the range of 50-80% depending on the temperature and fibre type. The increase in torrefaction temperature led to a decrease in mass yield. Moreover, the presence of polyester or viscose fibres within acrylic blends decreased the torrefaction mass yield. The HHV (higher heating value) of the resultant biochars were found to be quite high (between 24-30 MJ/kg depending on the blend) and it was found that torrefaction temperature did not have a considerable effect on higher heating value. As shown in Figure 1, O:C and H:C atomic ratios showed that the resultant biochars had similar properties to bituminous coal. On the other hand, acrylic based biochars were high in nitrogen content (12-24%) which will lead NOx emissions during combustion. Therefore, the use of wet scrubbers is recommended to remove NOx emissions from flue gas. On the other hand, nitrogen-rich biochars might be advantageous for the use as soil amendment applications.

Table 1. Properties of acrylic based textile wastes and resultant biochars

	Elemental analysis					HHV, MJ/kg	MY,%	ED	EY,%
	C	H	N	S	O*				
PAC	65.13	6.21	23.83	0	4.83	29.19	-	-	-
PAC-C300	67.84	4.86	21.88	0	5.42	28.52	79.80	0.98	77.95
PAC-C350	70.67	4.38	20.98	0	3.97	29.10	63.46	1.00	63.26
PAC-C400	70.45	3.87	19.83	0	5.85	28.82	57.89	0.97	56.02
PAC/PES	63.02	5.10	12.19	0	19.69	25.79	-	-	-
PAC/PES-C300	69.29	4.23	13.91	0	12.57	27.66	60.02	1.07	64.38
PAC/CV	55.77	6.02	14.40	0	23.81	23.88	-	-	-
PAC/CV-C300	66.40	4.63	14.95	0	14.02	26.96	49.50	1.13	55.88

*calculated by difference

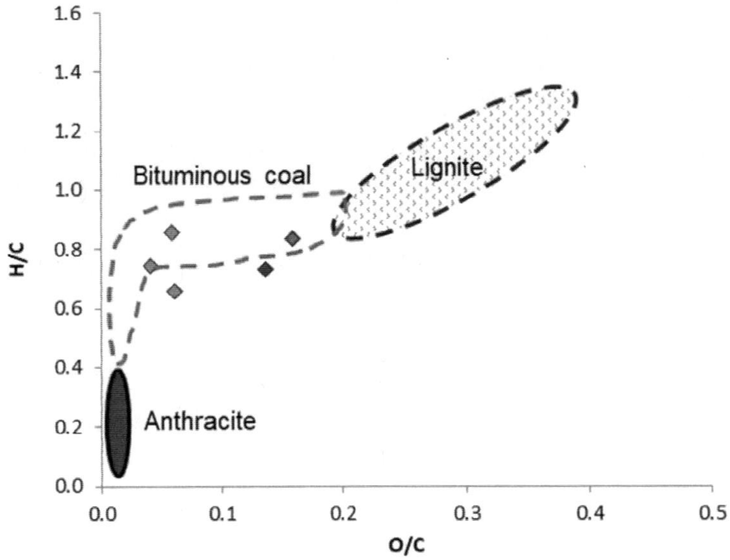

Figure 1. van Krevelen diagram of acrylic based biochars

4. Conclusion

Acrylic based textile wastes were converted into biochars by torrefaction with high mass yield. Resultant biochars had high HHV values and had similar properties to bituminous coal. Due to their high nitrogen content, a flue gas scrubber is recommended for combustion applications to remove NOx emissions.

Acknowledgment

The authors would like to gratefully acknowledge the financial support for this research received through the project (No: 216M406) of The Scientific and Technical Research Council of Turkey (TUBITAK); and the projects (No: 16-TKUAM-002 and 17-FEN-063) of Ege University Scientific Research Projects Coordination Unit.

References

[1] Fletcher K 2008 *Sustainable Fashion and Textiles* (London: Design Journeys, Earth Scan)
[2] Wang Y 2006 *Introduction* (Recycling in Textiles) ed Y Wang (Cambridge: Woodhead Publishing Ltd.)
[3] Clancy G, Fröling M, Peters G 2015 *J. Clean. Prod.* **99** 345-353
[4] Tag AT, Duman G, Ucar S, Yanik J 2016 *J. Anal. Appl. Pyrol.* **120** 200-206
[5] Channiwala SA, Parikh PP 2002 *Fuel* **30** 7746-7754

Gait analysis by using Smart Socks system

P Eizentals[1], **A Katashev**[1] **and A Oks**[2]

[1] Institute of Biomedical Engineering and Nanotechnologies, Riga Technical University, 1 Kalku Street, Riga, LV-1658, Latvia
[2] Institute of Design and Technology, Riga Technical University, 1 Kalku Street, Riga, LV-1658, Latvia

peteris.eizentals@gmail.com

Abstract. Feet plantar pressure measurement and gait pattern analysis can be utilized for different disease detection as well as post-operation rehabilitation monitoring. Increasingly popular for gait analysis in research have become smart sock systems as an alternative to the more expensive commercial analogs, however, analysis with such systems are complicated due to the limited sensor count and other inherit limitations of textile pressure sensors. This paper proposes a new method for gait pattern recognition for use with smart sock systems. According to this method, steps are classified depending on the sum of Manhattan distances (SMD) between two high-dimension vectors, from which one describes the step to analyze and the other is a predefined vector representing a specific gait pattern. The gait pattern with the lowest SMD value to the particular step vector is assigned to each step. An experiment was performed to verify the proposed method, where three volunteers simulated three gait types, normal, pronation and supination, and the simulated steps were compared to references. The experimental results confirmed that typically the lowest SMD value was obtained by the reference corresponding to the tested gait pattern type. In one of the demonstrated cases, the mean SMD value for the same set of steps with the reference for normal step was 1.26, 1.77, and 1.78, with reference for pronation it was 1.75, 1.57, and 2.59, and reference for supination, 1.75, 2.15 and 1.46, for normal gait, gait with pronation and gait with supination, respectively. It can be seen that the lowest mean SMD value for each gait type is when the correct reference is applied.

1. Introduction

The human gate monitoring by means of foot plantar pressure measurements could have numerous applications, such as post-operation rehabilitation and recovery assessment, abnormal gait condition detection, foot surface pressure monitoring for diabetic patients, and many more [1, 2]. Pressure plates are the typical plantar pressure monitoring systems used in clinics and are considered the golden standard for such measurements. They, however, cannot be used for continuous patient monitoring due to the limited surface area. Pressure-sensitive insoles are, generally, more suitable for out-of-lab measurements, but such devices are sensitive to folds and thus could not be used with orthopedic insoles or shoes. On top of that, both of these systems are rather expensive [3], which limits their usage outside of clinics. In recent years, smart textile based plantar pressure measurement systems have been studied worldwide as a possible alternative [4]. Such smart textile socks are rather inexpensive, they can be made to fit any foot size, can be used for a prolonged period, are washable and do not require a sophisticated data acquisition hardware [5]. Despite the above advantages, smart socks still are not widely used for gait analysis, as there is no well-established framework for employing the information from a limited number of sensors for detection of differences in gait pattern.

This research proposes a new method for gait pattern evaluation and classification for measurements using a smart sock system. For each step, a high dimension vector is constructed, which is then compared to reference vectors by using sum of Manhattan distances (SMD). The reference vectors were obtained from a set of recordings corresponding to particular step patterns. Such an approach allows evaluating the gait similarity to a certain abnormal gait pattern as well as deviation from a normal gait. The proposed method for step analysis was verified by performing test-walks by three volunteers, wearing custom-designed smart textile socks while walking normally or simulating one of the two analyzed abnormal gait patterns, excessive pronation or excessive supination. The preliminary results demonstrated that the proposed method could be used for gait pattern analysis. A significantly lower SMD value was observed when using the proper reference thus allowing gait type classification.

2. Materials and Methods

A novel gait analysis method for smart sock measurement system is proposed in this paper. According to this method, foot plantar pressure measurement is obtained by textile pressure sensors and then analyzed to evaluate the gait type. The evaluation is performed by extracting a set of step parameters from the measurement and combining them in a single multi-dimensional vector, which is used for step characterization. This vector then is compared to one of the pre-determined vectors by using Manhattan distance, which allows evaluating the difference between two high dimension vectors [6]. Finally, the total sum of Manhattan distances over all vector dimensions is used to determine the difference between a particular step and the reference. Using such approach, any step can be quickly classified by calculating to which of the pre-determined step types it belongs as long as sufficient reference data exists.

2.1. Smart Sock system

The method described in this paper was developed specially for application with a custom-made smart sock system that consists of a five textile pressure sensors incorporated directly in each sock (see figure 1). Smart socks used in this research were knitted from combed cotton yarn platted with elastane/polyamide yarn. Five pressure sensors were directly knitted in each sock, and silver-coated thread Shieldex 117/17 (Statex Produktions und Vertriebs GmbH) was used for electro-conductive lines which were also knitted in the socks during the manufacturing process. The sensors were distributed on the sock sole part in a configuration 2-1-2 with two sensors in the front, one on the lateral side in the middle, and two under the heel (see figure 2). Measurements of sensors output signals were performed with a dedicated circuit attached to electro-conductive lines by snap fasteners. The sampling rate of measurement was 33 SPS. The measurement was transferred to PC through Bluetooth connection, where it was processed as described below.

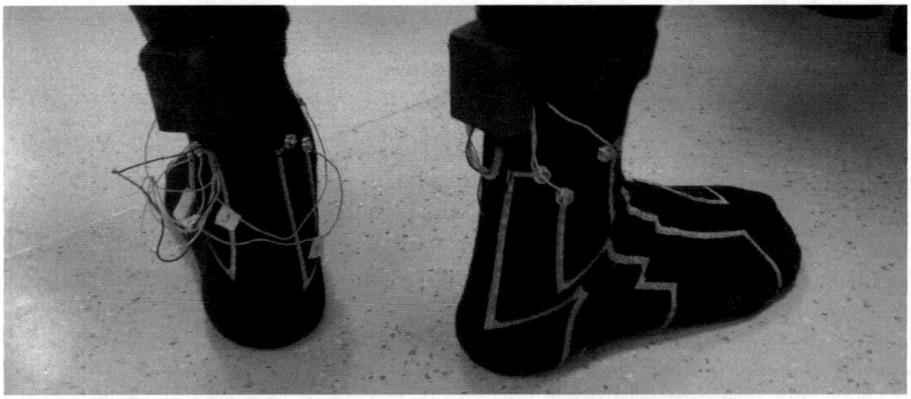

Figure 1. Smart Sock system used in the research.

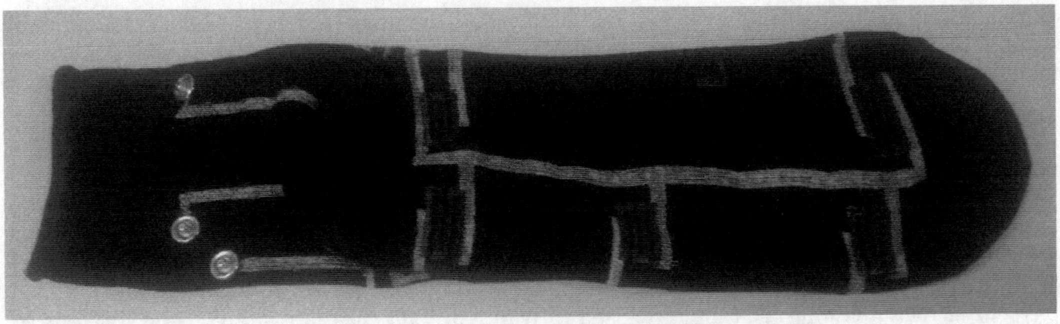

Figure 2. Sensor positions on the left sock.

2.2. Data Processing

The data processing consists of three major parts. First, the signal is filtered and individual steps are recognized. Afterward, step parameters are calculated and a high-dimension vector is constructed for each of the steps. Finally, the constructed vectors are compared to a reference to classify the step.

The individual step recognition is performed by combining together signals of all sensors, which are first inverted and normalized, according to the equations (1):

$$u_i' = \frac{u_i^{-1} - \min\limits_{a \le i \le b} u_i^{-1}}{\max\limits_{a \le i \le b} u_i^{-1} - \min\limits_{a \le i \le b} u_i^{-1}} \tag{1}$$

where u_i is the corresponding measurement and $\{a;b\}$ is a window around i-th measurement to which to normalize the signal. Such local normalization was chosen to solve the issue of creep, which is typical for textile sensors, while at the same time maintaining the local amplitude variations within a few steps. Separate sensor measurements and the combined signal can be seen in figure 3.

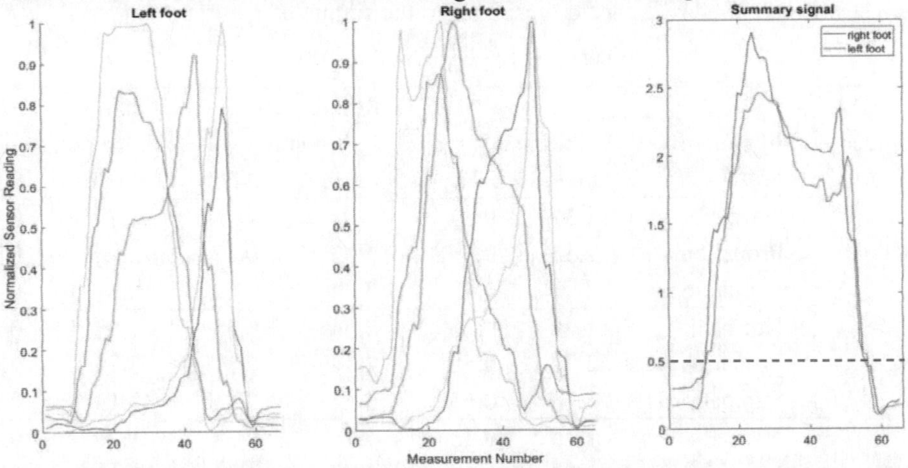

Figure 3. Separate sensor signals and the summary signal for a single step, step detection threshold shown with dashed line.

The beginning and the end of a step are determined by applying a level threshold of 0.5 to the total signal, where the first measurement above the threshold is considered the beginning of a step, and the first measurement under the threshold and after the start of a step is considered the end of the step. Afterward, detected steps are verified by a minimum length requirement to exclude false detections, where the signal could go above the threshold briefly due to a noise or misstep.

The high-dimension vector consists of 22 different parameters that describe both time and amplitude related characteristics of a step. These parameters are not discussed in this paper in details due to space limitation.

The final step of the algorithm is to use the SMD value to evaluate how close a step is to predefined vector values. The SMD value is calculated according to equation (2):

$$SMD = \sum \frac{|V_{left} - V'| + |V_{right} - V'|}{2} \tag{2}$$

where V_i is the high-dimension vector for the step to evaluate, and V' is the predefined reference vector for comparison. The step is considered to belong to the gait pattern with the lowest SMD value.

2.3. Experimental Setup

Following experiment was performed to verify the applicability of the developed method to gait pattern classification. Three volunteers were requested to walk several times in a straight line for approx. 10 meters in one of the following gait types: normal gait, gait with excessive pronation and gait with excessive supination. Gait with excessive pronation signifies increased pressure on the medial part of the foot, while excessive supination is the opposite with increased pressure on the lateral part of the foot. Measurements were performed with the smart sock system and data was analyzed as described above, and all measurements were compared to the predefined reference vectors to verify variations in SMD value depending on the reference gait type. The mean measurements of one walk per gait type were utilized as references for step type classification.

3. Results and Discussion

An experiment was performed to evaluate the proposed step type classification algorithm for use with smart socks. Three volunteers were requested to walk while simulating one of the gait types, normal gait, gait with excessive supination and gait with excessive pronation. The obtained measurement was processed as described above. In total 616 steps were recognized and used for this study, from those 304 were normal gait steps, 152 steps with pronation and 160 steps with supination. As it can be seen from the summarized data in table 1, the mean SMD between the test data and the reference in most (7/9) cases is lowest when the used reference corresponds to the simulated gait type.

Table 1. Experimental results

		Reference type					
Volunteer	Gait pattern	Normal		Pronation		Supination	
		Value	STD	Value	STD	Value	STD
V1	Normal	**1.39**	0.34	1.97	0.5	1.77	0.54
	Pronation	2.33	0.59	**1.83**	0.67	2.96	0.68
	Supination	2.73	0.49	3.34	0.62	**2.4**	0.48
V2	Normal	**1.46**	0.26	2.34	0.35	1.83	0.33
	Pronation	2.92	0.55	**2.14**	0.51	3.75	0.69
	Supination	2.18	0.61	3.12	0.61	**2.19**	0.55
V3	Normal	**1.26**	0.33	1.75	0.38	1.79	0.39
	Pronation	1.77	0.27	**1.57**	0.35	2.15	0.36
	Supination	1.78	0.38	2.59	0.51	**1.46**	0.3

Figure 4 demonstrates an example of the effect of different references on SMD. When a normal gait reference is used, most of the normal steps are with a significantly lower SMD value than other gait types (see figure 4a). By changing the reference to pronation gait (figure 4b), the mean SMD for normal gait increased from 1.26 to 1.75 while the mean SMD for pronation decreased from 1.77 to 1.57. The same trend for gait with supination can be seen in figure 4c. Moreover, as supination and pronation are somewhat opposites, so each of them can be identified by using the opposite reference and setting a high threshold instead of low. This can be seen in the example given in figures 4b and 4c, where the mean

SMD for supination increases from 2.19 with supination reference to 3.12 with pronation reference, and for pronation from 2.14 to 3.75.

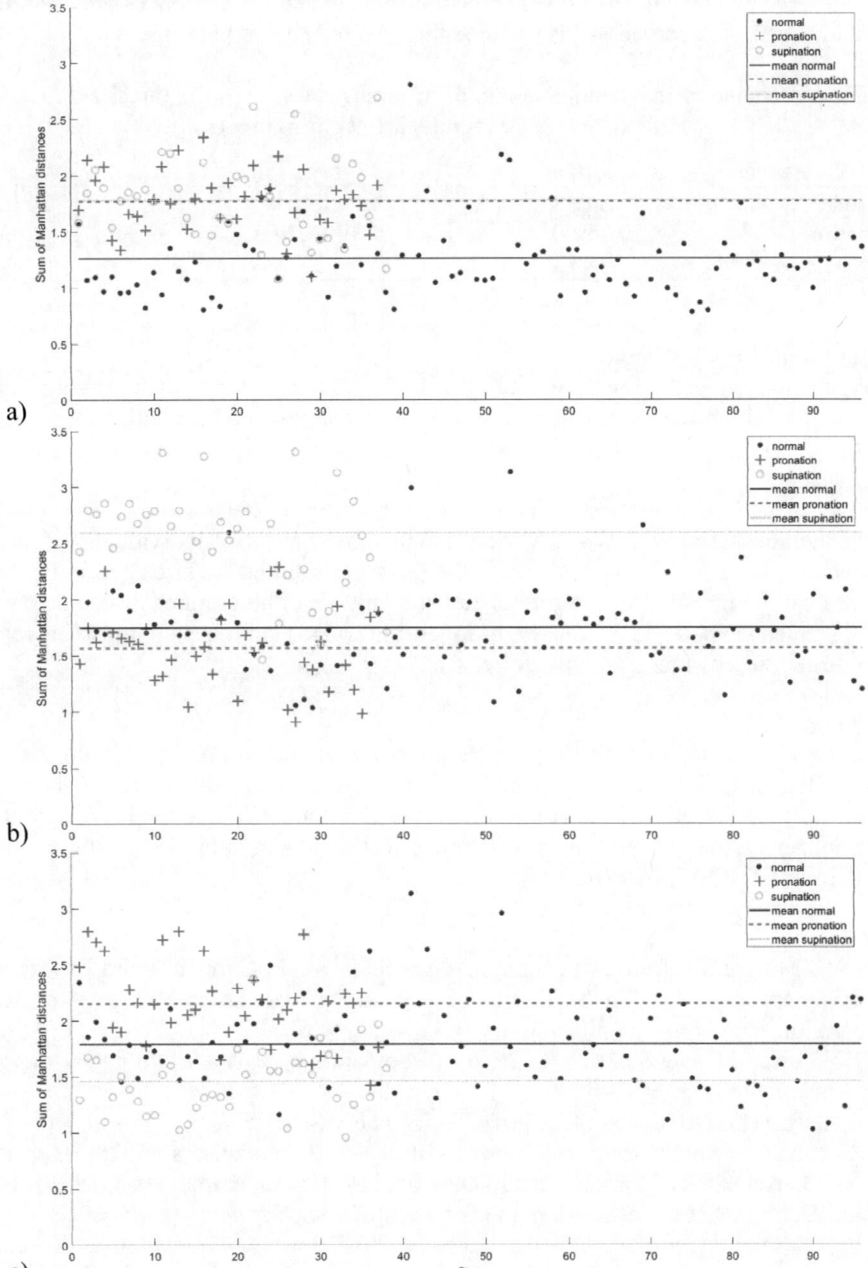

Figure 4. SMD values for each step for volunteer V3 when (a) normal, (b) pronation, and (c) supination reference is used.

Significant changes in the mean SMD can be observed in figure 4 despite the variation and overlap in the result of different gait types for individual steps. Although this overlap complicates setting a detection threshold, in practice, however, there is no necessity to classify each step correctly as the overall gait pattern can be concluded from the majority of steps. Table 2 provides the percentage of steps with SMD below the threshold of 1.5 for data in figure 4 for each of the gait pattern. As it can be seen, in case of normal gait, using normal gait pattern reference more than 80% of steps have SMD below the threshold, while only 13.51% and 26.32% if pronation and supination references are used respectively. A similar situation can be observed in case of the other two tested gait patterns.

Table 2. The percentage of steps with the sum of differences less than the threshold of 1.5 for each reference type (example data from figure 4).

Gait type / Reference	normal	pronation	supination
normal	**84.38**	13.51	26.32
pronation	25.00	**43.24**	2.63
supination	28.13	2.70	**50.00**

4. Conclusions and Future Work

A new gait type detection method for application with smart socks system was proposed, and three gait patterns were observed: normal gait, gait with excessive pronation and gait with excessive supination. The applicability of the proposed method for gait pattern type detection was demonstrated with experimental data.

For future work, additional tests are necessary to further improve the method. Considering that the clinical gait conditions used in this study were simulated by healthy individuals, clinical tests are necessary to acquire a true reference and verify the proposed method with real patients. Additionally, more gait types can be included in the reference list to increase the number of detectable gait types. Finally, the parameters included in the high-dimension vector of the step must be reviewed and optimized to further improve the gait type detection.

Acknowledgments

This work has been supported by the European Regional Development Fund within the Activity 1.1.1.2 “Post-doctoral Research Aid” of the Specific Aid Objective 1.1.1 “To increase the research and innovative capacity of scientific institutions of Latvia and the ability to attract external financing, investing in human resources and infrastructure” of the Operational Programme “Growth and Employment” (No. 1.1.1.2/VIAA/1/16/153).

References

[1] Eskofier B M, Lee S I, Baron M., Simon A, Martindale C F, Gaßner H and Klucken J 2017 An overview of smart shoes in the internet of health things: gait and mobility assessment in health promotion and disease monitoring *Appl. Sci.* **7**(10) 986.

[2] Hegde N, Bries M and Sazonov E 2016 A comparative review of footwear-based wearable systems *Electronics* **5**(3) 48

[3] Chen S, Lach J, Lo B and Yang G Z 2016 Toward pervasive gait analysis with wearable sensors: a systematic review *Journal of Biomedical and Health Informatics* vol **20**(6), pp 1521 – 1537

[4] Preece S J, Kenney L P, Major M J, Dias T, Lay E and Fernandes B T 2011 Automatic identification of gait events using an instrumented sock *J Neuroeng Rehabil* **8** 32

[5] Oks A, Katashev A, Bernans E and Abolins V 2017 Smart socks system as an equipment to analyze temporal parameters of human gait and running *Proceedings of the 11th Inter-national Scientific and Practical Conference* **3** pp 238 – 241

[6] Aggarwal C C, Hinneburg A, and Keim D A 2001 On the surprising behavior of distance metrics in high dimensional space *In: Van den Bussche J., Vianu V. (eds) Database Theory — ICDT 2001. ICDT 2001. Lecture Notes in Computer Science*, vol **1973** pp 420 – 434

An evaluation of usability of warp knitted spacer fabrics as hip protector

G Ertekin and A Marmaralı

Ege University, Faculty of Engineering, Department of Textile Engineering, 35100, Bornova, İzmir, Turkey

gozde.damci@ege.edu.tr

Abstract. The aim of this study is to examine the usability of warp knitted spacer fabrics as a hip protector, in order to eliminate the complaints frequently encountered in the market, due to their higher air and water vapor permeability, thermal conductivity, good compressibility and lightweight. For this purpose, the physical and performance properties of silicone and latex coated warp-knitted spacer fabrics were determined. With an attempt to discuss the effect of coating and fabric structural parameters on the physical and performance characteristics, six warp knitted spacer fabrics were produced by varying thickness (12.5 and 15 mm) and mesh structure (small and large hexagonal mesh). Then, fabrics were coated with silicone and latex by using vacuum infusion process in order to improve the force attenuation capacity of the fabrics. The fabrics were coated with silicone and latex substrates in a 1:1 and 1:½ weight ratios, respectively. Mass per unit area, thickness, air permeability, compressibility, dimensional stability and compression set properties were measured and evaluated.

1. Introduction

As the people are getting older, the incidence of falling, fracture and vulnerability to injury increases due to their decreased physical capabilities. The populations of many countries are both ageing and feminizing, meaning hip fractures now occur in around one in four women and one in eight men. More than 90% of hip fractures result from a fall [1-3]. A sideways fall produces a high-energy impact, which increases six times of the risk of a hip fracture compared to a backward or a forward fall, even in young subjects [4, 5] (Figure 1). Therefore, reducing the rate of falls and the resulting hip fractures is an increasingly important part of healthy ageing.

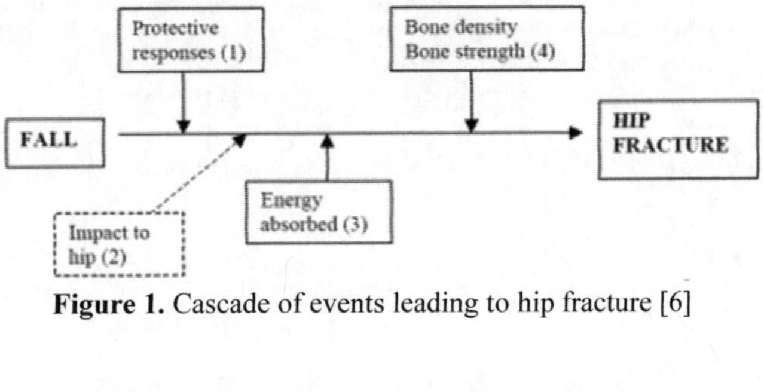

Figure 1. Cascade of events leading to hip fracture [6]

Hip protectors are hard or soft plastic shields embedded into pants or undergarments in proximity of the outside of the hip, just over the trochanter. The aim of hip protectors is to reduce the energy transmitted to the trochanter during a fall below the force causing a fracture, referred to as hip fracture threshold [4, 7]. Mainly, hip protectors can be either "rigid pads" designed for spreading the energy of the impact to the soft tissue of surrounding of the hip or "energy absorbing" soft materials that decrease the force of the impact ultimately transmitted to the bone [8, 9].

Several studies have been indicated that hip protectors are not often used for reasons such as uncomfortable (too tight/poor fit), wearing difficulties, urinary incontinence and physical difficulties/illnesses [10-15]. Attention has been paid to the features of the hip protector garment concerning comfort, appearance and fit. Changes in the design of hip protectors making them more comfortable and acceptable to users justify new studies addressing the effectiveness of hip protectors [16]. The aim of this study is to examine the usability of warp knitted spacer fabrics as a hip protector, in order to eliminate the complaints frequently encountered in the market, due to their higher air and water vapour permeability, thermal conductivity, good compressibility and lightweight.

2. Material and method

2.1. Production of spacer fabrics

Warp knitted spacer fabrics were produced by using a Karl Mayer RD 6 raschel machine with double-needle bars and six yarn guide bars (GB). A 167f48x4 and 334f72 polyester PTY multifilaments were used to create the top outer and bottom outer layers. Polyester monofilament yarns of 0.243, 0.200 and 0.170 mm in diameters were used as spacer yarn to connect two outer layers. The spacer monofilament inclination (shifting one needle distance) was kept constant.

The distance between the needle bars of the machine was set to 12.5 and 15 mm for varying the thickness values of the samples.

Brückner (Vn-Sfp-24/6-Q99) heat-setting machine was used in order to increase the structural stability and to achieve a determined final width (stretching). The samples were horizontally (course-wise) pinned on the sliding aluminum frame pins and heat set at the temperature of 180 °C for time duration of 6 min. The tension applied to the samples during heat-setting was arranged so that the final width of the samples will be 110 and 160 cm. After heat-setting, the fabric samples were allowed to cool down at room temperature for 24 h. The characteristics of the fabrics are given in Table 1.

Table 1. Specifications of the fabrics

Fabric type	Yarn thickness (dtex/mm/dtex)	Distance between the needle bars (mm)	Mesh size (Finished width of the samples)	Courses /cm	Fabric pattern
1	167f48x4-334f72x3 /0.243/ 167f48x4-334f72x3	12.5	Large (160 cm)	5.25	Two side open structure
2		15.0	Large (160 cm)		
3		12.5	Small (110 cm)		
4		15.0	Small (110 cm)		
5	167f48x4/0.200/167f48x4	15.0	130 cm	6.5	Two side close structure
6	334f72x3/0.170/334f72x3	15.0	130 cm	5	Face side open/back side close structure

2.2. Coating of spacer fabrics

Fabrics were coated with silicone and latex by using vacuum infusion process in order to improve the force attenuation capacity of the fabrics. Silicone substrate with two-components (10:1) from ACC Silicons Company mixed with a laboratory mixer and the mixture was put under vacuum before use to avoid air bubbles [17]. The fabrics were coated with this substrate in a 1:1 weight ratio. Latex substrate were applied in 1: ½ weight ratio. The photographs of the coated fabrics are given in Figure 2.

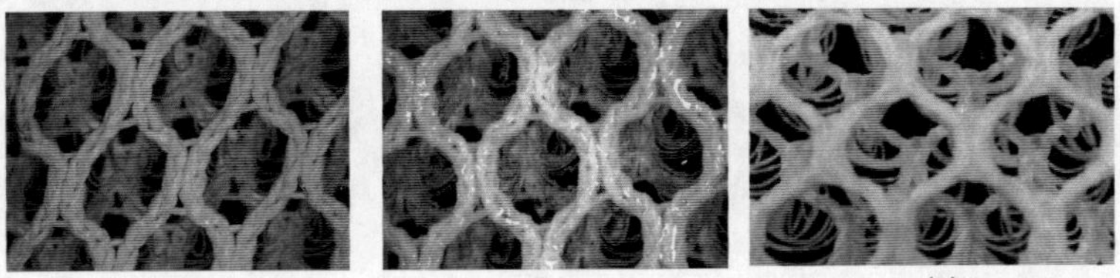

(a) (b) (c)

Figure 2. Surface photos of (a) non-coated, (b) silicone coated, (c) latex coated fabrics.

2.3. Measurements of spacer fabrics

The physical properties such as fabric thickness (measured with a caliper) and mass per unit area (according to EN 12127:1997) of the spacer fabric samples before and after coating process are measured and shown in Table 2.

Air permeability values were obtained by using Textest FX 3300 instrument according to TS 391 EN ISO 9237. The results of the measurements are averages from the values of 10 readings. The performance characteristics of the spacer fabrics such as compressibility, dimensional stability, compression set and air permeability were tested. Compressibility tests were carried out by Zwick Z010 Instrument based on the ISO 3386 standard. 950N was applied to the samples and the set compression rate is 25%. The measurement was performed by using two layers of fabric samples in both course and wale direction. The results are averages of 3 readings in N. Dimensional stability was tested according to the DIN 53377 standard. The samples are cut into square form of 10 x10 cm and are placed in an oven maintained at 90°C for 1h. The dimensional stability value was calculated according to the Eq. (1):

$$Dimensional\ \ stability\ \ (\%) = \frac{(l_O - l_F)}{l_O} x100 \tag{1}$$

where l_o: The original dimensions of the specimen, l_F: The dimensions of the specimen after treatment. The compression set is the difference between the initial thickness and the final thickness of a test piece of the material after compression for a given time at a given temperature and after a given recovery time and is determined according to TS 2013 EN ISO 1856 standard. The samples (50x50x25 mm) are compressed at a rate of 75% and are placed in an oven maintained at 70°C for 22h. The compression set value was calculated according to the Eq. (2):

$$Compression\ \ set\ \ (\%) = \frac{(d_O - d_r)}{d_O} x100 \tag{2}$$

where d_o: The original thickness of the specimen, d_r: The thickness value of the specimen waited for ½ h after experiment.

3. Result and Discussion

The mean values of the physical and performance properties are given in Table 2 and 3, respectively.

Table 2. Physical and air permeability properties of the fabrics

Surface structure	Space distance (mm)	Mesh size	Type of coating	Mass per unit area (g/m^2)	Thickness (mm)	Air permeability (l/m^2/s)
Two sides open	12.5	Large (160 cm)	Non-coated	525.12	10.51	6998.00
			Silicone coated	1125.33	10.02	6170.67
			Latex coated	864.00	11.57	6113.33
	15.0	Large (160 cm)	Non-coated	555.73	12.81	7124.00
			Silicone coated	1168.00	11.61	5939.33
			Latex coated	901.33	13.13	5909.33
	12.5	Small (110 cm)	Non-coated	683.57	10.80	6024.00
			Silicone coated	1402.67	10.70	5691.33
			Latex coated	922.67	11.49	5665.33
	15.0	Small (110 cm)	Non-coated	737.57	13.02	6078.00
			Silicone coated	1370.67	12.52	5654.00
			Latex coated	1125.33	14.15	5634.67
Two sides close	15.0	130 cm	Non-coated	1171.81	10.49	2254.00
			Silicone coated	2304.00	10.40	1135.87
			Latex coated	1616.00	11.00	2183.33
Face side open/back side close	15.0	130 cm	Non-coated	514.45	9.84	4614.00
			Silicone coated	1072.00	9.39	4354.00
			Latex coated	730.67	10.54	4644,67

Table 3. Performance properties of the fabrics

Surface structure	Space distance (mm)	Mesh size	Type of coating	Compressibility (N)		Dimensional stability (%)		Compression set (%)
				Course-wise	Wale-wise	Course-wise	Wale-wise	
Two sides open	12.5	Large (160 cm)	Non-coated	201.12	193.25	-2	0	33.16
			Silicone coated	436	427	-1	0	25.34
			Latex coated	520	518	-1	0	21.38
	15.0	Large (160 cm)	Non-coated	139.04	118.11	-3	0	38.12
			Silicone coated	368	341	-1	0	31.80
			Latex coated	421	409	-1	0	29.80
	12.5	Small (110 cm)	Non-coated	279.84	219.15	-2	0	32.51
			Silicone coated	627	539	-1	0	23.08
			Latex coated	647	604	-1	0	20.95
	15.0	Small (110 cm)	Non-coated	236.64	218.62	-3	0	37.47
			Silicone coated	440	413	-1	0	27.10
			Latex coated	513	508	-1	0	26.65
Two sides close	15.0	130 cm	Non-coated	306.72	300.28	-2	0	36.48
			Silicone coated	>950	>950	0	0	28.14
			Latex coated	>950	>950	0	0	25.14
Face side open/back side close	15.0	130 cm	Non-coated	271.68	249.63	-3	0	24.97
			Silicone coated	Not measured		-1	0	21.20
			Latex coated	465	443	-1	0	19.20

3.1. Air permeability

The air permeability results indicated that, the highest and lowest air permeability values were observed in the non-coated spacer fabric with a large hexagonal mesh structure and a distance between the needle bars of 15 mm and silicone coated spacer fabric having two side closed structure, respectively. For determination of air permeability properties, it was observed that, fabric density is the most effective parameter rather than coating, a distance between the needle bars and hexagonal mesh size. Generally, it was examined that, fabrics having a distance between the needle bars of 15 mm and/or large hexagonal mesh structure and/or non-coated have lower density resulting in higher air permeability.

When the effect of surface structure on the air permeability of the fabrics is evaluated, it was determined that the surface structure of the fabric has a dominant effect than the fabric density. As it is expected, fabrics having two sides open structures have higher air permeability followed by face side open/back side close structure and two sides close structure.

3.2. Compressibility

The force necessary to compress a fabric has to overcome the internal stresses of the fibres and the inter-fibre frictional force. A fabric that compresses easily is likely to be judged as soft, possessing a low compression modulus or high compression [18]. Factors affecting compressibility of three-dimensional fabrics are fabric thickness, stitch density, bending behaviour and inclination angle of spacer yarn.

According to Table 3, it was observed that as the thickness values increased, the compressibility of the fabrics decreased. As the fabric becomes thicker, it was thought that the fabric becomes softer due to the extension of the spacer yarn and facilitates its deformation.

The compressibility results indicated that, the lower compressibility values in both direction were observed in the spacer fabric with large hexagonal mesh structures and a distance between the needle bars of 15 mm due to its lower stitch density. When the effect of coating on compressibility is examined, it was noted that the coating increased the compressibility of the fabrics and latex coated spacer fabrics had higher compressibility values following by silicone coated and non-coated spacer fabrics.

When the effect of surface structure is investigated, it was examined that, the highest compressibility value was observed in the spacer fabric with two sides close structure, followed by face side open/back side close structure and two sides open structures. The compressibility of the spacer fabric having two sides close structures was higher due to its more stable and rigid structure. As the inclination angle of the spacer yarn decreases, spacer yarn provides better compressibility property. The reduction of compressibility is attributed to the inclination angle of monofilament yarn that can be calculated by image processing of the cross-sectional view of fabrics. The monofilament yarns are placed at nearly right angles in the spacer fabrics having two sides open structure. This led to more easily exposed to bending resulting in a decrease of compressibility.

3.3. Dimensional stability

The dimensional stability characteristics of the fabrics were measured in both directions: wale-wise and course-wise. The results indicated that, no dimensional change in wale direction was observed in all fabrics. The results indicated that, non-coated spacer fabrics with two sides open structures having a distance between the needle bars of 15 mm and non-coated spacer fabric with two sides close structures have 3% shrinkage in course direction. According to DIN 53 377, the acceptable limit value of dimensional stability for both directions is ± 3 %. Additionally, for the non-coated fabrics, as the fabric thickness increased, the shrinkage increased as well. The shrinkage values of the fabrics were 1% in course direction after coating process except fabrics with two sides close structure. The hexagonal mesh size didn't any effect on the dimensional stability of the fabrics.

When the effect of surface structure is investigated, it was observed that, fabrics with two sides close structure had lower shrinkage values than fabrics having two sides open and face side open/back side closed structures.

3.4. Compression set

Compression set value known as aging test is one of the important factors determining the durability and lifetime of spacer fabrics. There is an inverse relationship between the durability and compression set characteristic of the fabrics. The higher the compression set, the lower is the durability of the fabrics [18]. It can be seen from Table 3 that, non-coated fabrics with higher compression set values offered a shorter lifetime of the fabrics, whereas coated fabrics especially latex coated fabrics had lower compression set values. Additionally, an increase in fabric thickness and/or hexagonal mesh size led to an increase in compression set values resulting in shorter lifetimes of the fabrics. Fabric density has major effect on the compression set properties of the fabrics. The compression set values of the fabrics increase due to the decrease of fibre content per unit area. Spacer fabrics with face side open, back side close structure had lower compression set values than the other fabrics. It was observed that fabric density and surface structure have a mutual effect on compression set characteristic of fabrics.

4. Conclusion

The aim of this study is to examine the usability of warp knitted spacer fabrics as a hip protector. For this purpose, the physical and performance properties of silicone and latex coated warp-knitted spacer fabrics were determined. The results clearly demonstrated that:

– Coating process improved the compressibility property and decreased dimensional stability and compression set. Although coating reduced the air permeability values, no significant decrease in air permeability property was observed after coating. Latex coated spacer fabrics due to their higher compressibility and lower compression set properties provided longer duration and lifetime of the spacer fabrics.

– An increase in thickness values led to decrease in compressibility and increase in compression set values resulting in shorter lifetime of the fabrics. Additionally, for non-coated fabrics, as the thickness increased, shrinkage in % increased as well.

– Fabrics with small hexagonal mesh structure had higher compressibility, and lower air permeability and compression set.

– Fabrics with face side open/back side close structures had moderate air permeability and compressibility properties, whereas their compression set values were lower than the other fabric structures.

To sum up, parameters such as small hexagonal mesh size, lower thickness, face side open/back side close surface structure and latex coating can be preferable for better comfort, longer durability and life time of the warp knitted spacer fabrics which are intended to be used as hip protector.

Acknowledgments

The authors would like to acknowledge the support of TUBITAK 3001 "Starting R&D Projects Funding Program" within project no 117M824.

References

[1] Järvinen TL, Sievänen H, Khan KM, Heinonen A and Kannus P 2008 *BMJ* **336**124–126.
[2] Järvinen TL, Sievänen H, Kannus P, Jokihaara J and Khan KM 2011 *BMJ* **342** 2175.
[3] van Schoor NM, van der Veen AJ, Schaap LA, Smit TH and Lips P 2006 *Bone* **39** 401–407.
[4] Cianferotti L, Fossi C and Brandi M L 2015 *Calcif Tissue Int* **97**(1) 1-11.
[5] Kannus P, Leiponen P, Parkkari J, Palvanen M and Järvinen M 2006 *Bone* **39** 383–384.
[6] Sveistrup H., Lockett D., Hip Protectors and Community-Living Seniors: A Review of the Literature, 2004.
[7] Parker M and Johansen A 2006 *BMJ* **333** 27–30.
[8] Derler S, Spierings AB and Schmitt KU 2005 *Med Eng Phys* **27** 475-485.
[9] Laing AC, Feldman F, Jalili M, Tsai CM and Robinovitch SN 2011 *J Biomech* **44** 2627–2635.
[10] Bentzen H, Forsen L, Becker C and Bergland A 2008 *Osteoporos Int* **19** 101–111.
[11] Parker MJ, Gillespie WJ and Gillespie LD 2006 ***BMJ*** 1-4.
[12] van Schoor NM, Deville WL, Bouter LM and Lips P 2002 *Osteoporos Int* **13**(12) 917–924.
[13] Becker C, Kron M, Lindemann U, Sturm E, Eichner B, Walter-Jung B and Nikolaus T 2003 *J Am Geriatr Soc* **51**(3) 306-13.
[14] Cox H, Puffer S, Morton V, Cooper C, Hodson J, Masud T, Oliver D, Preedy D, Selby P, Stone M, Sutcliffe A, Torgerson D 2008 *Age and Ageing* **37**(2) 167-72.

[15] Cryer C, Knox A, Martin D and Barlow J 2002 *Injury Prevention* **8**(3) 202-6.
[16] Ertekin G and Marmaralı A 2012 *International Congress on Healthcare and Medical Textiles* May 17-18.
[17] Dow Corning Textile & Leather Solutions 2017 How to apply and cure silicone coatings. Retrieved from http://www.dowcorning.com/content/textiles/26-1256-01a.pdf (Accessed on July 02, 2018).
[18] Ertekin G and Marmaralı A 2016 *J Eng Fiber Fabr* **11**(3) 64-71.

IOP Conf. Series: Materials Science and Engineering **459** (2019) 012039 doi:10.1088/1757-899X/459/1/012039

Thermal comfort properties of nonwoven fabrics used in surgical gowns

S H Eryuruk[1], B Karaguzel Kayaoglu[1] and P Altay[1]
[1]Istanbul Technical University, Textile Technologies and Design Faculty, Textile Engineering Department, 34437, Istanbul, Turkey

eryuruk@itu.edu.tr

Abstract. Nonwoven fabrics are used for surgical gown applications due to their low cost, lightweight, durability, breathability, low hairiness and disposability. Multi-layered nonwovens with membranes are used in disposable surgical gowns to enhance protection against fluids, aerosol, toxins and microorganisms to meet the needs of surgical applications. In this study, we have analysed thermal comfort properties of SMS (spunbond-meltblown-spunbond) fabrics with three different weights (35 g/m^2, 50 g/m^2 and 80 g/m^2) and also evaluated the layered nonwoven structures with a breathable and water impermeable membrane. As a result of the study, fabric weight and layered structure had been found to have an important effect on the comfort properties of nonwoven fabrics.

1. Introduction

Multi-layered nonwovens with membranes are used in disposable surgical gowns to enhance protection against fluids, aerosol, toxins and microorganisms to meet the needs of surgical applications. Nonwoven fabrics are used for surgical gown applications due to their low cost, lightweight, durability, breathability, low hairiness and disposability. Clothing comfort is another aspect that need to be considered besides protection.

There are limited number of studies concerning the thermal comfort properties of nonwoven surgical gowns [1-6]. Pamuk et. al. evaluated the thermal comfort properties of different disposable surgical gowns (Spunlace-normal, Spunlace-reinforced, and SMS-normal) by using the thermal manikin [1]. Woo et. al. developed a theoretical model providing a reliable prediction of thermal transmission through nonwoven structures. They measured the thermal conductivity of various nonwoven barrier fabrics for the validity the model. The model explains fiber and fabric variables in determining the thermal insulation values of nonwoven barrier materials [2]. Issa et. al. evaluated the effects of the sterilisation process on their thermal comfort so they analysed some of the thermal properties of disposable surgical gowns before and after different sterilisation methods [3]. They concluded that the disposable materials used in the hospital sector (laminate,nonwoven, PE) are affected by the sterilisation process. Bogdan et. al. analysed thermal insulation of the modern materials used in medical clothing in the context of thermal comfort experienced by surgeons wearing a full clothing ensemble [4]. They stated that medical clothing ensembles made of modern materials bring about the risk of thermal stress. Aslan et. al. investigated comfort and microbial protection performances of two reusable and two disposable surgical gowns by subjective wear trials [5]. They found that thermal comfort performance of the microfiber polyester woven gown was the best according to subjective wear trial and microbial resistance test results. Disposable nonwoven gowns had lower comfort performances despite their higher permeability and lower resistance values. Zwolinska and Bogdan studied the thermal stress of surgeons using medical coverings [6].

In this study, we have analysed thermal comfort properties of SMS (spunbond-meltblown-spunbond) fabrics with three different weights (35 g/m^2, 50 g/m^2 and 80 g/m^2) and also evaluated the layered nonwoven structures with a breathable and water impermeable membrane.

2. Materials and Methods

35, 50 and 80 g/m^2 SMS (spunbond-meltblown-spunbond) fabrics were supplied from Mogul Textile Ltd., Turkey which were the most commercially used fabric types for surgical gowns. An elastic nonporous PU membrane with a weight of 80 g/m^2 were used that is pre-laminated on to a polyester raschel knitted fabric.

Alambeta test instrument was used to test thermal conductivity, thermal absorptivity, and thermal resistance of the fabrics [7]. Permetest instrument was used to measure water vapour resistance and water vapour permeability according to the ISO 11092:2014 (sweating guarded-hotplate test) standard [8]. The air permeability of the samples was measured by using Pro-white test apparatus according to standard EN ISO 9237:199519 by applying 100 Pa constant air pressure and 20 cm^2 sample size [9].

3. Results

Comfort properties of the surgical gowns are very important in determining the overall preference. Thermal conductivity values of fabrics are presented in Figure 1. As it is seen from the figure, as the fabric weight increases, thermal conductivity of fabrics increased due to the increased amount of fibres inside the nonwoven fabrics. Membrane has not a significant effect on thermal conductivity. All three fabric structures with membrane had nearly the same thermal conductivity values which is greater than the thermal conductivity of nonwoven fabrics.

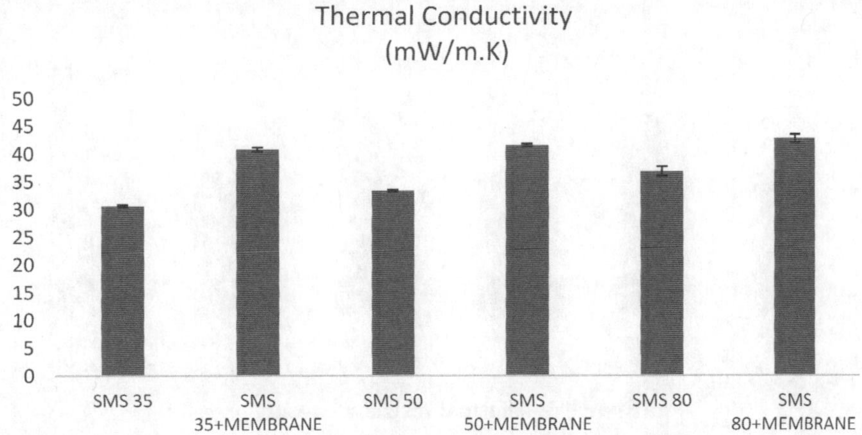

Figure 1. Thermal conductivity results

Thermal diffusion is related to the heat flow through the air in the fabric structure, higher thermal diffusion value is mainly related to bulky fabric structure and large amount of air inside the fabric. As the fabric weight increased, thermal diffusion of nonwoven fabrics decreased because of more compact structure and less air inside the fabric structure (Figure 2). Membrane increased the thermal diffusion because of more air inside the layered fabric structure.

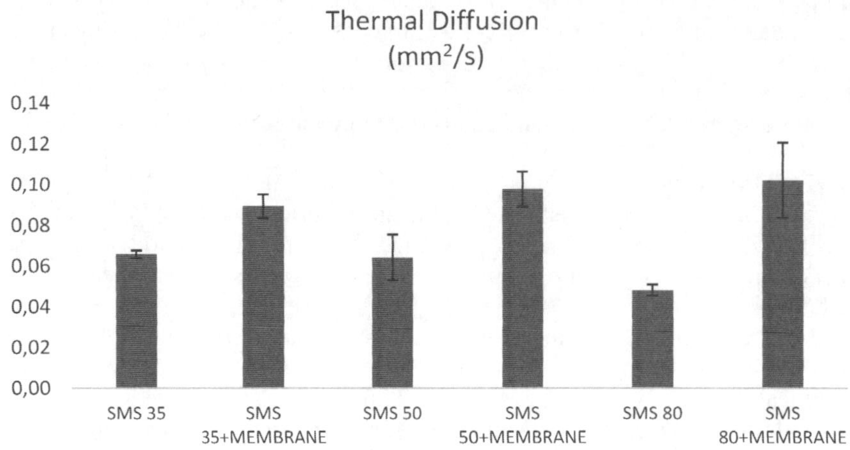

Figure 2. Thermal diffusion results

Thermal resistance represents the thermal insulation property of a fabric. According to the results, the highest thermal resistance values were obtained fabrics with membrane. Fabric weight has also an increasing effect on thermal resistance values of fabrics.

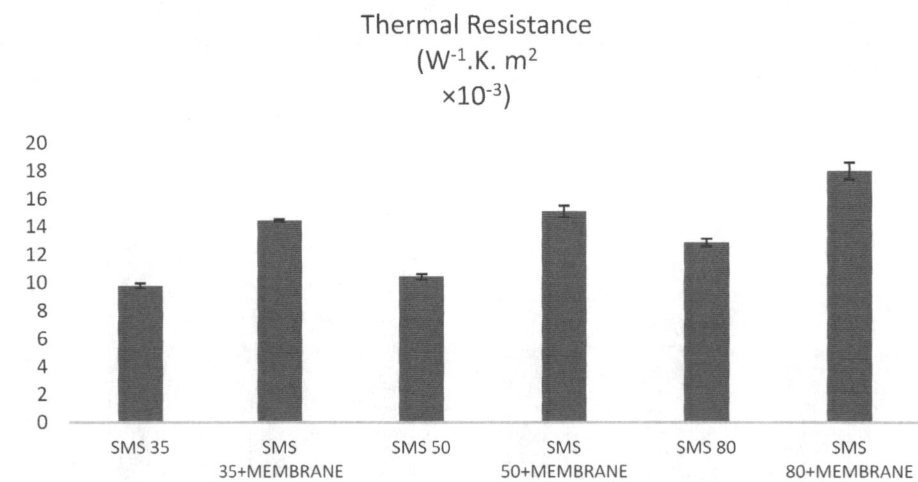

Figure 3. Thermal resistance results

Thermal absorptivity depends on the contact area between the skin and fabric surface. As it is seen from the Figure 4, thermal absorptivity increased as the fabric weight increased. Especially fabric weight 80 g/m^2 had the greatest value for thermal absorptivity because more contact area is obtained between the fabric and skin resulting to a cooler feeling. Membrane had no significant effect on the thermal absorptivity values of fabrics.

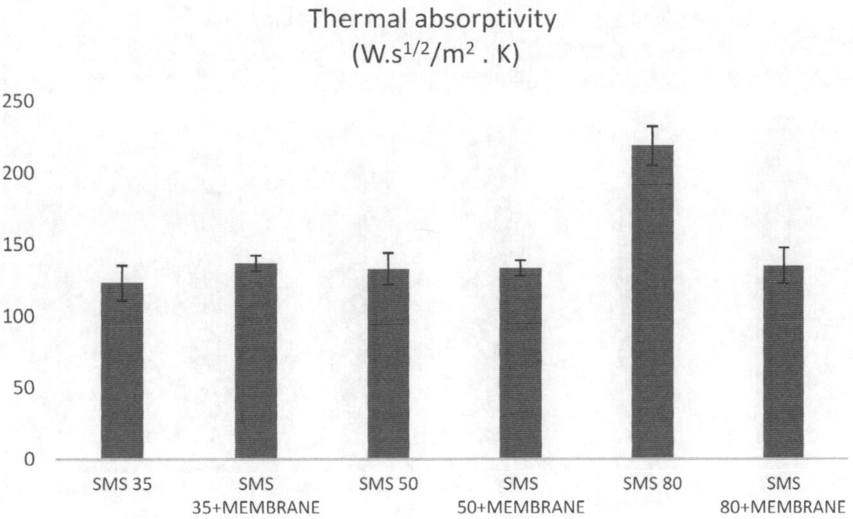

Figure 4. Thermal absorptivity results

Water vapor resistance is the ability of a fabric to allow moisture vapor to be transmitted through it. Water vapor resistance (R_{et}) values has increased as the fabric weight increased (Figure 5). Membrane had a significant effect on the R_{et} values which increased significantly. Relative water vapor permeability index of nonwoven fabrics with membrane has decreased for the nonwoven fabrics with membrane (Figure 6). Moreover, air permeability of nonwoven fabrics decreased as the fabric weight increased (Figure 7).

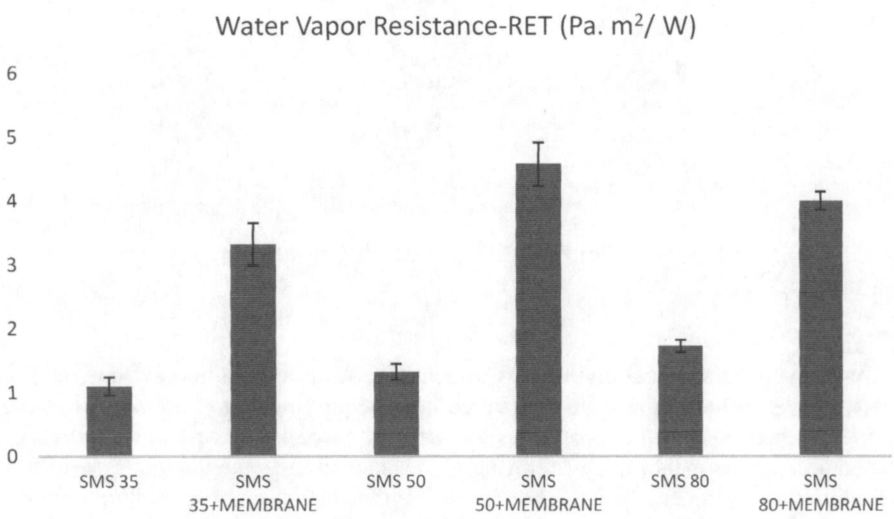

Figure 5. Water vapour resistance (Ret) results

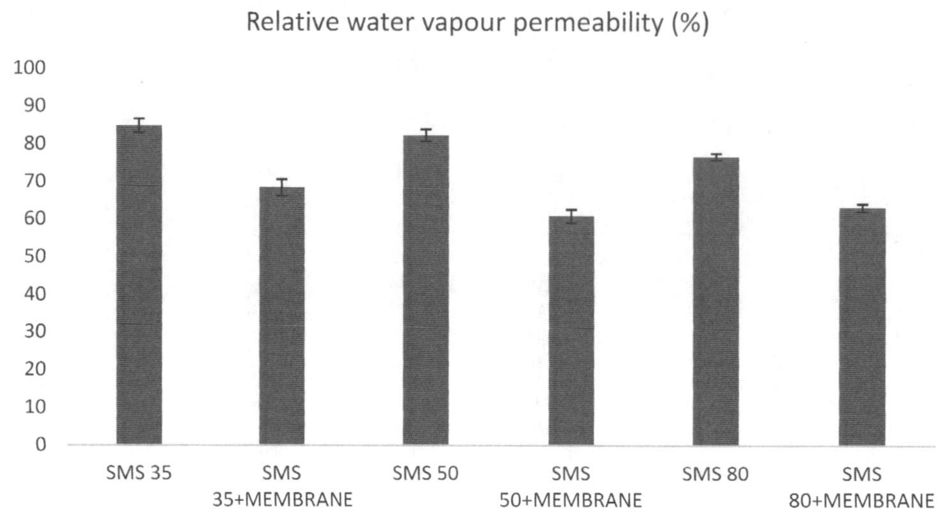

Figure 6. Relative water vapour permeability results

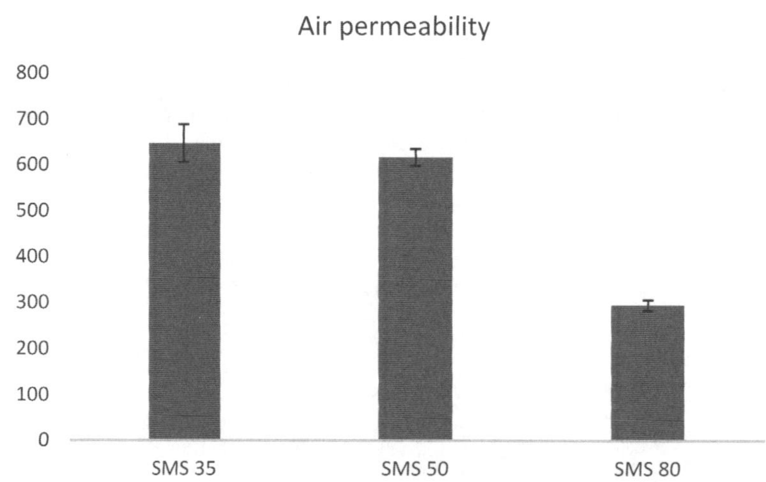

Figure 7. Air permeability results

4. Conclusions

It is very important for a surgical clothing to protect the health personnel and patient from infecting each other. Moreover, it has to meet the thermal comfort requirements to minimise physiological stress on the wearer so thermal comfort properties are needed to be evaluated like thermal conductivity, thermal resistance, air permeability, water vapour resistance and water vapour permeability values. Results of this study showed that as the fabric weight increased, thermal conductivity, thermal resistance, thermal absorptivity and water vapour resistance values of nonwoven fabrics increased whereas air permeability and water vapour permeability values decreased. The layered nonwoven structures with a breathable and water impermeable membrane had greater thermal conductivity, thermal resistance, thermal, diffusion, water vapour resistance values and lower relative water vapour permeability results. Especially membrane had a significant effect on the thermal resistance and R_{et} values. Results of this study showed that thermal comfort performances of nonwoven disposable gowns had decreased as the fabric weight increased and membrane added.

Acknowledgments
The present work is supported by ITU BAP (Istanbul Technical University Scientific Research Projects) with project number MGA-2018-41026 for the period 06.03.2018-06.03.2019.

References

[1] Pamuk O, Abreu M J, Ondogan Z, 2008 An Investigation on The Comfort Properties for Different Disposable Surgical Gowns by Using Thermal Manikin Tekstil ve Konfeksiyon **3** pp 236-239

[2] Woo S S, Shalev I and Barker R L, 1994 Heat and Moisture Transfer Through Nonwoven Fabrics
Part I: Heat Transfer Textile Research Journal **64 (3)** pp 149-162

[3] Issa M, Abreu M J, Schacher L, Adolphe D, Silva M E C, 2004 The Influence of The Sterilisation Process on Certain Thermal Properties Eur J Appl Physiol **92** pp 673–678

[4] Bogdan A, Sudoł-Szopińska I, Szopiński T, 2011 Assessment of Textiles for Use in Operating Theatres with Respect to the Thermal Comfort of Surgeons Fibres & Textiles in Eastern Europe **19** 2 (85) pp 65-69

[5] Aslan S, Kaplan S & Cetin C, 2013 An Investigation About Comfort And Protection Performances Of Disposable And Reusable Surgical Gowns By Objective And Subjective Measurements The Journal of The Textile Institute **104** 8 pp 870–882

[6] Zwolinska M and Bogdan A 2012 Impact of The Medical Clothing on The Thermal Stress of Surgeons Applied Ergonomics **43** pp 1096-1104

[7] Hes L, De Araujo M, Djulay V, 1996 Effect of Mutual Bonding of Textile Layers on Thermal Insulation and Thermal Contact Properties of Fabric Assemblies Textile Research Journal **66** pp 245–250

[8] ISO 11092:2014 Textiles -- Physiological effects -- Measurement of thermal and water-vapour resistance under steady-state conditions (sweating guarded-hotplate test)

[9] ISO 9237:1995. Textiles-determination of permeability of fabrics to air

Thermal comfort properties of firefighters' clothing with underwear

S H Eryuruk[1], V Koncar[2], F Kalaoglu[1] , H Gidik[3] and X Tao[2]
[1]Istanbul Technical University, Textile Technologies and Design Faculty, Textile Engineering Department, 34437, Istanbul, Turkey
[2]ENSAIT, Ecole Nationale Supérieure des Arts et Industries Textiles, 59056, Roubaix Cedex 1, France
[3]UCLille, HEI, GEMTEX, F-59046 Lille, France

eryuruk@itu.edu.tr

Abstract. Optimal comfort will enable the wearer to work efficiently over long periods of time and help to protect the body from imminent overheating. The thermal performance of fire fighters' protective clothing is primarily based on the thermophysical properties of the materials that are used to construct the clothing. In this study, we have used new fire resistant underwear. The main purpose of this study is to evaluate the single fabric layer thermal comfort behaviours of underwear, outer shell, moisture barrier and thermal barrier fabrics together with their three-layered and four-layered combinations to understand their multilayer fabric thermal comfort performances.

1. Introduction

Protection together with comfort is a very important subject for firefighters' performance and health. Heat and moisture transfer from the skin through the multilayer textile structures has a direct influence on the performance and safety of the firefighters. Thermal protective garments and firefighters uniforms are produced using special multilayered fabrics which are bulky and heavy. Thermal protection from fire and metabolic heat stress generated by the human body due to metabolic activities must be balanced [1-7].

Chung and Lee studied comfort of protective clothing for fire fighters and suggested on choosing system of clothing designs and material layers to balance protection and comfort [8]. Mah and Song outlined factors affecting firefighting clothing's heat and moisture transfer capacity material properties, the style, fit, size and drape of garments [9]. Li et al. research work evaluated the effects of material components and design features on heat transfer properties of firefighter turnout clothings [10]. By using a sweating manikin, heat and moisture transfer performances of firefighter turnout clothings including outer shell, moisture barrier and thermal liner were evaluated considering clothing material, design, size, accessories (design details in clothing) and clothing design. Lawson and Vettori suggested that thermal performance of firefighters clothings must be evaluated while dry, when wet, in full loft and when fully compressed [11]. Wanga et al. concluded that the thermal and moisture comfort of firefighters' ensembles when combined with the polyester inner clothing was worse than the other types of inner clothing [12]. He et. al. studied heat and moisture transfer in a multilayer protective fabric system under various ambient conditions[13].

The firefighter garments typically consist of three single layers of technical fabrics; outer shell fabric, moisture barrier fabric and thermal barrier fabric. Besides firefighter uniforms, firefighters also wear underwear garments inside their clothings. In this study, we have used new fire resistant underwear. The main purpose of this study was to evaluate thermal comfort behaviours of the single fabric layers underwear, outer shell, moisture barrier and thermal barrier fabrics together with their three-layered and four-layered combinations to understand their multilayer fabric thermal comfort performances. One firefighter garment type was selected for the study.

2. Materials and Methods

Fabric properties are presented in Table 1.

Table 1. Fabric properties

Fabric Code	**Fabric Type**	**Mass per unit area (g/m^2)**
Outer Shell	75% Meta Aramid-23% Paraaramid-2% Antistatic	200
Moisture Barrier	85% Metaaramid-15% Paraaramid PU membrane laminated to FR nonwoven Fabric	120
Thermal Barrier	Aramid felt quilted to Aramid/Viscose FR nonwoven fabric	115
Underwear	78% FR Viscose, 20% Paraaramid, 2% Antistatic	220

Thermal comfort properties of the firefighter clothing and underwear were evaluated considering three comfort properties, thermal resistance, water vapour permeability and moisture management properties of fabrics. The sweating guarded hot plate apparatus was used to measure the thermal resistance of clothings (R_{ct}) (m^{2}K/W), under steady-state conditions according to ISO 11092 (ISO, 1993) [14]. The temperature of the guarded hot plate was kept at 35°C (like human skin) and for the determination of R_{ct} of the fabrics, the standard atmospheric conditions were set as 65% R.H and 20°C. Water vapour permeability is the ability of a material to allow water vapour to pass through it. Water vapor permeability values of the samples were measured using the dish method, according to ISO 15496 (ISO, 2004) standard [15]. This method involves determination of weight losses with the evaporation time (24 h) of water contained in a cup, the top of which is covered by the cover ring. The difference in water loss between a cup covered with the reference fabric and one with the test fabric enables to study the relative rates of moisture movement through the test fabrics, so that the moisture vapor permeability of the test specimen can be calculated. Moisture management tester (MMT) was used to evaluate moisture management properties [16]. This method quantitatively measured the liquid moisture transfer in one step in a fabric in multi directions, where liquid moisture spreads on both surfaces of the fabric and transfers from one surface to the opposite.

Figure 1. The sweating guarded hot plate apparatus

Figure 2. Water vapour permeability test

Figure 3. MMT Instrument

3. Results

3.1. Thermal Resistance Results

Thermal resistance represents thermal insulation property of a fabric, the higher the thermal resistance, the higher the thermal insulation capacity of a fabric. As the number of fabric layers increased, thermal resistance of fabrics increased (Figure 4). Also it was found high R^2 values between thermal resistance & fabric thickness (0.978) and thermal resistance & fabric weight (0.724) (Figure 5 and Figure 6).

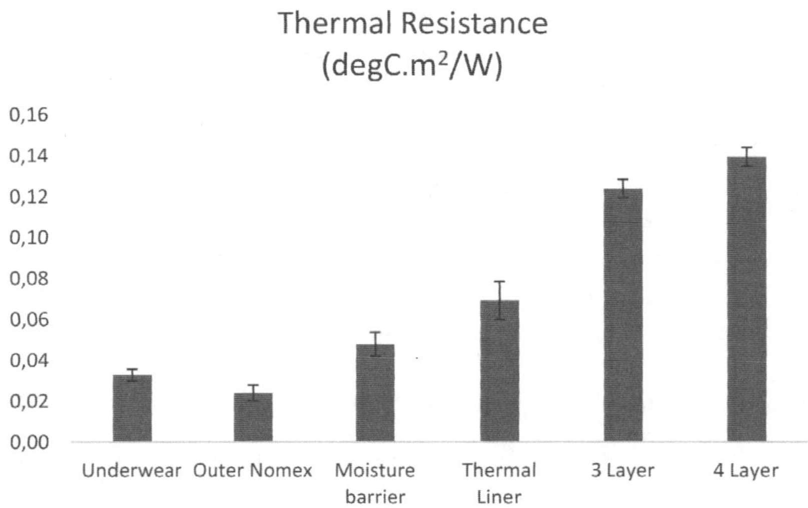

Figure 4. Thermal resistance results

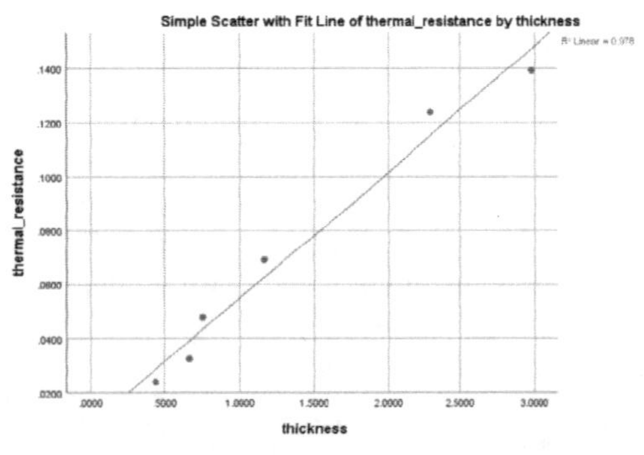

Figure 5. Thermal resistance & thickness linear regression graph

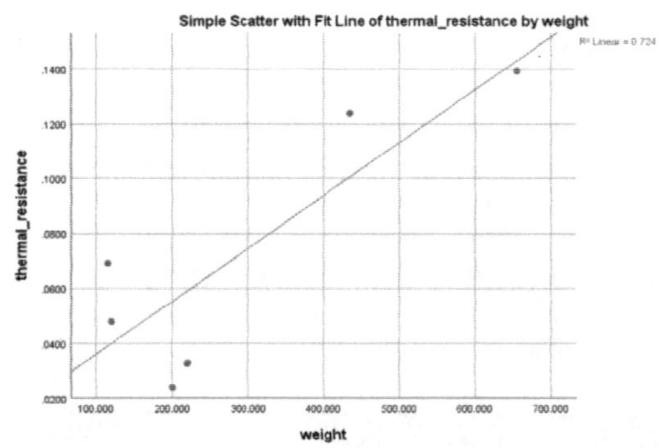

Figure 6. Thermal resistance&weight linear regression graph

3.2. Water Vapour Permeability Results

Water vapour permeability is the ability to transmit vapour from the body. Water vapour permeability and water vapour permeability index of fabrics are seen in Figure 7 and 8. As it is seen from the figures, underwear and outer shell fabrics had the highest water vapour permeability values. As the number of fabric layers increased, the water vapour permeability of fabrics decreased and this means that the uncomfortable sensation increases. Moreover it was found a negative correlation between water vapour permeability and water vapour permeability index and fabric thickness (Figure 9 and Figure 10).

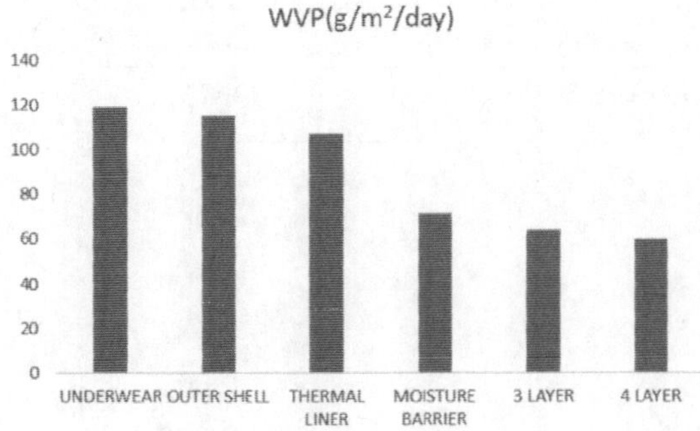

Figure 7. Water vapour permeability of fabrics

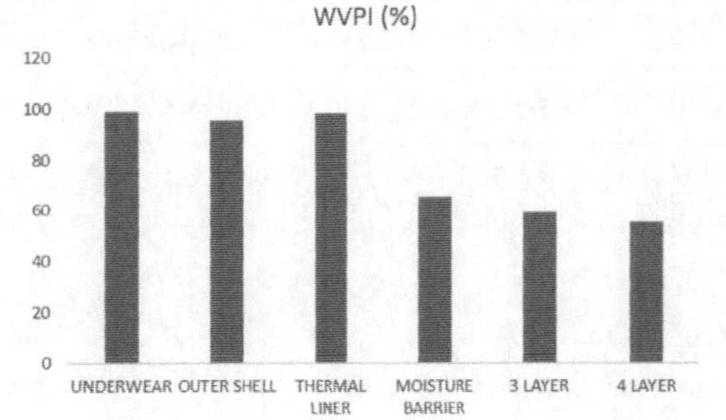

Figure 8. Water vapour permeability index of fabrics

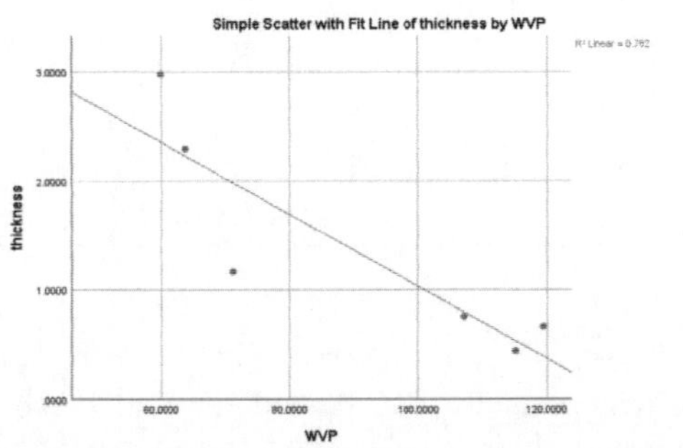

Figure 9. Water vapour permeability &thickness linear regression graph

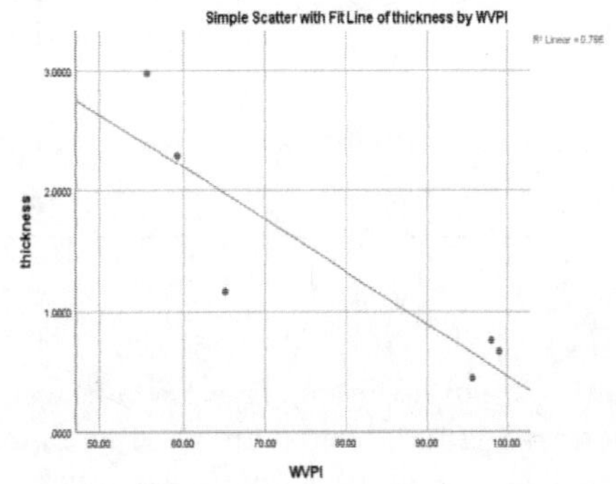

Figure 10. Water vapour permeability index &thickness linear regression graph

3.3. MMT Results

Liquid moisture transfer in clothing significantly influences the wearer's perception of moisture comfort sensations. Dynamic liquid transfer values of clothing materials were measured using MMT test instrument. Cumulative one-way transport capacity (OWTC) is the difference in the cumulative moisture content between the two surfaces of the fabric in the unit testing time period. Overall moisture management capacity (OMMC) is an index to indicate the overall ability of the fabric to manage the transport of liquid moisture, which includes three aspects of performance, moisture absorption rate of the bottom side, one-way liquid transport ability, and moisture drying speed of the bottom side, which is represented by the maximum spreading speed.

Table 2 shows cumulative one-way transport capacity (OWTC) and overall moisture management capacity (OMMC) obtained from the MMT test device. As it is seen underwear fabric and thermal liner fabric had the highest liquid moisture management capacities (OMMC) and one-way transfer capacities (OWTC), showing that liquid sweat can be transferred from next to the skin to the outer surface to keep the skin dry. Moisture management and outer shell fabrics face side had poor liquid moisture management properties and negative one-way transport capacities (OWTC) indicating that the liquid (sweat) cannot diffuse easily from the next-to-skin surface to the opposite side and will accumulate on the top surface of the fabric. These fabrics were found as a water repellent fabrics according to MMT results. As the number of layers were increased, fabrics were getting more water proof. 3 layered and 4 layered fabric structures had zero OMMC values and very high and negative accumulative one-way transport index values. Back sides had lower one way transport index values which were nearly half of the front sides. According to the MMT results, these fabrics have been found as water proof fabrics. Underwear fabric had also a positive effect by increasing accumulative one way transport index on the back side.

Table 2. MMT Results of fabrics

Fabric Code	Accumulative one-way transport index (%)	OMMC
Underwear front	97.168	0.447
Underwear back	114.776	0.502
Outer shell front	-717.590	0.000
Outer shell back	-243.534	0.083
Moisture barrier front	-981.595	0.000
Moisture barrier back	-997.663	0.000
Thermal liner front	66.970	0.455
Thermal liner back	453.438	0.757
3 layer front	-1015.405	0.000
3 layer back	-568.148	0.000
4 layer front	-1036.212	0.000
4 layer back	-417.720	0.000

4. Conclusions

In this study the single fabric layer' (underwear, outer shell, moisture barrier and thermal barrier fabrics) thermal comfort behaviours were measured together with their three-layered and four-layered combinations. As a result of study, it was found that as the number of fabric layers increased, thermal comfort properties of firefighter clothing are getting worse. As the number of fabric layers increased, thermal resistance of fabrics increased and the water vapour permeability of fabrics decreased meaning that the uncomfortable sensation increases. MMT results showed that underwear fabric and thermal liner fabric had the highest liquid moisture management capacity (OMMC) and one-way transfer capacity (OWTC), showing that liquid sweat can be transferred from next to the skin to the outer surface to keep the skin dry. Moisture management and outer fabric face side had poor liquid moisture management properties that the liquid (sweat) cannot diffuse easily from the next-to-skin surface to the opposite side and will accumulate on the top surface of the fabric. As a general conclusion for MMT results is that, as the number of layers increased, fabrics were becoming more water proof. On the other hand, underwear fabric had a positive effect by increasing accumulative one way transport index from back to front side.

Acknowledgments

This project has received funding from the European Union's Horizon 2020 research and innovation programme under the Marie Skłodowska-Curie grant agreement No 644268.

References

[1] Teunissena L P J , Wang L C, Chou, S N 2014 Evaluation Of Two Cooling Systems Under a Firefighter Coverall Applied Ergonomics **45** 6 pp 1433–1438

[2] Levels D, Koning K, Mol J J, Foster E, Hein C, Daanen A M 2014 The Effect Of Pre-Warming On Performance During Simulated Firefighting Exercise Applied Ergonomics **45** 6 pp 1504–1509

[3] Chung G S, Lee D H 2005 A Study On Comfort Of Protective Clothing For Firefighters Elsevier Ergonomics Book Series **3** pp 375-378

[4] Jiang Y Y, Yanai E, Nishimura K, Zhang H, Abe N, Shinohara M, Wakatsuki K, 2010 An Integrated Numerical Simulator For Thermal Performance Assessments Of Firefighters' Protective Clothing Fire Safety Journal **45** pp 314-326

[5] Kim J H, Williams W J, Coca A, Yokota M, 2013 Application Of Thermoregulatory Modeling To Predict Core And Skin Temperatures In Firefighters International Journal of Industrial Ergonomics **43** 1 pp 115-120

[6] Eryuruk S H 2016 Analysis of Thermal Properties of Firefighter's Protective Clothings Tekstil ve Konfeksiyon **26** 3 pp 270-279

[7] Gidik H, Bedek G, Dupont D, Codau C, 2015 Impact Of The Textile Substrate On The Heat Transfer Of A Textile Heat Flux Sensor Sensors and Actuators A: Physical A230 pp 25-32

[8] Chung G-S and Lee D H, 2005 A study on comfort of protective clothing for firefighters Environmental Ergonomics **3** pp 375-378

[9] Mah T. and Song G. W, 2010 Investigation of the contribution of garment design to thermal protection. Part 2:Instrumented female mannequin flash-fire evaluation system Textile Research Journal **80** 14 pp 1473–1487

[10] Li J, Barker R L and Deaton A S, 2007 Evaluating the Effects of Material Component and Design Feature on Heat Transfer in Firefighter Turnout Clothing by a Sweating Manikin Textile Research Journal **77** 2 pp 59-66

[11] Lawson R and Vettori R L, 2002 Thermal Measurements For Fire Fighters Protectıve Clothing. Building and Fire Research Laboratory National Institute of Standards and Technology (NIST) Gaithersburg MD 20899 pp 1-15

[12] Wanga Y, Zhang Z, Li J and Zhua G, 2013 Effects of inner and outer clothing combinations on firefighter ensembles' thermal- and moisture-related comfort levels The Journal of The Textile Institute **104** 5 pp 530–540

[13] He J, Li J and Kim E, 2015 Assessment of the heat and moisture transfer in a multilayer protective fabric system under various ambient conditions Textile Research Journal **85** 3 pp 227-237

[14] ISO 11092 Textiles-physiological effects-measurements of thermal and water-vapour resistance under steady-state conditions (sweating guarded hotplatetest) (ISO, 1993), Genève

[15] ISO 15496 Textiles – measurement of water vapour permeability of textiles forthe purpose of quality control (ISO, 2004), Genève

[16] Li Y, Xu W and Yeung K W, Moisture Management of Textiles U.S. patent 6,499,338 B2 2000

Textile-based pressure sensors for step detection: a preliminary assessment

A G Ferreira[1,2], A P Catarino[2], J L Monteiro[1] and A M Rocha[2]

[1]University of Minho, School of Engineering, Centro Algoritmi, Guimarães, Portugal
[2]University of Minho, School of Engineering, Centre of Textile Science and Technology, Guimarães, Portugal

Email: id4541@alunos.uminho.pt

Abstract. This paper presents the development and performance assessment of textile-based sensor based on a three layer architecture for the step detection. Two different transducing elements (EeonTex™ LG-SLPA and velostat) and electrodes (Satatex Techniktex P-130 and Elitex yarns) were selected for the construction of the sensors. The performance of the resulting sensors was assessed based on a dynamometer cyclic compression/decompression test with different compressions loads and at different speeds. Additionally, a real-life experiment was conducted to evaluate the sensor response during walking. The results show that all sensors configurations have a non-linear resistance-force relation. The best sensor configuration for the step detection was the combination of EeonTex™ LG-SLPA as a transducing element and the Elitex yarns for the electrodes. In this configuration, the resistance magnitude varies in an order of hundreds of kohms between the stance and the swing phases.

1. Introduction

Position estimation based on the Pedestrian Dead Reckoning (PDR) technique suffer from cumulative error due to the drift on the inertial sensors measurements. Typically, to prevent the exponential growth of the error on the position estimation, the PDR-based systems apply the zero-velocity update (ZUPT) technique. This technique takes advantage of the human gait to minimize the drift impact and correct the position estimate. I.e., every time the foot is in touch with the ground (stance phase), the inertial sensors readings are considered drift and the pedestrian position is corrected based on the magnitude of those readings. However, the ZUPT technique relies on the definition of thresholds on sensors' readings (e.g., stance duration, acceleration and angular velocity magnitudes) [1]. These thresholds make the performance user- and motion-dependent, characteristics that are undesired in Indoor Positioning Systems (IPSs) for emergency responders [2].

In this paper, we study the viability of embedding textile-based pressure sensors on a sock to detect the stance phase. To do that, different materials for both the electrodes and the sensing element were used and the performance of the resulting sensors was assessed based on two different experiments.

Besides improving the stance detection of the conventional ZUPT methods, the proposed solution is wearable, comfortable, easy-to-use, and cheaper than solutions with pressure sensors embedded on the insole [3].

2. Materials and Methods

In this section, the assembly details and materials used for the construction of the pressure sensor prototypes as well as the equipment and experimental setup used/designed to assess their performance are described.

2.1. Pressure Sensor Prototype

Figure 1 shows the sketch and the fabrication details of the pressure sensors developed. The sensors are produced in three layers: the top and bottom layer are made with conductive materials (fabric and yarn knitted on the sock), and the middle layer is a piezoresistive substrate. The sensing element is placed on the sock's heel and conductive leads connect the sensor to the acquisition device (placed on the user's ankle).

The dimensions of the sensing element and electrodes are 10x10 mm and 7.5x7.5 mm, respectively. The dimension of the electrodes is smaller than the sensing element to prevent any short-circuit between the two conductive layers.

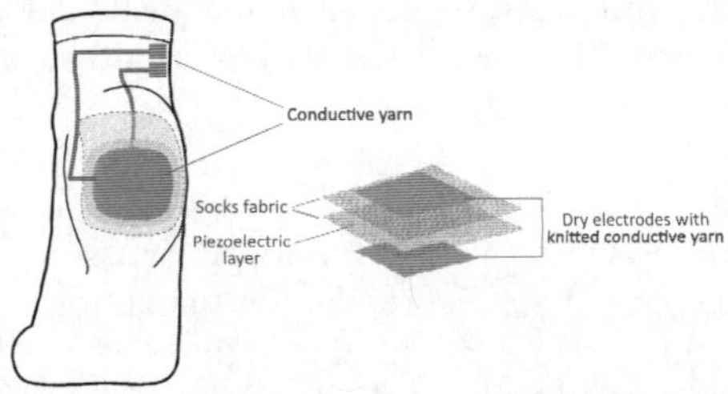

Figure 1. A sketch of the three layer textile-based pressure sensor with the conductive leads embedded on a sock.

2.2. Configuration of the Sensors' Prototypes

To assess the best combination of materials (electrodes and sensing element), three different sensor configurations were tested. They differ on the piezoresistive element (EeonTex™ LG-SLPA and velostat) and the conductive materials (Satatex Techniktex P-130 and a silver coated yarn from Elitex) used. Table 1 shows the materials used for the construction of each pressure sensor prototype, which are labeled as S1, S2 and S3.

Table 1. Description of the materials used for the construction of each pressure sensor prototype.

Sensor Label	Conductive Materials	Piezoresistive Sensing Element
S1	Elitex yarns	EeonTex™ LG-SLPA
S2	Satatex Techniktex P-130	EeonTex™ LG-SLPA
S3	Satatex Techniktex P-130	Velostat

2.3. Experimental Setups

Two types of experiments were conducted to assess the viability of using the pressure sensors for step detection. In the first experiment, a uniaxial dynamometer (Houndsfield, H100KS) and a digital multimeter (Agilent, 34410A) were used to characterize the electrical behavior of the pressure sensors during a 10 cycle compression/decompression test. This setting aims to study the sensor's response, in terms of electrical resistance, when it is submitted to different pressures. Therefore, for each sensor configuration, three different compression cycles (2-136N, 2-173N, and 2-210N) were applied. These compression forces corresponds, based on the ratio between the heel and dynamometer probe areas, to a body weight of 55, 70, and 85 kg, respectively. The application of different compression forces aims to study how the sensor respond to different body weights and if it possible to detect steps for all body weights and determine the wearer's weight based on the sensor output. Additionally, three different speeds (5, 25, and 50 mm/min) were set to evaluate the dynamic sensor response and to assess if the sensor is able to detect steps at different walking speeds. For the dynamometer experiment, a total of 27 individual tests were performed, 9 per each sensor configuration.

On the other hand, the second experiment consisted on asking a volunteer (male, 30 years old, 70kg) to wear a sock with the pressure sensor embedded, perform five consecutive steps and then stopped with the heel on the ground. This experiment was repeated five times for each sensor configuration and the sensor output was recorded using the same digital mutimeter (Agilent, 34410A). This experiment aims to evaluate the sensor response on a real scenario, since the dynamometer cannot replicate the velocities experienced during walking (≈1m/s).

3. Results and Discussion

During the stance phase, the heel presses the piezoresistive material and the sensor's resistance decrease. In the swing phase, the pressure between the heel and the footwear decreases and the resistance of the sensor increase. This resistance variation allows the step detection and higher its variation better the stance phase can be detected.

3.1. Dynamometer Experiment

Figure 2 shows the typical response of the developed pressure sensors during a dynamometer compression cycle. As can be seen in the figure, the resistance-force relation is non-linear (resembles to an exponential function), which is in line with the findings of similar research [4,5]. Since the goal of this work is to detect steps, this behavior of the sensor is beneficial as it allows to easily detect the step, the difference on the sensors resistance between the stance and swing phases can be in the order of hundreds of kilo ohms (figure 2). As expected, when the sensor is compressed (stance phase) its resistance decrease and increases during the decompression phase (swing phase). Additionally, the sensor response during the 10 compression/decompression cycles is almost the same, meaning that the sensor output is repeatable.

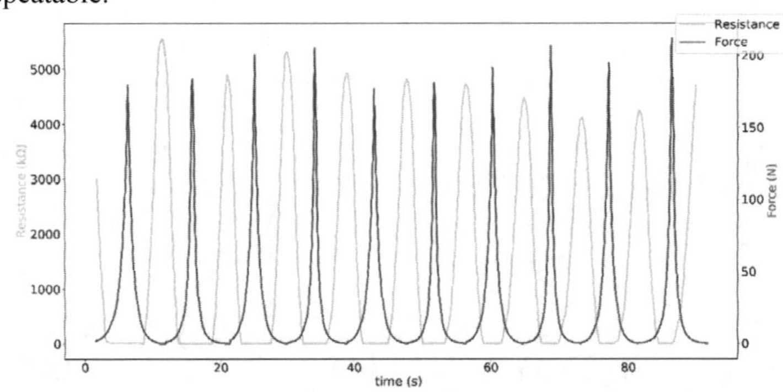

Figure 2. Example of a sensor output for the 10 cycles of compression/decompression for the dynamometer experiment. This graph illustrates the response of the S1 sensor configuration, for a simulated user weight of 70 kg (173 N) and a test velocity of 25 mm/min.

Figure 3 shows the resistance versus time percentage plots for all sensors configurations, weights and speeds. The time of the plots was converted to a percentage for an easier comparison between the experiments performed under different speeds.

Based on the results obtained, the S1 sensor configuration stand out as the best option to detect the steps as the resistance magnitude changes in the order of thousands of kilo ohms between the compression and decompression cycle. Whereas the sensor configuration S2 has a variation magnitude in the order of hundreds of kilo ohms and the S3 configuration in the order of hundreds of ohms. Additionally, unlike S2 and S3 configurations, the response of the S1 configuration is almost the same

IOP Conf. Series: Materials Science and Engineering **459** (2019) 012041 doi:10.1088/1757-899X/459/1/012041

for different speeds. For the S2 and S3 configurations the magnitude of the sensor response at lower speed is lower than for higher speeds. This phenomenon can be related with the limitations of the dynamometer. I.e., for higher speeds the applied force is higher than the threshold defined (e.g., see figure 2).

Although the resistance magnitude seems to decrease during the 10 compression/decompression cycles, it is not possible to state that the sensor is not repeatable since the dynamometer is not capable of applying the same force for each compression/decompression cycle. For lower velocities (5 mm/min) the sensors' magnitude is almost the same. Regarding the different forces applied, by analysing the graphs of figure 3 does not seem that they have a significant impact of the sensor output. However, more tests with better equipment should be conducted to assess if it is possible to determine the wearer's body weight based on the sensor output.

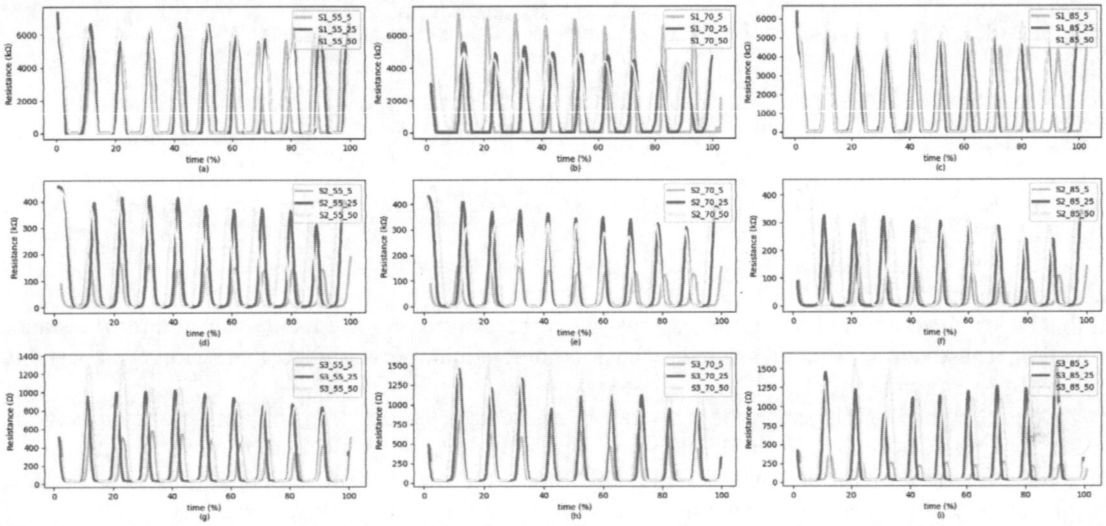

Figure 3. Resistance versus time percentage plots for the different sensors configurations, simulated users' weight, and test velocities.

3.2. Walking Experiment

Figure 4 shows the results of the tests for the walking experiment. As can be seen in the figure, the best results were obtained for the S1 configuration, followed by the S2 and S3 configurations. Compared with the results obtained from the dynamometer tests, it is possible to see that the resistance magnitude for all sensor configurations decreased significantly. This phenomenon can be explained by the existence of a continuous force applied to the sensor due to its placement between the heel and shoe. Nevertheless, the stance and the swing phases can be clearly distinguished, validating the use of these pressure sensors for step detection.

Additionally, we can see that a peak appears in the beginning of the swing phase. This peak refers to the heel off moment and can be explained by the diminution of the pressure applied by the heel on the sensor, which is later restored during the swing phase. Moreover, for all the sensor configurations, the resistance when the heel is pressing the sensor is very stable and its magnitude after the five steps is very close to the initial value. This means that the sensor is capable of providing reliable results for stance detection.

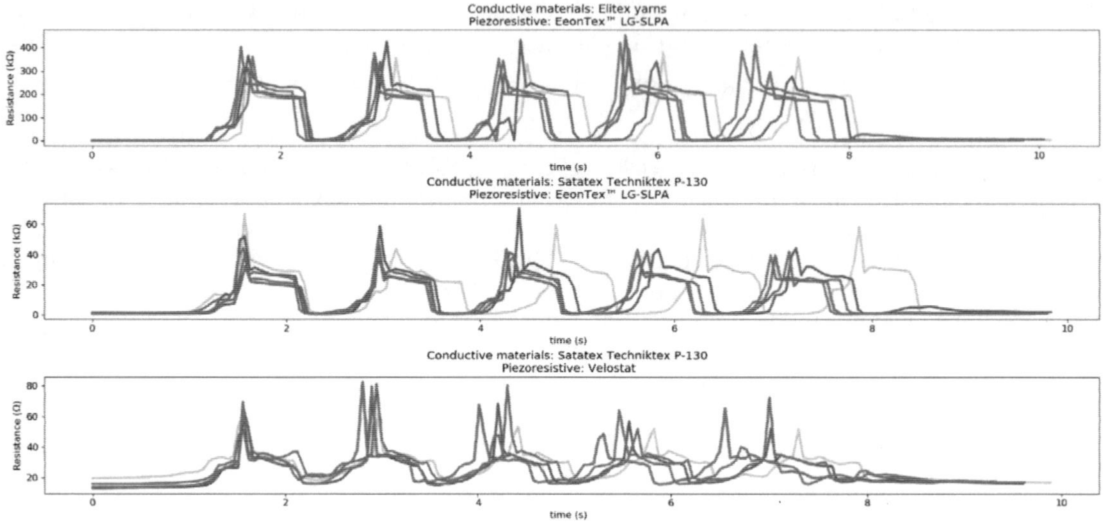

Figure 4. Resistance versus time plots for each sensor configuration while a volunteer performs five consecutive steps.

4. Conclusions

In this paper, three types of textile-based pressure sensors were proposed and evaluated for the stance detection. These sensors were evaluated on a dynamometer compression/decompression cycle test and by a real-life experiment.

As demonstrated in this paper, the steps can be easily detected by means of a textile-based pressure sensor integrated on a sock. From the three configurations evaluated, the combination of Elitex yarns as conductive materials and EeonTexTM LG-SLPA as the piezoresistive element provided the best results. In this first assessment, many relevant aspects of sensor construction have been investigated and a detailed knowledge about the sensor behavior has been acquired.

As future work, the sensors will be fully characterized in a more advanced dynamometer and more sensor configurations will be tested. Namely, the creation of 3D structures with conductive yarn to replace the three layer architecture and simulate the piezoresistive behavior of the sensing elements used in this work. Additionally, the textile-based pressure sensor will be integrated into PDR system [6] to assess its viability in the drift control of the inertial sensors.

Acknowledgments

This work was partially financed by FEDER funds through the Competitively Factors Operational Programme—COMPETE and by national founds through FCT -Foundation for Science and Technology within the scope of Project POCI-01-0145-FEDER-007136 and Project PEst-OE/EEI/UI0319/2014. Authors would like to thanks André Paiva and Sérgio Branco for their collaboration in sensors production and data acquisition, respectively.

References

[1] Jimenez AR, Seco F, Prieto JC, Guevara J. Indoor pedestrian navigation using an INS/EKF framework for yaw drift reduction and a foot-mounted IMU. 2010 In: *2010 7th Workshop on Positioning, Navigation and Communication* p. 135

[2] Ferreira AFG, Fernandes DMA, Catarino AP, Monteiro JL. Localization and Positioning Systems for Emergency Responders: A Survey. 2017 *IEEE Communications Surveys & Tutorials* **19** 2836

[3] Truong P, Lee J, Kwon A-R, Jeong G-M. Stride Counting in Human Walking and Walking Distance Estimation Using Insole Sensors. 2016 *Sensors* **16** 823

[4] Dinparast Tohidi S, Zille A, Catarino AP, Rocha AM. Effects of Base Fabric Parameters on the Electro-Mechanical Behavior of Piezoresistive Knitted Sensors. 2018 *IEEE Sensors Journal* **18** 4529
[5] Carvalho H, Yao Y, Gonçalves LM. Flexible force sensors for e-textiles. 2017 In: *IOP Conference Series: Materials Science and Engineering* **254** p. 072007
[6] Ferreira AG, Fernandes D, Monteiro JL, Catarino AP, Rocha AM. A Pedestrian Positioning System integrated into a Cyber-Physical System for Emergency Responders' Monitoring. 2018 In: *2018 International Conference on Indoor Positioning and Indoor Navigation (IPIN)*

Yarn Deformation with view of its structural structure

J K Gafurov[1], B Mardonov[1], K Gofurov[1] , O Sh Dushamov[1], O O Ergashev[1] and H T Bobajonov[2]
[1]Tashkent Institute of Textile and Light Industry, Technology of Textile Materials, Technology of Silk and Spinning, 5 Shohjahon, Yakkasaroy dist., Tashkent, UZBEKISTAN
[2]Namangan Engineering Technological Institute, Department of The Chair of Constructing Textile Productions and Technology, 7 Kosonsoy, Namangan, UZBEKISTAN
jgafurov@mail.ru

Abstract. The aim of this research is to implement suitable model for estimation mechanical properties of spun yarn. The behavior of fibers in the initial stage of their loading, which are oriented in the direction of the action of the load, is considered. The location of the fibers along the spiral lines has been studied, which leads to reduction in the cross section of the yarn as a result of the approach of the fibers. It is revealed that when the axial force is removed in the fibers, irreversible deformations remain, and in some fibers, residual stresses. This leads to an uneven distribution of strain across the section and it is one of the reasons for the structural unevenness of the yarn. It is established that the time for the realization of the internal resource of the yarn depends on the properties of the raw materials used, the way the yarn is formed, the condition of its winding and other factors. Two types of yarn are considered - compact and conventional, and the change in their deformation in time has been experimentally investigated. Various rheological models were used to develop the patterns of yarn deformation, taking into account the change in its structure during the loading process. The evaluation of yarn deformation taking into account its structure was performed using the Kelvin model with time-variable parameters, which take into account the instant and long modulus of elasticity of the material and the increase in strength characteristics of compact yarn is proved.

1. Introduction

Textile fiber in the form of yarns is a structural formation of fibers of certain length and thickness. The resulting tension is connected to interacting force, which is realized both through the friction of each fiber in the bundle and through the resilient contact between the ends of the fibers in the spinning triangle [1, 2] and in the yarns. Thus, the mechanical characteristics of the yarns formed depend both on the difference in fiber length and fiber diameter (linear density) and on spinning conditions. For example, mechanical behaviour of yarns under tension [3-6] presented as follows. Supposing that each stretched fiber has a fixed geometric position, determined by the angle of its inclination with respect to the yarns axis, and the tension applied here causes a fiber stress that depends on its position. Thus, at the maximum stress of fibers bundles as a whole (yarns), each fiber demonstrates an effective resistance to the actual load. Moreover, the fiber stiffness can affect the slope of yarns stretch curve. In this case, depending on geometric configuration, the medium tension of fiber bundle converts into the local tension. And the main index is the lateral pressure, which increases from zero on the outer surface of the yarns to its center, according to the law close to parabolic one [7]. In the absence of pressure then two neighboring fibers would easily slide relative to each other. However, if pressure between them is too high, then at full tension transfer the fibers will often tear before sliding and their axis tension in end points will go to zero.

2. Theoretical part

The foregoing description of the yarns mechanical behaviour during stretching indicates the possibility of developing and improving the representations of mechanical yarns models. In the initial stage of loading, the fibers oriented along the direction of the load begin to stretch simultaneously in the presence of frictional force, and the fibers located at an angle to the direction of this force are partially stretched, slide and moved together with parallel oriented fibers. Under the action of twisting, the fibers begin to be arranged along the helical lines, and the cross section of the yarns as a result of fibers convergence. When removing the axial force in the fibers the irreversible deformations persist, and in some fibers - the residual stresses. This leads to an uneven tension across the cross-section and is one of the reasons for the structural unevenness of the yarns. With repeated axial loading, there is the stress first along the yarn cross section and then yarns length redistributed, partial loading of some fibers, figuratively speaking, the yarns spend some time testing the load, thus "rolling" to achieve its final technological indicators for the further implementation of the internal resistance. The magnitude of this time depends on properties of the raw materials used, the method used for the construction of the yarn, the condition of winding and other factors. Moreover, for a sufficiently long time of load application, the yarn reaches the maximum stress resistance of the structure and the fibers. It becomes evident that the study of the processes occurring in the yarns up to the loss of the internal stress resistance as well as the determination of the time needed for the loss of yarn integrity (in connection with the magnitude of the load applied) are of great importance.

The tear time under different load values is presented in Table 1, for a ring-spun (20 Tex) tested for two different numbers of turns per unit length (740 tpm and 860 tpm). Yarns samples were obtained at a spindle speed =17000 min^{-1}. It is noticeable that as the tensile force increases, the time before yarns tears falls significantly. This indicates that at high loading values, the residual stresses responsible for the formation of the internal stress resistance.

Table 1. Tear time of yarns samples with linear density of 20 tex under different force.

Variants	$K_p = 740кр/м$								$K_p = 860кр/м$							
Yarns	compact				usual				compact				usual			
F, H	2.6	2.5	2.4	2.3	2.1	2.0	1.9	1.8	2.8	2.7	2.6	2.5	2.6	2.5	2.4	2.3
t_{fl}, c	3.1	6.0	26	51	4.1	6.1	8.6	38	3.7	15	29	46	7.6	9.4	12	24

Thus, the structural changes leads to the conclusion that the properties of all yarns made from a fibre mass depend on the time interval in the initial stage of loading and unloading.

Figure 1 shows the experimental curves of deformation dependence on time of constant tensile force F = 0,5 H for two kinds ringspun yarns (usual, compact) with linear density of 20 tex [8]. The first sample of yarns is obtained on common flyer frame and the second one is obtained by means of mechanical compaction unit installed on outlet cylinder of flyer frame's drafting roller.

As a result of tests, the curves were obtained. And the analysis of such curves revealed the following mathematic model. In initial stage of deformation change the curve has a concave section facing downwards. Therefore, in this section there is a change of deformation by time. Further, with increasing time, the concave section of curve passes to the convex one, and at the inflection point, the deformation change is by time, and the rate of yarns deformation rapidly decreases and becomes equal to zero. Such a change to the deformation curve can be explained physically as follows. At the time of force application in the yarn structure the straightening of fibers occurs and their convergence, as a result of which the radial stress values in the cross section of the yarn gradually increase, which increases the stiffness of the product under tension. At the same time, viscous properties first appear more vividly, and then, with increasing force, these properties pass from single to a compact system of fibers, with a high modulus of elasticity and a smaller viscous parameter. Such an approach of the explanation of the mechanical behaviour presented by experimental deformation curves makes it possible to use various rheological models to understand the mechanism of formation and analysis of the deformation low of yarns by taking into account the change in its structure in the process of loading.

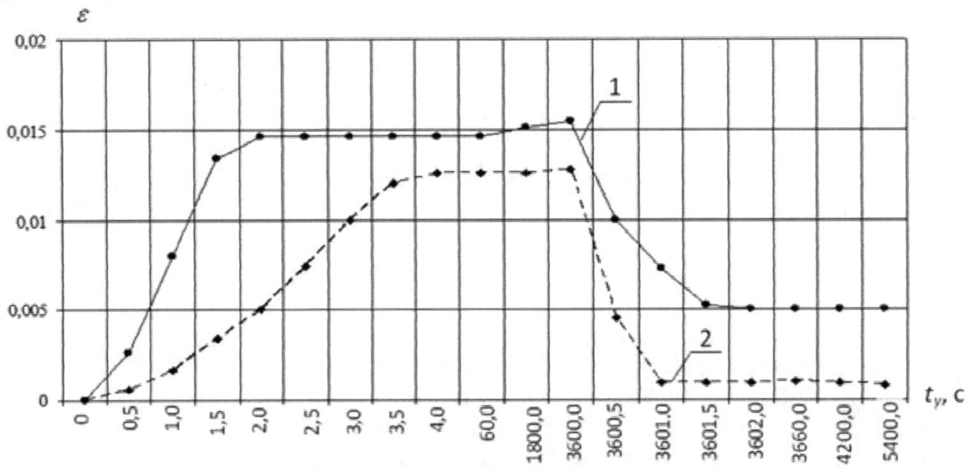

1 – usual yarns; 2 – compact yarns.
Figure 1. Change of deformation ε in time t

Let us consider the first cycle of yarn loading with a constant axial tensile force and let us determine the change in strain for the initial moments in time at which an internal redistribution occurs. This stress redistribution is related to the yarn structural change.
To describe this process, it is acceptable to use different models with elastic viscous and plastic elements. Consider some of them under the action of a constant axial load F.

1. The Kelvin rheological model with constant parameters ($\sigma_i = F/S$):

$$E\varepsilon(t) + \eta\dot{\varepsilon}(t) = \sigma_i \tag{1}$$

Integrating equation (1) with the initial condition $\varepsilon(0) = 0$, we find

$$\varepsilon = \frac{\sigma_i}{E}\left[1 - \exp(-E_1 t/\eta)\right] \tag{2}$$

From the formula (2) we establish

$$\ddot{\varepsilon} = -\frac{\sigma_i}{E\eta^2}\exp(-E_1 t/\eta) < 0$$

Thus, when using a simple Kelvin model, the tension in the yarns cross section remains constant and the rate of deformation in time decreases.

2. Generalized Kelvin model with parameters E_1, η_1, E_2, η_2:

$$\varepsilon = \frac{\sigma_i}{E_1}[1-\exp(-E_1 t/\eta_1)] + \frac{\sigma_i}{E_2}[1-\exp(-E_2 t/\eta_2)] \quad (3)$$

$$\ddot{\varepsilon} = -\sigma_i\left[\frac{\exp(-E_1 t/\eta_1)}{E_1} + \frac{\exp(-E_2 t/\eta_2)}{E_2}\right]$$

It can be seen that here also the rate of deformation decreases monotonically in time. Thus, the Kelvin model with constant parameters is not always suitable for describing the deformation process. Now, we consider the Kelvin model with time-varying parameters. According to this model, the change in the deformation with respect to time can be represented as follows

$$\eta(t)\dot{\varepsilon} + E(t)\varepsilon = \sigma_i \quad (4)$$

where $\eta(t) = H\eta(t)$ - parameter characterizing the viscous properties of the material, $\sigma_i = F/S_y$ With known types of functions $\eta(t)$ and $E(t)$ the equation (4) is easy to integrate under the zero initial condition $\varepsilon = 0$ at $t = 0$, and its solution presented as quadrature:

$$\varepsilon = \frac{\sigma_i}{H} f(t)\Phi(t) \quad (5)$$

where $f(t) = \exp\left[-\int \frac{E(t)dt}{H\eta(t)}\right]$, $\Phi(t) = \int_0^t \frac{dz}{\eta(z)f(z)}$

From formula (5) we note that yarns regularity of deformation in time is finally represented as a product of functions $f(t)$ and $\Phi(t)$.

Since $\eta(t) > 0$, $E(t) > 0$, then the functions $f(t)$ and $\Phi(t)$ in formula (5) are positive. Herewith the function $f(t)$ will be monotonically decreasing. As shown by the experimental curves (see Figure 3), the deformation for the initial time increases monotonically. It means that the function must be monotonous with increasing function.

We consider the cases of functions $\eta(t)$ and $E(t)$ as follows:

$$\eta = \eta_0 \exp(-\beta_c t), \quad E = E_0[1 - k\exp(-\alpha_c t)] \quad (6)$$

where $\eta_0 > 0$, α_c, E_0, $0 \le k \le 1$ и $\beta \ge 0$ are constant values.
In this case we have

$$f = \exp[-f_0(t)], \quad \Phi = \int_0^t \frac{\exp[f_0(t) + \beta_c t]}{n_0} dt, \quad f_0 = \frac{E_0 \exp(\beta_c t)}{Hn_0\beta_c}\left[1 - \frac{k\beta_c \exp(-\alpha_c t)}{(\beta_c - \alpha_c)}\right].$$

Let us denote relation $\gamma = E_0/H$ and analyze the function $\exp = \varepsilon(t)H/\sigma_i$ for various values of the dimensionless parameters k, γ and values of η_0, α_c and β_c with dimensions $сек^{-1}$ construct its graph with respect to the dimension of time t in seconds. Values k, γ и η_0 we call initial and α_c with β_c are structural parameters of yarns stiffness and viscosity. On Figure 2 shown the diagrams of function change $\exp(t)$ in time for $\gamma = 1$, $\beta_c = 0{,}5$ and different values of parameters k, α_c, η_0.

IOP Conf. Series: Materials Science and Engineering **459** (2019) 012042 doi:10.1088/1757-899X/459/1/012042

Since $E(0) = E_0(1-k)$, then, as can be seen, the curves with the approximation of the parameter k to 1 assume large values indicating a decrease in resistance of the yarn to stretching. Moreover, with increase of initial viscosity η_0, the deformation can be significantly decreased, which is the influence of the total frictional force that forms between the fibers in the yarn structure during its deformation. Easing of the yarn occurs either with small twists or weak interaction of the fibers with each other before applying the load. The increase in the value of η_0 parameter also leads to an increase in the time to reach the deformation of the limiting value and, depending on the value k, the inflection points occur (see Figure 2 curves 2, 3, 4 at $k = 0,9$). The deformation regularity of yarn is significantly influenced by the structural parameter α_c. From a comparison of the curves presented in Figure 2, it can be seen that the curves under the condition $\alpha_c = 0$ do not reach the maximum value in a finite period of time (see Figure 2.a). If you take $\alpha_c = 1$ (see Figure 2.б), then, as can be seen, the curves have maximum points in their determination area, and they have a pronounced character at small values η_0 и $k \approx 1$ (see Figure 2.б at $k = 0,9$).

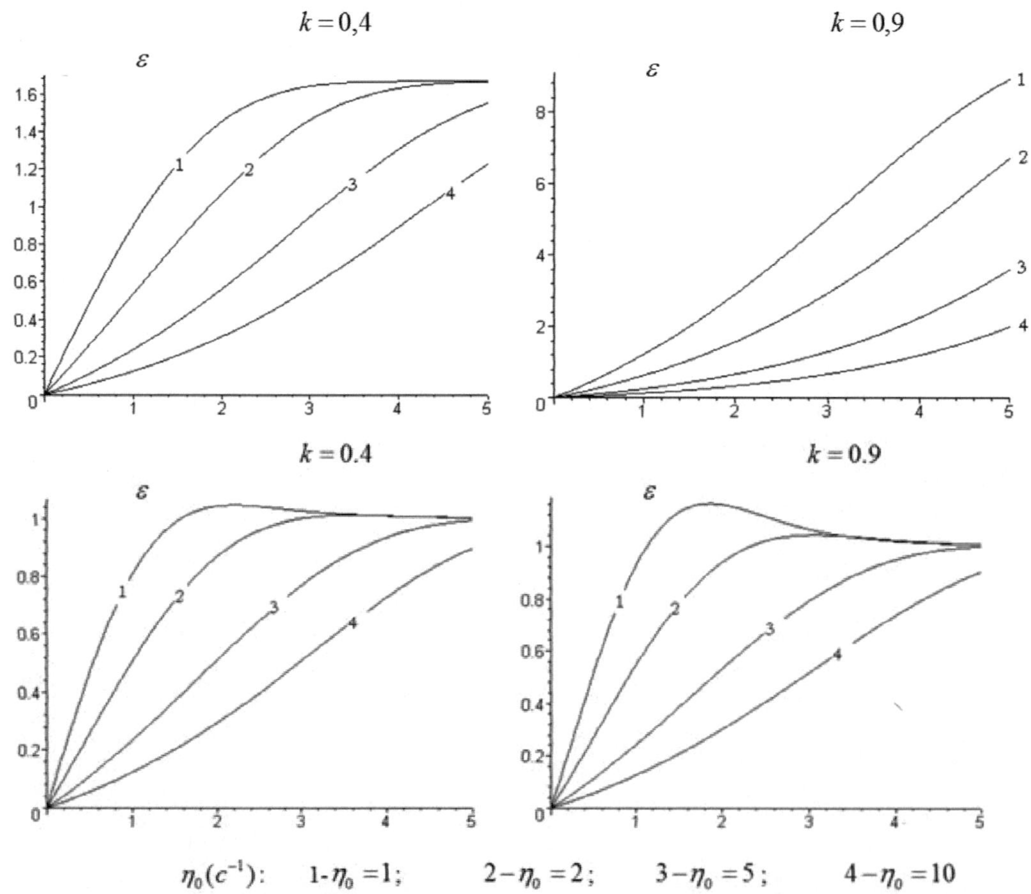

Figure 2. Deformation change $\varepsilon = \exp$ (related to σ_i / H) in time t at $\alpha_c = 0$ a), $\alpha_c = 1\,(c^{-1})$ б), $\beta_c = 0{,}5\,(c^{-1})$, $\gamma = 1$ and different values of k

The appearance of the maximum point is due to the fact that when $\alpha_c > 0$, according to dependence (3) the Young's modulus $E(t)$ also increases with increasing time and in certain combinations of parameters α_c and β_c the curve $\exp(t)$ reaches a maximum in a finite time interval.

A comparative table of the theoretical data obtained by the formula (2) with the experimental data presented in Fig. 1. In this case, the values $H, \alpha_c, \beta_c, \eta_0, k$ are calculated by minimizing the functional of the quadratic deviation of theoretical and experimental data, where the relative error between them is no more than 9%.

Table 2. Theoretical and experimental values of yarn deformation

Yarns indications	Kinds of yarns and initial conditions											
	Compact: $H = 400Па;\ \alpha_c = 0{,}7(c^{-1});\ \beta_c = 1{,}02($ $E_0 = 580Па;\ \eta_0 = 0{,}24(c^{-1});\ k = 0{,}1$						Usual: $H = 100Па;\ \alpha_c = 0{,}6(c^{-1});\ \beta_c = 1{,}95(c^{-1})$ $E_0 = 121Па;\ \eta_0 = 0{,}115(c^{-1});\ k = 0{,}11$					
t, c	0,5	1	1,5	2	2,5	3	0,5	1	1,5	2	2,5	3
$10^4 \varepsilon_{экс}$	0,4	1,6	3,2	5,3	7,4	10,0	2,24	8,1	13,2	14,6	14,6	14,6
$10^4 \varepsilon_{теор}$	0,37	1,52	3,48	5,72	7,77	9,2	2,22	7,6	12,1	14,3	15,3	15,7
Deformation error, (%)	3,5	4,95	8,86	8,02	4,95	7,68	5,1	5,75	7,89	1,59	5,17	7,89

From the analysis of tabulated data it follows that the elastic moduli H and E_0 of compact yarn slightly exceeds the moduli of conventional yarn, and the viscosity parameter β_c, characterizing the decrease in the modulus of elasticity has low values, therefore the compact yarn has higher strength characteristics. In general, it can be concluded that if the yarn deformation with respect to time starts with a positive acceleration, i.e. $\ddot{\varepsilon}(0) > 0$, then under the action of tensile force, the yarn can further acquire a structure with better mechanical properties.

Thus, to assess the deformation of yarns, taking into account its structure, the Kelvin model with time-variable parameters is validly adopted, where the instantaneous and long-term elasticity moduli of the material are taken into account. By means of a comparative study of the tensile strength of conventional and compact yarns, it has been established that for compact yarns, the instantaneous and long modulus of elasticity slightly exceed those of conventional yarns, and the viscosity parameter characterizing the decrease in the modulus of elasticity has low values (see Figure 3). This explains the increase in the strength characteristics of compact yarns.

IOP Conf. Series: Materials Science and Engineering **459** (2019) 012042 doi:10.1088/1757-899X/459/1/012042

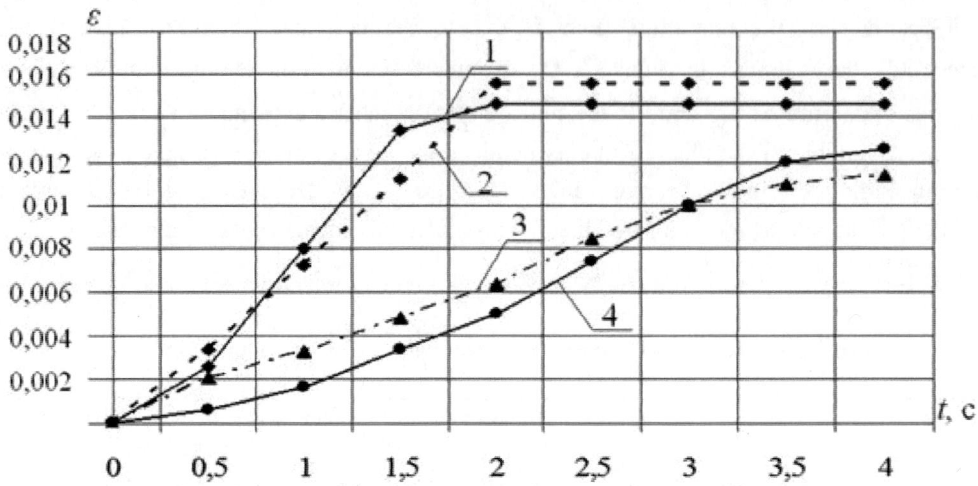

1-experimental curve of usual yarns, 2 – theoretical curve of usual yarns;
3- experimental curve of compact yarns; 4- theoretical curve of compact yarns

Figure 3. Deformation change in time at initial lading moments (spindle speed 13000 min^{-1} twist 740 tpm.)

Appendix Symbols Used

A	–	amplitude of natural vibrations of the spinning chamber, mm
A_1,	–	constant
A_n,	–	constant
C_1,	–	constant
C_2	–	constant
D	–	dispersion
E	–	Young's modulus, Pa
E_f	–	Young's modulus for fiber, Pa
E_{1f}	–	fiber hardening module, Pa
$E(t)$	–	long modulus of elasticity, Pa
E_e	–	energy density of yarn deformation
F	–	tensile force, N
F_c	–	centrifugal force, N
F_p	–	stretching force, N
F_{fr}	–	frictional force, N
G	–	shear modulus, Pa
G_t	–	lateral pressure, Pa
H	–	instantaneous modulus of elasticity, Pa
$H(t)$	–	Heaviside function
K	–	yarns twist, twm
K_f	–	actual yarns twist, twm

K_n – rated twist of yarn, twm
L_s – bundle length, mm
M_t – rotation moment, Nm
N – normal pressure force, N
P – force applied to the end of yarn piece, N
P_y – tensile strength of yarn, cN, N
R – yarns radius, mm
R_y – specific tensile strength of yarn, cN /tex
$R_{км}$ – yarn strength index, gs/tex
S_f – reduced cross-sectional area of the fiber, mm^2
S_y – yarn cross-sectional area, mm^2
T – линейная плотность пряжи, текс
U – fiber movement, mm
Q the immersive force, guiding the fiber into the tray of the chamber
a_1 – complex quantity
a_2 – mathematical expectation
a_3 – acceleration, m/sec^2
a_s – complex variable
b_s – length of slide zone, mm
b_t – width of the ribbon, providing a dense yarn structure, mm
d_f – diameter of fiber, mm
d_r – diameter of the spinning chamber, mm
d_y – yarns diameter, mm
e_0 – yield strength of yarn material, Pa
f – frequency of spinning chamber vibrations, Hz
g – reduced radial stress, Pa
h – winding height, mm
k – Coulomb friction switch
l_1 – triangle twisting height, mm
l_e – length of fiber end, detached from the matrix, mm
l_c – part of fiber length on which the disturbance propagates towards its center, mm
l_f – fiber length, mm
l_y – length of a piece of yarn, mm
m_f – mass of fiber, mg
n_r – rotational speed of the spinning chamber, min^{-1}
n_s – spindle speed, min^{-1}

r_0 – reduced fiber radius, mm
r_1 – radius to the fiber under investigation, mm
r_f – distance from the center of the spinning chamber to the fiber moving on its inclined surface, mm
r^* – relation of ρ to R
r_* – Relation of ρ_* to R
t – variable
t_f – the moment at which the fiber stops movement
t_y – time of yarns deformation, min
t_{fl} – yarns tear time, min
v – speed of fiber movement, m/min
v_1 – speed of shock wave propagation, m/min
v_y – yarns outlet speed, m/min
x – variable, mm
x_0 – length of rest zone, mm
x_1 – slide length, mm
x_2 – the distance between the fibers in bundle, mm
β – torsion angle, degrees
β_i – initial value of the torsion angle, degrees
β_* – torsion angle, limiting the sliding zone from the stretching zone, degrees
β_r – angle of inclination of the collecting surface of the spinning chamber, degrees
β_n – coefficient of thread nonlinearity
ε – breaking elongation, %
ε_1, – constant
ε_f – fiber deformation
ε_y – yarns deformation
ε_s – ultimate deformation upon reaching the zone of plastic deformation
η – viscosity modulus, Pa
θ – angle of orientation of the fiber in the yarn, degrees
λ – constant
μ – coefficient of friction
v_{12} – Poisson's ratio for longitudinal deformation of yarns
v_{12f} – Poisson's ratio for longitudinal deformation of fiber
ρ – distance from the center of the yarn to the fiber in question, mm
ρ_d – density of the material, mg / mm ^{2}
ρ_* – distance limiting the sliding zone from the stretching zone, mm
σ – axial stress of fiber, Pa
σ_a – axial tension of yarns, Pa
σ_1 – fiber tension in the sliding zone, Pa

σ_f – the stress acting on the fiber in the axial direction, Pa
σ_f^* – fiber tension in the rest zone, Pa
σ_i – initial tension, Pa
σ_r – radial stress between fibers, Pa
σ_s – ultimate fiber tension, Pa
σ_t – tangential tension, Pa
$\overline{\sigma}$ – average tension in fibers, Pa
ω_r – angular speed of the spinning chamber, sec $^{-1}$
τ – the dry friction force of a single fiber, N

References

[1] Xuzhong Su, Weidong Gao, Xinjin Liu, Chunping Xie, Bojun Xu. (2014), Theoretical Study of Yarn Torque Caused by Fibre Tension in the Spinning Triangle, *Fibres & Textiles in Eastern Europe*; 2014, Vol. **22**, № 6(108): pp. 41-50.
[2] Dehghan Mahdiyeha & Ekhtiyari Esfandiyar. (2014), Review of distribution of fiber tension at the spinning triangle. *Journal of the Textile Institute*, 2014, Vol. **105**, № 11, pp. 1167-1177.
[3] Platt MM. (1950), Mechanics of elastic performance of textile materials. *Textile Research Journal* 1950, Vol. **20**, №10, pp. 665-667.
[4] You Huh, Hyung Jin Kim, Jong Seong Kim. (2008) Tension dynamics of the fibers in a triangular plane flow of bundle. *Fibers and Polymers*, 2008; Vol. **9,** № 3, pp. 349-354.
[5] C. M. Deng, L. J. Wang, X. G. Wang. (2007), Modeling the Tensile Strains of Non-uniform Fibers. *Fibers and Polymers*, May 2007, Vol. **8**, № 3, pp. 289-294.
[6] Wenyi Yan1, Zhi Li, Xungai Wang and Lijing Wang. (2009), Dynamic Responses of Irregular Fibers under Axial Tension. *Journal of Applied Polymer Science*, 2009, Vol. **113**, № 4, pp. 2561-2568.
[7] J.K. Gafurov, B.M. Mardonov, M.H. Peerzada, K. Gafurov (2015), Investigating the Strain State of Fibres Located on the Helical Line in Extended Yarn. *Fibres & Textiles in Eastern Europe*, 2015; Vol. **23**, №2(110), pp. 19-24.
[8] J.K. Gafurov, A. Sobirov, H.T. Bobojanov, K. Gafurov (2013), Evaluation of yarn structure based on its deformation, *Problems of textiles*, 2013; № **4**, P. 46-51.

Fashion designer, fashion consumer, fashion learner.

M G Guedes[1] and A Buest[2]
[1, 2] University of Minho, School of Engineering, Textile Engineering, Campus de Azurém 4804 - 533, Guimarães, Portugal
[1] mgg@det.uminho.pt, [2] id3983@alunos.uminho.pt

Abstract. Fashion Design is a complex sector, formed by scattered consumers and companies, by non-linear models of production and by dynamic information flow. Sustained by technology pervasiveness fashion designers and fashion consumers create content around fashion brands, beyond the formality of products. New models of consumption fashion push new models of fashion design and should be pushing new models of learning. What kind of learning model can better prepare future professionals to contribute to the sector? This paper hypothesizes that online collaborative environments represent a valid alternative to recognize fashion designers, as fashion consumers, as 'lifelong' fashion learners. We confronted the results obtained in previous studies with recent data gathered from providers about consumer behavior on global online and mobile usage and penetration, fashion design sector in Europe and fashion design 'e-commerce' worldwide. The results demonstrated that although fashion figured as the most significant eCommerce segment in 2017, fashion design higher education programs offered online are either non-existent or merely transpose the traditional lecture-based education models to online settings. This paper also presents heutagogical principles for fashion design higher education, demonstrating how they could collaborate to bring together fashion consumption, fashion design development, and fashion design learning.

1. Fashion Design sector contextualization and the research problem.

Fashion Design belongs to a complex, demanding and technologically saturated sector, critical to the global economy, formed by scattered consumers and companies, and by non-linear models of production, distributed in an extended supply chain ([1]). Characterized by planned obsolescence, fashion products present short life cycles and vast differentiation, so brand loyalty is a necessity. In order to retain consumers loyalty, fashion brands need to captivate their consumers continually. In reality fashion is characterized by a high information flow, where the immaterial level has a greater significance and consumers buy 'access to ideas' ([2]) embodied in fashion products, or share the fashion brands values and lifestyle through meaningful experiences and points of interactions. The most influencing fashion brands communicate with hyper-connected, dispersed fashion consumers while exploring new business models. Benefiting from the mobile penetration and social media reviews, fashion brands capitalize on co-creation, personalization through data analytics, artificial intelligence (AI) or virtual reality (VR), on product quality and sustainability through 3D technology, on creating loyalty throughout the purchase journey integrating in-store and online services, on built communities around brand's culture, history. So, while fashion designers research in trend forecast online platforms, and create through specialized software, they also resort to social media and fashion consumers, lifestyle bloggers, vloggers, 'Instagrammers' that generate content around fashion, through fashion and for fashion. This 'fluid' landscape, where fashion brands, fashion designers, and fashion consumers interact in a seamless and symbolic dimension, beyond the materiality of products, as a global community engaged online. Non-traditional models of consuming fashion push new models of fashion design, new models of fashion retail, and should be pushing new models of fashion design learning. So, if we assume that fashion design higher education is immersed in this complex, technologically integrated, albeit dispersed landscape, what kind of learning model can better prepare future professionals to contribute to the sector? This paper hypothesizes that, if fashion designers, fashion consumers and fashion learners 'navigate' in a technology-immersed, online, global and fast-changing network, then a more evident fluidity between the development, the consuming and the learning of fashion design should be studied. We also hypothesize that fashion designers and fashion consumers are more inclined to become fashion learners and vice-versa. Therefore, this paper aims to demonstrate the need of a different approach for fashion design higher education, and that online collaborative, hybridized environments, represent a valid and innovative alternative for fashion designers/consumers/'lifelong' learners.

2. Theoretical Framework and methods

The theoretical framework and the methods adopted in this paper are presented in five phases, elaborated to confront the results obtained in previous studies about fashion design higher education in Europe, offered in online settings, with recent data gathered from providers of market and consumer research, in order to demonstrate how the online, digital and mobile worldwide usage and penetration, the fashion sector in Europe and the fashion 'e-commerce' worldwide relate to fashion design education online offer. Although focused in Europe, the comprehensiveness of the data researched is also worldwide oriented, since fashion brands, companies, designers, consumers, and learners share a global, borderless, timeless environment. This paper applied a quantitative approach to understanding whether and how technological development and adoption (online, mobile, social networking, AI, and VR) were influencing fashion sector and would consequently influence fashion learning contexts.

The first phase presents data from statistical sources that covers the period between 2015 and 2018, with projections to 2019 up to 2025. It focuses on world population, on internet, social media and mobile penetration and adoption worldwide ([3], [4], [5], [6]), on digital economy and eCommerce market, emerging trends, consumer behavior and business models ([7], [8], [9]). Next, to understand how the digital, online technology affects the education sector, the second phase of the study resorted to data centered on learning technology markets ([10], [11]) covering a period from 2016 - 2018, with projections to 2019 up to 2025. In the third phase, the study sought to understand how technology is affecting fashion design practice, fashion market trends, and consumer behavior. This study analysed information about the fashion industry in Europe ([12]), and more specifically about fashion eCommerce ([13]) with sectoral reports that provided a broad understanding of fashion and design trends, retail, consumption, as well as behavior changes in fashion consumer habits, desires and needs in relation to technology. The fourth phase meant to confront the fashion design education offer, with the scenario formed by the data gathered in the previous phases, in order to understand to what extent, it met market demands and professional/personal needs. Since this paper is part of a broader research, the perspective remained on fashion design programs, delivered online, in Europe, anchored in previous studies ([14], [15]) that had identified two programs, accordingly to the criteria stipulated: Architecture and Design, Fashion direction (*Architettura e Design Industriale, Indirizzo Moda*), provided by Università Telematica San Raffaele Roma and Design and Fashion Education (*Design e Discipline Della Moda*), offered by Università Telematica eCampus. From the relations previously established between heutagogical principles ([16]) and fashion design learning principles, a final phase is presented to demonstrate how the results from the previous four phases are relevant for the fashion sector as a whole and the fashion design higher education specifically. This fifth phase presents five recommendations to be discussed in the next section: *Reflective Thinking; Research and Interpretation; Creativity and Imagination; Collaboration and Communication; Complexity and Uncertainty.* The relations presented in the recommendations are neither direct, definitive, closed, exclusive, nor hierarchically organized. Resort to the heutagogical principles as a theoretical foundation for fashion design learning, requires understanding that they work systemically and can relate differently, in different contexts and for different purposes.

3. Results and discussion

This section presents the four 'themes' (paragraphs) that synthesize the results obtained from recent data gathered from providers of the market and consumer behavior, confronting it with the fashion design programs offered online. The first 'theme' presents results on global online and mobile usage and penetration. It dialogues with the second 'theme,' that presents data about the fashion sector in Europe and fashion 'e-commerce' worldwide. The third 'theme' presents data about the learning market in technological contexts, and the fourth theme presents the fashion design programs selected, in the light of the information gathered in the previous 'themes.' Finally, we discuss five recommendations.

From the 7 billion of the world population, more than half are the internet, mobile and social users and almost 2 billion are digital buyers ([3], [5]). In fact, eCommerce market revenue (€1,612,984m in 2018) will reach €2,353,428m by 2022 ([3], [5]). Fashion industry market value worldwide is US$3bn, (2018 statistics [12], [13]), and Western Europe is €333 million (US$380m). Interestingly, eCommerce market's largest segment is precisely Fashion (€472,245m, 2018), accounting for 28% of the total revenue ([13]) and 'Apparel' is Fashion's largest segment (€308,256m, 2018). Furthermore 'Fashion and Beauty' is also the common category in which consumers spend more ([3], [4]). In Europe, clothing segment accounted for revenue of €75,828m in 2016 and will remain among the most important digital markets by 2020. In 2017, Clothing and Footwear accounted for 40% online and 51% In-Store buying preference (from 24,471 respondents worldwide). Similarly, in a 2016 survey, products related to fashion ranked as the most popular online shopping category among 30,000 internet users, with 58% share ([13]).

The purchase behavior of fashion buyers is more participative when based on community inspiration and influence ([6]), as in recommendations and reviews. It is notorious how buyers and their peers engage and share views and opinions among them or with the brands ([7]). Social media integration, higher mobile internet connectivity, accessible smartphones and data tariffs, 5G technology (which will be commercially available by 2019), indicate a global shift to mobile ([4]). As a consequence, the mobile commerce is projected to reach $250bn in global sales by 2020 and mobile phone users are expected to surpass 5 billion in 2019 and 6 billion in 2025 ([4]). This trend reveals a new dynamic between creators, brands, companies, and consumers, increasing meaningful connections around the content, through digital platforms and services ([6]). The shift to mobile also resonates with the eCommerce Fashion segment, where unlimited mobile access and social media (especially Instagram or Snapchat) seamlessly integrate online and offline channels and services, from virtual fitting rooms to visual data analysis, creating a much more connected and personalized approach to consumer preferences.

In five to ten years global changes in demographics (35% growth of the +65y population), in the job market (instability, the constant need for professional improvement, career shifts) and disruptive digital technologies, accessible through mobile devices, will demand and promote more flexible, engaged and interactive learning experiences. Student-centered learning will also mean game-based and simulation-based learning, augmented reality (contextualization), virtual reality (personalization), m-learning, cognitive learning, micro-learning, high-velocity training, training through problem-discovery.

When confronting the structure, functionality, overall presentation and delivery model of the two fashion design programs offered online it became evident their distance from the digital, connected, flexible, mobile, collaborative, user-centered global scenario. The journey and the process are not straightforward, logical or quick and the interaction touchpoints, like links (repeated or deficient), buttons, search tools (non-existent or inefficient), the travel between screens, or the visual inconsistency, do not facilitate the information access and do not promote interaction or a smooth experience. The platforms are structured to support the physical campus, classes, and contacts, they are presented mainly in Italian, and the language options are not transversal to the whole content. This practice goes against English as the #1 of the 'most common languages for web content' ([3]). Both platforms present static pages, with large texts and links that open more static pages with more texts, and the Fashion Design curriculum does not informs about the adopted syllabus of learning units per academic year (e.g., 2017/2018 or 2018/2019). Specific lectures and final exams are structured in the traditional face-to-face model, in one of their physical campuses and following a fixed calendar. The social media presence is irrelevant or nearly non-existent, and it is a transposition from the desktop version, there is no space for engaged or interactive experiences. UT San Raffaele di Roma has 8.582 followers on Facebook, 147 on Twitter and 860 in Instagram and UT eCampus has 46.521 followers on Facebook, 1.954 in Instagram and 432 Youtube subscribers. UT eCampus app in Google Play has four users and a review of 1.5. UT San Raffaele di Roma does not have an app in Google Play.

3.1. Fashion design learning recommendations based on fashion design learning and heutagogical principles.

Digital connectivity, population aging, fewer jobs ([18], [19], [20]) are promoting the demand for self-improvement and a desire to keep learning, in a lifelong perspective. Under these circumstances, the learning experience must be accessible, flexible, mobile and integrated through different channels. In collaborative online learning environments, fashion design can harness deeper *reflective thinking*, while doing, mixing craftsmanship and technology. 3D design and printing, for instance, are possible ways to develop, evaluate, adapt concepts and ideas while prototyping, allowing the learner to reflect creatively upon his/her process. Fashion design learners could even share their proposals with colleagues and potential consumers, receiving feedback and learning to create collaboratively.

Especially in the fashion sector, production has reached the full potential globally, placing 'new designs' available to consumers in a few weeks. Fashion 'fast' and soulless cycle of change ruined Fashion's characteristics: originality, creativity, and fun. Ironically, digital technology, AI, VR, simulation, gaming can potentially promote *research and interpretation* in a more human approach to fashion design. They can 'zoom in' products, materials, people stories, diversity, intriguing imperfections, and singularities. Those technologies can potentially prompt fashion design learners to 'navigate' into an otherwise hidden world of textile fibers, a garment origin, product design and manufacture process and know more about the people who produce the clothes in different countries, their history, their culture, their knowledge. This path will create a sense of connectivity, purpose, and relevance that will impact fashion designers understanding about what Fashion is and their role in it. It a learning process focused on fashion products as a part of a global system formed by multiple perspectives, knowledge, and experiences ([8]).

The social network global expansion and penetration, as well as users high adoption, harnessed by digital tools and services promotes a collaborative and shared behavior, in which *creativity and imagination* are not individually make, but shared and explored together ([3], [8], [18], [19], [21]). However, what capacities a fashion designer is learning currently in order to create spaces, moments of co-creation, cooperation with their future consumers? Are they learning to develop fashion collections based on collective inspirations? Fashion design collaborative online learning should happen within a community of shared values and multiple viewpoints, so the learner acknowledges the contribution of colleagues and understands Fashion Design from a participatory model perspective.

The development of DNA, data science, and analytics as economic forces behind industries, systems, and processes will potentially change *collaboration and communication* within the fashion supply chain and the product life-cycle. This will impact one of Fashion's principles - planned obsolescence, since revive, redesign, extends a fashion product durability and reintroduction into the productive process. Fashion designers must understand materials, processes, and creation from both a molecular and systemic perspective. A much more efficient approach to fashion supply chain and fashion design is created since products will be developed to small segments instead of being stocked, and it will alter fashion traditional segmentation process. Fashion designers, resorting to 'emotional data analytics' technologies ([8]), should be learning to identify personal tastes and improve product design by knowing the full (re)living potential of a product or a collection.

Globalization and worldwide internet penetration also brought *complexity and uncertainty*, mistrust, misinformation. Under this intricated scenario, it is necessary to develop multiple perspectives and rely on trustworthy sources. We seek globally for information, but we validate it with our community, locally. Online social platforms are built around shared values, reliance, clarity. Especially in the case of fashion sector, which supply chain suffers from lack of transparency and trust, digital connectivity can potentially be employed to disclose fashion activities and create a much more aware and reliable community of fashion creators, consumers, and learners. Fashion design learning is to be structured around collaborative online environments, where connections and interactions are stimulated, and small (micro) networks or communities are created to contribute for knowledge and learning through the transference of their experiences, preferences and ideals in a personal and more meaningful level.

4. Conclusion

Technology quick changes are affecting perspectives on consumption, work, and education globally. AI, VR, 3D, 5G technologies will increase internet integration, prompt more flexible, mobile, social, collaborative business models, and change behavior patterns in different sectors, including in fashion and fashion education. It is necessary, then to regard fashion design learning under a holistic approach in which fashion designers, fashion consumers, fashion learners belong to a community connected researching, sharing, creating, buying, blogging, following, promoting, reviewing fashion, using a variety of apps, web services, and social platforms, enjoying the integration between physical and online channels. Confronting the data collected about foreseeing changes in the next five to ten years with data about fashion design higher education in Europe, it became clear that the education sector is missing the technological ubiquity and mobility, the flexibility and collaborative scenarios, the innovation and personalization of learning, the mixed role of the designer-consumer-learner altogether. Despite fashion figures among the most significant eCommerce segments worldwide, with an active and interactive presence of fashion brands, professionals, consumers and indeed students, the online, mobile, digital, innovative offer in fashion design higher education in Europe is limited, non-existent, or the ones existent are merely transpositions of the traditional 'onsite', lecture-based education models. There is a a constant resistance to consider alternative routes for fashion design higher education, especially in online, collaborative, innovative, digitally immersed settings. This resistance is coherent if we choose to hold on to the traditional roles of the fashion designer, to traditional manners of creating, manufacturing and selling fashion, to traditional ways of communicating and interacting with the fashion consumer. A new generation of fashion designers will relate and interact with their sector differently, so they will demand new ways of learning. Therefore, this paper intended to collaborate to present possible, but certainly not unique, educational pathways.

Acknowledgments

This work is financed by FEDER funds through the Competitivity Factors Operational Programme - COMPETE and by national funds through FCT – Foundation for Science and Technology within the scope of the project POCI-01-0145-FEDER-007136.

References

[1] Bohemia E, Harman K 2008 Globalization and product design education: the global studio. *Design Management Journal*, **3** issue 2 pp53-68. doi: 10.1111/j.1948-7177.2008.tb00014.x.

[2] Rifkin J 2000 Chapter three: the weightless economy. *The age of access: The new culture of hypercapitalism, where all of life is a paid-for experience.* [Kindle version] Retrieved from Amazon. New York: J.P. Tarcher/Putnam.

[3] Digital in 2018, essential insights into internet, social media, mobile and ecommerce use around the world [Internet]. We are social ltd and Hootsuite Inc [org]; 2018 [cited 2018 Jun 5] Available from: https://digitalreport.wearesocial.com/

[4] The mobile economy 2018 [Internet]. GSMA Association; 2018 [cited 2018 Aug 4] Available from: https://drive.google.com/file/d/1UeSn3p6Q9wF6KlSie5zECz1z4InOjJGA/view.

[5] eCommerce worldwide [Internet]. Statista Digital Market Outlook; 2018 [cited 2018 Aug 9] Available from: https://www.statista.com/outlook/243/100/ecommerce/worldwide.

[6] Digital Economy Compass 2018 [Internet]. Statista; 2018 [cited 2018 Jul 12] Available from: https://www.statista.com/page/compass.

[7] Commerce Global Web Index's biannual report on the last trends in online commerce [Internet]. Global Web Index; 2018 [cited 2018 Jun 12] Available from: https://www.globalwebindex.com/reports/commerce.

[8] Zero Hour Trend report 2018 [Internet]. We are Social; 2018 [cited 2018 Jul 17] Available from: https://thinkforward.wearesocial.com/

[9] Megatrend Analysis: Putting the Consumer at the Heart of Business [Internet]. Euromonitor; 2018 [cited 2018 Ago 3] Available from: http://go.euromonitor.com/white-paper-2017-megatrend-analysis.html

[10] Adkins, S S The 2016-2021 Worldwide Self-paced eLearning Market, Global self-paced eLearning Market in Steep Decline [Internet]. Metaari Advanced Learning Technology Research; 2016 [cited 2018 Ago 3] Available from: http://www.ambientinsight.com/Reports/eLearning.aspx

[11] Current Trends in the Global Education Sector [Internet]. Euromonitor; 2017 [cited 2018 Ago 3] Available from: http://go.euromonitor.com/strategy-briefing-2017-Trends-in-the-Global-Education-Sector.html

[12] Global Fashion Industry Statistics [Internet]. © Fashion United Business Intelligence; 2018 [cited 2018 Jul 5] Available from: https://fashionunited.com/global-fashion-industry-statistics.

[13] eCommerce Report 2018 – Fashion [Internet]. Statista Digital Market Outlook – Segment Report; 2018 [cited 2018 Jun 24] Available from: https://www.statista.com/study/38340/eco mmerce-report-fashion/

[14] Guedes M G, Buest, A 2018 Heutagogical principles for fashion design higher education. *Proc 10th Int Conf on Edu and New Learn Tech,* July, 2-4 pp. 8603-12 Palma, Spain ISBN: 978-84-09-02709-5 / ISSN: 2340-1117 doi: 10.21125/edulearn.2018 Publisher: IATED.

[15] Guedes M G, Buest, A 2018 Semiotic perspectives for a global fashion design higher education. *Proc 10th Int Conf on Edu and New Learn Tech,* July, 2-4 pp. 8613-23 Palma, Spain ISBN: 978-84-09-02709-5 / ISSN: 2340-1117 doi: 10.21125/edulearn.2018 Publisher: IATED.

[16] Hase S, Kenyon C 2007 Heutagogy: a child of complexity theory. *Complicity: An International Journal of Complexity and Education*, **vol 4** issue 1, pp 111–118.

[17] Apparel Market in Europe [Internet]. Statista; 2018 [cited 2018 Jul 30] Available from: https://www.statista.com/study/27600/apparel-market-in-europe-statista-dossier/

[18] WGSN Vision Team [Internet]. The Vision A/W18/19: The Thinker; 2016 [cited 2018 Jul 9] Available from: https://www.wgsn.com/content/reports/#/Future+Trends/w/The_Vision/26209

[19] WGSN Vision Team [Internet]. The Message 2020, Part 1: Fix the Future; 2018 [cited 2018 Jul 20] Available from: https://www.wgsn.com/content/reports/#/Future+Trends/w/The_Vision/27534

[20] WGSN Vision Team [Internet]. The Vision 2019: Part 2 – Purpose Full; 2017 [cited 2018 Jul 20] Available from: https://www.wgsn.com/content/reports/#/Future+Trends/w/The_Vision/27150

[21] Social Media Usage Worldwide [Internet]. Statista; 2018 [cited 2018 Ago 5] Available from: https://www.statista.com/study/12393/social-networks-statista-dossier/19

How Industry 4.0 concepts are applied in the Portuguese clothing industry: some evidences.

L T Ha[1], A D Marques[2] and F Ferreira[2]
[1]Industrial University of Ho Chi Minh, Department of Garment Technology – Fashion, 12 Nguyen VAN BAO, Ward 4, GO VAP DISTRICT, HO CHI MINH, VIETNAM
[2]University of Minho, School of Engineering, Department of Textile Engineering, Campus of Azurém, 4800-058 Guimarães, PORTUGAL
haluusgn@gmail.com

Abstract. Industry 4.0 is the current trend of automation and data exchange in manufacturing technologies, including textile and clothing industry. The concept of Industry 4.0 is based on the integration of information and communication technologies, industrial technology and is mainly dependent on building a Cyber-Physical System (CPS) to prepare a digital and intelligent factory, to become more digital, customized and (why not?) also green. The purpose of Industry 4.0 is to build a highly flexible production model of personalized and digital products and services, with real-time interactions between people, products and devices during the production process. This paper will show how a Portuguese clothing company (GUIMA) is preparing their processes, skills and resources to be in the Industry 4.0. The research team is following the industrial process for four months, starting from the reception and quality control of raw materials and accessories, until the orders expedition to different customers. Some results already obtained shows how the digitalization is yet in a preliminary stage in the clothing industry and is necessary to prepare the entire resources from the clothing company to achieve the horizontal and vertical integration in the manufacturing I4.0.

1. Introduction

The Portuguese clothing industry is a traditional, important and mature industrial sector. The textile and clothing industry (ITV) include 6,000 societies and 5,600 individual companies, employing a total of about 130,000 direct workers, producing 6.4 billion Euros and generating € 6.8 billion in turnover, in which € 4.8 billion are due to export activities [1].

The Fourth Industrial Revolution is taking place in this new millennium, also called the Digital Revolution, through technologies such as the Internet of Thing (IoT); Artificial Intelligence (AI); Virtual Reality (VR); Social, Mobile, Analytics and Cloud (SMAC) interaction, etc., to transform the real world into a digital world. In 2013, the new keyword "Industry 4.0" starts to emerge from a German government report that mentions this phrase in terms of high-tech strategy. Industry 4.0 is the current trend of automation and data exchange in manufacturing technologies that based on the integration of information and communication technologies, industrial technology and is mainly dependent on building a Cyber-Physical System (CPS) to prepare a digital and intelligent factory [2]. The purpose of Industry 4.0 is to build a highly flexible production model of personalized and digital products and services, with real-time interactions between people, products and devices during the production process [3].

Industry 4.0 is the intelligent interaction between machine-to-machine and human-to-machine through sensors, RFID, barcodes and tags, all of it used to collect data in the real production process. Industry 4.0 is driven by digitization and integration of vertical and horizontal production chains and end-to-end value chains. Vertical and horizontal integration allows the company to share real-time data, product specificity of each location, real-time productivity of workers, units produced, and data are quickly accessed and shared. The use of sensors and RFID makes the access and retrieval process fast and accurate. Finally, end-to-end use to the product development process [4]. Through IoT and CPS, the information is shared, monitored, made decentralized and self-optimizing. Real-time communication Machine-to-Machine (M2M) and Machine-to-Person (M2P) are connected in standard protocols [5],[6].

Industry 4.0 facilitates the creation of "smart factories" or "digital factories." With IoT, these virtual physics systems interact with each other and with people in real time, and through IoS the user will be involved in the value chain using these services. Industrial control systems will become far more complex and widely distributed, enabling flexible, fine-grained process control. Smart and connected embedded devices will be everywhere. Sensors and radio frequency identification technologies (RFID) will be assisted in every stage during the manufacturing process, providing the raw data as well as the feedback that is required by control systems. Industry 4.0 will be the virtualization and modularization of manufacturing and supply chain processes, achieving flexibility and personalization based on CPS and IoT along with ERP, MES, PLM, SCM [7].

According the last data of Digital Economy and Society Index 2017 [8], the Portuguese companies have integrated the Digital Technology in their processes: 44% of them are using Electronic Information Sharing (EIS) and 8% RFID.

2. Methodology Research

The research method used in this paper is the case study and documental analysis [9], from that using evaluated, estimated, tested standards and to predict trends of evolution in the future for throughout Bayesian process [10]. This is a method which can give some indications and allow further elaboration and hypothesis creation for companies that have similar conditions [11],[12].

The way to be performed this method is gathering data of the Portuguese clothing company GUIMA which is preparing their processes, new technologies, resources and professional skills to be in the Industry 4.0 revolution. This enterprise is a clothing company firm belongs to a bigger textile group (SOMELOS SGPS), with vertical production (spinning, warping, sizing, weaving, dyeing and finishing) and own brand in their fabrics. SOMELOS Fabrics is the main supplier of raw materials of GUIMA (fabrics to produce the GUIMA's shirts).

The research team followed the industrial process of this company for four months, with targets and check points to be followed and analysed such as the reception of raw materials and other items for the shirts, quality control of raw materials and accessories, production processes (quantities, efficiencies and lean times) and closes with the orders expedition to different customers.

3. Case study: GUIMA

The clothing production process starts with yarn, fabric, accessories and apparel manufacturing. Fashion accessories industry is known as one of the parts of the global fashion industry, which is the place that logistic providers are using many supply chain management (SCM) and used in retail stores.

In 2016, GUIMA was beginning to deploy a developed solution called FluxoConf. The aim of this solution is to support the process of storage, production, monitoring of production, efficiency analysis in order to make a controlling decision timely. This solution is focused in some respects as:

- Orders are ready to go into production;
- Collect average real time in production (product unit/ time unit);
- Evaluate individual average levels of productivity;
- Positioning of parts in production;
- Production efficiency of company.

The case study research method is applied in context of GUIMA that will be categorized into three steps (figure 1): Preproduction processes, Production processes and Postproduction processes (Classification in basic steps in the apparel production [13][14]), they are going to be described for each of steps follow by:

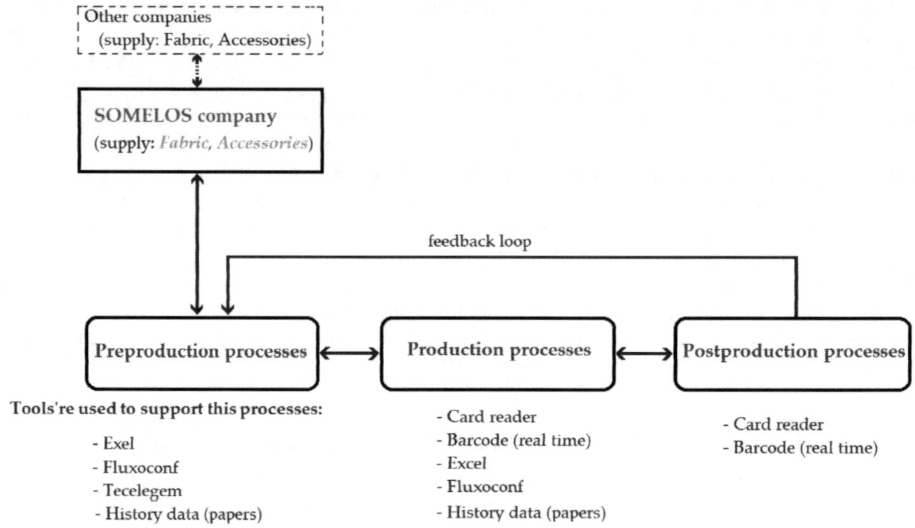

Figure 1. GUIMA processes system

3.1. Preproduction processes

The GUIMA uses Tecelagem software to create the production planning program.

-First, they are sent to the involved sections to find out the price of the items; initial estimate consumption of fabrics, trims, threads and materials for each unit of product as well as for all products in the orders respectively.

-Second, they will analyze the real productive capacity available (i.e. these capacities are depended on aspects of FluxoConf solution) to give a confidence interval for established order deadline and to reply to customers about delivery time.

-Third, using Tecelagem software to buy fabrics from Somelos in the shortest time. This software is supplied by Somelos Company.

Moreover, this department is also responsible for preparing all information for GUIMA's production such as patterns, cutting plans, technical assembly and the consumption of materials. They use CAD/CAM provided by the Gerber software. However, on the same time, the consumption of materials still uses an excel template and has not been integrated into the Fluxoconf software. Assembly steps are created without the help of the software and then go to the production department to join specific work steps.

The next step is designing pattern and preparation of the cutting plan (marker). GUIMA works with a CAD/CAM system provided by Gerber software. Patterns are created and then they are used to prepare the cutting plan. Patterns are scaled in the various sizes according the commercial order from clients in obtain the best fabric consumption. Cutting plan must be carefully considered because the most significant part of the cost of a clothing item is imputed to the material, and so any reduction in consumption leads to an increase in expected profit. The plans are made according to the orders of the various models and customers.

Fluxoconf is used in the preparation of the work phases and the documents have still been given to the production department by hand to ensure that the production will be done on time.

3.2. Production processes

First, this department receives, checks and counts the materials, trims, threads and accessories from preproduction processes, to ensure they are enough to complete the order before to go to the production process. In this period, they create and describe very specific operations step by step as well as time to imply them (the document will define the sum of the duration of operations of each part and the cost to be assigned to the product/minute), timekeeping management and always coordinates with other departments to handle any arising problems (if any) in the production process. Warehouse management, productivity and improvement of sewing performance are also based on software data to monitor.

The fabrics and accessories delivered to GUIMA are stored in the warehouse on shelves for future cutting and using operation. The cutting part is only performed after this department was received all the information related to the needs of fabric and accessories, the quantities to be produced and the deadline of delivery to the customer, and then material requirements are forwarded to their warehouses.

The storage process was also restructured by a program developed by Fluxoconf that allows assigning locations to the shelves by "street", "letter" and "number". This program allows in the entrance of fabric to be assigned a location that identifies its position in the warehouse. This process allows extremely fast location to meet the needs for a order. Another advantage is that it is possible to control all the entrances and exits of fabrics and other items, thus maintaining always updated the material stocks. The rack-to-rack location process was also applied to the accessories.

Fabrics are spread in flat tables and cut by a knife cutter. After cutting, the components are classified separately, bundled and separated the bundle with the instruction of documents, that include barcode, and continue forward to the sewing operations. This document use barcode to read production data and follow up to the finishing.

The process of sewing starts with the distribution of production batches by the machines compatible with the operations to be performed to finish the shirt. At each workstation, the operator has access to detailed information, step-by-step with a description of what to do as well as how to do it, and all information are showed for workers in the next step before they will do it. All this data is collected by barcode.

The program collects the quantity of products that each worker can produce in a day and provides it to the data having production monitoring on real time. The information flow system is created by using a JAVA application after they are collected by barcode. Before beginning any operation, the worker must identify the machine in which is working using the accurate function to show all information and to identify about status of every corresponding product.

At this stage, a barcode reader is attached to each machine. At the beginning of the task, the barcode entry to know exactly what to do and identifying the work. This period ends with the barcode scanner to complete the work. Then, the data is updated automatically into the system and managers can keep track of worker's productivity. The computer analysis done by the algorithm that has been installed, calculates several information for reporting and the degree of completion of work is displayed by different colors. At the end, workers are going to scan with a bar code reader for final product confirmation and the computer report the final output with the corresponding real-time.

3.3. Postproduction processes

Processes include thread trimming, pressing, inspection and quality control, folding, packaging and expedition. A barcode is randomly generated in the process of cutting and used for the finishing process. This code is also used in logistic and commercial.

3.4. Bayesian process for GUIMA

It is proposed to build a mathematical model supported by a Bayesian process and programed languages to GUIMA's production system achieve higher results (Figure 2). Bayesian process in production system is applied from preproduction processes to postproduction processes is having a purpose to estimate work efficiency as well as detecting for possible errors in each of the smallest and most specific details.

Moreover, this process also provides forecasts of product's productivity or any factor in system that can influence to production efficiency and is helping to have better decisions and controlling on time. Bayesian processed model integrated into one mechanism that is learned from data of previous process will give out important experiences and data for next process - this is one of automated applications in machine learning and Artificial Intelligence fields.

Tools are intended to support this proposal that are programmed languages (Java, Pythons or Matlab) to simulate the Bayesian process.

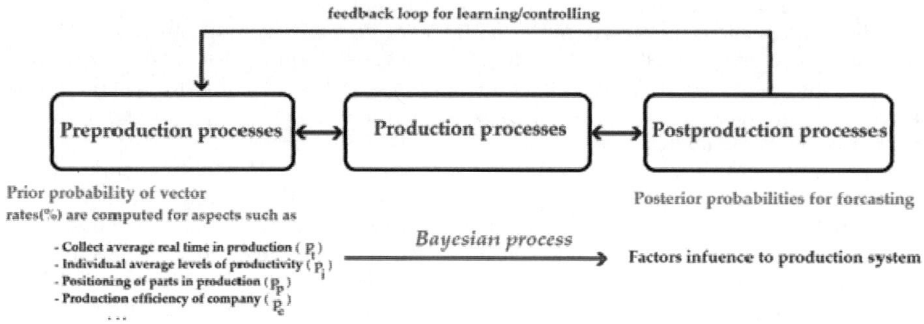

Figure 2. Bayesian process for production system

4. Conclusions

Throughout some evidences in the status of GUIMA that has already surveyed and characterized, shows how the digitalization is yet in a preliminary stage in the clothing industry. Some months ago, even the stocks of raw material were without a suitable organization to advance to digitalization.

The Industry 4.0 is a very important oriented approach for many fields and also for the competitiveness of the clothing industry in particular.

Industry 4.0 has two important components in manufacturing processes: the horizontal and vertical integration [15]. Horizontal integration is the connection between devices, workstations or separate plants, and it is also the cooperation between the different companies involved. Vertical integration in manufacturing is involving the actuators, sensors, controllers, production management, production and business planning starting from the suppliers until the customers. So, these two factors are necessary prepare the entire resources from the clothing companies (both human and technology) as well as organizing systems and processes to achieve this preliminary stage in GUIMA clothing company.

Acknowledgments

"This work is financed by FEDER funds through the Competitivity Factors Operational Programme - COMPETE and by national funds through FCT – Foundation for Science and Technology within the scope of the project POCI-01-0145-FEDER-007136".

COMPETE PROGRAMA OPERACIONAL FACTORES DE COMPETITIVIDADE
QREN QUADRO DE REFERÊNCIA ESTRATÉGICO NACIONAL PORTUGAL 2007.2013
UNIÃO EUROPEIA Fundo Europeu de Desenvolvimento Regional
FCT Fundação para a Ciência e a Tecnologia

References

[1] ATP 2017 *Estatísticas da ITV 2016* (Porto: Edições da ATP)

[2] Kagermann H, Wahlster W and Helbig J 2013 *Recommendations for implementing the strategic initiative INDUSTRIE 4.0.* Final report of the Industry 4.0 working group. (Berlin: National Academy of Science and Engineering)

[3] Zhou K, Liu, T and Zhou, L 2015 Industry 4.0: Towards future industrial opportunities and challenges, *Fuzzy Systems and Knowledge Discovery (FSKD),* 12th International Conference on Fuzzy Systems IEEE, pp. 2147-52

[4] Wang, S, Wan, J, Zhang, D, Li, D and Zhang, C 2016 Towards smart factory for industry 4.0: a self-organized multi-agent system with big data based feedback and coordination *Computer Networks*, **101**, pp.158-168.

[5] Hermann, M, Pentek, T and Otto, B 2016 Design principles for industrie 4.0 scenarios. In *System Sciences (HICSS), 2016 49th Hawaii International Conference on* IEEE, pp. 3928-37

[6] Bücker, I, Hermann, M, Pentek, T. and Otto, B 2016 Towards a methodology for Industrie 4.0 transformation. In *International Conference on Business Information Systems* (Cham: Springer), pp. 209-221

[7] Brettel, M, Friederichsen, N, Keller, M, and Rosenberg, M 2014 How virtualization, decentralization and network building change the manufacturing landscape: An industry 4.0 perspective. *International Journal of Mechanical, Industrial Science and Engineering*, **8** (1), pp. 37-44.

[8] European Commission 2017 *Digital Economy and Society Index 2016 – Portugal* https://ec.europa.eu/digital-single-market/en/scoreboard/portugal (accessed on 2 of May 2018)

[9] Saunders M, Lewis P and Thornhill A 2009 *Research Methods for Business Students* (London: Financial Times Prentice Hall)

[10] Sebastien T. 2010 Sales forecasts in clothing industry: The key success factor of the supply chain management. *International Journal of Production Economics*, **128** (2) pp.470–483

[11] Orum, A.M., Feagin, J.R. and Sjoberg, G., 1991 Introduction: The nature of the case study. In *A case for the case study*, (North Columbia: UNC Press Books), pp.1-26

[12] Feagin, J R, Orum, A M and Sjoberg, G 1991 *A case for the case study* (North Columbia: UNC Press Books)

[13] Nayak, R and Padhye, R (Eds.) 2015 *Garment manufacturing technology* (London: Elsevier)

[14] Dissanayake, G and Sinha, P 2015 An examination of the product development process for fashion remanufacturing. *Resources, Conservation and Recycling*, **104**, pp. 94-102

[15] McGuire, T W and Staelin, R 1983 An industry equilibrium analysis of downstream vertical integration. *Marketing science,* **2** (2), pp.161–191

IOP Conf. Series: Materials Science and Engineering **459** (2019) 012045 doi:10.1088/1757-899X/459/1/012045

Sensory analysis: approach for total handle evaluation of wool-type fabrics

R Harpa[1], C Piroi[1] and M Blaga[1]

[1]"Gheorghe Asachi" Technical University of Iasi, Faculty of Textiles, Leather and Industrial Management 29 D. Mangeron Blvd., 700050, Iasi, Romania

rharpa@tex.tuiasi.ro

Abstract. This research investigated the tactile sensory analysis of total handle in a range of wool-type fabrics. Considering the interaction of textile materials from garments with the consumer's senses as an essential performance property, seven wool-type fabrics selected as suitable for the final use, which is women office trousers for the cold season, were evaluated by a panel with non-experts for assessing the Total Handle Value. Experiments were conducted in two successive stages: by the individual panelists (as blind handling) and by the panel of assessors for consensus (visualization-with-handling). Previously developed ICT-based learning tools, STAT-Hand and STAT-ConCor, were used for both, the sensory analysis trials accomplishment and the processing of the total handle results. One outcome was that a panel of non-experts can get a good enough degree of agreement among individuals after the blind handling of a range of fabrics and, statistically, this cannot be considered accidental. The result of the consensus reaching on a panel level, through visualization and handling, about the Total Handle Value was not as good as expected and this was argued in the paper. After the blind handling, within the panel of individuals and visualization with handling as a panel consensus, the most proper fabric for women office trousers for the cold season was emphasized from the selected range.

1. Introduction

Considering the well-known definitions for fabric handle, the approach of fabric quality by sensory analysis is a widespread practice, due to the physical interaction with the textile product [1-7]. Regarding the total handle, this is not a property in itself but it was defined as a summary of all the feelings of a person when handling the fabric so, the total handle subjective evaluated could be assessed because the overall sensory analysis allows emphasizing the overall feelings of the fabrics properties linked to the sensorial comfort as handle attributes. Consequently, an overall fabric quality measure associated with fabric sensory properties, used by panelists as subjective judgment criterion, can be defined as Total Handle Value subjective evaluated (THVs) [1, 3, 5-11] .

An earlier work approached the subject of wool-type fabrics selection for ensuring sensorial comfort in women's clothing, using the multi-criteria decision analysis based on a rating scale delivered by customers (participants in a survey) with knowledge in the textile field but non-experts in the fabrics hand evaluation topic [12].

The current study aims to investigate the results of subjective evaluation sessions for THVs and to highlight the links with the virtual perception of potential buyers, expressed in the survey mentioned above. The sensory analysis will be conducted by a panel of master students for the same collection of wool-type fabrics and carried out in two successive stages, considering just the subjective assessment of Total Handle Value : first by "handling" (as blind individual evaluation of each panel member) and afterwards, by "visualization-and-handling" (as visualization and consensus evaluation from the entire panel).

2. Materials and Methods

2.1. Fabrics

Seven wool-type fabrics were chosen for this study for their similarities as a range of materials suitable for the final use that is women office trousers for the cold season [12]. Table 1 summarizes the main description of the selected fabrics with codifications F1 to F7 by including: pattern, weave and fibre content, yarn count size for warp and weft, warp, and weft density.

The seven wool-type fabrics where tested for properties linked to the sensorial comfort (flexibility, thickness, and weight) under controlled environmental conditions.

Table 1. Characteristics of the selected wool-type fabrics.

Fabrics	F1	F2	F3	F4	F5	F6	F7
Pattern	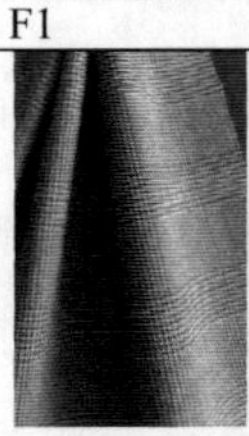		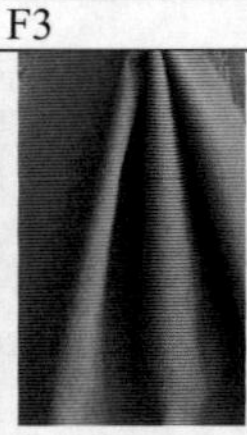			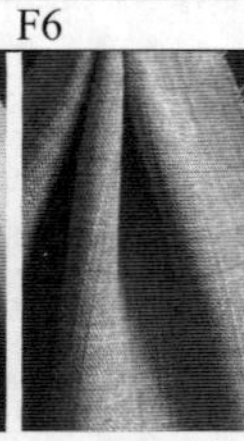	
Weave type	plain	plain	plain	twill 2/2	plain	twill 2/2	twill 2/1
Fibre content	PES 64% /CV 34% /EL 2%	Wool 45% / PES 55%	PES75%/ CV23%/EL 2%	Wool 100%	Wool 44% /PES 54% / EL 2%	Wool 45% /PES 55%	Wool 44% /PES 54% / EL 2%
Yarn count size (warp & weft)	Nm 64/2	Nm 64/2	Nm 52/2	Nm 72/2	Nm 64/2	Nm 64/2	Nm 56/2
Warp density /10cm	336±10	254±10	284±10	350±10	234±10	338±10	313±10
Weft density/10cm	280±10	230±10	244±10	316±10	222±10	288±10	194±10
Fabrics properties linked to the sensorial comfort							
Flexibility, %	64	51	67	53	63	58	61
Thickness, mm	0.43	0.35	0.34	0.36	0.36	0.49	0.48
Weight, g/m^2	216	167	180	182	158	210	220

2.2. Previous work

A former paper advanced the subject of wool-type fabrics selection for ensuring sensorial comfort in women's clothing by using the previously developed software for multi-criteria decision analysis STAT-ADM, based on a rating scale with established criteria for rating given not by experts, but by a large group of consumers as decision makers [12]. The analysis criteria for STAT-ADM were three fabrics properties (flexibility, thickness, and heaviness) tested in the textile laboratory (see table 1), selected due to the estimated links with the sensorial comfort defined by three groups of bipolar attributes (flexible/stiff, thin/thick, and light/heavy). The ranking has been achieved from a survey with nineteen questions about preferences manifested in buying clothes, conducted on a sample of 187 consumers with background in the textile field (women aged between 18 and 60 years). One aim of the survey was to ask women to make a hierarchy for the virtually feeling of sensory comfort, in the idea of purchasing office trousers for the cold season, made of wool-type fabrics. The survey result was the "niche reaction to the market" and the virtual feeling of sensory comfort was characterized by a hierarchy of bipolar attributes in the following order: thin/thick, flexible/stiff, and light/heavy.

Therefore, the criteria grading of the seven wool-type fabrics for running STAT-ADM was carried out according to the ranked options reached from the survey. Hence, a hierarchy of the seven wool-type fabrics was obtained in descending order of preference (from the best to the worst choice) for office trousers suitable for the cold season, as follows F7, F1, F6, F3, F5, F4 and F2.

2.3. Subjective evaluation of fabrics' total handle

Sensory analysis techniques are widely used for subjective assessing of the fabrics hand, mostly for garments design [2-4, 7-9]. In pursuit of making a subjective evaluation that is characterized by accuracy and reliability, a greatest attention should be paid to the following factors.

Selection of the evaluators for the panel configuration

Because the sensory analysis uses by default the human senses, the panel must be composed of six assessors at least, hence, for this work, the panel of evaluators included a homogeneous group consisting of eight master students as experts, having similar preferences in fashion style.

Subjective assessment procedure for the tactile properties perception

The panel has been trained to evaluate a wide range of woven fabrics by quantifying their individual feelings when blind handling for the evaluation of the mechanical and surface properties by means of bipolar physical attributes of fabric handle. However, the assessment technique applied for this paper was just a free fabric manipulation for total handle evaluation, having in view only the definition of total handle and the proposed fabric end-use. The assessment was carried in two successive stages: by the individual panelists (as a "blind handling") and afterwards, by the panel of assessors (as a "visualization-and-handling" for panel consensus).

Ranking the Total Hand Value during the subjective evaluation trials

After the total handle sensory analysis, the ultimate goal was to correlate the evaluators' feelings with the fabrics' end use. Therefore, it was assumed that grades should be given according to the evaluators' purchase preferences for each fabric item as if they were wearing office trousers for the cold season made of each of the evaluated fabric. For quantifying their sensorial feelings during fabrics handling, panelists were taught to use a rating scale for THVs, with grades from 1 to 5 (the most appropriate to the most inappropriate for the end use).

Analysis' strategy of the evaluation results

The panel performance checking is mandatory and needs specific techniques for data analysis, to establish the reliability and validity of the sensory analysis accomplished by the panel of assessors. For the selected case study, previously developed interactive ICT-based learning tools STAT-Hand and STAT-ConCor [10, 11].

3. Results and Discussion

3.1. On the subjective evaluation of Total Handle Values

After using STAT–Hand software for the blind evaluation trials, the THV profiles of the seven wool-type fabrics, based on the assigned grades awarded by each of the eight evaluators, are quite different, as are the fabrics (figure 1).

For each of the three fabrics (F2, F3 and F6) the evaluations were more heterogeneous and different grades for the same fabric were obtained, which rated it from the most appropriate to the most inappropriate material for the designated destination.

For each of the other four fabrics (F1, F4, F5, and F7), some individuals ranked similar or remarkably close the fabric suitability to a specific end-use. Evaluators E01, E05 and E06 awarded the grade 1 to fabric F1 (considered to be the most appropriate for the intended end use) and evaluators E02, E03, E04, E07 and E08 awarded the grade 2 (they rated the fabric as appropriate for the end use). For the fabric F4, evaluators E01, E02, E05, E06 and E07 awarded the grade 3 (as moderately proper for the intended end use) and evaluators E03, E04 and E08 awarded the grade 2 (they rated the fabric as right for the end use). For the fabrics F5 and F7 the assessors judged also in an analogous way.

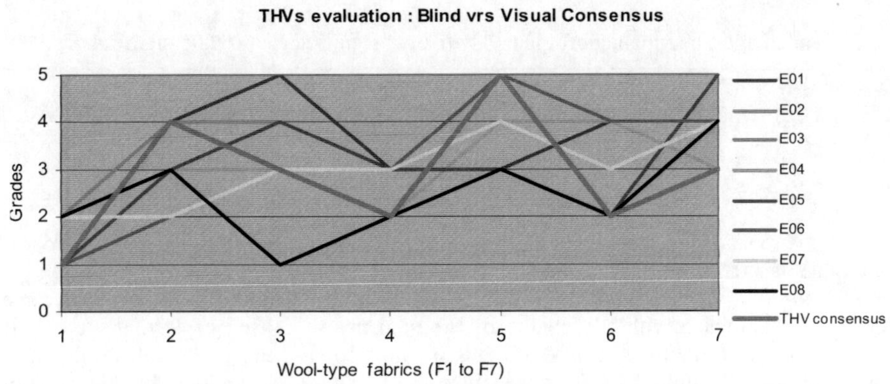

Figure 1. Total Handle Values subjective evaluated: "blind handling" and "visualization with handling".

In the figure 1, the red line stands for the Total Handle Values subjective assessed by the panel of assessors for consensus after "visualization-and-handling". In this case, obviously the "consumer perception" about the overall sensorial comfort potentially assured by each of the seven wool-type fabrics was influenced not only by the fabric properties linked to the overall sensorial comfort perceived when handling, but also by the aesthetic properties perceived visually. The consensus in the panel had the following results: for the fabrics F1 and F4, the grades were very close to the blind evaluation and for the fabrics F6 and F7 an evaluation very different from the blind handling was achieved. For the fabrics F2, F3 and F5, the feelings' meaning after "visualization-and-handling" had only a few tangents with the perceptions of some assessors at the blind handling. Altogether, according to the consensus after "visualization-and-handling", the seven wool-type fabrics have the following potential for the designed end use as women office trousers for the cold season: fabric F1 was chosen as the most appropriate for the end use (grade 1); fabrics F4 and F6- appropriate (grade 2); F3 and F7 -moderately appropriate (grade 3); F2 -inappropriate (grade 4) and, fabric F5 was chosen as the most inappropriate for the end use(grade 5).

In terms of common results, the blind subjective evaluation of Total Handle Values can be considered as moderate compared to the panel evaluation, achieved after the visualization and handling.

However, only by studying the THV profiles after both, individual blind handling, and visualization-with-handling at the level of the entire panel, it is hard to predict whether the panel structure with non-expert evaluators may be endorsed as a disturbing factor for the data accuracy.

3.2. On the rating consistency after blind evaluation vs. visual and handling evaluation

The outcomes from the overall processing with STAT-ConCor facilities of the subjective evaluation results after both, blind and visualization for consensus are presented in table 2. This summary table includes overall results by respecting the usual calculation relationships for each of coefficients (Kendall and Spearman), and the statistical significance tests to prove coefficients' statistical significance, considering the "null hypothesis" as a general statement that there is no relationship / agreement among the rankers.

Regarding the reliability of the panel with non-experts, the Kendall's coefficient of concordance value was 0.393, which means that, during the blind handling the eight evaluators (master students) were in poor to medium agreement on the suitability of the seven wool-type fabrics in women office trousers for cold season. However, following the results of the statistical significance test, the coefficient of concordance was significant at a level of 0.05. This situation strengthens the hypothesis that there was agreement between non-expert assessors during the blind handling evaluation but with a lower than a middle one degree.

Table 2. Kendall and Spearman coefficients' statistical significance (at 0.05 significance level).

Chi-Squared Test (*Kendall's coefficient of concordance*, W=0.393)		Student's t Test (*Spearman's correlation coefficient*, r_S=0.571)	
Computed value	Critical value χ^2 0.05,6	Computed value	Critical value t 0.05, 5
18.86	12.6	1.56	2.01

Regarding the connection between the two ways of evaluation the Spearman's correlation coefficient value was 0.571 so, it was emphasized a better than medium level of correlation between the blind evaluations from individuals and evaluations with visualization from the panel consensus. Nevertheless, in accordance with the results of the statistical significance test, the coefficient of correlation was insignificant at a level of 0.05. This situation strengthens the hypothesis that there was no agreement between the blind handling evaluation and visualization handling for this panel of non-expert assessors.

Table 3 summarizes an overall hierarchy (after three approaches) of the seven wool-type fabrics in descending order of preference for destination. The greatest ranking for THV could be regarded as the best fabric that would be designed and manufactured for specific end use.

Table 3. Overall hierarchy of seven wool-type fabrics after sensory and virtual analysis.

Ranking for the intended end use / Fabric evaluation	Most proper – appropriate	Moderately appropriate	Inappropriate-most inappropriate
Blind handling (panel individuals)	F1, F4	F2, F6, F3	F5, F7
Visualization and handling (panel consensus)	F1, F4,	F6, F3, F7	F2, F5
Virtually (without visualization and / or handling) [12]	F7, F1	F6, F3, F5	F4, F2

Altogether, from the above range of wool-type fabrics, the fabric with the best evaluated overall comfort seems to be F1.

4. Conclusions

In this paper, sensory analysis techniques were applied for a collection of seven wool-type fabrics, considering just the subjective assessment of Total Handle Value (THVs) by both, "blind touch" and "visualization-with-touch". To be consistent with the reality when it comes to the consumer niche demands, a reliable sensory analysis of the THVs was considered to have a key position in fabrics' choice for the specific end-use, which is women office trousers wool-type designed for the cold season.

Altogether, an interesting outcome was that a panel of non-experts with a low degree of unanimity among the individual responses after the blind handling of each fabric is good enough and, statistically, cannot be considered accidental. The reaching of consensus regarding the Total Handle Values at the panel level through fabrics visualization and manipulation had a less good result than expected although it was supposed that the visual and handling perception to provide a better assessment of fabrics than the blind handling. It seems that when different evaluators are looking at a particular fabric and handle it, they are influenced in evaluations do to the aesthetic aspects of the fabric, depending on the personality of each one, and subconsciously can create links with the final destination of the material, altering the objective manner on which a consensus can be achieved. Thus, for the selected fabrics, the sensory analysis trials provided better results when the evaluation was with blind handling than with visualization and handling.

The current study also aimed to analyse the results of subjective evaluation sessions for Total Handle Value and to highlight the links with the virtual perception of potential buyers, expressed in the survey and the outcome was the following: from the seven wool-type fabrics after blind handling (panel individuals), visualization with handling (panel consensus) and in a partial consent with virtually assessment for multi-criteria decision making (without visualization and handling), the most appropriate wool-type fabric designed for women office trousers for cold season, was the plain fabric F1 made of PES 64%/CV 34% /EL 2% .

The obtained results highlighted the sensitive aspects of the textile value chain, which can turn from strengths to weaknesses due to the influence of the potential buyers' preferences and it should be basic to consider the subjective perception of the consumer niche on the sensorial comfort properties expectations.

References

[1] Stearn A E, D'Arcy R, Postle R and Mahar T J 1988 A Statistical Analysis of Subjective and Objective Methods of Evaluating Fabric Handle. Part I: Analysis of Subjective Assessments *J. Tex. Mach. Soc. Japan* **38** (8) pp 62-68

[2] Naima H A, Mehdi S and Morched C H 2013Training and Control of Performance of a Tactile Sensory Panel *Journal of Applied Sciences* **13** (3) pp 366-376

[3] AATCC EP 5-2011 Fabric Hand: Guidelines for the Subjective Evaluation 2012 *AATCC Technical Manual* (American Association of Textile Chemists and Colourists) pp 388-390

[4] Schacher L, Bensaid S, El-Ghezal Jeguirim and Adolphe D 2011 Sensory and Physiological Issues *Advances in Modern Woven Fabrics Technology* ed Dr Savvas Vassiliadis (InTech)

[5] Valatkienė L and Strazdienė E 2006 Accuracy and Reliability of Fabric's Hand Subjective Evaluation *Mater. Sci. (Medžiagotyra)* **12** (3) pp 253 – 257

[6] Sülar V and Okur A 2007 Sensory Evaluation Methods for Tactile Properties of Fabrics, *J Sens. Stud.* **22** 1-16

[7] Kanai H, Morishima M, Nasu K, Nishimatsu T et all 2011 Identification of principal factors of fabric aesthetics by the evaluation from experts on textiles and from untrained consumers *Text. Res. J* **81** (12) pp 1216–1225

[8] Tester D, McGregor B A and Staynes L 2014 Ultrafine wools: comfort and handle properties for next-to-skin knitwear and manufacturing performance *Text. Res. J* **85** (11) pp 1181 – 1189

[9] Xue Z, Zeng X, Koehl L and Shen L 2015 Consistency and reliability of untrained consumers' perceptions of fabric hand of men's suiting *Text. Res. J* **86** (13) pp 1425–1442

[10] Harpa R 2016 Fabric Hand: Approach by Means of Software Package for Teaching Students. Part I. Software Development *12th Int. Scientific Conf. on eLearning and Software for Education* (Bucharest: Romania) ed. Roceanu I vol. **3** pp 512-517

[11] Harpa R, Cristian I and Piroi C 2017 STAT-CONCOR: Software for Validating the Panel of Assessors, Subsequently to the Fabric Hand Evaluation *13th Int. Scientific Conf. on eLearning and Software for Education* (Bucharest: Romania) ed. Roceanu I vol. **3** pp 431-438

[12] Harpa R 2017 Selecting wool-type fabrics for sensorial comfort in women office clothing for the cold season, using the multi-criteria decision analysis. *IOP Conf. Ser.: Mater. Sci. Eng.* **254** 142009

Air permeability characterization of glass fiber nonwoven fabric for liquid composite molding applications

M E Ince
Textile Engineering Department, Gaziantep University, Gaziantep, Turkey
Email: eince@gantep.edu.tr; meince@ncsu.edu

Abstract
Due to tailorable and superior mechanical properties normalized by their density; the use of fiber reinforced polymers (aka composites) is increasing steadily. Nonwoven surfaces from randomly oriented glass fibers are commonly used reinforcement fabrics in composite industry, and their resin infusion and final composite performances are largely dictated by their porosity that is quantified by air permeability. This research dealt with air permeability characterization of nonwoven fabric from chopped and randomly oriented glass fibers. Number of layers and fabric test face direction were selected as product input variables, while pressure drop and test head area were selected as process input variables. Front and back faces of the fabric stacks with single, double and triple layers were subjected to air permeability test with test head areas of 20 and 30 cm^2, and across a pressure drop of 100 and 200 Pa. As expected, multiple fabric layers exhibited lower air permeability. While test face of fabric and test head area did not affect the air permeability; measurements performed under 200 Pa pressure drop resulted in higher air permeability than that of 100 Pa at statistically significant level, which was explained by Bernoulli Equation.

Keywords: glass fiber, nonwoven fabric, air permeability, liquid composite molding

1. Introduction
Affordable price, good mechanical properties, and resistance against harsh environmental conditions render glass fibers attractive in technical textile applications. They are extensively consumed in protective clothing against flame and chemicals, filter media in severe circumstances, and reinforcement fiber in polymer composites [1]. Among textile fabric manufacturing techniques; nonwoven fabrics eliminate yarn production and following conversion of yarn into fabrics processes that enhance the production speed and lower costs dramatically. Controlled placement of fibers enhances the isotropy of the nonwoven fabrics that cannot be achieved easily in woven and knitted fabrics. Nonwoven fabrics from randomly oriented short glass fibers are consumed in large quantities by liquid composite molding industry. Resin infusion performances of these fabrics in single- and multiple-layer form are largely dictated by their permeability in thickness direction. Completely wetted and void-free composite production depends on the accurate air permeability characterization of reinforcement fabric stacks [2].

In this study, air permeability of single-, double-, and triple-layer randomly oriented glass fiber nonwoven fabrics were measured from both front and back faces of the fabric stack, along a pressure drop of 100 and 200 Pa, with test head areas of 20 and 30 cm^2. The results of this study will guide people working on liquid composite molding and its modeling.

2. Material ve methods
Emulsion bonded chopped strand mat provided by Şişecam Cam Elyaf Sanayi A.Ş. was used to characterize air permeability behavior of nonwoven fabric from glass fiber. Specification of the fabric is given in Table1.

Because reinforcement fabrics are stacked to achieve target thickness during composite production; having a sound understanding about the influence of number of fabric layers on air permeability behavior is valuable. Therefore, we varied number of layers as single, two and three layers within the experimental plan of this study. Fabrics can display different air permeability response depending on the fabric face from which the flowing fluid strikes. Thus, another input variable of this study is test face of the fabric. Air permeability devices are marketed by different test head areas depending on the tightness of the test fabric and power of the device. We also varied test head area as 20 and 30 cm^2 to reveal the effect of test head area on air permeability. Another input test parameter is the pressure drop of flowing fluid (air in our case) during test, and commonly used pressure drops are 100 and 200 Pa. We also tried to understand the effect of pressure drop on air permeability performance of nonwoven fabric from glass fiber. Experimental plan of this study is shown in Table 2. Five air permeability measurements from different surface spots were performed for each different combination of independent input parameters, thus in total 120 (5x3x2x2x2) air permeability measurements were completed in this study.

Table 1. Specifications of nonwoven fabric from glass fiber

Property	**Value**
Type of glass	E
Weight per unit area, g/m^2	450 ± 7%
Nominal filament diameter, micron	13
Split tex count, g/1000m	35
Length of chopped fibers, mm	50
Coupling agent	Silane
Resin compatibility	Polyester

Table 2. Experimental plan

Variables:	**Number of layers**	**Test face of fabric**	**Test head area,** cm^2	**Pressure drop,** Pa
Levels:	One layer	Front	20	100
	Two layers	Back	38	200
	Three layers			

Fabric unwinded from the roll was laid on the table. While top face of the fabric was assumed as front face, bottom face that is in contact with the table was assumed as back face of the fabric. In half set of the measurements the flowing air fluid hit the front face of the fabrics, while in other half set; it hit from the back face of the fabrics. Specimens with an area of 20x25 cm^2 were cut from the fabric lying on the table. The specimens with different combination of number of layers were used for air permeability test on SDL ATLAS M021A (Figure 1) according to ASTM D737-04. Air permeability unit was selected as $cm^3/cm^2/s$ - the volume of air (in cm^3) passing through per unit area (in cm^2) of fabric per unit time (in a second).

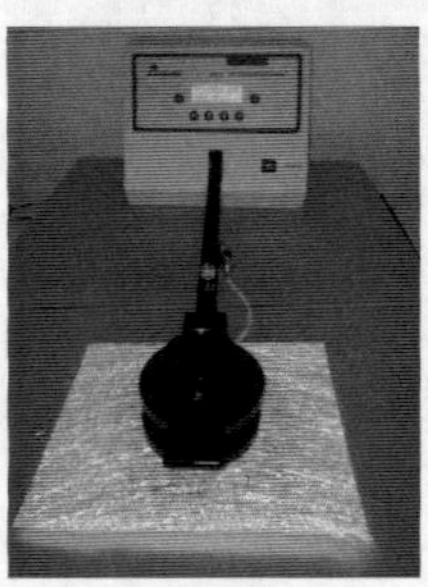

Figure 1. SDL ATLAS M021A air permeability tester (left), air permeability measurement of nonwoven glass fabric (right)

3. Results and discussion

3.1. The effect of number of fabric layers on air permeability

Data analysis indicated that increasing number of fabric layers resulted in statistically significant decrease in air permeability (Figure 2). Each layer level showed statistically significant air permeability than the others (Table 3). Increasing number of layers from one to two layers lowered air permeability at 34 %, while increasing it from two to three layers lowered air permeability at 22 %. Two- and three-layer fabric stacks exhibited less air permeability variation than single layer that was attributed to formation of more uniform air flow channels by closure of big and non-uniform pores in one layer by the adjacent layer.

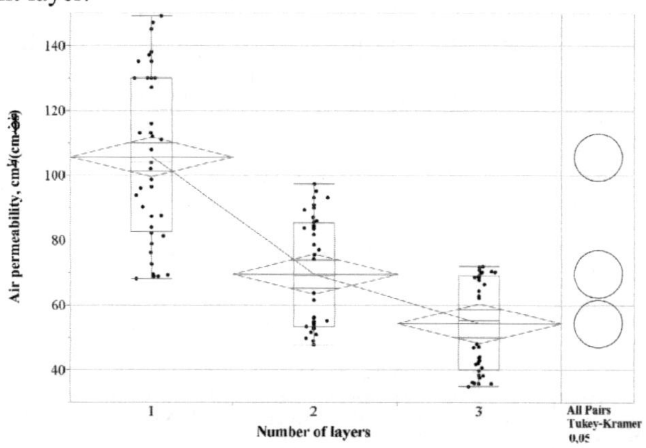

Figure 2. The effect of number of layers on air permeability

Note: Mean of each layer level is connected through blue line. The top and bottom of each green diamond represent the 95% confidence interval for each layer level. Comparison circles (given on the right column) for means those are significantly different either do not intersect, or intersect slightly.

Table 3. Number of layers versus air permeability report

# of layers		**mean** cm³/(cm²×s)	**sd** cm³/(cm²×s)	**LL** cm³/(cm²×s)	**UL** cm³/(cm²×s)
1	A	105,78	25,74	97,55	114,01
2	B	69,70	16,86	64,31	75,09
3	C	54,44	14,46	49,82	59,07

Note: levels not connected by the same capital letter are significantly different. sd: standard deviation, LL: lower limit, UL: upper limit. Limits are based on 95% confidence level.

3.2. The effect of test fabric face on air permeability

Results indicated that whether flowing fluid (air) strike on front or back face of fabric did not create a statistically significant change (p-value of 0,94) in air permeability, and this result is valid for all number of fabric layer levels (Figure 3, Table 4). This study reveals that during stacking of fabric layers before resin infusion, a specific preference for fabric face orientation does not create an advantage to speed up the permeation of resin through the thickness of the fabric stack.

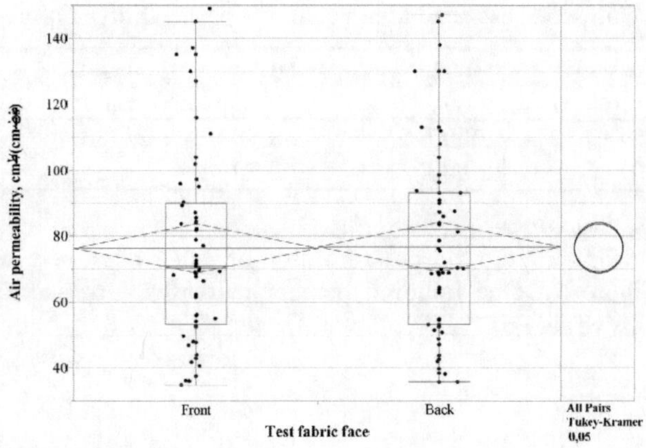

Figure 3. Effect of test fabric face on air permeability

Note: If the comparison circles intersect by an angle of more than 90 degrees, or if they are nested (as shown in the figure), the means are not significantly different.

Table 4. Test fabric face versus air permeability report

Test face	mean, $cm^3/(cm^2 \times s)$	sd $cm^3/(cm^2 \times s)$	LL $cm^3/(cm^2 \times s)$	UL $cm^3/(cm^2 \times s)$	p-value
Front	76,44	29,43	68,83	84,04	0,94
Back	76,84	28,98	69,36	84,33	

Note: p-values greater than 0,05 indicate lack of statistical significance.

3.3. The effect of test head area on air permeability

Data analysis indicated there was no statistically significant difference between the air permeability test results of 20 cm^2 test head area and that of 38 cm^2. Once steady flow state conditions were achieved at given pressure drop, same volume of air passing through per unit fabric area per unit time for both 20 and 38 cm^2 test head areas (Figure 4 and Table 5).

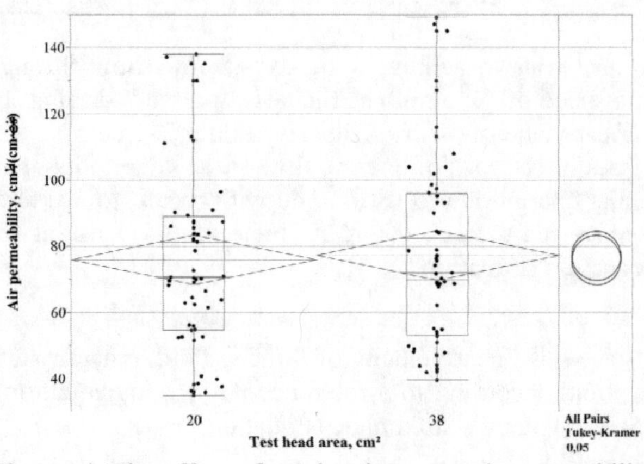

Figure 4. The effect of test head area on air permeability

Table 5. Test head area versus air permeability report

Test head area, cm^2	**mean,** cm^3/(cm^2×s)	**sd,** cm^3/(cm^2×s)	**LL,** cm^3/(cm^2×s)	**UL,** cm^3/(cm^2×s)	**p-value**
20	75,98	28,97	68,50	83,46	0,80
38	77,30	29,43	69,70	84,90	

3.4. The effect of pressure drop on air permeability

Air permeability tests performed across a pressure drop of 200 Pa resulted in higher air permeability than that of 100 Pa (Figure 5). One hundred percent increase in pressure drop resulted in 57 % increase in air permeability (Table 6).

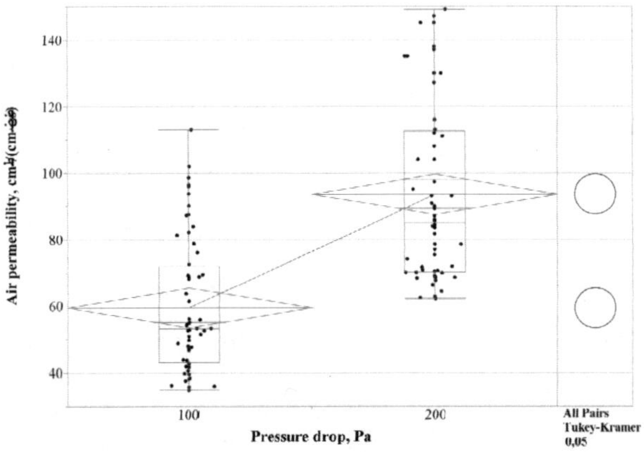

Figure 5. Effect of pressure drop on air permeability

Table 6. Pressure drop versus air permeability report

Pressure drop, Pa	**mean,** cm^3/(cm^2×s)	**sd,** cm^3/(cm^2×s)	**LL,** cm^3/(cm^2×s)	**UL,** cm^3/(cm^2×s)	**p-value**
100	59,59	20,09	54,40	64,77	<0,0001
200	93,70	26,67	86,80	100,59	

Once the device is run, it tries to achieve a steady state air flow through the test area at given pressure drop by adjusting speed of the air. Once the target pressure drop is achieved, air permeability (in other words average linear velocity of air) is displayed on the screen.

A flowing fluid possesses three types of energy; flow energy due to its pressure, kinetic energy due to its velocity, and potential energy due to its altitude with regard to a reference points. These three types of energy can be converted to each other, but their sum is constant along a streamline that is quantified by Bernoulli equation (Equation 1) [3].

$$\frac{P}{\rho}+\frac{V^2}{2}+g \quad = c \tag{1}$$

Where P is fluid pressure, V is linear velocity of flowing fluid, g is gravitational acceleration, and z is the altitude of flowing fluid according to a reference point. This equation can be written between two points (point 1 and point 2) along a streamline (Equation 2).

$$\frac{P_1}{\rho}+\frac{V_1^2}{2}+gz_1=\frac{P_2}{\rho}+\frac{V_2^2}{2}+gz_2 \tag{2}$$

Point one can be regarded as air flow before fabric, while point two is that after the fabric. Velocity of air at point 1 is zero, because it is the air (stationary air) in laboratory atmosphere, while velocity of air at point 2 is the air permeability of the test fabric (Figure 6). Pressure drop (*ΔP)* is the difference between pressures at points 2 and 1 ($\Delta P = P_2 - P_1$). As pressure drop is increased, flow energy lost by pressure drop is converted into kinetic energy that results in air velocity (air permeability) increase at point 2. Therefore, measurements performed under higher pressure drop results in higher air permeability results. If air permeability of two fabrics is compared, pressure drop must be kept constant.

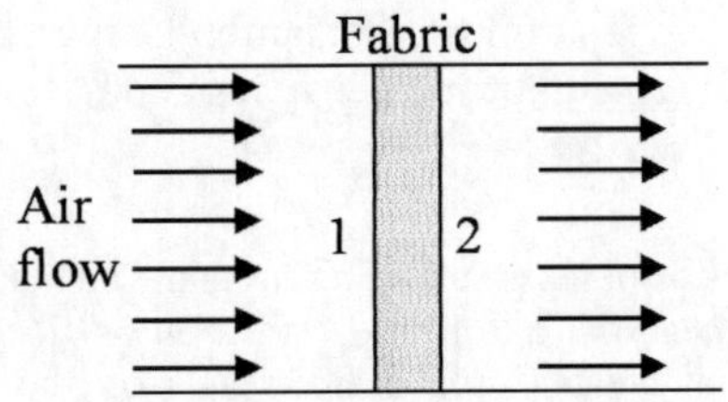

Figure 6. Schematic drawing of air permeability measurement

4. Conclusion

This study investigated the effect of number of fabric layers, test face of fabric, test head area of the device, and the pressure drop along the flow of the fluid on air permeability of nonwoven fabrics from randomly oriented chopped fibers. Due to closure of open straight channels with enhanced tortuosity in flow channels; increase in number of fabric layers lowered air permeability. The striking face of air over front or back face of the fabric did not affect the air permeability. Similarly, comparable air permeability measurements were obtained at both test head areas of 20 and 30 cm^2. Air permeability measurements obtained at pressure drop of 200 Pa resulted in higher values than those at 100 Pa that was explained by Bernoulli equation where the total energy of flowing fluid is constant along a streamline of the fluid.

References

[1] Chawla K K 1998 *Composite materials: science and engineering.* 2nd ed. (New York: Springer)
[2] Mazumdar S K 2002 *Composites manufacturing: materials, product, and process engineering.* (Boca Raton Fla.: CRC Press)
[3] Çengel Y A, Cimbala J and Turner R H 2017 *Fundamentals of Thermal-Fluid Sciences.* (New York: McGraw-Hill Education)

IOP Conf. Series: Materials Science and Engineering **459** (2019) 012047 doi:10.1088/1757-899X/459/1/012047

Strength loss of glass yarn during weft knitting

M E Ince and H Yıldırım
Textile Engineering Department, Gaziantep University, Gaziantep, Turkey
E-mail: eince@gantep.edu.tr; meince@ncsu.edu

Abstract

Weft-knitted fabrics from glass yarns present many advantages in technical textile applications. However, due to their stiff and brittle nature; glass fibers can easily be broken during weft knitting process that creates broken fiber ends on the fabric surface, and deteriorates the performance of resultant fabric. In this study, weft-knitted glass fabrics with four different knit architectures, and two different cam settings were produced. Number of yarn ply was also varied as 2- and 3-ply. The effect of all these input parameters on strength loss of glass yarn was examined via measuring virgin and unravelled yarn strengths in loop form. Results showed that knit pattern, cam setting and number of yarn ply did not affect the unravelled loop yarn strength at statistically significant level. On the other hand; unravelled yarns exhibited better loop strength than virgin yarns that was attributed to uniform distribution of fibers over cross sectional area of yarn as a result of knitting process.

Keywords: glass fiber, technical textiles, weft-knitted fabric, loop yarn strength

1. Introduction

Weft-knitted fabrics have stretchable, three dimensional, intermeshed, and porous structures that make them distinctive as compared with woven and nonwoven fabrics. Due to their snug fit comfortable nature, they are widely used in apparel production. They are also used in many technical textile applications. Once high performance fibers are converted into weft knitted fabrics, they offer unique advantages: comfortable protective clothing against flame or chemicals, reinforcements for impact resistant composites, and filter media for filtration of hazardous materials [1, 2]. Due to their low price, good mechanical properties, and high resistance against flame and chemicals; glass fibers are widely used in technical textiles applications. Stiff and brittle nature of glass fibers make their conversion into weft knitted fabrics challenging. Glass fibers can easily be broken, as they are forced to take loop shape, and broken fibers deteriorate the surface appearance and performance of knitted fabrics in the following service life. Quantification of this fiber breakage (damage) and determination of the most influential process and product input parameters on strength loss of the yarn after knitting process is crucial [3, 4]. Therefore, weft-knitted fabrics from glass yarn were produced within the context of this study. Knit pattern, number of yarn ply, and cam setting were varied and their effects on yarns strength loss of glass yarn were studied via measuring virgin and unravelled yarn strengths in loop form.

2. Material and methods

E-glass multi filament yarn with single-end count of 133 tex was consumed to manufacture fabrics on Brother KH-864 manual, flat weft knitting machine with a fineness of 5 gauge. Table 1 indicates the experimental plan of our study. Three input variables; knit architecture, cam setting and number of yarn ply were considered; thus sixteen (4x2x2) different samples were produced within the context of full factorial experimental design. Figure 1 indicates technical notations (needle diagrams) of the weft-knitted fabrics; and Figure 2 illustrates hand-draft drawings, and actual images of knit architectures. To quantify the damage level of various loop formation mechanisms over yarn strength; knit architectures including different loop types (plain, skip, tuck, and weft-inlay yarn) were selected.

IOP Conf. Series: Materials Science and Engineering **459** (2019) 012047 doi:10.1088/1757-899X/459/1/012047

Table 1. Experimental plan

Variables:	Knit architecture	Cam setting	Number of yarn ply
Levels:	Plain loop Skip loop Tuck loop Weft-inlay	Loose fabric Tight fabric	2-ply 3-ply

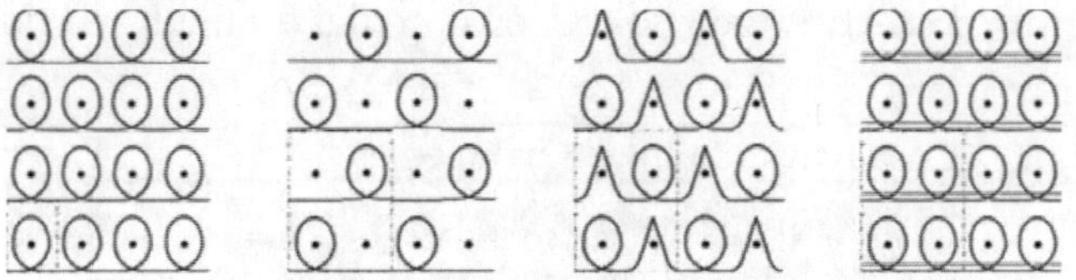

: plain loop, : skip loop, : tuck loop, : weft-inlay, : knit repeat boundary

Figure 1. Technical notations of the knit architectures

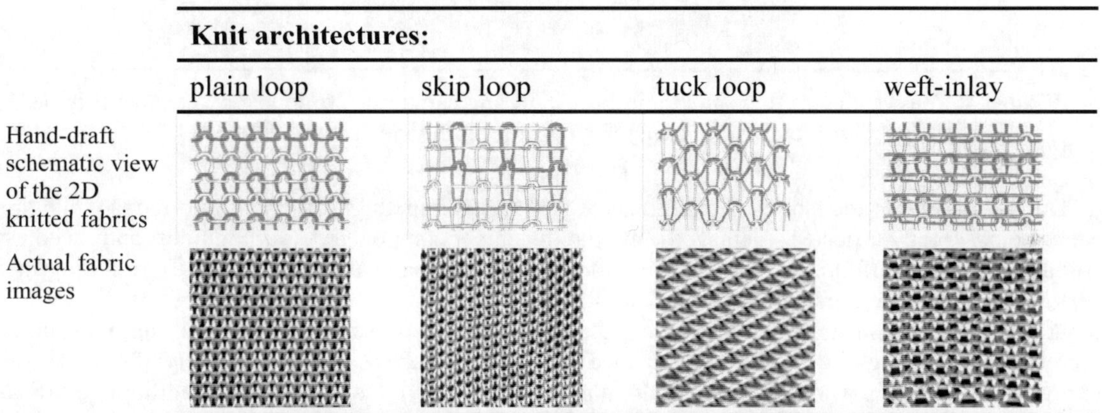

Figure 2. Hand-draft drawings and actual images of knit architectures

After the needle passed the hooked yarn through the formerly formed loop; descending distance of the needle on the cam track is controlled by cam setting that directly specifies the size of the loop. The longer the needle descending distance is, the greater the loop size is. Cam setting is adjusted by a dial called “Tension Dial” (Figure 3). This dial is scaled from 0 to 10, and each scale being subdivided into three parts. While 0 is the shortest descending distance of the needle that corresponds the smallest loop size with the highest internal yarn tension within the loop; 10 is the longest descending distance of the needle that corresponds the largest loop size with the lowest internal yarn tension within the loop. Thus, the cam setting indirectly determines internal yarn tension within the loop; and hence the dial for cam setting is called as “Tension Dial”. Figure 4 illustrates cam tracks followed by needles and fabric structures manufactured by different tension dial numbers. We decided on two different tension dial settings; number 3 and 8, where number 3 created a tight fabric structure (with smaller loops), and number 8 created a loose fabric structure (with larger loops). The last input variable of this study was number of yarn ply that was fed to the machine through the guides. Half of the samples were produced from 2-ply yarn, while the other half from 3-ply yarn.

Figure 3. K-carriage and the tension dial on the K-carriage

The dial setting	Cam tracks followed by needles	Fabric structure
# 10		
# 0		

Figure 4. Tension dial settings and their corresponding cam tracks followed by needles and the resultant fabric structures (Brother Industries Ltd., 1990)

During knitting of the fabrics; 2 or 3 adjacent yarns acted together to form various loops while they were being jointly affected (damaged) by the machine components (yarn guides and needles). Additionally, yarns of knitted fabrics respond to external mechanical forces in loop form during their performance (tensile, bursting) tests, and service life. Therefore; to reveal the effect of the selected input parameters (knit architecture, cam setting and number of yarn ply) on yarn strength and to correlate yarn strength with the performance of resultant fabrics more accurately; 2-, and 3-ply unravelled and virgin yarn strengths in loop form (Figure 5) were measured according to ASTM D2256 with a gauge length of 250 mm, pretension of 0.500 cN/tex, and extension rate of 50 mm/min. Load cell with a maximum capacity of 600 N was selected. To prevent yarn slippages inside the jaws during tests; fibers at both ends of the yarns where they are grabbed by the jaws were glued by two-part epoxy (Figure 6). The crossing point of the yarns was positioned into middle of the gauge section (Figure 7). Maximum load (Newton) and the extension (mm) at maximum load were obtained from the raw data file of the machine software. Maximum load (cN) was divided by the total tex between the jaws to calculate the loop yarn strength in cN/tex unit.

Figure 5. Preparation of 2-ply and 3-ply yarns for loop strength test

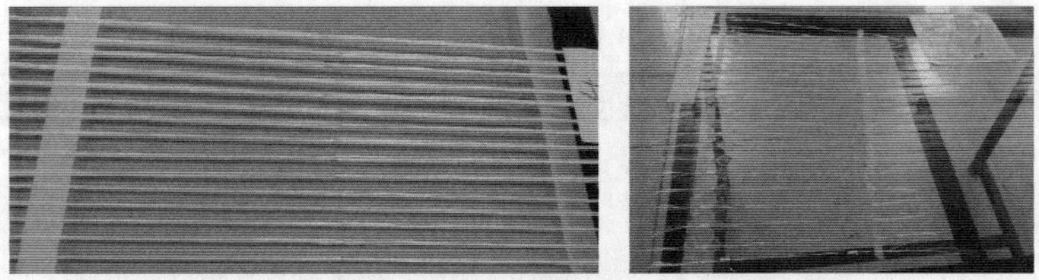

Figure 6. Gluing of yarn ends to prevent yarn slippage at jaws during test

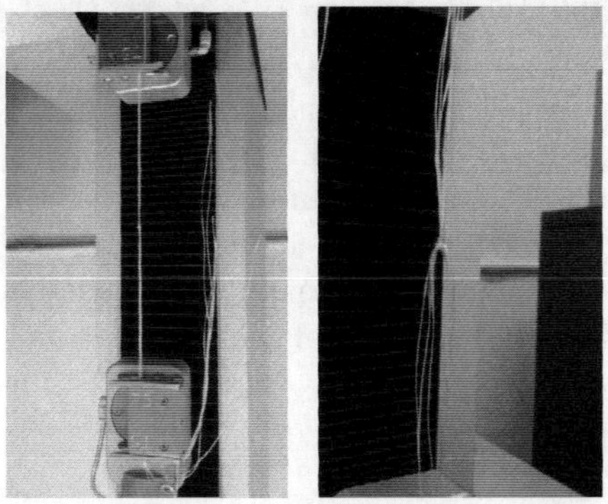

Figure 7. Running of loop yarn strength test (left), failure at crossing point

3. Results and discussion

3. 1. The effect of knit pattern on loop yarn strength

While unraveled yarns from skip stitch patterned fabric exhibited the highest yarn strength and unraveled yarn from weft-inlay showed the lowest unraveled yarn strength, no statistically significant difference was observed among the unraveled yarn strength of all knit pattern that proved lack of effect of knit pattern on yarn strength. Contrary to expected, unravelled yarns exhibited better strength than virgin yarn at statistically significant level (Figure 8 and Table 2). Knitting process improved loop yarn strength. Conversion of virgin yarn into the loop shape in the course of knitting process enhanced the uniform distribution of fibers in the yarn cross section that enhance the load carrying capability of unravelled yarns.

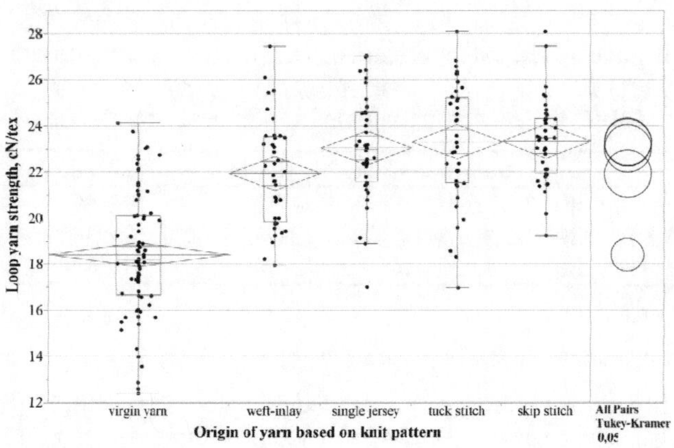

Figure 8. The effect of knit pattern on loop yarn strength

Note: The distance between top and bottom spikes of each green diamond represent the 95% confidence interval for the related origin of yarn level. Comparison circles (given on the right column) for means those are significantly different either do not intersect, or intersect slightly.

Table 2. Knit pattern versus loop yarn strength report

Knit pattern			**n**	**mean,** cN/tex	**sd,** cN/tex	**LL,** cN/tex	**UL,** cN/tex
skip stitch	A		40	23,34	1,76	22,77	23,90
tuck stitch	A		40	23,31	2,58	22,49	24,14
single jersey	A		40	23,05	2,22	22,34	23,76
weft-inlay	A		40	21,95	2,26	21,23	22,68
virgin yarn		B	80	18,43	2,64	17,85	19,02

Note: Levels not connected by the same capital letter are significantly different. n:number of measurements, sd: standard deviation, LL: lower limit, UL: upper limit. Limits are based on 95% confidence level.

3.2. The effect of cam setting on loop yarn strength

Yarns unravelled from tightly knitted fabrics exhibited slightly higher loop strength than yarns from loosely knitted fabrics; however this difference did not reach statistically significant level (at $\alpha = 0{,}05$ significance level) (Figure 9, Table 3). Yarns in tight fabrics have shorter loop length with smaller size and greater internal tension than yarns in loose fabrics. Short loop length and higher internal tension enhanced uniform distribution of fibers in the yarn cross section that slightly improved loop strength of unravelled yarn. Virgin yarns displayed lower loop strength than unravelled yarns that again was attributed the uniform spread and distribution of fibers over yarn cross section in case of unravelled yarns from knitted fabrics (Figure 9, Table 3).

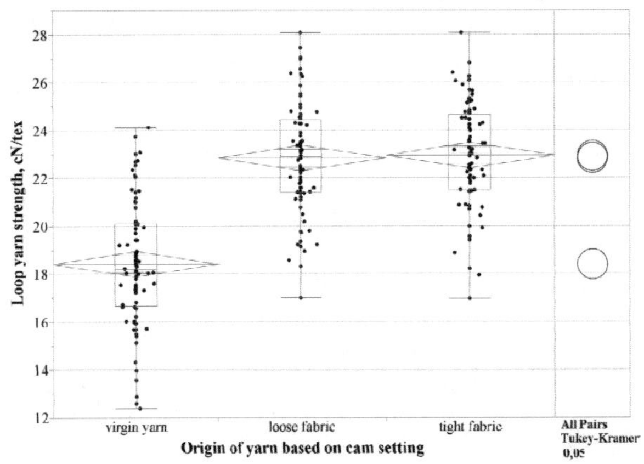

Figure 9. The effect of cam setting on loop yarn strength

Table 3. Cam setting versus loop yarn strength report

Cam setting			**n**	**mean,** cN/tex	**sd,** cN/tex	**LL,** cN/tex	**UL,** cN/tex
tight fabric	A		80	22,96	2,21	22,47	23,46
loose fabric	A		80	22,86	2,35	22,34	23,38
virgin yarn		B	80	18,43	2,64	17,85	19,02

3.3. The effect of number of yarn ply on loop yarn strength

While number of yarn ply did not affect the unravelled yarn loop strength; it affected virgin yarn loop strength at statistically significant level. 3-ply virgin yarns exhibited better strength than 2-ply yarns. Both 2-, and 3-ply unravelled yarns displayed better loop strength than their counterpart virgin yarns (Figure 10 and Table 4). Knitting process enhanced the distribution of the fibers and promoted their load bearing capability.

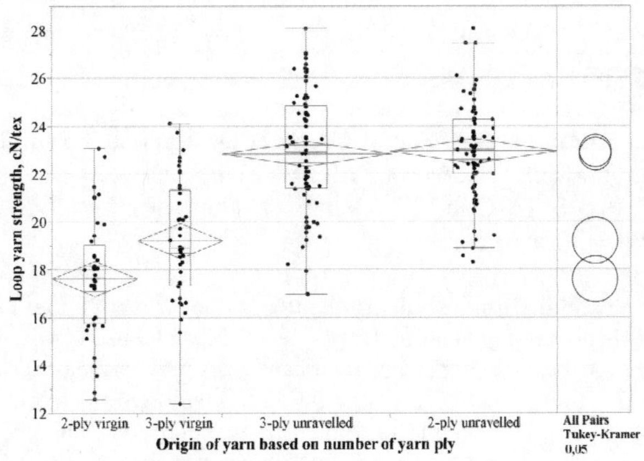

Figure 10. The effects of number of yarn ply on loop yarn strength

Table 4. Number of yarn ply versus loop yarn strength report

Number of yarn ply				n	mean, cN/tex	sd, cN/tex	LL, cN/tex	UL, cN/tex
2-ply unravelled	A			80	22,97	2,00	22,53	23,42
3-ply unravelled	A			80	22,85	2,53	22,29	23,42
3-ply virgin		B		40	19,22	2,57	18,40	20,05
2-ply virgin			C	40	17,65	2,51	16,84	18,45

4. Conclusion

This study investigated the effects knit pattern, cam setting, and number of yarn ply on loop strengths of unraveled yarns from weft knitted fabrics. Contrary to expectations; knit pattern and other independent input variables (cam setting and number of yarn ply) did not show statistically significant effect on unraveled yarn strength (at $\alpha = 0{,}05$ significance level). On the other hand, again contrary to expectations; conversion of virgin yarns into weft knitted fabrics improved loop yarn strength at statistically significant level that was attributed to uniform distribution of fibers within the yarn cross section during the knitting process. For future studies; cross sectional views of virgin and unravelled yarns will be investigated, and the comparison of other mechanical properties such as extension at maximum force (mm), time to break (s), elastic modulus (MPa), breaking work (J) will be revealed in view of the selected input parameters plus unravelled and virgin yarn perspective.

References

[1] Chawla K K 1998 *Composite materials: science and engineering.* 2nd ed. (New York: Springer)
[2] Mazumdar S K 2002 *Composites manufacturing: materials, product, and process engineering.* (Boca Raton Fla.: CRC Press)
[3] Hong H, Mei Z 2005 A study of the degree of breakage of glass filament yarns during the weft knitting process, *AUTEX Research Journal* **5** no 5 pp 141-148
[4] Xiao-Ming Liu, Nan-Liang Chen, Xun-Wei Feng 2009 Investigation on the knittability of glass yarn, *The Journal of The Textile Institute* **100** no 5 pp 440-450

IOP Conf. Series: Materials Science and Engineering **459** (2019) 012048 doi:10.1088/1757-899X/459/1/012048

Sewability interdependence on rigid structures

M Karypidis[1, 2]
[1]University of Applied Sciences of Central Macedonia
Faculty of Kilkis, Department of Garment Design and Technology,
3rd km Kilkis-Metaliko, 61100 Kilkis, Greece
[2]A.KARYPIDIS SA-Clinitex, 2 Vilara st. Thessaloniki 54625, Greece
mkarypidis@hotmail.com

Abstract. Sewability is the ability of the fabric to be sewn without faults and represents the ease of fabric processing at a satisfactory level. It can be used to predict problems which may occur in the garment sewing stage and it can be accurately measured using the L&M sewability tester as a function of that needle penetration force though the fabric. Sewability is attributed to fabric properties, pretreatment, finishing, sewing conditions and mishandling. Fabric softeners affect the ease of sewing processing of garments, which is mostly reflected by the needle penetration force. The current work was driven by previous study focused on the analysis of factors affecting the sewability of cellulosic woven fabrics of a wide range of designs and constructions, where the influence of several fabric properties on the needle penetration force had been investigated, revealing the existence of a relationship between the inherent properties of mass per square metre, breaking strength and bending length. However, the in-depth statistical analysis had shown that rigid structures follow a different trend in comparison to the light ones. In the current work, thin and light structured fabrics were excluded from the investigation and focus was given exclusively to the rigid structures of higher mass per square metre fabrics which are known to be problematic in sewing stage. Although the investigation was confined to the main three inherent properties which had proven to influence needle penetration, it included fabric treatments of washing and softening on the samples. Results show an increase in the interdependency of the studied properties while treatments show a minimal effect probably to low level of application.

Keyword(s). Needle penetration force, sewability, rigid woven structures.

1. Introduction

The ultimate and most critical stage in the chain process of textile construction is the garment making up process. Much of the commercial success of the final product depends on this last stage. Faulty products deriving from this stage do not have the chance to be corrected as it is not followed by any subsequent. Additionally, being the last stage, it bears the total cost from all previous stages. This underlines the importance of the area to investigate. Sewability has always been a major aspect of textile processing but quite often has not been given the correct attention by the textile manufacturers.

The ability of a fabric to be converted into a garment without faults through stitching process defines its sewability. It is attributed to fabric properties, pretreatment, finishing, sewing conditions and mishandling [1]. The finishing treatments influence fabric properties as well as the stitching quality of resulting garments. Studies have demonstrated that customer product selection is strongly related to the handle and softeners can significantly help meet the customer demands [2, 3, 4].

Sewability of fabric composes of merely two basic aspects, namely tailorability and sewing damage. Several studies have focused in the aspect of the sewing damage as seam puckering of the processed fabric (ASTM D1683 and ISO 7770), while fewer concentrated in the tailorability aspect. Tailorability reflects the ease of process during stitching of fabric at a satisfactory level and it can be used as a prediction tool to avoid processing problems which may result in poor performance of the garment sewing stage [5]. Several studies aimed to quantify the previously mentioned aspects of sewability as an expression of needle penetrating force through fabrics using sewing or overlock machines equipped with sensors or cameras. Although they help in building knowledge and understanding, they were mainly used for research purposes [6]. Woven fabrics were investigated with the use of an Instron tester with some modifications [5,7,8]. The L & M Sewability tester, devised by Leeming & Munden, served the same scope and was widely used mainly for the investigation of knitted fabrics [9]. Finally, other researchers devised modelling methods for predicting the sewability of cotton fabrics by establishing of a ratio between the strength of the original and the strength of the seamed fabric, as a more reproducible and sensitive test [10].

2. Methods and Materials

2.1 Scope

The scope of the current was to investigate dependence of the inherent properties such as strength, bending and mass per unit area with the needle penetration force through rigid fabrics of mid-high weight per square metre, in their natural, washed and softened states.

2.2 Analysed Properties

Studies have explained that needle penetration causes the yarn to move away from it but friction against it is inevitable [11]. As friction is much dependant of the vertical force and the coefficient of friction, fabric properties related to these factors such as fabric bending, yarn strength and fabric mass per unit area were investigated for the following reasons.

1. Fabric mass per unit area which represents the mass density of the material to come in contact with the needle. Fabric with higher mass would normally impose more impedance to the moving needle. Additionally higher mass could make the structure more rigid affecting fabric deformation and flexibility during needle penetration. Yildiz et al [12] claimed that the penetrating needle force increases exponentially with the increasing mass.
2. Fabric bending length representing the ability of the fabric to deform perpendicular to its plane before or during needle penetration. Deformation takes place at the moment of the needle contact with the warp and weft yarn found on its path. Fabric bending length is directly related to the seam puckering formation [13]. Additionally, flexural rigidity which is derived from bending length, and is the factor which Stylios [11] has based his mathematical model for prognosis of sewability damage.
3. Fabric strength representing the force which the fabric yarns are able to withstand when needle force penetration displacement occurs in in two planes [14]:
    a. Perpendicular (z-displacement) to the plane of the fabric when a number of yarns come in contact with the penetrating needle. The yarns are tensioned until they slide aside of the point of the needle or stay on top of it with the pressure building until fail.
    b. Within the plane (x-y displacement) of the fabric when the yarns are pushed away by the penetrating needle as its diameter increases between the point and the bud, which characterises the needle title.

2.3 Samples

Fabric samples from rigid and heavier cellulosic fabrics of a more than 220g/m² of weight per square metre were selected as previous work had shown to follow different trend to lighter structures. The fabrics covered a range of weave designs, since fabric construction play important role in the needle penetration [15]. Weave design diversity reassured taking this factor well in to account. The sample fabrics used are presented with the same sequence throughout the study and can be seen in table 1:

IOP Conf. Series: Materials Science and Engineering **459** (2019) 012048 doi:10.1088/1757-899X/459/1/012048

Table 1. Sample Fabrics.

No	Name	Weave	GSM
1	Twill	3/1	220
2	Panama (Rib)	2/1-1/1	500
3	Double Face	1/1 and 2/1	850
4	Denim	2/1	270
5	Dobby	Dobby design	370

2.4 Specimens and Sample Treatment

The fabrics in the in table 1, were used to create samples for testing. Three sets of samples were created, namely, control, washed and softened samples.

- Control samples were derived from fabrics in their normal finished state and were named as control.
- Samples derived from fabrics that have undergone a washing cycle at 90°C under Wascator with ECE Detergent (5A program) (BS EN ISO 2633) were named as washed. Washing treatment acts as a de-lubricating process by removing of any fatty substances on its surfaces and hence intent to increase the coefficient of friction.
- Finally samples resulting from fabrics that were processed with a cationic softener, known for their strong affinity to cellulosic substrate, were named as softened. The presence of softener on the fabric's surface decreases the coefficient of friction and improves abrasion performance[16]. The softener application was performed at 10g/l applied by mechanical deposition with a Benz pad mangle at wet pick-up 75% and dried evenly in a Mathis Steam Drier fabric-stenter at 85°C under the absence of any tension for a period of 2-5 minutes, depending on their dry weight.

2.5 Testing Machinery and Procedure

The previous mentioned properties have been measured in warp and weft direction and their average was taken in to account, while in the case of bending an additional set of samples was prepared and examined at a 45° to warp (or weft). Merely the readings are expressed as:

a) Bending Length for Warp, Weft and 45° (mm). Measurements conformed the BS 3356:1990 procedure with 5 samples for each direction

b) Strength for Warp and Weft (kg). Measurements conformed the ISO 13934-1 procedure with 5 samples for each direction

c) Mass per unit area (gr/m²). Measurements conformed the ISO 3801:1977 procedure with 5 samples

d) Needle Penetration Force (known as Sewability Force). Measurements were conducted using L&M sewability test apparatus. The instrument enables consecutive readings of force build during the penetration of a fabric by the selected needle (90s ball point) to be measured on a sample fabric at a rate of 100 penetrations/min [17]. The average force of 100 perforations is measure over a specimen length of about 350mm wide.

3. Results and Discussion

The SPSS software ver.13.0 was used for the statistical analysis of the correlation between the above mentioned variables. The dependency of the needle penetration force as a function of sewability value of the fabric was compared against the independent variables of the bending length, strength and mass per unit area. The degree of the dependence was expressed by Pearson R correlation value. The scatter diagrams were used to check the spread of the data which is also reflected by the coefficient of determination R^2 when a linear fit line passes through them, indicating a linear relationship. The analysis was performed under the confidence level of 95% using a 2-tailed model, while the data population was 75 reading for each case. The analysed data derived from the average values of the readings taken in the warp and weft direction for all samples for the needle penetration force and for all the above properties. In case of the bending a third set of data was analysed at 45° to the warp in order to detect best fit.

The resulting Pearson R values from the correlation statistical analysis are presented in the table 2 and their scatter diagrams of the figures 1, 2 and 3. In case of bending the resulting values of Pearson R from the data set of the average of the warp and weft data exhibit closer relationship than those taken at 45° to the warp. The R^2 values of 0.767, 0.784 and 0.732 reveal a high degree of goodness of fit and a good linear relationship, as presented by the corresponding scatter diagrams.

Table 2. Results from correlations analysis between needle penetration force and fabric property

Dependent Value Needle Penetration Force		N	Correlation	Sig.**
1	Bending	75	,876	,000
2	Bending (45°)	75	,838	,000
3	Strength	75	,885	,000
4	Mass per unit area	75	,856	,000

**Correlation is significant at the 0.01 level (2-tailed).

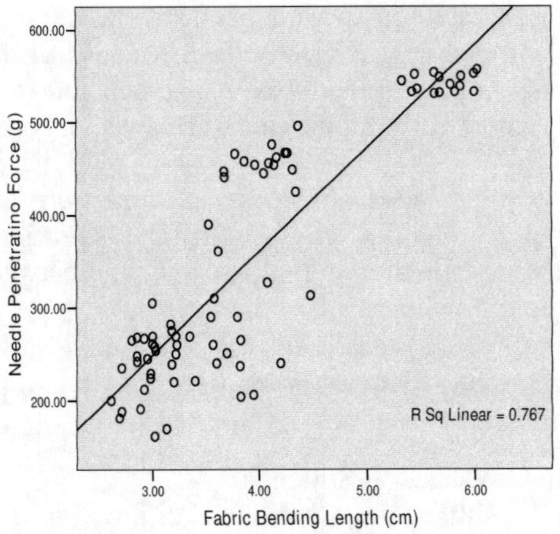

Figure 1. Scatter diagram of needle penetration force (g) against bending length (cm)

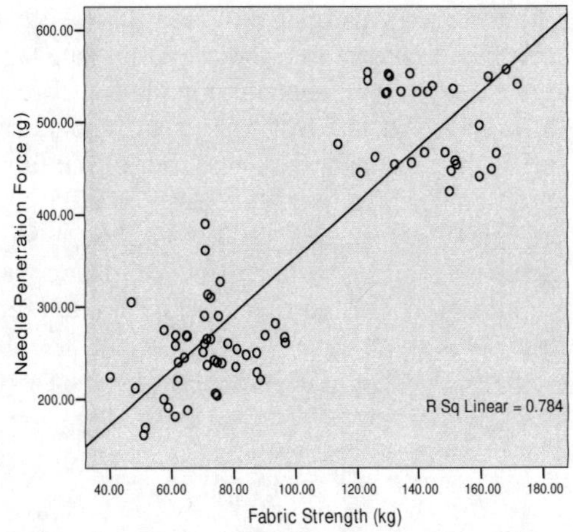

Figure 2. Scatter diagram of needle penetration force (g) against fabric strength (kg)

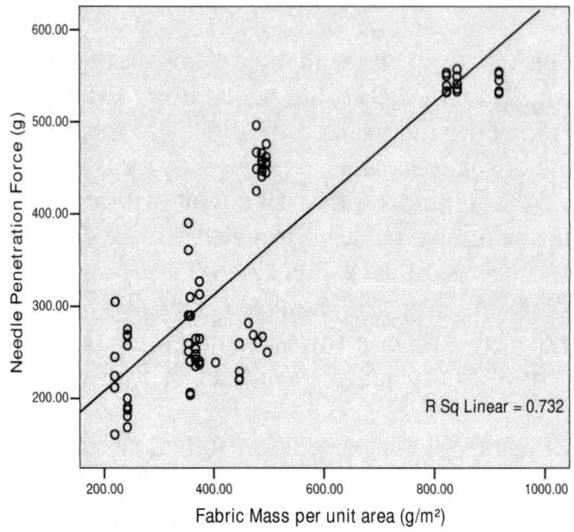

Figure 3. Scatter diagram of needle penetration force (g) against mass per unit area (g/m²)

The significance values of below 0.01 in table 2, validate the Pearson R values which demonstrate the existence of a strong relationship between the penetration force and all three properties, fabric strength 0.885, bending length 0.876 for average of weft and warp and 0.838 for 45° to warp, and mass of the fabric per unit area 0.856. Bending and mass are measurements which compose the flexural rigidity of the fabric, a property which exerted a similar behaviour to the model of Stylios [11] with similar accuracy, about 85%, confirming the above results. Additionally, the correlation values found are significantly higher than that observed in previous work [18], were light and heavy fabric structures were analysed together, underlining that the fact the more rigid the structure is the stronger becomes the interdependence of the sewability, in form of penetration force, and fabric strength, mass per unit area and bending length.

High needle penetration values may alert problems in fabric processing when they are over a certain threshold. L&M sewability empirically quantifies this limit in single or double jersey knitted structures [17]. Currently work is in progress to model this threshold on woven structures.

In order to investigate and identify the effect of the fabric treatments, washing and softening, on the needle penetration force the previous data were analysed by one way ANOVA. In the means of that the average value of needle penetration force for weft and warp of every fabric was compared in its control, washed and softened state and checked for its statistically significant difference in order confirm alterations in the behaviour of the fabric.

The confidence level used was a=0.05 with population of 15 cases (15 averages of all warp and weft tests), while the LSD post hoc test was selected to identify group statistical significant differences. Values for significance of one way ANOVA were tabulated in the first row, and between group differences (treatments) LSD sig. are presented in the following rows, of the table 3.

Table 3: ANOVA and L.S.D. significance values for treated fabric

Fabric	ANOVA Sig.	L.S.D. Sig. Control-Washed	Washed-Softened	Softened-Control
Twill	0.005	0.047	0.001	0.075
Panama	0.781	0.503	0.633	0.845
Double Face	0.953	0.805	0.975	0.782
Denim	0.217	0.432	0.316	0.088
Dobby	0.009	0.103	0.003	0.068

Table 4: Needle Penetration Force Values for treated fabric

	Penetration Force		
Fabric	Control	Washed	Soften
Twill	229.40	185.80	267.80
Panama	460.80	453.20	458.60
Double Face	541.40	543.00	543.20
Denim	250.00	276.40	310.40
Dobby	248.80	233.80	265.80

The ANOVA Sig values and LSD Sig values are over 0.05 in almost all the cases. Therefore the null hypothesis cannot be rejected and any difference in the Table 4 is statistically insignificant. This can be attributed to the low level of softener application in comparison to the course fabric structure. As a consequence, alterations, due washing or softening, of the needle penetration force through rigid fabrics could not be claimed. In the case of Twill fabric, which has the least rigid structure bearing the least mass per unit area of the fabrics and most flexible construction, the values were on border line of acceptance with the washed sample to differ from the rest as the LSD pair values reveal. Previous work in thin and light structures had confirmed a significant drop in needle penetration force [18] similarly to Alime's findings [9] were silicone softeners were used as lubricants on knitted structures.

4. Conclusions

Bending length, mass per unit area (grams/m²) and strength are properties of rigid fabrics which are closely relate to the force needed for a needle to penetrate those fabrics which reflects their sewability. This relationship seems to get stronger as the fabric structure becomes heavier and more rigid in a proportional manner. However, washing and softening fabric treatments of rigid structures seems to have a negligible effect on the needle force needed to penetrate through them, on low level application.

Acknowledgement

The author expresses his gratitude to the University of Applied Sciences of Central Macedonia Research Committee and A.Karypidis SA Textiles-Clinitex for their support on materials, apparatus and knowledge base.

References

[1] Grancaric, A. M., Lima, M., Vasconcelos, R., Tarbuk, A., Handle of Cotton Knitted Fabrics-Influence of Pretreatments, 5th World Textile Conference AUTEX, Portoroz, Slovenia, 43-47, (2005).

[2] El-Dessouki, H. A., The Thermal Comfort Properties of Certain Egyptian Stretched Knitted Fabrics, International Design Journal, 5 (1), 69, (2015).

[3] Sand, C., Brückmann, R., Zyschka, R., Fashionable Trends in Textile Finishing, Melliand International, 7 (1), 71, (2001).

[4] Bernd Wahle.and Falkowski Jurgen, Softeners in Textile Processing. Part 1: An Overview, Coloration Technology 32 (1), 118, (2010).

[5] Stylios, G., The Principles of Intelligent Textile and Garment Manufacturing Systems, Assembly Automation, 16 (3), 40, (1996).

[6] Chmielowiec R., Sewing Machine, Fabric and Thread Dynamics, PhD Thesis, University of Leeds, 6-26, (1993).

[7] Ezzatollah Haghighat, Seyed Mohammad Etrati, Saeed Shaikhzadeh Najar, Evaluation of Woven Denim Fabric Sewability based on Needle Penetration Force, Journal of Engineered Fibers and Fabrics, 9, (2), 47, (2014)

[8] Stepan V. Lomov, A Predictive Model for the Penetration Force of a Woven Fabric by a Needle, International Journal of Clothing Science and Technology, 10 (2), 91, (1998)

[9] Alime Ashli Illeez, Eylen Sema Dalbasi, Gonca Özcelik Kayseri, Improving of Sewability Properties of Various Knitted Fabrics with the Softeners, World Conference on Technology, Innovation and Entrepreneurship, Procedia - Social and Behavioral Sciences 195, 2786– 2795, (2015).

[10] Frederick, E.B., Development of a Sewability Test for Cotton Fabrics, Textile Research Journal, 22 (10), 687, (1952).

[11] Stylios, G., Prognosis of Sewability Problems in Garment Manufacture Using Computer Based Techniques, Proceedings of the IEEE International Conference on Systems Engineering, Pittsburgh, PA, USA, 371-373, (1990)

[12] Yildiz E. Z., Pamuk O., Ondogan Z., A Study about the Effects of Interlinings to Sewability of the Woven Fabrics, Tekstil ve Konfeksiyon, 21 (1), 87, (2011).

[13] Stylios, G., Lloyd D. W., The Mechanism of Seam Pucker in Structurally Jammed Woven Fabrics, International Journal of Clothing Science and Technology, 1 (1), 5, (1989).

[14] Mallet E. and Du R., Finite Element Analysis of Sewing Process, International Journal of Clothing Science and Technology, 11 (1), 19, (1999).

[15] Khan R. A, Hersh S.P. and Grady P.L., Simulation of Needle-Fabric Interactions in Sewing Operations, Textile Research Journal, 40 (6), 489, (1970).

[16] Karypidis M., Wilding M. A. & Carr C. M. and Lewis D. M, The Effect of Crosslinking Agents and Reactive Dyes on the Fibrillation of Lyocell, AATCC Review, 1(8), 40-44, (2001).

[17] L&M Sewability Tester Manual, John Godrich, (2010).

[18] Karypidis M., Savvidis G, Analysis of Factors Influencing Needle Penetration Force through Woven Fabrics, Journal of the Textile Association of India, 80, Preprint Jul-Aug, (2018).

Bending force of LLDPE monofilaments at high temperatures measured in DMA (Dynamic Mechanical Analyzer)

I Kola[1], B Kolgjini[1], G Guxho[1] and Paul Kiekens[2]
[1] Polytechnic University of Tirana, Faculty of Mechanical Engineering, Department of Textile and Fashion, Sheshi "Nënë Tereza", Nr. 1, Tirana, Albania
[2] Ghent University, Department of Materials, Textile and Chemical Engineering, Technologiepark 907, 9052 Zwijnaarde, Ghent, Belgium

ikola@fim.edu.al

Abstract
The bending force of monofilaments used as pile layer in artificial turf football fields is a very important property due to the strong relation with the performance of the entire system being the force that mostly influences the deformation of monofilaments. The existing test methods for measuring this force can only be used to perform tests at room temperature. Nowadays, artificial turf surfaces are being installed worldwide, even in regions with very high values of air temperature, leading to the need of evaluating the performance of monofilaments and thus, of testing their bending force at elevated temperatures.
The aim of this paper is to use the DMA's advantage of testing at high temperatures, by using a new test method developed for measuring the bending force of monofilaments at room temperature and to evaluate the results obtained. For this purpose, six different LLDPE (Low Linear Density Polyethylene) monofilaments are tested at temperatures 25°C, 50°C and 80°C and the bending force is measured. The DMA is used successfully on performing tests at high temperatures. Considering the results, an analysis of the bending behavior of LLDPE monofilaments depending on the temperature is done. All fibers show the same trends regarding this influence, with decreasing of the bending force while increasing the temperature.

1. Introduction

The use of artificial turf fields in sports and other surfaces has been increased recently [1]. There have been great and continuous improvements over the years on these fields [2, 3]. So, although their first installations were made in regions with low air temperatures, nowadays synthetic turf surfaces are being installed more and more, even in locations where the air temperature is qualified high.

Artificial turf systems are mostly used in football fields and with football being the most popular sport in the world, the geography of installing and using these fields is worldwide. High temperatures are one of today's problems of the world of football. They have an impact not only on footballers in form of health problems and their playing performance, but also in the artificial turf field, by affecting the performance of the fibers and the whole system [4]. Values up to 50°C, are reached in many regions where artificial grass field systems are already, or aim to be installed. The air temperature in these locations can sometimes reach up to 70°C in the sun. With artificial turf fields being mainly produced from linear low density polyethylene (LLDPE) monofilament, and with this polymer being sensitive to high temperatures [5-7], it is strongly needed to test and evaluate their behavior in high temperatures. The force that mostly influences the deformation of the fibres is the bending force [8].

A new method consisting on testing the bending force of monofilaments on the DMA, was developed successfully [9]. The DMA has a great advantage compared to other test methods used to measure the bending force of the fibers, because the tests can be performed at elevated temperatures [10].

The bending force of all samples is measured on the DMA at two high temperatures, besides the room temperature. Based on the results obtained from the tests, an analysis of the bending behavior of LLDPE monofilaments depending on the temperature and the distance of the applied force is done.

2. Materials

For this study were used six different Linear Low Density Polyethylene (LLDPE) monofilaments fibers (denoted A-F) The linear density is measured for the six fibers and the cross section of each fiber is observed in the microscope. The measurements of both linear density and cross section were performed at the University of Gent laboratories, Department of Textile. The values of the linear density (Tex) for each fiber are respectively: A=205, B=225, C=149, D=208, E=225, F=206. The cross sections of the fibers were of different shapes. There are two diamond shape fibers, A and E, but slightly different in the extension of the sides, and they also differ in the linear density. Two other fibers, B and F, are almost "c" section. The "c" shape has two sides: the concave side denoted as (l) and the convex side denoted as (m). The last two fibers are of different shapes, where fiber C is almost rectangular shaped, while fiber D has a small ball in the center.

3. Test method

DMA TA Instrument was used to perform the measurements of the bending force. The testing is performed at the Department of Textile at University of Gent. The Single Cantilever is used for clamping the fiber in one side and the DMA is set to Controlled Force Mode [10]. The distance of applying the force is set to 2 mm and 3 mm. The temperature is set at the required value (25°C, 50°C and 80°C). The DMA device is equipped with a closed oven, which not only achieves temperature set, but also keeps this temperature unchanged during the testing. Fibers B and F are tested in both sides of the "c" shape, because they perform differently in each of the sides [11]. The TA Instruments Universal Analysis (UA) Program is used for data analyses [12]. The static bending force (N) is plotted versus the displacement (μm) and The Onset Point (OP) 1 and Onset Point (OP) 2 are found in each graphic, through the UA Program (Figure 1). The Onset Point determines the force at which a change in the curve occurs [13], where the Onset Point 1 represents the bending force where the deformation starts but it can still be reversed. From the Onset Point 1 to the Onset Point 2 the deformation continues and on the Onset Point 2 can be considered as the breaking point. The same plot as below is obtained for each repetition of each sample.

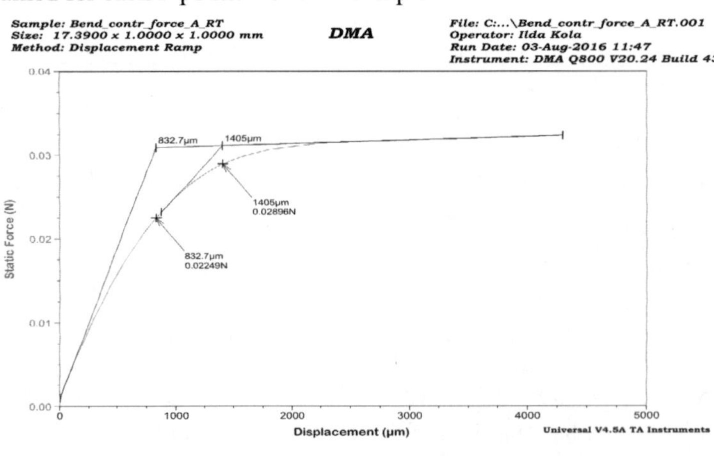

Figure 1. Bending force of fiber A versus displacement and the Onset Points on the DMA.

IOP Conf. Series: Materials Science and Engineering **459** (2019) 012049 doi:10.1088/1757-899X/459/1/012049

4. Results and discussions

The testing of all fibers was performed at 25°C, corresponding to the room temperature, and two high temperatures corresponding to 50°C and 80°C on two distances of the applied force (2 and 3mm).
Fibers A and B (both sides) were tested at three temperatures at both distances, while other fibers (C, D, E, F) were tested at three temperatures only on the distance of 2 mm, while on 3mm, the testing was carried out at 25°C and 50°C, because the changing values trend was the same as on the 2 mm distance and similar to fibers A and B. For each fiber 3 repetitions are performed.
For each measurement the bending force is plotted versus the displacement as showed above. After each plot was obtained, the bending force values on the Onset Point 1 and 2 are located (see Figure 1) and the mean value was calculated for each fiber, at three temperatures and both distances.

In order to evaluate the changing of the bending force at elevated temperatures, the plots of each fiber per each temperature are overlaid separately for each distance. This was accomplished for fiber A and B, for which the overlaid plots at three temperatures on both distances of the application force are obtained (Figure 2, 3).

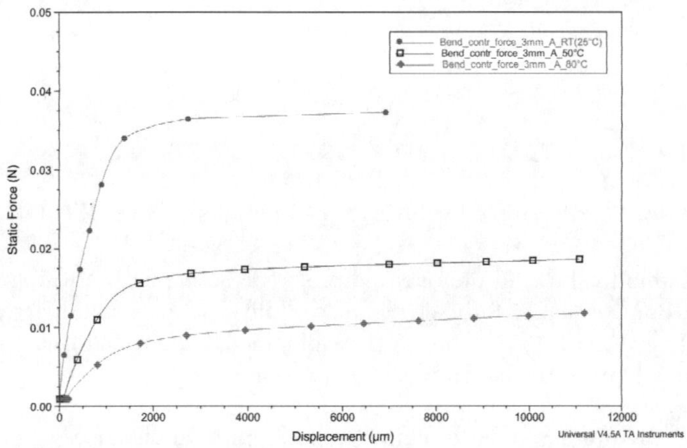

Figure 2. Bending force for fiber A, on 3 mm distance, at RT, 50°C and 80°C.

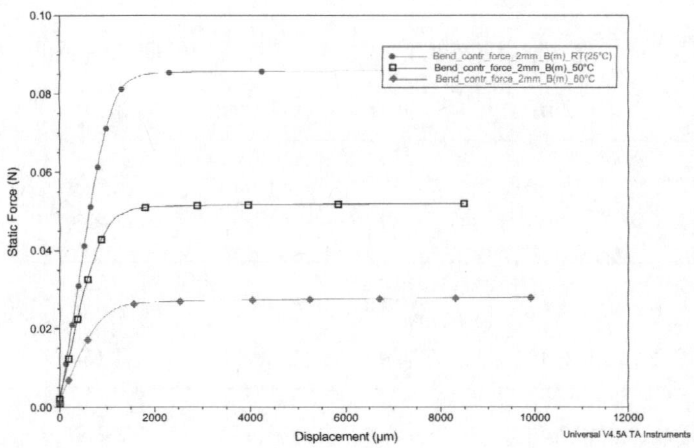

Figure 3. Bending force for fiber B(m), on 2 mm distance, at RT, 50°C and 80°C.

For the remaining fibers (C÷D) the overlaid plots of two temperatures on the 3mm distance were obtained (Figure 4).

Regarding the overlaid plots, it is obvious that the performance of each fiber is different at each temperature. So, with the increasing of temperature, the bending force was reduced significantly and at 80°C, the fibers exhibited a really poor bending performance (Figure 2, 3, 4). This applies to all fibers on both distances.

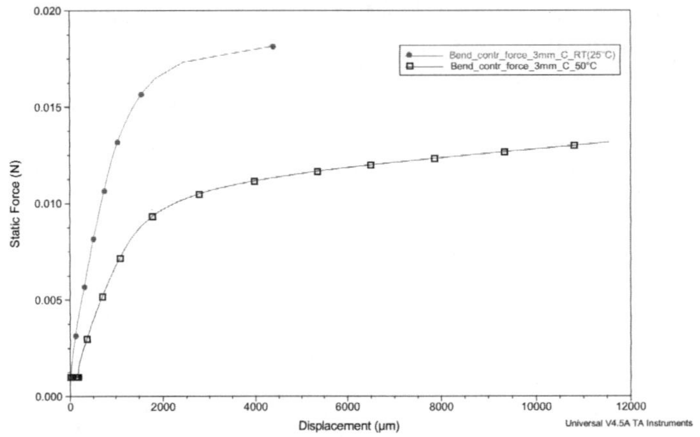

Figure 4. Bending force for fiber C, on 3 mm distance, at RT and 50°C .

Since it is impossible to show all the overlaid plots on the paper, the results of the bending force values on the Onset Points 1 and 2 for each fiber, for both distances and different temperatures were represented on Table 1. Based on the force values obtained, a better and more accurate analysis on how the bending force depends on the temperature was done.

Table 1. Bending force values at Onset Points 1 and 2, for both distances and three temperatures.

Fiber sample	Onset Points	Bending force (cN)					
		RT (25°C)		50°C		80°C	
		2 mm	*3mm*	2 mm	*3mm*	2 mm	*3mm*
Fiber A	OP 1	3.361	***2.272***	1.944	***1.119***	1.178	***0.61***
	OP 2	4.243	***3.213***	2.586	***1.466***	1.43	***0.83***
Fiber B(l)	OP 1	4.669	***2.759***	3.055	***1.636***	1.571	***1.057***
	OP 2	6.115	***3.543***	4.058	***2.184***	2.024	***1.332***
Fiber B(m)	OP 1	7.433	***3.465***	3.919	***2.268***	2.083	***1.448***
	OP 2	8.278	***4.291***	5.090	***2.980***	2.568	***1.866***
Fiber C	OP 1	2.21	***1.101***	1.188	***0.768***	0.559	***NA***

	OP 2	2.8076	***1.401***	1.453	***1.044***	0.785	***NA***
Fiber D	OP 1	3.125	***1.423***	1.162	***0.636***	0.599	***NA***
	OP 2	3.869	***1.764***	1.5086	***0.858***	0.827	***NA***
Fiber E	OP 1	4.717	***2.299***	2.159	***1.064***	1.124	***NA***
	OP 2	6.121	***2.809***	2.8496	***1.588***	1.438	***NA***
Fiber F(l)	OP 1	2.835	***1.672***	1.39	***0.863***	0.836	***NA***
	OP 2	4.474	***2.2556***	1.984	***1.228***	1.153	***NA***
Fiber F(m)	OP 1	6.124	***3.625***	3.028	***3.028***	1.649	***NA***
	OP 2	6.69	***4.168***	3.517	***3.518***	1.904	***NA***

The results show that the reduction of bending force values was quite high when passing from room temperature to 50°C. As it can be noticed from the plots and the table, for all fibers, and for both distances, there is a decrease by almost 50% on the values obtained at 50°C temperature compare to those obtained at 25°C. This can be explained with the changes that occur in the molecular structure of the polymer. So, the LLDPE polymer is stable and no changes occur on its internal structure, when exposed to temperatures up to 40°C [6, 7]. At temperatures higher than 50°C, internal molecular movements start, causing changes in the inside structure of the polymer, and consequently also in its mechanical properties [7]. This is reflected on the significant changes of the bending force values at 50°C temperature.

Regarding the bending force values at 80°C (Table1), to be noticed is another decrease of almost 50%, compared to those obtained at 50°C and a high reduction on these values while compared to the values obtained at 25°C. This is due to the increasing of the internal molecular movements causing changes on the polymer chains and increasing the disorder on the internal structure of the material [6]. The LLDPE fibers are estimated working up to temperatures 65-70°C resulting on a very high loss of their properties when exposed to higher temperatures [7]. This is the reason why they show a significant drop in performance, reflected in big changes of the bending force values at 80°C.

When comparing the performance of the fibers at high temperatures, is noticed that the better performing fiber is B on the convex side followed by fiber F on the same side and fiber E, while the worst performing fibers are C and D. For both distances, the same conclusion applies. This is explained with the cross section shape of the fibers [11].

Also, the highest values of the bending force are encountered while testing the fibers at high temperatures on the 2mm distance of the applied force, in analogy with the room temperatures measurements [13].

So, when tested at high temperatures, the fibers show the same tends of performance as they show when tested at 25°C [11, 13]. This results to a successful testing on the DMA at elevated temperatures.

5. Conclusions

The testing of the LLDPE monofilaments on the DMA at elevated temperatures, other than room temperature and the evaluation of the bending force measured depending on the temperature was the purpose of this paper. Six different monofilaments of LLDPE were tested at 25°C, 50°C and 80°C and in 2 and 3 mm distance of applying the force.

The challenge of using DMA for testing at elevated temperatures was successfully accomplished.

From the analysis of the bending force values obtained, resulted that fibers exhibited significantly lower performance at higher values of temperature than 25°C. Thus, the bending force at 50°C temperature was reduced by almost 50% compare to 25°C temperature, and at 80°C the force values were also reduced by almost 50% compare to those at 50°C. The same principle applies to all fibers on both distances. This is due to the influence of the temperature on the inside molecular fiber movements and changes occurring in the polymer chains, which increases the disorder in the internal structure, resulting in loss of fiber properties.

Considering the performance of each fiber for the same high temperature, the results show that the better performing fibers were the "c" shaped fibers, B and F on the convex side and the worst performing were fibers D and C. This is in full analogy with the fibers performance at 25°C.

References

[1] Young C 2009 *Maintenance: Cost Benefits* (Sport SURF 7th workshop Loughborough) http://www.sportsurf.org/workshops/7/CY.pdf

[2] ASGi *Association Turf Installers* (valid form webpage http://www.asgi.us/public-library)

[3] *Artificial Grass Market Study* 2007 (valid from webpage http://www.asgi.us/59.htm)

[4] Joosten T 2003 *Players experiences of artificial turf* (ISSS Stadia turf summit Amsterdam) http://www.isss.de/conferences/Amsterdam2003/Joosten.pdf

[5] Sperling L 1993 *Introduction to Physical Polymer Science* (Academic Press NY)

[6] Ferry J and Wiley J 1980 *Viscoelastic Properties of Polymers* (New Work)

[7] Peacock A 2000 *Handbook of Polyethylene Structures Properties and Applications* (New York)

[8] Schoukens G 2009 Development in textile sports surface In Goswami KK *Advanced in Carpet Manufacture Cambridge* (Woodhead Publishing in Textiles ltd.)

[9] Kola I, Kolgjini B, D'Hooge D, Guxho G and Kiekens P 2017 Comparison between new developed test method (using DMA) and the existing one for measurement of the bending force of monofilaments *International Journal of Innovative Research in Science, Engineering and Technology (IJIRSET)* **Vol 6** Issue 6 (ISSN 2347-6710) DOI:10.15680/IJIRSET.2017.0606001

[10] Menard K P 2008 *Dynamic Mechanical Analysis* (CRC Press Taylor & Francis Group NW 33487-2742)

[11] Kola I, Kolgjini B, Rambour S, Guxho G and Kiekens P 2017 (2) Influence of linear density and cross section shape of the monofilaments on their bending behavior *Journal of Natural and Technical Sciences (JNTS)* **Vol XXII (44)** (ISSN 2074-0867 pp. 147-158)

[12] TA-Instruments 2000 *Advantage Software v5.5.22* (Universal Analysis)

[13] Universal Analysis 2000 Issued September 2001 *Operator's Manual (*PN 925809.002 Rev J)

[14] Kola I, Daelemans L, Kolgjini B, Rambour S, Guxho G and Kiekens P 2017 Bending force of LLDPE monofilaments in relation with the distance of applied force *International Journal for Science Technics and Innovations for the Industry/ Machines Technology Materials* (Year XI ISSN **1313-0226**) pp. 221-224

IOP Conf. Series: Materials Science and Engineering **459** (2019) 012050 doi:10.1088/1757-899X/459/1/012050

Properties of goat fibres from selected Albanian regions

B Kolgjini[1], A M Grancaric[2], L Botteri[2], I Kola[1], E Shehi[1] and G Guxho[1]
[1]Polytechnic University of Tirana, Faculty of Mechanical Engineering, Department of Textile and Fashion, Sheshi Nënë Tereza, N01, Tirana, Albania.
[2] University of Zagreb, Faculty of Textile Technology, Prilaz baruna Filipovića 28a, 10000 Zagreb, Croatia
bkolgjini@fim.edu.al / bkolgjini@yahoo.co.uk

Abstract
The focus of this paper is to study the quality of goat fibres /hair from various Albanian regions for possible use in the textile industry. In Albania there are several regions where the collection of these fibres is possible but interest to employ them is lacking. This dearth of interest could be due to the paucity of information regarding the quality of such fibres. In the not-so-distant past Albania had a tradition of using these fibres for various handmade textile products, especially for products where strength, durability and dimensional stability are required.
In this investigation the physical mechanical and thermal properties of this material are studied. Also, given that wool fibres are the main attraction as a component in high performance thermal protection applications to provide protection from fire, high limitation oxygen index (LOI) tests are performed. Samples are taken from five regions and from the obtained results the finesses of the fibres are at very high values, approximately 78µ; however, the values of LOI are rather promising, 41%, and comparable values concerning the special fibres used for fire protection.

1. Introduction

Throughout the last 20 years, although some farmers in Albania have expressed interest to increase the number of the goats [1] to be used for meat and milk, the same level of enthusiasm has not been shown for using their fibres in textile products. This lack of interest could be because there is a dearth of information regarding the quality of fibres; there are no published papers or studies regarding such an issue. Although there are several papers and studies providing information about the quality of wool fibres from local sheep herds in Albania, surprisingly not much is known about the quality of goat hair.

In addition, in the recent past Albania had a tradition [2] in using these fibres for several textile products, especially for products where strength, durability and dimensional stability were required. It should be mentioned here that "m'taf", a kind of rug produced by 100% from goat fibres, usually covered the carpeted area of a guest's room, which used to be the largest room in the house; "duq" a kind of satchel used to transport goods in the villages, "norta", amongst numerous other handmade items, were also made of this material. Several such traditional products could be enumerated here, which these days rarely can be seen or used. Thus, this paper can be used like an "emergency call" to assist in the passing down of the tradition so that it does not die out! The objective of this paper is to investigate the properties of goat fibres to increase awareness and draw attention to matters concerning the possibility to improve the industrial interest for utilizing such fibres, and indirectly to create sustainable business opportunities for local farmers. When it comes to the definition of the quality of the fibres, it is not as simple as just characterising the strands. As Khan et al [3] explains there are several factors that influence their properties, including environmental, genetic, physiological state, hormones, parasite and diseases, amongst others.

Regarding these factors, conducting research, in particular concerning evaluating the quality of the goat fibres in Albania is reasonable, especially considering that in recent years there has been a gap concerning information about the race of goats and conditions for properly treating them. Physical and mechanical properties are the classical properties to define in order to assess the quality of fibres, but it might be of interest to also study: Limitation Oxygen Index (LOI), Thermogravimetric Analyses (TGA) and Microscale Combulsion Calorimeter (MCC) of local goat fibres as rather important parameters, especially concerning floor covering textiles. Wool is regarded as a naturally flame-resistant fibre, and thus wool fabrics were the traditional materials for flame and heat protection [4, 5]. These fibres have been found to yield high performance thermal protection, i.e. to provide protection from fire. In general wool is known to have a high ignition temperature, high LOI and a low heat release from the combustion, but it varies depending on the type of wool as well as the region of the collected samples [6].

2. Materials and Methods

Fibre samples; samples are randomly selected from five different regions in Albania. The chosen areas include Hasi, Tropoja, Zagore, Vorfë (Shkodër) and Përmeti. Per each region, 500g of hair were considered for analyses. The hair samples were sheared directly from the goats, specifically from the midrib area of the body, to be processed for the quality parameters. Sample preparation for this test was according to ISO 1130:1975 Textile fibres -- Some methods of sampling for testing [7].

Physical and Mechanical tests were carried out at the accredited laboratory at the Department of Textiles and Fashion (Polytechnic University of Tirana, Albania) according to the ISO standards for determining fibre characteristics. The fibre length was calculated according to ISO 6989:1981Textile fibres - Determination of length and length distribution of staple fibres (by measurement of single fibres) [8]. Breaking strength and fibre elasticity were performed according to ISO 5079:1995 Textile fibres -- Determination of breaking force and elongation at break of individual fibres [9].

LOI, MCC and TGA tests were performed at the Textile Department Laboratory in Zagreb Croatia.

The burning behaviour of the fibre was determined with Limiting Oxygen Index (LOI) in a LOI chamber (Dynisco), according to the ISO 4589:1996– Plastics – Determination of Burning behaviour by Oxygen Index [10].

Fabric was able to be produced from goat fibres (figure 1) taken from the Hasi region. The structure of the textile that was consisted of 100% goat fibres, plain weave; the density of warp and weft yarns are respectively 23y/10cm and 34y/10cm.

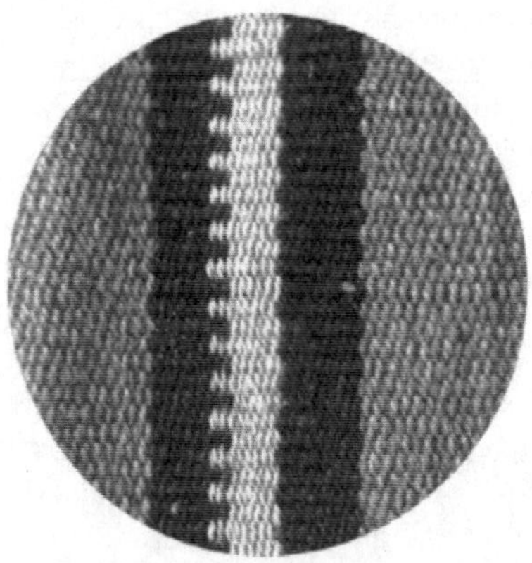

Figure 1. Picture of 100% of goat fabric from the Has region in Albania.

Thermogravimetric analyses (TGA) was performed on TGA Pyris 1 (PerkinElmer). The mass of the sample was 5 mg of goat fibres.

Microscale Combustion Calorimeter (MCC) tests were performed according to ASTM D 7309 Standard Test Method for Determining Flammability Characteristics of Plastics and Other Solid Materials Using Microscale Combustion Calorimetry [11], on MCC-2, Govmark, USA. The 5 mg of test sample (three samples per each system) were placed within the sample cup and the test was performed under the following conditions: pyrolysis operating temperature range of min. 25°C to 600°C at a heating rate of 1 K/s in an inert gas stream (nitrogen, 80 ml/min), with a detection sensitivity limit of min. 5 mW, repeatability of ± 2 %.

For samples taken from the Hasi region, the fibre diameter was measured by using SEM. SEM pictures were made on FE–SEM MIRA/LMU (Field Emission Scanning Electron Microscope).

3. Results and discussion

In table 1 the mean length, fibre diameter, fibre elasticity and elongation per each sample taken from five different regions in Albania are presented.

Table 1. Values of mechanical properties of goat fibre samples.

Sample	Length[mm]	Diameter [m]	Elongation [%]	Max Force [cN]
Tropoj	118.6 ± 13.5	78.79	68.4	106.03
Hasi	95.5 ± 7.93	77.62	69.4	95.28
Zagore	134 ± 17.8	82.79	56.38	149.87
Vorfë	103 ± 8.73	84.91	47.51	103.28
Përmet	109.1± 14.01	74.73	63.6	78.58

From the obtained results, regarding the length, the samples taken from Zagor area show the highest value (134mm), followed by Tropoja and Përmeti. Samples taken from two other regions have almost the same value. The differences in the samples could be related to several factors, which should be considered: not much is known about the race of the goats, in that different regions have different races; in addition to this genetic influence, the environment itself could serve as a strong indication. These values, however, are almost the same as those reported by other studies conducted in Turkey; Deger et al. (2008) [12] the mean length value was 136.3mm, Della et al. 2014 [13] and Soylezmezoglu et al. (2002) [14], where the mean value lengths were 118mm and 122mm, respectively.

Regarding the values concerning the diameter of the tested samples, the situation is completely different from the length of the fibres; all the samples have nearly the same value, approximately 78 . These values are comparable with studies conducted by Dellal et al which reported that the diameter of the goat hair was between 64-93 m [13]. Only samples from Përmet have the lowest value 74.73 m, almost at the same value as those reported by Deger et al (2008), where the mean value was 76.70 m [12].

Similar to the length of the fibres, the diameter is very much influenced by the conditions where the goats are treated [3].

Regarding the evaluation of the fibres in term of industrial use, the length and diameter are rather important parameters. The length does show comparable values with the wool fibres from sheep, but the diameter is far away from the values of the angora goats (approximately 17.2 m).

The burning behaviour of the textiles of 100% of goat fabric was performed through a LOI test according to the ISO 4589 standard. The results are presented in table 2. According to the results, it can be seen that the textile burns was a presence of 41%. The same results were found for three repetitions. Compared with other studies [6, 15], the values of LOI for wool was 25% ; the results seem to be promising (41%) and yield comparable values with the special synthetic fibres used for fire protection after special treatment [16].

Table 2. LOI values of 100% goat fabric.

Sample	LOI	$t_{100\ mm}$ [s]	t_{glow} [s]
Goat fabric	41	113	120
	41	118	105
	41	115	111

By performing the TGA, the evaluation of the thermal stability of the textile was possible. In figure 2 and table 3, the curve and the values for the thermal degradation of goat fabric, which takes place at 300–400 °C through two competitive processes, namely depolymerisation and dehydration, are presented. The goat fabric started the degradation process at 92 °C (Tonset5%), where the maximum temperature of first weight loss (Tmax1) is 247 °C and the maximum temperature of the second weight loss (Tmax2) is 526 °C. The residue at the end of TGA measurement is 4.9%.

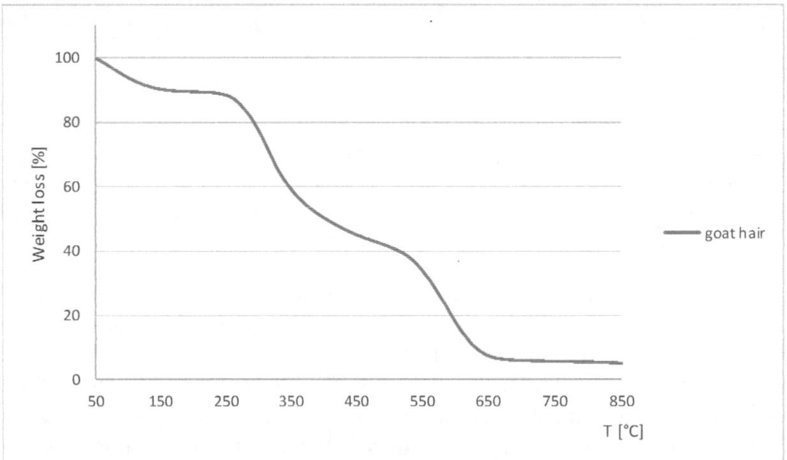

Figure 2. TGA curve of goat fabric in the air, heating rate 10 °C/min.

Table 3. Thermogravimetric data of goat fabric measured in the air condition.

Sample	$T_{onset\ 5\%}$ [°C]	T_{max1} [°C]	T_{max2} [°C]	Residue at 800 °C [%]
Goat hair	92	247	526	4,9

For samples taken from the Hasi region, it has been performed the measurement of MCC, as shown in figure 3 and table 4. The results of the micro combustion calorimeter (MCC) analyses show that goat fabric has a rather low Heat Rate (80,9 W/g), evidence that these samples exhibit good thermal properties. A yield of pyrolysis residue at the end of the measurements is 24%. For the samples taken from the Hasi region it was possible to realize some measurements of the fibre diameter by using SEM. (see figure 4). The SEM pictures were made on FE–SEM MIRA/LMU (Field Emission Scanning Electron Microscope). It can be seen some impurities.

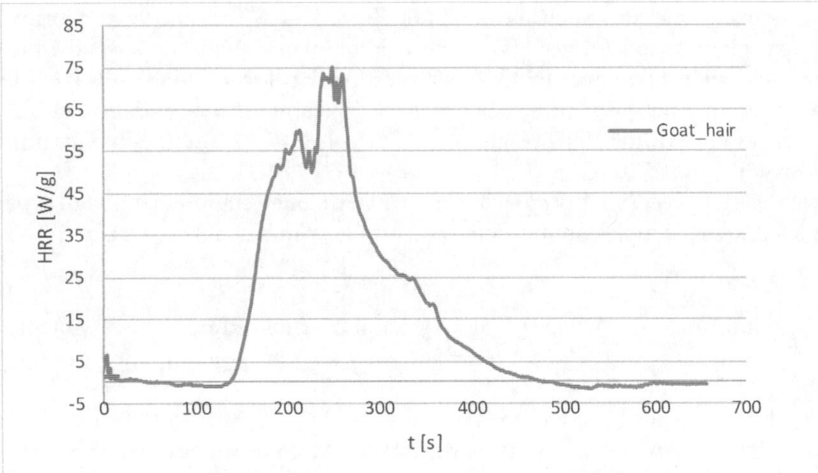

Figure 3. MCC curve of goat's fabric.

Table 4. MCC data of goat hair.

Sample	Heat release temperature T_{max1} (°C)	Maximum specific heat release Q_{max1} (W/g)	Specific Heat release hc (kJ/g)	Heat release capacity ηc (J/g-K)	Yield of pyrolysis residue Yp (g/g)	Spec. heat of fuel gases combustion hc.gas(kJ/g)
Goat hair	351,1	80,9	9,6	92	0,24	12,7

Figure 4. SEM images of goat fabric.

4. Conclusion

This paper includes information describing the physical and mechanical properties of goat fibres. The values of length vary from 95.5 to 134 mm, and the diameter varies from 74.7 to 84.91 m. The values differ depending on the region, thus suggesting that the race and environment have a strong influence on the goat hair properties.

Regarding thermal properties, such as the LOI, TGA and MCC, the goat fabric from the Hasi region is rather promising; the LOI was 41 %, comparable values with the special fibres used for fire protection. The MCC shows that goat fabric has rather low Heat Rate (80, 9 W/g), evidence that these samples exhibit good thermal properties. The goat fabric started the degradation at 92 °C (T_{onset}5%); the maximum temperature of the first weight loss (T_{max1}) is 247 °C and the maximum temperature of second weight loss (T_{max2}) is 526 °C.

In terms of industrial processing of goat fibres from Albanian regions, although a part of the work has been completed; more samples are needed to be tested, which is the next stage of the research.

References

[1] Source of information: Ministry of Agriculture and Rural Development, available at http://www.instat.gov.al/al/temat/bujq%C3%ABsia-dhe-peshkimi/blegtoria/#tab2, accessed on 2018, 08.10.

[2] Gjergji A., Ligjerata per Etnologjine Shqiptare, ISBN 99927-36-25-92001

[3] Khan M., Abbas A., Ayaz M., Naeem M., Akhter M. and Soomro M., *Factors affecting wool quality and quantity in sheep. African journal of Biotechnology* Vol. **11** (73), pp. 13761-66, 2012.

[4] Guxho G., Shehi Y., Shehi E., Shkenca dhe teknologjia e materialeve tekstile, ,ISBN: 978-99943-49-99-9; 2011.

[5] N. A. G. Johnsons, I. Russell, Advances in Wool Technology, Woodhead Publishin, ISBN 978-1-84569-5469-0 (e-book).

[6] IWTO Fact Sheet: Wool and Flame resistance international wool Textile Organization aisbl.2018.

[7] ISO 1130:1975 Textile fibres -- Some methods of sampling for testing.

[8] ISO 6989:1981Textile fibres - Determination of length and length distribution of staple fibres (by measurement of single fibres).

[9] ISO 5079:1995 Textile fibres - Determination of breaking force and elongation at break of individual fibres.

[10] ISO 4589:1996– Plastics – Determination of Burning behaviour by Oxygen Index.

[11] ASTM D7309- Standard Test Method for Determining Flammability Characteristics of Plastics and Other Solid Materials Using Microscale Combustion Calorimetry.

[12] Deger H, Toplu O and Altinel A, 2008. Some production traits of indigenous Hair goats bred under extensive conditions in Turkey. 1st communication: reproduction, milk yield and hair production traits of does. *Arch Tierz* **51**:498-506.

[13] Dellal G, Soylemezoglu F, Erdogan Z, et al., 2014. Present situation and future of animal fiber production in Turkey: A Review. *J Life Sci* **8**: 192-200.

[14] Soylemezoglu F, Erdogan Z, Dellal G and Tatar AM, 2002. The Relationships between colourfullness and physical traits of down and coarse fibers obtained from goats raised in Turkey. *Turk J Vet Anim Sci* 26:1395-1400.

[15] Edited by N.A.G. Jahnson and I.M.Russell, Woodhead Publishing 2009, Isbn 978-1-84569-546-0-e-book.

[16] Horrocks A. and Anand S., Handbook of Technical Textiles, Woodhead Publishing ISBN 1 85573 385 4.

IOP Conf. Series: Materials Science and Engineering **459** (2019) 012051 doi:10.1088/1757-899X/459/1/012051

Evaluating the ultraviolet protection factor (UPF) of various knit fabric structures

E Louris[1], E Sfiroera[1], G Priniotakis[1], R Makris[1], H Siemos[1], C Efthymiou[2] and M N Assimakopoulos[2]

[1] University of West Attica, Faculty of Engineering, Department of Industrial Design and Production, University Campus 2, 250 Thivon & P. Ralli Str., 12244, Egaleo, Greece

[2] National and Kapodistrian University of Athens, Faculty of Physics, Division of Environmental Physics and Meteorology, University Campus Zografou, Building PHYS-V, 15784, Athens, Greece

elour@puas.gr

Abstract. Public awareness regarding the risks of prolonged skin exposure to the sun light and more specifically to the UV spectrum part is increased during the last decades. Clothing is the most natural and suitable way of protecting the human body, thus the market interest in clothes that can offer adequate UV protection is growing continuously. Previous research works have revealed the main factors that influence the ability of fabrics to block harmful UV radiation. However, the variability of these factors and the versatility of their combined effect make UV protection factor prediction difficult and hence the design of fabrics with high performance against UV radiation becomes a complicated task. Hopefully, the most critical and predictable among all factors is the fabric structure itself. Expectedly, closer and tighter structures offer higher UV protection. Due to this fact, the majority of previous research concern woven structures which generally are less porous and offer a higher UV protection. However, the possibility to obtain knitted fabrics with adequate UV protection factor is of great interest, since knitted fabrics are more appropriate for sports as well as for casual summer fashion garments. Current literature regarding the UV protection factor of knitted fabrics is very limited and concerns mostly fabrics produced in machines of relative large gauges. In the present work the UV protection factor of various typical weft-knitted structures, produced in a flat knitting machine with 7 gauge and by using grey 100% Organic Cotton yarns, Ne 30/2, 330 TPM (twists per meter) is studied. The yarn has been selected due to the increasing market interest for Organic Cotton products.

1. Introduction

Since clothing is the most natural and convenient way of protecting the human body from external environmental factors, the market interest in clothes that can offer adequate protection against UVR is growing continuously. Depending on the radiation intensity, the penetration degree and the respective harm that the various wavelengths can cause to human skin significantly increases. . Generally, skin damage risk increases exponentially as wavelength decreases, thus radiation at 280 nm is estimated to be 1000 times more damaging than radiation at 340 nm [1]. Prolonged exposure to UVA causes ageing of the skin and some risk of skin cancer, while much shorter exposure to UVB causes erythema (skin reddening and burn) and increased risk of cancer [1,2]. The most dangerous region is UVC, which is related to extensive skin damages and extremely high risks of skin cancer. Fortunately, the earth's stratospheric ozone layer absorbs radiation with wavelengths between 100 nm and 290 nm, therefore it blocks the UVC and some part of the UVB radiation [1-3]. Thus, protection from sun-light refers to protection mostly from the harmful UVB, but for longer exposure times also from the UVA region of the solar spectrum. In the previous years, in a wide range of UV shield products, the protection from the UVA used to be neglected, but nowadays it is considered as essential.

In order to rate the UV protection that is provided by various protection shields, like sunscreens, fabrics or colored glasses, the international scientific community has established the Ultraviolet Protection Factor (UPF), which is calculated by the following equation (1):

$$\mathrm{UPF} = \frac{\sum_{290}^{400} E_\lambda \cdot S_\lambda \cdot \Delta_\lambda}{\sum_{290}^{400} E_\lambda \cdot S_\lambda \cdot T_\lambda \cdot \Delta_\lambda} \qquad (1)$$

where,
E_λ is the solar UVR spectral irradiance in $W \cdot m^{-2} \cdot nm^{-1}$
S_λ is the relative erythemal effectiveness according to the CIE [4,5]
Δ_λ is the wavelength interval of the measurements
T_λ is the spectral transmittance at wavelength λ
λ is the wavelength (in nm)

Based on their measured UPF, UV protective shields are classified into specific UPF rating categories in accordance to international standards (Table 1). The practical meaning of a UPF rate can be explained with the following example. If an unprotected normal fair skin starts exhibiting signs of erythema (reddening) after 10 minutes of continuous exposure to UV solar light, a protective shield with UPF=15 can extend this exposure time by 15 times, i.e. 150 minutes [2]. Higher UPF values above 40 (rating 40+) or 50 (rating 50+) are considered as "totally" safe "sun-blocks".

Table 1. UPF rating categories according AS/NZS, ASTM and European standards [1-4,6,7]

	AS/NZS 4399:1996 and ASTM D6603			EN 13758-2:2003		
Protection	% UVB Transmittance	UPF Range	UPF Ratings	% UVB Transmittance	UPF Range	UPF Ratings
Excellent	less than 2.5	above 40	40, 45, 50, 50+	less than 2.5	above 40	40+
Very good	4.1–2.6	25–39	25, 30,35	3.3–2.5	30-40	30
Good	6.7–4.2	15-24	15, 20	5.0–3.4	20-29	no value

In order to express more precisely the UV shielding performance of a material in each part of the UVR spectrum, the UVA and UVB transmittances are calculated as the arithmetic mean of all intermediate transmittance measurements in each region, according to equations (2) and (3) [5].

$$\text{UVA}=\frac{1}{m}\sum_{290}^{315} T_\lambda \quad (2)$$

$$\text{UVB}=\frac{1}{k}\sum_{315}^{400} T_\lambda \quad (3)$$

where

T_λ is the spectral transmittance at wavelength λ
m and k are the number of measurement points between 315 nm and 400 nm and between 290 nm and 315 nm respectively, i.e. depending

Previous research has revealed the main factors that influence the ability of fabrics to block harmful UV radiation. A summary of these factors can be found in the next section. However, the variability of these factors and the versatility of their combined effect makes any a priori estimation of the UV protection performance of a fabric, difficult. Thus, the design and engineering of a fabric with high UV protection performance becomes a complicated task, since the only way to secure its performance is to measure the final incident UV transmittance and calculate its actual UPF. Since closer and tighter structures offer higher UV protection, the majority of previous research concern woven structures which generally are less porous and offer a higher UV protection. However, the possibility to obtain knitted fabrics with adequate UV protection factor is of great interest, since knitted fabrics are more appropriate for sports as well as for casual summer fashion garments. Current literature regarding the UV protection factor of weft-knitted fabrics is very limited and concerns only fabrics produced in machines of relative large gauges, like T-Shirts. At the same time, there is a strong market interest in fashion garments made of natural fibres, such as organic cotton, which can offer high comfort levels.

The target of the present research is to evaluate the UPF of fabrics produced in a small gauge flat knitting machine (7 gauge) by using organic cotton yarns and to define the structures which can serve as a basis for further developing and constructing heavier gauge knitwear with good UV protection capabilities.

2. Factors influencing the UPF of fabrics

2.1 Fabric structure (construction)

The fabric structure is the most critical factor in the UV protection performance of a fabric. Closer and tighter structures exhibit less open spaces between the yarns that form the fabric, thus allowing less UV radiation to be transmitted through the fabric. In this respect, the actual structural parameter that determines the UV protection performance of a fabric is the porosity or otherwise the cover factor (CF) of the fabric, which can be defined as the percentage area occupied by yarns in a given fabric area. Previous research has shown the relation between some other constructional parameters, such as the fabric weight or the fabric thickness, with the resulting UPF. For a given fabric structure, an increase in the weight per unit area or in thickness, normally decreases the fabric's porosity and results to an increase in the UPF. However, a different lighter and thinner structure made of the same material may exhibit higher UPF, indicating that the dominant parameter in the UV protection performance is the fabric structure itself and its respective porosity. [7,11,13]

2.2 Material (fiber type)

Fiber type is the second most critical factor that determines UV protection performance of a fabric. Generally, synthetic yarns have superior UV performance than natural fibers. Fabrics made of 100% polyester yarns exhibit the highest UPF ratings and thus are widely used for technical sports garments.

IOP Conf. Series: Materials Science and Engineering **459** (2019) 012051 doi:10.1088/1757-899X/459/1/012051

However, natural fabrics due to their hydrophilic properties offer superior comfort levels. Blending of natural with synthetic fibers may lead to more comfortable garments with increased UPF ratings [11,13]

2.3 Color
The fabric color influences significantly the UPF. As a general rule, darker colors absorb more UVR resulting higher UPFs [11]. Dark colors, like black, blues, dark greens, improve significantly the fabric UPF, while light colors show little improvement when compared with the undyed. Since different dyes maintains a different absorption spectrum, the UPF improvement exhibits a large variability which depends not only on the shade but on the dye type as well [7,11,13]. A UPF increase from 12 to 34 has achieved with the use of red or blue dyes [13]. In the case of 100% polyester fabrics, which already have high UPFs, the effect of color is not significant [13]. Naturally-colored cotton that contains color pigments has also found having high UPF values (green UPF = 30 to 50+ , tan UPF = 20 to 45 and brown UPF=40 to 50+) while conventional unbleached cotton has UPF = 8 [12].

2.4 Bleaching (whitening)
Bleaching of fabric reduces the UPF. Natural fibers contain substances which act as UV absorbers. With bleaching these natural UV absorbers are removed and the UPF drops. Conventional unbleached cotton has a UPF = 8 because of its content in natural pigments, pectin, and waxes, while bleached cotton exhibits UPF < 4 [12,18].

2.5 Stretch (fabric deformation)
When a fabric is stretched, structural deformation tends to move yarns apart and create more open spaces between them resulting in an increase in fabric porosity. By applying higher tensions and increasing fabric stretch, significantly larger drop in the UPF values is observed [11]. Especially weft-knit fabrics which are inherently susceptible to dimensional changes and structural deformation exhibit dramatic changes in the UPF when stretched. In order to get the maximum UV protection from a knitwear it should be worn in a loose and un-stretched fit.

2.6 Wetness (water absorption)
The influence of wetness on the UPF depends primarily upon the fabric's material, namely the type of the containing fibers. Cotton fabrics provide significantly less UVR protection in their wet state than in their dry state. However, study in linen, viscose and polyester fabrics has shown that UPF significantly increases with wetting [11]. It is worth mentioning that the same study gives important evidences that, in almost all fabric cases saturation with tap water or salt water have the same effect on the UPF, and the moisture of the skin which is in contact with the fabric does not alter the fabric UPF value [11]. Alteration of the UPF due to water absorption can be explained partially due to the swell or shrinkage effect on particular fibers, which results in change of the fabric's porosity, and partially due to the alteration of the optical properties of the fabric, i.e. scattering effect can be reduced, leading to an increase in UV radiation penetration [11].

2.7 Washing
Generally, washing increases the fabric UPF value. The effect of washing is particularly significant for cotton and polycotton fabrics, which exhibit a permanent increase in their UPF after the first washing cycle [11,16]. On the contrary, the UPF of bleached cotton fabrics does not significantly change with washing [12]. The increase of the UPF after the first laundering may be explained by the shrinkage that occurs in the fabric structure and the slip of some fiber ends outside the yarn as a result of the mechanical stresses that undergoes the fabric during laundering. Both of these phenomena will decrease the fabric porosity and consequently increase the UPF.

2.8 UVR absorbing additives (agents)
Special fabric treatments with chemical additives that absorb UVR can be used in order to increase the UV protection performance of garments. UV absorbers are substances, such as substituted benzothiazoles, which selectively absorb UV radiation and convert it to heat [11]. For permanent results, UV absorbers should be attached to the fibers with strong covalent bonds. Special treatments of Cotton with nanoparticles of Zinc oxide (ZnO) or Titanium dioxide (TiO2) can also be applied resulting higher levels of UV blocking [18-19]. In the case of synthetic fibers, special pigments such as Titanium dioxide (TiO2) can be incorporated into the fiber mass during the manufacturing process, offering permanent UV protection performance [17-19]. Finally as already mentioned, special dyes with an absorption spectrum within the UV range may also be used as UV absorbers.

2.9 Yarn morphology
According to older considerations, the UV transmission through yarns is practically zero. However, more recent research showed that the yarn morphology has an impact in fabric UPF. Yarns made of more wavy filaments have a structure that allows higher UV transmission levels. Higher twists make yarns more compact but create a more irregular surface, which prevents the closer packing of the yarns during the weaving process and results in an increased fabric porosity [14]. The yarns that are used in knitting are less twisted and their less compact structure may allow higher UV transmission levels. Thus, in order to be more accurate regarding the porosity of knitted fabrics the bulk density of the knits, ρk (g/cm3) is taken into account, and the total porosity P (%) is defined as the portion of all air spaces in the fabric both between yarns and inside them [3].

3. Materials and methods
For the purposes of this research, 14 typical weft-knitted structures were considered, as presented in Table 2. For each one of them two samples were constructed, one by using a 2-ply yarn and another one by using a 3-ply yarn. For the Interlock structure, only one sample with a 2-ply yarn was constructed. In total 27 different knitted fabric samples have been constructed in a flat knitting machine STOLL CMS 411.6 - 7 Gauge.

The material used was grey yarns made of 100% Organic Cotton, Ne 30/2, 330 TPM, which have been offered by the Greek cotton industry VARVARESSOS S.A. The selection of the particular material has been made due to the increasing market interest for Organic Cotton products. In order to achieve the main target of our research, which is to verify the knit structures that can offer the highest possible UV protection, undyed grey yarns was chosen as the most proper option, since a different dyeing in any shade will increase even more the fabric UPF.

Table 2. The evaluated knitted structures

Knit Structure	**Technical Front**	**Technical Back**	**Knitting Notation**
Single Jersey	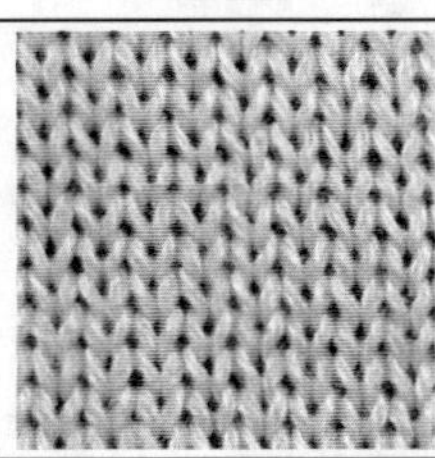	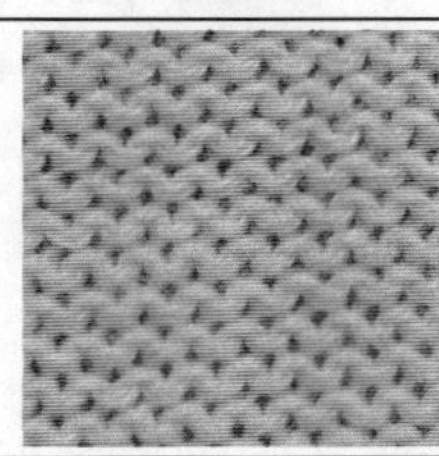	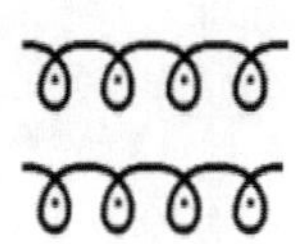

Lacoste	
Full Cardigan	
Half Cardigan	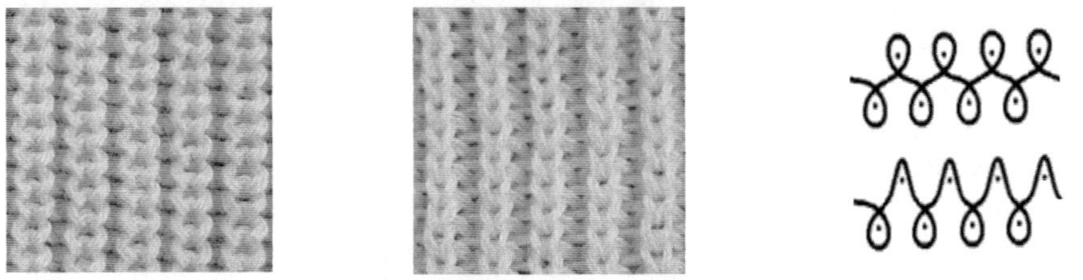
Full Milano	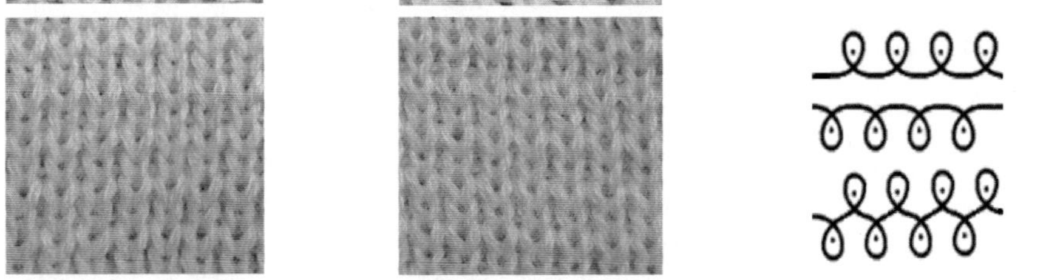
Half Milano	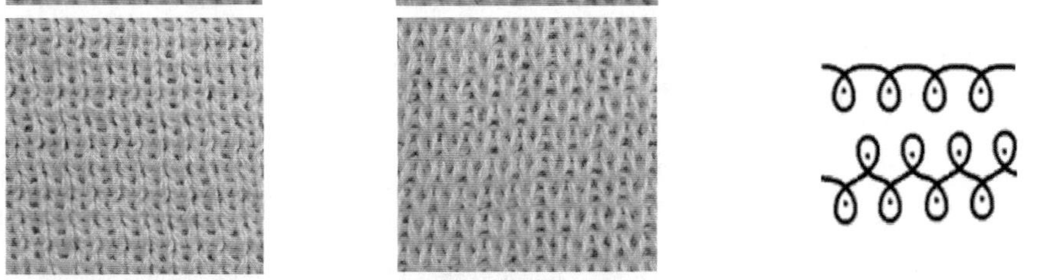
Pineapple	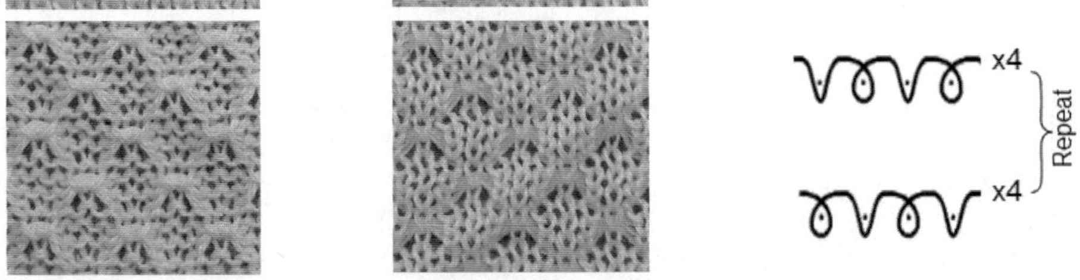
KM22C	

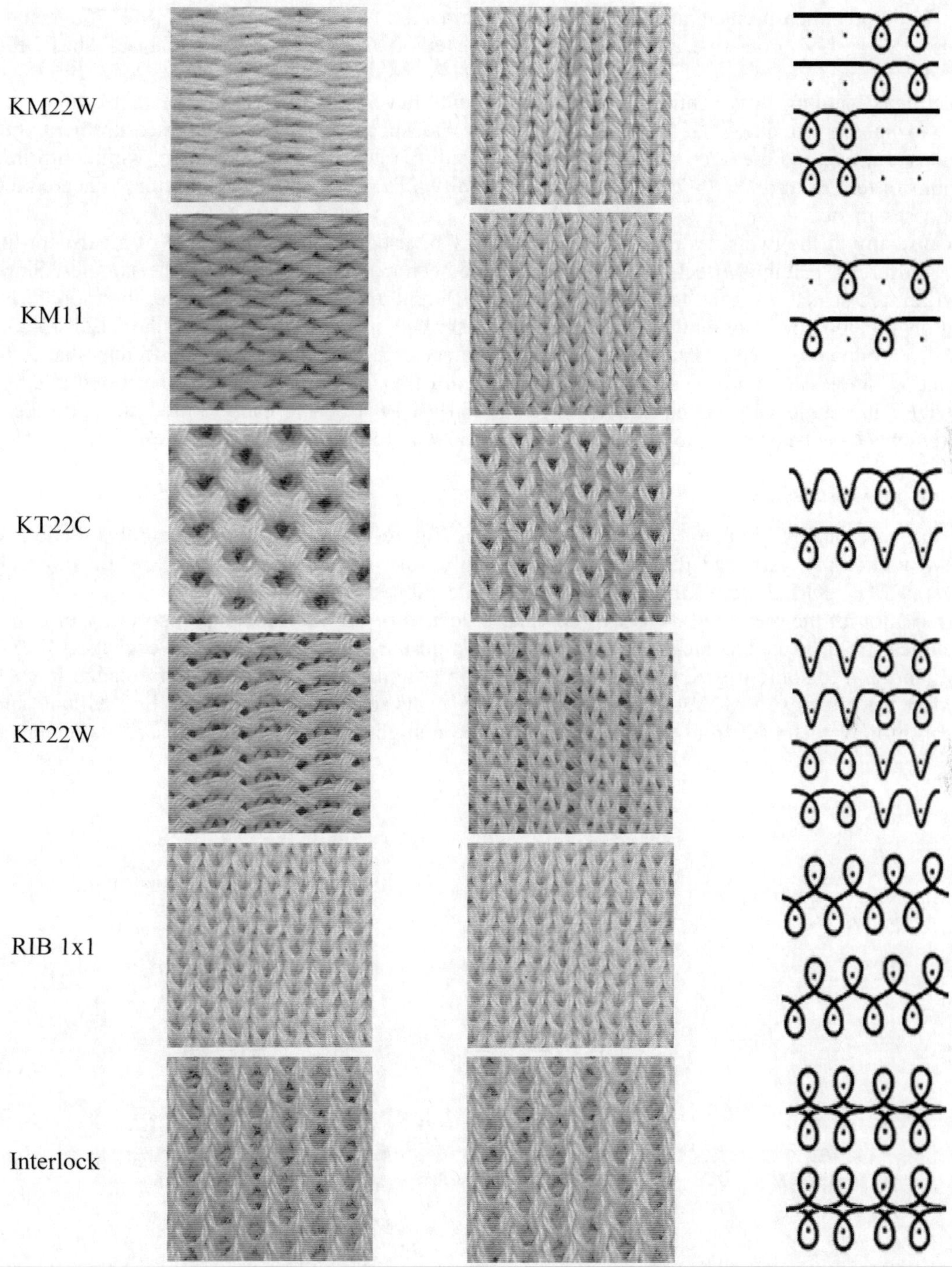

The in vitro UV transmittance measurements have been realized in a Varian Cary 5000 UV-Vis-NIR spectrophotometer, equipped with integrating sphere. Measurements made in the range from λ_{min}= 290 nm to λ_{max}= 400 nm with a step of Δ_{λ} = 5 nm. Since measurements concern non-bleached and non-dyed samples, no band-pass filter UG-11 for fluorescent materials has been used.

For the sample measurement and evaluation procedures the "European Standard EN 13758-1:2001 Textiles - Solar UV protective properties - Part 1: Method of test for apparel fabrics" has been followed.

Specimens handling was gentle and all measurements have been made with the fabric under no stress. No special conditioning has been applied to the specimens before the measurement, but specimens' storage and measurements have been realized in a laboratory environment with controlled conditions of temperature 25 +/- 2°C and relative humidity 55 +/-5 % RH. The laboratory is accredited for UV assessments.

Because any of the two sides of a knitted fabric may be used in the external face of a garment for creating different "fashion" effects, measurements have been made in both sides of the samples. Since in most knitted structures, each technical side has a different appearance and texture, the more strict evaluation method for "Non-uniform samples" as described in the European Standard EN 13758-1:2001 was followed. According to this method, in materials that contain areas of different shades or construction, at least two specimens of each colour and of each texture area shall be tested and the lowest UPF measured value is considered the material's UPF [5]. For each knitted structure eight measurements have been made, four in the technical front and four in the technical back.

4. Results and discussion

Seven structures have been found with high UPF values (figure 1). Three of them, the Full Milano and KM22W with 3-ply yarn and the Interlock with 2-ply yarn exhibited UPF > 50, while the Half Cardigan, KM11, KM22C and Rib 1x1 with 3-ply yarn exhibited UPF > 40.

Observation of the measured data gave no clear evidence of the UV blocking superiority of either the technical front or the technical back side for almost all the evaluated structures, except KM22W, Full Milano and to some degree Cardigan, which exhibit significantly different performance in each side. More specifically, KM22W exhibits higher UPF in its technical back, while Full Milano and Cardigan in their technical front. Single Jersey showed a slightly better performance in its technical back.

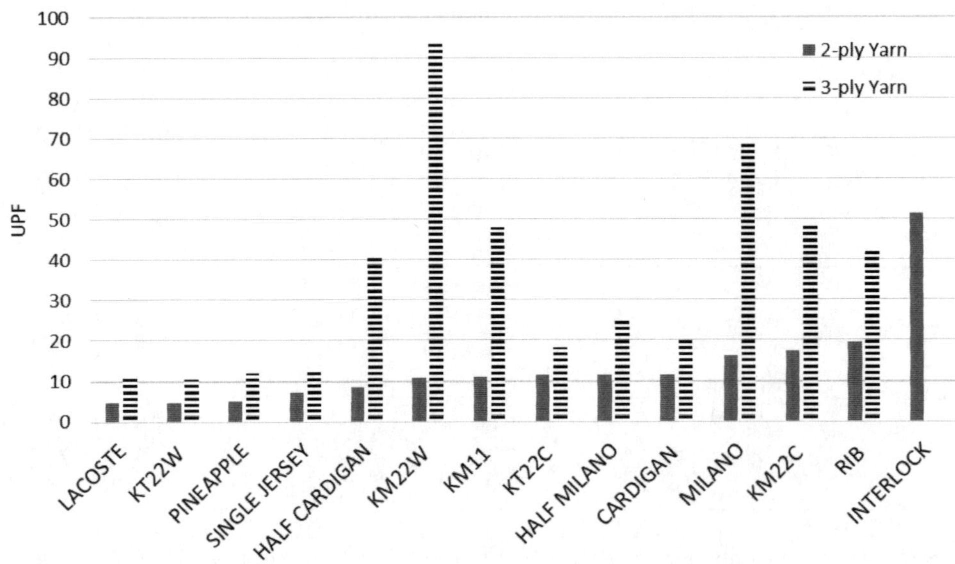

Figure 1. Minimum UPF values calculated for each knitted structure with 2x or 3x yarn

The maximum UVA and UVB, which correspond to the minimum UPF, of each knitted structure are depicted in the next figures 2 and 3 respectively.

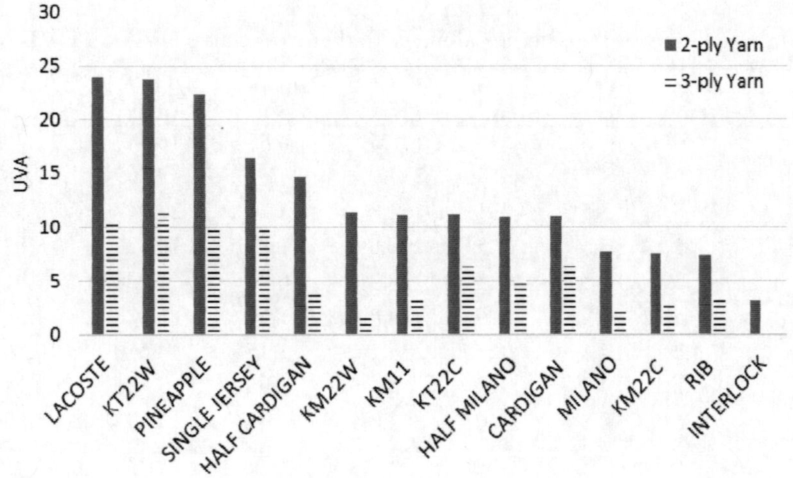

Figure 2. Maximum UVA values calculated for each knitted structure with 2x or 3x yarn

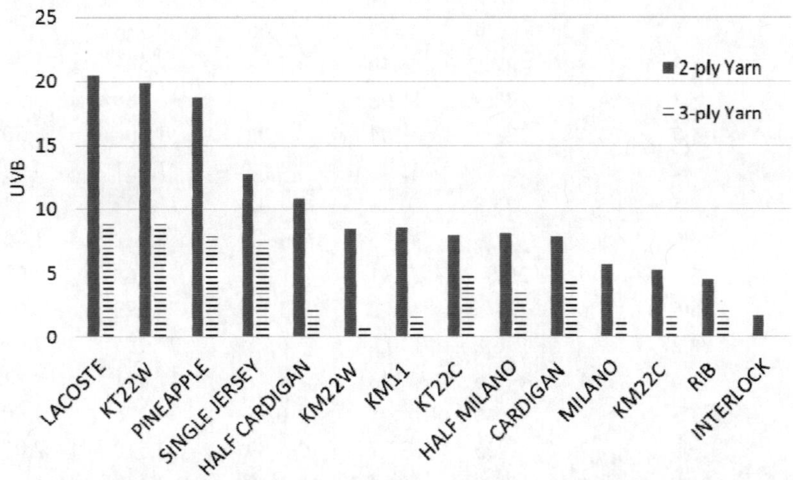

Figure 3. Maximum UVB values calculated for each knitted structure with 2x or 3x yarn

The technical textile parameters of the evaluated samples along with their respective UVA, UVB and UPF values are presented in table 3. The use of thicker yarns (3-ply instead of 2-ply) increases dramatically UV protection performance of all knitted structures. Thicker yarns produce thicker and heavier fabrics with improved UV blocking capacity. However, this improvement depends upon the specific knitted structure that is used. For instance, the Half Cardigan structure, despite its lower loop density, weight and thickness in comparison with the Half Milano structure, presents a four times higher UPF when using 3-ply instead of 2-ply yarns, while the Half Milano presents a twice higher UPF. Another example indicating the structure configuration as the dominant parameter that affects the UPF, is the case of the lighter Rib 1x1 structure with a 2-ply yarn, which provides a higher UPF than the heavier Lacoste structure with a 3-ply yarn.

The presence of miss stitches in the knit structure seems to be related with an increased UV blocking effect. All evaluated structures which contain miss stitches, i.e. KM11, KM22C and KM22W, exhibited high UPFs. On the other hand, tuck stitches certainly affect the UPF, but their positive or negative effect depends on how they are combined and repeated within the structure. From all the evaluated structures which contain tuck stitches, only Half Cardigan reached a UPF above 40.

Table 3. Textile parameters of the samples along with their respective UVA_{max}, UVB_{max} and UPF_{min}

KNIT TYPE	Course /cm	Wales /cm	Loops /cm^2	Weight gr/m^2	Thickness mm	UVA (max)	UVB (max)	UPF (min)
Plain Jersey x2	6	5	30	178,5	0,85	16,447	12,705	7,342
Plain Jersey x3	5,9	4,5	26,55	254,9	1,1	9,809	7,487	12,429
Lacoste x2	6	3,5	21	162,8	1,76	23,947	20,526	4,640
Lacoste x3	5	3,5	17,5	299,6	1,85	10,629	8,838	10,657
Full Cardigan x2	5	5	25	228,3	0,96	11,028	7,917	11,603
Full Cardigan x3	5	4	20	329,4	1,66	6,828	4,519	19,899
Half Cardigan x2	4,2	4,5	18,9	236,1	1,15	14,650	10,861	8,557
Half Cardigan x3	4,2	4,5	18,9	344	1,95	3,836	2,143	41,002
Full Milano x2	5	8	40	304,8	1,45	7,693	5,690	16,167
Full Milano x3	6	8	48	442,6	1,96	2,296	1,277	68,807
Half Milano x2	7	8	56	265,9	1,45	10,942	8,098	11,435
Half Milano x3	7	8	56	419,5	2,17	5,044	3,685	24,941
Pineapple x2	5	4	20	165,8	0,91	22,437	18,718	5,061
Pineapple x3	5	4	20	314	1,65	9,718	7,874	11,946
KM22C x2	5,5	5	27,5	240	1,08	7,525	5,216	17,457
KM22C x3	7	4,7	32,9	412,2	1,51	2,768	1,944	48,595
KM22W x2	5,5	5,9	32,45	204	1,02	11,404	8,507	10,866
KM22W x3	7	5	35	423,7	1,54	1,761	0,927	93,448
KM11 x2	5	5,3	26,5	207,8	1,02	11,142	8,534	10,904
KM11 x3	6	5,5	33	370,5	1,42	3,132	1,844	47,902
KT22C x2	5,8	3,6	20,88	222,1	1,5	11,268	8,015	11,404
KT22C x3	6,5	3,5	22,75	367,2	1,89	6,572	5,070	18,335
KT22W x2	4,2	4	16,8	140,2	0,77	23,755	19,914	4,759
KT22W x3	4	4	16	243	1,45	11,411	8,847	10,525
RIB 1:1 x2	5	8	40	178,5	0,82	7,376	4,601	19,499
RIB 1:1 x3	5	7	35	383,7	1,89	3,510	2,059	42,729
Interlock x2	5,3	6	31,8	361,4	1,69	3,140	1,667	51,359

The above mentioned structures with UPF > 40 are expected to exhibit relative high UPFs also when used in finer gauge fabrics. This assumption is also supported by previous research on finer gauge knitted fabrics [8, 9]. Although the thinner samples knitted with 2-ply yarn showed significantly lower UPFs, a selection among these structures can be considered as the optimum starting point in the design of finer gauge knitwear with a relative high UPF which then can be further improved by optimizing other fabric engineering parameters. By using one of these structures, it is expected that the final knitwear will have a high UPF, since coloration and post-knitting wet treatments will further increase the UPF. The final finishing stage of ironing is anticipated to increase even more the UPF, but it has to be confirmed in a future research.

To understand better the protection level that is offered by the various knitted structures across the entire UVR spectrum, the measured Transmittance (T%) over evenly incremented wavelength intervals is presented in the following table 4 and figure 4 for the knitted samples with 2-ply yarn and in table 5 and figure 5 for the knitted samples with 3-ply yarn. All the knit structures show a transmittance increase as the wavelength increases, especially above 350 nm, resulting lower protection levels in the UVA region.

Table 4. UVR Transmittance (T%) of the knit structures with 2-ply yarn

λ (nm)	KT22W	Lacoste	Pineapple	Single Jersey	Half Cardigan	KM11	KM22W	KT22C	Cardigan	Half Milano	KM22C	Full Milano	RIB 1:1	Interlock
290	17,19	17,21	14,82	10,60	8,21	7,12	7,05	6,44	5,80	5,51	4,38	3,98	2,54	1,06
295	17,77	17,76	15,33	11,06	8,56	7,44	7,38	6,78	6,13	5,77	4,62	4,20	2,72	1,16
300	18,33	18,28	15,84	11,51	8,91	7,79	7,73	7,12	6,44	6,05	4,86	4,40	2,93	1,27
305	18,87	18,80	16,31	11,90	9,24	8,10	8,04	7,45	6,75	6,31	5,11	4,63	3,10	1,38
310	19,35	19,27	16,74	12,32	9,57	8,41	8,37	7,78	7,06	6,56	5,34	4,82	3,28	1,49
315	19,83	19,72	17,19	12,72	9,89	8,72	8,67	8,10	7,36	6,82	5,57	5,03	3,47	1,62
320	20,31	20,17	17,59	13,11	10,18	8,96	8,96	8,43	7,66	7,06	5,78	5,22	3,69	1,73
325	20,74	20,62	18,02	13,48	10,50	9,20	9,20	8,75	7,96	7,30	5,98	5,37	3,87	1,84
330	21,22	21,06	18,45	13,88	10,84	9,45	9,48	9,06	8,28	7,56	6,20	5,56	4,06	1,97
335	21,70	21,49	18,86	14,26	11,16	9,76	9,78	9,41	8,62	7,82	6,43	5,77	4,29	2,11
340	22,03	21,83	19,20	14,60	11,45	10,02	10,07	9,68	8,88	8,06	6,69	5,97	4,51	2,25
345	22,52	22,27	19,59	15,01	11,82	10,35	10,42	10,08	9,24	8,37	6,95	6,22	4,75	2,39
350	22,76	21,93	20,12	15,21	12,16	10,68	10,27	10,26	9,43	8,62	7,09	6,31	5,06	2,48
355	23,21	22,36	20,55	15,62	12,52	11,02	10,63	10,63	9,79	8,93	7,39	6,56	5,33	2,66
360	23,66	22,78	20,97	16,04	12,88	11,37	11,00	11,02	10,17	9,24	7,70	6,83	5,62	2,85
365	24,10	23,21	21,38	16,45	13,25	11,72	11,38	11,40	10,54	9,55	8,01	7,09	5,90	3,04
370	24,54	23,63	21,80	16,87	13,63	12,08	11,75	11,79	10,92	9,87	8,32	7,36	6,20	3,24
375	24,96	24,04	22,21	17,28	14,01	12,43	12,12	12,17	11,30	10,19	8,64	7,63	6,50	3,45
380	25,38	24,46	22,62	17,70	14,40	12,78	12,48	12,55	11,69	10,51	8,95	7,91	6,81	3,67
385	25,81	24,87	23,03	18,12	14,79	13,15	12,88	12,95	12,08	10,85	9,29	8,20	7,14	3,90
390	26,21	25,26	23,43	18,54	15,17	13,53	13,27	13,34	12,47	11,20	9,64	8,50	7,47	4,14
395	26,59	25,65	23,81	18,94	15,55	13,91	13,67	13,72	12,86	11,53	9,99	8,80	7,80	4,39
400	26,96	26,02	24,18	19,34	15,92	14,29	14,06	14,10	13,24	11,87	10,35	9,12	8,14	4,64

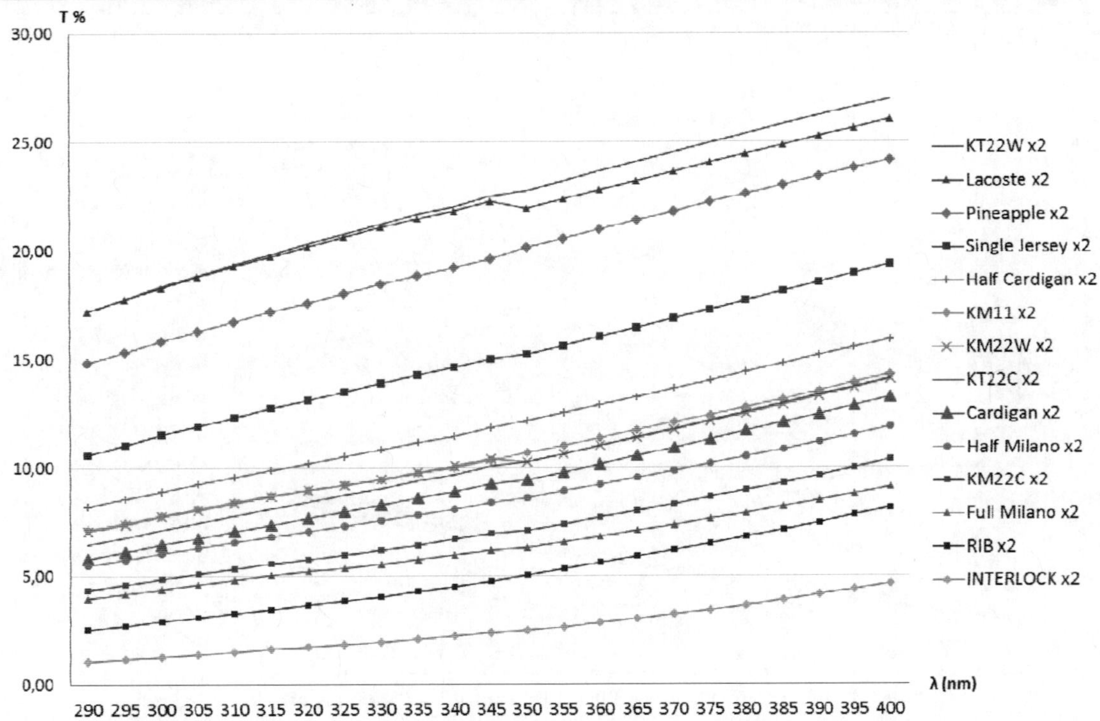

Figure 4. Transmittance T% in the UVR spectrum of the knitted structures with 2-ply yarn

Table 5. UVR Transmittance (T%) of the knit structures with 3-ply yarn

λ (nm)	KT22W	Pineapple	Lacoste	Single Jersey	KT22C	Cardigan	Half Milano	Half Cardigan	RIB 1:1	KM11	KM22C	Full Milano	KM22W
290	6,51	6,97	6,26	6,03	3,88	2,83	2,29	1,19	1,18	1,43	1,40	1,05	0,58
295	6,84	7,24	6,56	6,33	4,07	3,04	2,40	1,30	1,28	1,54	1,48	1,12	0,64
300	7,14	7,51	6,84	6,61	4,28	3,22	2,54	1,40	1,38	1,65	1,59	1,20	0,70
305	7,44	7,76	7,12	6,86	4,47	3,42	2,67	1,50	1,49	1,75	1,68	1,28	0,74
310	7,75	7,99	7,37	7,10	4,66	3,61	2,77	1,62	1,59	1,83	1,77	1,36	0,81
315	8,02	8,22	7,62	7,36	4,84	3,79	2,90	1,73	1,71	1,93	1,86	1,44	0,85
320	8,31	8,42	7,86	7,60	5,01	3,97	3,01	1,84	1,81	2,02	1,96	1,51	0,93
325	8,56	8,62	8,06	7,85	5,17	4,15	3,11	1,96	1,92	2,11	2,03	1,59	0,97
330	8,83	8,83	8,29	8,09	5,36	4,35	3,23	2,08	2,01	2,20	2,13	1,66	1,05
335	9,13	9,05	8,57	8,36	5,56	4,59	3,35	2,21	2,16	2,31	2,21	1,75	1,12
340	9,36	9,25	8,79	8,56	5,73	4,75	3,49	2,36	2,28	2,40	2,34	1,86	1,16
345	9,71	9,48	9,08	8,84	5,89	4,99	3,62	2,51	2,43	2,53	2,44	1,94	1,26
350	9,70	9,39	8,95	8,81	5,86	5,07	3,63	2,60	2,53	2,68	2,49	2,03	1,36
355	10,02	9,63	9,23	9,08	6,08	5,33	3,77	2,77	2,70	2,82	2,62	2,14	1,46
360	10,36	9,87	9,51	9,36	6,30	5,59	3,92	2,95	2,86	2,96	2,76	2,27	1,56
365	10,69	10,12	9,80	9,65	6,52	5,85	4,08	3,13	3,03	3,11	2,90	2,39	1,67
370	11,02	10,36	10,07	9,94	6,75	6,12	4,23	3,32	3,21	3,26	3,05	2,51	1,78
375	11,36	10,60	10,35	10,22	6,98	6,39	4,39	3,52	3,39	3,41	3,20	2,64	1,89
380	11,69	10,86	10,64	10,52	7,20	6,67	4,56	3,73	3,58	3,57	3,36	2,78	2,02
385	12,04	11,11	10,93	10,82	7,45	6,97	4,74	3,95	3,78	3,75	3,53	2,93	2,15
390	12,38	11,38	11,24	11,12	7,70	7,27	4,92	4,17	3,99	3,92	3,71	3,08	2,29
395	12,74	11,64	11,53	11,42	7,95	7,57	5,11	4,40	4,21	4,11	3,89	3,24	2,44
400	13,08	11,91	11,84	11,71	8,20	7,88	5,30	4,64	4,43	4,31	4,08	3,41	2,60

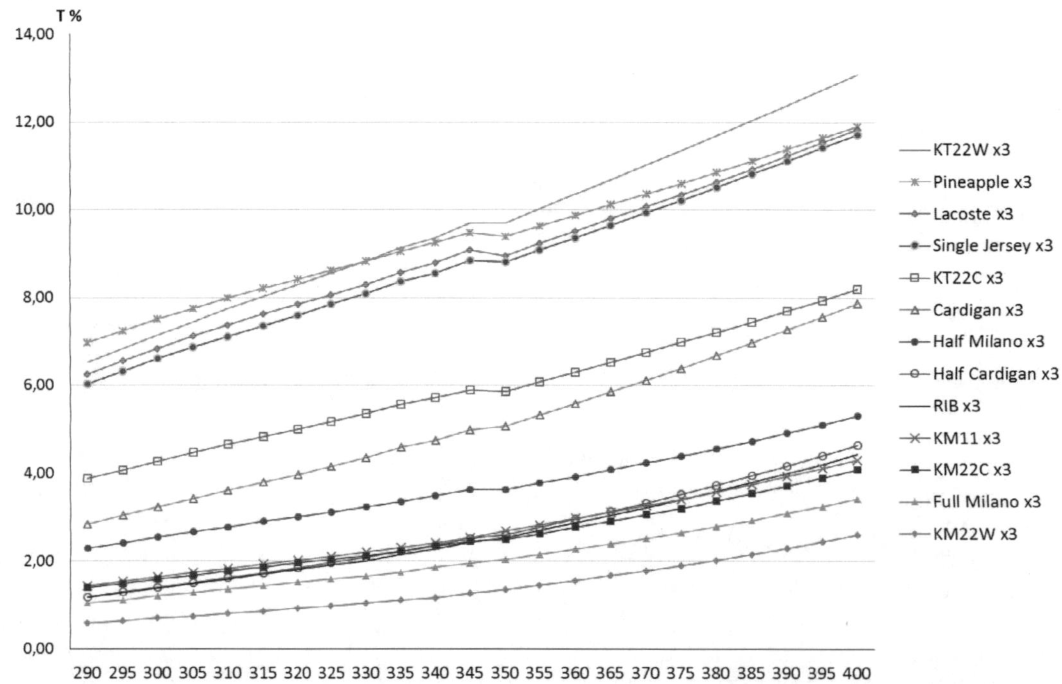

Figure 5. Transmittance T% in the UVR spectrum of the knitted structures with 3-ply yarn

5. Concluding Remarks

As expected, the use of thicker yarns (3-ply instead of 2-ply) results to increased UV protection performance of the fabric. However, the comparison between different samples leads to the following conclusions:

- Knit fabric structure has the greatest influence in the UPF. The presence of miss stitches seems to be related with an increased UV blocking performance, since all the evaluated structures with miss stitches exhibited high UPFs.
- Tuck stitches, although are generally associated with low UPFs, may have a positive or negative effect depending on how they are combined and repeated within the knit structure.
- Seven knit structures have been found with UPF > 40, namely, Half Cardigan. Full Milano, Rib 1x1, KM11, KM22C, KM22W with 3-ply yarn and Interlock with 2-ply yarn, with three of them reaching UPF values above 50 (Full Milano, KM22W and Interlock). By using one of these structures we can be pretty sure that the final knitwear will have a high UPF, taking into consideration that coloration and post-knitting wet treatments, like washing, will further increase the UPF.
- The final finishing stage of ironing is expected to increase even more the UPF, but it has to be confirmed in a future research. The above structures is expected to have high UPF also in higher gauges, which is in accordance with the results of previous research works.

Finally, it is worth mentioning that in all knit structures a transmittance increase is observed as the wavelength increases, especially above 350 nm, resulting in lower protection levels in the UVA region.

Acknowledgments

We thank the European Spinning Mills Industry VARVARESSOS S.A. and in particular Mr George Kostopoulos, the Sales Manager in VARVARESSOS S.A., for the support provided during the course of the present research.

VARVARESSOS

References

[1] Allen M W and Bain G, Measuring the UV Protection Factor of Fabrics, *Thermo Fisher Scientific, Madison, WI, USA*, retrieved on 18/05/2018 from www.analiticaweb.com.br/newsletter/13/51664_UV_fator_protecao_solar_tecidos.pdf

[2] Measurement Standards Laboratory of New Zealand, Ultraviolet protection rating of fabrics, retrieved on 18/05/2018 from www.msl.irl.cri.nz

[3] Stankovic S B, Popovic D, Poparic G B and Bizjak M 2009 Ultraviolet Protection Factor of Gray-state Plain Cotton Knitted Fabrics *Textile Research Journal* **79** (11) pp 1034–1042

[4] Gies P, Roy C, McLennan A, Pailthorpe M, Hilfiker R, Osterwalder U, Monard B, Moseley H, Sliney D, Wengraitis S, Wong J, Human S, Bilimis Z and Holmes G 2003 Ultraviolet Protection Factors for Clothing: An Intercomparison of Measurement Systems *Photochemistry and Photobiology* **77**(1) pp 58–67

[5] The European Standard EN 13758-1:2001 Textiles - Solar UV protective properties - Part 1: Method of test for apparel fabrics

[6] The European Standard EN 13758-2:2003 Textiles - Solar UV protective properties - Part 2: Classification and marking of apparel

[7] Grifoni D, Bacci L, Zipoli G, Carreras G, Baronti S and Sabatini F 2009 Laboratory and Outdoor Assessment of UV Protection Offered by Flax and Hemp Fabrics Dyed with Natural Dyes, Photochemistry and Photobiology **85** pp 313–320

[8] Kan C W 2014 A Study on Ultraviolet Protection of 100% Cotton Knitted Fabric: Effect of Fabric Parameters *The Scientific World Journal* Volume **2014** doi.10.1155/2014/506049

[9] Wong W-Y, Lam J K-C, Kan C-W and Postle R 2013 Influence of knitted fabric construction on the ultraviolet protection factor of greige and bleached cotton fabrics *Textile Research Journal* **83**(7) pp 683–699

[10] Ghazi S, Couteau C and Coiffard LJM 2010 What level of protection can be obtained using sun protective clothing? Determining effectiveness using an in vitro method *International Journal of Pharmaceutics* **397** pp 144–146

[11] Gies P H, Roy C R, Toomey S and McLennan A 1998 Protection against solar ultraviolet radiation Mutation Research **422** pp 15–22

[12] Hustvedt G and Cox Crews P 2005 The Ultraviolet Protection Factor of Naturally-pigmented Cotton *The Journal of Cotton Science* **9** pp 47–55

[13] Aguilera J, de Galvez M V, Sanchez-Rold C and Herrera-Ceballos E 2014 New Advances in Protection Against Solar Ultraviolet Radiation in Textiles for Summer Clothing *Photochemistry and Photobiology* **90** pp 1199–1206

[14] Singh M K and Singh A 2013 Ultraviolet Protection by Fabric Engineering *Journal of Textiles* Volume **2013**, Article ID 579129

[15] Gambichler T, Hatch K L, Avermaete A, Altmeyer P and Hoffmann P 2002 Influence of wetness on the ultraviolet protection factor (UPF) of textiles: in vitro and in vivo measurements *Photodermatology Photoimmunology Photomedicine* **18** pp 29–35

[16] Stanford D G, Georgouras K E and Pailthorpe M T 1995 The effect of laundering on the sun protection afforded by a summer weight garment *Journal of the European Academy of Dermatology and Venereology* **5** pp 28-30

[17] Hoffmann K, Kaspar K, Gambichler T and Altmeyer P 2000 In vitro and in vivo determination of the UV protection factor for lightweight cotton and viscose summer fabrics: A preliminary study J American Academy of Dermatology **Dec 2000**

[18] Saravanan D 2007 UV Protection Textile Materials *AUTEX Research Journal* Vol. **7**, No 1, pp 53-62

[19] Tsuzuki T and Wang X 2010 Nanoparticle Coatings for UV Protective Textiles *Research Journal of Textile and Apparel*, Vol. **14** iss 2 pp. 9 – 20

IOP Conf. Series: Materials Science and Engineering **459** (2019) 012052 doi:10.1088/1757-899X/459/1/012052

Pressure comfort of sock welts produced on single cylinder sock machine

A Marmaralı[1] and B Cüreklibatır Encan[2]
[1]Ege University, Engineering Faculty, Textile Engineering Department, 35100, İzmir, Turkey
[2]Ege University, Emel Akın Vocational Training School, 35100, İzmir, Turkey

arzu.marmarali@ege.edu.tr

1. Introduction
Consumers have become more conscious about every piece of clothing they buy due to easy access to information with the help of technology. Therefore, they are demanding high quality, more comfort and low price. Recently pressure comfort, which is one of the components of clothing comfort, has prominently been researched. Pressure exerted on the body by garment depends on shape of the body, fiber type and fabric structure [1]. Wear trials conducted by Momota et al. [2] and Tsujisaka et al. [3] revealed that pressure values of 10 mm Hg and 2.02 ± 0.29 kPa (~15 mm Hg) respectively are felt as comfortable. Wu and Li [4] stated that clothing pressure affects integumentary, circulatory, respiratory and other physiological systems of human body. Therefore, in this study pressure comfort of sock welts produced on single cylinder sock machines were investigated.

2. Materials and Method
All samples were knitted on Da Kong Sock machine (E14, 156 needles, $3^{3/4}$ ") using Nm 50/1 acrylic yarn. Two types of welts, two types of knit structures and two types of welt heights were used (Table 1). Welt types and knit structures are given in Figure 1. Mock rib and mock pique structures of accordion welts were produced by laying elastomeric yarn in the same or alternate needles. Single jersey welt has been sold in market recently and was added to our study to test whether it provides lower pressure.

a b c

Figure 1. Welt types and knit structures (a. accordion welt – mock pique structure, b. accordion welt – mock rib structure, c. single jersey welt

Welt pressures were measured with Kikuhime pressure monitoring device and MST Professional 2 – Medical Stocking Tester. As in TSE CEN/TR 15831:2010 standard [5], leg circumference was taken as 30 cm for point B_1 (calf area) for both devices.

PASW Statistics 18 programme was used to evaluate test results.

3. Results and Discussion

Table 1. Welt characteristics and mean pressure values.

Welt Type	Knit Structure	Welt Height	Mean Pressure with Kikuhime device	Mean Pressure with MST Professional 2 device
Accordion	Mock Pique	3 cm	21,1 mm Hg	9,1 mm Hg
Accordion	Mock Pique	6 cm	20,5 mm Hg	12,6 mm Hg
Accordion	Mock Rib	3 cm	19,2 mm Hg	10,5 mm Hg
Accordion	Mock Rib	6 cm	20,1 mm Hg	13,6 mm Hg
Single jersey	-	-	6,8 mm Hg	5,3 mm Hg

To determine the statistical importance of variations of both devices' measurements, Independent Samples T Test method was applied. According to the results there was a significant difference between measured mean pressure values (p=0,000). Additionally, a strong positive correlation (p=0,822) was found between the results of both devices.

4. Conclusion

Results indicated that the difference between mean pressure values measured by these devices was significant. In order to do measurements with the Kikuhime pressure monitoring device, the socks were placed on a special metal ring that does not have the flexibility like a human skin. Therefore, despite the existence of a strong correlation between the two devices, it was seen that MST Professional 2 Medical Stocking Tester device could simulate human body more correctly.

When the test results and statistical analysis were evaluated, single jersey welt was found as the most comfortable in terms of pressure comfort. However, this type of welt can easily get loose due to its weaker structure resulting in a short lifetime. The strength and laundry performance of this type of welt may be investigated in a further research.

If MST Professional 2 – Medical Stocking Tester's results are considered, it can be seen that mock pique welt with 3 cm height provides more comfort, when single jersey welt is put aside.

Acknowledgement

This study was supported by Ege University Scientific Research Projects Coordination Unit (Project number 2017-EAMYO-001).

References

[1] Pazireh E, Gharehaghaji AA and Haghighat E 2014 Study on the comfort of knitted elastic fabrics based on compressive deformation behavior, *JEFF,* **9(4)** 83-89

[2] Momota H, Makabe H, Mitsuno T and Ueda K 1993 A study of clothing pressure caused by Japanese men's socks, *Japan Res. Association for Text. End-uses,* **34 (4)** 175-186

[3] Tsujisaka T, Azuma Y, Matsumoto Y and Morooka H 2004 Comfort pressure of the top part of men's socks, *Textile Research Journal* **74 (7)** 598-602

[4] Wu X and Li Y 2015 Testing methods and effect of clothing pressure on human body, *International Journal of Business and Social Science* **6 (4-1)** 126-129

[5] TSE CEN/TR 15831:2010 – Method for Testing Compression in Medical Hosiery

IOP Conf. Series: Materials Science and Engineering **459** (2019) 012053 doi:10.1088/1757-899X/459/1/012053

Photochromic textile materials

S Morsümbül[1] and E P Akçakoca Kumbasar[1]

[1]Ege University, Faculty of Engineering, Department of Textile Engineering, 35100, Bornova İzmir Turkey

perrin.akcakoca@ege.edu.tr

Abstract. Smart textiles are the materials that can perceive and respond to the changes in environmental conditions. Photochromic textiles, one of the smart textile products, can reversibly change their color by UV irradiation. Photochromic textiles have become more attractive with the increasing interest in functional textile materials. In this paper, the applications of photochromic dyes in textile industry, the problems experienced in the processes and general information about the solution possibilities of these problems are given.

1. Introduction

People's expectations from textile materials are changing with the development of technology and accordingly, functional properties of textile materials have also become important beside their properties such as aesthetics, design and suitability to fashion. In this context, functional textile products increase the competitive power of the companies in the sector. The chromic materials, which are one of the functional textiles and can change colors with external factors, can be used to obtain smart textiles and also to get fashion effect.

Chromism is a general term used for color-changing materials and is used as a suffix. Chromic materials are also called "*chameleon materials*" due to their color changing properties caused by the effect of an external stimulus. There are various types of chromism according to the external stimulus (Table 1) [1].

Table 1. Chromism types [1]

Chromism type	External stimulus for changing of color
Thermochromism	Heat
Photochromism	UV radiation
Electrochromism	Electrical potential
Solvatochromism	Solvent polarity
Piezochromism	Mechanical pressure
Cathodochromism	Electron beam irradiation
Halochromism	Solution pH

Chromism has been explored since 1900's and its main applications are in the field of photochromism, thermochromism and electrochromism, such as dye, cosmetics, plastics and many optical applications. In the textile industry, photochromism and thermochromism applications are generally used among these types.

2. Photochromism and its applications

Photochromic materials change from colorless when indoors to color when outdoors. Especially, photochromics change color in response to UV light. When the UV light source is removed, their color returns to their original state (Figure 1) [2].

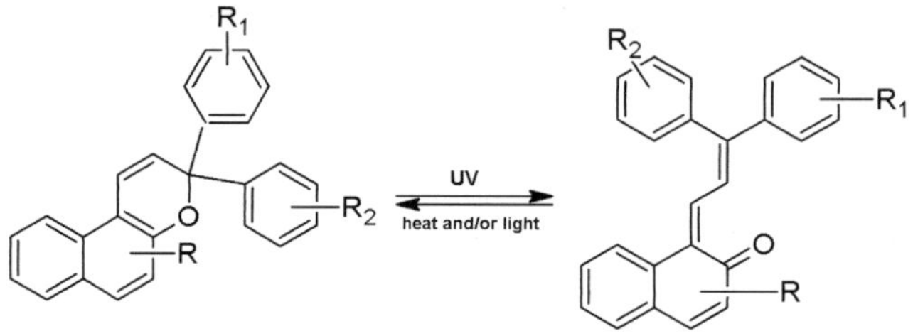

R: H, amino, etc.; R1, R2: H, alkoxy, amino, etc.

Figure 1. Photochromic compound (naphthopyran) and its photochromic reaction [2].

The most widely used class of photochromic dyes, along with many different classes, exhibit photochromism based on pericyclic reactions. These compounds, which show photochromism effect by the ring opening/closing reaction, are separated into 5 groups, spiropyran, spirooxazine, naphthopyran, diarylethene and fulgide [2, 3].

Photochromic eyeglasses are the most known photochromic products. These glasses get darker with the increase in the intensity of the UV light and thus the amount of light passing through the glass is reduced (Figure 2). Photochromic dyes can also be used at window glasses and, like the photochromic eyeglasses, with the increase in UV light intensity, the glass get darker and reduce the sunlight entering the building [4].

Figure 2. The photochromic eyeglass at indoor (picture on the left) and outdoor (pictures on the right) [5].

The use of photochromic dyes in the optical industry is widespread, and the applications of these dyes are also available in the plastic and cosmetics industries. In addition, one of the commercial uses of these dyes is printing ink, which can be transferred to the materials such as cloth or paper by different processes such as screen printing, flexography, dry offset etc. with different effects [4].

The use of photochromic dyes in the textile field is based on the 1990's, and today there are some examples of commercial use [4, 6, 7]. Photochromic dyes can be added in the polymer matrix during the production of synthetic fibers and so, photochromic yarns can be obtained (Figure 3) [6]. These dyes are also used by the fashion industry to obtain different effects such as photochromic t-shirts produced by printing method (Figure 4) [7].

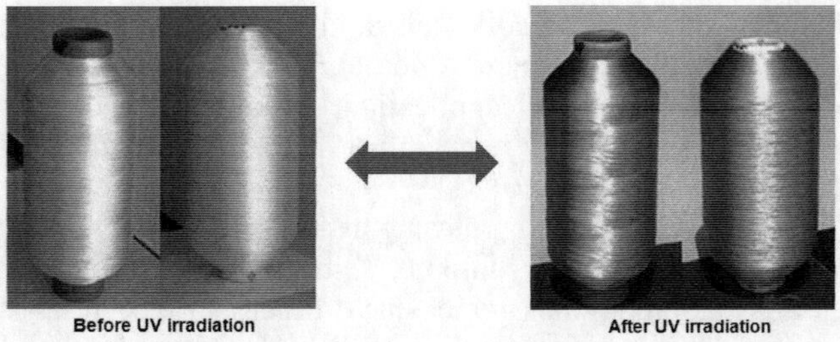

Figure 3. Photochromic yarns before and after UV irradiation [6].

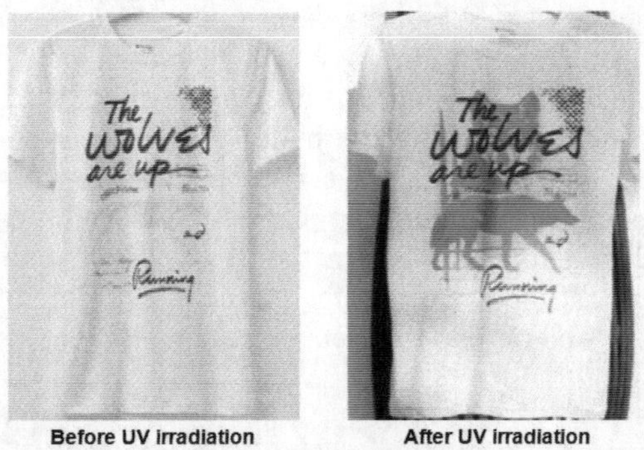

Figure 4. Photochromic t-shirt before and after UV irradiation [7].

Photochromic textile materials can be used as a UV sensor by changing the color depending on the amount of UV light in the environment. Thus, the person using the material would be warned that UV protection is required [4].

Photochromic dyes have also been used in the nanofiber production to obtain functional nanofibers. These dyes are incorporated into the polymer solution and then the nanofiber surface shows photochromic effect. Thus, sensitivity of photochromic dyes increase and the time necessary to respond to the UV irradiation reduce due to large surface area of the nanofibers. Photochromic nanofibers could find applications in fields such as optical data storage devices, optical sensors, processing media and functional components for smart surfaces [2].

3. The problems and solutions in the textile applications of photochromic dyes

In the studies on photochromic textiles, several problems have been encountered due to the sensitive structure of these dyes and their low water solubility [8-13]. The technologies such as encapsulation, sol-gel and electrospinning can be used to solve the problems in the textile applications of photochromic dyes [14-16]. The aims of these technologies are to provide the homogeneous distribution of the photochromic dyes in the solvent or carry the dyes in the polymer matrix. Thus, the use of photochromic dyes in textile materials can be improved with these alternative methods.

Encapsulation is a coating process of a core material with a shell material [17]. The encapsulation process provides advantages such as protecting the core from atmospheric conditions, increasing stability, improving processability, extending the shelf life of core material, etc. Along with many different encapsulation methods, methods such as in-situ polymerization, interfacial polymerization, emulsion-solvent evaporation and spray drying have generally been used to encapsulate photochromic dyes [15, 18-23]. Photochromic dyes based on spirooxazine, naphthopyran and diarylethene were used as core materials and polymers such as ethyl cellulose, polystyrene, polyurethane, polymethyl methacrylate, melamine and chitosan were used as shell materials in these studies.

Sol-gel technology can be used as another application method for photochromic textiles [16, 24, 25]. Sol is the stabilized suspension of colloidal solid particles in liquid and gel refers to a web structure with a form between solid and liquid. Photochromic dyes can be applied on textile materials with sol-gel method which consists of application, drying and condensation steps. However, photochromic dyes are sensitive to high temperature and therefore the curing temperature of the photochromic sol-gel matrices is limited [26].

Photochromic dyes exhibit ring opening/closing reaction when exposed to UV light, and color change reactions may not occur due to their fatigue. Spiropyrans have relatively low fatigue resistance and the spirooxazines and naphthopyrans which have higher fatigue resistance have become more important than spiropyrans. The use of different stabilizers such as hindered amine light stabilizers (HALS), antioxidants, UV absorbers, etc. can also improve the fatigue resistance of photochromic dyes [27].

4. Conclusion

The competition in textile industry is increasing and the production of high-value added materials is getting more importance. Smart textiles are the materials which are high-value added and therefore highly competitive. Photochromic materials have also become one of the remarkable products in this area, but the problems experienced in their applications restrict the use of these dyes in the textile industry. However, many different application methods have been studied to solve these problems. In this context, it is considered that the use of photochromic dyes in the textile industry will continue to be of interest by solving the problems experienced in the application.

References

[1] Kelly F M and Cochrane C 2015 Color-Changing Textiles and Electrochromism *Handbook of smart textiles* ed X Tao (Singapore: Springer Science+Business Media) chapter 32 pp 859-890

[2] Akçakoca Kumbasar E P, Morsunbul S and Alır S 2018 Photchromic Nanofibers *Novel Aspects of Nanofibers* ed T Lin (London: Intech Open) chapter 4 pp 69-85

[3] Crano J C and Guglielmetti R 1999 *Organic Photochromic and Thermochromic Compounds* vol 1 (New York: Plenum Press) p 378

[4] Ferrara M and Bengisu M 2014 *Materials that Change Color* (Singapore: Springer) p 139

[5] Aliexpress 2016 *New photochromic lenses http://www.aliexpress.com/item/2015-new-photochromic-lenses-colored-lenses-for-astigmatic-myopia-optical-glasses-lens-color-from-one-end/32352939794.html* (Date accessed: 25 February 2016)

[6] Diytrade 2017 *https://www.diytrade.com/china/pd/19055265/Photochromic_Pigment_for_Yarn_Making_Ultraviolet_UV_photosensitive_color_yarn.html* (Date accessed: 17 December 2017)

[7] DeFacto 2017 *http://www.defacto.com.tr/renk-degistiren-kurt-baskili-genc-erkek-tisort-344729* (Date accessed: 27 July 2017)

[8] Little A F and Christie R M 2016 *Color. Technol.* **132(4)** pp 304-309

[9] Fan F and Wu Y 2017 *J Appl. Polym. Sci.* **134(15)** pp 1-13

[10] Ayazi-Yazdi S, Karimi L, Mirjalili M and Karimnejad M 2017 *The J. of The Text. I.* **108(5)** pp 856-863.

[11] Little A F and Christie R M 2011 *Color. Technol* **127** pp 275–281

[12] Little A F and Christie R M 2010a *Color. Technol.***(126)** pp 157–163

[13] Little A F and Christie R M 2010b *Color. Technol.***(126)** pp 164–170

[14] Ali S, Ahmed F, Khatri Z and Kim S H 2015 Text. Sci.Eng. **52(2)** pp 126-131

[15] Feczkó T, Samu K, Wenzel K, Neral B and Voncina B 2012 *Color. Technol.* **129** pp 18–23

[16] Mahltig B, Textor T and Akcakoca Kumbasar E P 2015 *CBU J. Sci.* **11(3)** pp 306-315

[17] Bakry A M, Abbas S, Ali B, Majeed H, Abouelwafa M Y, Mousa A and Liang L 2016 *Comprehensive Reviews in Food Science and Food Safety* **15(1)** pp 143-182
[18] Han M, Lee E and Kim E 2002 *Optical Materials* **21** pp 579–583
[19] Feczko T,Varga O, Kovacs M, Vidoczy T and Voncina B 2011 *Journal of Photochemistry And Photobiology A: Chemistry* **22** pp 293-298
[20] Zhou Y, Yan Y, Du Y, Chen J, Hou X and Meng J 2013 *Sensors and Actuators B* **188** 502 – 512
[21] Fan F, Zhang W and Wang C 2015 Cellulose **22(2)** pp 427-438
[22] Akçakoca Kumbasar E P, Çay A, Morsunbul S, Voncina B 2016 *AATCC Journal of Researc*h **3(2)** pp 1 – 7
[23] Morsunbul S, Akçakoca Kumbasar E P and Çay A 2016 *Int. Conf. of Applied Research on Textile (Hammamet),* November 9-12
[24] Cheng T, Lin T, Fang J and Brady R 2007 *Textile Research Journal* **77(12)** pp 923-928
[25] Cheng T, Lin T, Brady R and Wang X 2008 *Fibers and Polymers* **9(3)** pp 301-306
[26] Akçakoca Kumbasar E P and Morsunbul S 2017 Photochromic Textiles *Textiles: Advances in Research and Applications* ed B Mahltig, (New York: Nova Science Publishers) chapter 18 pp 359-380
[27] Feczkó T, Kovács M and Voncina B 2012 *Journal of Photochemistry and Photobiology A: Chemistry* **247** pp 1-7

Corresponding author:
Emriye Perrin AKÇAKOCA KUMBASAR
Ege University, Faculty of Engineering, Department of Textile Engineering, 35100, Bornova İzmir Turkey
E-mail: perrin.akcakoca@ege.edu.tr

IOP Conf. Series: Materials Science and Engineering **459** (2019) 012054 doi:10.1088/1757-899X/459/1/012054

The perception of tactile feeling and corresponding textile attributes worldwide

R N Nagamatsu[12] , M J Abreu[1] and C D Santiago[2]

[1]University of Minho, 2C2T-Center for Textile Science and Technology, Campus de Azurém, Guimaraes, Portugal
[2]Federal Technology University of Paraná, Apucarana, Paraná, Brasil

josi@det.uminho.pt

Abstract: The textile industry has been exploring sensory analysis as means to evaluate different products as a strategy to improve product comfort. The sensory analysis uses a set of attributes standardized for sensory evaluation: the lexicons. This paper compares three lexicons developed to assess the sensory comfort of textile fabrics, by researchers from France, Portugal and Brazil. Quantitative Descriptive Analysis was conducted for the development of these lexicons. The study showed that France and Portugal have thirteen attributes in common and seven attributes are common in France, Portugal and Brazil: Light- Heavy, Gross- Fine, Fresh- Hot, Soft, Plushy, Elasticity and Falling.

1. Introduction

Different types of fabrics are usually classified according to their physical characteristics that influence the quality of the textile products. Among several attributes of textile quality one of the properties that motivates the acceptability of the consumer and repeat their purchase, is comfort. The sensorial comfort is the result of a complex series of sensory stimuli between the fabric and human skin when wearing or touches a garment.

The textile comfort has been studied by instrumental and sensorial methods. The sensory comfort influences by the feeling that the clothing provides the user when in contact (mechanical or thermal) with the skin is the touch. Descriptive sensory analysis, much explored by the food and cosmetic industries, can characterize the attributes of different product types qualitatively as quantitatively [1,2]. Textile hand properties evaluate quantitatively through adaptations of methods of sensory analysis. This assessment is the result of descriptive psychological and physiological responses of individuals.

The sensory analysis is a technique where the sensations, when a material is touched, are quantified, and the human hand is the only instrument applied.

Therefore, the sensory analysis is a new and significant tool for general textile materials where the tactile feeling is of paramount importance for the consumer. The development of a global lexicons for tactile sensory analysis is an important tool for tactile evaluation of products. Before the evaluation process of the textile comfort is necessary to develop descriptive terminologies attributes that best define the characteristics of a product- Lexicons.

Among the methods of sensory analysis, researchers from France [3] and Portugal [4] adapted the Quantitative Descriptive Analysis (QDA) [5], with the purpose of quantifying the quality of touch in textiles. The researchers used this method to develop lexicons for the evaluation of the tactile sensorial attributes in textiles. Each national panel creates its own consumer-based (lexicon) word for the evaluations of a product type.

IOP Conf. Series: Materials Science and Engineering **459** (2019) 012054 doi:10.1088/1757-899X/459/1/012054

A Brazilian lexicon was developed by Nagamatsu et al. They used the same method as the France and Portuguese researchers, resulting in a lexicon of 11 attributes [6].

After the three lexicons were compared, seven terms reveled to be common among the three countries: three bipolar (light & heavy; thin & thick; Cold & warm) two describing the surface (soft; plush) and two describe the material (elasticity; falling) [7].

2. Materials and methods

The procedure of selection and identification of the descriptors was adapted from ISO 11035 [8]. A panel of Brazilian textile sensorial evaluators were invited to participate in the development of the Brazilian lexicon. They are 14 assessors (five men and nine women) from the city of Apucarana/ Brazil, with different personal taste, education level and expertise participated in this study.
20 samples with different textile structures and fiber composition, were cut into 20x20 centimeters pieces. Sensor Assessors touched the fabric samples behind a cabin and described the sensations when touching the samples without seeing them and using free vocabulary (Figure 1).

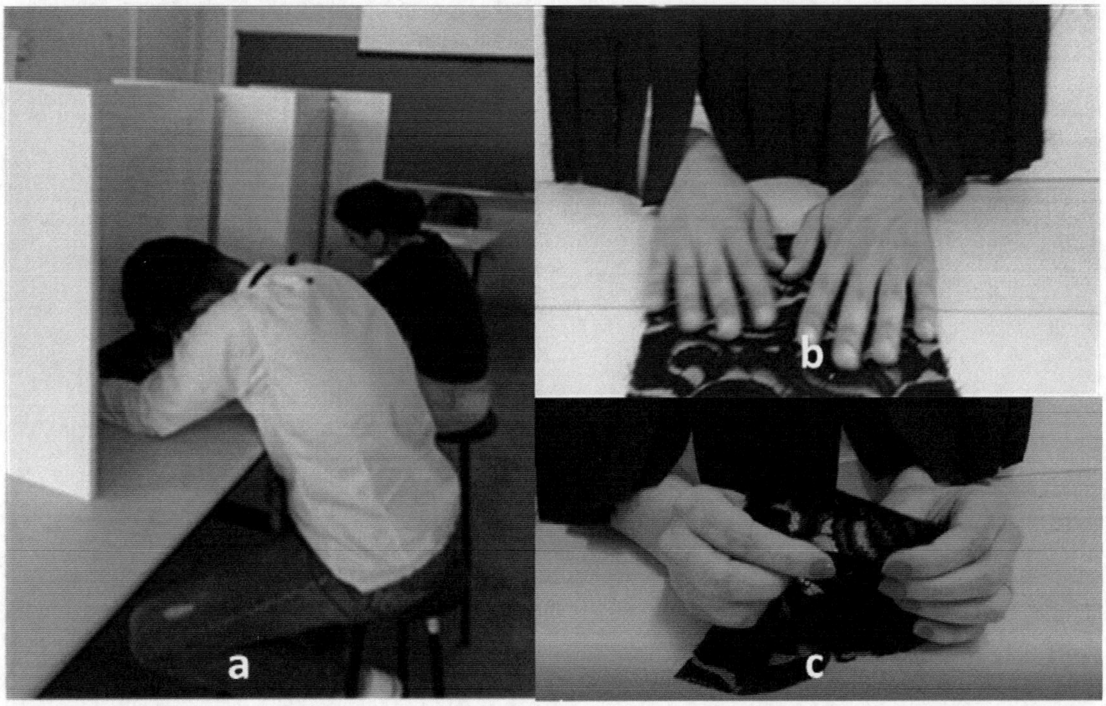

Figure 1. (a) Sensory evaluation cabin; (b) Touching one side of textile sample (c) Touching with both hands both sides

They generated 299 terms that were reduced in four steps, the first three by the panel during 3 meetings. First eliminating hedonic terms (128); Second reduction was rejecting descriptors with same meaning (122); Third reduction was combining singular terms (28); and in the fourth and last reduction, the panel quantified the perceived intensity of the attributes and these data were treated by means of statistical methods obtaining the last eleven attributes (figure 2).

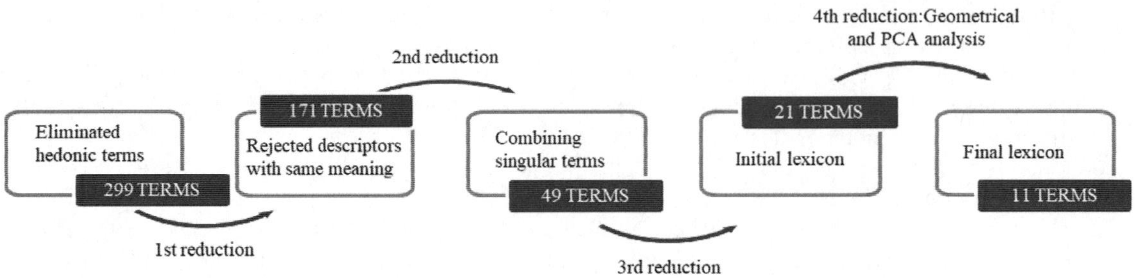

Figure 2. Process to reduce terms for the formation of Brazilian lexicon

For the development of the tactile sensory comfort lexicon for Brazilian textile products, four statistical models were applied. In Geometric Analysis it was possible to classify the order of perceived intensity. The interpretation of Correlation Analysis and the map of Principal Component Analysis - PCA (figure 3) allowed the identification of some differences and similarities between the attributes (positive correlation) as Itches and Rugged; Relief & Rough. Soft & Smooth and Fit and Flowing, as well as identifying the opposing attributes (negative correlation) that contributed to the classification of the bipolar terms: Light – Heavy; Thick – Thin; Fresh – Hot; Dry – Humid; Rough – Smooth.

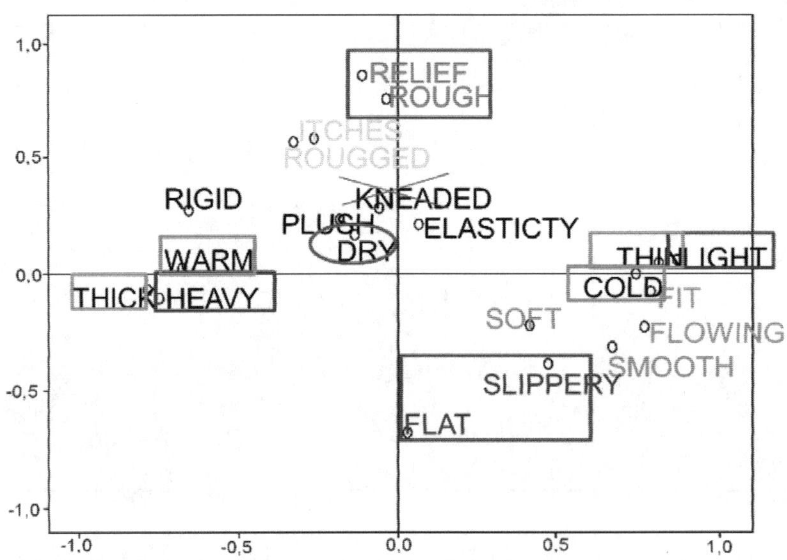

Figure 3. PCA of 21 attributes to evaluation sensory hand textile

The test-t was conducted to verify if there are significant differences between the arithmetic mean of the attributes. The attributes Kneaded obtained the p-value > 0,05. The attribute Kneaded was excluded because it does not contribute to differentiate the hand sensorial qualities between the textile samples. (table 1).

Table 1. Test-t of means againt reference constant p > 0,05

VARIABLE	MEAN	N	STD. ERR	REFERENCE CONSTANT	T-VALUE	DF	P
LIGHT	2,268519	108	0,167757	0,00	13,52266	107	0,000000
HEAVY	1,583333	108	0,153490	0,00	10,31554	107	0,000000
PLUSHY	1,027778	108	0,147940	0,00	6,94727	107	0,000000
SMOOTH	4,425926	108	1,584813	0,00	2,79271	107	0,006193
ITCHES	2,018519	108	0,934232	0,00	2,16062	107	0,032956
SOFT	2,768519	108	0,929182	0,00	2,97952	107	0,003573
ELASTICITY	0,962963	108	0,148796	0,00	6,47171	107	0,000000
KNEADED	**1,453704**	**108**	**0,934640**	**0,00**	**1,55536**	**107**	**0,122813**
ROUGH	2,064815	108	0,141857	0,00	14,55560	107	0,000000
FLAT	1,333333	108	0,147084	0,00	9,06510	107	0,000000
THICK	2,768519	108	0,931600	0,00	2,97179	107	0,003657
THIN	1,527778	108	0,161369	0,00	9,46762	107	0,000000
FIT	2,259259	108	0,148472	0,00	15,21671	107	0,000000
COLD	1,564815	108	0,158564	0,00	9,86868	107	0,000000
WARM	2,259259	108	0,142525	0,00	15,85169	107	0,000000
RIGID	1,759259	108	0,155587	0,00	11,30722	107	0,000000
FLOWING	1,879630	108	0,162576	0,00	11,56156	107	0,000000
DRY	3,287037	108	0,924181	0,00	3,55670	107	0,000561
SLIPPERY	1,675926	108	0,157224	0,00	10,65948	107	0,000000
RUGGED	1,537037	108	0,157002	0,00	9,78991	107	0,000000
RELIEF	2,314815	108	0,177795	0,00	13,01959	107	0,000000

The analysis of the results of the application of the statistical models in the 21 attributes allowed decision making for a more precise formation of the Brazilian lexicon. The final list is composed of 11 most significant attributes (table 2).

The final list of Brazilian lexicons was composed with 11 descriptors. The panel also determined the definition, valuation techniques and references for each attribute. The lexicon in this study was made to assess the comfort of textile materials.

Table 2. Brazilian lexicons

BIPOLAR ATTRIBUTES	SURFACE ATTRIBUTES	MATERIAL ATTRIBUTES
LIGHT-HEAVY	SOFT	ELASTICITY
ROUGH - FLAT	PLUSHY	RIGID
THICK - THIN	RUGGED	FIT
FRESH - HOT		
DRY - HUMID		

IOP Conf. Series: Materials Science and Engineering **459** (2019) 012054 doi:10.1088/1757-899X/459/1/012054

3. Results and discussion

Comparison between consumers of the three countries/different continents was made and obtained a sensory profile, as introduced in table 3. The French and Portuguese lexicons have fourteen terms and Brazilians developed eleven terms.

The study showed that France and Portugal have thirteen attributes in common and seven attributes are common in France, Portugal and Brazil: Light- Heavy, Gross- Fine, Fresh- Hot, Soft, Plushy, Elasticity and Falling.

Table 3. Lexicons French, Portuguese and Brazilian

	FRANCE	PORTUGAL	BRAZIL
BIPOLAR ATRITUTES	Light- Heavy	Light- Heavy	Light- Heavy
	Thin- Thick	Thin- Thick	Gross- Fine
	Cold- Warm	Cold- Warm	Fresh- Hot
	Supple- Rigid	Supple- Rigid	
		Sleek – Rugoss	Rough- Flat
		Rough – Smooth	
			Dry- Humid
SURFACE ATRIBUTES	Soft	Fluffy	Soft
	Pilous	Pilous	Plushy
	Granulous	Granulous	
	Sticky	Sticky	
	Slippery	Slippery	
	Greasy		
	Grooved		
			Rugged
MATERIALS ATRIBUTES	Elastic	Elastic	Elasticity
	Falling	Falling	Falling
	Responsive	Shape Recovery	
	Crumple-Like	Crumple	
			Rigid

The rigid and supple bipolar attribute were not considered by Brazilians. Brazilians grouped Portuguese bipolar terms Sleek - Rough and Smooth Rough differently.

Only the French introduced terms such as Greasy and Grooved and the Brazilians presented Dry - Humid.

Although these three lexicons were developed for textile hand assessment of textile fabrics, Nagamatsu et al. [9] adapted three attributes to sports cap wear trial evaluation: tridimensional products (Light-Heavy; Gross-Fine; Fresh-Hot). The authors compared the results of the Fresh-Hot attribute between objective evaluation with wear trial and objective evaluation with thermal/humid sensors.

4. Conclusion

In this work, the three developed lexicons, by France, Portuguese and Brazilian panels, were developed in different countries and compared in this reearch. We can consider, that the seven common attributes are the most mentioned among panelists and offer higher improvement in the development of innovative clothing, resulting in greater comfort to the consumer.

Furthermore, we can consider these seven attributes form the lexicon of tactile comfort and they can be exploited worldwide for hand textile assessment.

Key words: Tactile Feeling, Thermal comfort, Textile Lexicon, Attributes

Acknowledgments
This work was supported by FEDER funds through the Competitivity Factors Operational Program - COMPETE and by national funds through FCT – Foundation for Science and Technology within the scope of the project POCI-01-0145-FEDER-007136.

References

[1] AFNOR, "ISO 8586 - Sensory analysis - General guidelines for the assessors and expert sensory assessors, 2014, Association Française de Normalisation, p. 28.
[2] Lawless, L. J. R. and Civille, G. V., 2013, Developing Lexicons: a review in *J. Sens. Stud.*, vol. **20**, pp. 270–281.
[3] Philippe, F., Schacher, L., Adolphe, D. C. and Dacremont, C., 2004, Tactile feeling: Sensory analysis applied to textile goods, *Text. Res. J.*, **vol. 74**, no. 12, pp. 1066–1072,
[4] Nogueira, C., 2011, Análise sensorial de produtos têxteis, p. 218.
[5] ISO, 2003, ISO 13299 – Sensory analysis – Methodology – General guidance for establishing a sensory profile. International Organization for Satandardization, Geneve, p.24.
[6] Nagamatsu, R. N., Marques Abreu, M. J. and Santiago, D., 2016, Lexicon Development for Brazilian Textile Sensory Analysis in *CIMODE - International Fashion an Design Congress*, pp. 1–6.
[7] Nagamatsu, R. N., Marques Abreu, M. J. and Santiago, D., 2017, Tactile feeling of textile : a comparative study between textiles comfortable attributes of France , Portugal and Brazil," in Fiber Society Spring Conference 2017.
[8] ISO, 1994, ISO 11035 - Sensory analysis - Identification and selection of descriptors for establishing a sensory profile by a multidemensional approch. International Organization for Satandardization, Geneve, p. 31.
[9] Nagamatsu, R. N., Marques Abreu, M. J. and Santiago, D., 2018, Determination of Total Comfort of Sport Caps using Wear Trials. in Fiber Society Spring Conference 2018.

IOP Conf. Series: Materials Science and Engineering **459** (2019) 012055 doi:10.1088/1757-899X/459/1/012055

Buckling analysis of plain-woven fabric structure using shell element and a one cell-based integration scheme in smoothed finite element method

Q T Nguyen[1], **A J P Gomes**[2] and **F B N Ferreira**[1]

[1]Universidade do Minho, 2C2T - Centro de Ciência e Tecnologia Têxtil, Campus de Azurém, 4800-058 Guimarães, Portugal

[2]Instituto de Telecomunicações and Universidade da Beira Interior, R. Marquês de Ávila e Bolama, 6201-001 Covilhã, Portugal

E-mail: quyenum@gmail.com

Abstract. A one smoothing cell integration scheme in the strain smoothing technique in finite elements (referred as SFEM) was proposed to evaluate the nodal train fields of a four-node quadrilateral (Q4) shell element, which is based on the first-order shear deformation theory of plate (FSDT). A mixed interpolation of tensorial components (MITC) approaches for Q4 transverse shear strains also applied to eliminate a shear locking phenomenon that may occur when the thin plate/shell elements are geometrically distorted in curved geometries of fabric sheet. The numerical eigenvalues of buckling analysis of a plain-woven fabric sample, of which physical and mechanical parameters extracted from Kawabata evaluation system for fabrics (KES-FB), obtained a higher efficiency in numerical computation and approximated to Q4 shell element implemented in the finite element method (FEM).

1. Introduction and kinematics of shells

Analysis of buckling behaviour of textile fabric using thin plate/shell finite elements helps understand the basic practical problem of residual curvature in the fabric to understand the behaviour of fabric during garment manufacturing as well as during its use as a garment.[1, 2] An excellent review in the past development of plate/shell finite elements can be found in the works of Yang et al..[3] This paper presents an Q4 shell element based on the FSDT [4-6] and the cell-based smoothed finite element in SFEM [7, 8] for buckling analysis of thin and moderately thick plain-woven fabric structure. The numerical implementation and results indicated a better efficiency of numerical computation but accuracy results compared to the same Q4 shell elements without integrating cell-based models. The strain smoothing operator and the summarized formulation of Q4 shell plate/shell elements are respectively presented in the following sections.

1.1. A strain smoothing method in finite elements

The strain smoothing technique recently proposed by Liu et al. [9], referred as SFEM, which improved the accuracy and convergence rate of the existing conventional finite element finite element method (FEM) of elastic solid mechanics problems.[10-12] This technique avoids evaluating derivatives of mesh-free shape functions at nodes and therefore eliminates defective modes. The major techniques used in smoothed finite element methods appear summarized in references [8, 13, 14].

The strain smoothing operator over an element Ω^e bounded by Γ^e referred as a smoothing domain (or cell) Ω_k^s, which is defined as

$$\bar{\nabla}\boldsymbol{u}(x_k) = \boldsymbol{\varepsilon}(\mathrm{x}) = \int_{\Omega_i^s} \nabla\boldsymbol{u}(x)\Phi(x - x_k)d\Omega \quad (1)$$

and has to satisfy conditions:

$$\Phi \geq 0 \; and \int_{\Omega_i^s} \Phi d\Omega = 1, \Phi(x - x_k) = \begin{cases} \frac{1}{A_k^s}, & x \in \Omega_k^s \\ 0, & x \notin \Omega_k^s \end{cases}, \quad (2)$$

where Φ is a smoothing or weight function in Ω_k^s and $A_k = \int_{\Omega_k^s} d\Omega$ is the area of smoothing domain.

The strain fields in Ω_k^s can be further assumed to be a constant and equals $\bar{\boldsymbol{\varepsilon}}(\mathrm{x}_k)$, which gives:

$$\bar{\boldsymbol{\varepsilon}}_k = \bar{\boldsymbol{\varepsilon}}_k(x) = \bar{\boldsymbol{\varepsilon}}(x_k) = \frac{1}{A_k^s}\int_{\Omega_k^s} \boldsymbol{\varepsilon}(x)d\Omega. \quad (3)$$

Substituting function Φ into Eq. (1), the smoothed gradient of displacement can be written as

$$\bar{\boldsymbol{\varepsilon}}(\mathrm{x}_k) = \frac{1}{A_k^s}\int_{\Gamma_k^s} \boldsymbol{n}(\mathrm{x}).\,\hat{\boldsymbol{u}}(\mathrm{x})d\Gamma, \quad (4)$$

in which $\boldsymbol{n}(\mathrm{x})$ is the outward unit normal matrix containing the components of the outward unit normal vector to the boundary Γ_k^s.

1.2. Kinematics of shells

Based on the FSDT, the displacement components $\boldsymbol{u} = [u, v, w]$ at a local coordinate system (x, y, z) within problem domain Ω bounded by Γ are defined as follows:

$$\begin{aligned} u(x,y,z) &= u_0(x,y) - z\theta_x(x,y) \\ v(x,y,z) &= v_0(x,y) - z\theta_y(x,y) \\ w(x,y,z) &= w_0(x,y) \end{aligned} \quad (5)$$

where u_0, v_0 and w_0 be the translation displacements, θ_x and θ_y be the rotations about the yz and xz planes in the Cartesian coordinate system.

In terms of the mid-plane deformations, the strain vector $\boldsymbol{\varepsilon}$ can be written using Eq. (5), which gives:

$$\boldsymbol{\varepsilon} = \{\varepsilon_x \quad \varepsilon_y \quad \varepsilon_{xy} \quad \varepsilon_{xz} \quad \varepsilon_{yz}\}^T = \begin{Bmatrix} \boldsymbol{\varepsilon}^m \\ 0 \end{Bmatrix} + \begin{Bmatrix} z\boldsymbol{\varepsilon}^b \\ 0 \end{Bmatrix} + \begin{Bmatrix} 0 \\ \boldsymbol{\varepsilon}^s \end{Bmatrix} \quad (6)$$

in which superscripts m, b and s are respectively the membrane, bending and the shear terms. The generalized strain vector $\hat{\boldsymbol{\varepsilon}}$ can be written as

$$\hat{\boldsymbol{\varepsilon}} = \begin{Bmatrix} \boldsymbol{\varepsilon}^m \\ \boldsymbol{\varepsilon}^b \\ \boldsymbol{\varepsilon}^s \end{Bmatrix}, \quad (7)$$

where

$$\boldsymbol{\varepsilon}^m = \begin{Bmatrix} \frac{\partial u_0}{\partial x} \\ \frac{\partial v_0}{\partial x} \\ \frac{\partial u_0}{\partial x} + \frac{\partial v_0}{\partial x} \end{Bmatrix}, \boldsymbol{\varepsilon}^b = \begin{Bmatrix} \frac{\partial \theta_x}{\partial x} \\ \frac{\partial \theta_y}{\partial y} \\ \frac{\partial \theta_x}{\partial x} + \frac{\partial \theta_y}{\partial y} \end{Bmatrix}, \boldsymbol{\varepsilon}^s = \begin{Bmatrix} \frac{\partial w_0}{\partial x} - \theta_x \\ \frac{\partial w_0}{\partial y} - \theta_y \end{Bmatrix}. \quad (8)$$

The constitutive equations can be expressed as

$$\hat{\boldsymbol{\sigma}} = \widehat{\boldsymbol{D}}\hat{\boldsymbol{\varepsilon}}, \tag{9}$$

where

$$\hat{\boldsymbol{\sigma}} = \begin{Bmatrix} \widehat{\boldsymbol{N}} \\ \widehat{\boldsymbol{M}} \\ \widehat{\boldsymbol{Q}} \end{Bmatrix}, \widehat{\boldsymbol{D}} = \begin{bmatrix} \boldsymbol{D}^m & \mathbf{0} & \mathbf{0} \\ \mathbf{0} & \boldsymbol{D}^b & \mathbf{0} \\ \mathbf{0} & \mathbf{0} & \boldsymbol{D}^s \end{bmatrix}, \tag{10}$$

in which $\widehat{\boldsymbol{N}} = \{N_x \quad N_y \quad N_{xy}\}^T$ is the vector of membrane force, $\widehat{\boldsymbol{M}} = \{M_x \quad M_y \quad M_{xy}\}^T$ is the vector of bending moment, $\widehat{\boldsymbol{Q}} = \{Q_x \quad Q_y\}^T$ is the vector of transverse shear force and $\boldsymbol{D}$ are stiffness constitutive coefficients matrix, which given as

$$\boldsymbol{D}^m = \int_{-\frac{h}{2}}^{\frac{h}{2}} \begin{bmatrix} \frac{E_1}{1-v_1v_2} & \frac{v_2E_1}{1-v_2v_1} & 0 \\ \frac{v_1E_2}{1-v_1v_2} & \frac{E_2}{1-v_2v_1} & 0 \\ 0 & 0 & G \end{bmatrix} dz\,, \boldsymbol{D}^b = \int_{-\frac{h}{2}}^{\frac{h}{2}} z^2 \begin{bmatrix} B_1 & 0 & 0 \\ 0 & B_2 & 0 \\ 0 & 0 & H \end{bmatrix} dz, \boldsymbol{D}^s = \int_{-\frac{h}{2}}^{\frac{h}{2}} \frac{5}{6} G \begin{bmatrix} 1 & 0 \\ 0 & 1 \end{bmatrix} dz, \tag{11}$$

where h is the thickness of the plate or shell, G is shear modulus, E_* is Young's modulus and v_* is Poisson's ratio corresponding to the warp and weft direction of yarns, B_* and H stand for flexural moduli and torsional rigidity.

According to FEM procedures for buckling analysis, the problem domain Ω is discretized into a set of four-node quadrilateral flat shell elements Ω^e with boundary Γ^e, the generalized mid-plane displacement vector $\hat{\boldsymbol{u}}$ can be then defined as

$$\hat{\boldsymbol{u}} = \sum_{I=1}^{4} \begin{bmatrix} N_I(\mathbf{x}) & 0 & 0 & 0 & 0 \\ 0 & N_I(\mathbf{x}) & 0 & 0 & 0 \\ 0 & 0 & N_I(\mathbf{x}) & 0 & 0 \\ 0 & 0 & 0 & N_I(\mathbf{x}) & 0 \\ 0 & 0 & 0 & 0 & N_I(\mathbf{x}) \end{bmatrix} \begin{Bmatrix} u_I \\ v_I \\ w_I \\ \theta_{xI} \\ \theta_{yI} \end{Bmatrix} = \sum_{I=1}^{4} \boldsymbol{N}_I(\mathbf{x})\boldsymbol{d}_I, \tag{12}$$

where $N_I(\mathbf{x})$ is the basis function associated to node I of a four-node quadrilateral shell element.

The eigenvalue equation for buckling analysis can be expressed through the direct application of variational principles and using Eq. (5 to 12), which is given as

$$(\boldsymbol{K} - \lambda\boldsymbol{K}_g)\boldsymbol{d} = 0, \tag{13}$$

where λ is the critical buckling load, $\boldsymbol{K}$ and $\boldsymbol{K}_g$ are the global stiffness matrix and the geometric stiffness matrix, respectively, which express in [7, 15]

The shear locking phenomenon may appear due to incorrect transverse forces under bending, or in the case of the thickness of the plate tends to zero. To overcome the shear locking phenomena, the approximation of the shear strain $\boldsymbol{\varepsilon}^s$ in Eq. (8) can be formulated with MITC4 element as in [4].

Applying Eq. (1, 3 and 4) with $k = 1$ in order to evaluate the membrane and bending trains of a Q4 shell element being formulated in this section using the following shape functions:

$$N_1 = (1,0,0,0), N_2 = (0,1,0,0), N_3 = (0,0,1,0), N_4 = (0,0,0,1). \tag{14}$$

2. Numerical implementation and results

The span-to-thickness ratio l/h of a square woven fabric sample was taken to be 23.5849 in the numerical example. Two types of boundary conditions, simply supported (S) and clamped (C) edges applied for meshes such as 5x5, 10x10, 15x15 and 20x20. Both the Q4 shell elements with and without being integrated smoothing cells were programmed.

Mechanical and physical parameters of a plain-woven fabric sample were measured with Kawabata evaluation system for fabrics (KES-FB) [16, 17] and derived as: elastic modulus [gf/cm], $E_1 = 3823.7993$, $E_2 = 14092.4464$ and $E_{12} = 6896.5517$, Poisson's ratio $v_1 = 0.0211$ and $v_2 = 0.0778$, bending rigidity $[gf.cm^2/cm]$ of $B_1 = 0.1237$, $B_2 = 0.1333$ and $B_{12} = 0.0880$, transverse shear modulus $[gf.cm^2]$ of $G = 217.3100$.

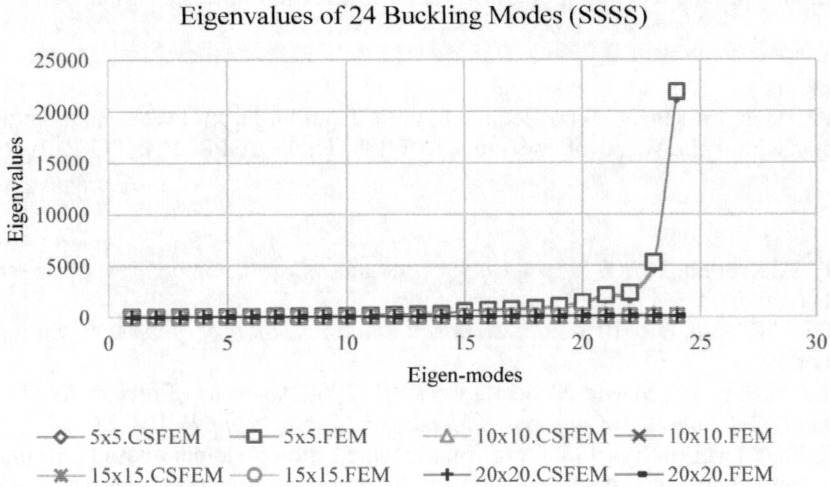

Figure 1. Compare the eigenvalues of 24 buckling modes between the implemented CS-FEM model and FEM with simply supported edges and different mesh density

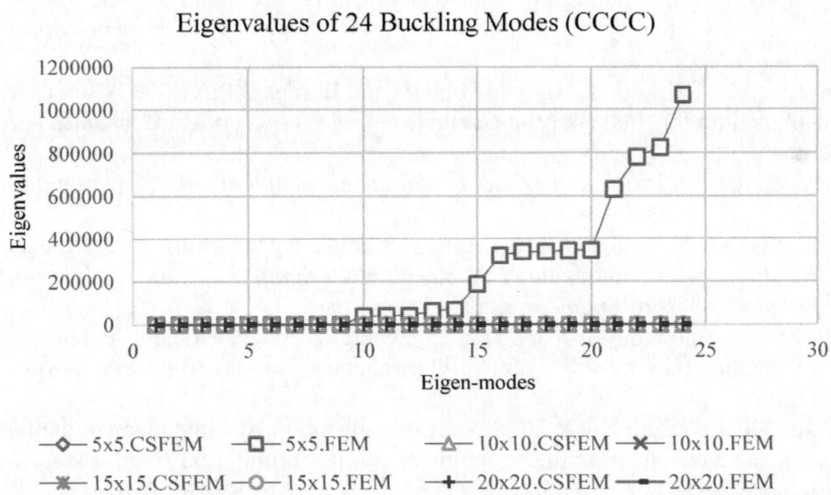

Figure 2. Compare the eigenvalues of 24 buckling modes between the implemented CS-FEM model and FEM with clamped edges and different mesh density

Figures 1 and 2 indicated that the eigenvalues of 24 modes of the critical buckling load λ computed by CSFEM model is approximate and coincided with that one of FEM on the same boundary conditions and mesh configurations. The shape functions in SFEM model are constants as presented in Eq. 14. Vice versa, the corresponding shape functions of a Q4 element in FEM are bilinear shape functions interpolated in natural coordinates (ξ, η, ζ). Thus, the strain smoothing technique reduces the numerical implementation and computation time in terms of the central processing unit (CPU) time.

3. Conclusions

A one cell-base integration scheme in SFEM for a Q4 element, which based on the FSDT, presented an effective computation in terms of CPU time for the buckling analysis of plain-woven fabric structure, produced an accurate numerical result and also reduced tasks in numerical implementation compared with FEM.

Acknowledgments
The authors wish to express their acknowledgment to FCT funding from FCT – Foundation for Science and Technology within the scope of the project "PEST UID/CTM/00264; POCI-01-0145-FEDER-007136".

References

[1] Oñate E and Kröplin B 2006 *Textile Composites and Inflatable Structures*: Springer Netherlands)
[2] Oñate E and Kröplin B 2010 *Textile Composites and Inflatable Structures II*: Springer Netherlands)
[3] Yang H T Y, Saigal S, Masud A and Kapania R K 2000 A survey of recent shell finite elements *International Journal for Numerical Methods in Engineering* **47** 101-27
[4] Bathe K-J and Dvorkin E N 1985 A four-node plate bending element based on Mindlin/Reissner plate theory and a mixed interpolation *International Journal for Numerical Methods in Engineering* **21** 367-83
[5] Duan H and Ma J 2018 Continuous finite element methods for Reissner-Mindlin plate problem *Acta Mathematica Scientia* **38** 450-70
[6] Wu F, Zeng W, Yao L Y and Liu G R 2018 A generalized probabilistic edge-based smoothed finite element method for elastostatic analysis of Reissner–Mindlin plates *Applied Mathematical Modelling* **53** 333-52
[7] Nguyen-Thanh N, Rabczuk T, Nguyen-Xuan H and Bordas S P A 2008 A smoothed finite element method for shell analysis *Computer Methods in Applied Mechanics and Engineering* **198** 165-77
[8] Liu G-R and Nguyen-Thoi T 2010 *Smoothed finite element methods*: Taylor and Francis Group, LLC)
[9] Chen J-S, Wu C-T, Yoon S and You Y 2001 A stabilized conforming nodal integration for Galerkin mesh-free methods *Int. J. Numer. Meth. Engng. International Journal for Numerical Methods in Engineering* **50** 435-66
[10] Liu G R, Zeng W and Nguyen-Xuan H 2013 Generalized stochastic cell-based smoothed finite element method (GS_CS-FEM) for solid mechanics *Finite Elements in Analysis and Design* **63** 51-61
[11] Yue J, Liu G-R, Li M and Niu R 2018 A cell-based smoothed finite element method for multi-body contact analysis using linear complementarity formulation *International Journal of Solids and Structures*
[12] Tootoonchi A, Khoshghalb A, Liu G R and Khalili N 2016 A cell-based smoothed point interpolation method for flow-deformation analysis of saturated porous media *Computers and Geotechnics* **75** 159-73
[13] Liu G, Dai K and Nguyen T 2007 A Smoothed Finite Element Method for Mechanics Problems *Computational Mechanics* **39** 859-77
[14] Liu G R, Nguyen T T, Dai K Y and Lam K Y 2007 Theoretical aspects of the smoothed finite element method (SFEM) *International Journal for Numerical Methods in Engineering* **71** 902-30
[15] Nguyen-Van H, Mai-Duy N, Karunasena W and Tran-Cong T 2011 Buckling and vibration analysis of laminated composite plate/shell structures via a smoothed quadrilateral flat shell element with in-plane rotations *Computers & Structures* **89** 612-25
[16] Hu J 2008 *Fabric Testing*: Woodhead Publishing Ltd.)
[17] Fan J, Yu W and Hunter L 2004 *Clothing appearance and fit: Science and technology*: CRC)

Moisture Management and Drying Properties of Double Face Knitted Fabrics

N Oğlakcıoğlu, A Çay and B Sarı

Ege University, Department of Textile Engineering, İzmir, Turkey

E-mail: nida.gulsevin@ege.edu.tr

Abstract. This study presents the drying time and moisture spreading characteristics of double face interlock fabrics, which were knitted by the combination of polyester, cotton and viscose yarns. The results revealed that polyester-cotton samples showed the lowest drying time and better moisture transfer capability among all tested fabric, when the polyester face is placed inside of the garment. Also, a relationship was determined between water spreading and drying properties.

1. Introduction

The human body is in constant interaction with its surroundings and it tries to maintain a balanced temperature around 37°C. If the thermal equilibrium condition deteriorates, the body will attempt to lose or gain heat. The increment or decrement of the body temperature due to environmental changes or activity levels is controlled by hypothalamus. In the case of heat increase, the basic responses are to accelerate the blood flow in skin vessels (vasodilation) or the output of water vapour and liquid sweat on skin surface for reducing inner temperature of body (sweating) [1]. Although the sweating process is one of most important weapons for thermal regulation system, it is also a potential cause of discomfort feeling. Particularly, textile garments allow very little microclimate areas on skin, which prevents quickly dispersing to atmosphere. Also wet textile surface by absorbing liquid cools faster than skin and causes irritating cold feeling at the time of contact. Therefore, the properties expected from a comfortable textile product are to allow efficiently the transportation of heat and moisture in the condition of thermal imbalance, to remove quickly the liquid sweat with capillary effect to outer surface and to spread it to a wide area on fabric [1, 2].

It is very difficult for a conventional textile product to fulfil these complex functions. For example, hydrophobic fibres quickly spreads the moisture which leads to faster drying, however wet feeling can not be prevented. On the other hand, highly hygroscopic fibres have good moisture absorbency, but weak drying efficiency [3]. Therefore, the thermoregulation effect of double face fabric structures formed by combining different layers with different properties is thought to be better than single layer structure. Due to this synergistic effect of different moisture transfer mechanisms of each type of fibres within double face composite fabric structure, highly comfortable products are expected to be obtained, especially for sportswear and inner garments. From this point of view, this study aims to reveal the effects of the combination of different yarn types on moisture management and drying characteristics of double face knitted fabrics.

2. Experimental

Double face interlock fabrics were knitted using polyester, cotton and viscose yarns (Ne 30/1, αe 3.5). Moisture management properties were tested by MMT (SDL Atlas) according to AATCC TM 195. Drying properties were investigated through water drop method [4] in a laboratory scale climate chamber (BINDER) in which the ambient air conditions were set to 35 °C and 65% RH.

3. Results and Discussion

According to the MMT test results, moisture spreading speed and spreading radius of PES-CO and PES-CV samples were found to be higher compared to other fabrics composed of one type of yarn.

As expected PES-PES fabrics dried faster than CO-CO and CV-CV fabrics. The results of the fabrics including different yarns on both sides (PES-CO, PES-CV) revealed that, when wetting occur on the hydrophobic surface, a faster drying was achieved compared to wetting on hydrophilic surfaces (CO or CV). Moreover, it was observed that PES-CO samples showed excellent drying behaviour which has the lowest drying time among all tested fabric, even PES-PES fabrics. This situation showed that there is a strict correlation between moisture spreading capability and drying properties.

Table 1. Drying and water spreading properties of double face knitted fabrics

Notation	Face side	Back side	Mean drying time (min)	Water spreading speed (mm/s) top/bottom	Water spreading radius (mm) top/bottom
PES-PES	Polyester	Polyester	22.5	2.7 / 2.8	25 / 25
CO-CO	Cotton	Cotton	31.5	0.8 / 0.9	15 / 10
CV-CV	Viscose	Viscose	40.5	1.3 / 1.3	10 / 10
PES-CO	Polyester	Cotton	21.0	4.7 / 3.8	30 / 30
PES-CV	Polyester	Viscose	30.0	4.8 / 3.8	30 / 30

4. Conclusion

In this study, moisture management and drying properties of double face knitted fabrics were investigated. The results showed that drying behaviour and moisture transfer capacity can be improved by the combination of polyester and cotton yarns in the double face structure, which will provide the rapid transmission of the sweat.

Acknowledgment

The authors acknowledge the Research Foundation of Ege University for the financial support given to this study (project number: 17-MÜH-054).

References

[1] Angelova R A 2016 Textiles and Human Thermophysiological Comfort in the Indoor Environment (Florida: Taylor & Francis Group) 328
[2] Song G 2011 Improving Comfort in Clothing (Cambridge: The Textile Institute/Woodhead Publishing) 459
[3] Uttam D 2013 *Engineering and Applied Sciences Research* **4** 34-40
[4] Cay A, Gurlek G, Oglakcioglu N 2017 *Drying Technol* **35** 509-521

IOP Conf. Series: Materials Science and Engineering **459** (2019) 012057 doi:10.1088/1757-899X/459/1/012057

The effect of cotton fibre characteristic on yarn properties

E Oner[1], **S Topcuoglu**[1] **and O Kutlu**[1]

[1]Usak University, Textile Engineering Department, Usak, Turkey

eren.oner@usak.edu.tr

Abstract: In this study, the effects of cotton fibre properties on the tenacity, elongation, unevenness and hairiness of the compact yarns were investigated in detail. For this purpose, compacted yarns having four different yarn counts as Ne 30/1, 40/1, 50/1 and 60/1 with similar twist coefficient between α_e 4.2-4.4 were produced systematically. The fibre, sliver, roving and yarn, that in cops and package form, properties were measured step by step from the beginning to the end of yarn production. According to the findings, unevenness, imperfections, hairiness, breaking strength and elongation results of the compact yarns were evaluated taking into account unevenness results of sliver and roving, and fibre properties tested by High Volume Instruments (HVI).
Keywords: Cotton, compact yarn, HVI, unevenness, hairiness, tenacity.

1. Introduction

In order to produce high quality yarn, choosing the appropriate cotton for the field of use, quality of fibre properties and efficient use of fibre blends are priority requirements. Fibre cost constitutes more than 50% of the producing cost of cotton yarn, and quality properties of yarn change depending on the fibre quality. Considering this fact, many researchers have studied the effect of raw materials and spinning condition on yarn properties [1-8]. In this study, the quality properties of the materials were measured at each stage of production and very thin cotton compact yarns at different yarn count were investigated differently from the literature.

2. Materials and methods

In this study, four different types of cotton compact yarn at different linear densities were produced on the Saurer Zinser 35 2Impact FX compact spinning machine. The cotton bales that used in the study were mixed at the same blow room. Trützschler TC11 type carding machine was used. Two passages drawing process was used with Trützschler TD7 and Trützschler TD8. Rieter E 80 combing machine was used for combing processes and Saurer Zinser 670 machine was used to produce roving. The cotton compact yarns that produced in the study are presented in Table 1.

Table 1. The yarns produced in the study

Sample Code	Yarn Count (Ne)	Twist Coefficient (α_e)
1	30.15	4.20
2	39.92	4.21
3	49.88	4.31
4	59.74	4.39

The fibres, slivers and yarns were preconditioned in a conditioning room at standard atmospheric conditions (20 ± 2°C, 65 ± 2% RH) for 24 hours The fibre properties were measured on High Volume Instruments (HVI) in accordance with ASTM D4605. By using Uster Tester 4 and Uster Tensorapid-3, the measurements of breaking strength, elongation, hairiness and unevenness properties were performed in accordance with Uster and ASTM D2256 standards. The obtained results were evaluated in 95% confidence interval (α=0.05) using SPSS 23.0 statistical software and by applying variance and Pearson's correlation and regression analyses.

3. Results and discussion

The fibre properties, carding, draw and roving sliver properties, unevenness, imperfections, hairiness, breaking strength and elongation properties of the yarns were measured in the study.

Uster HVI Testing System was used to determine all the physical and quality properties of the Aegean and South East Anatolian regions of Turkey's cotton fibres used in the study. The cotton samples were collected from cotton bales of the blends, which were prepared for each of the yarn count. The mean properties of the cotton fibre blends formed for each yarn count are presented in Table 2.

Table 2. The quality properties of the cotton fibres used in the study

Property	Blend for Ne 30	Blend for Ne 40	Blend for Ne 50	Blend for Ne 60
Neps	222.38	250.67	99.91	82.72
Spinning Consistency Index (SCI)	135.17	119.00	148.66	164.95
Micronaire	4.50	4.56	4.74	4.66
Maturity Index	0.86	0.87	0.88	0.91
Length (mm)	29.18	29.19	29.86	30.19
Uniformity	81.87	81.37	83.77	83.94
Short Fibre Index	8.89	8.33	6.58	6.32
Strength (g/tex)	31.16	30.30	33.62	38.23
Elongation (%)	10.96	10.93	11.10	10.89
Moisture Content (%)	6.08	6.57	7.49	6.96
Reflectance (Rd)	77.08	65.90	69.53	70.55
Yellowness (+b)	8.09	8.20	7.94	7.97
Trash Area (%)	0.59	0.34	1.07	1.07
Trash Grade	4.00	2.67	5.78	5.95

As seen in the HVI results, the fibre properties improve in finer yarn counts, in general. This state is also confirmed by SCI, which is a calculation from micronaire, uniformity strength and reflection and uses for predicting the overall quality and spinnability of the cotton fibre. Besides, the higher SCI value shows that the better quality yarn can be produced.

The unevenness (CV%) measured by Uster Tester of the carding, draw and roving slivers are shown in Table 3. As expected, the unevenness of the slivers is decreased from first to second passage drawing. The highest unevenness value is seen for Ne 40 slivers in roving form.

Table 3. The unevenness properties of the slivers

Property	Sliver for Ne 30	Sliver for Ne 40	Sliver for Ne 50	Sliver for Ne 60
CV% of the Carding	2.55	2.66	2.51	2.45
CV% of the I. Drawing	3.23	2.57	2.98	2.47
CV% of the II. Drawing	2.46	1.84	2.23	2.01
CV% of the Roving	3.44	5.34	3.02	3.21

The results of the tenacity (RKM) and elongation at break of the yarns are given in Figure 1 and Figure 2, respectively. As understood from the tensile strength results, Tenacity values of the yarns decrease with the increase of the yarn count. Similarly, elongation ability of the yarns decreases when yarns become thinner. Although fibres used for Ne 60 have the highest strength values, the tenacity of these yarns is lower. The reduction of the number of fibres in the cross-section of the yarn when yarn becomes thinner may be reason for this result.

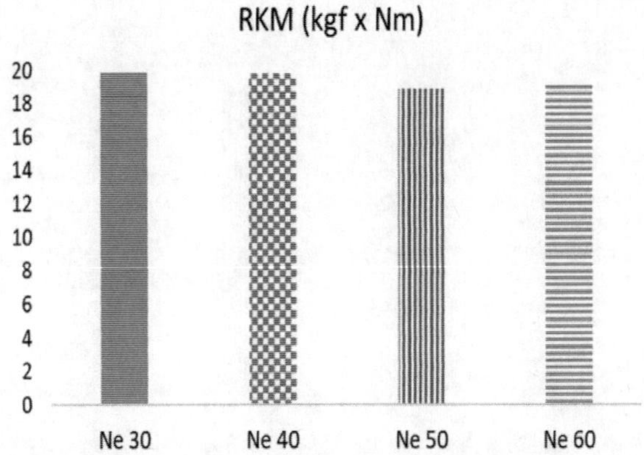

Figure 1. Tenacity values of the yarns produced in the study.

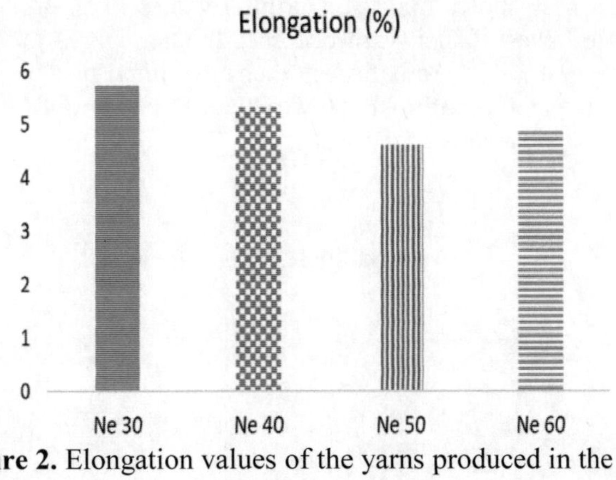

Figure 2. Elongation values of the yarns produced in the study.

The results of the imperfections as thin places (-50%/km), thick places (+50%/km) and neps (+200%/km) of the yarns are given in Table 4. As seen from the results, the imperfections of the yarns increase with the increase of the yarn count. In spite of cotton fibres blended for making thinner yarns have low neps content, they have higher neps values in yarn form.

Table 4. The imperfection values of the yarns

Yarn Count	Thin (-50%/km)	Thick (+50%/km)	Neps (+200%/km)
Ne 30	0.43	16.21	16.43
Ne 40	0.52	19.46	47.59
Ne 50	5.16	33.80	50.12
Ne 60	13.03	38.38	80.15

The results of the unevenness (CV%) of the yarns are illustrated in Figure 3. It is observed that the unevenness of the yarns increases with the increase of yarn count. In terms of tensile strength results, there was an opposite condition to the results of unevenness. Besides CV% values of the yarns change between 12.19-13.74, which are good values in terms of yarn regularity.

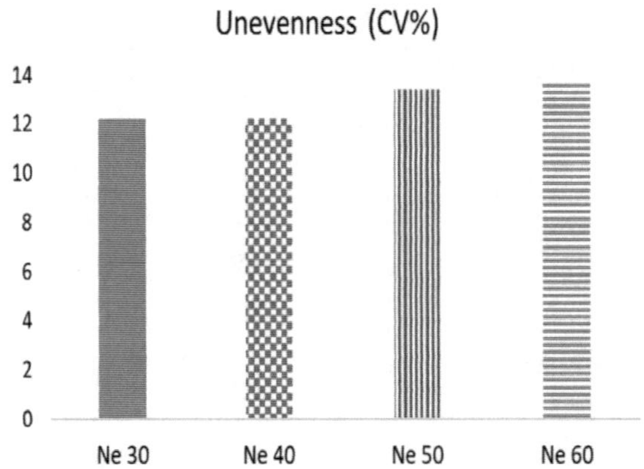

Figure 3. Unevenness values of the yarns produced in the study.

The hairiness values of the yarns produced in the study are shown in Figure 4. An evaluation of the effect of yarn count on hairiness shows that the hairiness values of the Ne 60 yarns are the lowest whereas Ne 30 yarns are the lowest. When yarn becomes thinner, fibres join the structure better and less prone to go out from the yarn axis. Actually, another most important fibre parameters that affects hairiness after yarn count is the amount of trash. However, trash content of the fibres has less effect on the yarn hairiness according to the results of the study.

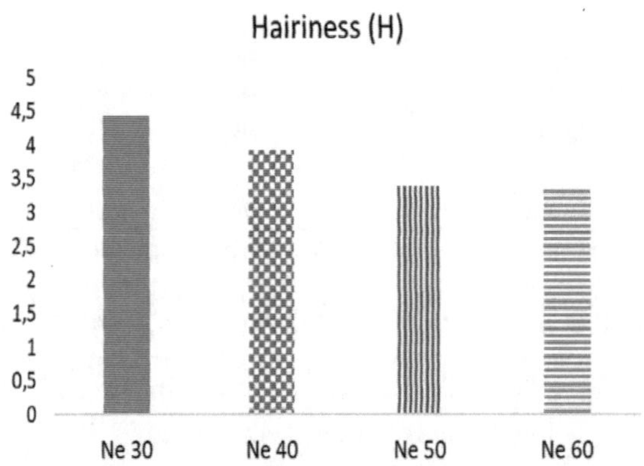

Figure 4. Hairiness values of the yarns produced in the study.

In order to investigate the relationship between fibre properties and yarn properties, Pearson correlation coefficients were calculated and the results are given in Table 5. According to the correlation analyses results, generally the correlation coefficients between fibre and yarn parameters are statistically significant at significance level of α=0.05. This finding shows that, fibre properties measured in this study may be used for estimating yarn properties. As expected, yarn tenacity has statistically significant correlations with fibre length, uniformity and strength in positive direction, whereas it has statistically significant correlations with neps, short fibre index and trash content in negative direction. Elongation ability of the yarns has positive correlation with fibre fineness and moisture content, and has negative correlation with short fibre index. As understood from the correlation analysis, when neps, short fibre index and trash content in the blend increase, unevenness of the yarn increases. Besides, the increase of fibre fineness, maturity, length and uniformity causes the reduction of unevenness of the yarn. According to the yarn hairiness correlation analysis it is found that, neps, short fibre index and trash in the fibre blend make high hairiness on the yarn surface.

Table 5. Correlations between fibre and yarn properties

		Yarn Properties			
		Tenacity	**Elongation**	**Unevenness**	**Hairiness**
Fibre Properties	**Neps**	-0.952*	0.832	0.983*	0.941*
	Micronaire	0.972*	0.983*	-0.861	-0.932*
	Maturity	0.511	0.728	-0.954*	-0.861
	Length	0.940*	0.815	-0.995*	-0.948*
	Uniformity	0.954*	0.831	-0.977*	-0.830
	Short Fibre Index	-0.965*	-0.934*	0.983*	0.967*
	Strength	0.943*	0.604	-0.819	-0.743
	Elongation	0.412	0.842	0.106	-0.161
	Moisture Content	0.436	0.933*	0.803	-0.216
	Reflectance	0.254	0.515	-0.183	0.571
	Yellowness	0.903	0.740	-0.787	0.680
	Trash Area	-0.937*	-0.757	0.936*	0.941*
	Trash Grade	-0.894	-0.776	0.932*	0.904

* Correlation coefficients are significant at 95% confidence level

In other respects, linear regression models were established to estimate the yarn tenacity, elongation unevenness and hairiness from fibre parameters. The calculated linear regression models for the yarn properties are presented below:

Tenacity= -10.121 + 0.811 Strength + 0.257 Length – 0.411 Short Fibre Index (R=0.861)

Elongation= 23.562 – 7.943 Micronaire + 1.815 Elongation – 0.22 Short Fibre Index (R=0.907)

Unevenness= -15.829 + 10.789 Neps + 0.253 Short Fibre Index – 0.215 Uniformity (R=0.836)

Hairiness= 4.354 + 0.731 Trash Area + 0.496 Short Fibre Index – 1.056 Micronaire (R=0.886)

The linear models were selected from different equations with take an attention adjusted R square values. This shows that the linear regression models which contain the parameters calculated by the method used in this study provide a valuable estimation of yarn properties.

4. Conclusion

This study focuses at examining and evaluating the relation among fibre properties, process controlling and yarn properties. The test results have revealed the relation between cotton fibre properties, slivers and yarn properties, and the performance of thin cotton compact yarns.

The results show that the unevenness of the yarns increases with the increase of yarn count, while there was an opposite condition in tensile strength values according to the results of unevenness. As understood from the findings, yarn tenacity increases with the increase of fibre length, uniformity and strength, whereas it decreases with neps, short fibre index and trash content. Extensibility of the yarns increase with increase of the fibre fineness and moisture content, and it decreases with short fibre index in the fibre blend. On the other hand, unevenness and hairiness of the yarns increase with increase of neps, short fibre index and trash content in the blend. Besides, the increase of fibre fineness, maturity, length and uniformity causes the reduction of unevenness and hairiness of the yarn. Moreover, linear regression models were established to estimate yarn properties from other fibre parameters and valuable estimation models were obtained.

The findings of this study may be helpful for predicting yarn properties from fibre characteristic studies and cotton yarn manufacturers.

References

[1] Atlas S and Kadoglu H 2006 Determining fibre properties and linear density effect on cotton yarn hairiness in ring spinning *Fibres and Textiles in Eastern Europe* **14** 48-51

[2] Pan N, Hua T and Qiu Y 2001 Relationship between fiber and yarn strength *Textile Research Journal* **71(11)** 960-964

[3] Zeidman M and Sawhney PS 2002 Influence of fiber length distribution on strength efficiency of fibers in yarn *Textile Research Journal* **72(3)** 216-220

[4] Ureyen M E and Kadoglu H 2006 Regressional estimation of ring cotton yarn properties from HVI fiber properties *Textile Research Journal* **72(3)** 216-220

[5] Krifa, M and Ethridge, M D H 2006 Compact spinning effect on cotton yarn quality: Interactions with fiber characteristics *Textile Research Journal* **76(5)** 388-399

[6] Tyagi, G K, Bhowmick, M, Bhattacharyya, S and Kumar, R 2010 Effect of spinning conditions on mechanical and performance characteristics of cotton ring-and compact-spun yarns *Indian Journal of Fibre & Textile Research* **35** 21-30

[7] Baykal, P D, Babaarslan, O and Erol, R 2006 Prediction of strength and elongation properties of cotton polyester-blended OE rotor yarns *Fibres and Textiles in Eastern Europe* **14(1)**, 18-21

[8] Karapinar, B O and Erdem, N. 2003 Comparison of quality characteristics of yarns spun from Aegean cotton fibres and their mixtures with South-east Anatolian cotton fibres. *Fibres and Textiles in Eastern Europe* **11(4),** 26-29

IOP Conf. Series: Materials Science and Engineering **459** (2019) 012058 doi:10.1088/1757-899X/459/1/012058

Design and characterization of a textile extension sensor for sports and health applications

A Paiva, F Ferreira, A Catarino, M Carvalho and H Carvalho

University of Minho, School of Engineering, Centre for Textile Science and Technology (2C2T), Campus de Azurém, 4800-058 Guimarães, Portugal

dmpaiva.s@gmail.com

Abstract. Electronic textiles (e-textiles) are a form of textiles that, integrating technology, can add new functionalities or forms of expression. The fabric can, for instance, have sensing properties, emerging as a textile sensor, which is a subfield of the e-textiles, widely reported in literature. One such kind of sensors are the extension sensors, usually studied for breathing monitoring, posture detection or characterization of movements. There are several challenges associated with the production of a textile extension sensor with high precision and repeatability. This paper reports the design and development of a knit-based sensor that uses a thermoplastic polyurethane (TPU) adhesive and film to either bond or insulate the sensing area (respectively), in order to improve the sensor's performance. The sensors were found to be appropriate for measuring breathing rate, but not for absolute extension measurements. The use of elastic tape or TPU produced the most stable results.

1. Introduction

There has been an increased interest in exploring e-textiles to build sensors that are able to read physiological and biometrical data in real-time, with more products appearing on the market. What seems to be the most appealing characteristic of this kind of sensors is the fact that we are in constant contact with textiles, almost as if they are an intrinsic part of our bodies.

Wearable sensors may play a fundamental part making instruments for healthcare widely available, allowing better healthcare management. They are expected to allow continuous monitoring of the person, in real time. But they can also be of significant value in sports, leisure or daily activities.

Textile-based strain sensors are an example of the many textile based sensors that have been studied and reported in literature, mostly focusing on the monitoring of breathing rate or posture, and characterization of movements. Most textile-based stretch sensors reported in literature are based on the piezoresistive effect – changes in electrical resistance due to deformations of the material – however, other properties can also vary when stretching a material, and have been reported.

Gioberto and Dunne [1] have studied the possibility of using a four-thread overlock stitch, in which a conductive yarn forms the upper looper thread, as a stretch sensor. They have found that, when the fabric stretches to a certain dimension, the contact points between the loops decrease, resulting in an increase of the electrical resistance, but after some point, the conductivity slightly increases again. In another study [2], the authors have explored the properties of a two-needle coverstitch with top and bottom cover, in which the top thread is replaced by a conductive yarn (silver coated polyamide). The sensor is fairly linear in the extension range in which it is used and, although it produces a small resistance variation (approximately 9 ohms), it is enough to detect breathing rate.

Ramos-Garcia et al. [3] has used a coverstitch to create a sensor to monitor breathing rate. He has also used a bonding tape to insulate the sensor. Tests were carried out and the measurements were compared to a commercially available belt (SleepSense Inductive Plethysmography). Although the sensor has shown some potential, its precision still needs to be improved, with the breathing rate error increasing in each cycle.

Stewart and Skach [4] have compared three conductive knitted fabrics as potential candidates for strain sensors. The authors have produced two of the sensors, one in a flatbed knitting machine and the other by hand knitting, with a stainless steel coated polyester yarn. The third substrate was a commercially available conductive knitting fabric (silver coated polyamide/elastane fabric, by Eeonyx), produced in a circular knitting machine. The tests showed that the difference between the first two knitted fabrics, in terms of signal acquisition, is not significant, with these two fabrics presenting low stability and producing much noise, unlike the latter, which seems to be a better candidate for strain measurement.

Atalay et al. [5] studied the effect of knitting structure compactness on electrical resistance and sensing characteristics. Three knitted strain sensors were compared. The samples were produced in a flat-bed machine, differing in either the elastomeric yarn input tension (thus varying the structure density) or elastomeric yarn thickness – one sample varies in yarn tension and the other differs in yarn thickness, both in relation to the first sample. The conductive yarn (silver coated polyamide) was embedded in a cotton interlock knitted structure. Tests were done to determine the change in resistance as a function of fabrics extension. The authors found that the sample with the thicker elastomeric yarn, which is also the most compact structure, is almost linear in all its extension, while the other sensors have a smaller linear region.

Zhang [6] has compared jersey-based knitted structures (single jersey, combination of jersey and float loops, and combination of jersey and tuck loops) as possible candidates for strain sensing. The samples were all produced in seamless knitting machine. The sensing yarn is a stainless steel polyester coated yarn (Bekaert Bekintex). The tests indicated that the sensors would become more stable after 50% elongation, thus suggesting that a pre-tension would allow the sensor to work in a more linear region. Also, the single jersey fabric seems to produce more stable results. The author also ran some tests, coating the samples with silicone, to understand the effect of a silicone layer on the sensor's performance, but it didn't improve the sensor's response, as initially expected.

Not all textile sensors are based on the piezoresistive effect. Guay [7] has introduced a spiral antenna in the chest of a t-shirt, to monitor breathing rate. The spiral antenna is composed of silver coated polyamide/glass fibre. The expansion of the chest stretches the antenna, varying its resonant frequency, which is proportional to the breathing rate. Krehel et al. [8] has used optical fiber to create a sensor that detects changes in light intensity as a response to strain, to monitor breathing rate. An LED emits light into one side of the optical fiber, while a photodiode detects the remaining light that reaches the other end of the fiber.

Disregarding the method, producing a sensor with high precision seems to be a great challenge when using e-textiles. This paper reports the study of knit-based extension sensors. Knitted extension sensors have been widely reported in the literature, but in this project, thermoplastic polyurethane (TPU) adhesive and film, which is claimed to have a good elastic recovery, were used, either to bond the conductive knit fabric to a non-conductive substrate or to create a protective layer. This material should improve the elastic recovery, thus providing a better signal response.

2. Materials and Methods

Three samples of extension sensors were produced and tested (figure 1). In all samples, conductive knitted strips were cut in the weft direction and bonded to a textile substrate, using a polyurethane based thermoplastic adhesive (Bemis 5256), which improves elastic recovery. The conductive knitted fabric is a jersey fabric composed of cotton (83%) and silver coated polyamide (17%) – Statex Silverell. The strips are of 13 cm length and 1 cm width. They are bonded to a knit fabric strip of 20 cm length. In samples 1 and 2, the sensors are bonded to a jersey PA/EL fabric and in sample 3, it is bonded to an elastic band. In sample 2, the strip was covered with a TPU film (Bemis Exoflex 3900) that slightly decreases elasticity, while improving the fabric recovery. The strips and TPU film were bonded to the substrates using a flat press machine, at 130° and 5.5 Bar, for 20s.

Figure 1. Sensor samples: 1- conductive knit thermobonded to a jersey fabric; 2- equal to sample 1, but reinforced with a TPU layer; 3- conductive knit thermobonded to an elastic band.

Preliminary tests were executed to determine the change in voltage over time. A dynamometer was used to stretch the textile sensor, during 10 cycles, and obtain the variation of force in relation to elongation. A Fluke 45 multimeter connected to a PC was used to measure the electrical resistance during stretching and relaxing. The multimeter only allows a sample rate of two samples per second, which is not enough, especially for trials with higher extension speed. For this reason, a signal conditioning circuit was built to produce a voltage according to the extension and thus electrical resistance of the sensor. The signal conditioning circuit is connected to a National Instruments NI-6210 data acquisition board and sampled at 1000 Hz.

Knowing the voltage in response to extension, it is also possible to determine the electrical resistance variation. The relation between sensor resistance and output voltage, as reported in [9], is given by

$$Vo = Voff\left(1 + \frac{R}{Rs}\right)$$

where V_o is the output voltage, V_{off} is the offset voltage, R is the feedback resistance and Rs is the sensor's resistance. As can be seen, the output voltage is reciprocal to the resistance. Previous experiments have shown, however, that resistance variations varies similarly with extension. The circuit's transfer function thus provides some linearization of the voltage versus extension function.

The elongation versus force data are obtained from the dynamometer. Then, for each trial, two data sets are combined: voltage over time, and elongation versus force at a constant speed. The combination provides voltage versus elongation.

The various tests were executed with different values of speed, elongation and pre-tension. Three trials are reported: test A starts with the samples relaxed and stretches 48 mm, at a speed of 200 mm/min; B and C were done with a small pre-tension (20 mm) and the samples were stretched 18 mm, at 200 mm/min and 400 mm/min, respectively (the latter was done to simulate breathing). Adding some pre-tension is expected to set the resistance and voltage behavior in a more linear region [6]. Moreover, it would be more similar to a slim fitted shirt.

3. Results and Discussion

3.1. *Trial A*

The variation of resistance and voltage over time is represented in Figure 2 (note that resistance and voltage cannot be measured during the same test, they are obtained in separate runs). As expected, resistance varies reciprocally to extension (Figure 2 a), and it can be observed that the voltage waveforms vary almost linearly with extension (Figure 2 b), as previously discussed.

Samples 2 and 3 are the one showing the most stable maximum voltage levels (corresponding to the lowest resistance values, i.e. the highest elongations). Sample 1, not stabilized by a TPU tape or the elastic band, shows a more variable behavior. At minimum extension, all sensors present some instability, despite the stabilization through the TPU or elastic band, but it is worse with sample 1.

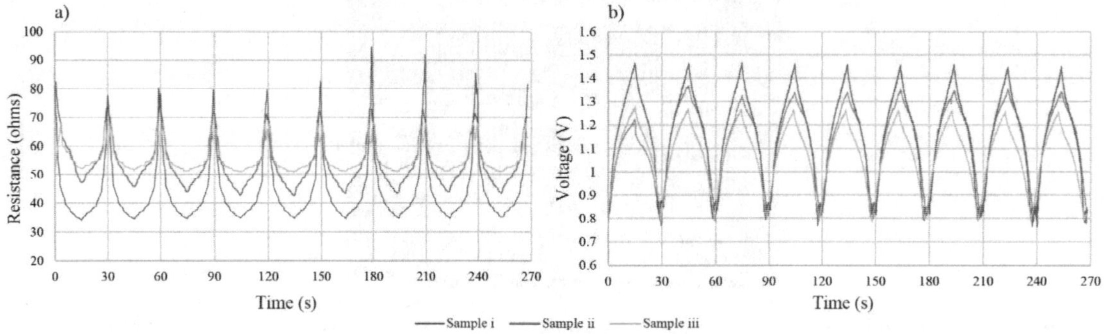

Figure 2. Resistance and voltage over time, for the three samples (two separate tests for resistance and voltage)

In Figure 3, voltage output is plotted against elongation of the sensor. It can be seen that after some elongation value, different from sample to sample, the relation becomes almost linear. Overall, the behavior of the sensors is more stable with the samples using TPU or set on elastic tape.

With increasing elongation, the variations become smaller, as if the sensor is saturating. This is particularly noted for sample 1 and 3.

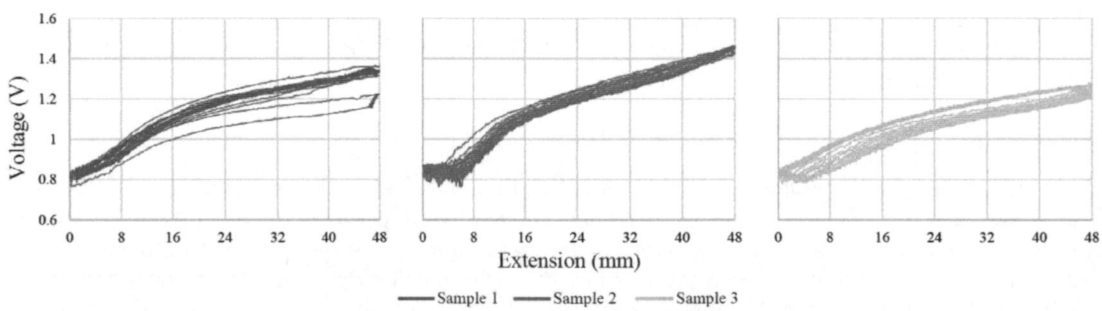

Figure 3. Variation of voltage versus elongation

3.2. Trials B and C

Figure 4 shows the results of trials B and C (for simplicity, only the voltage output is represented). In both graphs, sample 3 shows the lowest voltage offset and amplitude. Sample 2 has the highest voltage, as in previous results. However, from one trial to the other, both voltage offsets and amplitudes vary significantly. Sample 2 produces an increase of amplitude from B to C, whilst sample 1 shows a decrease.

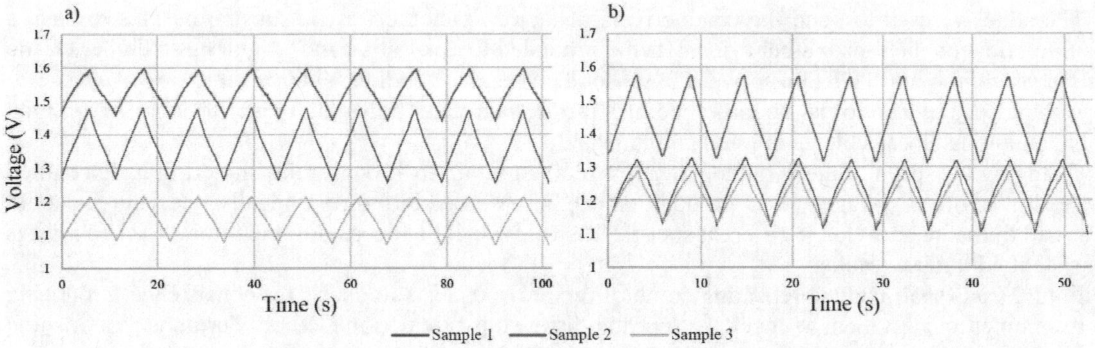

Figure 4. Voltage variation over time: a) trial B; b) trial C

It can thus be observed that the signals have different offsets and amplitudes from one trial to the other, although the extension they are subjected to are the same. Only sample 3 presents approximately the same amplitude at the two speeds, but varies in offset. Besides that, the voltage signal varies almost linearly with elongation for all samples.

The fact that offset and amplitudes of the signals are unstable from one trial to another means that these sensors are not appropriate to make absolute measurements of extension or force. However, the signal is clean and constant regarding the calculation of the movement frequency.

Figure 5 presents the voltage versus elongation graphs, once again confirming the almost linear behavior, some hysteresis and variability between cycles, and varying offsets and amplitudes.

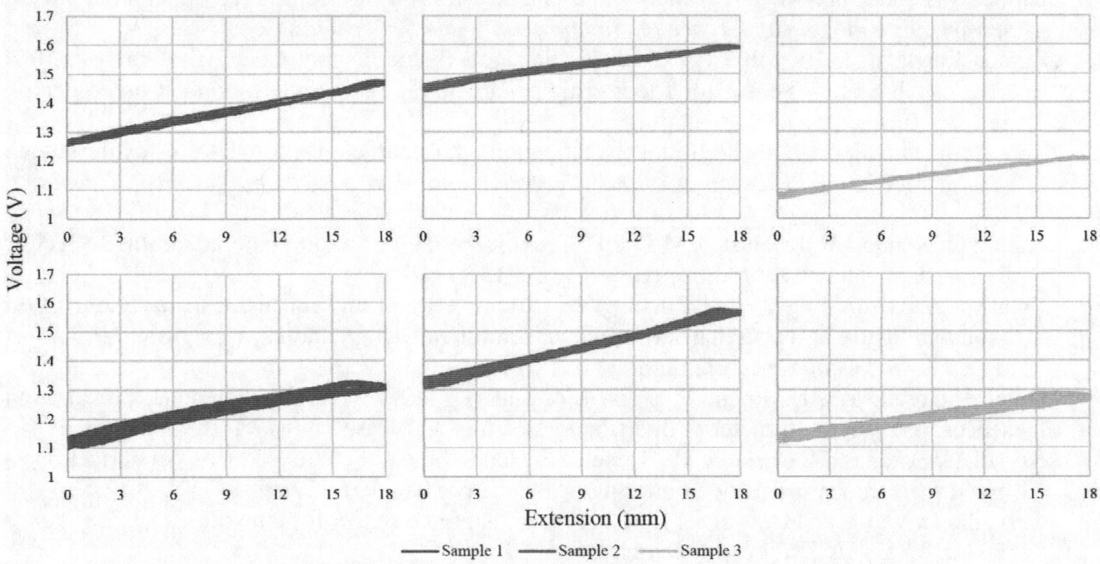

Figure 5. Voltage variation in response to elongation – trials B (top) and C (bottom)

4. Conclusion

Knit-based extension sensors, using thermo bonding materials (TPU film and adhesive) to bond the sensing material (conductive knit fabric) to a knitting substrate, were produced and tested. Three samples were produced and the results were reported.

All three samples show some potential for breathing rate monitoring, considering that the variation rate is clearly depicted by the voltage signals produced. For parameters where an absolute measurement is required, such as posture or joint angles, higher precision is required.

The adhesive used to bond the conductive knit fabric to another textile substrate seems to have a positive effect on the sensor's behavior. Having it bonded to an elastic band, which has a higher elastic recovery, or applying a TPU layer over the sensor also seem to produce better results.

Another positive factor is the existence of a pre-tension that allows the sensor to work in a region where an almost linear voltage output is obtained.

Changing the speed of the dynamometer (from 200mm/min to 400mm/min) shouldn't have resulted in the shift in offset voltage and amplitude that can be observed in figure 4. Ideally, the sample should have had the same behavior at different speeds. The trials shall be repeated in the future, to check if it is consistent with these results.

Future trials shall include repeating some of the previous tests to check for consistence, producing more samples and test them to check the repeatability of this production method. Furthermore, washing tests will be run to understand the sensor's behaviour with washing.

Acknowledgements
This work is financed by Project "Deus ex Machina", NORTE-01-0145-FEDER-000026, funded by CCDRN, through Sistema de Apoio à Investigação Cientifica e Tecnológica (Projetos Estruturados I&D&I) of Programa Operacional Regional do Norte, from Portugal 2020 and by FEDER funds through the Competitivity Factors Operational Programme - COMPETE and by national funds through FCT – Foundation for Science and Technology within the scope of the project POCI-01-0145-FEDER-007136.

References

[1] Gioberto G and Dunne L 2013 Overlock-stitched stretch sensors: characterizations and effect of fabric property *Journal of Textile Apparel, Technology and Management* **8**

[2] Gioberto G and Dunne L 2012 Theory and characterization of a top-thread coverstitch stretch sensor *International Conference on Systems, Man, and Cybernetics* Seoul

[3] Ramos-Garcia R I, Da Silva F, Kondi Y, Sazonov E and Dunny L E 2016 Analysis of a Coverstitch Stretch Sensor for Monitoring of Breathing *Tenth International Conference on Sensing Technology (ICST)* Nanjing

[4] Stewart R and Skach S 2017 Initial investigation into characterization DIY e-textile stretch sensors *MOCO'17 Proceedings of the 4th International Conference on Movement Computing* 1-4

[5] Atalay O, Kennon W R, Husain M D 2013 Textile-based weft knitted strain sensors: effect of fabric parameters on sensor properties *Sensors* **13** 11114-27

[6] Catarino A P, Yin ning Z, and Rocha A M 2016 Electrical and mechanical behaviour of an extention textile based stretch sensor whem submitted to high and slow extension *16th Autex World Textile Conference* Ljubljana

[7] Guay P, Gorgutsa S, La Rochelle S and Messaddeq Y 2017 Wearable contactless respiration sensor based on multi-material fibers integrated into textile *Sensors* **17** 1-13

[8] Kerhel M, Schemid M, Rossi, R M, Boesel L F, Bona G L and Schere L J 2014 An optical fiber-based sensor for respiratory monitoring *Sensors* **14** 13088-101

[9] Carvalho H, Yao Y and Gonçalves L M 2017 Flexible force sensor for e-textiles *IOP Conference Series: Materials Science and Engineering* **254**

Regarding the effect of finishing processes on some properties of stretch denim fabrics

C Piroi, R Harpa and M Oprea
"Gheorghe Asachi" Technical University of Iasi, Faculty of Textile, Leather and Industrial Management, 29 D. Mangeron Blvd. 700050, Iasi, Romania
cpiroi@tex.tuiasi.ro

Abstract. The present study aimed to evaluate the effect of industrial finishing and washing processes on some properties of stretch denim fabrics. The investigations have been carried out using stretch denim fabrics with a composition of 98% CO/2% EL and 3/1 Z twill weave. Starting from a greige denim fabric, through the application of various finishing treatments, six variants of finished denim fabrics were obtained. Subsequently, the finished fabrics have been subjected to washing processes commonly used for washing denim garments, in order to attain the appearance and properties as close as possible to those required by the garment products to be made of these materials. The effect of finishing treatments and washing processes on the denim fabrics was assessed by measuring certain structural and physical-mechanical properties. Have been selected and determined by usual laboratory testing methods - as objective assessment - those fabric properties that can be also subjectively assessed by the customers: fabric weight and thickness, fabric extensibility and flexural rigidity, coefficient of friction. The obtained results showed significant changes of fabrics properties both after industrial finishing treatments and after washing processes. These confirm the possibility of obtaining a diverse range of denim fabrics featuring various properties, starting from a single type of greige fabric, by using various finishing treatments and washing processes.

1. Introduction

Traditionally, denim is a strong and durable cotton fabric, produced with 2/1 or 3/1 twill weave constructions, in which the warp yarns are dyed with blue pigment obtained from indigo dye and the weft yarns are white. It is used for making jeans, jackets, shirts, skirts and many other garments for men and women of any age, as well as for children. Denim fabrics are widely used in the clothing industry because they bring together several features highly appreciated by the consumers. The denim garments are comfortable, durable, versatile, affordable, and always fashionable, being an appropriate choice for all age groups [1, 2, 4].

Since the people prefer comfortable and fashionable clothing, denim brands have become more interested in improving both the comfort and functional properties of their products, giving customers the opportunity to feel better while wearing these products. Consequently, a major area of innovation and development is devoted to diversifying the range of denim fabrics, by using varied types of fibres, yarns and finishing treatments [3,4]. Among these approaches, one important trend is represented by the manufacture of stretch denim fabrics. These fabrics, in which the weft system incorporates elastic core-spun yarns, show a high elasticity compared to other types of denim and ensure valuable comfort properties of the garment such as formability, fitting to the human body and shape retention after wearing [5, 6, 7].

The supply chain for denim products includes, among other stages, the necessary steps required for production of denim fabrics, garments manufacturing and finishing of denim garments. Generally, after weaving, the denim fabric is subjected to mechanical and chemical treatments designed to ensure dimensional stability, soft surface and to prevent contraction in subsequent finishing and washing processes. Furthermore, by applying various fabric finishing processes, denim manufacturers have the opportunity to diversify their production and get a varied range of denim fabrics. Additionally, the finished denim fabrics may be subjected to diverse type of washings in order to obtain those aesthetic features expected by the customers for the garment products [8, 9]. This is a marketing strategy that results in a number of competitive benefits such as quicker response to market and fashion demands, higher flexibility in production, more design options available already in fabric stage, large variety of possible wash-down effects and increased added value. For these reasons, the denim companies make substantial investments in machines and equipment as well as in the development of processes and techniques allowing them to achieve a flexible production and to quickly meet the continuous changing in fashion trends.

2. Materials and methods

After manufacturing, the denim garments are often subjected to washing processes to obtain a wide variety of aesthetic and tactile effects. Therefore, for denim manufacturers, not only the characteristics of the finished fabrics provided to the garment factories are important, but also the way the washing processes can amend these. Considering this aspect, the present study aims to evaluate the effect of industrial finishing and washing processes on some properties of stretch denim fabrics.

The research has been carried out using stretch denim fabrics with a composition of 98% CO/2% EL and 3/1 Z twill weave, produced on PICANOL OMNI PLUS 800- air jet weaving machines. Structural characteristics of greige denim fabric are displayed in table 1.

Table 1. Structural characteristics of greige denim fabric.

Fabric code		V0
Composition		98% CO/ 2% EL
Weave		Twill 3/1
Yarn density	warp	265
(yarns/10 cm)	weft	200
Weight (g/m^2)		328
Yarn linear	warp	57
density (tex)	weft	63

Starting from one type of greige fabric, through the application of four types of finishing treatments (F1÷F4), six variants of finished denim fabrics was obtained. The finished fabrics were subsequently washed by three different washing processes (W1÷W3), commonly used for washing denim garments. The finishing processes applied to greige fabric as well as the washing processes used for finished fabrics are briefly described in table 2.

To assess the effect of finishing treatments and washing processes on denim fabrics, have been selected and measured by the usual laboratory testing methods - as objective assessment - those fabric properties that can be subjectively assessed by the customers: fabric weight and thickness, fabric extensibility and flexural rigidity, fabric coefficient of friction. These properties are important because contribute to creating a certain sensory feeling to customers when they manipulate the fabric and may have a considerable influence on their decision regarding the buying a specific garment item [10].

Physical-mechanical properties of the resulted denim fabrics variants were evaluated using specific testing equipment and methodologies, according to the applicable standards, as follows:

- Fabric weight, M (g/m^2) was measured using a Digital Electronic Balance-MESDAN, following the SR EN 12127:2003 standard.
- Fabric thickness, g_p (mm) was measured with a Digital Thickness Tester Lab-MESDAN, according to the SR EN ISO 5084:2001 standard.

- Fabric extensibility in warp direction, E (%) was determined with the Fryma Extensometer-SDL ATLAS, according to the BS EN 14704-1:2005 standard.
- Fabric flexural rigidity in warp direction, R (mg·cm) was measured using the Fabric Stiffness Tester MTFEEM FF20-METRIMPEX, following the ASTM D1388:1996 standard.
- Dynamic coefficient of friction, μ_D was determined using the Shirley Fabric Friction Tester-SDL ATLAS, according to ISO 8295:1995 standard.

Before testing, all denim fabric samples were conditioned for 24 h at 65% relative humidity and 20°C, according to the SR EN ISO 139:2005 standard.

Table 2. Description of finishing and washing processes

Industrial finishing process	Expected effect on the fabric surface appearance	Washing process	Fabric code
F1: Brushing → Singeing → Caustification →Drying →Sanforizing → *Acrylic coating** → Softening → Drying →Sanforizing	- Lustrous surface, leather effect	*W1: Light rinse* - protects the coating and gives the fabric an extra shine effect.	V1
	- Mat surface effect		V2
	- Semi-lustrous surface effect		V3
F2: Brushing → Singeing → Caustification → Drying → Sanforizing → *Resin treatment* → Softening → Drying →Sanforizing	- Anti-crease and soft surface effect	*W2: Replica of the stonewashing process* - used to achieve the worn and faded look of the fabric	V4
F3: Brushing → Singeing → Caustification → Drying → Sanforizing	- Soft surface effect		V5
F4: Brushing → Singeing → *Overdyeing* → Softening → Drying → Sanforizing →Heat setting	- Various shades and soft surface effect	*W3: Rinse for over dyed articles* - used to remove the sizing agents and unfixed dyes	V6

**Acrylic coating* allows obtaining different fabric surface effects by changing the composition of coating component

3. Results and discussion

Table 3 summarizes the values of physical-mechanical properties for both finished and washed denim fabrics, obtained as an average of five tests. The results show differences, sometimes considerable, between the properties of various finished denim fabrics. Changes of these properties were also observed after performing the washing process.

Table 3. Physical-mechanical properties of finished and washed denim fabrics.

Fabric code	Fabric weight M (g/m^2)		Fabric thickness g_p (mm)		Fabric extensibility E (%)		Flexural rigidity R (mg·cm)		Dynamic coefficient of friction μ_D	
	Finished	Washed	Finished	Washed	Finished	Washed	Finished	Washed	Finished	Washed
V1	405	412	0.67	0.79	2.5	3.6	815.4	525.1	0.5546	0.6791
V2	361	376	0.69	0.82	3.4	3.7	493.4	433.7	0.7129	0.8102
V3	351	376	0.68	0.84	3.3	4.3	512.4	426.9	0.6809	0.7605
V4	341	358	0.72	0.88	4.2	4.3	385.4	349.9	0.7540	0.8497
V5	343	360	0.70	0.89	5.5	6.0	316.8	301.2	0.7534	0.8388
V6	332	373	0.72	0.91	6.0	6.5	279.1	413.4	0.7530	0.8385

3.1. Fabric weight

As a result of the finishing processes, all variants of finished denim fabrics exhibit an increase in weight compared to the greige fabric (figure 1). This effect can be explained by the normal changes of fabric structural parameters, such as increasing the yarn density due to the fabric shrinkage, but also by the addition of finishing products such as coating components, resins, softeners. The highest increase of fabric weight was observed for the fabrics V1 (23.4%), V2 (10%) and V3 (7%), which were finished by acrylic coating, while for the others fabric variants the percentage increase was under 5%.

Following the washing process, the fabric weight increases again, but to a lesser extent (<7%), except for V6 variant that show a growth of weight by 12%.

3.2. Fabric thickness

Finished denim fabrics do not show significant differences in thickness (0.67 mm - 0.72 mm), the smallest values being obtained for acrylic coated fabrics V1 - V3.

After washing processes, the thickness increases at all fabric variants, but the trend is more significant for V4-V6 (figure 2). These fabric variants have undergone other finishing processes than the acrylic coatings - treatment with resin, overdyeing - and have been washed in a more aggressive manner to obtain the desired surface effect.

Based on these results it can be concluded that the acrylic coating has contributed to reducing the tendency of the fabric to shrink (by fixing the yarns in the fabric structure) thereby diminishing the effect of increasing the thickness.

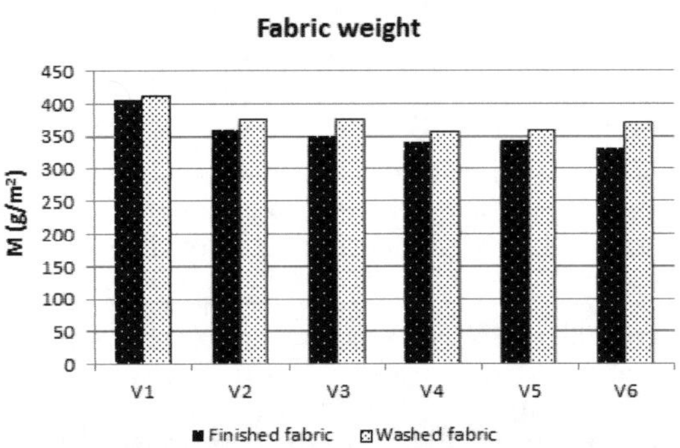

Figure 1. Effect of the applied treatments on the fabric weight.

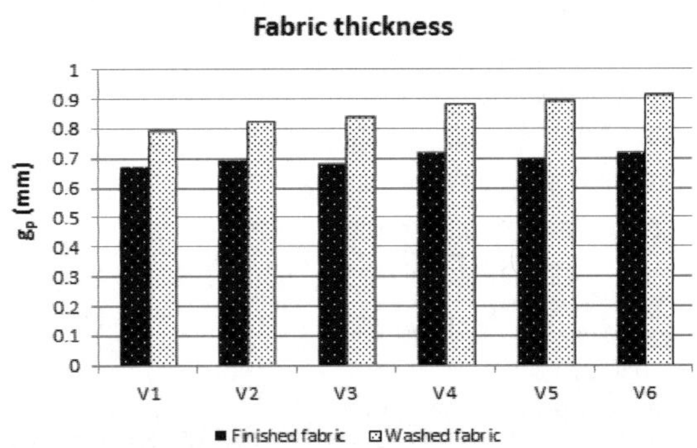

Figure 2. Effect of the applied treatments on the fabric thickness.

3.3. Fabric extensibility

Fabric extensibility measures the increase in fabric dimensions when it is subjected to an applied load and is a significant mechanical parameter that influences the proper fit of the garment and the wearing comfort. From this point of view, the results revealed a different behaviour of the examined denim fabrics (figure 3). In the case of finished denim fabrics, the variants V1 - V4 exhibit lower values of extensibility (2.5% - 4.2%), while for the V5 and V6 variants this parameter has higher values (5.5% - 6%). Again, the particularities of the finishing processes can explain this behaviour. For V1-V4 variants, acrylic coatings and resin treatment limit the mobility of the yarns and hence reduce the fabric's ability to increase its size when subjected to loading, while for the other two finishes this effect does not occur.

Following the washing process, the extensibility of all fabric variants examined has increased, but in a different manner. The trend is more pronounced for variants V1 and V3 with 44% and respectively 30% increase, while for the other variants the increase in extensibility is under 10%.

3.4. Fabric flexural rigidity

Bending properties of the fabrics are determined by an assembly of factors such as yarn bending behaviour, fabric weave, finishing treatments and have a major influence on fabric's hand value and draping behaviour.

In terms of flexural rigidity, the analysed fabrics showed a quite different behaviour, but in agreement to the trends amending extensibility (figure 4). Acrylic coating imparts stiffness to finished denim fabric and this effect is confirmed by the high flexural rigidity values obtained especially for fabric variants V1 but also for V2 and V3. The other variants of denim fabrics are less rigid, but they are also different from each other. After washing processes, the rigidity of fabrics decreased, with percentages varying between 35% (V1) and 5% (V5); only for fabric variant (V6) it was seen that the washed fabric is stiffer than the finished fabric.

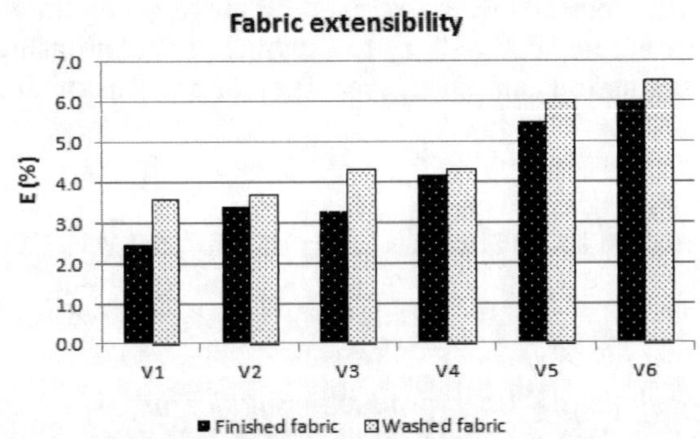

Figure 3. Effect of the applied treatments on the fabric extensibility.

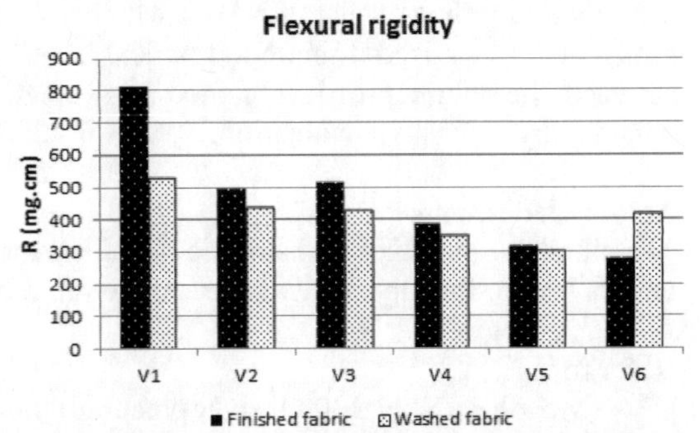

Figure 4. Effect of the applied treatments on the fabric rigidity.

3.5. Coefficient of friction

Coefficient of friction is a parameter that can be used to evaluate the fabric's surface appearance in terms of smoothness or roughness, the smaller values showing a smoother surface.

The obtained results confirm that both the finishing and washing treatments have influenced the surface properties of denim fabrics (figure 5). Analysing the coefficient of friction's values, one can say that the variants of acrylic coated denim fabrics (V1-V3) have a smoother surface compared to the other variants. The same situation is maintained after the washing process, but in this case the coefficient of friction has higher values and the differences between fabric variants become smaller, except the variant V1 which remain in first place, with the smoothest surface.

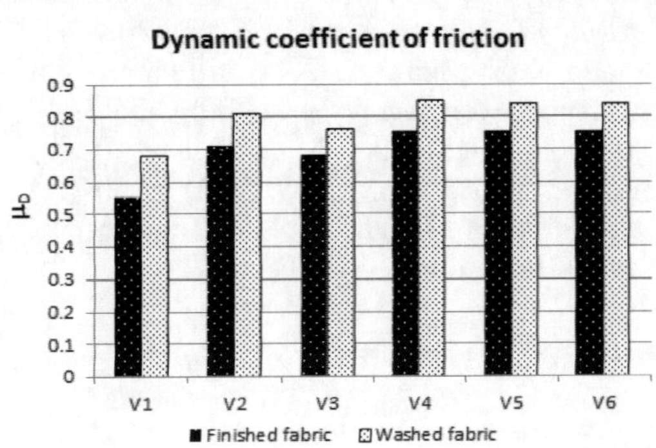

Figure 5. Effect of the applied treatments on the fabric coefficient of friction.

4. Conclusions

This study aimed to evaluate the effect of industrial finishing and washing processes on the properties of denim stretch fabrics by assessing some physical-mechanical properties important from the point of view of the sensory perception that customers have when handling the material.

The results showed substantial changes in the fabric properties both after the industrial finishing treatments and after the washing processes, confirming the possibility of obtaining a range of denim fabrics with different properties, starting from a single type of greige fabric. For denim factories, this is an intelligent strategy that allows them to be more flexible and respond quickly to the changing demands of the market and customers.

Complementing the classic strategy for testing finished denim fabrics with testing of washed denim fabrics give a clearer picture of the possibilities for further diversification, providing designers and garment manufacturers with more information about how the final products will look like.

It is important, however, to find out whether these differences can be perceived by customers who do not have a specialized training in textiles. In our future studies, we will try to establish the relationship between the fabric profile obtained by objective evaluation and the perception of the customers, obtained by subjective evaluation.

Acknowledgments

The authors would like to thank the management of UCO TESATURA Co., part of UCO RAYMOND Denim group, for supplying the stretch denim fabrics.

References

[1] Mogahzy Y E 2009 Development of traditional textile fiber products *Engineering Textiles: Integrating the Design and Manufacture of Textile Products*, Woodhead Publishing p. 362
[2] Annapoorani S G 2017 Introduction to denim *Sustainability in Denim,* Ed. Muthu S S *The Textile Institute Book Series* Woodhead Publishing
[3] Gokarneshan N, Sandip k R, Malathi R, Aathira 2018 Advances in denim research. *Res Dev Material Sci.* 3(1). RDMS.000551
[4] Kumar S, Chatterjee K, Padhye R and Nayak R 2016 Designing and Development of Denim Fabrics: Part 1 - Study the effect of fabric parameters on the fabric characteristics for women's wear. *J Textile Sci Eng* **6(4)** p 265
[5] Kaynak H K 2017 Optimization of stretch and recovery properties of woven stretch fabrics *Text. Res. J* **87(5)** 582–592
[6] Choudhary A K and Sheena B 2018 Influences of elastane content, aesthetic finishes and fabric weight on mechanical and comfort properties of denim fabrics. *J Text Eng Fashion Technol* **4(1)**
[7] Özdil N 2008 Stretch and bagging properties of denim fabrics containing different rates of elastane *Fibres & Textile in Eastern Europe* 1(66) pp 63-67
[8] Jucienė M, Dobilaitė V and Kazlauskaitė G 2006 Influence of Industrial Washing on Denim Properties *Materials science (Medžiagotyra)* **12(4)** pp 355-359
[9] Halleb N A, Sahnoun M and Cheikhrouhou M 2015 The effect of washing treatments on the sensory properties of denim fabric *Text. Res. J* **85(2)** p 150-159
[10] Kayseri G Ö, Özdil N and Süpüren Mengüç G S 2012 Sensorial Comfort of Textile Materials, *Woven Fabrics*, Ed. Prof. Han-Yong Jeon InTech,

IOP Conf. Series: Materials Science and Engineering **459** (2019) 012060 doi:10.1088/1757-899X/459/1/012060

Inventory management concepts and techniques

G Priniotakis[1] and P Argyropoulos[2]
[1] Department of Industrial Design and Production Engineering,
School of Engineering - University of West Attica, Athens, Greece
[2] Prosent S.A.,Technical Fabrics, Production Planning and Control (PPC)Department,
84 Archimidous str., 19400, Athens, Greece

gprin@uniwa.gr

Abstract: Inventory management has become one of the key elements of the supply chain management and can greatly affect the performance of a business. The textile industry is no exception. Traditional approaches in decision making based on manager instincts and hunches are no longer enough in the today's increasingly competitive environment. Small to medium sized family owned textile businesses are usually prone to this way of thinking.
This paper discusses some basic concepts and techniques for classifying inventory, controlling inventory levels, avoiding stock outs and increasing customer satisfaction. It also discusses the importance of forecasting demand and uses the Root Mean Square Error (RMSE) as an effective measure of the forecast error, which later becomes a basic driver for inventory management. It addresses the Service Level (SL) as a performance metric and emphasizes on the importance of Safety Stock (SS). Finally, it discusses the use of the Reorder Point (ROP) as an efficient indicator for triggering production replenishment and proposes a simple technique for prioritizing production orders.

1. Introduction

Inventory management is the process of monitoring and controlling inventory level and ensuring adequate replenishment in order to meet customer demand. Determining the appropriate inventory level is crucial since inventory ties up money and affects performance. Having too much inventory reduces the working capital and impacts the company's liquidity. On the contrary, having too little inventory leads to stock outs and missed sales which leads to less profit. It becomes clear that management attention should be focused on keeping inventory level somewhere in between, striving for increased customer satisfaction and minimum stock outs while keeping inventory costs as low as possible.

2. Service Level

The Service Level (SL) is an important performance indicator which in a simplified manner, measures a company's ability to service customer demand and is expressed as a percentage. In inventory management, service level is the probability that the customer demand is met or that the customer demand does not exceed the inventory. A service level of 95% means that there is 95% probability that demand will be met and customer orders will be fulfilled on time, while the probability that a stock out will occur, resulting in missed sales, is 5%.The higher the Service Level, the higher the customer satisfaction but also the higher the inventory level. Since the future demand is uncertain, achieving a 100% Service Level would require an infinite amount of inventory which is clearly unachievable. Management should understand the trade-off between the cost of inventory and the cost of stock-outs and position against inventory levels based on specific criteria.

IOP Conf. Series: Materials Science and Engineering **459** (2019) 012060 doi:10.1088/1757-899X/459/1/012060

3. Inventory Classification

ABC inventory classification is a very popular inventory control technique that follows the Pareto Principle which states that, for many events, roughly 80% of the effects come from 20% of the causes. In a case of a business, it could be stated that roughly the 20% of the end products generates the 80% of the income. In ABC analysis, a company reviews its inventory and sorts all items into three categories, called "A" items, "B" items and "C" items. A typical breakdown would possibly describe "A" items as those that produce 70% of income, "B" items as those that produce 25% of income and "C" items, as those that produce 5% of income. This classification might be different from company to company but managers should be able to find the pattern that suits best their needs.

Clearly, "A" items require closer attention and should be handled differently. Assigning higher Service Levels for those items is a wise choice. The higher service level will lead to higher inventory but will also decrease the probability of a stock out. A 5% probability of a stock out for an "A" item will result to much higher losses than of a 5% probability for a "C" item. A 99% Service Level for an "A" item could have about the same impact as an 85% Service Level for a "C" item and managers should position against each category accordingly.

4. Forecast Error and Safety Stock

It has been already stated, that customer demand is uncertain. Managers should try to predict future demand based on statistical data and taking into account multiple criteria. It is highly desirable to try to predict the future demand and get properly prepared even with a certain degree of uncertainty than having no expectation of what is about to happen. The methods and tools used for forecasting are not under the scope of this paper. What is really important, is to find a way to calculate how closely the prediction of the demand meets the actual demand, thus how accurate a forecast is. The difference between the actual and the forecasted data is the Forecast Error.

Since the Forecast Error can be calculated and not just predicted, it can be quite a safe driver for inventory management. The goal is to keep just as much inventory as it is really needed and by knowing that there is a fixed error in our estimation, we can safely take it into account and add up a bit of extra stock to our inventory to compensate for this misalignment. This is called Safety Stock and its purpose of existence is to absorb the error of the estimation and to protect the company against an unexpected and unwanted stock out. In a simplified scenario, a 30% variation of the forecasted demand should result in a 30% increase of the inventory level.

5. Root Mean Square Error (RMSE)

In order to calculate the errorof a forecast the RMSE method is preferred among others, as it calculates the standard deviation of the residuals between the actual and the forecasted data. As the name suggests, the difference between the forecasted demand and the actual demand is squared and then it is expressed as the square root of the average squared residuals. In the table below, real data are presented.

Table 1. Calculation of the RMSE of the Forecasted Demand for SKU X

Month	Actual Demand	Forecasted Demand	Squared Error	Total Squared Error	Average Squared Error	RMSE
1	1.508	1.533	625	1.658.065	138.172	372
2	1.884	1.867	289			
3	1.024	1.234	44.100			
4	1.458	1.523	4.225			
5	2.433	2.763	108.900			
6	3.523	2.707	665.856			
7	2.322	1.998	104.976			

8	818	444	139.876
9	1.753	1.015	544.644
10	1.889	1.767	14.884
11	1.438	1.509	5.041
12	1.283	1.440	24.649

The reader will quickly realize that the forecasted data are not highly accurate. Although forecast accuracy is important, it will soon become clear that calculating the forecast error and taking itinto account when positioning against inventory levels is more significant than the accuracy of the forecast itself, when our primary concern is to protect against stock outs.

6. Mixing Service Level, Safety Stock and RMSE

When the desired Service Level for a stock keeping unit (SKU) is selected and the RMSE of the forecast has been defined then the Safety Stock can be accurately calculated. The SKU in table 1 was classified as an "A" item and the Service Level was set to 97%. Under the assumption that our data are following a normal distribution curve, then in 97% of the cases, the actual demand will be less than 1,88 standard deviations above the mean demand (see appendix A).

Any sample value will lie within $\mu \pm \sigma \times Z$. Since we are looking only for extra stock we actually calculate only for $\mu + \sigma \times Z$. In addition, we know our estimation's variance but we need to calculate the variance during the lead time. The safety stock needed during the manufacturing lead time can be calculated using the following formula:

$$Safety\ Stock = Z \times \sqrt{LT} \times RMSE$$

Where:

Z =Z-score for the desired Service Level (see appendix A)
LT = The Lead Time using in the same time period as in the forecast
RMSE = The Root Mean Square Error between the actual and the forecasted demand

Table 2. Parameters for SKU X

RMSE	372
Service Level	97%
Working Days per Month	22
Manufacturing Lead Time (Days)	5
Z Statistic	1,88

Given the data from table 2, the Safety Stock for SKU X can be calculated as below:

$$Safety\ Stock = 1{,}88 \times \sqrt{\frac{5}{22}} \times 372 = 333\ units$$

Thus, keeping 333 units of extra stock for SKU X, will prevent a stock out for 97% of the time.

One could note that in the textile industry, the manufacturing lead time is also been subjected to a certain variance. This can be due to the irregular machine cycle time or due to an unexpected downtime of the machinery. For instance, the setup time of a stenter machine is greatly affected by the process it previously underwent resulting in extra time for heating it up or cooling it down. It is wise to

IOP Conf. Series: Materials Science and Engineering **459** (2019) 012060 doi:10.1088/1757-899X/459/1/012060

take this variance under consideration and adjust the lead time in the equation accordingly. If the variance of the demand and the variance of the lead time are independent of each other,then the safety stock is Z-score times the square root of the sum of the squares of the individual variabilities, as in the following equation:

$$Safety\ Stock = Z \times \sqrt{(LT \times RMSE_D)^2 + (RMSE_{LT} \times \bar{D})^2}$$

But if both the variance of the demand and the variance of the lead time are dependent of each other then safety stock can be computed as below:

$$Safety\ Stock = Z \times \sqrt{LT} \times RMSE_D + Z \times RMSE_{LT} \times \bar{D}$$

Where:

Z = Z-score for the desired Service Level (see appendix A)
LT = The Lead Time using in the same time period as in the forecast
$RMSE_D$ = The Root Mean Square Error between the actual and the forecasted demand
$RMSE_{LT}$=The Root Mean Square Error between the actual and the standardlead time
$\bar{D}$= The average forecasted demand

Table 3. Parameters for SKU X

$RMSE_D$	372
Service Level	97%
Working Days per Month	22
Manufacturing Lead Time (Days)	5
Z Statistic	1,88
$RMSE_{LT}$ (Days)	0,25 (6 hrs)
Average Forecasted Demand (monthly)	1650

Given the data from table 3, the safety stock can be calculated. It was decided that the variance of the lead time and the variance of the demand was dependent of each other. This is quite typical in textile factories since the life expectancy of the industrial machinery is very high (many decades) and as a result, new machinery with higher output rate is added before old machinery is regarded obsolete and its use is been discontinued. Thus, when demand rises and extra capacity is needed, managers decide to add shifts and make use of old and new machines simultaneously which produce the same products at different output rates resulting in higher variance in the manufacturing lead time. Taking this into account, the Safety Stock is calculated below:

$$Safety\ Stock = 1{,}88 \times \sqrt{\frac{5}{22}} \times 372 + 1{,}88 \times \frac{0{,}25}{22} \times 1650 = 369\ units$$

Thus, keeping 369 units of extra stock for SKU X, will prevent a stock out for 97% of the time.

7. The Re-Order Point

The Re-Order Point (ROP) technique is a process of triggering inventory replenishment based on inventory level. The ROP is set at a level which is calculated based on the forecasted demand during the replenishment time plus a safety stock, as in the following equation:

$$ROP = \overline{D} \times \overline{LT} + SS$$

Where:

$\overline{D}$= The average forecasted demand
$\overline{LT}$= The average Lead Time
SS = The Safety Stock

Keeping the same data as before, the ROP for SKU X is calculated below:

$$ROP = \frac{1650}{22} \times 5 + 369 = 744\, units$$

Thus, the inventory level of SKU X should be closely monitored and as soon as it becomes equal or less than 744 units, then production should be reinitiated. There should be just enough time to reproduce the item before its inventory level reaches the safety stock level but even if it does there will be 97% chance that the safety stock will be enough to protect against an unexpected customer demand or a production delay and only 3% chance that the inventory level will fall to zero and a sales order will not be satisfied on time.

There will also be times when the accumulated pending orders will exceed the inventory level even before the ROP is reached. Thus, the future fulfilment of the pending orders will probably lead to a stock out. Clearly, the Projected Inventory level is of concern and it is calculated according to the formula bellow:

$$Projected\ Inventory = Stock\ On\ Hand + WIP - Pending\ Orders$$

Where:

Stock On Hand = The available inventory
WIP = The inventory that is already in process but not yet finished (Work in Process)
Pending Orders = The pending orders that are due today or already past due

Based on the above approach, production should be initiated whenever the Projected Inventory level falls below the ROP.

8. Prioritizing Production Orders

In real life, the inventory or the projected inventory level of multiple stock keeping units will simultaneously reach or fall under their ROP. There are also capacity constraints or other constraints that force managers to perform some kind of prioritization of the production. When avoiding a stock out is the primary concern, then some of the key elements for deciding how to prioritize production are: the ABC classification, the Inventory Turnover Ratio and the Projected Inventory Level.

The Inventory Turnover ratio is metric that calculates the number of times that inventory cycles (turns) during a given period. It is calculated as the Cost Of Goods sold over the average inventory but can also be expressed as the number of units sold over the average number of units. The importance of the ABC classification and the Projected Inventory Level has already been discussed.

The following excel sheet depicts a real life situation:

Table 4. Excel Inventory Report

Code	WIP	Stock	SS	ROP	Pending Orders	ABC Rating	Inv. Turnover	Projected Stock	Status
SKU2463		260	102	194	756	C	4	< 0	●
SKU2794		296	88	167	32	C	4	>ROP	●
SKU2854		300	75	143	280	C	2	<=SS	●

Code	WIP	Stock	SS	ROP	Pending Orders	ABC Rating	Inv. Turnover	Projected Stock	Status
SKU1282		910	370	681	51	B	6	>ROP	
SKU1362	600	800	502	871	900	A	7	<=SS	
SKU2341		778	88	167	350	C	2	>ROP	
SKU2361		1.001	292	537	450	B	5	>ROP	
SKU2461	600	1.082	419	725	960	A	8	<=ROP	
SKU2791		1.618	923	1.600	253	A	12	<=ROP	
SKU2851		780	641	1.111	858	A	7	< 0	

The Status becomes green and signifies the need for production when the projected inventory level falls below the ROP. SKU2851 & 2463 have not enough stock to fulfil the pending orders thus their production should be prioritized. SKU1362 & 2854 will soon fall under their safety stock level thus they should be produced right after. SKU2461 & 2791 will soon fall under their ROP but SKU2791 cycles a bit faster than 2461 thus it would be wiser to produce it first.

The idea behind it is to prioritize those items that their projected stock falls below zero. If more than one item meets this rule then prioritize based on their ABC classification. If again more than one item meets this rule then prioritize based on their turnover ratio. Then check for all the items that their projected inventory will fall under their safety stock level prioritizing them the same way and finally check for those items that their projected inventory level that will fall under their ROP.

9. Conclusions

This paper emphasizes on the importance of inventory management. The methods and tools that are used can add great business value. Holding the right amount of inventory can increase business performance by reducing the response time to customer demand which results to higher customer satisfaction. The textile managers and business owners should weigh and balance the trade-offs when deciding how much inventory to hold and strategically decide based on the proven concepts and techniques that are described.

IOP Conf. Series: Materials Science and Engineering **459** (2019) 012060 doi:10.1088/1757-899X/459/1/012060

Appendix A

Normal Curve Areas
Standard normal probability
in right-hand tail

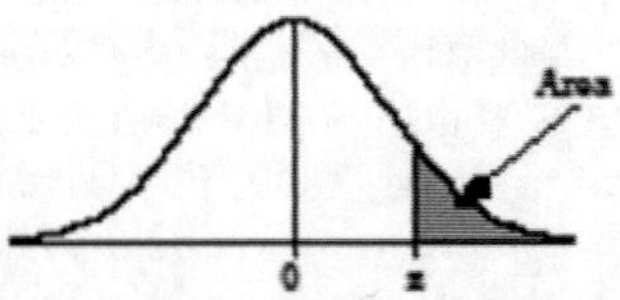

	Second decimal place of z									
z	.00	.01	.02	.03	.04	.05	.06	.07	.08	.09
0.0	.5000	.4960	.4920	.4880	.4840	.4801	.4761	.4721	.4681	.4641
0.1	.4602	.4562	.4522	.4483	.4443	.4404	.4364	.4325	.4286	.4247
0.2	.4207	.4168	.4129	.4090	.4052	.4013	.3974	.3936	.3897	.3859
0.3	.3821	.3783	.3745	.3707	.3669	.3632	.3594	.3557	.3520	.3483
0.4	.3446	.3409	.3372	.3336	.3300	.3264	.3228	.3192	.3156	.3121
0.5	.3085	.3050	.3015	.2981	.2946	.2912	.2877	.2843	.2810	.2776
0.6	.2743	.2709	.2676	.2643	.2611	.2578	.2546	.2514	.2483	.2451
0.7	.2420	.2389	.2358	.2327	.2296	.2266	.2236	.2206	.2177	.2148
0.8	.2119	.2090	.2061	.2033	.2005	.1977	.1949	.1922	.1894	.1867
0.9	.1841	.1814	.1788	.1762	.1736	.1711	.1685	.1660	.1635	.1611
1.0	.1587	.1562	.1539	.1515	.1492	.1469	.1446	.1423	.1401	.1379
1.1	.1357	.1335	.1314	.1292	.1271	.1251	.1230	.1210	.1190	.1170
1.2	.1151	.1131	.1112	.1093	.1075	.1056	.1038	.1020	.1003	.0985
1.3	.0968	.0951	.0934	.0918	.0901	.0885	.0869	.0853	.0838	.0823
1.4	.0808	.0793	.0778	.0764	.0749	.0735	.0721	.0708	.0694	.0681
1.5	.0668	.0655	.0643	.0630	.0618	.0606	.0594	.0582	.0571	.0559
1.6	.0548	.0537	.0526	.0516	.0505	.0495	.0485	.0475	.0465	.0455
1.7	.0446	.0436	.0427	.0418	.0409	.0401	.0392	.0384	.0375	.0367
1.8	.0359	.0351	.0344	.0336	.0329	.0322	.0314	.0307	.0301	.0294
1.9	.0287	.0281	.0274	.0268	.0262	.0256	.0250	.0244	.0239	.0233
2.0	.0228	.0222	.0217	.0212	.0207	.0202	.0197	.0192	.0188	.0183
2.1	.0179	.0174	.0170	.0166	.0162	.0158	.0154	.0150	.0146	.0143
2.2	.0139	.0136	.0132	.0129	.0125	.0122	.0119	.0116	.0113	.0110
2.3	.0107	.0104	.0102	.0099	.0096	.0094	.0091	.0089	.0087	.0084
2.4	.0082	.0080	.0078	.0075	.0073	.0071	.0069	.0068	.0066	.0064
2.5	.0062	.0060	.0059	.0057	.0055	.0054	.0052	.0051	.0049	.0048
2.6	.0047	.0045	.0044	.0043	.0041	.0040	.0039	.0038	.0037	.0036
2.7	.0035	.0034	.0033	.0032	.0031	.0030	.0029	.0028	.0027	.0026
2.8	.0026	.0025	.0024	.0023	.0023	.0022	.0021	.0021	.0020	.0019
2.9	.0019	.0018	.0018	.0017	.0016	.0016	.0015	.0015	.0014	.0014

IOP Conf. Series: Materials Science and Engineering **459** (2019) 012060 doi:10.1088/1757-899X/459/1/012060

References

[1] Plossl G, 1994, *Orlicky's Material Requirements Planning*
[2] King L P (2011): *Crack the code: Understanding safety stock and mastering its equations*, APICS Magazine, July/August 2011, p. 33-36
[3] Chockalingam M, 2009, *Forecast Accuracy and Inventory Strategies*
[4] Curwin J and Slater R, 2002, *Quantitative Methods for Business Decisions*
[5] Slack N, Chambers S and Johnston R, 2004, *Operations Management*

IOP Conf. Series: Materials Science and Engineering **459** (2019) 012061 doi:10.1088/1757-899X/459/1/012061

Effects of different conductive yarns' knitting structure on electromagnetic shielding effectiveness

G Priniotakis[1], E Sfyroera[1], S Symeonidis[2], S A Mitilineos[2], S Vassiliadis[2], C Zafeiri[1] and S Moudatsou[1]
[1] Department of Industrial Design and Production Engineering
[2] Department of Electrical and Electronic Engineering
School of Engineering - University of West Attica, Athens, GREECE

gprin@uniwa.gr

Abstract. Recently, increasing number of studies are performed on protective fabrics containing conductive yarns for electromagnetic shielding purposes. In this paper, several weft knitted fabrics were developed with conductive yarns of the same yarn count and their conductivity and electromagnetic shielding effectiveness (EMSE) were measured. The variations in EMSE as well as reflection, transmission and other physical properties of knitted conductive fabrics were investigated considering conductive yarns and knit structure. It is observed that an increase in the conductive yarns content significantly alters conductivity while all samples demonstrate sufficient EMSE in high frequency bands; Measurements results are presented herein and future research directions are provided.
Keywords: electromagnetic shielding effectiveness; knitted fabrics; conductive fabrics.

1. Introduction

Wireless devices, Wi-Fi, and electronic security systems, that are some of the newer devices, which are implementations of electromagnetic (EM) fields, have been common in recent years. These applications, use industrial, scientific and medical (ISM) band of 2400 MHz. In this regard, various research teams have examined if the EM fields have possibly effects on human health. So, it is important to create public concern about the safety of these EM fields (1). Conductive fabrics can easily be the shielding material in addition to a number of other advantages. Fabric is the friendliest and most widely used material by humans. Conductive fabrics are lightweight, flexible and inexpensive.

For a few years conductive fabrics have been used for various shielding applications by the electrical and electronic industries, instead of electrically conductive metal sheet or wire mesh. Previous research that used selected copper as a conductive filler to produce copper core yarns with cotton fibre as sheath material to make plain and twill woven fabrics, observed an increase in shielding effectiveness with an increase in the number of conductive fabric layers, finer yarn count, warp density, weft density, cover factors and a decrease in shielding effectiveness with copper wire diameter (2). It measured the electromagnetic shielding effectiveness of these fabrics in the frequency range of 20 –18,000 MHz with coaxial transmission equipment.

Roh et al. (3) used metal composite yarns, which were used in the construction of plain woven metal composite fabrics, and were produced with commercially available metal filaments and polyester (PET) filaments. Planewave shielding properties of the composite fabrics were measured between 30-1500 MHz using the coaxial transmission line method. They observed that while the overall EM shielding effectiveness increased with metal content, different frequency dependence related to the aspect ratio of metal grid structure.

Duran et al. (4) investigated the electromagnetic shielding effectiveness of 3/1 twill woven fabrics with electromagnetic competence test device in 5 different frequencies; 200MHz, 400MHz, 600MHz, 800MHz and 1GHz. While 100% cotton yarns was used as warp yarns, two different weft yarns, copper/cotton core yarns and 100% cotton yarns were used in two different weft densities. It was observed that fabrics woven with copper/cotton core yarns had considerably higher electromagnetic shielding effectiveness values than the fabrics woven with 100% cotton wefts. The results showed that weft density had only a slight effect on the electromagnetic shielding property of the fabric produced with core yarns, whereas it did not have any effect on the electromagnetic shielding property of 100% cotton fabrics.

Palamutçu et al. (5) developed an electromagnetic shielding efficiency measurement set and tested its reliability with two cases, the electrically conductive plain knitted and plain woven fabrics. In these fabrics, Co/Cu and Co/Cu/Ag yarns with different ratios were used. For woven specimens at the 860MHz-960MHz frequency range the highest level of average EMSE value was found to belong to the specimen which had the lowest level (finest conductive fiber) of conductive fiber content. For knitted specimens at the 860 MHz -960 MHz and 1750 MHz - 1850 MHz frequency the level of average EMSE value was found to be similar among all four knitted specimens. Average attenuation values of all knitted specimens were found to be lower than those of the woven specimens for both frequency ranges. It was also observed that the highest attenuation was obtained at 1790 MHz with the woven specimen, which had the finest conductive fiber.

Sandrolini and Reggiani (6) tested five electrically conductive woven and non-woven fabrics by circular coaxial transmission line holder in the frequency range 300 kHz-3.6 GHz. The results showed that a low surface resistivity value was a requirement for a higher shielding effectiveness especially for woven materials.

Özdekir et al. (7) were examined the electromagnetic shielding effectiveness of twill and cellular woven conductive fabrics made with stainless steel core yarns at different weft densities by free space measurement technique at horizontal polarization of the antenna. It is observed that woven fabric samples, which have been positioned so that the weft yarns have been parallel to the antenna polarization, have shown good electromagnetic shielding performances in high frequency band, namely industrial, scientific and medical band.

In this study it is investigate the EM shielding effectiveness of single jersey fabrics and to compare those with different conductive yarn density in the structure.

2. Materials and Methods

2.1. Materials

The yarns used in the tests (conductive and cotton) are shown in the Table 1.

Table 1. Yarns Specifications

Materials	Yarn count	Composition	Conductivity
Conductive yarn	12/1x275/100z	Bekinox VN Stainless steel	1 Ω/cm
Background yarn	30/2Nm	100% Cotton	-

The experimental samples were still knitted on a STOLL CMS computerized flat knitting machine (H. Stoll GmbH & Co.KG). To make the sample specifications uniform, the knitting parameters for producing the samples were as listed in Table 2.

Table 2. Sample specifications

Sample	Knit structure	Composition
1	Single jersey	2 /2 (Cotton/SS)
2	Single jersey	4/2 (Cotton/SS)
3	Single jersey	8/2 (Cotton/SS)
4	Single jersey	16/2 (Cotton/SS)
5	Single jersey	32/2 (Cotton/SS)

In this research, 5 types of knitted fabric samples on five types have been produced in STOLL CMS computerized flat knitting machine (H. Stoll GmbH & Co.KG) 100% using cotton and stainless-steel yarns. The specifications of yarns are given in Table 1. Knitted patterns are shown in Figure 1. The conductive yarns have inserted in certain intervals to obtain different open grid structures of conductive yarn within the fabrics, which resulted in different conductive yarn densities. In the sample No 1 there a repetition of 2 cotton yarns after 2 conductive yarns, in the sample No 2 there are a repetition of 4 cotton yarns after 2 conductive yarns, in the sample No 3 there are a repetition of 8 cotton yarns after 2 conductive yarns, in the sample No 4 there are a repetition of 16 cotton yarns after 2 conductive yarns and in the sample No 5 there are a repetition of 32 cotton yarns after 2 conductive yarns. The characteristics of the conductive fabrics are shown in Table 2.

2.2 Measurement of Transmission Coefficient and Electromagnetic Shielding Effectiveness (EMSE)

Electromagnetic shielding effectiveness (EMSE) is defined as the ratio of the incident to transmitted power of the electromagnetic wave [8]:

$$EMSE\big|_{dB} = 10\log\left(\frac{P_i}{P_t}\right) = 10\log\left(\frac{E_i}{E_t}\right)^2 = 20\log\left(\frac{E_i}{E_t}\right), \tag{1}$$

where P_i, P_t are the incident and transmitted power in W/m^2, and E_i, E_t are the incident and transmitted electric field in V/m, respectively.

The transmittance or S_{21} scattering parameter of the material is defined by [9]

$$S_{21} = \frac{E_t}{E_i}. \tag{2}$$

As such, the transmittance of the material in dB is equal to

$$S_{21}\big|_{dB} = 10\log\left(\frac{E_t}{E_i}\right)^2 = 20\log\left(\frac{E_t}{E_i}\right). \tag{3}$$

Therefore, the EMSE and the transmittance of the material are related with the following equation:

$$EMSE\big|_{dB} = -S_{21}\big|_{dB}. \tag{4}$$

From the above discussion it can be deduced that a large EMSE corresponds to a low S_{21} and, hence, better electromagnetic shielding.

Furthermore, the well-known transmission line technique (with a waveguide) was used to measure the EMSE of the provided fabric samples. We used a Vector Network Analyzer (VNA), model Rohde & Schwarz ZVA24 together with a WR430 (WG9) waveguide set of components (straight waveguides and coaxial to waveguide adaptors). The technique consists in launching a radiofrequency (RF) signal at the one end of the waveguide apparatus and measuring the output signal at the other end, as illustrated in Figure 1(a). The fabric sample is placed in between two straight waveguides. The transmitted signal is measured both with and without the fabric sample in place and the EMSE is calculated as the difference between S_{21} for the two different measurements.

A photo of the testbed with a fabric sample placed between two straight waveguides is displayed in Figure 1(b). It should be noted that according to widely accepted guidelines, the metallic part of the waveguides should be electrically insulated vs. the fabric sample. For this reason, the inner part of the waveguide flanges attached to the samples were covered with a thin plastic insulator film.

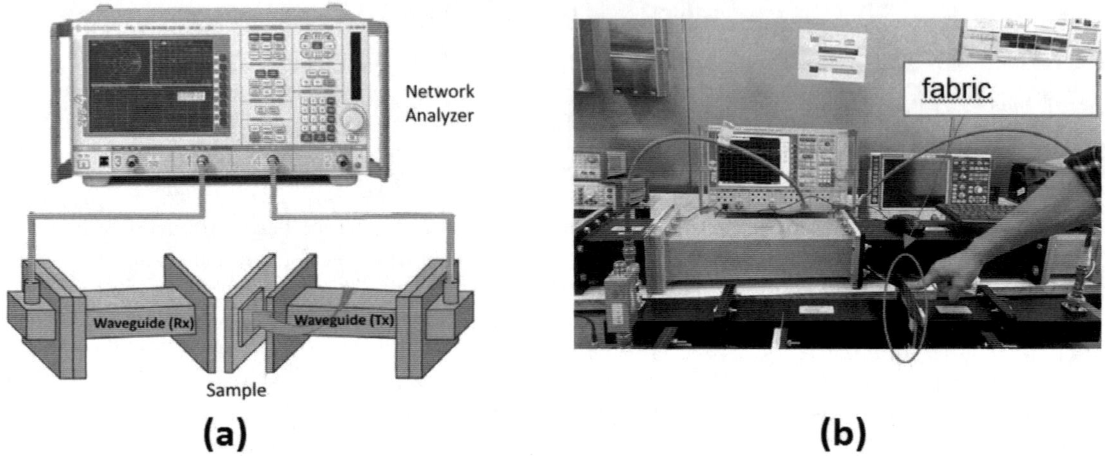

Figure 1: (a) Schematic of the testbed used for EMSE measurements and (b) photo of the testbed with the fabric sample placed and secured between two straight waveguides

3. Results and discussion

The transmittance for the 2-3.5 GHz frequency range of fabric samples according to the orientation of conducting threads in parallel or in perpendicular to the large dimension of the waveguide cross section has been examined.

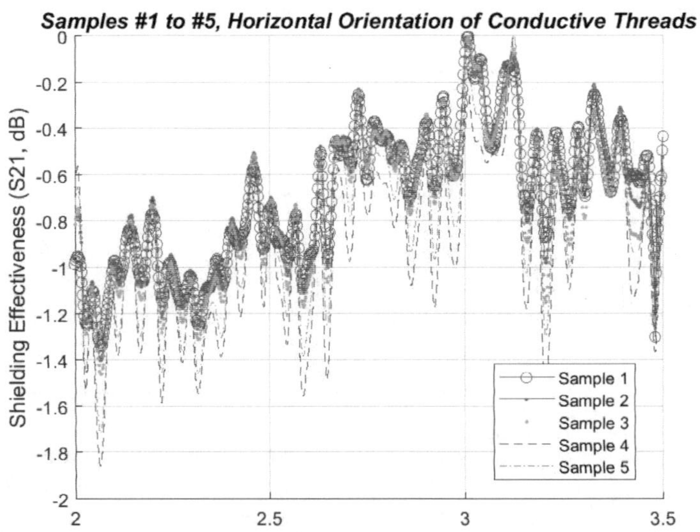

Figure 2: Transmittance of fabric samples #1 to #5 for the 2-3.5 GHz frequency range; orientation of conducting threads parallel to the large dimension of the waveguide cross section

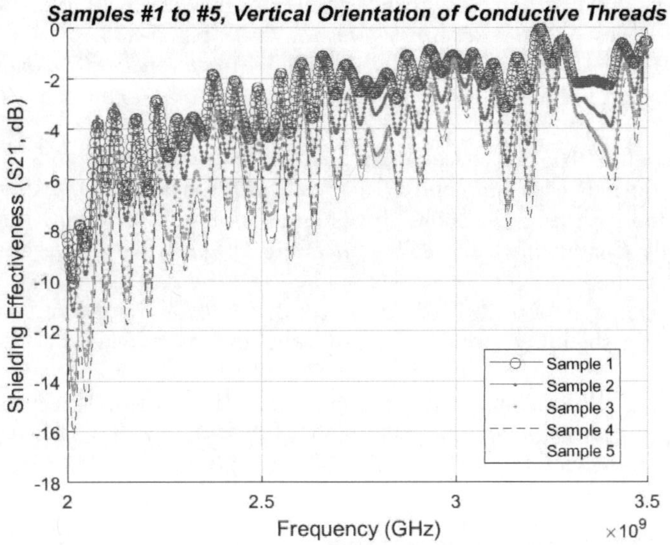

Figure 3: Transmittance of fabric samples #1 to #5 for the 2-3.5 GHz frequency range; orientation of conducting threads perpendicular to the large dimension of the waveguide cross section

Figures 2 and 3 demonstrate the transmittance or, equivalently, the EMSE of the five provided fabric samples for the 2-3.5 GHz frequency range with the transmission line technique. Since the conductive threads in the provided samples are forming parallel lines, it was decided to measure the transmittance for both parallel and perpendicular orientation of the lines with respect to the large dimension of the waveguide cross section. This choice was proven to be valuable since there is a significant effect of orientation to the resulting EMSE.

Indeed, Figure 2 illustrates that the maximum EMSE is equal to ~1.9 dB for a frequency close to 2 GHz and sample #4. It can be seen that EMSE diminishes with increasing frequency; this should be expected since for larger frequencies the corresponding wavelength is smaller and, thus, it can be easier transmitted for a given conducting threads grid.

On the other hand, Figure 3 illustrates the maximum EMSE is equal to ~-16 dB for the same frequency and sample #. Even more, all samples exhibit larger EMSE for this orientation.

4. Conclusions

In this paper, the electromagnetic shielding effectiveness of single jersey fabrics produced on STOLL CMS computerized flat knitting machine (H. Stoll GmbH & Co.KG) 100% using cotton and stainless-steel yarns. have been investigated. The setup includes the positioning of the fabric samples according to the orientation of conducting yarns in parallel or in perpendicular to the large dimension of the waveguide cross section.

It has been proved that the orientation of the conductive yarns has play a significant role in the EMSE

Nevertheless, the EM shielding properties of these fabrics should be improved especially in low and medium frequency bands. Moreover, it is supposed to model the EMSE of conductive knitted fabrics with simulation programs to clear up the circumstances referred above.

References

[1] Gandhi, O. P., 2002, "Electromagnetic Fields: Human Safety Issues", *Annual Review of Biomedical Engineering*, No.**4**, p.211-234.

[2] Perumalraj, R. et al., 2009, "Electromagnetic Shielding Effectiveness of Copper Core-Woven Fabrics", *The Journal of the Textile Institute*, **100(6)**, p. 512-524.

[3]. Roh, J.S. et al., 2008, "Electromagnetic Shielding Effectiveness of Multifunctional Metal Composite Fabrics", *Textile Research Journal*, **78**(9), p.825-835.

[4]. Duran, D. et al., 2009, "Electromagnetic Shielding Effectiveness of Woven Fabrics Produced with Conductive Core Yarns", *AUTEX World Textile Conference*, May 26-28, İzmir, Turkey, p.779-786.

[5]. Palamutcu, S. et al., 2010, "Electrically Conductive Textile Surfaces and Their Electromagnetic Shielding Efficiency Measurement", *Tekstil ve Konfeksiyon*, **20**(3), p.199-207.

[6]. Sandrolini, L. and Reggiani, U., 2008, "Investigation on the Shielding Effectiveness Properties of Electrically Conductive Textiles", *Asia Pacific Microwave Conference*, December16-19, Hong Kong, China.

[7] Hakan Özdekir, Ahmet Özkurt, 2013, The effects of weave and conductive yarn density on the electromagnetic shielding effectiveness of cellular woven fabrics, *Tekstil ve Konfeksiyon*, **23**(2)

[8] Maity, S., Singha, K., Debnath, K., Singha, M., "Textiles in Electromagnetic Radiation Protection", *Journal of Safety Engineering* 2013, **Vol. 2**, No. 2, pp. 11-19, 2013.

[9] Pozar, D. M., Microwave Engineering, Wiley, 4th Ed., 2011.

IOP Conf. Series: Materials Science and Engineering **459** (2019) 012062 doi:10.1088/1757-899X/459/1/012062

Studies on manufacturing of pajamas for patients with extensive burn surfaces

C D Radu[1], M Blaga[1], C Tamas [2], V C Macarel[1], V Rotaru[1] and R M Ivanus[1]
[1] Technical University "Gheorghe Asachi" Iasi/ Department of Textile, Leather and Industrial Management, 12 Mangeron street, Iasi zip code 700050
[2] University of Medicine and Pharmacy " Gr.T.Popa" Iasi, Clinic of Plastic Surgery, 16 Universitatii street, Iasi, Romania

rcezar2010@yahoo.com

Abstract. Paper carries out the manufacturing of pajamas made of different knitting structures. One emphasized on criteria necessary to fulfill the medical requirements according with patient critical condition. Considering burns wounds of 3rd degree one has in view the need of thermal insulation (the remaining skin surface does not cover up the ability of thermal insulation and patient is confronted with hypothermia), keeping sepsis condition liquid feeding and a specific pharmacodynamics therapy. The paper has in view two lines one is to obtain a pajamas based on high comfort for hyper sensitized and damaged skin made by means of specific knitting systems and secondly to promote on inner surface of pajamas a specific biomaterial able to speed up healing and fast recovery of burn wound. There presented detailed on progress research obtained on a specific biomaterial and on main lines to obtaining a clothing system able to promote a medical therapy.

1. General background

The state of a patient with deep burns is critical and there is enough room in the therapy field for improvement and innovation. Clinicians consider that the main elements to influence the patient evolution are the depth and surface of the lesions, the sepsis and the promptness of medical intervention. The ideal textile cloth specific for a patient with burns, has to meet the following requirements:

i) the main biological systems of the body are severely damaged;
ii) the pain associated with burn injuries is so high that an optimal function of the nervous sistem cannot be maintained;
iii) the absence of skin barrier allows continuous heat loss leading to an unbalanced methabolysme caused by hypothermia;
iv) the body loses a significant volume of liquids and proteins (electrolytes, enzymes, blood, lymph, etc) whith a marked effect the cells methabolysme,on the circulatory and excretory systems;
v) the protective function of skin against patogen agents is damaged, the wound becoming a large gate for bacterias; on the other hand, the excretory function of the teguments is altered and body toxins are not anymore neutralised; there is a huge pathogenic agression towards the whole body and the patient is not able to fight against it;
vi) because of the severe changes, the functions of the nervous sistem are near colapse.

Under these circumstances the clinician needs to form a strategy that covers all the demands of the patient.

2. Principles of treatment for burn injuries

The treatment of patients with burns is complex, with two main complementary aspects: general and local. The burn injury is defined by a continuous evolution, mainly during the first 72 hours after the incident, with a high risk of extension, in both surface and depth. A correct fluid ressuscitation is decisive in the further evolution of the wounds.

The surgical treatment of a burn injury depends on the general status of the patient and the local aspect of the wound.

For the superficial lesions, with good resources for spontaneous regeneration, conservative treatment is indicated. Clinical studies stated that a correct fluid resuscitation and the maintenance of a moist medium arround the wound, can limit the agravation af the local injury. The use of ointements on the burned surface has the following purposses: to calm the continuous pain, to hidrate the tissues, to stimulate the deterssion and to optimally control the bacterial collonisation (density and composition). The most paintful moment of the day for a burned patient is the bandage replacement. Any applicable solution that can minimize the pain (special gauses with lipids or oils to avoid the adherence to the wounds, notpaintful ointments) are useful. On the same note, any substance that stimulates the skin regeneration is welcome (skin growth factor, vitamins). Chemical substances with antiseptic properties, but not agressive towards the fragile newly formed skin layer, are very usefull in the fight against bacterias (silver, iodine)[1-4].The pH of the wound is very important, too and can be influenced by substances added to the bandage.

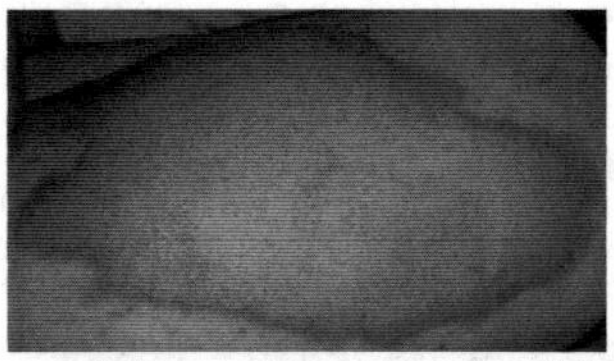

Figure 1. Second degree burn on the thigh (first day)

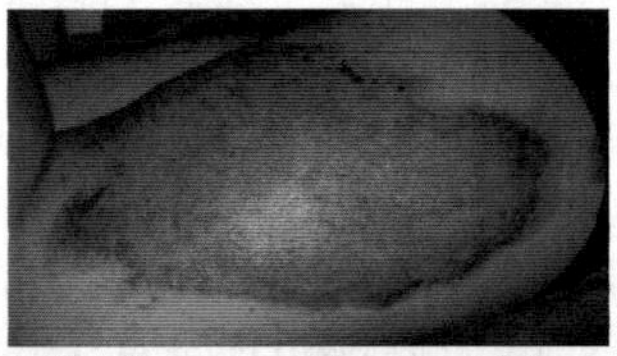

Figure 2. Conservative local treatment

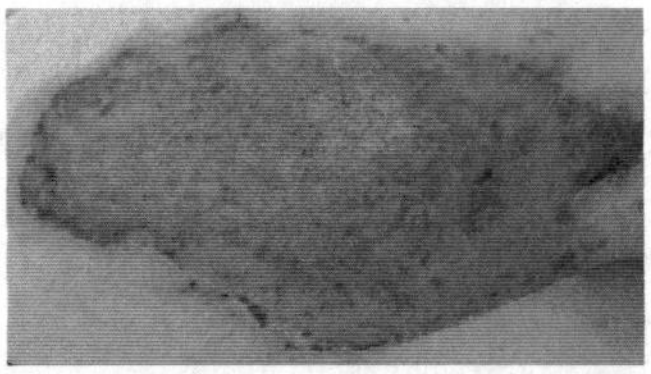

Figure 3. Same injury ,complete healing (day 20)

The ideal approach for the deep wounds is: early excision and grafting. This option is not always available, mainly when the burned aria occupies more than 20% of the body surface. There are many advantages of early excision of the damaged skin layers in deep burns (within the first 3 to 5 days after burn injury): less bacterial contamination of the tissues, functional dermal vascular network, less bleeding after excision. For circumferential deep burns with limbs localisation,this technique manages to avoid the compartment sindrome. Early excision and grafting will always guarantee the rapid wound healing and will avoid general complications and death.

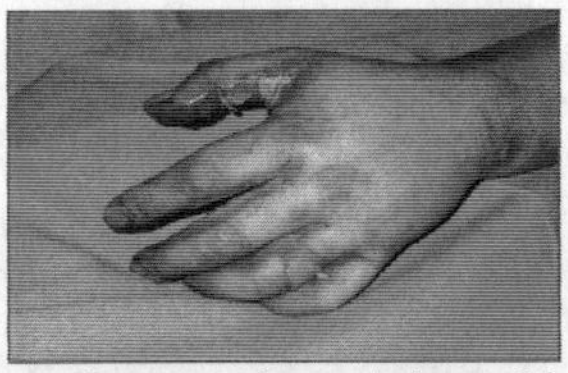

Figure 4. Deep burn injury (day 1)

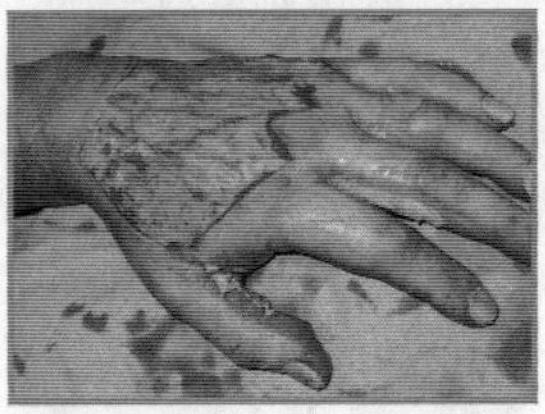

Figure 5. Early excision

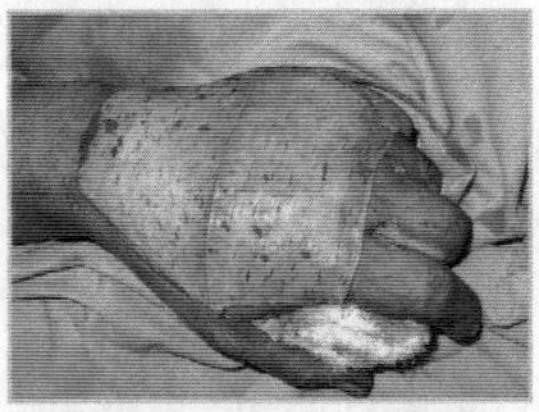

Figure 6. Early grafting

The grafted aria and the donor zones (surfaces where grafts are harvested from) have particular bandage requierements.Within the first 5 postoperatory days ,the grafts need to be covered with nonadherent gauses (grasolind) and moist, compressive bandages.The donor zones,with active,dermal bleeding,need special haemosthatic gauses.The mobile segments of the body (as hand and fingers) need bandages that allow all the active movements,in order to avoid complications like muscular atrophy and joints blockage [5-8]. The burned patient usually presents arias with different types of injuries,so a special bandage or „pajamas" destinated to cover the body surface has to meet the varied requierements corresponding to each leasion and to offer thermal protection.

3. Textile solutions to improve the condition of a patient with burn wounds

To built up a garment able to be worn on wound is a sensible task but in the near future textile garment will provide a solution for some acutely requirements neccesary to accomplish in the therapy od skin damaged by burn. Our research still in progress has to solve obtaining of a garment in the shape of pajamas made by a blouse and trousers in a new concept.

According with our design the blouse and trousers has to be performed under the folowing general shape illustrated in figure 7 shown below.

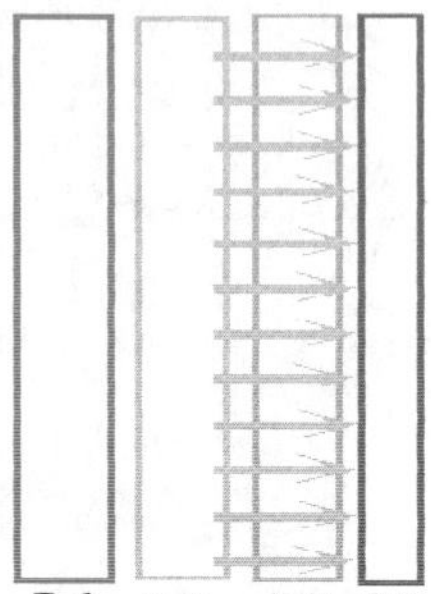

Figure 7. Principle outline of a pajamas fabric

The general concept of a pajamas is made by three layers: L1, L2, and L3. The pajamas fabric covers the whole surface of body skin and on the wound location the structure is shown in the figure 1 from above. The pajamas is customised according with disposal the wound location. The garments has to be ready in a proper period of time to assist the skin damaged. The three layers are ready before the patient entrance in clinic. The clinician provides information on the map and surface of wound skin in an electronic shape to textile company. By using a specific software the company already knows the dimensions and the way of linked different parts in the structure shown in figure 7.

Type of fabric inserted in each layer is the result of the research still in progress. But layer 1 is an polyester woven structure; the layer 2 is a knitted structure and layer 3 is a nonwoven fabric.

4. Conclusions

The goal of research is to help the work of surgery tasks and to ameliorate the patient comfort. The main lines exposed have to be obtaining by some reasonable textile solutions. On the other hand a tendency consider the need of customisation the main parameters according with specific aspects of each patient and if the cost of such pajamas. Up to now we have to determine the criteria which have much influence on obtaining an even diffusion of both drug and solution which need more accuracy. The research is in progress.

References

[1] Proksch E, Brandner JM, Jensen JM.,The skin: an indispensable barrier. Exp. Dermatology. **17**(12), 2008, p.1063-72

[2] Shores JT, Gabriel A, Gupta S. Skin substitutes and alternatives: a review. Adv Skin wound Care. Sep; **20**(9), 2007, p.493-508. Review. PubMed PMID: 17762218

[3] Sheridan RL, Tompkins RG. Alternative wound coverings. Total Bum Care, Elsevier, 2007, Cap 16, Pag 239-245;

[4] Kagan RJ, Robb E, Plessinger RT. The skin Bank. Total Burn Care, Elsevier 2007,Chapter 15, p. 229-239.

[5] Shores JT, Gabriel A, Gupta S. Skin substitutes and alternatives:a review. Adv Skin wound Care. 2007 Sep; **20**(9ptl):493-508. Review. PubMed PMID: 17762218

[6] Burn Statistics | Burn Injury Guide burninjuryguide.com/burn-statistics

[7] https://evidencebasedpractice.osumc.edu/.../ABLSProviderManual_20101018.pdf

[8] ameriburn.org/wp-content/uploads/2017/05/burncenterreferralcriteria.pdf

IOP Conf. Series: Materials Science and Engineering **459** (2019) 012063 doi:10.1088/1757-899X/459/1/012063

On the Thermal and Pressure Characteristics of Sport Compression Garments

B Sarı and N Oğlakcıoğlu

Ege University, Department of Textile Engineering, Izmir, Turkey

E-mail: nida.gulsevin@ege.edu.tr

Abstract. In this study, various knitted fabric structures were designed to help preventing muscular difficulties in sports. The samples, containing channels in widthwise direction, were produced by jacquard knitting technique. The effects of channel size and quilted inlay design were investigated in terms of pressure behaviour, thermal and water vapour resistance. Increasing of channel size and adding inlay yarns provided higher thermal resistance, as well as an increase in water vapour resistance, they also significantly affected pressure characteristics of samples.

1. Introduction

Sport is one of indispensable activities of daily life and it is becoming a more common living standard with increasing worldwide interest day by day. The sportive activities that constitute the basic building block of the healthy lifestyle dominated in our time have triggered the development of a unique production market. Textile products that are used in sports need to carry many features apart from casual wear. Generally, desired properties are comfort, freedom of movement, breathability, easy care, durability, body fit and lightness [1].

In addition to the “required performance characteristics” in sport textiles, the other important topic is “athlete's health”. As it is known, the human body is influenced by both the environmental factors and the risks in sport activity. Environmental factors occur usually with the effects of nature, such as weather conditions, cold and hot climates and sport field [2]. Risks of sportive activity are the extreme forces applied to the body during sports, such as various muscle and skeletal soreness, collision and falling. Especially in sports such as athletics, football and basketball, which require a high level of energy, the muscles are exposed to various difficulties.

Products with compression effect have both a performance boosting function and a protective function that allows the muscles to work more regularly. Compression products support the muscles with extra forces that are created from outside, tightly wrap the muscles to prevent more energy from vibrating and accelerate blood flow [3]. In addition to these effects, the use of compression products during pre-sports or during sports also increase muscle temperature. The warming of the muscles prolongs the muscular length for making the desired movement easier and the muscular reaction time is shortened in the nerves that control the muscles. Thus, the risk of muscle injury reduces.

During sport, the body produces excessive heat that will affect athlete’s health and durability. Therefore, the body should be in thermal balance to eliminate heat-load. In terms of thermal comfort, a successful product should transport heat, vapour and liquid sweat to the outside as fast as possible and help to control the body's thermal balance. In addition, the sport product must perform different tasks depending on the ambient temperature [4].

This study aimed to investigate thermal comfort and pressure behaviour of special quilted knitted fabrics, which were designed to use in compression garments. The goals were compression effect that could support muscular systems and thermal insulation to warm muscles before or during sports.

2. Experimental

Fabric design (channel size) and inlay design are selected parameters for this investigation. Experimental design is given on Table 1

Table 1. Experimental design

Fabric Type	Fabric design (channel size)	Inlay design
A1	5 mm	Empty
A2	7.5 mm	Empty
A3	12.5 mm	Empty
B1	7.5 mm	1x1
B2	7.5 mm	3x1

In the scope of study, the transverse channelled fabric samples are knitted as raw material 150 denier PET, 70/40 Spandex yarns and as quilted inlay yarn 600 denier PET at Monarch VLEC 6SC E20 38" electronic circular knitting machine. Water vapour and thermal resistance of samples were tested in accordance with related standards by using Permetest. The pressure behaviour properties were determined by using Kikuhime pressure sensor with 25% reduction factor of pressure samples on a nonflexible cylinder (12 cm diameter).

3. Results and Discussion

Test results reporting the fabric characteristics, water vapour resistance, thermal resistance and pressure values are given in Table 2. Effects of channel size and inlay design on results were statistically evaluated and significant differences were determined ($p<0.05$). Analysis of variance tests (ANOVA) were applied by using SPSS software.

Table 2. Test results of fabric samples

Fabric Type	Thickness (mm)	Mass per area (g/m^2)	Pressure value (mmHg)	Water vapour resistance (m^2Pa/W)	Thermal resistance (m^2k/W)
A1	1.52	312.17	12.36	3.55	0.0156
A2	2.03	321.62	12.90	6.13	0.0298
A3	2.58	340.36	10.75	12.95	0.0597
B1	2.46	456.68	19.40	9.68	0.0409
B2	2.62	520.45	21.70	11.12	0.0332

Pressure is defined as force per unit area. According to the analytical results, while the large channelled samples showed the lowest pressure value, similar pressure properties were observed for other channel sizes. The inlay yarn design caused a systematic difference on pressure characteristic. As the number of inlay yarn increased, higher pressure values were measured.

Water vapour resistance is a measure of the water vapour transfer through fabric. The results stated that the samples with smaller channel size and without inlay yarn provided better water vapour permeability among all samples, since thicker and compact fabrics constituted a handicap for permeability.

Thermal resistance is a barrier effect for heat flow from one to another surface of fabric. Similar to water vapour resistance, thermal resistance of the fabrics with larger channel sizes increased as well. Also inlay yarns improved thermal resistance effect compared to empty fabric design.

4. Conclusion

This study aims to investigate the effects of knitting structure (channelled effect) and inlay design on pressure characteristics, water vapour permeability and thermal resistance in order to help preventing muscular difficulties in sports. The results revealed that bigger channel size constituted both higher water vapour and thermal resistance, but they had lower pressure values. Increasing of inlay yarn amount provided higher pressure values. While the thermal resistance of the fabrics with inlay yarns was increased, the water vapor resistance values were also increased.

Acknowledgment

The authors acknowledge the Research Foundation of Ege University for the financial support given to this study (Project Number: 16-MÜH-089).

References

[1] Uttam D 2013 Active sportswear fabrics *International Journal of IT, Engineering and Applied Sciences Research (IJIEASR)* **Vol 2** No 1 pp 34-40
[2] Au K F ed. 2011 *Advances in knitting technology* (Woodhead Publishing Limited)
[3] Sarı B, Oğlakcıoğlu N 2018 Analysis of the parameters affecting pressure characteristics of medical stockings, *Journal of Industrial Textiles,* **Vol 47** No 6 pp 1083-1096
[4] Song G 2011 *Improving Comfort in Clothing* (Woodhead Publishing Limited)

Development of fibrous materials for shielding of electromagnetic radiation

L A Shcherbina[1], G Priniotakis[2], I S Haradniakova[1], P V Chvirou[1], Y V Bibikau[1] and I A Budkute[1]

[1]Mogilev State University of Food-stuff, Chemico-Technological Faculty, Department of Chemical Technology of High-Molecular Compounds, Av. Shmidt, 3, Mogilev 212027, Belarus
[2]University of West Attica, School of Engineering, Department of Textile Engineering, 250, Thivon & P. Ralli Str. 12244 Egaleo, Athens, Greece

E-mail: htvms@tut.by

Abstract. To analyze the possibility of creating polyacrylonitrile fibrous materials modified with carbon nanomaterials (CNM), the effect of CNM on the process of synthesis of a copolymer of acrylonitrile (AN), methyl acrylate (MA) and 2-acrylamide-2-methylpropane sulfonic acid (AMPS) in dimethylformamide (DMF) was studied. The deterioration of the kinetic characteristics of the process and the molecular mass properties of the obtained copolymer is established. The rheological properties of solutions of poly [AN-co-MA-co-AMPS] in DMF containing CNM were investigated. The effect of CNM on the spinning capacity of AN copolymer spinning solutions, the ability of the spun fibers to plasticization stretching and thermo-orientation one, as well as the physico-mechanical properties of the obtained fibers is shown.

1. Introduction

At present materials (including textile ones) with electrophysical properties that allow to screen electromagnetic radiation are of great interest. Screening of electromagnetic fields is an urgent task of protecting health, information security, electromagnetic compatibility and electromagnetic ecology of living quarters, protecting premises for servers and / or electronic equipment.

Materials, coatings, structures perform the function of shielding electromagnetic radiation, mainly due to its reflection. If radiation is reflected almost completely from metallic materials, only about 80% of the energy is reflected from the screens of conductive polymer materials, and 20% is absorbed. The ability of a medium to absorb electromagnetic radiation is determined by its electrical and magnetic properties which include specific electrical conductivity, dielectric and magnetic permeabilities [1].

At present for shielding electromagnetic radiation by means of reflection and absorption of electromagnetic energy the following items are used:

- polymeric materials with electrically conductive and magnetic components;
- polymeric electrically conductive coatings;
- structures using reflective and absorbing electromagnetic radiation of polymeric materials and coatings.

IOP Conf. Series: Materials Science and Engineering **459** (2019) 012064 doi:10.1088/1757-899X/459/1/012064

The necessary effect can be achieved by various methods. One of them is to introduce electrically conductive fillers such as carbon materials (carbon black, graphite, soot, fullerenes, carbon fibers and their various textile forms (yarns, strands, fabrics, tapes)) into the polymer base. Fibrous materials are being widely used to develop products with radio-absorbing and radio-shielding properties. When designing and manufacturing screens from fibrous materials, high technological efficiency of such products is an undoubted advantage, which, together with high productivity of modern textile equipment and relatively low cost of both raw materials and production process, makes the use of fabrics or knitted fabrics promising to create flexible screen designs and absorbers of electromagnetic radiation [2].

The use of carbon black in the form of soot is due to high dispersity of such materials. Electrically conductive carbon black can be represented as a substance formed from separate blocks. The smallest blocks are carbon particles in the form of microspheres with a diameter of approximately 30 nm, formed by layers of carbon atoms. They can be either whole or hollow. A large number of carbon particles forms the primary aggregates responsible for the structure of the substance. Their size varies from 100 to 1000 nm. Increasing the degree of filling leads to a sharp decrease in electrical resistance. However, at certain filling levels, the conductive network is saturated and further soot filling practically does not affect the electrical conductivity. As the particle sizes are reduced (down to nanoscale), their content in the compositions decreases to achieve the required level of electrical conductivity. In this connection, it is promising to use carbon nanomaterials (fulleroid multilayer carbon nanoparticles, carbon nanotubes, nanofibres, etc.) in radio absorbing materials [1].

2. Materials and methods

Carbon nanotubes (CNT) and technical carbon (TC) were used as additives in the study. Carbon nanotubes consisted of nanocones with a diameter of 10 to 60 nm, connected in chains up to 1000 nm in length. Technical carbon contained up to 50 % of particles 30-60 nm in size, had an average particle size of 30-35 nm and had a specific adsorption surface of 84 ± 4 m^2/g. The structure of the TC was of spherical formations, presented by graphene layers as a "cabbage head".

Two methods were considered for the preparation of polyacrylonitrile fibers modified with carbon nanomaterials (CNM):

– introduction of highly disperse carbon additives at the stage of synthesis of the fiber-forming copolymer of acrylonitrile (AN), methyl acrylate (MA) and 2-acrylamide-2-methylpropane sulfonic acid (AMPS);

– introduction of additives into the spinning polymer solution.

Synthesis of model copolymers of AN was carried out by the homophase free-radical synthesis method under conditions simulating a standard production process for obtaining in the reactor an ideal mixing of a continuous type of fiber-forming copolymer in dimethylformamide (DMF). The reaction mixture contained 90 % AN, 9 % MA and 1% AMPS (based on the weight of the monomers). As the initiator of free radical synthesis, azobisisobutyronitrile (AIBN) was used. As the CNM, TC was used, which was introduced into the reaction mixture after preliminary dispersion into DMF by ultrasonic action with a frequency of 22 and 44 kHz. The synthesis temperature was 70 °C and 80 °C. Dynamics of the synthesis of poly [AN-co-MA-co-AMPS] was estimated by the gravimetric method. The molecular weight characteristics of the polymer samples obtained were evaluated by the viscometric method.

To assess the effect of the content of carbon additives of different morphology on the physic- mechanical properties of fibrous materials, model fibers were obtained by wet-molding from solutions in DMF. For this purpose, a specially developed laboratory pilot plant was used, which allows modeling various production processes for obtaining fibers and threads and studying the influence of various technological factors on the properties of finished fibrous materials. Spinning solutions were prepared on the basis of poly [AN (90%) -co-MA (9%) -co-AMPS (1%)]. The polymer content in the spinning solution was 21%. The path length of the fiber in the precipitation bath was 0.9 m, the temperature of the precipitation bath being about 12 °C. The precipitation bath consisted of 55 % (by weight) of DMF and 45 % (by weight) of water. Further, the fibers were stretched in a plasticizer bath, dried and stretched again. The plasticizing stretching temperature in a 5% aqueous solution of DMF was 95 °C.

The drying temperature and subsequent thermal orientation stretching of the fibers was 95 and 135 °C. respectively. As a result, at each stage of the technological process, fiber samples were obtained, which made it possible to evaluate the effect of highly dispersed carbon additives on the formation of the structure and properties of modified fibers. The tensile deformability, strength, elongation at break of model fibers were determined. The evaluation of the physico-mechanical properties of the fiber was carried out in accordance with the current standards. The spin-purity of solutions of poly [AN-co-MA-co-AMPS] in DMF was evaluated by determining the maximum rate of reception of fiber on the take-up tray from the precipitation bath before fiber breakage.

3. Results and discussion

From a practical point of view, the introduction of carbon modifiers into the fiber-forming polymer at the stage of synthesis could reduce the number of technological operations, and possibly increase the degree of its dispersion, stability and uniformity of the suspension. To assess this possibility, the synthesis of fiber-forming copolymers of AN with the introduction of CNM into the initial reaction medium was studied. The results of the study of the effect of (CNM) on the dynamics of the copolymerization of AN, MA, and AMPS in DMF are presented in figure 1. The quantitative analysis of the kinetic parameters of the synthesis of poly [AN-co-MA-co-AMPS] is shown in table 1.

The experimental data characterizing the dynamics of synthesis (figure 1) indicate, first of all, the extensification of the process of obtaining fiber-forming poly [AN-co-MA-co-AMPS] in the presence of CNM at all the synthesis temperatures. Based on the data of table 1, the extensification of the synthesis process is observed, primarily, as a result of an increase in the duration of the induction period of the polymerization process in the presence of CNM. Simultaneously, there is a slight decrease in the rate and its temperature dependence, as well as the overall conversion of monomers in the presence of CNM in the reaction mixture. The possible reason for the negative effect of CNM on the synthesis of poly [AN-co-MA-co-AMPS] can be that the CNM with its graphene (polynuclear aromatic) structure can act as a "trap" of free-radical particles generated as a result of the initiator decay, which reduces the efficiency of initiation.

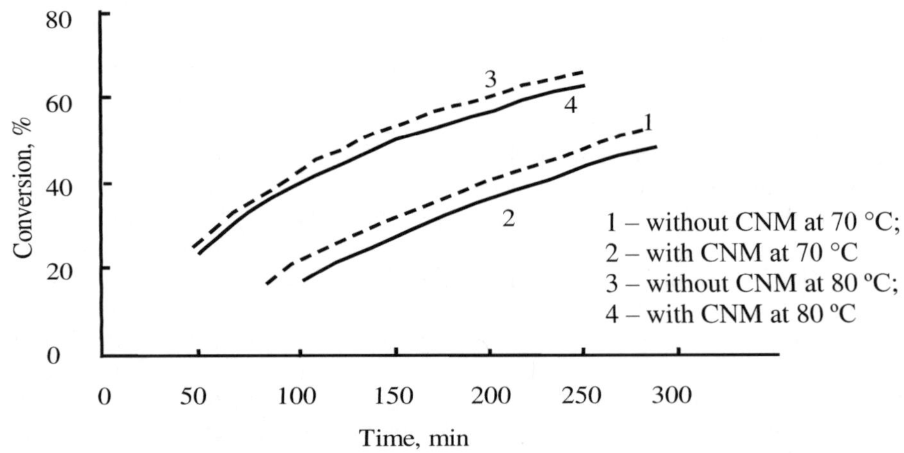

Figure 1. Dynamics of the synthesis of poly[AN-co-MA-co-AMPS].

This assumption is confirmed by the results of studying the rheological properties of dilute solutions of poly [AN-co-MA-co-AMPS] samples synthesized in DMF. Thus, from the data presented in table 1, it follows that in the presence of CNM in the reaction mixture, a significant increase in such an index of the molecular weight characteristics of the polymer as the intrinsic viscosity is observed. This may be the reason that the effective concentration of free radicals in solution in the presence of CNM decreases, and the trapped by CNM radicals may be the cause of the formation of branched macromolecular structures. Such structures should contain particles of CNM at the branch nodes. The significant growth of molecular weight and the presence of branched macromolecular formations make it difficult to process the polymer into fiber and do not provide the possibility of obtaining fibers of the required quality.

Table 1. Kinetic parameters of synthesis of poly[AN-co-MA-co-AMPS].

Indicator	without CNM		in the presence of CNM	
	70 °C	80 °C	70 °C	80 °C
Induction period, min	19,6	7,4	40,9	13,6
Copolymerization rate, $mol \cdot dm^{-3} \cdot s^{-1}$	$3{,}23 \cdot 10^{-4}$	$1{,}21 \cdot 10^{-3}$	$3{,}11 \cdot 10^{-4}$	$1{,}10 \cdot 10^{-3}$
Conversion of monomers over 300 min, %	57	64	54	60
Apparent activation energy, kJ/mol	140		120	
Characteristic viscosity, dl/g	0,91	0,76	2,43	1,81

One of the methods for modifying fibers with highly disperse materials is to introduce them into the spinning solution at the stage of preparation of the latter. However, the high tendency of CNM to self-agglomeration is a significant difficulty for their dispersion and reduces the stability of their dispersion. In this paper, this problem was solved in a complex manner by gradually increasing the viscosity of the concentrated dispersion of CNM in DMF. To this purpose the dispersion of CNM in DMF was subjected to alternate long-term action of ultrasonic oscillations of different frequencies, which was accompanied by the introduction of poly [AN-co-MA-co-AMPS] into the slurry of the CNM. The obtained CNM dispersion concentrate (CCNM) containing a certain amount of polymer was used to prepare spinning solutions intended for obtaining fibers with a specified content of CNM. The results of systematic microscopy of films cast from CCNM confirmed the high degree of dispersion of CNM in CCNM and the stability of this dispersion.

In preparing spinning solutions for the production of fibers modified with CNM, a significant increase in viscosity and anomalies in the rheological behavior of the spinning liquid was observed even when small amounts of additives were introduced (figure 2). This fact is very important, since it leads to high shear stresses during extrusion through the holes of the spinnerets and changes the behavior of the spinning-liquid jet drawn in the spin bath during fiber spinning.

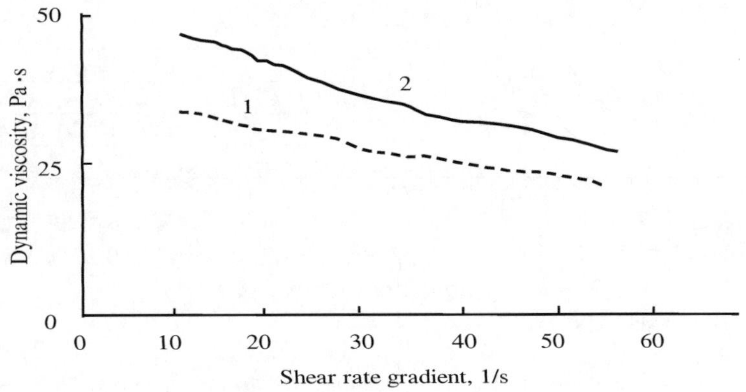

Figure 2. Rheograms of solutions of poly [AN-co-MA-co-AMPS] in DMF at 20 °C without CNM (1) and with 0.1 % of CNM (by weight of polymer) (2).

An increase in the content of CNM leads to a decrease in the ability of the jet of the spinning liquid to deform the stretching (figure 3). However, when the content of the modifier in the spinning solution was up to 0.1 % (by weight of the polymer), no disturbance in the filament stability was observed. Reduction of spinnability when introducing a modifier appears is due to the fact that with an increase in the number of particles of CNM and in their total surface area, the proportion of the polymer phase to which the filler effect extends increases. As a rule, the near-surface phase has a high packing density of polymer macromolecules, which reduces its ability for high viscous-flowing and highly elastic deformations. This is most revealed when the spinning liquid of a CNT is modified. It can be explained by much higher anisotropy of the CNT form, compared with the spherical particles of the TC. On the one hand, this creates more hydrodynamic resistance for deformation, and on the other hand, it also significantly increases the contact area between CNT particles and the total energy of their interaction with the polymer phase in the spinning liquid.

Despite the decrease in the spinnability of solutions of poly[AN-co-MA-co-AMPS] in DMF as the content of CNM increases and when the morphology of the carbon modifier is varied, practically there was no revealed the effect of these factors (table 2) on the maximum possible values of drawing ratio and basic physico-mechanical properties of fibers even with an increase in the content of CNM in the fiber to 1 %.

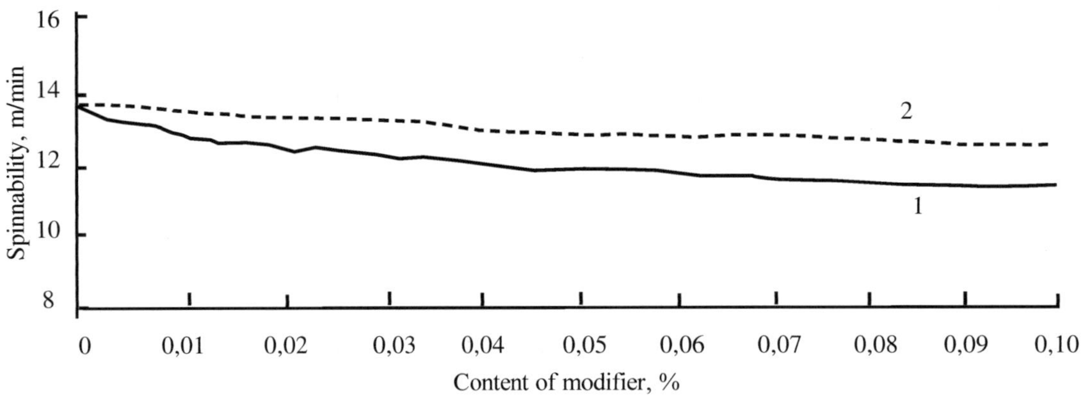

Figure 3. Spinnability of solutions of poly [AN-co-MA-co-AMPS] in DMF with CNT (1) and TC (2).

Table 2. Parameters for the production of modified fibers and their physical and mechanical properties of fibers.

Indicator	Modifier				
	–	CNT	TC	CNT	TC
		Content of CNM,% of polymer			
		0,1		1,0	
Maximum fold of drawing, times					
plasticization drawing	9,0±0,3	8,8±0,3	8,9±0,3	no data	no data
thermal orientation drawing	2,2±0,2	2,2±0,2	2,2±0,2	no data	no data
Plasticization drawing by 5 times					
Specific breaking load of fiber, cN / tex	38±2	44±2	43±2	45±2	44±2
Elongation at break,%	19,8±1	22,0±1	21,8±1	21,9±1	21,7±1
Plasticization drawing by 5 times, thermal orientation drawing by 2 times					
Specific breaking load of fiber, cN / tex	68±3	75±3	72±3	77±3	73±3
Elongation at break,%	14,3±0,5	13,9±0,5	13,5±0,5	13,6±0,5	13,4±0,5
Fiber color	white	light-gray		dark-gray	

At the same time, in order to make fibrous materials capable of shielding electromagnetic radiation, it is necessary to create conditions for having at least tunnel electrical conductivity. This requires the introduction of a larger amount of CNM into the polymer base.

Analysis of the possibility of increasing the content of carbon filler in the spinning solution by the example of the use of TC revealed the possibility of increasing this parameter to 15-20 %. In this case, the shrinkage of the filaments formed in the precipitation bath is substantially reduced, as well as spinning (figure 4) and drawing ratio of the fiber. The properties of such fibrous materials will be presented in other papers.

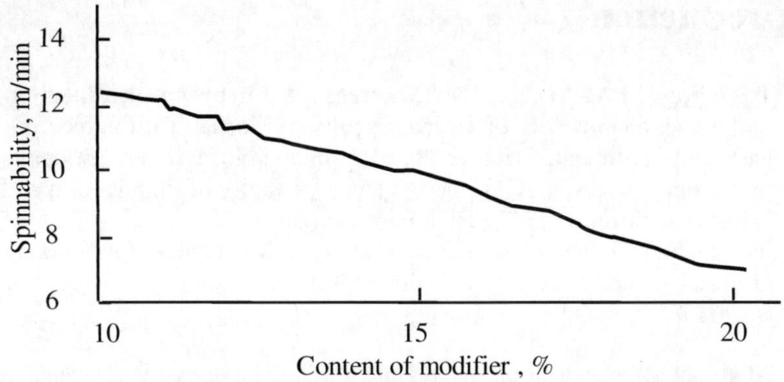

Figure 4. Influence of the TC content in the fiber on the spinner.

4. Conclusions

The study of the possibility of modifying fibers obtained from solutions of poly[AN-co-MA-co-AMPS] in DMF with the addition of CNM, with a view to further developing of the technology for the production of textile materials for shielding electromagnetic radiation, showed:

– the introduction of CNM into the polymer structure of the fiber at the stage of obtaining fiber-forming copolymers of AN by the method of free radical synthesis is not technologically advisable, since it substantially extensifies this process and impairs the technological characteristics of the fiber-forming copolymer;

– modification of the fiber by CNM is more expedient to carry out by dispersion in spinning liquids;

– the most effective way of dispersion of CNM in liquids is ultrasound treatment;

– the effective viscosity and anomaly of the viscous properties of spinning solutions containing CNM in the investigated range of the shear rate gradient and the shear stress are much higher as compared with unmodified solutions;

– the introduction of up to 1 % (by mass) carbon materials of nanometric sizes into the fiber and their morphology do not significantly affect the plasticization and thermo-orientation stretching of modified polyacrylonitrile fibers, as well as their specific tensile load and elongation at break.

– the maximum possible content of CNM in a fiber is actually limited by the spinnability of solutions of polymers, while the physico-mechanical properties of the fibers are also reduced.

References

[1] Mikhailin U A 2009 *Special polymeric composite materials* (St. Petersburg: Scientific bases and technologies) p 660

[2] Lynkov L M, Bogush V A, Borbotko T V, Ukrainets Ye A and Kolbun N V, 2004, New materials for electromagnetic radiation screens, *Reports of the Belarusian State University of Informatics and Radioelectronics*, **3**, 152-167

[3] Malik P, Sharma A, Gianender and Sharma J P, 2018, Textiles for protection against Electromagnetic Radiations: A review *J. of Engineering Research and Application*, **8**, **6**, 32-37

Adhesion of 3D printing polymers on textile fabrics for garment production

T Spahiu[1], M Al-Arabiyat[2], Y Martens[2], A Ehrmann[2], E Piperi[3] and E Shehi[1]
[1]Polytechnic University of Tirana, Faculty of Mechanical Engineering, Textile and Fashion Department, "Mother Tereza" Square, No. 1, Tirana, Albania
[2]Bielefeld University of Applied Sciences, Faculty of Engineering and Mathematics, ITES, Interaktion 1, 33619 Bielefeld, Germany
[3]Polytechnic University of Tirana, Faculty of Mechanical Engineering, Department of Production and Management, "Mother Tereza" Square, No. 1, Tirana, Albania
tspahiu@fim.edu.al

Abstract. Applications of 3D printing in fashion industry are numerous, including 3D printing geometries on textile fabrics, printing parts of the garments, or even the entire garment. A crucial point of making such polymer/textile combinations usable is the adhesion between both materials. Investigations related to this adhesion, however, are still scarce. Similarly, the influence of 3D printed patterns on the geometries and the mechanical parameters of textile fabrics are only rarely examined. Continuing our previous research, the experiments shown here illustrate the influence of different printing parameters on the adhesion between fabrics and 3D printing polymer. These results will support other researchers as well as fashion designers in developing new 3D garment shapes by combining common textile fabrics with the relatively new possibility of 3D printing.

1. Introduction

3D printing is an innovative technology with applications in different areas of production. These applications include the fashion industry, where examples can be found in various case related to 3D printing geometries on textile fabrics, printing parts of the garments or whole garments. Even though combinations of 3D printed objects and textile fabrics are innovative for the fashion industry, they are still rare compared to the traditional fabrics. Applications for footwear products are more often found than garments. This is related to the polymers which are more often used in footwear products for sole production [1-3].

Applications of 3D printing on textile fabrics strongly depend on a high adhesion of the 3D printed polymers on the textile fabrics. Investigations related to this adhesion, however, are still scarce [4-6]. Only few research groups have conducted investigations on the impact of textile and printing parameters on the adhesion between both materials. Especially the distance between nozzle and printing bed was found to significantly influence the adhesion [6], while other researchers underlined chemical influences or pretreatments of the textile fabrics [5,7,8]. In another study, several printing parameters were found to weakly influence the adhesion of an imprinted polymer [9]. More comprehensive investigations, however, correlating the different parameters, are still missing.

Similarly, the influence of 3D printed patterns on the geometries and the mechanical parameters of textile fabrics are only rarely examined [10]. This is why this paper depicts results of adhesion tests on different textile fabrics typically used for garments and shows some possibilities to create new 3D forms by adding 3D printed patterns to common textile fabrics.

2. Methods and Materials

In this work printing was performed using an FDM (fused deposition modelling) printer Orcabot XXL (Prodim International). The following printing parameters were used: nozzle diameter 0.4 mm, layer thickness 0.2 mm, nozzle temperature 200 °C or 230 °C, and printing bed temperature 60 °C or 100 °C. The distance between nozzle and printing bed was varied.

As printing materials, was used PLA, which has shown to have the highest adhesion in textile fabrics in previous investigations [6].

Different cotton woven fabrics of varying thickness and areal weight were used as the base for printing. Table 1 depicts the fabric parameters. Since for 3D printing with a fine nozzle on the fabric, the thickness measured by a micrometer caliper is more important than the value measured with the usual textile thickness tester, both values are given here. Opposite to the tests performed earlier [8], the fabrics are raw and were not chemically or mechanically pre-treated.

Table 1. Parameters of the fabrics used in this study. d_{micro} and $d_{textile}$ depict the thicknesses measured with a micrometer caliper and with a textile testing instrument (according to DIN EN ISO 5084), respectively. The areal weight is measured according to ASTM D3776.

No.	Weave	d_{micro} / mm	$d_{textile}$ / mm	Areal weight / (g/m²)
1	Plain	0.32	0.49	144
2	Plain	0.27	0.37	127
3	Plain	0.31	0.39	154
4	Twill 2/1	0.56	0.74	276

Rectangles of 250 mm x 25 mm area with a nominal height of 0.4 mm, i.e. two layers, were printed on the fabrics for the adhesion tests, using three printed samples per parameter set. Autodesk Inventor was used for 3D designing of the rectangles. The exported .stl files were imported in the repetier host for slicing them.

Adhesion tests were performed according to DIN 53530 and results were evaluated according to ISO 6133 defining the analysis of multi-peak results obtained in adhesion force investigations. The method for more than 20 peaks was used due to our experimental results.

Microscopic images were taken using a digital optical microscope VHX-600D (Keyence, Neu-Isenburg, Germany) with a nominal magnification of 50 x.

3. Results & discussion

Fig. 1 depicts one z-distance dependent measurement series (samples printed with a nozzle temperature of 200 °C and a printing bed temperature of 60 °C). The shape of the curve is typical for printing PLA on diverse woven fabrics [6]. With decreasing distance, the adhesion force increases, until in a certain distance (here: -0.2 mm) when the filament is blocked too strongly and cannot be transported to the nozzle anymore, so that printing stops. It should be mentioned that the z-distance of 0 mm means that nozzle and printing bed are in contact before the textile fabric is placed on the printing bed, i.e. the nozzle is printing *inside* the textile fabric to press the polymer with enough force into the open pores of the woven fabric.

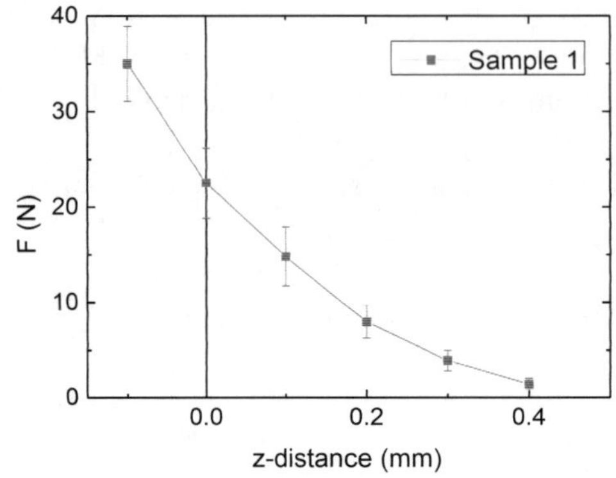

Figure 1. z-distance dependence of the adhesion force of PLA on sample 1.

To investigate the influence of the textile fabric thickness and weave, the same test series was performed with the other three woven fabrics under examination, using the same nozzle and printing bed temperatures. The results are depicted in Fig. 2, showing the z-distance dependent adhesion force (Fig. 2a) as well as the derivative, i.e. the slope of the curve (Fig. 2b).

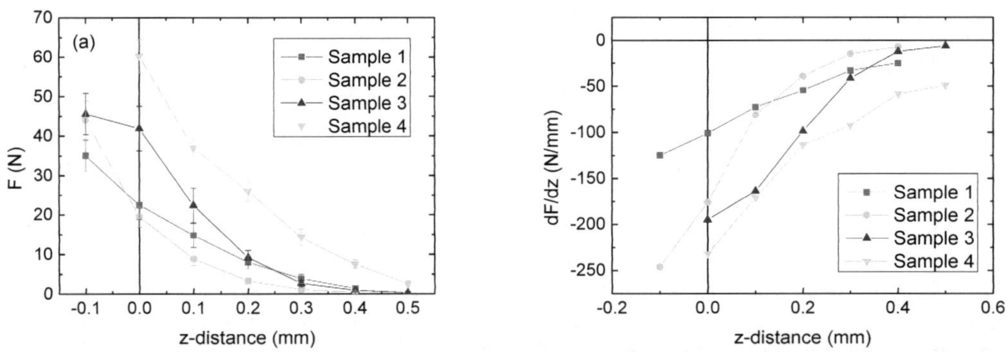

Figure 2. z-distance dependence of the adhesion force of PLA on the different textiles samples under examination (a) and corresponding derivatives dF/dz (b). For the calculation of the derivative of sample 3, the smallest z-distance (showing problems in the first printed layer) was ignored.

Firstly, it becomes visible that the graph of sample 4 – which is 0.25-0.3 mm thicker than the other fabrics – is shifted to higher z-distances, what could be expected. On the other hand, the slopes of the graphs differ. This is best visible by calculating the derivatives dF/dz (Fig. 2b).

For sample 2 – with the smallest areal weight – and sample 4 – with the highest areal weight, the largest absolute values are reached, i.e. the strongest changes of the adhesion force with the z-distance. Apparently, there is no correlation between the maximum of the derivative and the areal weight or the thickness.

Another possibly interesting value can be calculated by taking into account the different thicknesses, measured by the textile testing instrument and the micrometer caliper. If the ratio $d_{micro}/d_{textile}$ is defined as compression, the following compressions can be calculated for the samples used in this study: 0.65 (sample 1), 0.73 (sample 2), 0.79 (sample 3), and 0.76 (sample 4). Obviously, for sample 1 which can be compressed most strongly, the derivative has the smallest value. The less compressible samples have correspondingly higher maximum values of the derivatives. It should be mentioned that

printing on sample 3 with a z-distance of -0.1 mm resulted in clearly visible problems in the first layer. Here setting the z-distance to -0.05 mm may have led to a similarly high absolute value of the derivative as for samples 2 and 4.
Due to the relatively low amount of different compressibility values, it can only be estimated that the slope of the z-dependent adhesion force curve is correlated to the compressibility. This idea has to be investigated in more detail in the future.

Another interesting parameter is the temperature of the printing bed and the nozzle. The influence of increased temperatures was tested with sample 3 and is depicted in Fig. 3.

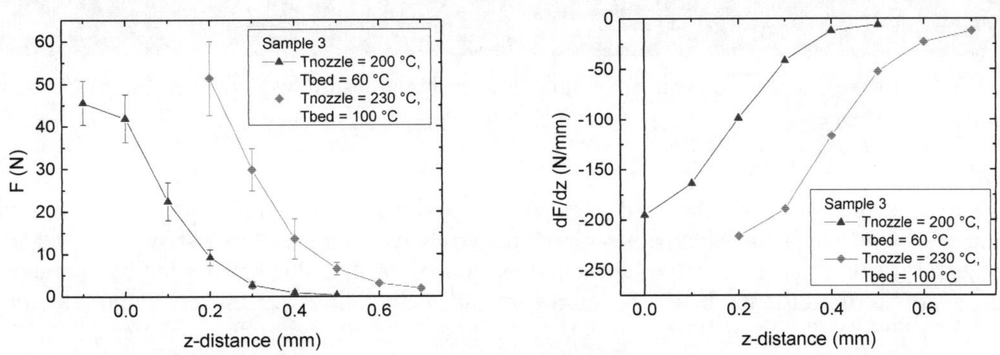

Figure 3. z-distance dependence of the adhesion force of PLA on sample 3, printed with different temperatures, (a) and corresponding derivatives dF/dz (b). For the calculation of the derivative, the smallest z-distance was ignored.

In an earlier investigation [11], we found a strong dependence of the adhesion force on the printing bed temperature as well as the nozzle temperature while letting all other parameters constant. Fig. 3 now reveals the reason for this finding. Apparently, the maximum possible adhesion force is not increased, and even the slopes of the curves of both temperature situations are nearly identical (Fig. 3b). Instead, the curves are shifted along the z-distance. It must be mentioned that after changing the temperatures, the distance between nozzle and printing bed was carefully re-adjusted to avoid erroneous modifications of this parameter.
This finding means that by increasing both temperatures and thus decreasing the viscosity of the molten PLA during printing, it is possible to print in an increased distance to the textile fabric. In this way, however, it is not possible to increase the maximum adhesion force since the z-distance in which the nozzle is clogged is also shifted to larger values.
Finally, Fig. 4 shows a comparison of the back of the 3D printed layer on sample 3 after the adhesion tests, comparing the maximum and the minimum distance. When printing in the maximum z-distance, the typical diagonal stripes of the printing process are clearly visible, with only a slight modification by the textile fabric below. Only a few fibers are stuck on the polymer. Oppositely, the back of the polymer printed in the smallest distance is fully covered with fibers, and in the polymer, the structure of the woven fabric is clearly visible. These images show clearly how strongly the z-distance influences the first printed layer.
It should be mentioned that the large amount of fibers visible in Fig. 4b suggests printing on a fabric woven from a filament yarn where – opposite to the cotton fabrics used here – the filaments cannot be pulled out of the fabric easily. In this way, it may be possible to further increase the adhesion between a 3D printed object and the textile fabric below.

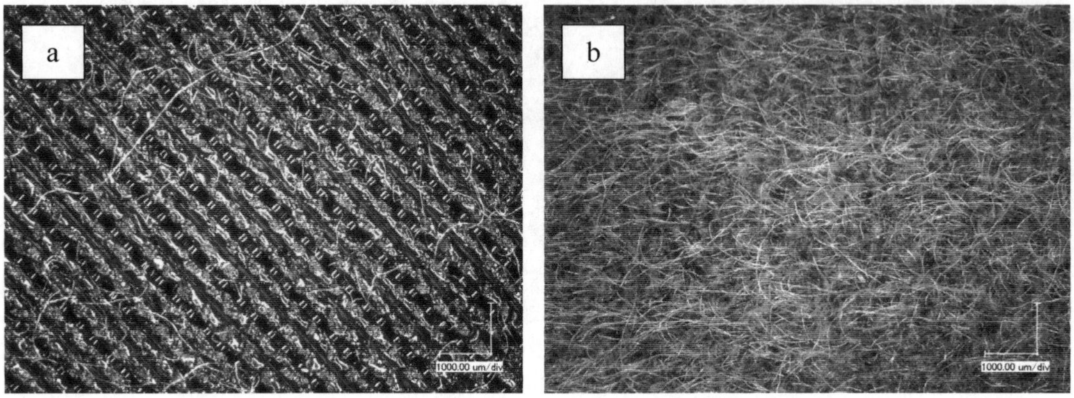

Figure 4. Back of the 3D printed layer on sample 3 after the adhesion tests. (a) maximum z-distance; (b) minimum z-distance.

Conclusions

3D printing on textile fabrics can be used to create composite garments in new 3D shapes. A crucial point is to ensure sufficient adhesion between both materials, making previous tests with the materials in use necessary. The results of our study give an overview of the most important printing parameters.

The z-distance, i.e. the distance between the nozzle and the printing bed, is very important for the adhesion on textile substrates. With decreasing distance, the adhesion force increases until the minimum distance is reached in which the filament does not clog the nozzle. By reducing the z-distance, the nozzle presses the 3D printing polymer with higher forces into the open pores of the woven fabric.

Other important parameters tested in this work are the temperatures of the printing bed and the nozzle. By increasing both temperatures and thus decreasing the viscosity of the molten PLA during printing, it was possible to print in an increased distance to the textile fabric. In this way, however, the maximum adhesion force could not be increased since the z-distance in which the nozzle was clogged was also shifted to larger values.

These results will support other researchers as well as fashion designers in developing new 3D garment shapes by combining common textile fabrics with the relatively new possibility of 3D printing.

References

[1] [Online]. Available: https://www.theverge.com/2017/4/7/15216724/adidas-3d-printed-sneaker-futurecraft. [Accessed August 2018]

[2] [Online]. Available: https://www.3dnatives.com/en/ecco-launches-3d-shoes-260320184/. [Accessed August 2018]

[3] [Online]. Available: https://all3dp.com/reebok-launches-latest-3d-printed-shoe-called-liquid-floatride-run/. [Accessed August 2018]

[4] Sabantina L, Kinzel F, Ehrmann A and Finsterbusch K 2015 IOP Conf. Series: Mat. Sci. Eng 87 012005

[5] Eujin Pei D, Pei E, Shen J and Watling J 2015 Rapid Prototyping J. 21 556-571

[6] Grimmelsmann N, Kreuziger M, Korger M, Meissner H, Ehrmann 2018 Rapid Prototyping J. 24 166-170

[7] Korger M, Bergschneider J, Lutz M, Mahltig B, Finsterbusch K, Rabe M 2016 IOP Conf. Series: Materials Science and Engineering 141 012011

[8] Kozior T, Döpke C, Grimmelsmann N, Juhász Junger I, and Ehrmann A 2018 Advances in Mechanical Engineering, *online first*

[9] Sanatgar R H, Campagne C, Nierstraz V 2017 Applied Surface Science 403 551-563

[10] Spahiu T, Fafenrot S, Grimmelsmann N, Piperi E, Shehi E and Ehrmann A 2017 IOP Conf. Series: Mat. Sci. Eng. 254 172023
[11] Spahiu T, Grimmelsmann N, Ehrmann A, Piperi E, and Shehi E 2017 Proc. of 1st International Conference Engineering and Entrepreneurship

An innovative method for reducing water use in process after dyeing

G Tandoğan and M Kertmen
İskur Tekstil Enerji Ticaret ve Sanayi A.Ş. Kahramanmaraş, Turkey

gokhan.tandogan@iskur.com

Abstract. Textile industries consume large volumes of water and chemicals for wet processing of textiles. Textiel wastewater damage the environment as they contain high concentrations of organic and inorganic chemicals. There is also a lot of waste water and energy consumption while processes. Reactive dyes react with cellulose to establish covalent bond and are water soluble. Because of these properties and low fix rates, they cause high levels of dyes in waste water. Reactive dyes which are not totaly fixed to the fabric throughout the dyeing process. Washing is required to remove un-fixed dye from the fabric, The control of water pollution has become of increasing importance in recent years. The control of water pollution has become of increasing importance in recent years. The release of dyes into the environment constitutes water pollution. Since conventional treatment methods are insufficient for purifying dyehouse wastewaters, new treatment technologies are developed and applied for its treatment. Nowadays, removal of dyes methods are often very costly and have many problems. For this reason, effective and economical alternative treatment methods are needed in removing large amounts of dyes from waste water. In this study, it is aimed to reduce the number of washing treatment after direct and reactive dyeing. It is planned to develop filters made from cationic fibers while the use for after treatment processes. When experiments are done, we observed that non-hydrolysis of the dyes trapped into the developed filters. As a result of these study, the wastewater likely to show better results such as; less energy and, less water consumption has been achieved.

1. Introduction

Textile industries consume large volumes of water and chemicals for wet processing of textiles [1]. Textile wastewater damages the environment as they contain high concentrations of organic and inorganic chemicals. There is also a lot of water and energy consumption while processes. Dyes are also one of the unwanted pollutants in the wastewater. These dyes are harmful and dangerous because of their carcinogenic properties [6]. Reactive dyes are one of the most preferred types of dyes in textiles. Reactive dyes react with cellulose to establish covalent bond and are water soluble [2]. Because of these properties and low fix rates, they cause high levels of dyes in waste water. Reactive dyes which are not totaly fixed to the fabric throughout the dyeing process. Washing is required to remove un-fixed dye from the fabric, the control of water pollution has become of increasing importance in recent years. The release of dyes into the environment constitutes water pollution [1]. The studies show that in the last few decades there has been a huge increase in the production and use of organic pollutants and as a result there is a major pollution threat. Although there are different methods for removing organic pollutants in wastewater, appropriate and cost effective methods should be preferred in terms of textile factories [5]. Since conventional treatment methods are insufficient for purifying dye from wastewaters so new treatment technologies are developed and applied for its treatment. Nowadays, methods of removaling dyes are often very costly and have many problems. For this reason, effective and economical alternative treatment methods are needed in removing large amounts of dyes from waste water [4].

2. Materials and Methods

Samples A, B, D, E, F, G, H and I were taken from rinsing water after standard washing at 55, 98, 98, 80 and 80 °C temperatures and the absorbance values of these samples were measured with UV-VIS Spectrophotometer. Samples of rinsing water are given in Figure 1.

Figure 1. Samples of rinsing water

3. Results and Discussion

As a result of the adsorption experiments, The COD test was carried out for samples A and sample I and found that 1640 mg / l and 176 mg / l values A and I, respectively. Accordingly, there was a decrease of 89% in COD value. The samples A, B, D, F, H were taken from rinsing water after standard washing at 55, 98, 98, 80 and 80 °C respectively without adding cationized fabric. Table 1 gives the samples and absorbance values. It was observed that there is a decrease in the absorbance value from the sample A to the sample H due to the washing process and the effect of the temperature, which means that the color intensity of the rinse water decreases. In addition, when the cationized fabric was added to the washing bath according to the same temperature and washing sequence, the absorbance values were lowered. Table 2 shows % reduction in absorbance values of samples.

Table 1. Samples and Absorbance Values

Sample Code	Sample Code	Absorbance Value	Sample Code	Absorbance Value
1. Washing at 55°C	A	>3.3		–
2. Washing at 98°C	B	>3.3	C	1.98
3. Washing at 98°C	D	0.937	E	0.292
4. Washing at 80°C	F	0.135	G	0.105
5. Washing at 80°C	H	0.109	I	0.056

Table 2. % Reduction in Absorbance of Samples

Sample Code	Sample Code	Sample Code	% Reduction in Absorbance Value Without Cationized Filters	% Reduction in Absorbance Value With Cationized Filters
1. Washing at 55°C	A		–	–
2. Washing at 98°C	B	C	0	40%
3. Washing at 98°C	D	E	71,6%	91,15%
4. Washing at 80°C	F	G	95,9%	96,82%
5. Washing at 80°C	H	I	96,7%	98,30%

Thus, for example, without the cationized fabric, the absorbance value after the third washing step decreased to 0.937 and the decrease in the absorbance value became 71,6%. On the other hand, when the cationized fabric was added to the washing water, this value was found to be 0.292 and the decrease in the absorbance value was 91.15%. Accordingly, it was possible to reduce the absorbance values at the beginning of washing steps by cationized fabric addition.

It is estimated that lowering the washing temperature or decreasing the number of washing steps after the third washing step will reduce the costs in terms of the factory and easier disposal of the waste water in terms of environmental health.

4. Conclusions

In this study, it is aimed to reduce the number of washing steps after reactive dyeing. It is planned to develop filters made from cationic fibers while the use for after treatment processes and washing processes. When experiments are done, non-hydrolysis of the dyes trapped into the developed filters. As a result of this study, the wastewater likely to show better results such as less energy and, less water consumption has been achieved.

References

[1] Robinson T and McMullan G and Marchant R and Nigam 2001 *Bioresource Technology,* Volume **77**, Issue 3, Pages 247-255
[2] Eren, H, Kurcan, P, Aniş, P 2007 *Tekstil ve Konfeksiyon,* **17** (2), 119-125.
[2] Chakraborty J N 2014 *in Fundamentals and Practices in Colouration of Textiles*
[3] Gümüş, D 2009 *Master of Applied Science Thesis,* Yıldız Technical University Istanbul
[4] K F O Alkan U *Uludağ Üniv. Müh. Mim. Fak. Dergisi*, No **1**, 47-55, 2002.
[5] Annadurai, G, Juang, R S, Lee, D J 2002 *Journal of Hazardous Materials,* Volume **92**, Issue 3, Pages 263–274.
[6] Asgher, M, Bhatti, H N 2012 Ecol Eng, Volume **38**, Pages 79-85.

IOP Conf. Series: Materials Science and Engineering **459** (2019) 012067 doi:10.1088/1757-899X/459/1/012067

Development of the calculation method of polymer compound mass to be applied onto the textile garment pieces

S Sh Tashpulatov[1], I V Cherunova[2], M K Rasulova[1], D D Inogamdjanov[3], M Yu Umarova [1], A D Daminov[1], U R Uzakova[1] and Sh G Jurayeva[1]

[1] Tashkent Institute of Textile and Light Industry, Design and Technology of Apparel Sewing Department, 5 Shokhjahon, Tashkent, Uzbekistan
[2] Institute of Service Sphere and Entrepreneurship (the Branch) of the Don State Technical University, Structuring, Technology, Design Department, 147 Shevchenko, Shakhty, Russian Federation
[3] Jizzakh Polytechnic Institute. Department of textile and light industry. Islom Karimov street #4, 130100. Jizzakh, Uzbekistan.
E-mail of the Presenting Author: weaving.uz@gmail.com

Abstract. The article is dedicated to the development of calculation method of a base line mass of a polymer compound to be applied onto the textiles with an added performance of friability relative to each other. The method of strengthening of the yarned joints of products from the range of fabrics of sparse structures is being considered. The obtained results of the research made possible to reveal the negative influence of such indicators as the friability and moving ability of the yarns on the quality of yarn crossections (move ability and friability of yarns in the fabric, it's tensile strength, stiffness, etc.) and the product as a whole, which make possible to use progressive processing methods during the selection of garment production technology. To ensure the required quality and reliability of yarned joints of garments made of highly expandable fabrics, the studies have been carried out. At the same time, as a control, a fabric was chosen without applying a polymer and another alternative variant was chosen as well, using the liquid-phase polymer compositions along the seams. The control and alternative variants of the samples were subject to comparative assessment according to the criteria of tensile strength, resistance to friability, flexural rigidity, air permeability and resistance to washing. As a result of the studies, the rational parameters of the technological conditions of applying the liquid-phase polymer composition onto the jointing stiches of the garment components were established to improve the performance properties.

1. Introduction.

It is known by history that within the territory of Uzbekistan, the amazingly beautiful, high-quality cotton, semi-silk, silk and other fabrics have been produced [1]. The traditions developed over several centuries had their rising ups and downs, but they never died. The manufacturing features have been passed from father to son, from teacher to student, and have been survived up to day. Nowadays, the traditional art weaving takes one of the leading places in the modern national art. National fabrics have not only historical and daily importance, but also great artistic value [2].

Despite all the positive aspects (appearance, air permeability, hygroscopicity, capillarity, hygiene, etc.), the classic national fabrics retain one negative feature - the yarns in the fabrics are prone to a high degree of moving and friability.

Resistance to friability and moving depends on the structure (interlacing) of the fabric formation, its fibrous composition, the structure of the yarn, the ratio of the linear densities of the warp and weft yarns, the structure phase of fabrics, finishes, etc., in short, on those factors that determine the strength of friction and mutual adhesion between warp yarns and weft in the process of manufacturing and finishing of fabrics [3]. For example, plain weave fabrics have less moving ability (compared to twill or satin weave fabrics) due to shorter and more frequent overlap between the warp and weft yarns. The stronger curvature of the warp yarns, compared with the wefts, leads to the fact that in the yarns the expansion occurs mainly by displacing the warp yarns against to the weft yarns. The value contributions to movement index of warp and weft and their total moving ability in cross-sections at all structure phases of fabric are summarized [13]. Therefore, the resistance to moving and friability can serve as an indirect indicator of the rationality of the structure of the produced fabric.

The works [4, 5, 6] have been devoted to solve the problems above, where the authors propose to decrease the moving and friability of yarns to increase the outsizes, to use additional operations on laying along the contour of sewing parts of adhesive glued materials, applying adhesive substances along the edges of the cut to prevent the friability of yarns.

However, the proposed methods for reducing the friability and moving of yarns in the fabric lead to the need to increase the outsizes for seams, as well as the application of additional technological operations of wet and heat treatment of parts in the sewing industry, which leads to an increase in the rate of yarns consumption due to additional parameters for seams and labor output ration of the technological process of manufacturing of sewing apparels.

The conducted studies [7], showed that the most common defects of sewing apparels made from textile fabrics are the friability and of the fabric yarns along the middle seam of the back, the relief seams of the back and the front, the seams of the hose connection with the armhole, the side and middle seams of the skirt. This circumstance makes necessary to conclude that there is a need for a deeper investigation of the friability of yarns in fabrics along the lines and development of more progressive ways to improve the stability of fabric structure and its strength properties.

The improvement of the strength indexes of yarned joints in fabrics of sparse structures due to the reduction of the friability and moving of yarns will improve the quality and extend the life of products made of these fabrics. Consequently, the problem of objective and qualitative assessment of the wear resistance of yarned joints in the fabrics of sparse structures, aimed at improving the quality of products and prolongation of life of products among them, is a very urgent task and is of practical and theoretical interest.

The application of a polymer composition onto the surface of textile fabrics will significantly improve the physical and mechanical properties of fabrics. In this case, it is especially important to increase the form stability of a garment. An important initial parameter when applying a polymer coating is the structure of the fabric, its surface density. The density of a fabric forms a certain thickness of the fabric.

The thickness of a fabric is an indicator that has a valuable influence on its purpose and processing in the garment manufacturing. The thickness of the fabric depends on the thickness of the yarn and its twist, the interlacing of yarns, the density, and the method of forming the fabric itself and the nature of the finish.

As thicker the yarn, the thicker is fabric, when all other parameters being equal. With the increase of the number of twist of the yarn, its diameter decreases, but to a certain limit, after, the yarn is shortened and, consequently, increase in its cross section.

Depending on the type of weave of the single layered fabric has been manufactured, its thickness may be different. The smallest thickness is a character of plain weave fabrics, and higher is for twill, satin and finely combined weave fabrics, the largest one for fabrics of complex weaves.

Fabric thickness depends on the degree of bending of the warp and weft yarns. If the warp and the weft are uniformly bending one another, floating in a space length of one diameter, the thickness of the fabric will correspond to the diameter of one weft and one warp yarn (Figure 1. a). There are fabrics in which the weft yarn is not bent at all, but the main yarn goes around it comletely. In this case, the thickness of the fabric will correspond to one diameter of the weft yarn and two diameters of the warp yarn (Figure 1. b).

If the weft is curved less, displacing in a surface by half a diameter, and the warp is larger, floating by one and a half diameter, then the thickness of the fabric will correspond to one diameter of weft and one and a half diameter of the warp yarn (Figure 1. c).

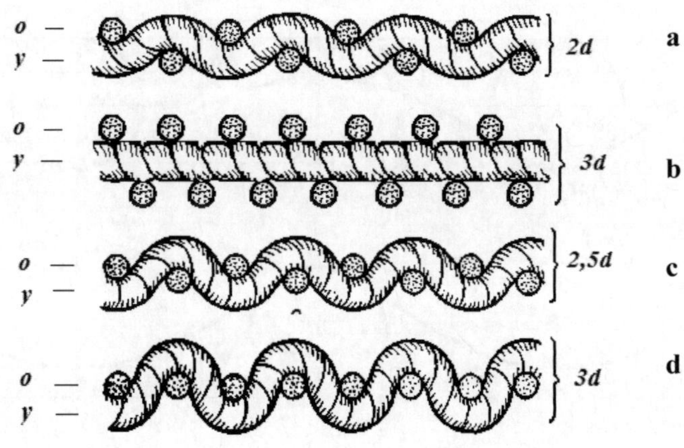

Figure 1. Scheme of the fabric structure with different bending of the warp (*o*) and weft (*y*) yarns

Thus, the thickness of single-layer fabrics can be in the range of two to three diameters of yarns from which fabric is produced.

To determine the mass of the polymer coating that has absorbed into the surface of the fabric, it is necessary to calculate the consumption and mass of the weft yarn and warp yarn for a single fabric structure. For a single layered fabric's structure, the total volume is determined from saw

$$V = t_1 \cdot t_2 \cdot h,$$

where, t_1 – distance between the warp yarns or the width of a single fabric structure;

t_2- distance between weft yarns or the width of the single-structure fabric in the transverse direction;

h –fabric thickness.

The Figure 2 shows the scheme of the single structure of the fabric according to the variant, Figure 1, a. From the calculation scheme, it is possible to determine the consumable length of the weft yarn for a single structure of fabric:

$$l_y = l_{KB} + l_{BC} + l_{CD} + l_{DM} + l_{ME} + l_{ET} \tag{1}$$

At that, the length of separate areas of a weft yarn will be as follows:

$$l_{KB} = \frac{l_{AB}}{2} = (r_1 + r_2)tg\frac{\alpha}{2};$$

$$l_{KB} = l_{CD} = l_{DM} = l_{ET}$$

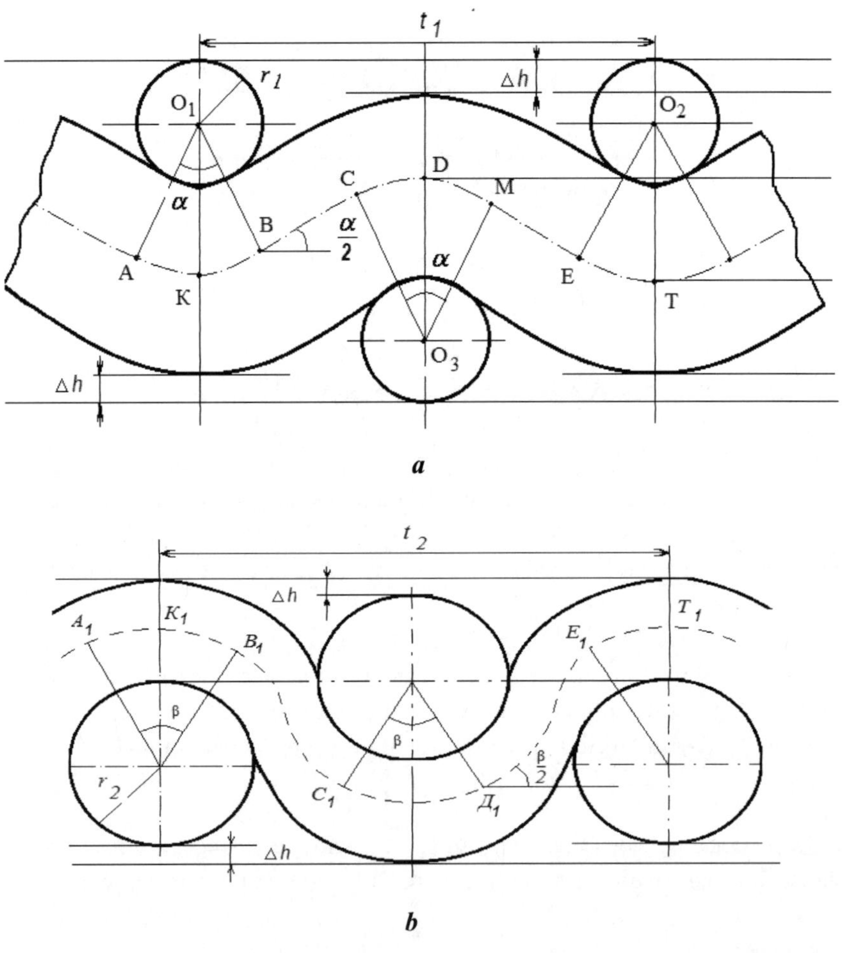

Figure 2.

The length of the weft yarn l_{BC} is defined in the following manner. Based on the Figure 2, it is seen that l_{BC} constitutes with a horizontal line the ground α/2, thus, its sides are mutually perpendicular, accordingly, with l_{O3D} *and* l_{O3C}. So, taking into account $l_{BC}=l_{DM}$, we have:

$$t_1 = 2l_{BC} \cdot \cos\frac{\alpha}{2} + 4(r_1 + r_2)\sin\frac{\alpha}{2} \tag{2}$$

From received (2) we will define l_{BC}:

$$l_{BC} = \frac{t_1}{2\cos\frac{\alpha}{2}} - 4(r_1 + r_2)\cdot tg\frac{\alpha}{2} \tag{3}$$

$$l_{BC} = \frac{t_1}{2\cos\frac{\alpha}{2}} - 2(r_1 + r_2)\cdot tg\frac{\alpha}{2}$$

where, r_1; r_2 – there after the radii of cross sections of the weft and warp yarns.
At that, the length of the weft yarn in the single structure of fabric will be as follows:

$$l_y = 4(r_1 + r_2)\cdot tg\frac{\alpha}{2} + \frac{t_1}{\cos\frac{\alpha}{2}} - 4(r_1 + r_2)\cdot tg\frac{\alpha}{2} \tag{4}$$

According to the Figure 2 *b,* in the same way we may calculate the length of the warp yarn of a single layered fabric:

$$l_o = 4(r_1 + r_2)\cdot tg\frac{\beta}{2} + \frac{t_2}{\cos\frac{\beta}{2}} - 4(r_1 + r_2)\cdot tg\frac{\beta}{2} \tag{5}$$

where, α – wrap angle by weft yarn to the warp yarn;
β – wrap angle by warp yarn to the weft one;
t_1–distance of a single cross section of fabric along the weft yarn length;
t_2 – distance of a single cross section of fabric along the warp yarn.

Total volume of a single cross section of fabric is defined from the expression according to the Figure 2.

$$V_E = t_1 \cdot t_2 \cdot (h + \Delta h) \tag{6}$$

where, h - total thickness of warp and weft yarns;
Δh–warp and weft yarns shifting in a single structure of fabric section.

It is known that when a polymer coating is applied to the surface of a fabric, the polymeric material fills the free space between the warp and weft yarns. In addition, depending on the density of the fabric yarns, the polymeric matter also penetrates between the fibers of yarns. Part of the polymer composition can also be absorbed into the fibers of the fabric yarn. To determine the equal amount of polymer matter when applied onto a single section of fabric, we calculate the volume of the polymer matter:

$$V_n = V_E - V_Y - V_O + \Delta V_n \tag{7}$$

where, V_y, V_o - are, accordingly, the volumes of weft and warp yarns of a single section of fabric;

V_n - is a volume of polymer matter soaked and penetrated into the structure of fiber. As per the data of experimental studies, it reaches up to (0.08 ÷0.11) V_n.

Taking into account of a cross section of yarns (6), will become as following:

$$V_n = t_1 \cdot t_2 [(r_1 + r_2)2 + \Delta h] - 2\pi(r_1 + r_2)\cdot(\alpha \cdot r_2^2 + \beta \cdot r_1^2) - \frac{\pi \cdot r_2^2 \cdot t_1}{\cos\frac{\alpha}{2}} - \frac{\pi \cdot r_1^2 \cdot t_2}{\cos\frac{\beta}{2}} + \\ 8\pi(r_1 + r_2)\cdot\left(r_2^2 \cos\frac{\alpha}{2} + r_1^2 \cos\frac{\beta}{2}\right) - \Delta V_n \tag{8}$$

$$V_n = t_1 \cdot t_2 [(r_1 + r_2)\cdot 2 + \Delta h] - 2\pi(r_1 + r_2)\cdot(\cos\alpha \cdot r_2^2 + \cos\beta \cdot r_1^2) - \frac{\pi \cdot r_2^2 \cdot t_1}{\cos\frac{\alpha}{2}} - \frac{\pi \cdot r_1^2 \cdot t_2}{\cos\frac{\beta}{2}} + \\ + 8\pi(r_1 + r_2)\cdot\left(r_2^2 \cos\frac{\alpha}{2} + r_1^2 \cos\frac{\beta}{2}\right) - \Delta V_n$$

Figure 3 provides the scheme of polymer composition application on to the surface of a single section of a fabric.

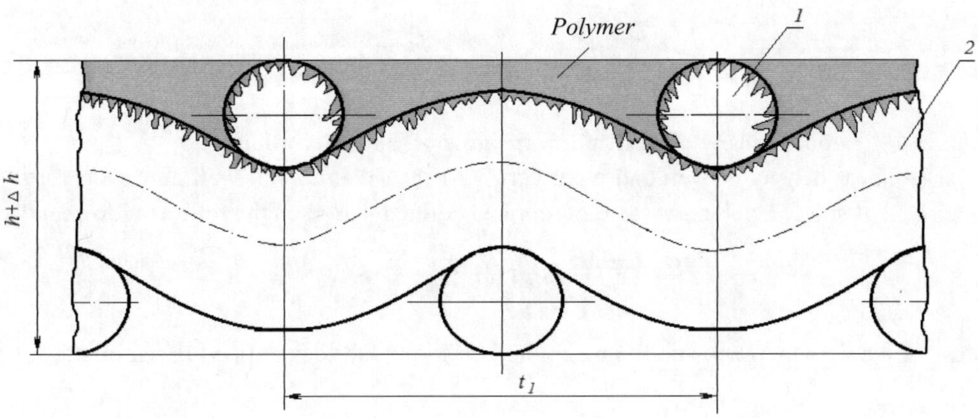

Figure 3. Scheme of polymer composition application onto the surface of a fabric

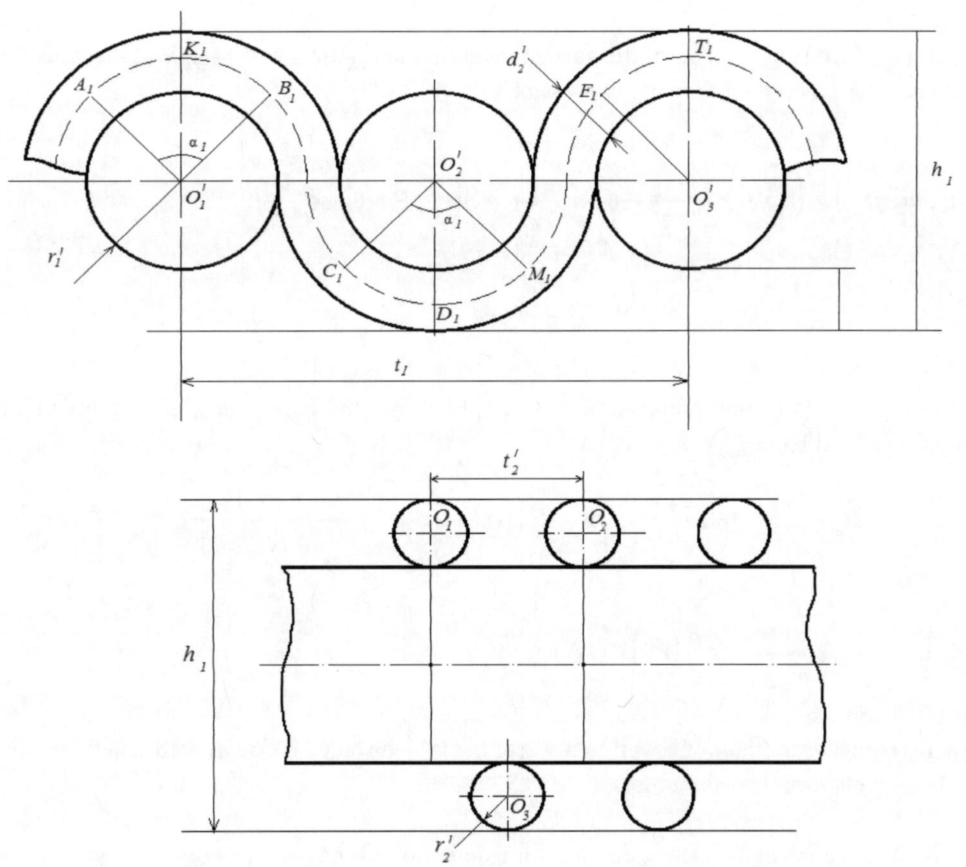

Figure 4. Structural view pattern of a single layered fabric

The relationship between the linear and bulk density of the weft and warp yarn, we can fix the following relationship:

$$\rho_y = \frac{\rho_{ул}}{\pi \cdot r_2^2}; \rho_0 = \frac{\rho_{ол}}{\pi \cdot r_1^2} \tag{9}$$

where, ρ_y, ρ_o – bulk density of weft and warp yarns of fabric accordingly;

$\rho_{ул}, \rho_{ол}$ – linear density of weft and warp yarns. At that, the mass of weft and warp yarns in the single layered structure of fabric is defined, based on the following formula:

$$m_y = \rho_{ул} \cdot l_y; m_o = \rho_{ол} \cdot l_o \tag{10}$$

Mass of a polymer composition to be applied on to the fabric is defined based on the following expression:

$$m_n = V_n \cdot \rho_n \tag{11}$$

where, ρ_n – specific density of polymer.

It is necessary to note that for the variant given in the Figure 1 *в*, the mass of the weft and warp yarn is defined, based on the following expressions:

$$m_{y1} = \rho_{ул}\left[4(r_1^1 + r_2^2)\frac{\alpha_1}{2} + \frac{t_i}{\cos\frac{\alpha_1}{2}} - 8(r_1^1 + r_2^1)\cdot tg\frac{\alpha_1}{2}\right]$$

$$m_{O1} = \rho_{ОЛ} \cdot t_2^1 \tag{12}$$

Volume of a polymer composition to be applied onto the fabric surface for the variant given in the Figure 1 *в*, will have:

$$V_{n1} = 2t_1^1 \cdot t_2^2(r_1^1 + 2r_2^1) - 2\pi r_2^{1\,2}\alpha_1^1(r_1^1 + r_2^1) - \frac{\pi \cdot t_1^1 \cdot (r_2^1)^2}{\cos\frac{\alpha_1}{2}} + 8\pi (r_2^1)^2 \cdot$$
$$\cdot (r_1^1 + r_2^1) tg\frac{\alpha_1}{2} - \pi t_2^1 (r_1^1)^2 - \Delta V_{n1} \tag{13}$$

where, $r^1{}_1$, $r^1{}_2$ - are radii of a weft and warp; $t^1{}_1$, $t^1{}_2$ – distance between weft and warp yarns; α_1- wrap angle by weft yarn the warp one.

At that, we may have the expression for calculation of the mass of polymer composition to be applied onto the surface of fabric for the variant given in the Figure 1, *в*:

$$m_{n1} = V_{n1} \cdot \rho_n \tag{14}$$

Conclusions. The results of experimental studies are presented in Figures 5-8. As can be seen from the given graphs, the mass of the applied polymer composition on the surface of the textile in the direction of the warp and weft yarns, as well as the width of the single parameters for the warp and weft yarns has the linear relationship.

Thus, the maximum mass (m) of the polymer composition at a maximum radius r1 of warp yarns (Figure 5). With a minimum radius of warp yarns of textile, the mass of the polymer composition is 55.4% from maximum value. At that, the average value of mass of the polymer composition at the range being considered of the warp yarn radius is 77.7% of the maximum value.

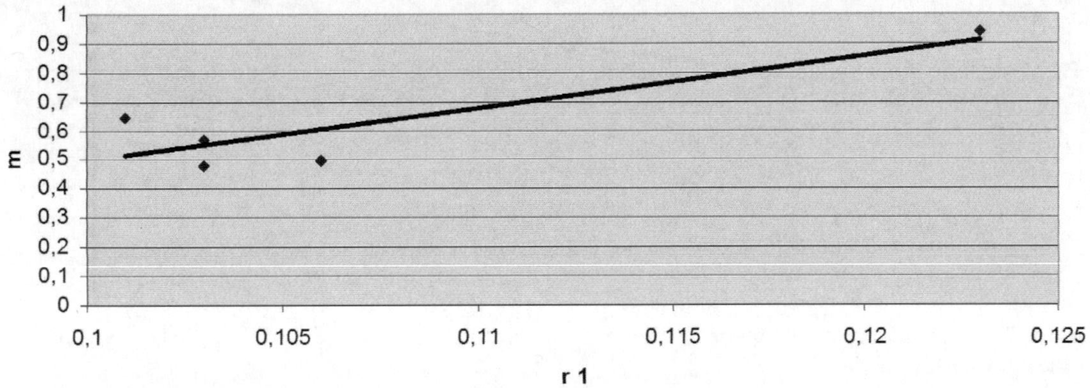

Figure 5. Diagram of relationship of a polymer composition with a radius of the warp yarn made of cotton

Maximum mass (m) of the polymer composition at a minimum radius r2 of the weft yarns (Figure 6). At the maximum radius of the textile warp yarns, the weight of the polymer composition is 70.5% of the maximum value. At that, the average value of mass of the polymer composition at the range being considered of the radius of the weft yarns is 85.25% of the maximum value.

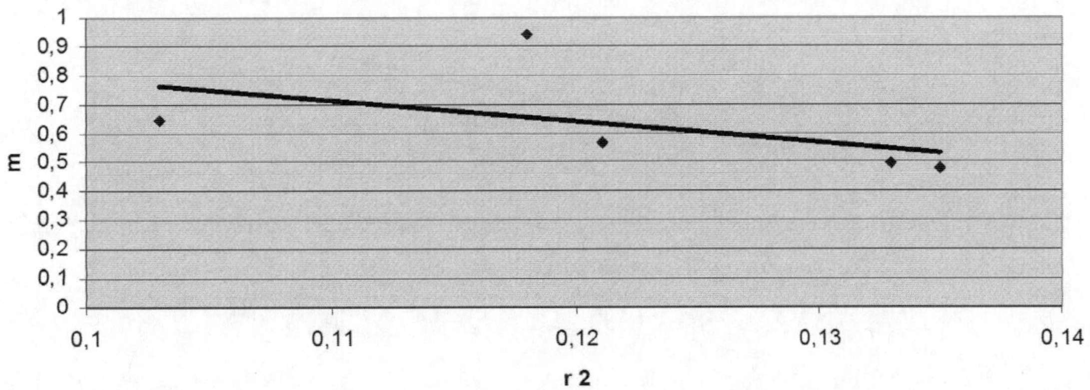

Figure 6. Diagram of relationship of a polymer composition with a radius of the weft yarn made of cotton

From the given schemes (Figure 7 and 8), the mass of the applied polymer composition on the surface of the textile from the width of single indices along the warp and weft yarns has the linear relationship.

Thus, the maximum mass (m) of the polymer composition at a maximum value of a single structure t1 for the warp yarns (Figure 7). At a minimum value of a single structure for the warp yarns, the mass of the polymer composition has 54.9% of the maximum value. At that, the average value of the mass of the polymer composition at the range being considered of the warp yarn radius is 77.45% of the maximum value.

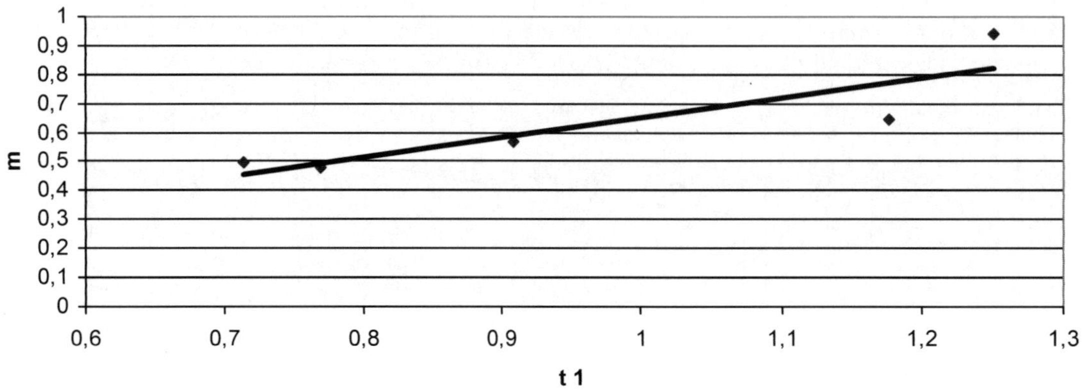

Figure 7. Diagram of relationship of a polymer composition from the width of a single layered structure fabric of warp yarn

The maximum mass (m) of the polymer composition at a maximum value of a single structure t1 for the weft yarns (Figure 7). At a minimum value of a single structure for the weft yarns, the mass of the polymer composition has 51.85% of the maximum value. At that, the average value of the mass of the polymer composition at the range being considered of the warp yarn radius is 75.9 % of the maximum value.

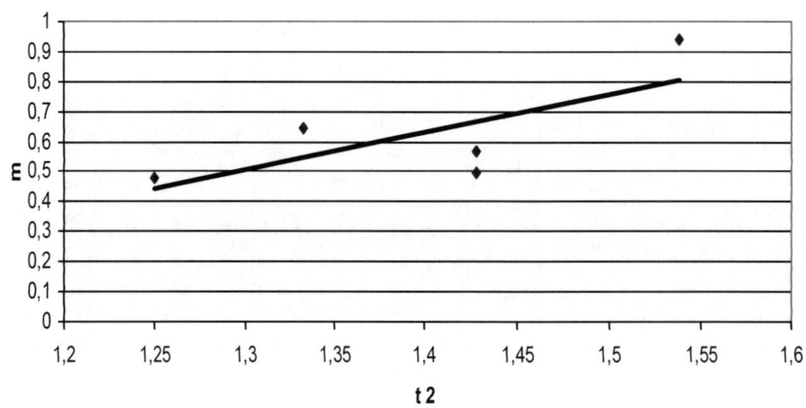

Figure 8. Diagram of relationship of a polymer composition from the width of a single layered fabric of weft yarn

References.

[1] Clark Ruby. Central Asian Ikats: From the Rau Collection. – London - New-York: V&A Publications; Distributed in North America by Harry N. Abrams, 2007. - **96** p.66 color plates.

[2] Hale Andy, Fitzgibbon Kate. Ikats: Woven Silks from Central Asia: The Rau Collection. - Oxford: Basil Blackwell, 1988. - 92 p.

[3] B.A. Buzov, Laboratory workshop on the material science of garment manufacture. – M.: Legpromizdat, 1991. – 322 p.

[4] V.V. Veselov, O.V. Metelyova, Roleof chemistry in the processes of garment manufacturing // The Russia Chemical Magazine – 2002. – XLVI, **No.1**. – p. 121–129.

[5] V.V. Veselov, G.V. Kolotilova, Chemicalization of the technological processes of garment manufacturers : textbook – Ivanovo: IGTA, 1999.

[6] S.Sh. Tashpulatov, Development of a high-efficiency sustainable technology of garment manufacturing. Abstract of a thesis, Doctor of engineering sciences, city of Tashkent, TITLI, 2008, 42 p.

[7] GOST 29104.1-91: Methods of definition of linear sizes, linear and surface densities in part of technical fabrics. – 15 p.

[8] Methods of intensification of processes of hygro-thermal treatment with the purpose of instilling the goods of a light industry the features of shape stability / V.V. Smirnov, L.V. Larina, I.V. Cherunova, A.V. Merkulova, E.A. Schenikova // Contemporary challenges of science and education. – 2012. – **No. 6.** – p. 133.

[9] Device for applying the polymer composition onto the close parts of garment apparels / S.Sh. Tashpulatov, B.G. Isroilova, Sh.Kh. Bekhbudov [and others] // License FAR No.00885 dd. February 7, 2014.

[10] Researches on influence of packetizing mechanisms on their physical and mechanical peculiarities /I.V.Cherunova, S.Sh. Tashpulatov, B.A.Rikhsiyev, L.N. Nutfullayeva, A.A. Kovaleva, T.Yu. Lesnikova // In the book: High technology in the service of a human's ecology: monography/under the general editorship of I.V. Cherunova. – Novocherkassk, 2015. – p. 36–40.

[11] of garment manufacturing/ B.G. Alimukhamedova, S.Sh. Tashpulatov, I.V. Cherunova // in the book: Technical regulation: baseline materials, goods and services [electronic resource]: collection of researches / under the editorship V.T. Prokhorov (prev.) [andothers]; institute of service and entrepreneurship (branch) DSTU, 2017. – p. 411-416.

[12] I.V. Cherunova, G.R. Milyutina, Research on development of nano-structural materials and compositions for a human's safe life // Fundamental researches. – 2014. – **No. 9–10**. – p. 2153–2156.

[13] D.Inogamdjanov, A. Daminov and O.Kasimov, The construction phase's influence to the moving ability of cross-sections of woven structure. 17th World Textile Conference AUTEX 2017 – Textiles – Shaping the Future. doi:10.1088/1757-899X/254/16/162003

Design development and parameters calculation methods of plastic diamond pattern bars on resilient supports in ginning machines

D S Tashpulatov[1], A J Muradov[1], A Juraev[1] J K Gafurov[1] and S Vassiliadis[2]
[1]Tashkent Institute of Textile and Light Industry, Technology of Textile Materials, Technology of Silk and Spinning, 5 Shohjahon, Yakkasaroy dist., Tashkent, UZBEKISTAN
[2] University of West Attica, Department of Electrical and Electronics Engineering, 250 Thivon & P. Ralli, Aigaleo 12244, GREECE
jgafurov@mail.ru

Abstract. The article presents effective construction scheme and operating principle of plastic diamond pattern on bars on resilient supports of raw cotton cleaners for large weeds. Basic parameters of cleaning zone of the proposed construction are based on theoretical and experimental research. Production testing confirmed effectiveness of plastic diamond pattern bars use.
Keywords. Cleaner, raw cotton, diamond bar, resilient support, plastic mass, vibration, amplitude, frequency, saw cylinder, optimization, effect, weeds.

1. Introduction

In the current stage of the development of the cotton ginning industry, it is important to intensify the cleaning of raw cotton, to develop more sophisticated designs, to find new, effective ways of raw cotton cleaning from small and large weeds, and to select appropriate modes for moving operating elements of cotton cleaners. Raw cotton, which comes for its primary processing, contains large and small weeds. Weeds of less than 0.8 mm penetrate deeply into the fiber and are removed by means of strong shaking of raw cotton. Large weeds, mainly located on raw cotton surface, have weak adhesion to fiber and are easily separated from it [1, 2]. In the cotton ginning industry, research is aimed at innovative techniques and technology development to provide effective application of modern science and technology achievements as well as modernization. In this industry, development and research of effective, resource-saving and technological designs of operating elements and bars of cotton cleaners are of great importance. From the above brief analysis of the research in the field of raw cotton and cotton fiber cleaning, it can be concluded that most researches are focused on the development and modernization of operating elements of cotton cleaners from large weeds, based on application of optimization techniques. The matter of use of such alternative materials for bars and their geometrical parameters as plastic diamond pattern bars on resilient supports is insufficiently investigated. Practically, there are no dynamic investigations that determine the basis for calculation of reliability, durability, endurance, etc. In addition, there are no investigations on vibration effect of operating elements on raw cotton, enabling efficient discharge of impurities [3, 4].

2. Theoretical part

In order to improve existing constructions and increase the cleaning effect, we developed a number of highly efficient designs of bar grilles that allow maximum possible preservation of natural qualities of cotton and seeds [5, 6].

According to design we recommend the plastic bars are installed in lateral segments by means of elastic sleeves, the thickness of which is selected as decreasing in the course of pulling the cotton. The construction works in the following way (fig. 1, a). Saw cylinder 4 grabs raw cotton with teeth, pulling them through the bars 2. At the same time the bars 2 fluctuate due to exposure of cotton pappus and deformation of elastic sleeves 3. Thickness of elastic sleeves 3 is selected as decreasing when pulling cotton and has a ratio of:

$$\Delta_1=r_1-R;\ \Delta_2=r_2-R;\ \Delta_n=r_n-R;\ \ \Delta_1>\Delta_2>\ldots>\Delta_n \tag{1}$$

where, R-bars radii; r_1, r_2,..., r_n external radii of elastic sleeves 3 of relevant bars 2; Δ_1, Δ_2,…,Δ_n- thickness of elastic sleeves 3 of relevant bars 2.

At the beginning of pulling zone raw cotton will be less friable and therefore due to greater thickness of elastic sleeves 3 the bars 2 in this zone will vibrate with greater amplitude and lower frequency, that allow not only to separate the weeds, but also to loosen the raw cotton. At the end of cotton pulling zone, due to less thickness of elastic sleeves 3 bars 2 vibrate with greater frequency and amplitude. This leads to separation of the weeds, located more deeply in stock.

Inter-bar gaps are also selected as decreasing in the course of pulling cotton and have a ratio of:

$$a_1-a_2=\Delta_1-\Delta_3;\ \ a_2-a_3=\Delta_2-\Delta_4;\ \ a_3-a_4=\Delta_3-\Delta_5;\ a_{n-1}-a_n=\Delta_{n-1}-\Delta_{n+1} \tag{2}$$

where, a_1, a_2,….,a_n- gaps between bars 2 in the course of cotton pulling.

Gaps difference between neighbouring bars 2 in the course of pulling cotton is selected according to (1, 2) equal to the difference of elastic sleeves 3 thickness respectively between even and odd neighbouring bars 2. In the course of pulling, cotton will be more divided on pappusses, and in order to reduce their loss through the gaps between bars 2, these gaps are made as decreasing according to displays (2). Recommended bar grid of fibrous material allows significantly increase the cleaning effect and reduce cotton pappus loss with discharged weeds.

The following bar grid of fibrous material cleaner is developed, containing diamond pattern bars with the different number of faces that in the course of pulling fibrous material are selected so that each subsequent bar has one more face in the number of faces than the previous bar [5].

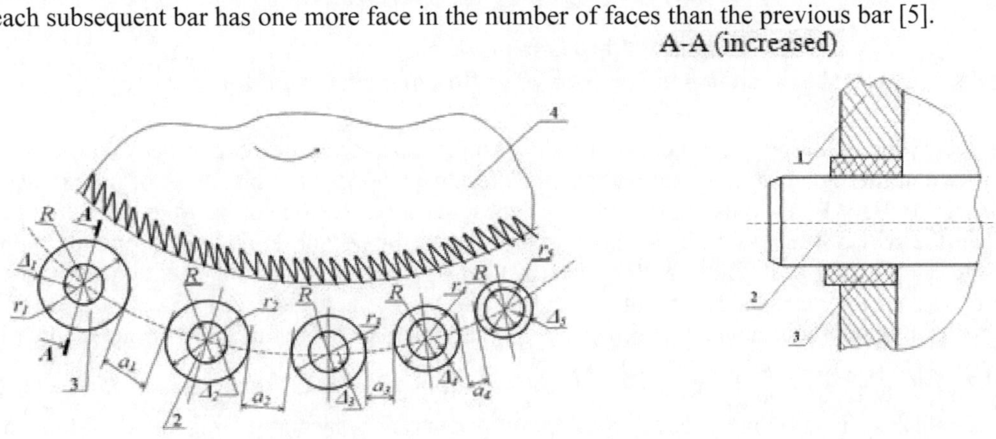

a- plastic bar grid of fibrous material cleaner with rubber bearings

IOP Conf. Series: Materials Science and Engineering **459** (2019) 012068 doi:10.1088/1757-899X/459/1/012068

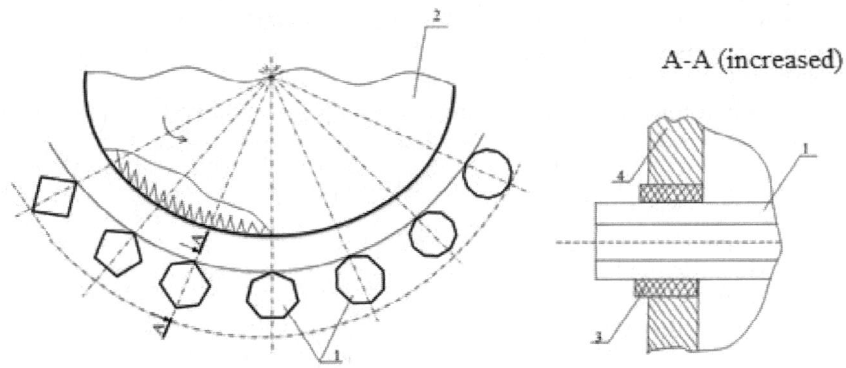

b- bar grid of fibrous material cleaner with plastic diamond pattern bars on resilient supports

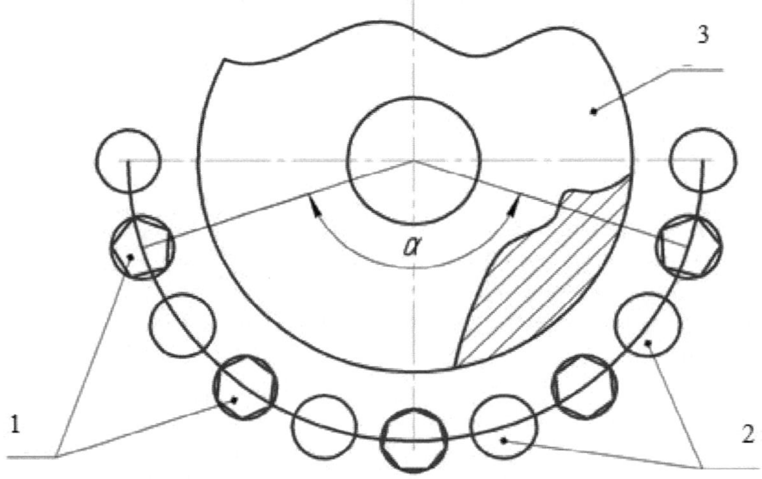

c- bar grid of fibrous material cleaner

Figure 1. Plastic bars scheme in the zone of cotton cleaning from large weeds

Construction consists of bars 1, which are rigidly installed, using elastic rubber sleeves 3, curved rods (not shown in the figure) 4 and rotating spiked roller 2 (included for explanation of operating of proposed bar grid). Bars 1 are of diamond kind. Each subsequent bar has one more edge in the number of edges than the previous bar 1 has. First bar 1 is four-sided, the second is five-sided one, the third bar 1, installed in the course of cotton pulling, is six-sided one and so on. In three sectional bar grid, each section has up to five bars 1, and then the last bar 1 will have 18-faces (fig. 1, б). During operation at the starting zone, raw cotton, clashing on the bars 1 with four faces, with great force is exposed to loosening and weeds separation. Further, when the number of bars 1 faces increases, interaction force between cotton and bars 1 faces reduces, but their frequency and interaction direction increase. This allows the effective separation of cotton and weeds, especially large weeds. Thus, the following relationship is maintained

$$K_i = K_{i-1} + 1 \tag{3}$$

where, K_i– number of faces of i – bar 1. K_{i-1} number of faces of i-1 – bar.

To improve productivity and cleaning effect, as well as to reduce cotton pappus loss, bar grid design is improved, using required sequence setup of round and diamond pattern bars in the course of pulling cotton. During operation, raw cotton (stock) is fed to spiked roller 3, teeth of which grab the raw cotton and pull it along the bar grid. In operating area of spiked roller 3, cotton impacts cyclically on diamond pattern bars 1 and round cross-section bars 2 (fig. 1 в). In this case force and direction of impacts along the rotation of spiked roller 3 will be different due to different number of bars' 1 faces. At the same time, with an increase in the number of faces of the bars 1, the momentum force of the cotton impacts on the face of the bar 1 decreases, and, with the number of faces of the bars 1 decreasing. Such interaction of cotton with diamond (of different number) bars 1 facilitates release of impurities of varying weight and varying depth in cotton. Alternation of round cross-section bars 2 with diamond pattern bars 1 when interacting with fibrous material dramatically reduces of fibres, and allows changing the momentum force of interaction and motion law of fibrous material in the area of their pulling.

3. Results of theoretical researches on argumentation of plastic bars parameters.

For investigation of vibration and analysis of main parameters, taking into account random influence of cotton on plastic bars of raw cotton cleaner from large weeds, bar vibration equation is obtained:

$$m\frac{d^2x}{dt^2} + в\frac{dx}{dt} + c_1 x + \frac{c_2}{\mu}x^3 = M(F_в) \pm \delta(F_в), \tag{4}$$

where m - reduced mass of plastic bar; μ - constant coefficient of nonlinearity; c_1, c_2 –stiffness coefficients of resilient support, в – dissipation factor, m (F) –mathematical expectation of random resistance component of raw cotton.

Taking into account the initial angles *t=0 да,* $\dot{X}=0$; $\ddot{X}=0$, the solution was obtained, using PC: the average value of plastic bar axis displacement is $X_{cp}=(1{,}4\text{-}1{,}6)\cdot10^{-3}$m, vibration amplitude will be $\Delta X=(1{,}8\text{-}2{,}1)\cdot10^{-3}$m. High-frequency vibrations of the bar match the frequency of resistance force δ ($F_в$).

Fig. 2. shows fragments of displacement, speed and acceleration of the plastic bar on resilient support with nonlinear restoring force when

$m = 1{,}0\,Hc^2/м$, $c_1 = 0{,}8\cdot10^4\,H/м$, $c_2 = 1{,}2\cdot10^4\,H/м$, $M(F_в) = 10{,}5\,H$,

$\delta F_в = (0{,}7 \div 1{,}0)H$. It should be noted that frequency of the bar vibrations is (40…55) Hz. In this case the high-frequency component of bar vibrations is (147÷178) Hz.

The low-frequency component corresponds to the frequency of forced oscillation speed serrate drum unit, and the high frequency component corresponds with the number in the grate section. Fig. 2 shows that, at forced vibrations, bar deviates in average by value $X_{cp} = (1{,}4 \div 1{,}6)\cdot10^{-3}\,м$, and vibrations amplitude at calculated values of parameters is $\Delta X = (1{,}8 \div 2{,}1)10^{-3}\,м$. For metal bars on resilient support under operation [4] the vibrations amplitude is $\Delta X = (2{,}2 \div 2{,}5)\cdot10^{-3}\,м$. Comparison of the results shows that in the proposed design of plastic bars the vibrations amplitude is increased by (10÷15)% due to low weight of the bar.

Values $\dot{X}$ and $\ddot{X}$ change accordingly. Amplitude of speed vibrations reaches the value of 0, 6 m/s - 1,25m/s, and amplitude of accelerations vibrations at the calculated parameters of the system changes within (6,5÷10)m/s^2. Frequencies of speed and accelerations vibrations correspond to high-frequency component of process load from cotton.

Fig. 2a graphically shows the dependences of changes of displacement, speed, and acceleration amplitude due to increase in bar grille mass. It is known that with mass increase in the vibration system the great strength is required for its agitation, i.e. with increase in mass, amplitude of the plastic bar oscillations reduces. With increase in bar mass from 0.7 Nc2/m to 1.5 Nc2/m the vibrations amplitude of the plastic bar decreases from $1{,}85\cdot10^{-3}$ м till $0{,}65\cdot10^{-3}$ м according to non-linear rule. Considering the vibration system, it should be noted that with increase in the bar mass the reducing speed and acceleration are also nonlinear. It is especially important that the intensity of the decrease in the amplitude ΔX, $\Delta\dot{X}$ and $\Delta\ddot{X}$ of the vibrations decreases with increasing mass. This is due to the nonlinear rigid characteristic of resilient support.

In the course the process of raw cotton cleaning from large weeds limiting of bars vibration amplitude is of great importance, as these vibrations have a direct impact on width of gap between the bars and saw cylinder. Researches have revealed that increase in rigidity coefficient C_1 of the resilient support leads to proportional reduction of vibration amplitude of the plastic bars. To provide plastic bars vibrations with amplitude $(0{,}6\div1{,}2)\cdot10^{-3}$ м the non-linear component of elastic rigidity coefficient of resilient support must have the value of $(0{,}8\div1{,}0)\cdot10^{4}$, and rigidity coefficient of $c_1=(1{,}2\div1{,}4)\cdot10^{4}H/м$. The change of thickness of the rubber bushing is up to $2{,}0\div2{,}5\cdot10^{-3}$ м (for NO-68 rubber).

Fig.2b graphically shows dependences of displacement, speed and acceleration change for plastic bar on resilient supports with nonlinear rigidity in variations of load from raw cotton. With increase in cotton resistance from 8,5 to 20 N (average value), the bar displacement increases from $0{,}65\cdot10^{-3}$ to $2{,}6\cdot10^{-3}$ m.

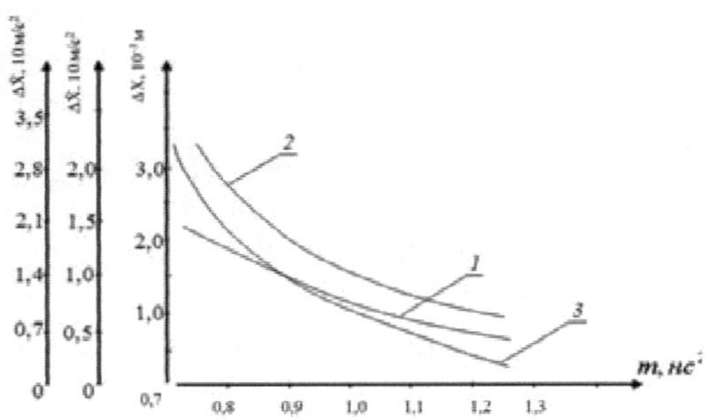

$1-\Delta X=f(m);2-\Delta\dot{X}=f(m);3-\Delta\ddot{X}=f(m)$;
when $c_1=2{,}5\cdot10^{4}H/м$, $c_2=1{,}2\cdot10^{4}H/м$,
deviations of $\delta\Delta X,\delta\Delta\dot{X},\delta\Delta\ddot{X}=(10\div12)\%$
a- Graphical dependence of vibrations amplitude changes of displacement, speed and acceleration of the plastic bar on bar mass change.

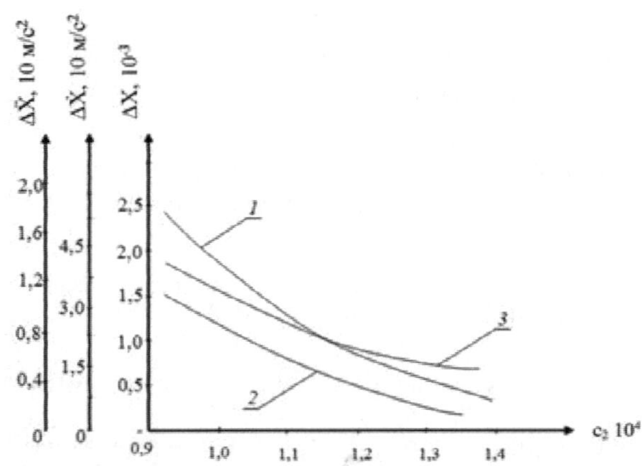

$1-\Delta X=f(c_2);2-\Delta\dot{X}=f(c_2);3-\Delta\ddot{X}=f(c_2)$;
deviations $(8{,}0\div10)\%$; $m=1{,}4нc^2/м$; $c_1=1{,}2\cdot10^{4}н/м$
b- Dependence of changes $\Delta x,\Delta\dot{x},\Delta\ddot{x}$ in function of rigidity coefficient c2 of of plastic bar resilient support.

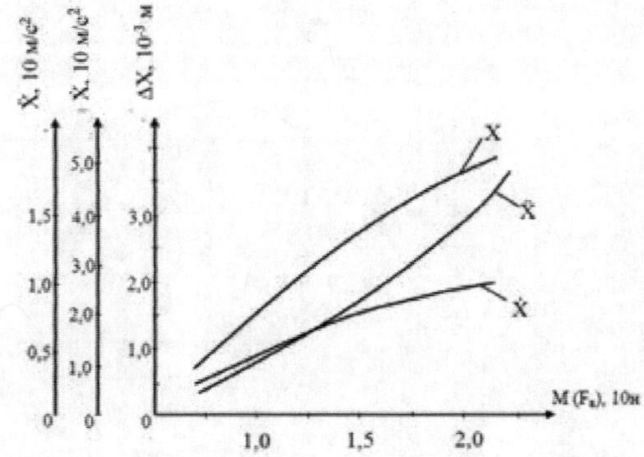

deviations $\delta x, \delta \dot{x}, \delta \ddot{x} = (8{,}0 \div 10)\%$ $c_1 = 1{,}0 \cdot 10^4 \, н/м$ $c_2 = 1{,}2 \cdot 10^4 \, н/м$

c- dependence of change of displacement, speed and acceleration of conical bars on resistance function of raw cotton.

Figure 2. Dependences of ΔX, $\Delta \dot{X}$ and $\Delta \ddot{X}$ changes

To prevent pappus loss between bars due to large vibrations amplitudes of the bars and shortening of the process gap between the spiked roller and bars, on the experiments results the plastic bars amplitude should not exceed $(1{,}0 \div 1{,}4) \cdot 10^{-3} м$. With increase in bar load the intensity of resilient supports deformation decreases that causes reduction of the bars vibrations amplitude (see figure.2c). Comparison of researches results with investigations of metal bar on resilient support shows that the amplitude of speed vibration and acceleration $\dot{X}$ and $\ddot{X}$ in recommended variant is (10÷13)% more than in cylindrical bars. This means that momentum force of plastic bars, influencing on raw cotton is (12÷15)% more than in existing cleaner bars. Recommended values for plastic bars mass are (1,1÷1,3) NC2/m. Thus, in order to provide vibrations amplitude less than $(1{,}0 \div 1{,}4) \cdot 10^{-3}$ m, it is recommended the parameters with values of $c_2=(0{,}8\text{-}1{,}0) \cdot 10^4$ Н/м, $c_1=(1{,}2\text{-}1{,}4) \cdot 10^4$ Н/м, F_0=(15-20)Н, Pr=(5,0-7,0) t/h. It is known that rotary overspeed of saw cylinder more than (31÷35) rad/s leads to raw cotton pappus dragging by saw cylinder and pappus inclination angle increases and without interaction with the bars, pappus passes through cleaning zone. In order to argumentate pappus exclusion, we shall consider calculation model (Figure. 3) and compose differential equation of seed cotton pappus vibration:

$$m_л l_1^2 \ddot{\beta} + m_1 (\omega_0 + \omega_1 \sin K_K t)^2 \cdot [\sqrt{l_1^2 + R^2 \sin^2 \varphi} + R\cos\varphi] \cdot \\ \cdot R\sin\varphi + m_л g h_1 + \sqrt{l_2 + R_2 \cos^2\varphi} \cdot F_л \theta_л \frac{\rho v^2}{2} l_1 = 0 \tag{5}$$

Solution (5), taking into account [11] will be.

$$\frac{d^2 h_1}{dt^2} + \frac{R h_1}{l_1}(\omega_o + \omega_1 \sin K_k t)^2 + \frac{g}{l_1} - h_1 = 0; \; \beta = \frac{h_1}{l_1}$$

$$h_1 = h_ю \cos f_{cb} t + \sqrt{\frac{{v_o}^2 l_1}{R\omega^2 + g}} \sin f_{cб} t \tag{6}$$

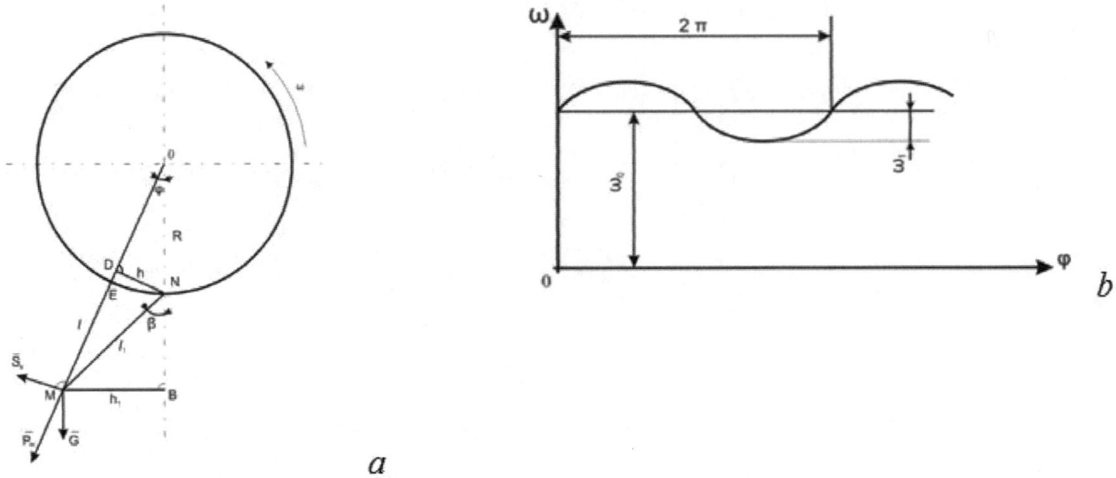

Figure 3. Analitical model (a) and diagram of spiked roller (б) rotary speed

Analysis (3) shows that at the starting moment when t=0, value h1 = h0 the deviation angle of pappus fibers strand deviates from the radius of spiked roller by 600...620. Taking into account the pappus strand length of 0,022 m, the h0 will be 0,019 m. Accordingly, to provide deviation angle of cotton pappus of less than 56÷63 and rotary speed of saw cylinder of 31,4 rad/s the design shutdown is selected within h1=(0,018÷0,621) м.

4. Experimental part.

Experiments were conducted in production environment of cotton gin. Measurement of load, bars vibrations was conducted both for plastic bars on resilient supports and for serial bars. When conducting experiments according to the rigidity coefficients values the following rubber were selected: SKF 32, SPF 26, SPF 260, 1847, NO-68, 1318, and their characteristics were also studied. It was chosen the SKF 32 rubber with rigidity coefficient of $(2\div 2,2)\cdot 10^4$ N/m and the sleeves were manufactured from it by vulcanization method. Accordingly were studied the properties of such plastics as PVC-polyvinylchloride, PP-polypropylene, PE-polyethylene, PA-polyamide. It was choosen the PA plastic consisting of antipyrene (fireproof), anti-static substance (not charged), glass fiber and basalt fiber of 3.5 mm to increase strength when impacts. Load displacements on axes X and Y were measured using strain gauge sensor, and recorded on a computer, using modulator, digital converter. Figure 4 shows diagram of measurements of bars displacement (loading) and overview of the measuring instrumentation [7].

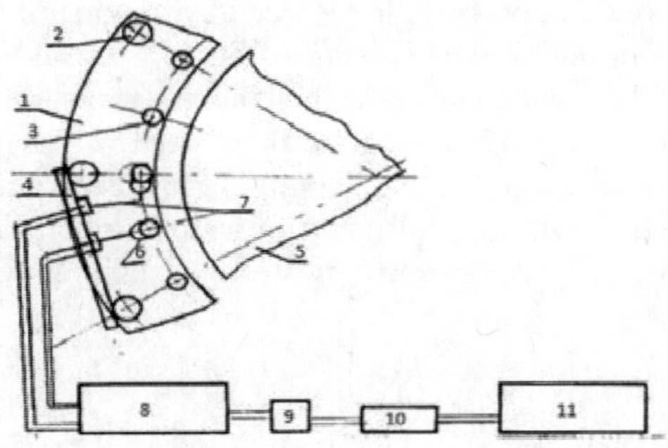

1- sidewall fastening of bar grid; 2- bars rigidity rods; 3- bars; 4- mounting bracket for dynamic sensor; 5- spiked roller; 6- YY axis dynamic sensor; 7- dynamic sensor of vibrations on XX axis; 8- high frequency amplifier; 9- Modulator; 10- ADC; 11- PC.

Figure 4. Measuring scheme for indications of bars displacements (vibrations) on two coordinates

Figure 5 presents oscillograms of plastic bar vibrations on axes X and Y. Thus, at spiked roller rotation speed of 300 rad/s, machine performance of 5.0 t/h and bars rubber bearings rigidity of $1.8 \cdot 10^4$ N/m the vibration amplitude of the plastic bar on Y axis reaches $0,55 \cdot 10^{-3}$ m, and on X-axis the bar does not actually vibrate. With increase in performance of cotton cleaner up to 7.0 t/h and rotation frequency of the spiked roller 310 rad/s and with stiffness of rubber supports $c=1,4 \cdot 10^4$ N/m, the vibrations amplitude of plastic bar increases up to $0,6 \cdot 10^{-3}$m on X axis, and on Y axis vibrations amplitude of plastic bar on resilient support reaches $3,0 \cdot 10^{-3}$ m.

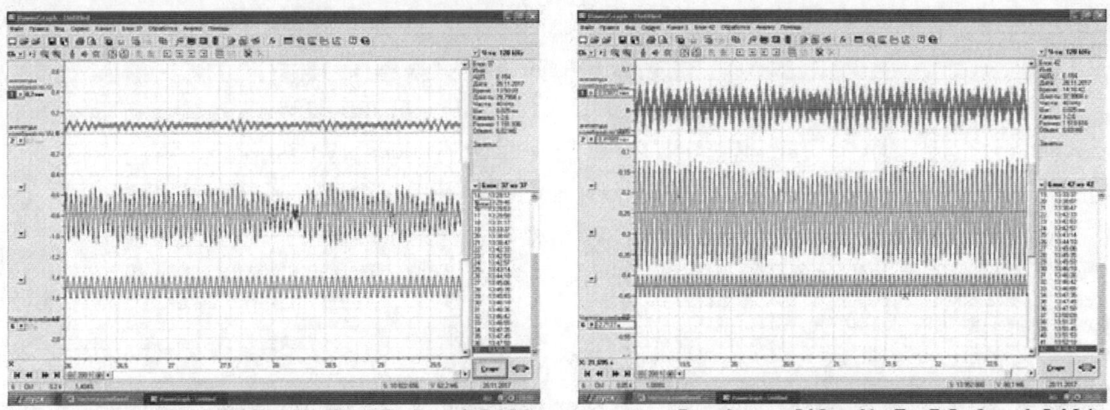

a - when n=300 rad/s; Pr=5,5 t/h c=1,6·104 N/m

б - when n=310 rad/s; Pr=7,5 t/h c=1,5·104 N/m

Figure 5. Oscillogram, characterizing vibrations of plastic bar on resilient supports

Obtained oscillograms were processed based on computer software oscillograms, and graphic dependences were built (fig. 6). On the basis of results analysis, it was revealed that the existing metal bars do not vibrate. Plastic bars on rubber bushes of X-axis vibrate with low amplitude (0,2÷6,3) mm, on Y-axis, bar vibration amplitude reaches (2,5÷3,1) mm, and vibration frequency is (6,5÷15,5) Hz, that is much less than rotation frequency of the spiked roller, and reaches (70÷80) Hz. When increasing rigidity coefficient of the rubber support from $1,2 \cdot 10^4$N/m to $2,8 \cdot 10^4$N/m, oscillation amplitude of plastic bar on X axis decreases from $0,67 \cdot 10^{-3}$ m to $0,18 \cdot 10^{-3}$ m, and on Y axis, it decreases from $2,25 \cdot 10^{-3}$ m to $0,58 \cdot 10^{-3}$ m (Fig.5).

Based on oscillograms processing, graphical dependences of the variation in the vibrations of bars of both variants along the X and Y axes from the change in load from the raw cotton to be cleaned, which are presented in Fig. 6. ΔX and ΔY are nonlinear. Value of ΔY is important for the plastic bar that

IOP Conf. Series: Materials Science and Engineering **459** (2019) 012068 doi:10.1088/1757-899X/459/1/012068

directly affects cleaning of raw cotton from large weeds. So, when increasing load from 5,0 N to 23,0 N, the vibration amplitude in the current version of bar ΔX and ΔY does not exceed $0{,}5\cdot 10^{-3}$ m. At that, vibration amplitude of the plastic bar on resilient supports on X-axis reaches $1{,}0\cdot 10^{-3}$ m, and on Y axis it reaches $3{,}15\cdot 10^{-3}$ m. At that, plastic bars vibrate vertically with an amplitude of $1{,}575\cdot 10^{-3}$ m with frequency for 6,5-15,5 times higher than rotation frequency of the spiked roller. It means that during operation of the cleaner with one turnover of spiked roller, the raw cotton pappus, pulled by saw cylinder, in average 6.5 -15.5 times clashes on the bar, i.e. frequency of forced vibrations of plastic bars reaches 70-80 Hz.

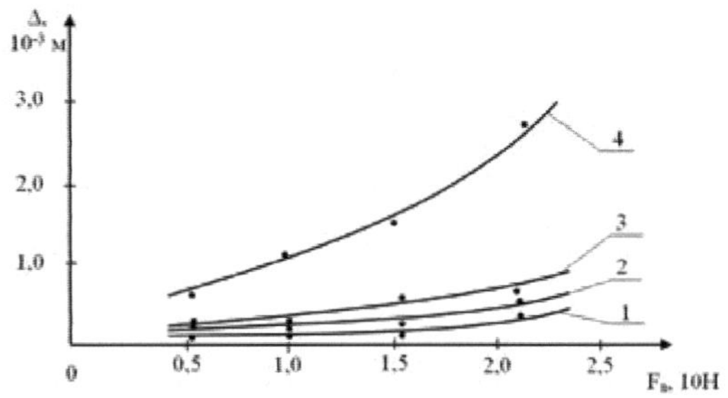

1,2 –for metal (serial) bar; 3,4- for plastic bar on resilient supports; 1-Δx and Δy - for metal bar; 3-Δx and 4-Δy – for plastic bar
Graphic dependences of bars vibrations amplitude changes on changes in load of raw cotton under cleaning

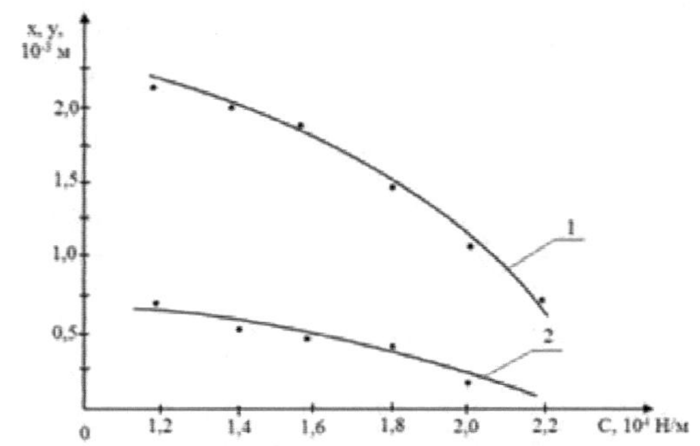

1-y=f(c); 2-x=f(c); when n=290 rad/s; nпр=7,0 t/h.
Dependence of changes in amplitude of the plastic bars on change in rigidity coefficient of resilient supports.

Figure 6. Regularity of change

It is necessary to manufacture the rubber support of the plastic bar from SKF 32 rubber with rigidity coefficient of $(1{,}6\div 2{,}0)\cdot 10^4$ N/m. Rotation frequency of saw cylinder is within $(3{,}0\div 3{,}15)\cdot 10^2$rad/s.

The following regression equations are based on planning of full-factorial experiments [8]:

for I-class cotton Namangan -77

$$Y= 84{,}08+0{,}96X_1 -1{,}28X_2-0{,}76X_3 -2X_1X_2 -0{,}34X_2X_3 + 0{,}27X_1X_2X_3$$

for II-class cotton Namangan -77

$$Y= 80{,}1 + 0{,}33X_1 -1{,}24X_2-1{,}87X_3 +0{,}4X_1X_2 -1{,}56\, X_1X_3 +0{,}73X_2\, X_3+0{,}37\, X_1X_2X_3$$

The input factors are: X_1-rubber rigidity of 10^4N/m, X_2- machine capacity Pr, ton/h, X_3-bar gap δ, 10^{-3} m; output parameter - effect of cotton cleaning. Analysis of results, obtained during full factorial experiment, allows to recommend the following values for the selected key factors: performance, t/h – 5,0; rubber rigidity - $2{,}0\cdot 10^4$ N/m; gap between the bar and spiked roller – 16 mm. At these values the efficient operation of raw cotton cleaner is observed, i.e. the cleaning effect is above 90%.

5. Results

Recommended diamond plastic bars on resilient supports were installed on the lines of universal cleaning complex of ginning factories in Tashkent region. Production test results are presented in Table 1.

Table 1. Production test results

Indications, %	After cleaning unit with the recommended plastic bars	After cleaning unit with serial plastic bars

Cotton, Humidity, (%)	8-9%	8-9%
Content of weeds, (%)	4,6/2,75 4,6/2,65 4,3/2,55	4,6/3,3 4,6/3,07 4,3/2,8
Cleaning Effect, (%)	$\frac{91,2}{93,4}$	$\frac{74,9}{77,1}$
Mechanical damage of seeds	$\frac{1,89}{2,17}$	$\frac{2,82}{3,22}$
Free fiber	$\frac{0,08}{0,11}$	$\frac{0,19}{0,21}$

6. Conclusions

Based on the analysis of the construction of cotton cleaner bar grilles for large weeds, new effective schemes for low-weight construction of plastic bars on resilient supports was developed. A mathematical model, of the vibrations of plastic bars on resilient supports with nonlinear rigidity characteristics, was made. It was shown that increase in mass of the plastic bar reduces displacement vibration amplitude, speed and acceleration of the bar according to non-linear regularity of (1,1÷1,3) $Hc^2/м$. Formulas were obtained to determine the distance and angle of cotton pappus deviation with variable rotation of the spiked roller and use of the plastic bar. During experiments the oscillograms were obtained, characterizing vibrating movement laws of the plastic bars on resilient supports when changing the load from raw cotton to be cleaned, rigidity coefficient of rubber bushings (supports) and rotation frequency of the spiked rollers. A mathematical model was developed that adequately describes cleaning process of raw cotton from large weeds in the form of regression equations. Use of plastic bars on resilient supports in raw cotton cleaners for large weeds helps to improve cleaning effect up to 17%, mechanical damage of seeds is reduced to 1%, and free fiber by 0.18% compared to serial variant.

References

[1] Cotton: World Statistics. Bulletin of the International Cotton Advisory Committee, NY, November 2015.http://www.ICAC.org.,http://www.USDA.gov.

[2] O. J. Murodov, A. Juraev. Groundation of the parameters of grate baronela stic support with non-line arhardness.// The European science review. № 7–8 2017. July–August. Vienna, Prague 2017. page. 109-111.

[3] A. Juraev, J. Mirakhmedov, A.V. Abdullayev. Theoretical bases and improvement of bars design of cotton cleaners for large weeds // Monograph. Fergana, 2013, p. 158

[4] Lugachev A.E. Development of theoretical bases of cotton supply and cleaning in relation to continuous processing. Doc. Diss. in engineering.- Tashkent, TITLI, 1998. -422 p.

[5] Djuraev A. J., Plekhanov A.F., Bitus E.I., Razumeev K.E., Tashpulatov D.S. Bar grille of fibrous material cleaner. Application to the Federal Service for Intellectual Property FIIP of the Russian Federation for issuing patent of invention №2017143328 from 12.12.2017.

[6] Djuraev, A. J. Murodov O. J., Tashpulatov D.S. Bar grille of fibrous material cleaner. Application to the Intellectual Property Agency of the Republic of Uzbekistan for issuing patent of invention № IAP 2017 0386 from 06/09/2017.

[7] Panovko Ya.G., Fundamentals of applied theory of vibrations and clashes, Engineering, L., 1976, 320p.

[8] Adler Y.P., Markova E.V. Planning of experiment when finding optimum conditions. Pub. M. 1976.

Stress relaxation of three-dimensional warp-knitted fabrics in compression

S Vassiliadis[1], D Matsouka[1], L Rešetárová[2], L Bouin[3] and A Marmarali[4]
[1]University of West Attica, Department of Electrical and Electronics Engineers, ,Egaleo, Greece
[2]Technical University of Liberec, Liberec, Czech Republic
[3]ENSAIT, Roubaix, France
[4]Ege University, Bornova/İzmir, Turkey

ptosky@gmail.com

Abstract. Spacer fabrics are a versatile textile construction that can be modified and adapted for a wide range of uses. While the behaviour of spacer fabrics in compression has been extensively studied, their time dependant properties such as creep and stress relaxation have not been investigated equally well. In this research project warp-knit spacer fabrics underwent consecutive compression cycles and the resulting stress degradation over time curves were analysed in order to develop a mathematical and mechanical model of the material behaviour. Based on this analysis it was found that the complexity of spacer structure results in the creep behaviour being connected to the creep bending of interconnecting yarns.

1. Introduction

The typical fabric structures (woven or knitted) in textiles can be considered essentially two-dimensional surfaces as the thickness of the fabric structure is negligible compared with the length and the width of the fabric piece. Nevertheless, there exist textile structures where the thickness is significant, and that thickness is the added value characteristic of the structure. These 3D knitted, or woven fabrics or spacer fabrics as they are also called, are utilized in applications where there is requirement for added "padding" or thickness and they tend to replace materials like polyurethane foam which has historically been used for such applications. These applications include automotive interiors, personal protective equipment, medical textiles, and construction. Combining satisfactory behavior in compression with advanced mechanical properties, breathability and air permeability 3D with acoustic properties spacer fabrics have been extensively studied [1–6]. 3D knitted spacer fabrics are three-dimensional knitted fabrics consisting of two layers of outer fabric that are connected by a layer of, usually, monofilament, synthetic fibres (Figure 1).

Figure 1. Side view of a 3D spacer fabric.

IOP Conf. Series: Materials Science and Engineering **459** (2019) 012069 doi:10.1088/1757-899X/459/1/012069

While the behavior of 3D warp-knitted spacer fabric in compression has been investigated and modeled so that the typical compression stress strain curve has been determined (Figure 2) [7–11], the mechanical behavior of the fabrics regarding time dependent deformations has not been yet investigated. The different areas on the typical compression graph have to do with the rearrangement of the interconnecting monofilaments during the compression among themselves and with the outer layers. The compression in the densification stage (stage III) shows a rapid increase of the force due to the swift densification of the entire fabric.

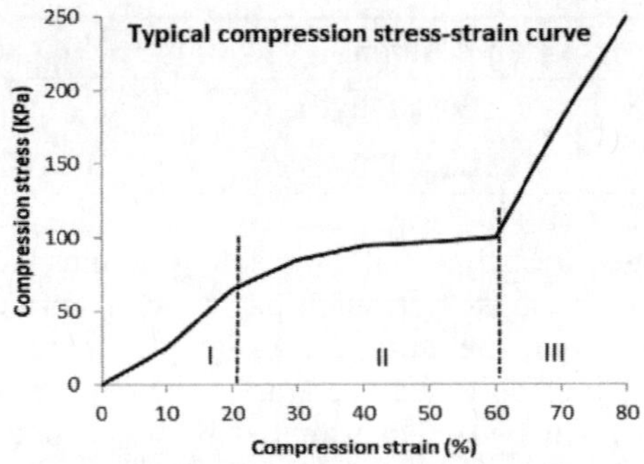

Figure 2. Typical compression stress strain curve of 3D spacer fabric where the three distinct areas are visible.

By comparison the time dependent properties of materials (creep/ stress relaxation) present a very different 2ehavior. Specifically stress relaxation is defined as the observed decrease in stress in response to the same amount of strain generated in the structure. This is primarily due to keeping the structure in a strained condition for some finite interval of time and hence causing some amount of plastic strain. The inverse phenomenon is called creep, which is defined as constant state of stress with an increasing amount of strain. The graphic illustration of stress relaxation can be seen in Figure 3.

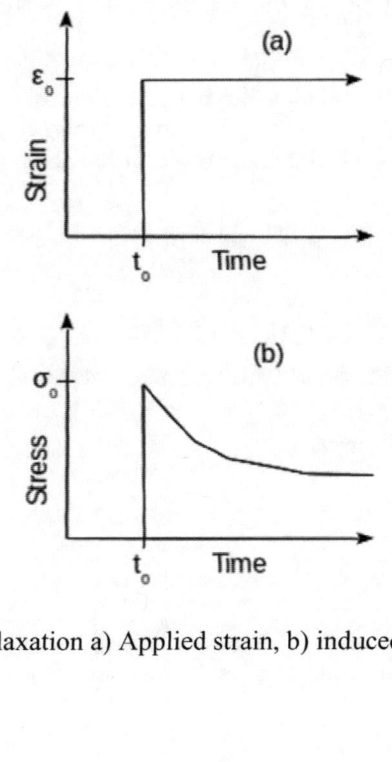

Figure 3. Stress relaxation a) Applied strain, b) induced stress as function of time.

2. Materials-methods

In this research project two samples of 3D warp knit fabrics (Table 1) with different weights where subjected to a compression deformation cycle of 1mm. The deformation was maintained for 4 mins. This interval was chosen since from preliminary tests it was observed that the phenomenon of stress degradation for these samples was moving towards equilibrium within this time. During compression the load that developed on the specimens was recorded (Figure 4).

Table 1. 3D sample characteristics

Sample	Composition	Weight (g/m^2)
51100120	100% Polyester	900
50100130	100% Polyester	1120

The load results were plotted over time and the resulting graphs were analysed with regards to the behaviour of the material. Firstly, stress relaxation phenomena were investigated in accordance with the exponential laws described in the literature. For all samples there was little agreement between the results and the behaviour described in the literature. Then the case of the creep in bending of viscoelastic beams, was investigated. This approach was applied to the behaviour of the interconnecting monofilaments during the deformation cycles and good agreement was achieved between the model and the experimental results.

3. Results and Discussion

From the literature the relationship between the applied strain and the induced stress can generally be described following Mawxell's model as follows, (Equation 1) [12].

$$\sigma = \sigma_0 e^{-Kt} \quad (1)$$

where K, is a constant representing the rate of spontaneous stress relaxation and σ and σ_0 are the stresses at times t and 0.

As can be seen by Equation 1, the relationship between time and stress is exponential in nature. When attempting to model the results of the experiments following the exponential relationship between the stress results (Figure 4) and time, there was little agreement between the results and the fitting curves produced.

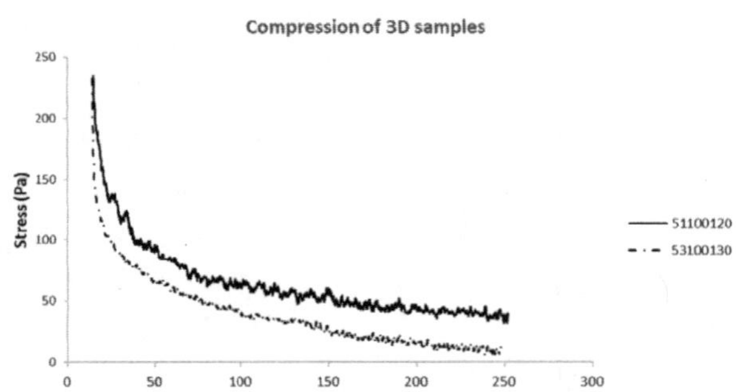

Figure 4. Compression of 3D samples.

In order to model the behaviour of the 3D spacer fabrics extensive review of the behaviour of viscoelastic materials under the application of stress/ strain, was carried out. During this research the laws describing the creep in bending behaviour of elastoviscoplastic beams [13] were located and fitting curves were produced based on this type of behaviour.

This approach showed good agreement with the behaviour of the 3D samples. In Figure 5 the results for sample 51100120 can be seen together with values obtained using the model that was produced. The curves follow a tendency of a logarithmic function, indicating the dominance of the secondary creep characterized by a stationary deformation rate as described in the work by Hadid et al. [13]. The behavior of the materials can be broadly described by Equation 2.

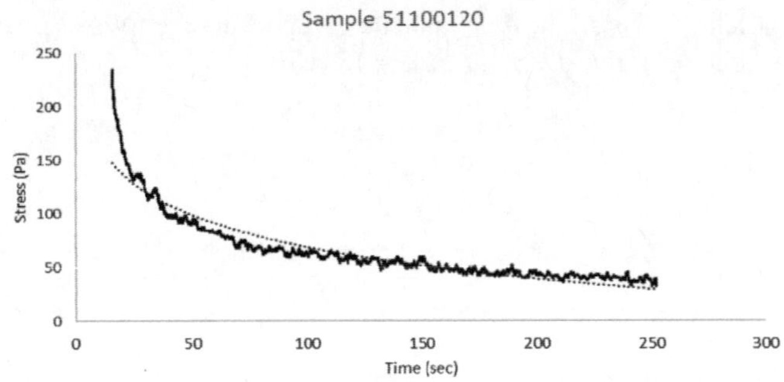

Figure 5. Graph of stress degradation over time with stress values obtained by model (dotted line).

$$\boldsymbol{\sigma = -A ln \sigma_0 + B} \qquad (2)$$

Where parameters A and B are connected to the structure of the 3D fabric and the composition of the interconnecting monofilaments.

4. Conclusions

Based on the analysis of the stress degradation curves produced by the compression of the spacer fabric specimens it was shown that the time dependant properties of the fabrics are connected to the creep bending of the interconnecting yarns. This observation underscores the complexity of 3D warp knit fabrics and the need for further investigation of time dependant properties of these materials.

References

[1] Yip J and Ng S-P Study of three-dimensional spacer fabrics: 2009 Molding properties for intimate apparel application *J. Mater. Process. Technol.* **209**(1) 58–62

[2] Armakan DM and Roye A 2009 A study on the compression behavior of spacer fabrics designed for concrete applications *Fibers Polym.* **10**(1) 116–23.

[3] Vassiliadis S, Kallivretaki A, Psilla N, Provatidis C, Mecit D and Roye A 2009 Numerical modelling of the compressional behaviour of warp-knitted spacer fabrics *Fibres and Textiles in Eastern Europe* **76**(5) 56–61.

[4] Umbach KH 2000 Physiological comfort on car seats *Kettenwirk-Praxis* **34**(1) E9-E13+34-40.

[5] Rajan TP, Souza LD, Ramakrishnan G and Zakriya GM 2016 Comfort properties of functional warp-knitted polyester spacer fabrics for shoe insole applications. *J. Ind. Tex.* **45**(6)1239–51.

[6] Arumugam V, Mishra R, Militky J and Novak J 2015 Thermo-acoustic behaviour of 3D knitted spacer fabrics *Fibers Polym* **16**(11) 2467–76.

[7] Mokhtari F, Vaghefi PM, Shamshirsaz M and Latifi M 2013 Analysis of compressibility behavior in warp knitted spacer fabrics: experiments and van wyk theory *J. Eng. Fibers and Fabrics* **8**(3):6.

[8] Hou X, Hu H and Silberschmidt VV 2012 A study of computational mechanics of 3D spacer fabric: factors affecting its compression deformation *J. Mater. Sc.* **47**(9) 3989–99.

[9] Li M 2015 Effect of structural parameters on compression performance of warp-knitted spacer fabric *Journal of Fiber Bioengineering and Informatics* 30 **8**(2) 267–76.
[10] Liu Y and Hu H 2016 Compressive mechanics of warp-knitted spacer fabrics. Part I: a constitutive model *Tex. Res. J.* **86**(1) 3–12.
[11] Mecit D and Roye A 2009 Investigation of a testing method for compression behavior of spacer fabrics designed for concrete applications *Tex. Res. J.* **79**(10) 867–75.
[12] Vorotnikov GS and Rovinskii BM 1966 Stress relaxation, creep, and uniaxial strain: General and special features *J. Appl. Mech. Tech. Phys* **7**(6) 13-7.
[13] Hadid, M., Saadallah, Y., Debilou, A. et al. The creep of an elastoviscoplastic beam under a bending loading *Mech Time-Depend Mater* **18**: 573-87.

A Research on the Use of Copper Core Yarns in Electromagnetic Shielding Application

I Yesildag[1] and H Kadoglu[2]

[1]Ege University, Institute of Science and Technology, Textile Engineering Department, Izmir, Turkey
[2]Ege University, Faculty of Engineering, Textile Engineering Department, Izmir, Turkey

E-mail: yesildagirem@hotmail.com

Abstract

In this work, possibilities for producing textile materials with copper core cotton yarns for the purpose of electromagnetic shielding were studied. Cu/Co core yarns were produced by using copper filaments as core material with two different diameters (0.05 and 0.07mm) and cotton as sheath material. All core yarns are produced in Ne 8 yarn count. Conductive Cu/Co yarns were integrated in to 3/1 twill woven fabric structure in the weft direction with 5 different weft densities (8, 13, 18, 23 and 28 per cm). According to the TS EN 50147-1 standard, the shielding effectiveness of these fabrics was measured in the frequency range 1 GHz – 6 GHz. The results have shown that EMSE of fabrics can be tailored by changing the weft density, core filament fineness and wave frequency parameters.

Keywords: Electromagnetic shielding textiles, Copper core yarns, Conductive woven fabrics.

1. Introduction

With technological progresses, electrical and electronic devices have reached a more important position in our daily life. The use of these devices makes our life easier but at same time they cause an important environmental pollution since they emit electromagnetic waves to their surroundings while they are working. Electromagnetic waves have harmful effects for both sensitive electronic equipment and also for human health. Electrical devices interfere with other electrical and electronic devices in that they generate the electromagnetic field. This interference may cause malfunctioning of other sensitive electronic devices, when specific limit is exceeded [1].

If an electromagnetic wave gets into an organism, it vibrates molecules to spread heat. In the same way, when an electromagnetic wave enters the human body, it will hinder a cell's regeneration of DNA and RNA. Moreover, it causes abnormal chemical activities to produce cancer cells [2].

Protection from electromagnetic waves is very significant because of these reasons. Shielding is one of the most efficient solutions for protection from electromagnetic waves. Electromagnetic shielding can be defined as prevention of electromagnetic radiation transmission by a material [1].

There are some previous studies conducted on this topic. Su and Chern, (2004), used stainless steel as a conductive filler to produce core, cover and plied yarns to make different types of woven fabrics. The electromagnetic shielding effectiveness (EMSE) of these fabrics was measured by coaxial transmission equipment. Electromagnetic frequencies range from 9KHz to 3GHz. The experimental results showed that a denser fabric structure had a higher EMSE. In respect to the influence of yarn structure, a fabric made from the core yarn has a higher EMSE than fabrics made from cover or plied yarn [2].

Cheng et al., (2006), produced twill copper fabrics (3/1) and studied the effects of varying weft density, warp density, wire diameter and lay-up angle on EMSE. They used a coaxial transmission line holder in the frequency range of 144- 3000 MHz. With an increase in the number of conductive fabric layers, warp and weft density, an increase in shielding effectiveness was observed, while with an increase in wire diameter, a decrease in shielding effectiveness occurred [3].

Das et al., (2009), studied the effect of material type, yarn count, pick density, type of mordant and layers of fabrics on the EMS properties of textile materials. The shielding effectiveness of the fabric was measured by estimating the insertion loss incurred to the signal when the sample is placed in the path of the signal at the frequency range 100 MHz- 3 GHz inside the coaxial transmission holder. They stated that there are some effects of type of material, yarn count, number of fabric layers and type of mordant on EMSE whereas, the number of apertures and thread density do not have significant effect on EMSE, particularly in case of metallic sheets [4].

Duran and Kadoglu, (2015), studied woven fabrics, produced with two different types of conductive yarns, namely silver containing (Ag/PA,Co) core yarns and silver-containing (Ag/PA-Co) blended yarns. The effect of various yarn and fabric properties on EMSE was investigated. They obtained results that the shielding effectiveness can be tailored by changing the yarn and fabric parameters and also there are significant differences between the electromagnetic shielding characteristics and performances of the fabrics produced with different types of yarns. Such fabrics are convenient for both daily and professional uses, since they have both high EMSE performance and comfort properties [5].

2. Material and Method

2.1. Production

Yarn production was carried out on a ring spinning machine. Copper core cotton yarns (Cu/Co) were produced by using copper filaments as core material and cotton as sheath matersial. A core yarn apparatus was used in the machine for feed the conductive filament into the centre of the cotton

IOP Conf. Series: Materials Science and Engineering **459** (2019) 012070 doi:10.1088/1757-899X/459/1/012070

sheath before twistings [1]. 0,05 mm and 0,07 mm copper wires were used as core material. All of copper core cotton yarns were produced in Ne 8 yarn count [6].

After the yarn production, copper core cotton yarns were integrated to the 3/1 twill woven fabric structure in the weft, with five different weft densities, namely 8, 13, 18, 23, 28 wefts/cm, for the evaluation of their electromagnetic shielding effectiveness (EMSE) properties [1]. Ne 20/2, 100% cotton yarns were used as warp yarns for all the fabrics. Fabric samples have been coded according to their wire diameters and weft densities [7]. For instance; Sample 5.28. has 0,05 mm copper core and 28 wires/ cm weft density.

2.2. Tests

Electromagnetic shielding properties of the woven fabrics were tested by using unechoic chamber test system according to EN50147-1 standard in 1GHz- 6GHz frequency range. The results were attained in decibels (dB) [5].

The system is consist of a signal generator which generates signals, an RF power amplifier which amplifies the signals before being sent to the sample, two antennas and two adjacent shielded rooms, each having one of the antennas inside (First antenna connected to the signal generator and second to the spectrum analyzer as signal receiver) and a spectrum analyzer which analyzes the signals obtained from the receiver antenna [1].

Fabric was placed the gap between two rooms (two antennas) during the measurements. Signals were produced and amplified and then sent onto the fabric by an antenna. The signals transmitted by the fabric were detected by the receiver antenna, situated on the other side of the fabric [1]. Results were read from spectrum analyzer. EMSE of the copper core conductive fabrics were measured at frequencies ranging from 1 GHz to 6 GHz [8]. Each fabric sample has been measured three times. So as to obtain the actual shielding effectiveness results, blank measurements (without fabrics/ shielding materials) are regarded at all frequencies [7]. The results were analyzed by using Excel tables, diagrams and SPSS software program.

3. Results and Discussion

The effects of wave frequency, fineness of the core filament and weft density on the electromagnetic shielding effectiveness of fabrics were investigated.

3.1. Effects of Wave Frequency

According to the ANOVA and SNK tests, wave frequency had a statistically significant effect on the shielding effectiveness [1,5].

Change of EMSE of fabric samples with wave frequency are given in Figure 1.

IOP Conf. Series: Materials Science and Engineering **459** (2019) 012070 doi:10.1088/1757-899X/459/1/012070

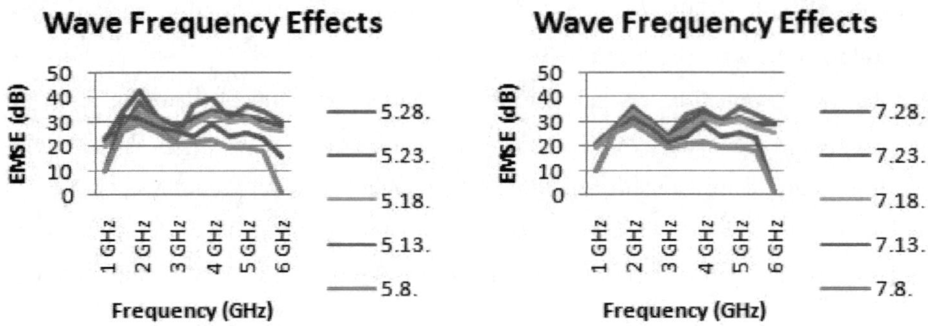

Figure 1. Change of EMSE of fabric samples with wave frequency.

Figure 1 shows the electromagnetic shielding effectiveness changes of fabrics according to wave frequency. When the wave frequency rise, (despite deviations) EMSE of three samples have 28, 23 and 18 wires/cm weft densities indicate slight increasing tendency but the samples have 13 and 8 wires/cm indicate slight decreasing tendency.

3.2. Effects of Core Fineness

According to the Paired Samples T Test, the core fineness has a statistically significant influence on the shielding effectiveness of the fabrics [3].

Figure 2 presents the variations of average EMSE for the fabrics with different core fineness in the frequency range 1 GHz – 6 GHz [9].

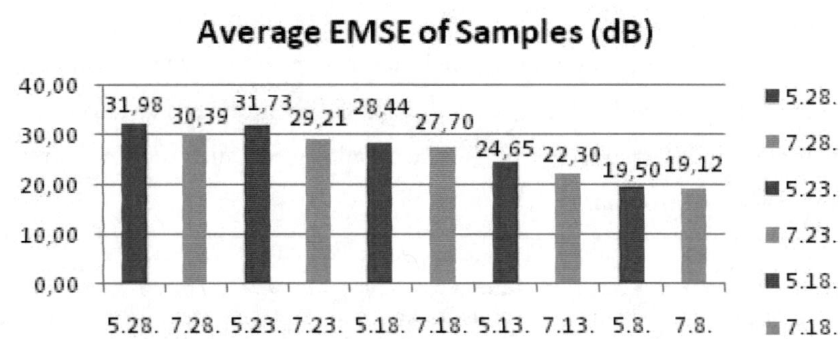

Figure 2. Variation of EMSE of samples with core fineness and weft density.

With an increase in wire diameter (wire becomes thick), a decrease in shielding effectiveness has been observed [3]. The reason of this, as copper filament becomes thick, its rigidity soars. Increased rigidity of the copper filament, makes it difficult to bent and take the form of copper. This can cause apertures on the fabric structure [1,8]. Therefore, EMSE of fabrics decrease. According to Figure 2, fabrics have 0,05 mm core fineness indicate higher EMSE than 0,07 mm at all frequencies.

IOP Conf. Series: Materials Science and Engineering **459** (2019) 012070 doi:10.1088/1757-899X/459/1/012070

3.3. Effects of Weft Density

In accordance with the results of ANOVA and SNK tests, weft density had a statistically important effect on the shielding effectiveness.

Figure 2 shows the changes in average EMSE of samples with different weft densities. As the weft density increase, the fabric comprises more conductive materials so EMSE values of the fabrics increase, as well. The highest average EMSE value obtained from the fabrics have 28 wires/cm weft density.

4. Conclusion

In this work, textile materials were developed for protecting from electromagnetic waves and their usage areas were investigated. For this purpose, Cu/Co core yarns were produced by using copper filaments as core material and cotton as sheath material on a ring spinning frame. Copper monofilaments of 0.05 mm and 0.07 mm were used as core materials. All core yarns are produced in Ne 8 yarn count. Conductive Cu/Co yarns were integrated in to 3/1 twill woven fabric structure in the weft direction with 5 different weft densities (8, 13, 18, 23 and 28 per cm).

The effects of wave frequency, core filament count and weft density on the EMSE were investigated. The shielding effectiveness of these fabrics were measured between 1GHz and 6GHz using the TS EN 50147-1 standard [10].

The results have shown that there are significant effects of fineness of the core filament and weft density on the EMSE. The fabrics that was produced with 0,05 mm core filament have represented higher EMSE than 0,07mm core fineness. EMSE values of the fabrics increased with increasing weft density [11]. Since the fabric comprises more conductive materials.

The highest results were obtained from the fabric that was produced with 0,05 mm core filament and 28 wefts/ cm weft density (sample 5.28) at 2 GHz. Average EMSE value of the sample 5.28 is 31.98 dB (Figure 2).

In conclusion, since satisfactory results were received, the fabrics produced in this research can be used for electromagnetic shielding. Especially against devices which emit between 1 and 6 GHz frequency as cell phones, computers, tablets, radios and household electrical appliances. For this purpose, the produced fabrics are thought to be used in various applications in both professional and daily life such as curtain, mosquito netting, apron, tent, awning and protective clothing.

IOP Conf. Series: Materials Science and Engineering **459** (2019) 012070 doi:10.1088/1757-899X/459/1/012070

References

[1] Duran D and Kadoğlu H 2012 A research on electromagnetic shielding with copper core yarns *Journal of Textiles and Engineer* **4** p 354-359

[2] Su C I and Chern J T 2004 Effect of stainless steel containing fabrics on electromagnetic shielding effectiveness *Textile Research Journal* **74** p 51-54

[3] Cheng K B, Cheng T W, Nadaraj R N, GiriDev V R and Neelakandan R 2006 Electromagnetic shielding effectiveness of the twill copper woven fabrics *Journal of Reinforced Pylastics and Composites* **25** No 7 p 699-709

[4] Das A, Kothari V K, Kothari A and Kumar A 2009 Effect of various parameters on electromagnetic shielding effectiveness of textile fabrics *Indian Journal of Fibre&Textile Research* **34** p 144-148

[5] Duran D and Kadoğlu H 2015 Electromagnetic shielding characterization of conductive woven fabrics produced with silver containing yarns *Textile Research Journal* **85 (10)** p 1009–1021

[6] Yesildag İ 2017 Researching of copper filaments availabilities on electromagnetic shielding textiles Master Thesis (Turkey: Ege University)

[7] Özdemir H and Özkurt A 2013 The effects of weave and conductive yarn density on the electromagnetic shielding effectiveness of cellular woven fabrics *Tekstil ve Konfeksiyon* **23(2)** p 124-135

[8] Perumalraj R, Dasaradan B S, Anbarasu R, Arokiaraj P and Harish S L 2009 Electromagnetic shielding effectiveness of copper core woven fabrics *The Journal of The Textile Institute* **100** No 6 p 512-524

[9] Chen H C, Lee K C, Lin J H and Koch M 2007 Fabrication of conductive woven fabric and analysis of electromagnetic shielding via measurement and empirical equation *Journal of Materials Processing Technology* **184** p 124- 130

[10] Safarova V and Militky J 2014 Electromagnetic shielding properties of woven fabrics made from high performance fibers *Textile Research Journal* **84(12)** p 1255–1267

[11] Chen M Y, Lai K, Sun R J, Wu H and Zha A X Electromagnetic shielding effectiveness of fabrics with metallized polyester filaments *Textile Research Journal* **77(4)** p 242-246

IOP Conf. Series: Materials Science and Engineering **459** (2019) 012071 doi:10.1088/1757-899X/459/1/012071

Market readiness of smart textile structures – reliability and washability

S U Zaman, X Tao, C Cochrane and V Koncar

ENSAIT/GEMTEX, 2 Allée Louis et Victor Champier, Roubaix, F-59100, France

Université of Lille, Cité Scientifique, Villeneuve d'Ascq, F-59650, France

Corresponding author email address: shahood-uz.zaman@ensait.fr

Abstract. Even if smart textiles gained a certain level of maturity, they are not yet ready for the market at large scale. The main issues explaining this are related to problems of reliability and difficulties to launder smart textile structures. Smart textiles are also facing problems for launching in market due to lack of standardizations. This paper is focused on washability standards of smart textiles. The objective is to co relate available textile mechanical tests with the washing of smart textiles and to observe the change in their performances. First, different washing functions and washing factors acting in laundering machines are analysed and discussed. Then Martindale abrasion tests and pilling box tests were performed in order to detect damages similar to those provoked by washing process. Two different types of silver coated conductive threads sewn on cotton plain woven fabric are used. Changes in threads resistance during these tests are analysed and correlated with modifications of electrical conductivity after washing of the similar samples. Therefore, the mechanical stresses underwent by conductive threads during washing cycle have been identified and reproduced by standardized testing equipment. More generally, this paper emphasizes all the problems, encompassing efforts that industry and research labs have to realize to make smart textile structures more robust and able to be cleaned or washed similarly to everyday textile products such as underwear, clothing, home textiles and even technical textiles. It gives indications on how the issues related to reliability and washability of smart textile structures have to be addressed in order to make them reliable and robust enough to be put on the market.

1. Introduction

Smart textile, also called e-textile, involves wearable electronic industry that is penetrating in our daily life now days. This new developing sector has captured numerous fields of applications including different trends in leisure and sports, some lifesavings in health care, security and military. In initial stages of this industry, main focus was on to use e-textile in healthcare applications. Although now, smart textile is being used widely in other fields like wireless communication, smart sporting, smart watches, monitoring etc. [1] Printed electronics and energy harvesting technologies are evolving to meet the demands of new, wearable formats. Next-generation wearables and smart technical textiles will rely on active fabrics conductor, insulator and semiconductor fibers. These products will integrate electronic functions with everyday comfortability of clothing and thus enhance the user experience [2]. Smart textiles can be defined as textiles that are able to sense and respond to changes in their environment. Conductive polymers and metals may be integrated in to e-textile structure for suppling conductivity. These conductive yarns should maintain the minimum requirement for conductivity and also physical characteristics [3]. In broader view, smart textile can have three types of components. Smart textile can have electronic modules equipped with a micro controller, memory and input output interfaces and it is dedicated to data processing and decision purposes. Smart textile can contain sensors, actuators and electrodes. These are normally used to detect external stimuli. Smart textile can contain connecting conductive yarn with different types of assembling techniques, used for connections among antennas, sensors and modules.

Even if smart textiles gained a certain level of maturity, they are not yet ready for the market at large scale. The main issues explaining this are related to problems of reliability and difficulties to launder smart textile structures. Washability of smart textile is one of the major hurdles to launch the product in market as user friendly and reduces the reliability of the product. These connecting yarns are more difficult to protect especially at connection points between yarn and sensors, actuators etc. Actually, interface between hard and soft point is the weakest point in smart textile structure increasing risks to be damaged within washing process. This is a big barrier to their readiness to market.

Washability of smart textile has gained attention now days and a lot of work is has been performed to make washable ready to market smart textile products. Washing itself is complex process as so many factors are involved in it. Numerous chemical and physical forces are acting during this process. Washing process is also sensitive to clothing type, quality of water, type of detergent and washing procedure adopted. Mechanical actions can be explained in term of abrasion among surfaces of textile structures, and with the wall of washing drum, hydrodynamic flow of solution and flexing action etc. [4].

Before analysing the effect of washing cycle on conductive yarns, it is important to analyse influences and stresses underwent by smart textiles during it. A washing cycle may be decomposed in four different damaging actions with possible interactions: mechanical stress (bending, torsion, friction, etc.), thermal stress (temperature), water stress and chemical influence (detergent) [5]. The mechanical stress is probably the most damaging factor on smart textile structures, particularly on conductive yarns dedicated to connect electronic modules with sensors & actuators and the interconnection between the conductive yarn and the electronic component in e-textile.

For washing of smart textile, there are two possibilities. One is laboratory standardized washing procedure and the other is washing in household washing machines. In order to be successful in smart textiles, products should be able to withstand washing in normal washing machines as ultimately they will be laundered. For these reasons, the focus is on commercially available washing machines. They have different washing cycles with difference in time, washing speed and drying capabilities.

2. Procedure & Experiment

To make smart textile reliable and washable, it is compulsory to understand exactly what is actually happening during laundering with lot of constraints that are playing role in the sample. These constraints can be explained as mechanical forces, chemical actions with detergent, water influence, effect of temperature and washing time. These washing factors are shortlisted in Figure 1. For reliability and washability, smart textile should withstand in all these circumstances. Smart textile can be performed in these pre-defined situations to predict their washability. However, all these constraints are interlinked to each other and interdependent. Their effect can be changed when performed independent to each other. One example is effect of water on smart textile that may be different in idle state as compared with effect during some mechanical forcing, such as abrasion; bending or frictions are acting on it mechanical forcing, such as abrasion; bending or frictions are acting on it.

Among all these washing factors, the mechanical stress during washing cycles is factor that is difficult to control. Standardized available mechanical tests textiles have been performed and then these test results were correlated with washability of smart textiles. During washing, samples are rubbed with other clothes and also with walls in washing drum of machine. These actions are supposed to have major effect on smart textiles durability. Martindale abrasion test can be considered as substitute to verify the effect on conductive yarns as the week point of smart textiles without washing the samples.

Two conductive yarns from different companies were used. These yarns are marked as yarn A and B. Both yarns were continuous filament polyamide silver coated yarns. They were two ply yarns consists of 117 fibres each having 20 μm diameters. These conductive yarns, of known length, were stitched on plain cotton fabric. Silver paste was pasted on stitched ends points of yarn to avoid wear off during mechanical abrasion. Martindale abrasion tests were performed on Martindale Abrasion Tester by James H. Heals & Co Ltd. Load on samples during testing was 9 KPa and woven felt was used. Unidirectional movement of Martindale top motion plate was selected. Linear resistance was measured after each 500 abrasion cycles up to 3000 cycles.

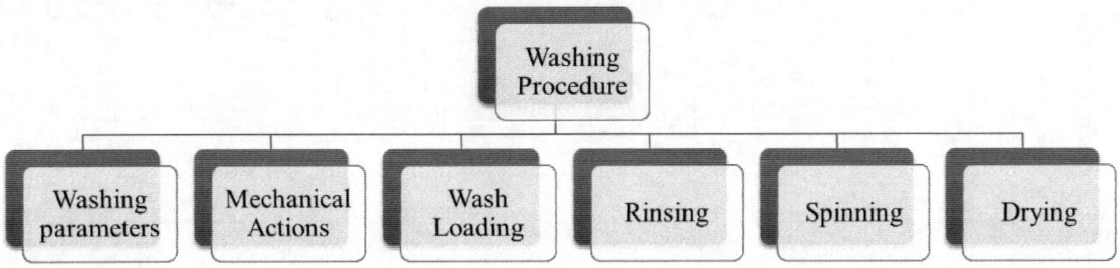

a)

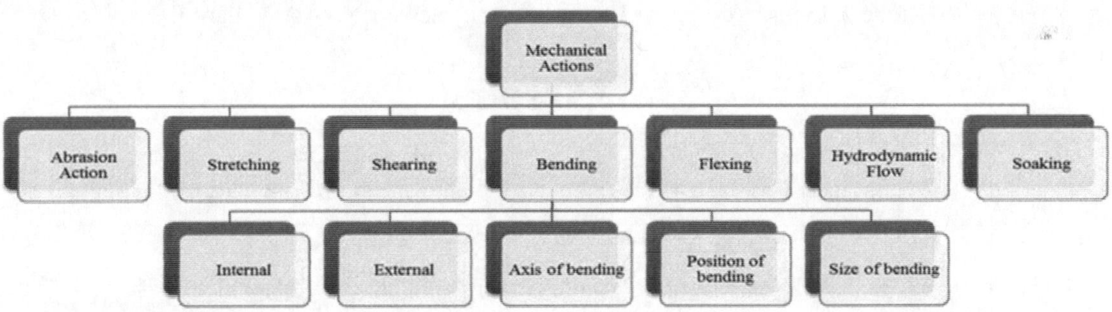

b)

Figure 1(a) Factors involved in washing process, **(b)** Mechanical factors involved in washing process.

For washing of samples "Silk" washing process was selected on local consumer used washing machine. Reason behind this selection was gentle washing and no interim tumbling during rinsing. Washing speed was 15 revolutions per minute (rpm), while tumbling speed was 400 rpm. Samples were dried after each washing at controlled room conditions (65%±5% R.H and 25°C± 2°C). Linear resistance was measured by Agilent digital multi meter using standard probes.

3. Results & Discussion

Abrasion test was performed in the group of 500 cycles for total 3000 abrasion cycles and linear resistance was measured after every 500 cycles. Average linear resistance against abrasion cycles is shown in Figure 2 for both yarns A and B. Resistance increases with increase in number of abrasion cycles. Thin black line shows fitted linear trend and thick blue line shows actual values. Fitted linear equation with R square values are in the graph. R square value for yarn A is better as compared to yarn

B. A sharp increase of electric resistance is observed at beginning of wash process. This initial damage of conductive layer on yarns maybe due to the damage of superficial silver, which are not strongly adhered onto the surface of thread.

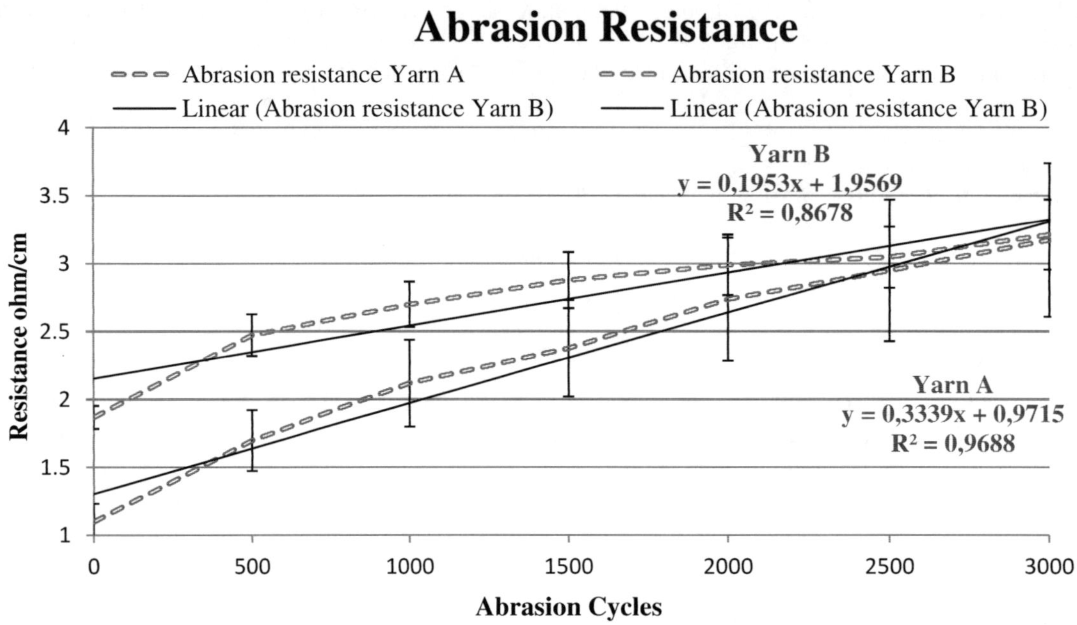

Figure 2. Linear Resistance after abrasion cycles for Yarn A & Yarn B

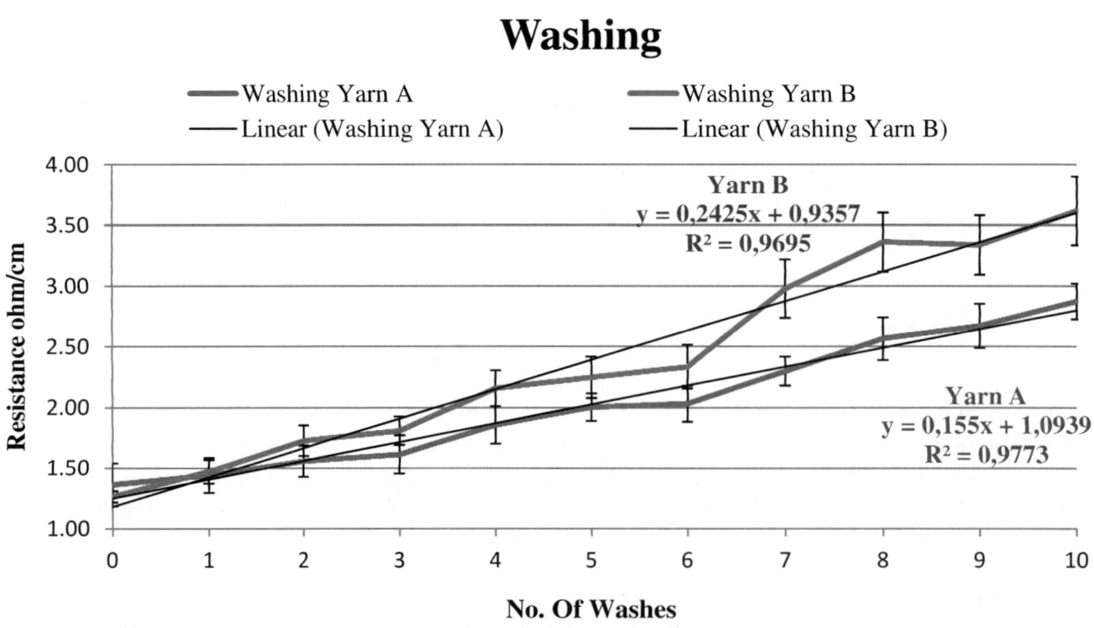

Figure 3. Linear resistance after washing for Yarn A & Yarn B

Also, at initial stage, the temperature of surface increases from low to high with the abrasion cycle. That's why at initial stage abrupt change is observed and afterward straight line was noted. For yarn A initial value of conductive thread was 1.10 Ω/cm and increased to 3.17 Ω/cm after 3000 abrasion cycles. For yarn B initial value was 1.87 Ω/cm and increased to 3.21 Ω/cm after 3000 abrasion cycles. In comparison yarn A has less initial value as compared to yarn B, but after 3000 cycles both yarns have almost same values. This means change in resistance is more important in yarn A as compared to yarn B.

Figure 3 shows the resistance per centimetre against number of washing cycles. This trend is also linear as in abrasion cycle experiment. Thin black line shows linear trend and thick blue line is actual values. Fitted linear equation and R square values for both yarn A and B are given in graph. R square value for both yarn A and yarn B are almost same and near to 1. This shows better linear trend for both yarns A and B. For yarn A resistance increased to 2.88 Ω/cm after ten washing cycles. For yarn B it was 3.6 Ω/cm after ten washing cycles. After washing, change in resistance in yarn B was more as compared by yarn A. Figure 4 shows the co relation between abrasion resistance effect and washing effect on smart textile proto types. For both yarns it can be predicted that 1500 abrasion cycles is equivalence to about 7 washing cycles (keeping in mind the washing type selected for this experiment). These experiments are initial outcomes for co-relation of mechanical tests with washing of smart textiles. Further experiments in future can give the final statement about the relation.

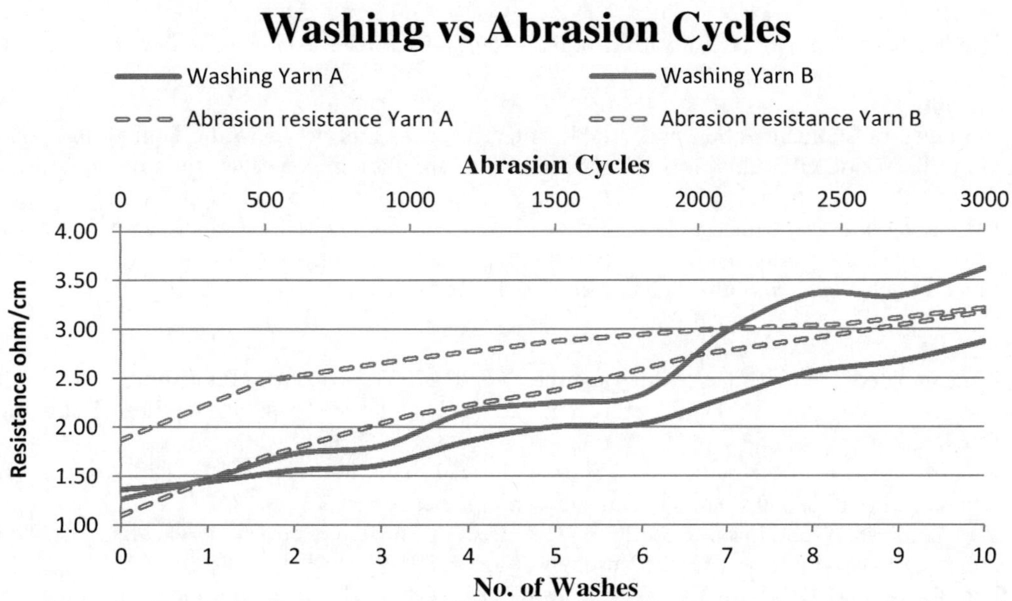

Figure 4. Co-relation between washing and abrasion resistance

Figure 5 shows the optical microscopic picture of conductive yarn at initial stage, after washing and after abrasion resistance. These pictures clearly point out the whitish damaged spots after washing and abrasion tests. These both tests involve mechanical movements and wear and tear that cause damage of surface. These spots are created due to damage of silver layer that is responsible for conductivity of yarn. These damaged parts caused the increase in resistance and decrease in conductivity of conductive yarn.

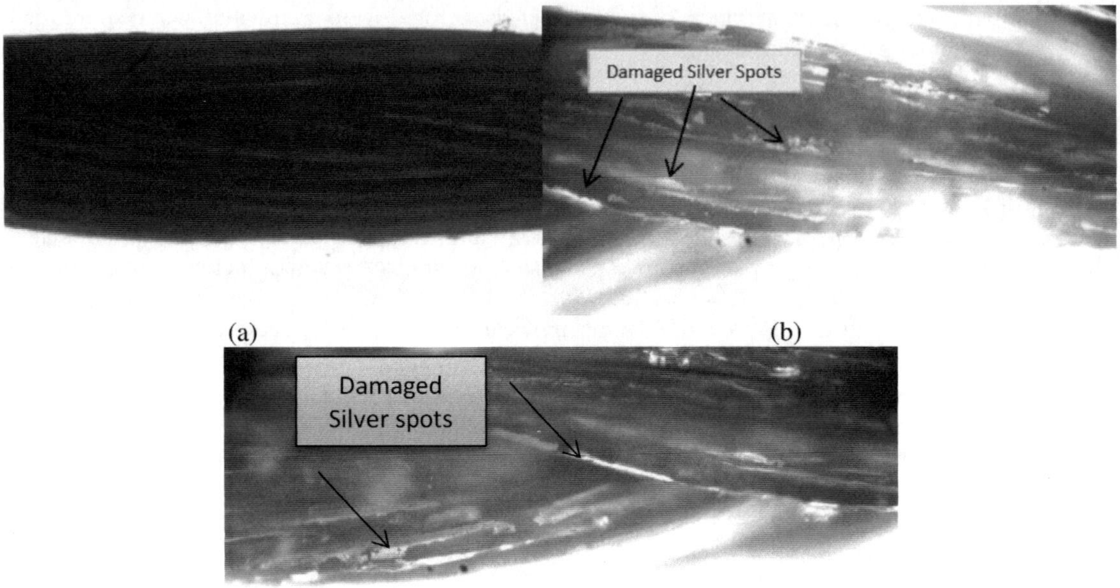

(c)

Figure 5. Microscopic picture of yarn (a) Initial stage, (b) After washing, (c) After abrasion test

4. Conclusion

The importance of standardization and establishment of tests able to fit to the laundering of smart textiles have been studied in this work. First approach to simulation of mechanical stresses underwent by conductive threads embedded to textile fabric, necessary to connect and supply sensors, actuators and electronic modules contained in smart textiles has been set up. The final objective of our research work is to establish standards and tests supposed to help companies involved in e-textiles development to meet market demands and realize reliable and washable products.

5. References

[1] Koncar, V. ed., 2016. Smart textiles and their applications. Woodhead Publishing.
[2] Xuyuan. T ed. and V. Koncar sec. ed., 2015, Handbook on smart textiles, Hong Kong: SPRINGER.
[3] Tao, X., Koncar, V., Huang, T.H., Shen, C.L., Ko, Y.C. and Jou, G.T., 2017. How to make reliable, washable, and wearable textronic devices. Sensors, 17(4), p.673.
[4] Yun, C., Park, S. and Park, C.H., 2013. The effect of fabric movement on washing performance in a front-loading washer. Textile Research Journal, 83(17), pp.1786-1795.
[5] Lee, A., Seo, M. H., Yang, S., Koh, J., & Kim, H. 2008. The effects of mechanical actions on washing efficiency. Fibers and Polymers, 9(1), pp. 101-106.

IOP Conf. Series: Materials Science and Engineering **459** (2019) 012072 doi:10.1088/1757-899X/459/1/012072

Integrating modern virtual engineering tools in footwear design and development

P Azariadis, P Papagiannis and Z Koutkalaki

University of the Aegean, School of Engineering, Department of Product & System Design Engineering, Ermoupolis, Syros, GR84100, Greece

azar@aegean.gr

Abstract. Footwear prototyping is constrained by significant volume of "sampling", time to market pressures, cost implications and the failure of the industry to attract and develop high quality professionals mastering modern computer-aided product development. Computer-Aided Engineering (CAE) and material selection are becoming increasingly important transforming factors of this process raising the question of how to best integrate them into the design function considering not only infrastructure issues but also human resources and financing. Their role in footwear development is presented in this paper, based on the view of design as a multistage process subject to Concurrent Engineering. Proposed interventions to the design process for exploiting modern CAE and material selection tools are discussed, taking into account the business Value Chain Model and modern Cloud Service approaches. Innovative approaches to harness the processing power of modern CAE tools are identified and related to business practices, considering, also, the role of product development professionals and financing constraints. In particular, the application of the cloud service model of managing Software as a Service is discussed along with relevant opportunities and limitations. The paper concludes with an overall assessment of the impact that this model may have on the footwear industry.

1. Modern Footwear Development

Footwear range from established traditional look-like products (e.g. classic leather shoes) to highly innovative complex structures utilizing modern materials and the state of the art in textile and additive technologies (e.g. wrap knitting, nanotechnology, 3D printing). They are assemblies of multi-material parts with extended supply chains, so formulating an effective product strategy dictates analysis of a complex global political, economic, social, technological, legal and environmental operating macro-environment [1], [2], [3]. Aligning with fashion and consumers' behavior are not the only challenges to address. Products have to comply with standardization requirements. They also need to achieve acceptable performance, comfort, durability and reliability for the intended use. Additionally, they have to comply with consumer safety, environmental and sustainability requirements. Operating globally is complicated due to a volatile political and economic environment that affects both demand and supply, driving material substitution and assembly location decisions. This is further affected by new trade wars and escalation of tariffs and other barriers to trade. In global trade, managing human resources and stakeholders means dealing with varied educational levels, different cultural and linguistic backgrounds and different approaches to problem solving and work organization. The members of footwear design teams may well be based at distant locations around the world.

Today, the effective and efficient development of new products has become a competitive advantage [3]. Although the contribution of Strategy, Financing, Human Resource Management, Purchasing, Logistics and Marketing cannot be ignored, Design is the heart of product development. Design is a process with inputs, outputs and performance metrics [1], [4] as depicted on Figure 1. There are five stages in the process: conceptualization, concept screening, preliminary design, evaluation and improvement, prototyping and final design [4]. These overlap, giving rise to the concept of concurrent engineering and its variants, which are enhanced through the use of Information and Communication Technology (ICT) tools, and Product Data Management (PDM) [5].

Figure 1. The Design Process (Adjusted from [4]).

In the footwear sector, prototyping usually requires a significant volume of "samples" prior to finalizing the details of the final product [5]. In addition, time to market is short, whereas overall product cost pressures are often significant. Furthermore, cost implications push product development budgets down, constraining design. Another issue is that the industry, generally, has failed to attract high quality professionals to staff critical development positions. Footwear designers have manufacturing or design-artistic background; however, most of their skill base development takes place through long practical industrial apprenticeships with limited exposure to modern scientific and engineering advances, in most cases.

Several frameworks to improve processes in the product assembly industries have been proposed ranging from incremental change ones (e.g. Total Quality Management) to radical Business Process Reengineering. These frameworks rely heavily on integrated ICT tools [6]. Modern Enterprise Resource Planning (ERP) systems incorporate PDM capabilities to support product development functions. Besides managing the extended supply chain, they even offer product style management capabilities, though they cannot address product visualization, pattern engineering, tooling development, as well as, product performance and sustainability assessment. With regard to product geometric modeling and rendering, pattern engineering and tooling (lasts, molds, cutting dies) there are already commercial CAD/CAM solutions, with increasingly improved interoperability. This, though still hindered by many complications [7], tends to increasingly satisfy the needs of several companies around the world. However, the industry has not managed to integrate CAE into their processes yet. In addition, during the last 20 years, the footwear sector has seen the emergence of several supplementary innovative aids to product design. It is the purpose of this paper, to revisit modern footwear design in the light of these advances and identify alternative practices that will facilitate increased effectiveness and efficiency of the process.

Technology has the potential to transform any process, but the ultimate success depends on the people involved. Access to knowledge and enhanced decision-making, both supported through improved training are of utmost importance [6]. This paper takes into account this view and also considers the third transforming resource input to the design process: financing. Human resource development and technology come at a cost that most enterprises cannot easily absorb into their budgeting. Fortunately, innovative cost-effective approaches to harness the processing power of modern CAE tools do exist and will be discussed along with their relation to business practices.

2. Computer-assisted material selection in the footwear industry

Material selection software addresses performance, consumer safety, environmental and sustainability aspects of design. With regard to performance, general engineering material-selection software cover common industrial materials used in footwear (e.g. wood, alloys, plastics, coatings, adhesives). Some interesting approaches to dealing with the far more challenging textiles, leather and footwear composites, including “shape memory” foams and membranes, also exist. Textile fabrics, in particular, are complex products and their final properties depend on fiber and yarn characteristics, fabric geometry, as well as any associated dyeing and finishing process. Within different design solutions for clothing, some applications and databases are also of interest to footwear. The LibTex database [8] based on analytical and regression models can be used to calculate fabric performance characteristics from fiber, yarn and textile design parameters. Though information on tensile characteristics of simple fabrics is useful, several textiles used in footwear manufacturing have been altered through special finishing processes, stiffened or used in composites. Also, several other critical problems related to the cyclic loading of 3D footwear structures are not easy to address. A database of textile, leather and footwear materials was also developed as part of the Envirotex Design project [9], in order to support the Interactive Matrix, a product performance evaluation module based on neural networks.

Regarding consumer safety and sustainability, both product and process aspects have to be considered along the supply chain. Relevant assessment methodologies can be facilitated through ICT aids. Most general-purpose material-selection applications offer environmental assessment capabilities that address substance restriction issues, carbon footprint, embodied energy, water usage, resource consumption, end of life and supply risk aspects. There are few applications specifically developed for the footwear industry. These have varied degree of coverage for materials and processes. Some well-known aids are the Higg Index Tools by the American Sustainability Apparel Coalition [10], the ShoeLaw tool [11] on legal requirements for footwear products and processes, the ShoeBAT tool [12] on Best Available Techniques for footwear production, the CO2SHOE estimator [13] for carbon footprint and the Envirotex Design platform [9], which addresses both process and product aspects. With the exception of the Higg Index, which is fully supported by an Association, all the other ones developed within cooperation programs have limited long term deployment and support.

3. Computer-Aided Engineering in the footwear industry

CAE in footwear design is limited to virtual performance evaluation (mechanical, thermal, comfort, etc) for an intended use and virtual optimization, in which key parameters (topological or material properties) are determined to achieve certain product performance. Footwear CAE takes place at either component or final product level. CAE for solid industrial components (e.g. heels, toe caps), that are normally inputs to the footwear industry, follows standard mechanical analyses practices and will be no further considered. With regard to textiles and leather, several analytical models, stochastic models, neural networks and other techniques exist [9]. Although useful for guiding material selection, these concern the performance of materials in flat form. However, footwear products have complex 3D topology and multi-material structure. Techniques such as multivariate analysis and Finite Element Analysis (FEA) have been applied on sole, upper and full footwear level for assessing performance of plantar pressure distribution, shock absorption, bending and torsion characteristics, upper pressure on the foot, upper fitting, plantar fitting, thermal and vapor exchange characteristics [9], [14].

FEA is suitable for problems, with complex loading and boundary conditions, involving the interaction of several multi-material solids with complex geometries, such as the footwear-foot-ground system. The availability of accurate biomechanical information on human motion and digital foot biomodels are fundamental to the application of FEA in footwear [15], [16]. Deploying FEA requires a high level of expertise; however, once the analyses are parameterized and standardized, then it is possible for non-experts to apply the techniques. This approach was used in the Optshoes project [17] for determining footwear comfort characteristics, where a simple to use web-based interface limits interactivity to parameterization of input files without any further involvement with the underlying FEA kernel [7]. The entire process is facilitated by using a specialized Material Database.

4. Integrating Computer-Aided Engineering and Material Selection in modern footwear design

Computer-aided engineering and material selection are transforming services that support engineering aspects of product development. These follow the stages of conceptualization, concept screening and preliminary design. The footwear industry is gradually moving towards the application of product engineering practices [9]. These are not just value adding but increasingly important elements of the value chain. Depending on the knowhow and infrastructure of the firm concerned and the economics of the process, design can either be an internal function or outsourced to specialists [18]. Large sports and technical footwear enterprises and smaller niche manufacturers of customized solutions often invest in specialized product engineering infrastructure. In these sectors the value of product development is significant. Owning the CAE functions ensures strict control of them, as well as quick response and flexibility of the CAE "service" to emerging market demands. When the intensity and value of product engineering activities are lower, then it makes business sense to outsource the whole function or part of it. In this case, the company has to avoid any adverse effects on the dependability on design. It has to ensure acceptable quality and timeliness of the outsourced design service and reasonable flexibility to emerging demands with all partners involved. In addition, any outsourcing has to be cost effective and meet budgetary requirements.

The strategic importance of design will continue to rise [3] and so will the need for dependability on its outsourced functions. A question of how to best deploy outsourced functions given the current state of technological progress, thus, arises. Inter-organizational deployment of CAE and material selection software that overcome limitations on design transforming resources is increasingly feasible through advances in engineering design technology and this facilitates outsourcing. Modern solutions now support collaboration, access to data on multiple sites and enhanced remote 3D visualization across different platforms. However, they require thorough and integrated development of both system architecture and software, with the latter exploiting advances in remote visualization, communication and distributed geometric computing algorithms [6], [7]. The aim of these is to overcome the key non-functional limitations of collaborative CAD/CAE systems: performance, usability, and scalability. Today, the realization of collaborative 3D visualization and access to multisite databases is taking place over the web. CAD/CAE inevitably follow wider trends in the IT Industry which is evolving around an Internet-based service model, according to which most end users receive services on demand, irrespectively of their location and means of delivery. Cloud computing seems to have evolved to the main tool for realizing this model, through external hosting of intensive processing and storage functions and delivery of software-based services to end users over the Internet [7]. It reduces client-side hardware and software requirements, as well as complexity and is appropriate for outsourcing processing power and storage. In the case of CAD/CAE, which is specialized service offered by specialist providers, the cloud service model of managing Software as a Service (SaaS) appears to be a promising manufacturing option. The concept of Software as a Service allows for quick deployment of necessary software updates and changes, access to specialized IT functions and takes the burden of hardware installations, system upgrades and maintenance off the client [7].

The question of whether to deploy fat or thin client configuration in order to deliver CAD/CAE as a Service is a critical one. Though fat client configurations may be feasible and cost effective for companies with extended and intensive footwear development activities, supported by all necessary hardware and the state of the art in design software, this is the exception rather than the rule. The reality for most manufacturers is that CAE is a peripheral activity and product development facilities and staff are not geared towards fully supporting it. Thin client visualization configuration, on the other hand, allows for resource intensive computations and rendering performed at remote specialized servers with only minimum processing taking place at the client's end. However, footwear manufacturers considering thin client options, have to take into account adverse effects on model quality, reduced speed of rendering, 3D content latency, limitations on interactivity, as well as issues regarding proprietary or open technologies. In fact, there are still several compatibility problems between different platforms utilizing various plug-ins and multimedia APIs, for some of which license fees are required. It is expected that the adoption of the new WebGL for graphics and HTML 5 for website applications will gradually leverage modern systems from such legacy issues [7]. With regard to improving 3D content latency, formats such as VRML, X3D, W3D and MPEG-4 have contributed to the emergence of innovative and faster CAD/CAE file exchange approaches [7]. In order to improve the performance of the thin client approach, the concept of the 'Balanced Client-Server' can be incorporated, where intensive data and visualization processing is taking place at the server close to the data, while final visualization and rendering are performed by the client. The key features of the aforementioned approaches are listed in Table 1.

A paradigm of CAD/CAE cloud-based deployment as a service, through the balanced client-server approach, for estimating the comfort parameters of footwear designs was developed recently [7], [17]. In this implementation, data management and processing take place at Cloud-based servers and the clients are only provided with simulation results. Through the development of a model-view-controller pattern, the end-users (typically footwear designers and specialists) do not need to engage in FEA tasks (i.e. pre-processing, boundary conditions), since the web-based interface allows for easy parameterization. In addition, licensing issues for CAD/CAE software are not the responsibility of the end-user but of the service provider. This minimizes service costs for the designer/manufacturer.

Table 1. Key features of fat, balanced and thin client-server configuration.

	Level of processing	Infrastructure	Interactivity
Fat Client	Most visualization, rendering and display data at client's side.	Heavy for graphics processing Increased network bandwidth.	Very high.
Balanced Client	Data intensive visualization and rendering on server. Final rendering and display on client.	Sufficient accelerated graphics units and software plug-ins.	Moderate.
Thin Client	Visualization and rendering processing on server. Only imagery sent to client.	The minimum of modern display arrangements.	Only pre-processed images displayed.

The deployment of CAD/CAE as a Service on demand, utilizing cloud computing, through ICT infrastructure typically found in modern footwear enterprises is therefore considered practically feasible. The end users do not need to be specialists to use key CAE and material selection services, heavy processing tasks are performed at the server side and the solutions have reasonable cost. In addition, with regard to material selection software, providers have come up with innovative modular offers, using cloud computing for managing queries taking the burden of operating extensive material databases and intensive query tools off the clients.

With regard to human resources, it is still important that key professionals understand the concepts and phenomena involved, are able to parameterize their design problem, understand the basic concepts of CAE tools and interpret the simulation results. The role of relevant training is essential in order to equip designers with new tools for developing competitive products in the footwear sector. International cooperation initiatives, such as [19], can be utilized to successfully provide training modules to the footwear industry in order to cope with the modern product design demands.

5. Conclusion

The options for integrating CAE and computer-assisted material selection in modern footwear design were investigated considering their Value Chain position and the state of the art in technology available for the sector. The key finding is that advances in cloud computing are making deployment of footwear CAD/CAE as a Service feasible. Further improvements of this approach are expected through disengagement from legacy plug-ins and APIs, as well as overcoming 3D content latency issues, as new development tools (WebGL, HTML5) and formats (VRML, X3D, W3D and MPEG-4) become widely adopted. The paradigm of managing CAD/CAE as a Service allows for companies to outsource intensive aspects of CAD/CAE functions with minimum costs. Cloud computing options, as well as modular options and flexible pricing offers have also contributed to the proliferation of computer-aided material selection in industry. For both CAD/CAE and material selection functions, footwear manufacturers can now deploy them as outsourced services, while maintaining high quality of the results, adherence to timeframes, flexibility to emerging demands and low cost, thus, ensuring high business dependability on the design process.

References

[1] Kotler P and Keller K L 2016 *Marketing Management* 15th edition (Pearson)

[2] Ease M 2009 *Fashion Marketing* 3rd edition (Wiley-Blackwell, Oxford)

[3] Muratovski G 2015 Paradigm Shift: The New Role of Design in Business and Society *She Ji: The Journal of Design, Economics, and Innovation* doi:10.1016/j.sheji.2015.11.002

[4] Slack N, Brandon-Jones A and Johnston R 2016 *Operations Management* 8th edition (Pearson)

[5] Eckert C and Demaid A 1997 Concurrent Engineering *Proc. of 78th International Conference of the Textile Institute* May 1997 Thessaloniki Greece

[6] Turban E, Volonino L and Wood G 2013 *Information Technology for Management: Advancing Sustainable, Profitable Business Growth* 9th edition (Wiley)

[7] Zissis D, Lekkas D, Azariadis Ph, Papanikos P and Xidias E 2016 Collaborative CAD/CAE as a Cloud Service *International Journal of Systems Science: Operations & Logistics* **4:4** 339-55 doi:10.1080/23302674.2016.1186237

[8] Křemenáková D, Mertová I and Kolčavová-Sirková B 2008 Computer-aided textile design 'LibTex' *Indian Journal of Fibre & Textile Research* **33** 400-4

[9] Azariadis P and Papagiannis P 2010 A new business model for integrating textile/clothing and footwear production *Proc. of the 3rd International Conference on Advanced Materials and Systems – ICAMS 2010* Bucharest September 16-18 2010 235-40

[10] Islam M and Khan M R 2014 Environmental Sustainability Evaluation of Apparel Product: A Case Study on Knitted T-Shirt *Journal of Textiles* http://dx.doi.org/10.1155/2014/643080

[11] SHOELAW LIFE08ENV/E/000147 project *Promotion of Environmental Legislation among European Footwear Industries* available at http://ec.europa.eu/environment/life/project/Projects/index.cfm?fuseaction=search.dspPage&n_proj_id=3438

[12] SHOEBAT LIFE12ENV/ES/000243 project *Promotion of best available techniques in the European footwear and tanning sectors* available at http://ec.europa.eu/environment/life/project/Projects/index.cfm?fuseaction=search.dspPage&n_proj_id=4621&docType=pdf

[13] CO2SHOE LIFE12ENV/ES/000315 project *Footwear Carbon Footprint* available at http://www.co2shoe.eu/en/

[14] Papagiannis P, Koutkalaki Z, Azariadis P and Papanikos P 2015 Definition and Evaluation of Plantar Mechanical Comfort for the Support of Footwear Design *Computer-Aided Design and Applications (Taylor & Francis) Special Issue: Technologies for Human Centric Free-form Products* http://dx.doi.org/10.1080/16864360.2015.1084189

[15] Papagiannis P, Koutkalaki Z and Azariadis P 2014 Footwear plantar mechanical comfort: Physical measures and modern approaches to their approximation *Proc. of the 5th International Conference on Advanced Materials and Systems* Bucharest October 23–25

[16] Azariadis P 2012 Finite Element Analysis in Footwear Design *The Science of Footwear* ed Goonetilleke R (Taylor & Francis Group) 321-37

[17] OPTSHOES project *Development of cost-effective and accurate computer-aided design and engineering (CAD/CAE) tools for the determination and optimization of footwear comfort parameters* available at http://optshoes.syros.aegean.gr

[18] Porter M 1998 *Competitive Advantage: Creating and Sustaining Superior Performance* (The Free Press, New York)

[19] Erasmus+ project 2015-1-RO01-KA203-015198 *Knowledge 4 foot - Knowledge Platform for Transferring Research and Innovation in Footwear Manufacturing* final report

Analysis of body differences for the design of children's clothing

G Bezerra[1], M Carvalho[2], A Araujo[3], M Rocha[4] and R Barboza[5]
[1]University Federal of Ceará/University of Fortaleza/University of Minho, www.ufc.br/www.unifor.edu.br/www.uminho.pt, Brazil/Portugal
[2]University of Minho, School of Engineering, Centre for Textile Science and Technology (2C2T), Campus de Azurem, 4800-058 Guimaraes, Portugal
[3]University Federal of Ceará/ www.dema.ufc.br, Brazil
[4]Federal Rural University of Pernambuco, www.ufrpe.br, Brazil
[5]University Federal of Ceará/University of Minho, www.ufc.br/www.uminho.pt, Brazil/Portugal

migcar@det.uminho.pt

Abstract: This paper discusses anthropometric data obtained through 3D body scanner technology, aiming the dimensioning of Portuguese children´s measurements for the design of clothing. The study comprises boys and girls from the Northern Region of Portugal aged between two and ten years old. The main objective of this part of the study was to evaluate the possibility of grouping children with six and seven years old in a same garment size, proposing the same body measurements in the final standard measurements table. In addition to the children's anthropometric measurements, it was checked the body conformation of the children and the possibility of identifying the main biotypes. As a preliminary result, through cluster analysis it was found that it is possible to combine these children in the same group in the proposed standard measurements table.

1. Introduction

Is very recent in the history of garments the perception of its construction as a design project. As soon as clothes started to be perceived that way, other concepts were being introduced, incorporating new scientific knowledge from Anthropometry and Ergonomics, combined with the science of comfort and safety. When it comes to children's clothing, the process requires more attention from the designer, considering the body characteristics of these users, not only accessing reliable and consistent measurements tables, but also being aware of their body shapes. Many authors have been dedicated to analyze the body, its shape and scale, but only a few study children's body aiming the design of clothing. Hence, the characteristics of the user must be observed and evaluated, studied to meet the requirements of functionality, usability and comfort combined with the aesthetics of the product [1].

The purpose of this paper is to identify the main body shapes of the Portuguese children at the ages of six and seven years old and investigate what steps can be taken in consideration in identifying the ways that published studies used to deal with clothing for adults. In addition, this study aims to identify whether it is advisable to set a medium size in the table of measurements, answering the body shapes of these children.

The adoption of engineering skills in the process of garment´s product development becomes necessary for the definition of the required measurements and consequently a production quantification associated with the new methodology, considering more rational approaches in the sense of three-dimensionality [2].

To do so, the use of 3D body scanner technology supported the creation of a database with statistical sampling of body segmentation, resulting in the definition of new measurement tables, with real measurements of the children under study, enabling the development of specific patterns and its subsequent optimization for garment production.

2. Materials and methods

Children's data was obtained through an anthropometric survey of children in the Northern Region of Portugal of both gender, using the technological capacities of the 3D body scanner KBI-Kinect Body Imaging. About 700 children from two to ten years of age were measured, in an ongoing larger PhD study in Fashion Design at University of Minho. Nevertheless, as mentioned before, for this paper, was only considered the data for children with six and seven years of age.

Thus, the final sample in this age group was 305 children, 161 aged six and 144 aged seven. Data was not separated by gender, because the published measurements tables do not differentiate boys and girls until the age of nine [3]. The existing studies for the analysis of biotypes of adults consider the following measurements: height and girths of bust, waist and hips [4] [5]. For this study, in addition to these measurements, it was also considered, the weight, in an attempt to evaluate how much the weight may influence the definition of the biotypes of these children. In this study, it was also analyzed the body images of the children obtained through the use of 3D body scanner. The scanning process was repeated three times for each child.

3. Anthropometric data collection with 3Dbody scanner technology

3D body scanners provide with great accuracy the necessary measurements for clothing design. These systems have a software for acquisition and processing of anthropometric data collected and all volumetric data in a few seconds, making the process efficient, effective and fairly fast, when compared with the traditional manual methods. When the study involves children, these technologies are even more important, as children have more dificulties in remaing colaborative during a long manual measurement process. Also, in these systems there is no phisical contact with the child during the process. A database with statistical sampling of body segmentation enables standardization of measurements tables, with accurate measurements of the human body, including a greater number of children, and thus enabling the design of more reliable patterns and the optimization of garments production.

The 3D image captured by the KBY System can display an error of 2 to 3 mm deep if the distance between the individual and the scanner is not in accordance with the parameters, or if the infrared camera is not calibrated correctly with the Kinect sensor. The calibration of the system is critical for the accuracy of the measurements and should be performed whenever a new study starts or if there is an inappropriate movement of the sensors [6]. The calibration process involves the use of a target, positioned in the middle of each pair of sensors, according to the instructions of the supplier [7]. The child must be positioned between the two groups of Kinect sensors, on a rug with a center mark, instructed not move for a few seconds, following the voice command at the end of the acquisition. Figure 1 illustrates the placement of each individual in the center of the two group of sensors.

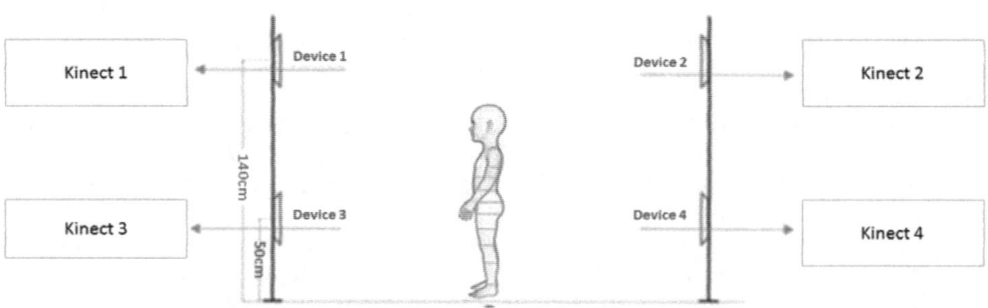

Figure1. Child positioned between the two groups of Kinectsensors - KBI System.

The 3D body scanner offers a range of variables and combinations of fundamental and complementary measurements of the human body, arranged according to the representation of Figure 2.

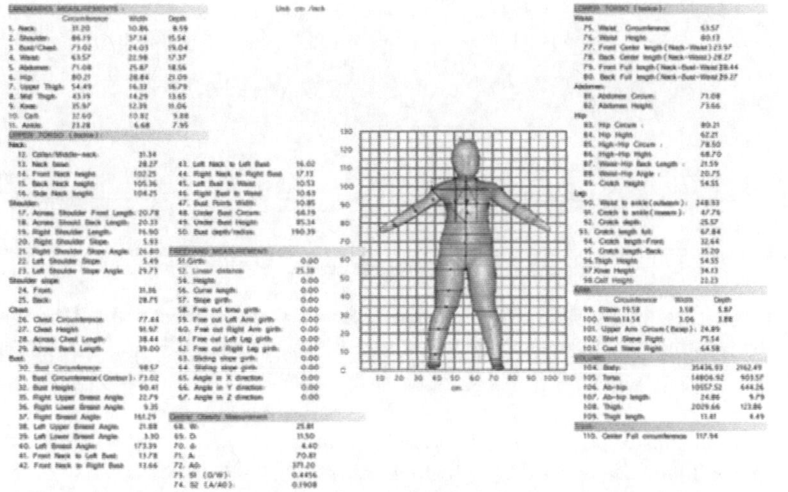

Figure 2. Measurementsand 3D body image obtained by the KBI-System.

The KBI System automatically provides 101 measurements of length and girth of the human body, as well as volume measurements of the main parts of the body.

4. Results

Most authors in this area work mainly with measurements of stature and bust girth and waist and hip girths. In this study, it was decided to add weight, considering that weight may interfere directly in the type of biotypes representing this population. Special attention was also given to the abdomen girth, usually more prominent in children, because of the body development, when compared with adults. By the age of ten, boys and girls begin to present the first traces of adolescence, changing significantly their body shapes. In this study was only considered the Portuguese children with six and seven years old. Therefore, the data in the tables bellow present descriptive measurements of this age range only.

Groups definition through cluster analysis

The multivariate cluster analysis technique was used to form homogeneous groups according to the variables studied. After a detailed analysis of all data, three main groups were considered. These groups are formed by children with similar body characteristics in the age ranges of 8-9 years (Group 1), 6-7 years (Group 2) and 4-5 years (Group 3). Nevertheless, the data used in these groups are not formed exclusively by children from a certain age group. There are children that have features of different age group and, therefore, belong to different groups. For example, a tall and heavy child of 6-7 years old, with high girth value scan be classified in Group 1. Initially six groups were formed, with children divided in these six groups as shown in Chart 1.

Chart 1 represent the distribution of children in their respective groups, defined by cluster analysis, according to the values of the Principal Components 1 and 2, generated by the method of principal components multivariate technique, which summarizes the information of the variables considered in the study in a small number of variables [8].

Observing the resulting distribution, it is possible to refer that the 3 groups are very clear, with each groupperfectly separated for the age ranges from four to nine years old. In Chart 1, group 2 is very clearly identified suggesting an adequate grouping of children with six and seven years old. With this distribution more than 70% of them were allocated to Group 2.

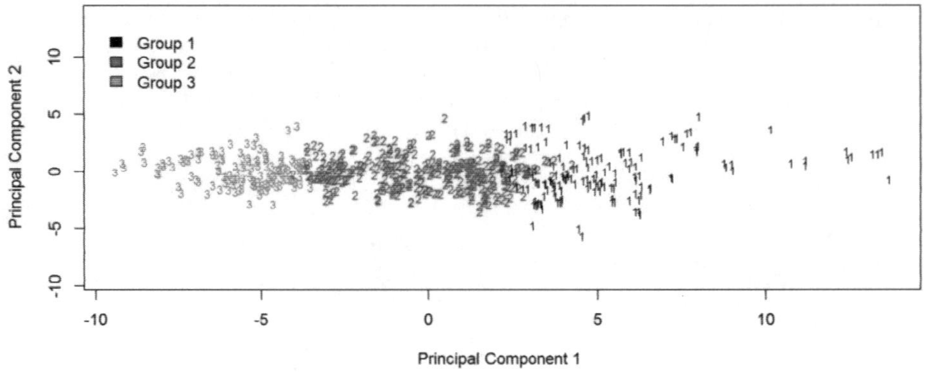

Chart 1. Distribution of children by age according to three groups defined by multivariate cluster analysis method, Legend: Group 1 (8-9y) - Black; Group 2 (6-7y) - Red; Group 3 (4-5y) – Green.

Tables 1, 2 and 3 present the descriptive measurements for each considered variable: minimum; maximum; first quartile; median; third quartile; and mean.

Table 1 presents data of 6-7 year old children as a group and Tables 2 and 3 present individually the measurementsof children with 6 and 7 years old, respectively.

The descriptive measurements considered for this study were: weight (kg), height (cm), bust girth (BG), waist girth(WG), abdomen girth(AG), hip girth (HG) of children in Group 2.

Table 1. Descriptive measurements of 6-7 year old children (Group 2).

Grupo 2 / Variables	Minimum	1st Quartile	Median	Mean	3rd Quartile	Maximum
Weight (Kg)	16.1	22.5	25.5	25.19	27.8	34.5
Height (cm)	104	118	124	123.17	129.00	140
BG	52.48	60.47	62.69	62.74	64.98	73.64
WG	49.33	56.90	59.57	59.85	62.28	72.69
AG	54.79	63.01	65.44	65.68	68.23	76.34
HG	58.46	66.75	70.20	69.95	72.53	78.73

Legend: Bust girth (BG); waist girth(WG); abdomen girth(AG); hip girth (HG).

The descriptive analysis of the measurements of the children with 6 and 7 years old individually is represented in Tables 2 and 3.

Through the percentiles of the measurements under study it was found that 50% of six year old children have weight up 23.90 kg; stature of up to 118 cm; bust girth up to 61.26; waist girth up to 58.14 cm; abdomen girth up to 63.29 cm and hip girth up to 67.32 cm.

Through the percentiles of the measurements under study it was found that 50% of seven year old children have weight up to 27.70kg; height up to 128cm; bust girth up to 64cm; waist girth up to 61.10 cm; abdomen girth up to 67.25 cm and hip girth up to 72.29 cm.

Through boxplot, represented in Chart 2, it is possible to observe how the measurements behave in terms of dispersionallowing the comparison of the two age groups. Three measurements build the central box: First quartile (bottom line of box); Median or second

IOP Conf. Series: Materials Science and Engineering **459** (2019) 012073 doi:10.1088/1757-899X/459/1/012073

quartile (internal line, thicker); and Third quartile (upper line of box). The upper and lower bars of the chart after a dotted line represent the minimum and maximum values of the data. The points below the bottom bar or above the top bar represent the outliers [7].

Table 2. Descriptive measurements for six year old children.

Measurements	Weight (Kg)	Height (cm)	BG	WG	AG	HG
Minimum	17.1	108	52.5	49.3	53.6	57.1
1st quartile	20.9	114	59.3	56.1	60.6	64.0
Median	23.9	118	61.3	58.1	63.3	67.3
Mean	24.1	119	62.0	59.1	64.1	68.0
3rd quartile	26.4	124	64.2	61.4	66.8	71.7
Maximum	36.0	130	78.2	77.7	82.0	83.6
Standard deviation	4.16	5.44	4.58	5.25	5.52	5.59

Table 3. Descriptive measurements for seven year old children.

Measures	Weight(Kg)	Height (cm)	BG	WG	AG	HG
Minimum	20.1	108	57.3	52.3	59.0	61.5
1st quartile	24.8	122	61.2	58.8	65.0	69.1
Median	27.7	128	64.0	61.1	67.2	72.3
Mean	27.5	127	64.6	61.5	68.0	72.7
3rd quartile	29.7	132	67.0	63.9	70.8	76.4
Maximum	37.0	140	77.7	73.4	81.6	86.8
Standard deviation	3.40	6.00	4.46	4.38	4.36	4.89

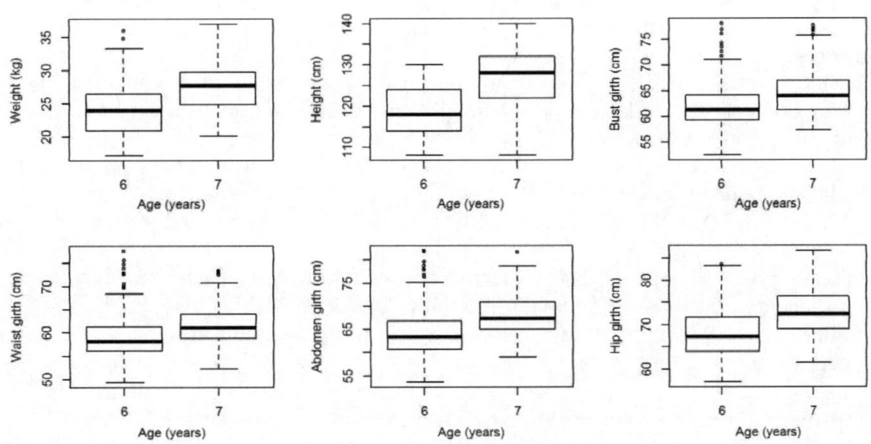

Chart 2. Boxplot of 6-7 years old children.

Through the Wilcoxon-Mann-Whitney test [9] significant differences were identified among children with six and seven years old for all measurements, requiring an individual treatment. There were clear differences in the measurements of their bodies. However, it was found that there is an intersection between points, i.e., children with six years more developed with the equivalent measurements of seven years, as well as children with seven years with measurements equivalent to those of children with six years. Hence, there are children that have features of different age groups.

From this statistical analysis it will be possible to define the most appropriate sizing for the Portuguese children and build tables of standard measurements that best meet the needs of a larger contingent of children, contributing to the improvement of quality and comfort of children'sclothing.

All the statiscal analysis were executed using the software R [9].

5. Conclusions

With a multivariate cluster analysis it was possible to confirme the possibility of grouping boys and girls from the Northern Region of Portugal aged between six and seven years old in a same garment size, proposing the same body measurements in the final standard measurements table. Through the measurement of height, weight, bust, waist, abdomen and hip girths of Portuguese children, combined with the analysis of their 3D body shape, provided by the KBI 3D body scanner, it will be possible to define their proportions and identify the main children biotypes. With this information compatible mannequins of the Portuguese children can be produced, considering the results of this anthropometric study, providing relevant information for fashion schools and for the garment industry. The definition of new measurement tables tailored to a specific audience, built with a wide coverage of consumers, can generate a smaller number of returned products and meet users expectations. This fact contributes to increasing the efficiency and sustainability of this industry and increase consumer satisfaction.

This paper is part of a broader study, including other regions of the country, with the purpose of presenting to the fashion industry a proposal of sizing and definition of children's standard measurement tables, respecting more effectively their needs. This initiative is expected to provide greater clarification for the garment producers and for the users.

Acknowledgments

We would like to acknowledge the Brazilian Institutions Federal University of Ceará and University of Fortaleza, and 2C2T-Science Center for Textile Technology from University of Minho. This work is financed by FEDER funds through the Competitive Factors Operational Program (COMPETE) POCI-01-0145-FEDER-007136 and by national funds through FCT-Portuguese Foundation for Science and Technology, under the project UID/CTM/000264.

References

[1] Bezerra G. Design de vestuário infantil com foco na ergonomia: desenvolvimento de uma metodologia de modelação a partir de um estudo antropométrico com base em digitalização 3D. PhD Thesis project, 2015. Supervisor: Carvalho, M.

[2] Gill, S. (2015). A review of research and innovation in garment sizing, prototyping and fitting.Textile Progress.

[3] Aldrich, W. 2009. Metric pattern cutting for children's wear and babywear. Fourth edition.W.B.

[4] Vuruskan, A., &Bulgun, E. (2011). Identification of female body shapes based on numerical evaluations. International Journal of Clothing Science and Technology, 23(1), 46-60.

[5] Simmons, K. P. (2003). Body shape analysis using three-dimensional body scanning technology.

[6] Barboza, R. B. D. M. G. (2016). Design de vestuário para jovens com Síndrome de Down, a partir de um estudo antropométrico com recurso à digitalização corporal 3D.

[7] KBI -Kinect Body Imaging - User's Manual.

[8]Zakaria, N. (2010). The development of body sizing system for school-aged children using the anthropometric data (Doctoral dissertation, Universiti Teknologi MARA).

[7] Bussab, W.O., &Morettin, P.A. (2013). Estatística Básica. Eighth Edition. Saraiva.

[8] Johnson, R.A., & Wichern, D.W. (2012). Applied Multivariate Statistical Analysis.Sixth Edition. Pearson.

[9] R Core Team (2018). R: A language and environment for statistical computing. R Foundation for Statistical Computing, Vienna, Austria.URL https://www.R-project.org/.

Initial designs of wheelchair rugby gloves

S Bragança[1], M Carvalho[2], J Steel[3], S Passman[3], S Gill[4], I Castellucci[5] and P Arezes[6]
[1]Research, Innovation and Enterprise, Solent University, Southampton, United Kingdom
[2]Department of Textile Engineering, University of Minho, Guimaraes, Portugal
[3]School of Art, Design and Fashion, Solent University, Southampton, United Kingdom
[4]School of Materials, University of Manchester, Manchester, United Kingdom
[5]Escuela de Kinesiología, Universidad de Valparaíso, Valparaíso, Chile
[6]ALGORITMI Centre, University of Minho, Guimarães, Portugal

migcar@det.uminho.pt

Abstract. Wheelchair rugby is a sport played by approximately 200 people in the United Kingdom. Despite being a small part of the population, research suggests that there are some aspects related to the sports-wear that need attention. This is especially relevant for gloves, as they have been identified as one of the main causes of performance losses. Until now, there are no gloves available on the market that were developed to respond to this sport's needs and requirements. The purpose of this paper was to present the concepts of the initial designs and first prototype of a pair of gloves created for wheelchair rugby. The designs were developed based on the feedback from one UK-based wheelchair rugby team. Of the designs produced only one was selected to be developed as a physical prototype. This particular design had in consideration the most important features identified by the athletes, namely the easiness of putting on and taking off the gloves, the padding and grip in specific areas and the elimination of the need for tape to secure.

1. Introduction

Wheelchair Rugby is a sport that was designed for male and female athletes with any type of physical disability. However, it has also gained popularity among abled-bodied athletes. Athletes are always striving for success and improved performance. The evaluation of performance is decisive for athletes, as it allows for a continuous improvement of the athlete and the wheelchair-user interface [1–3].

The very few studies available on wheelchair rugby sports-wear have found that the gloves are a major concern in this particular sport. These studies have found that there are no gloves specific for wheelchair rugby, which causes the athletes to modify existing ones in order to accommodate the performance requirements [4–8]. The International Wheelchair Rugby Federation states that "many different types of gloves are available to choose from", however, none are sport-specific. Further, they state that "Many athletes prefer using rubber coated cotton gloves because they are inexpensive and provide considerable grip to the player. Using tape to secure gloves to your wrist is a common practice and is highly recommended". However, this information is only available on their website [9]. The official document containing the International Rules for the Sport of Wheelchair Rugby [10] does not provide any information about gloves.

Mason et al. [4] tried to define the effectiveness of different types of gloves on mobility performance of wheelchair rugby. They concluded that the gloves performed significantly better when they were modified and adapted for the specific athletes' needs. The authors advise that further research is needed to evaluate the athletes' interaction with different types of gloves.

IOP Conf. Series: Materials Science and Engineering **459** (2019) 012074 doi:10.1088/1757-899X/459/1/012074

The work of Lutgendorf et al. [5] also evaluated the influence of the use of different types of gloves on standardized wheelchair rugby performance compared with not using gloves at all. This study lead to the conclusion that the use of different types of gloves significantly influenced performance. Nonetheless, some gloves helped the athletes to perform better than others (US National Football League gloves were the most favourable ones, whilst multipurpose were even worse than not wearing gloves at all).

In these two studies, it was clear that there is a real need to develop gloves specific to this sport that fulfil the athletes' specific needs and requirements. In other more recent studies, similar conclusions were withdrawn. In the study of Bragança et al. [11] a set of design recommendations for the development of wheelchair rugby sports-wear that responds to the users' particular needs and requirements was evaluated. The results of this study indicate that the athletes would be happy to have improvements introduced to their current sports-wear, particularly on the gloves.

Bragança et al. [12] analysed, by means of videos and photographs, the patterns of activity and movements performed by wheelchair rugby athletes during a set of wheelchair rugby training sessions. Their study served not only to inform on the movements and body postures but also on the problems and difficulties caused by the items of sports-wear worn by the athletes, concluding that gloves are causing the most problems with performance.

The purpose of the current paper is to present an initial prototype of a pair of gloves specifically designed to fit the needs of wheelchair rugby players.

Methodology

1.1. Sample
The sample used for this study was composed by a wheelchair rugby team of 25 athletes based in the United Kingdom (22 males and 3 females).

1.2. Data collection
The design of the glove started with the identification of the end-users' needs. This was done by means of both direct observation (with the capture of several videos and photographs) as well as of interviews to the end users. The videos, photos and interviews were recorded over the course of several training sessions of this particular team. The research team responsible for these observations was composed by a lead researcher and five product design students.

1.3. Data analysis
The information gathered in the unstructured interviews to the athletes was analysed and the users' needs were extracted. The videos and photographs allowed to understand how the athletes move and the main tasks performed during the training sessions. The areas of contact of the hand with the wheelchair and the ball were analysed in detail.

Once identified the needs and all the other necessary requirements, several potential designs were developed. These designs were then evaluated by a team of experts who selected the one that responded to the users' needs and the sports requirements better. The selected design was then transformed into a real physical prototype.

2. Results and Discussion
Due to the lack of availability of a specialised pair of gloves for wheelchair rugby, athletes use a variety of different types of gloves available on the market that have been designed to fulfil other requirements. These other types of gloves include cycling gloves; golf gloves; motorcycling gloves; football goal keeper gloves; and even industrial gloves. Table 1 shows a brief summary of the advantages and disadvantages of each one of these gloves.

Table 1. Characteristic of different types of gloves available on the market.

Type of glove	Pros and cons
Cycling gloves	Pros: Sweat absorbent; easy to get on and take off; breathable; enables grip. Cons: Not very durable.
Golf gloves	Pros: Easy to get on and take off; breathable; durable; has some padding. Cons: Does not provide enough grip.
Motorcycling gloves	Pros: Breathable; made for rough conditions; texturized grip areas; has protective patches; durable. Cons: Restricts movement on the areas with protective patches.
Football goal keeper gloves	Pros: Very protective; very padded; enables some grip; durable. Cons: Too padded; not breathable.
Industrial gloves	Pros: Easy to get on and take off; enables grip; durable; padded. Cons: Does not stay in place; not breathable.

However, gloves for wheelchair rugby need to have special features that none of these gloves have. Some of the gloves analysed present a few relevant features but there is not one pair that has all the necessary characteristics at the same time. Generally, according to the feedback from the athletes interviewed and to previous studies [8], gloves for wheelchair rugby need to be:

- durable;
- resistant to wear and tear;
- lightweight;
- sweat absorbent;
- allow movements of the fingers and wrists;
- enable grip on the areas of contact with the wheelchair and the ball;
- have padding on the areas of contact with the wheelchair for protection;
- able to be kept in place while manoeuvring the wheelchair;
- easy to put on and take off.

It is important to mention that each athlete has a very personal way of interacting with the wheelchair and the ball. As such, it is very difficult to develop a product that suits everyone's needs in the same way. The several designs created tried to respond to all of these needs and requirements in a general way. Nonetheless, the design selected to be developed into a prototype responded better to the needs and requirements of the majority of the athletes (Figure 1). Still, it is important to note that even though this has been designed for the majority of the athletes, it would be possible to modify this design to accommodate users with different needs by implementing the same changes and applying them to different areas of the hand.

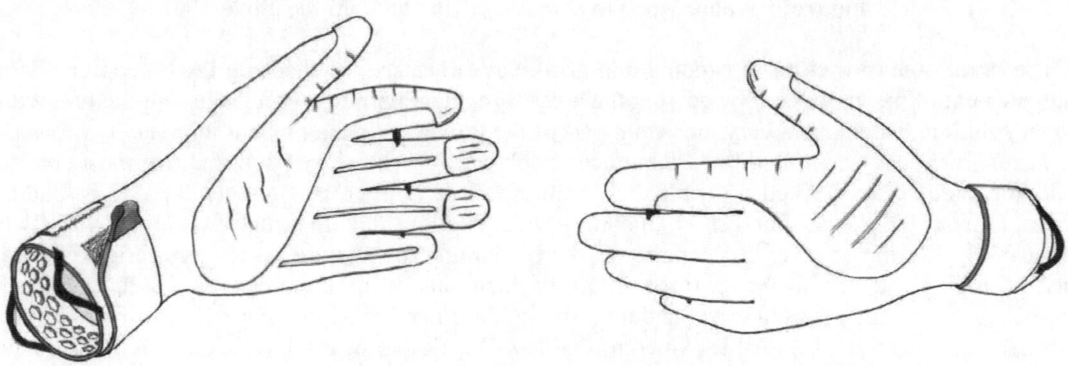

Figure 1. Draft of the prototype.

IOP Conf. Series: Materials Science and Engineering **459** (2019) 012074 doi:10.1088/1757-899X/459/1/012074

The design that was selected focused on the most important issues identified by the athletes: (i) the ability to secure the gloves in place; (ii) the ability to easily put the gloves on and take them off; and (iii) the provision of padding and grip in the areas that are in contact with the wheelchair.
The first issue (ability to secure the gloves in place) was addressed by exploring different ways of securing the gloves in the wrist area. It was decided that the best way of doing this would be by having a Velcro strap placed just above the wrist. This positioning of the strap would allow not only to ensure that the gloves do not come off while manoeuvring the wheelchair but it would also allow for an increase range of movement of the wrists (Figure 2).

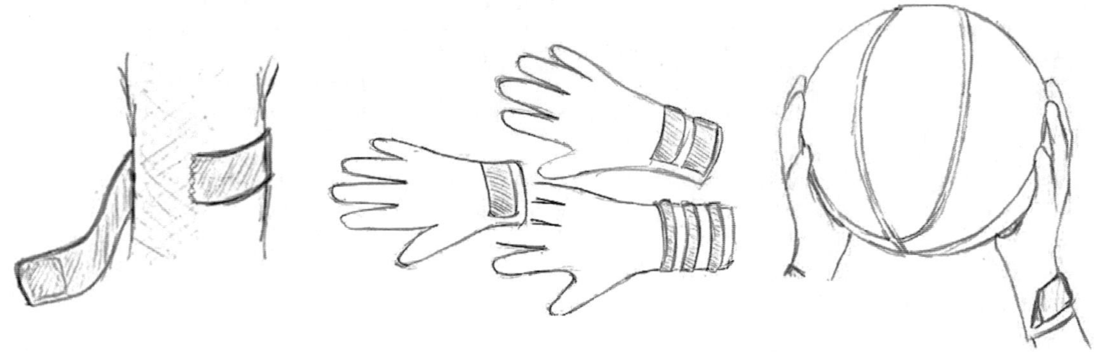

Figure 2. Testing on the wrist strap.

The second issue (ability to easily put the gloves on and take them off) was addressed by including webbed tabs in between the fingers to help take the gloves off with the opposite hand. To help in putting the gloves on, either with the opposite hand or with the teeth, the same type of tabs was added to the ends of the glove in the wrist area. These attachments can be seen in Figure 3.

Figure 3. Webbed tabs to help put on and take off the gloves.

The third issue (provision of padding and grip) was addressed by choosing thick neoprene as the main element of the gloves and by coating the areas of contact with the wheelchair and the ball with a rubber solution. Padding is a very important part of the gloves as it can prevent injuries, by protecting the hands from heavy contact with hard surfaces or objects. On the glove itself, the two most common areas for padding are located on the palm of the hand. It is important to select the right padding material for the gloves as, after a short amount of time, padding and other materials on the glove wear and tear. By looking at other contact sports where padding is usually used (e.g. American football, hockey, rugby and ice hockey), some different materials were classified as good options for wheelchair rugby: foam padding; gel padding; or wadding material. Neoprene can also be used as an option as it is a durable, breathable and comfortable material. Moreover, its thick and elastic structure can act as a good protection material to use between the hand and the objects. However, the construction of the glove using this neoprene as a base can become quite complex as it is somewhat difficult to work with when used in such an intricate manner. Figure 4 shows how the stitching of the neoprene was made (with the addition of a thinner neoprene layer to avoid the rubbing of the seams with the skin).

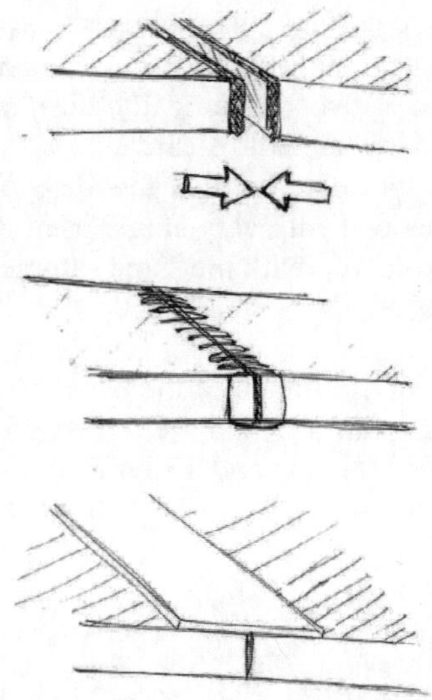

Figure 4. Stitching on neoprene breakdown.

As for the grip, there are some materials that are usually used for this purpose, such as rubber; silicone; or polyurethane. These materials are flexible and somewhat resistant so they represent good options for wheelchair rugby. In the prototype created, a rubber solution was used as it was the most inexpensive and easy way of obtaining the desired result. Nonetheless, a very good alternative would have been the application of a vulcanised rubber patch as this material is very durable and has a high resistance to wear and tear However, this would require the manufacture of a mould specific for this glove, which is a much more expensive solution. Figure 5 shows how the vulcanised rubber should be moulded to fit the parts of the hand that need more grip.

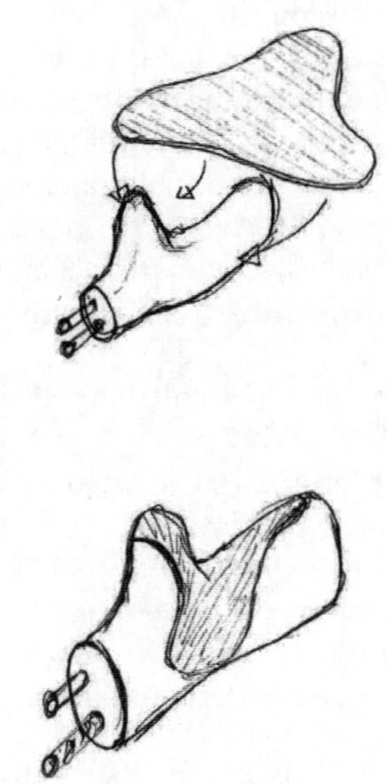

Figure 5. Vulcanised rubber moulding.

3. Conclusions

Several initial designs were created for a pair of gloves specific for wheelchair rugby. These designs were created based on the results of previous studies as well as on unstructured interviews conducted over the course of several training sessions of one UK based team. One of the designs was selected to be turned into a physical prototype as it focused on the most important features for the athletes, namely the ability to secure the gloves in place; the ability to easily put the gloves on and take them off; and the provision of padding and grip in the areas that are in contact with the wheelchair.

This first prototype of a pair of gloves specifically designed to accommodate the needs of wheelchair rugby players fills a gap in this area of research.

The major limitations of this study are the fact that this design has not yet been tested by the end users and that is might not satisfy the needs of every single user. As such, as future work, the authors intend to test and evaluate this first prototype with the same sample that provided the initial feedback.

Acknowledgements

The authors wish to acknowledge the commitment of the participants of this study. This study had the financial support of FCT - Fundação para a Ciência e a Tecnologia of Portugal under the projects UID/CEC/00319/2013 and UID/CTM/000264; and FEDER funds through the Competitive Factors Operational Program (COMPETE) POCI-01-0145-FEDER-007136.

References

[1] Chua J. A novel approach to identify and quantify activity and performance in wheelchair rugby. PhD. 2013;(May).
[2] Sasaki M, Kimura T, Matsuo K, Obinata G, Iwami T, Miyawaki K, et al. Simulator for optimal wheelchair design. *J Robot Mechatronics*. 2008;**20(6)**:854.
[3] Usma-Alvarez CC, Subic A, Burton M, Fuss FK. Identification of design requirements for rugby wheelchairs using the QFD method. Procedia Eng [Internet]. 2010 Jun;**2(2)**:2749–55. Available from: https://www.scopus.com/inward/record.uri?eid=2-s2.0-78149315665&partnerID=40&md5=7fa22758e0597fceca6b32c731b27d8c
[4] Mason BS, van der Woude LHV, Goosey-Tolfrey VL. Influence of Glove Type on Mobility Performance for Wheelchair Rugby Players. *Am J Phys Med Rehabil* [Internet]. 2009;**88(7)**:559–70. Available from: http://content.wkhealth.com/linkback/openurl?sid=WKPTLP:landingpage&an=00002060-200907000-00006
[5] Lutgendorf M, Mason BS, Van Der Woude LH V, Goosey-Tolfrey VL. The effect of glove type on wheelchair rugby sports performance. *Assist technol Res Ser.* 2009;**26(3)**:363–5.
[6] Churton E, Keogh JW. Constraints influencing sports wheelchair propulsion performance and injury risk. *Sport Med Arthrosc Rehabil Ther Technol* [Internet]. 2013;**5(1):3**. Available from: http://bmcsportsscimedrehabil.biomedcentral.com/articles/10.1186/2052-1847-5-3
[7] Braganca S, Steele J, Gill S, Carvalho M, Arezes P. Sports-Wear in Wheelchair Rugby: Establishing Design Needs. In: Di Bucchianico G, Kercher PF, editors. Advances in Design for Inclusion: Proceedings of the AHFE 2017 International Conference on Design for Inclusion, July 17–21, 2017, The Westin Bonaventure Hotel, Los Angeles, California, USA [Internet]. Cham: Springer International Publishing; 2018. p. 381–9. Available from: https://doi.org/10.1007/978-3-319-60597-5_36
[8] Bragança S, Castellucci I, Gill S, Matthias P, Carvalho M, Arezes P. Insights on the apparel needs and limitations for athletes with disabilities: The design of wheelchair rugby sports-wear. *Appl Ergon* [Internet]. 2018 Feb;**67**:9–25. Available from: http://linkinghub.elsevier.com/retrieve/pii/S0003687017302016
[9] IWRF. International Wheelchair Rugby Federation - Gloves in wheelchair rugby. 2017.
[10] IWRF. International Wheelchair Rugby Federation - International Rules for the Sport of Wheelchair Rugby. 2015;(January):1–44.
[11] Braganca S, Castellucci I, Gill S, Carvalho M, Arezes P. Evaluation of design recommendations for the development of wheelchair rugby sports-wear, *Advances in Intelligent Systems and*

Computing **Vol. 776**, 2019.
[12] Braganca S, Carvalho M, Gill S, Castellucci I, Arezes P. Image analysis as a basis to the design of wheelchair rugby sportswear. In: 91st Textile Institute World Conference. Leeds, United Kingdom; 2018.

Digital twin and men's underwear design

Z Cheng[1,2,3] and V E Kuzmichev[1,2,3]
[1] Fashion of School, Wuhan Textile University, Fangzhi Road, Wuhan, Hubei, China
[2] Textile Institute, Ivanovo State Polytechnic University, 21, Sheremetev Av., Ivanovo, Russian Federation
[3] Wuhan Textile and Clothing Digital Engineering Technology Research Center

zcheng@wtu.edu.cn, wkd37@list.ru

Abstract. Based on 3D body scanning technology, KES properties of real knitted materials, new body measurements for digital twin generating, we developed the virtual method of men's underwear try-on. The results show that by means of virtual software 3D CLO we can simulate well the body characteristic, the underwear structure, the knitted material elongation, and the compression pressure distribution. New underwear design can have a good presentation in customization and mass production.

Key words: men's underwear design; 3D simulation; digital twin; pressure; knitted material

1. Introduction

The current 3D simulation technology provides great convenience for apparel design and product development [1, 2]. Advances in 3D simulation technology can be used for clothing suitability assess. But there is only a small number of researches on tight-fitting clothing and few studies between male virtual body and knitted underwear. Most of the 3D simulation is about the design of clothing 2D pattern or the choice of material, few studies are concerned on "body-clothing" system. Many studies focus on some loose-fitting clothing that does not directly contact with human skin. Only a few studies on tight-fitting clothing, or between male virtual body and knitted underwear.

Most of 3D simulations are only about the common clothing pattern or the material simulation [3]. Few studies are concerned on system "body-clothing" and tight-fitting/compression clothing [4, 5]. The negative ease allowance is often designing required because the tight-fitting underwear fitness truly reflects the human body morphology [6]. Nowadays, men's compression underwear also needs to be customized due to different needs, comfortable satisfaction, and shape characteristics [7, 8]. For this purpose, we examine the system "body-underwear" from several points of view such as structure, wearing effect, material deformation, and pressure distribution. We also compared various underwear sizes with corresponding body to verify our new classification of body size.

2. Experimental methods and steps

This study includes 3D body scanning measurements, 2D&3D underwear structure design, virtual materials analysis, pattern, pressure evaluation etc.

IOP Conf. Series: Materials Science and Engineering **459** (2019) 012075 doi:10.1088/1757-899X/459/1/012075

We used VITUS Smart 3D body scanner measure 115 Chinese (18…28 age young males) and test 14 kinds knitted materials by KES. Some basic research has been done before to establish databases of reasonable pressure values, ease range etc. [9].

We cut 14 kinds of knitted materials "shell" and make underwear to measure pressure values (38 points per piece) on one hand. We use the software CLO 4.1 simulated these 14 kinds of knitted materials to check its pressure values accuracy and reliability, then build virtual underwear on another hand. After, we compare the difference of real and virtual pressure values to obtain the error value.
Next, we used new method of underwear design (for basic types), build virtual underwear in 12 sizes S^-, S, S^+, S^{++}, M…L^{++}, and change its eases from -20 to 0 %, the S, M, L is size to describe the hip girth, "+" and "–" are different type of waist girth and other details.

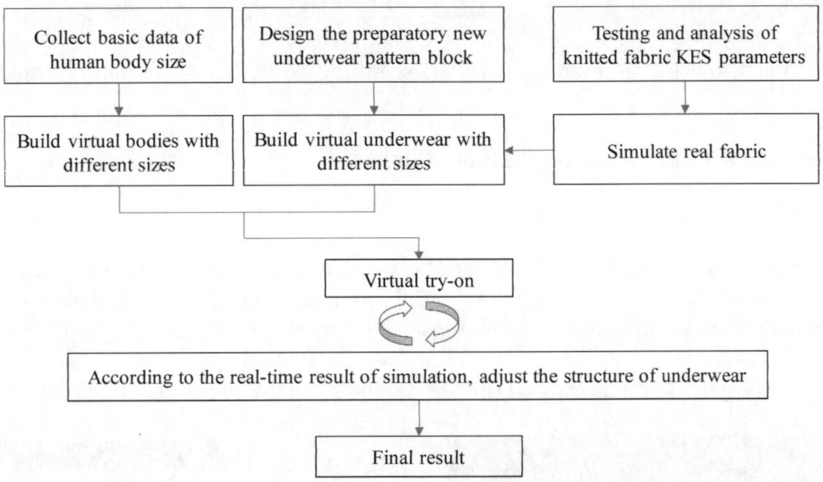

Figure 1. Flow chart

3. Reconstruction of human models

The 160 men scanned and reclassified according to their body features. Figure 2(a) shows three male profiles. We can see that the shape of the profile is very different, even at the same size M.

So, we used software of MakeHuman (see Figure 2b) to establish S, M, L sizes virtual male bodies (see Figure 2c), the male genitalia are set to the same at here. It can be clearly seen from the profiles that different waist sizes have different shape on the part of abdomen and the back of the body, when the hip circumference is the same. Therefore, a more detailed classification with the S, M, L number is necessary.

Then, we try on a variety of virtual bodies to test the pattern of underwear designed for more detailed body shape. Figure 2(c) are virtual models meets the average size of our classification, covers the genital area for lower torso of the body. Finally, the "obj" human model file is imported into the CLO software for subsequent simulation work.

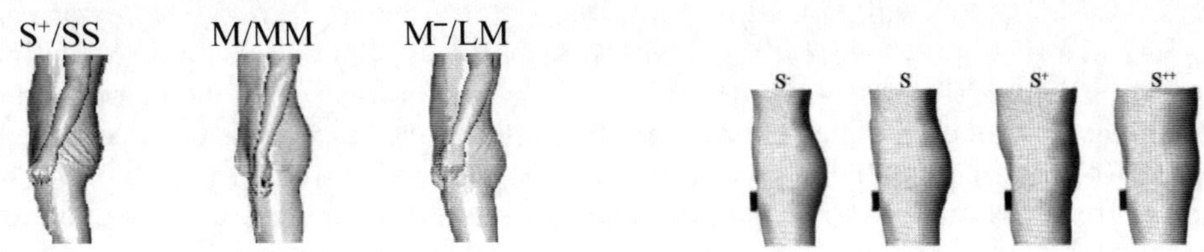

(a)

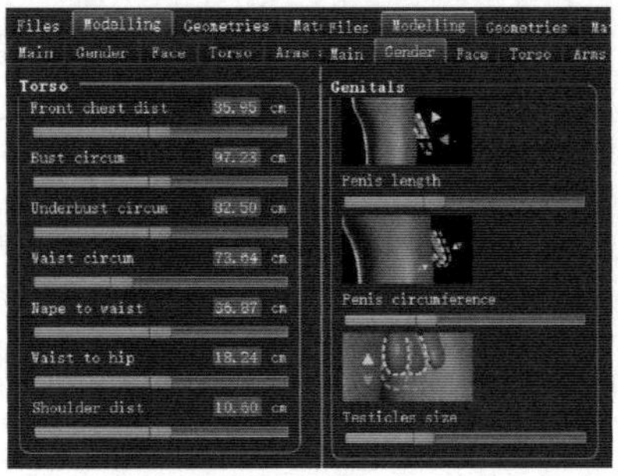

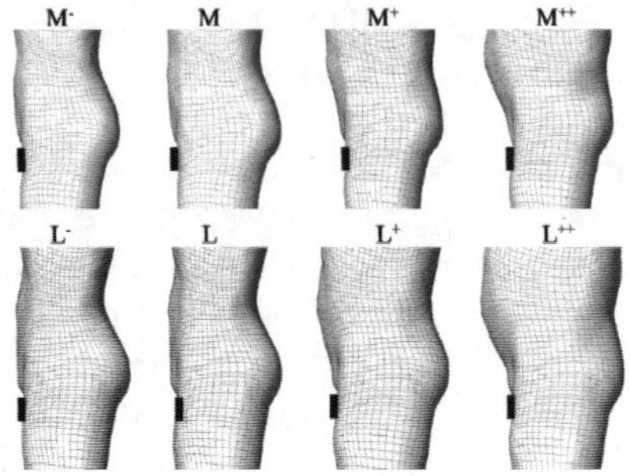

(b) (c)

Figure 2. 3D Human lower torso models: (a) – 3D scanning body; (b) – size modification; (c) – virtual body

4. Materials pressure error between real and virtual

4.1. Software parameter correction for material

Form the simulations of 14 kinds virtual materials, we adjust the parameters and put the KES results into CLO 4.1. Such as the options material's "detail", the best way is take the material parameters *G* (shear stiffness), thickness and density into these 3 options of "stretch", "shear", "thickness" and "density".

After this, we simulate the knitted material properties and measure pressure values (see Figure 3b).

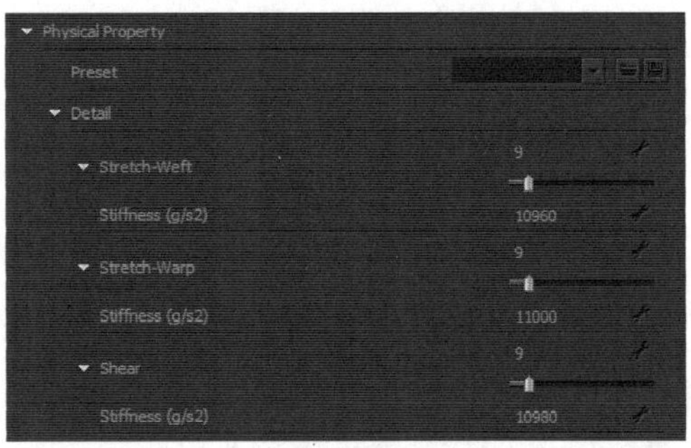

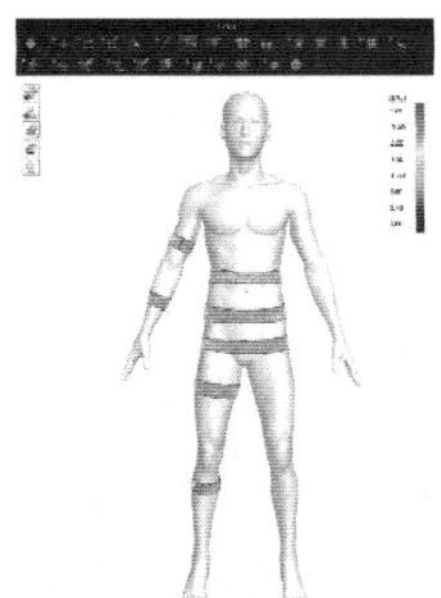

(a) (b)

Figure 3. Material simulation for virtual pressure test

4.2. Virtual pressure test

We have test nearly 5000 virtual pressure value data (PV) in knitted materials two yarn directions on 7 parts of virtual body. The simulated virtual pressure (PV) is 0.14–2.13 kPa (the real pressure on human body PB is 0.05–2.31 kPa) with knitted materials elongated from -20.0 to -5.0% in warp and pressure is 0.00–2.82 kPa (PB is 0.13–2.02 kPa) in weft. As shown in Figure 4(a), there are definite correlations between PB and PV of 14 knitted materials, which means that there are certain connections between them and the simulated pressure are feasible, but the error needs to be corrected.

As shown in Figure 4(b), we find that the knitted materials elongated value from -15.0 to -10.0% has the largest difference from -0.26 to 0.24 kPa, and lines intersect at -12.5%; in total, as for average is ± 0.19 kPa error.

Then, we build virtual underwear and measure virtual pressure PV', from ease -20.0 to -5.0%, the result as shown in Figure 4(b).

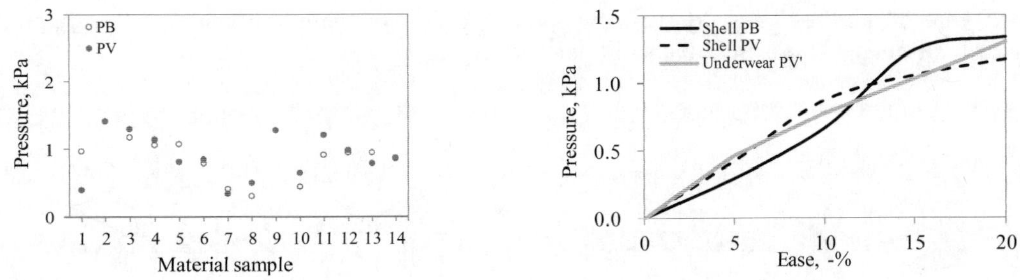

(a) (b)

Figure 4. Paired pressure values on body (PB) and virtual (PV): (a) – 14 knitted materials, (b) – elongated (ease) during -20.0…0.0%

4.3. Pattern adjustment and test

After, we checked the underwear pattern test the virtual pressure sensibility and underwear deformation at various parts. Figure 5(a) shows the virtual underwear try on in front view. We clearly see that there is an excessive pressure (red color) at the crotch part and front belt. We need to further modify this original pattern. Figure 5(a) left shows the modified front crotch part and belt length of the pattern (increased to the red), after this, we measured the pressure is lower than before and the deformation at crotch part is changed to 32.8…41.2%, see Figure 5(b) right. Figure 5(c, d) shows the back view. 2D pattern and 3D simulation are also shown. The crotch part of back pattern has also been modified.

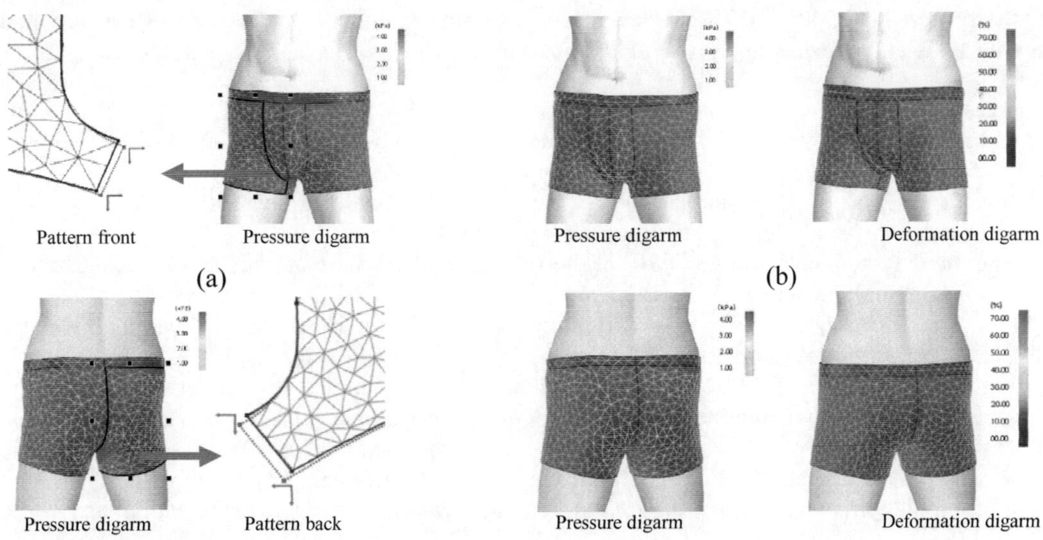

(a) (b)

(c) (d)

Figure 5. Basic type underwear simulations: (a) – front part modify; (b) – front part pressure and material defamation; (c) – back part modify; (d) – back part pressure and material defamation

4.4. Underwear comfort evaluation and grade

Figure 6 shows the several examples of pairing test of multiple sizes, pressure distributions and the rating evaluations. As for this, five grades are applicable to evaluate the compression underwear from low pressure to high. The range includes five grades: **1** (very low pressure/loose), **2** (low pressure/a little loose), **3** (comfort pressure/close-fit), **4** (high pressure/very close-fit), **5** (very high pressure/extra close-fit) to evaluate the final score which was equal to the average grades measured in six parts per underwear. During the **2-4** is fit, **4-5** also fit, just feeling tight.

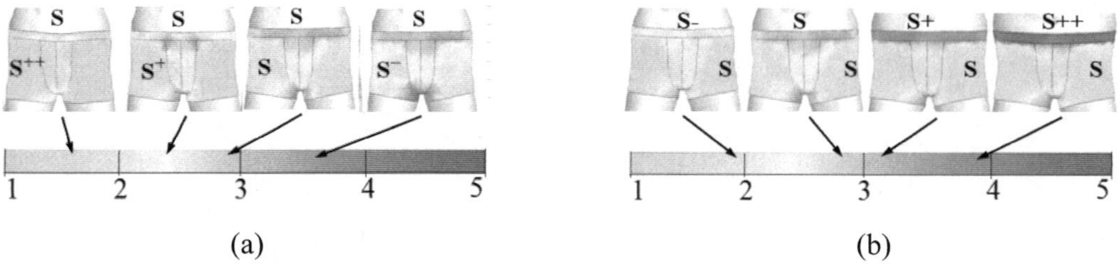

Figure 6. Virtual evaluation of pressure distribution when underwear has the ease = -19.5%: (a) different sizes underwear on S body; (b) same sizes S underwear on different bodies

5. Conclusion

By adjusting the virtual body dimensions to fit our new classification, we build a variety of virtual human models with different sizes and male characteristic. We use 3D simulation technology to test our new patterns on virtual bodies. After our correction, the CLO software can simulate well. The simulation experiment demonstrates the rationality of our design method and classifications. The body wears matching underwear size performed the best fitting and reasonable pressure distribution. Next, we will apply these results to the underwear factory for the verification of real products.

Acknowledgments

This work is supported by China Wuhan I'd Co., Ltd. and Russian Ministry of Science and Education under the project № 2.2425.2017/4.6 "Development of software for virtual design of system 'body-clothes' in static and dynamic and for virtual try-on FashionNet".

References

[1] Loker S and Ashdown S 2005 Size-specific analysis of body scan data to improve apparel fit *Journal of Textile and Apparel Technology and Management* vol **4** pp 1–15

[2] Hwalin S 2016 Fitting simulation evaluation on personalized avatars *Journal of Textile Engineering & Fashion Technology vol **4*** pp 125–130

[3] Ancutienė K and sinkevičiūtė D 2011 The influence of textile materials mechanical properties upon virtual garment fit *Materials science vol **17*** pp 160–167

[4] Fontana M and Rizzi C 2005 3D virtual apparel design for industrial applications *Computer-Aided Design vol **37*** pp 609–622

[5] Yeung K W and Li Y 2004 A 3D biomechanical human model for numerical simulation of garment-body dynamic mechanical interactions during wear *Journal of The Textile Institute vol **95*** pp 59–79

[6] Kuzmichev V E and Zhe C 2016 Improving men's underwear design by 3d body scanning technology *Proc. Int. Conf. on 7th 3D Body Scanning Technologies (Lugano, Switzerland)* p 328

[7] Kuzmichev V E, Zhe C and Adolph D C 2015 Development of male underwear compression designing *Int. Conf. on AUTEX 2015 World Textile (Bucharest, Romania)* (Gheorghe Asachi Technical University of Iasi), p 100

[8] Yeung K W and Li Y 2004 A 3D biomechanical human model for numerical simulation of garment-body dynamic mechanical interactions during wear *The Journal of The Textile Institute* vol **95** pp 59–79

[9] Zhe C, Kuzmichev V E and Adolphe D C 2017 Development of knitted materials selection for compression underwear *Autex Research Journal **vol 17*** pp 177–187

IOP Conf. Series: Materials Science and Engineering **459** (2019) 012076 doi:10.1088/1757-899X/459/1/012076

The influence of human body posture on the determination of total dimensions (morphological characteristics)

I Dabolina, E Lapkovska and A Vilumsone

Riga Technical University, Faculty of Material Science and Applied Chemistry, Institute of Design Technologies, Kipsalas street 6-220, Riga, LV-1048, Latvia

Inga.Dabolina@rtu.lv

Abstract. Most important human body measurements are used for sizing uniforms and work wear. If the measurements are gained and used inappropriate, the garments may be unsuitable for the end-user. The target group is soldiers which are provided with uniform jackets which appropriate fit is significant for successful accomplishment of their daily tasks. Anthropometry as a method studies total morphological characteristics which measurement procedure can be influenced by different factors – the chosen measuring tools and equipment, the influences of body posture and their changes. The aim of the research is investigation of chest circumference – factors affecting the measurement precision and differences between manual and 3D scanning anthropometrical methods. Different measurements of the chest circumference also in different body postures show that the manually obtained chest girth is closer to the chest girth perpendicular to the axis of the torso, rather than the horizontal chest girth which is most significant dimension in patternmaking of upper body garments.

1. Introduction

Size labelling rules support satisfaction ability of multiform anthropometric demand of clothes for different consumers allowing various height, girth and drop interval combinations if suppliers would comprehend complexity of the applicable demand [1]. For sizing and ensurance of uniformity and fit of uniforms and work wear most important body measurements (primary and secondary dimensions) are used. The absolute values of the dimensions used for developing sizing system must be as large as possible. Main body dimensions used for sizing protective clothing are chest girth, body height and waist girth [2].

Anthropometry studies the total dimensions of the body - total morphological characteristics which comprise the major anthropometric indicators: body height, chest circumference, and body mass. They reflect the external form of a human body and are the most significant indicators of physical development, therefore, their interrelationships, as well as body proportions and stature, external forms of individual parts of the body, sex and age characteristics, etc. are investigated. The constitution of a human body depends on many characteristics, for instance, the distribution of adipose tissue in the body, development of muscular system, person's age, the size and shape of the skeleton, the shape of the thorax and the thoracic region, as well as the shape of the abdomen and spine [3].

The horizontal chest circumference, which is the most significant body dimension in all methods of patternmaking of upper body garments and is used to calculate the total width of the garment, is used widely for setting the core aspect of sizing system. It is relatively arbitrarily possible to allocate any desired number of matching intervals, which may vary from 4 up to even 10 cm and are chosen

IOP Conf. Series: Materials Science and Engineering **459** (2019) 012076 doi:10.1088/1757-899X/459/1/012076

depending on garment type. Among the end users of garment standardized to the fit range, no one would have encountered the disproportion between the actual values of his unique body measurements and the average value of the standard intervals in proportion to the standardized dimensions of the designed apparel. If the measurement is made inappropriate to the patternmaker's method and/or practice, it may turn out to be in another fit range and the end product is inappropriate for the end-user, thus wearing comfort would not be provided [4].

There are two types of human body measurement acquiring methods: manual anthropometry methods (contact methods) and optical anthropometry methods (non-contact methods). Traditional methods use different manual tools (measuring tape, anthropometer, etc). As the technologies develop, new tools are being created and/or the existent ones are improved. The aim of this research is investigation of chest circumference – factors affecting the precision of measurements and differences between manual methods and 3D scanning anthropometrical methods.

2. Chest girth in pattern making

When using the methods of patternmaking of upper body garments, the chest circumference is used for calculation of the total width of the garment, whereas the total width of the pattern is divided into the back, armhole, and front in different variations. As soon as in beginning of the 20th century, the master-tailor Levitanus had suggested that the width of an armhole shall be defined as ¼ of chest circumference. The beginnings of the search for standard proportions date back to the mid-20th century. According to the most typical example, a half of chest circumference shall be divided in proportions as follows: 3/8 for the back, 2/8 for the armhole, and respectively 3/8 for the front. In some systems, this ratio is not so clearly defined and is used only to determine a pattern block. Also features of garment assortment reflected in form of ease allowances or different adjustment coefficients shall be considered. The most common methods were those according to which all body measurements were calculated only by means of body height and chest girth. The renowned M.Müller & Sohn method of patternmaking uses proportional body measurements which are calculated by taking into account four key dimensions: body height, chest, waist, and hip girths. The proportional methods do not, however, always provide clear distinctions between body measurements and sizes of pattern line segments of the garment or drawing.

The significance of the chest girth becomes apparent also when determining other relevant sizes of both horizontal and vertical drawings. Some proportional methods use specific empirical regularities to calculate the depth of the armhole. Levitanus suggests to allocate, along the centre line of the back, 1/4 of the sum total of the back length and chest half-girth. According to M.M&Sohn - 1/10 chest girth + 10.5 cm. In other it is calculated: chest girth/ 3+4cm. As for the proportional patternmaking methods, the width of the neck line, and the width of the sleeve are associated with the chest circumference. In all these empirical formulas, coefficients have great significance, whereas coherences are to be used for a limited number of figure types, a minor group of sizes, and for a particular garment assortment. Since the width of the back can be logically associated with the respective body measurement – the back width, later methods mostly use this coherence. It is rather difficult to associate the width of the armhole with the direct body measurement. The sagittal diameter of an arm is used by SEPP method, as other methods mostly use a part of the chest girth.

3. Body stature and posture characteristics

Factors influencing the accuracy of body measurements are the human's stature and body position during the measurement. The stature is the usual configuration of a natural standing position of the human body, i.e., without muscle tension, and is characterized by several features: the shape of the spine, the head and shoulder position, pelvic inclination, abdominal circumcision, limb position, muscle development, fat distribution. In turn, these characteristics are determined by heredity, sex and age, nature of a person's work, performance of central nervous system, and other causes [5].

Anthropological studies deal with different typological distinctions of body statures, and the identification of the main types allow usage of stature regularities in textile science. When drafting the patterns, the most convenient approach of classification is usage of the sagittal (profile) contour of the torso. Stature can be divided by 3 types – stooped, normal, and curved. Simple estimation of the stature can be done by some main measurements: – body position (Ks), – depth of lumbar lordosis (LVdz), – lumbo-pelvic lordosis (GVdz) [5].

The depth of lumbar lordosis (LVdz) has little impact on the fit of loose and semi-tight upper body garment. In textile science, greater attention is being paid to the body position (Ks). Solutions to prevent the deficiencies of sagittal imbalance have been sought using constructive technological methods. They can be reduced by patterning a larger neck line. In such case the garment depends less on the individual profile type of end-users.

A complex muscular system is responsible for the posture and movements of the human body. There are various groups of muscles: head, neck, upper extremity, trunk and lower extremity. Even small-scale body movements may involve several muscles and their groups, which contribute to changes in the external form of a body. Smaller muscles, for instance, act to move the scapula up and inwards. Other muscles contribute to the movements of the ribcage during breathing [6,7].

Factor affecting the measurement results is the breathing of the test person. In dynamics studies, which deal with the changes in body measurements caused by movements, a separate position is the dynamic indicators during the deep breathing. One of the studies shows that the chest girth dynamic effect of men was, on average, 2.61% of the indicator in a static position [8]. The results obtained in statics may be influenced by slight changes in thoracic form during normal breathing. When breathing in – the lower ribs are lifted up and out to the sides. The breastbone also moves out and contractions of the intercostal muscles causes the ribcage to expand both sideways and front to back [6,7].

Furthermore, fat percentage of the body and adipose tissue distribution depending on their location may both intensify and lessen a hump, waist, chest or hip protrusion, and other posture characteristics.

4. Methods and procedures

The target group of the performed research is Latvian National Armed Force soldiers, namely, a group comprised of 171 men with an average age of 30 years (aged 19-52), average body height of 180 cm (160.3 – 195.2 cm), and average body weight of 87kg (62.4 – 141.6 cm). Soldiers are supplied with uniform jackets, which are fitted with primary dimensions: chest girth and body height. Incorrect body measurements, especially in the case of a chest circumference, can result in an inappropriate size, which, for example, in the case of a smaller size, results in a poor fitness and movement limit during the soldier's daily tasks.

When using the traditional manual measurement methods, during the measurement of the chest girth, the person usually lowers the hands letting them hang down freely (schematically 'figure 1 a', 3D scanning reproduction of the front - 'figure 1 b'), whereas the laser scanning technology requires adjusted position with arms moved away from the body, thus not covering the sides and allowing laser rays to cover as much as possible (schematically 'figure 1 c', 3D scanning reproduction of the front - 'figure 1 d'). Combined images ('Figure 2') shows differences in postures.

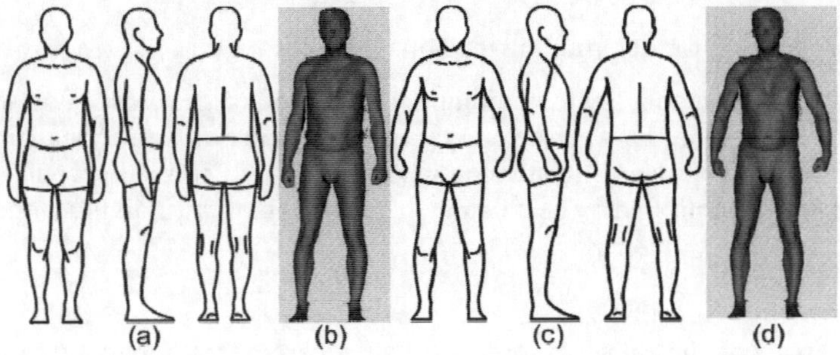

Figure 1. Measurement positions: schematic and 3D reproductions.

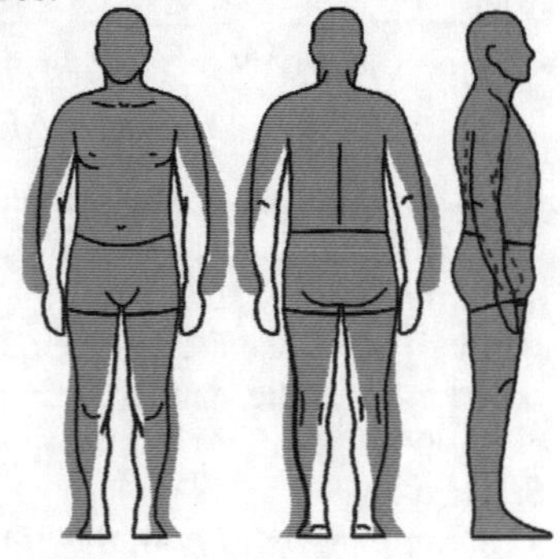

Figure 2. Combined images of postures.

The horizontal chest girth is defined as the greatest body measurement of the horizontal perimeter of the trunk at bust points level. As emphasized, during the measurement procedure, the measuring tape at all times must be positioned horizontally, which in contact method can be a condition rather difficult to control, in particular, when the measurements are taken by one person. The position of the measuring tape will most likely be subject to the location of torso axis of the subject.

The 3D system captures two types of chest circumferences, both of which are made across the bust point landmarks, while the first one "Bust/chest girth (horizontal)" is measured parallel to the standing surface ('figure 3 a'), but "Bust/chest girth" is perpendicular to the axis of the torso ('figure 3 b').

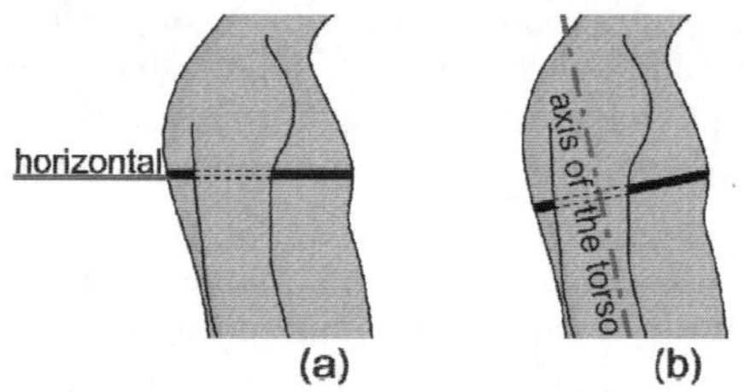

Figure 3. Chest circumference placement in 3D system (schematically): a - Bust/chest girth (horizontal) and b - Bust/chest girth.

End-users body measurements were obtained using both methods, for the contact method using measuring tape and anthropometer, but for the noncontact method using 3D anthropometrical scanner Vitus Smart XXL® (©Human Solutions GmbH and VITRONIC GmbH) with data processing system AnthroScan. Scanning measurements include 153 different body measurements, whereas the control measurements of chest girths were made manually. The research deals with the comparison of the both 3D and conventional manual chest girths by analyzing measurement differences.

Chest horizontal circumferences (gained manually and by 3D scanning) analysed for a separate group comprised of 22 men. For the pilot-experiment manual measurement precision, horizontal line (transversal plane at the level of bust points) is marked by placing additional measuring points on the back ('figure 4 a,b'). For more precision, the measurements were taken by two persons, one of which provides stable positioning of the measuring tape on the front, while the other – on back ('figure 4 c)'.

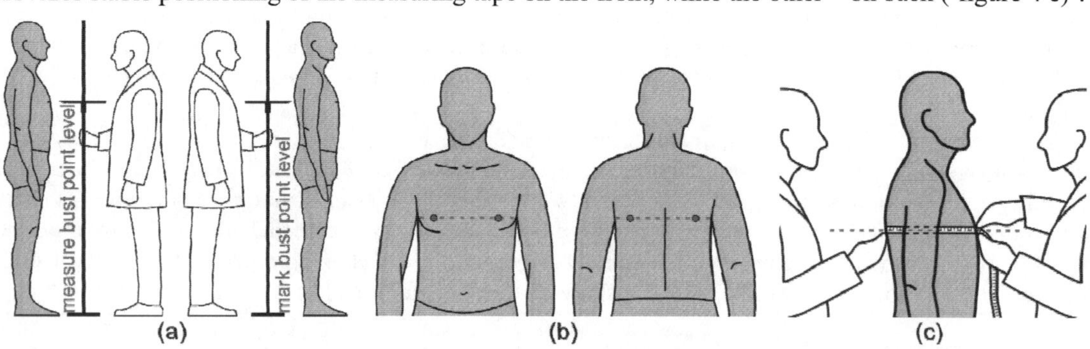

Figure 4. Manual method for determining the horizontal chest girth.

The postures of the subjects were analyzed, by determining stature indicators (Ks). For the calculation of the body position, the differences between two body measurements of the 3D scanning system, namely, the 'Distance 7CV' (7CV - the vertebra prominent) and the 'Distance scapula to vertical' were used. The type of the profile identified by calculating the deviation of the individual body position from the standard numerical value.

5. Results

The comparison of both the 3D and manual chest girths shows that, in the case of the horizontal girth, the differences from the manual chest girth vary from -3.9 to 8.8 cm (the average difference 1.8 cm), while from those of the girth perpendicular to the axis – from -5.4 to 6.1 cm (the average difference 0.4 cm). After having grouped the differences ('figure 5') with the frequency of 1 cm, it could be concluded, that the chest girth perpendicular to the axis of the torso is closer to the manually obtained chest girth, mostly exceeding the value by 0 to 1 cm. However, in the great majority of cases, the 3D horizontal chest girth difference from the manual chest girth is about 2 to 3 cm larger, which, in turn, confirms the assumption that, if during the manual measurement no additional body markers are used, the obtained measurement will not be the horizontal chest girth, but rather a measurement perpendicular to the axis of the torso.

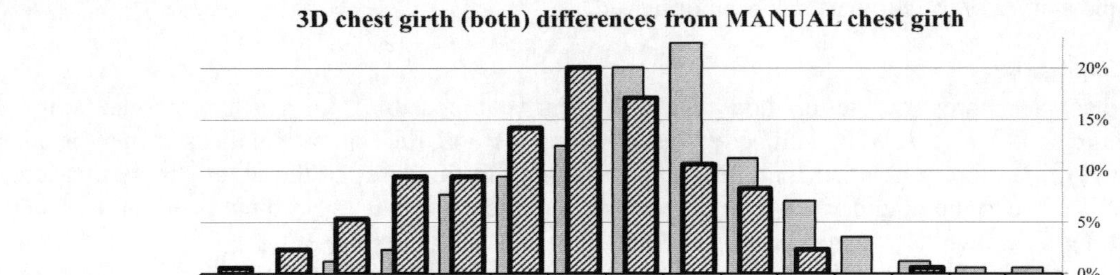

	(-6) - (-5)	(-5) - (-4)	(-4) - (-3)	(-3) - (-2)	(-2) - (-1)	(-1) - 0	0 - 1	1 - 2	2 - 3	3 - 4	4 - 5	5 - 6	6 - 7	7 - 8	8 - 9
Δ Chest girth (horiz.)	0,0%	0,0%	1,2%	2,4%	7,7%	9,5%	12,4%	20,1%	22,5%	11,2%	7,1%	3,6%	1,2%	0,6%	0,6%
Δ Chest girth	0,6%	2,4%	5,3%	9,5%	9,5%	14,2%	20,1%	17,2%	10,7%	8,3%	2,4%	0,0%	0,6%	0,0%	0,0%

Figure 5. 3D chest girth (both) differences from manual chest girth.

As for both types of 3D chest girth (horizontal and perpendicular to the axis of the torso), the frequency of occurrence was determined by selecting the chest girth intervals by means of the tables of control measurements of producer's jackets to be used for the sizing. The obtained distribution (the chart in 'figure 6'), indicates that the sizes of the garments to be produced or provided tend to differ from the selected chest girth type, which is either horizontal or perpendicular to the axis of the torso.

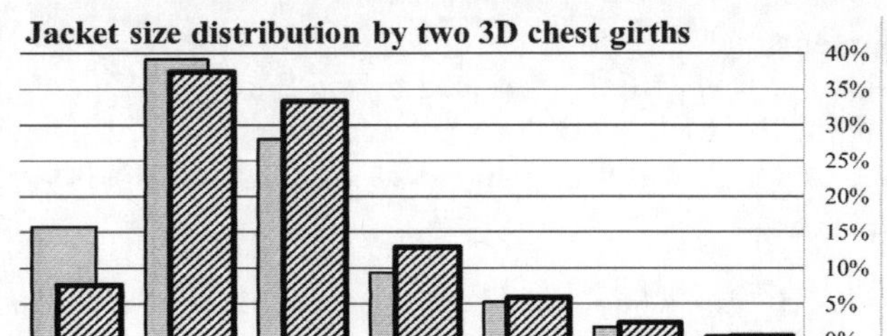

	M	L	XL	XXL	3XL	4XL	5XL
Chest girth (horizontal)	15,8%	39,2%	28,1%	9,4%	5,3%	1,8%	0,6%
Chest girth	7,6%	37,4%	33,3%	12,9%	5,8%	2,3%	0,6%

Figure 6. Jacket size distribution by two 3D chest girths.

As a result of a pilot experiment, it is observed that, compared with the manually measured chest circumference, the horizontally marked chest circumference values are less distinct from the 3D scan target when the difference varies from -2.7 to 5.1 cm. However, it is necessary to increase the research group so that it is possible to exclude erroneous measurements and to find reasonable regularities with other human body measurements.

It was concluded that soldiers within the target group mostly have a normal stature or body position. However, to great extent, the occurrence of subjects with stately and stooped posture can be observed. No correlation between the stature indicator (Ks) and differences between both 3D chest girths un manual chest girths have been observed, which indicates that the stature indicators of the individuals cannot be directly used within the particular group, in order to analyze the tendencies of the changes in measurements arising from the results of 3D scanning and manual measurement as a result of change of the body position (with arms down and moved away from the body). The differences in statures within the particular group could have been affected by such factors as heredity, physical development, and influence of physical activities. As from the very moment the soldiers join the military, their everyday activities involve carrying of heavy loads - backpacks, weapons, and equipment, which may occur in various conditions, as walking, running, overcoming barriers in different environments (meadows, forests, swamps, steep ascent areas), which, in turn, may contribute to the change in the posture, as the center of gravity needs to be shifted. On many occasions, changes in the posture are associated with person's age and ageing impacts. Given that the target group is comprised of men aged 19 to 52, the regularities between the age and body condition were analysed; however, no correlation between the age and stature of the subjects has been observed.

6. Conclusions

By manual anthropometric methods is hard (nearly impossible) to measure precise transversal measurement. By usage of human body laser scanning method, the transversal measurement is obtained precisely, therefore it is larger than manually measured. In the cases of the unique body measurement value is close to the boundary of the range of measurements, it causes problems with appropriate size choosing for individual wearer.

The comparison of the 3D and manual chest girths, as well as the analysis of the results within the particular group confirm the assumption that the manual chest girth would be closer to the chest girth perpendicular to the axis of the torso, rather than the horizontal chest girth, this being affirmed by smaller occurring differences.

The occurrence of chest girth differences as a result of the change of the position may be influenced by many different factors - in a way, the aforementioned posture, the amount of soft tissue, the impact on the body area to be measured, as well as the development of the muscles of the thorax, breathing indicators, the precision of measuring tools and skills of the person taking the measurements, the individual behavioral of the person to be measured, and the ability to precisely follow the given instructions, etc.

Acknowledgments

This research work was partially financed by the European Union's European Regional Development Fund, through the INTERREG BSR Programme, which awarded a grant to the SWW project (#R006). The authors gratefully acknowledge the received financial support.

References

[1] EN 13402-3:2017 (standard) *Size designation of clothes - Part 3: Size labelling based on body measurements and intervals*
[2] EN ISO 13688:2013 *Protective clothing - General requirements*
[3] Конопальцева Н М, Волкова Е Ю, Крылова И Ю 2014 *Антропометрия индивидуального потребителя.* Основы прикладной антропологии и биомеханики: лабораторный практикум (Москва, ФОРУМ-ИНФРА-М) 256 с
[4] Song G 2011 *Improving comfort in clothing* (Woodhead Publishing Limited) p 496
[5] Шершнева Л П, Пирязева Т В, Ларкина Л В 2004 *Основы прикладной антрпологии и биомеханики* (Москва, ФОРУМ-ИНФРА-М) 144 с
[6] Cohen B J, Hull K 2016 *Memmler's structure and function of the human body* (Wolters Kluwer) 11th edition p 512
[7] *"The human body". An illustrated guide to your body and how it works* 2017 ed P Abrahams, (London: Amber Books Ltd) pp 68-103
[8] Коблякова Е В, Куршакова Ю С, Зенкевич П И, Дунаевская Т Н 1974 *Размерная типология населения стран – членов СЭВ* (, Москва: Легкая индустрия) 440 с

Evaluation of clothing fit

I Dabolina[1], L Silina[1] and P Apse-Apsitis[2]

[1]Riga Technical University, Faculty of Material Science and Applied Chemistry, Institute of Design Technologies, Kipsalas street 6-220, Riga, LV-1048, Latvia

[2]Riga Technical University, Faculty of Power and Electrical Engineering, Institute of Industrial Electronics and Electrical Engineering, 12/1 Azenes street, Riga, LV-1048, Latvia

Inga.Dabolina@rtu.lv

Abstract. This paper focuses on analysis of clothing appearance and fit. Clothing fit ensures human body comfort in terms of microclimate, movements and the body well-being. Garments produced in serial production developed according sizing chart, which is suited to the individual measurements by indifference intervals only. Some, more elaborated systems use drop values. The fit is good for people whose size meets the middle of the table. If the measured values reach the end of the interval, fit is at risk. This is especially important for garments with integrated additional features. Clothing appearance and fit assessment methods are currently well developed - it is possible to assess the individual's 3D look and the distance from the body of the garment. Such an analysis can be done either through virtual simulation or by scanning a real person and/or sample. However, none of aforementioned methods assess the individual's senses in clothing. Therefore, a system is proposed which allows evaluating the analysis of clothing pressure.

1. Introduction

Nowadays different textile materials are developed – tendencies in the market are to produce textiles with smart applications. Smart applications in terms of garment include not only textile properties – it is complex system within human body and environment [1]. Clothing can increase the ability of body to function in different environments. Although all clothing is functional, the functional clothing design focuses on garments with special purposes. In order to meet all the determined functional and comfort requirements for clothing, patternmaking is one of the most significant steps. Pattern should be suitable for users' body type and its specifics, as well as for the specifics of the garment itself. Clothing is produced according to certain size charts (considering few main measurements – body height, bust, waist and hip horizontal girth) [2]. However, the figure of each individual is very different; therefore, it is hard to meet the classification due to the various proportions, which are rarely taken into account when designing clothes for production.

For a textile to meet the set requirements in terms of functionality and comfort, one of the most important and crucial factors is a pattern corresponding to human figure and the specific features of the particular garment [3]. Clothing is produced based on figure types [4], [5]. The main pre-condition for standardization and figure type indifference intervals is to serve as wide target group as possible so as to ensure that one size covers as large part of the population as possible. However, the figure of each individual is different, and even though it may be analyzed statistically, it is difficult to classify it due to different proportions (drop type), which in patternmaking for mass production are hardly ever taken into account.

Clothing designed for a specific function such as winter sportswear (providing protection against wind, moisture, and cold, pleasant microclimate, etc.) loses its functionality, if the pattern does not correspond to the specific features of the body – the interlining between the outer fabric and the lining of the garment becomes tightened, thus causing loss of air gap, which, in turn, results in considerable decrease in thermal protective properties of the garment. Moreover, discomfort is caused also by limited range of mobility [3].

Target of the research is to create a prototype for a system to experimentally determine the extent to which clothing clings to the wearer's body (or other clothing layers), how the clothing fit affects the wearer's comfort and ensures that garment meets all the functional requirements. Sensors elaborated in textile system can be used to analyze and determine these questions, also 3D prototyping and ease evaluation is useful.

2. Materials and methods

The fit of clothing and clothing systems greatly affect the provision of general garment performance, functionality, comfort, and mobility [3]. Clothing is an essential human need with several functions. The choice of clothing depends on several complex criteria, such as personal desires and needs, which are user dependent and therefore subjective. Lately more and more attention is focused on the concept and awareness of comfort, and the possibility to express the level of comfort by means of specific characteristics.

Thermophysiological comfort can be evaluated by performing the guarded-hotplate test (as referred to in ISO 11092) and by means of different thermal manikins (as referred to in ISO 15831) [6], [7]. Ergonomic comfort parameters can be defined by using a 3-D body scanning method to determine the air gap and its distribution between the clothing and human body [3]. Furthermore, different CAD/CAM systems [8] may be used in patternmaking, and virtual fitting when analyzing the respective color-coded bars regarding the fit [5].

Since the clothing covers a three-dimensional figure, the fit is hard to define due to the specific geometry of the human body, different characteristics of the material, and other complex factors (fashion, perception, purpose of wearing the specific garment, etc.). When designing garments in a 2-D environment, ease allowances are determined by linear body measures. However, they may not always correctly represent the fitness of the garment for the human figure [4], [5].

The 3-D scanner registers specific characteristics of human body and clothing proportions in three dimensions, which also allows determining deficiencies of 2-D patterns. Thus, it is possible to analyze the required ease allowances and the distance between the clothing and the human body. The air gap amount and the overall air gap distribution between the clothing and the human body are quantified in order to quantitatively characterize a three-dimensional garment fit.

2.1. Measurements

Measurements of subjects were obtained using two methods, for the contact method using measuring tape and anthropometer (device), for the noncontact method using 3D anthropometrical scanner Vitus Smart XXL® (©Human Solutions GmbH and VITRONIC GmbH) with data processing system AnthroScan. A human body surface reproduction dot cloud is created from the coordinate readings, which is to be used as a virtual mannequin or only the coordinates themselves can be used. A virtual reproduction of the human body for this research is used as virtual try-on object within the same pattern jacket for 6 different persons from point of view of drop values, although the size by sizing system is the same for each person – see values of total morphological parameters for each person used in the research in Table 1.

For the pattern size 40 as basic size is used. The intervals of main dimensions required are bust girth 90-94 cm, waist girth 74-78 cm and hip girth 98.5 – 101.5 cm. According to sizing system used in patternmaking method, indifference intervals for bust and waist measurements are 4 cm and 3 cm for hip girth, all involved test-persons are in size 40 at least in interval of bust girth.

Table 1. Measurement table for test-persons [cm]

#	Body height	Bust girth	Waist girth	Hip girth
1.	171.1	93.7	72.3	102.5
2.	180.6	90.5	66.8	96.7
3.	167.5	92.6	65.8	96.2
4.	166.7	93.0	75.9	99.0
5.	173.1	92.3	75.9	100.9
6.	156.8	91.2	77.4	99.4

2.2. Virtual try-on

For the virtual try-on and ease assessment, Lectra Modaris 3D Fit system is used. This system is useful for garment imitation – the systems allows performing a virtual fitting, evaluate the external appearance, shape, set, proportions of the garment (the garment is created in 3D by joining patterns constructed in a plane, creating an imitation of the garment with the intention to ascertain the conformity of the outer appearance to the expectations).

System allows choose material from built-in data by searching most important parameters. Model for simulation is one ply fleece jacket with zipper closure ('figure 1'). Fabric used is fleece interlock, Cotton 94%/ Elastane 6%, thickness 0.8 mm with good drape. For virtual prototyping side panels are gathered to 90 % and sleeves are gathered to 80 %.

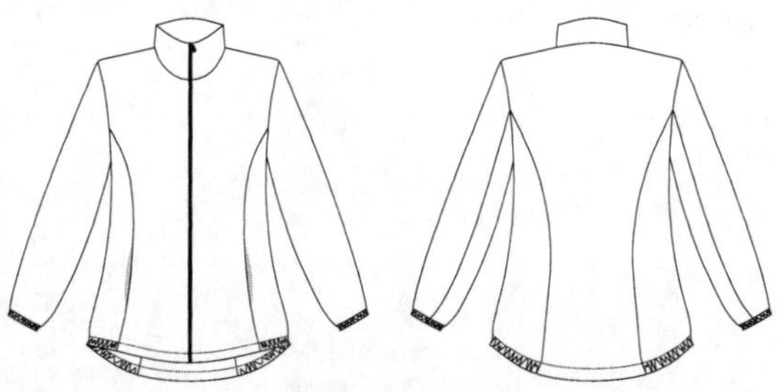

Figure 1. Jacket technical drawing.

3. Results

The results of virtual prototyping allow assess ease, fit and appearance of the model. As the chosen test-persons are different – three of them match the size only by bust girth, three of them fits into indifference intervals.

3.1. Assessment of fit

Fit assessment visually by appearance of simulated jacket displayed in Table 2.

Table 2. Fit of the virtual jacket to the test-person virtual scanatars

#	Front view	Right side	Back view	Left side
1.	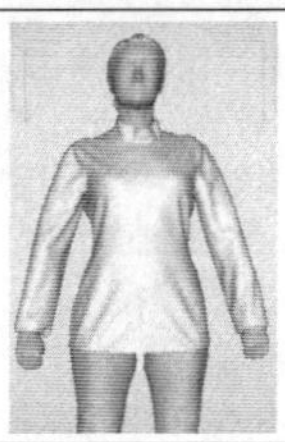		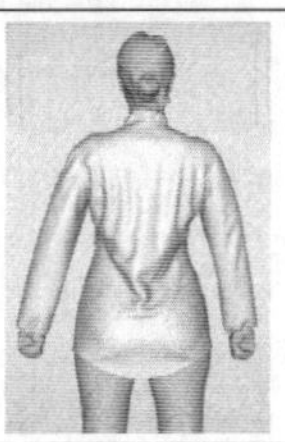	

IOP Conf. Series: Materials Science and Engineering **459** (2019) 012077 doi:10.1088/1757-899X/459/1/012077

Table 2. Fit of the virtual jacket to the test-person virtual scanatars

#	Front view	Right side	Back view	Left side
2.				
3.	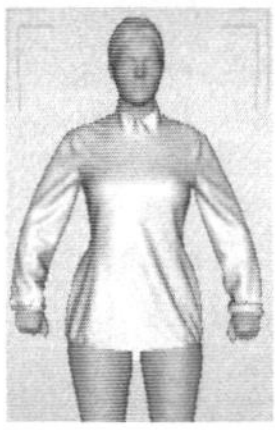			
4.			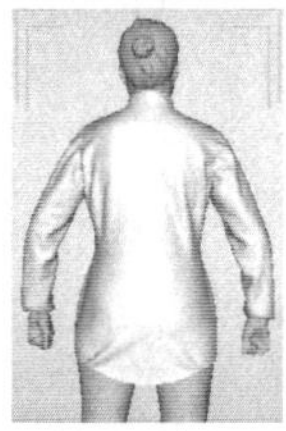	
5.	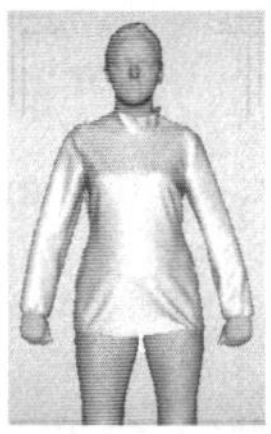	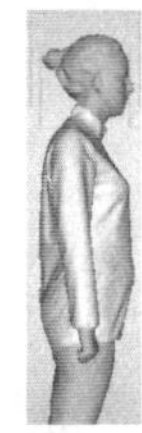	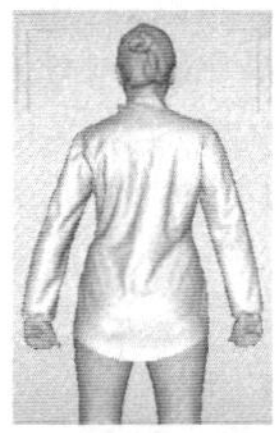	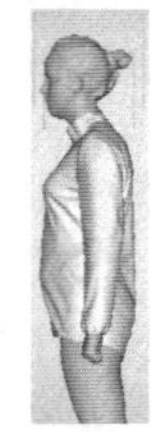
6.	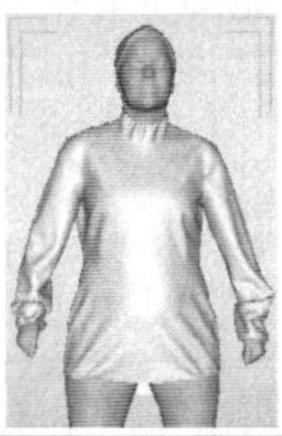	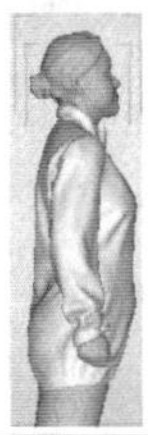	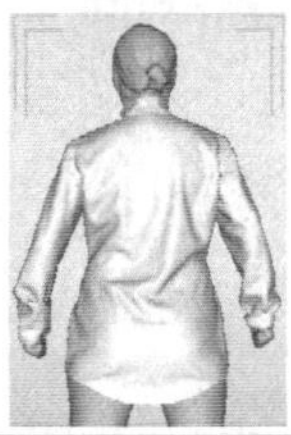	

Last three test-persons fit in the sizing table more proper (in terms of total morphological values) than first three and the appearance of the jacket is smooth and appropriate to the body. To evaluate ease of garment colour analysis is used where dark blue shows too loose garment and bright red is too tight garment. White shows appropriate fit, light blue loosened fabric and yellow – stretched fabric. Analysis of ease and fit in Table 3 is given.

Gaps between jacket and body shows spatial ease – one for free movements, other describe bad fit. As all six test-persons fit in sizing chart, ill fit is caused only if person's measurements are supreme or infima values in the indifference interval. For example, first test-person's bust circumference tends to reach supreme value – therefore folds around bust transversal plane are rather small. In addition, some folds are caused by posture and dedicated for free movement. Second person's jacket's appearance analysis shows large gaps between body and jacket – this is caused by smaller bust circumference.

IOP Conf. Series: Materials Science and Engineering **459** (2019) 012077 doi:10.1088/1757-899X/459/1/012077

Table 3. Analysis of ease and fit

#	1.	2.	3.	4.	5.	6.
Ease shadow	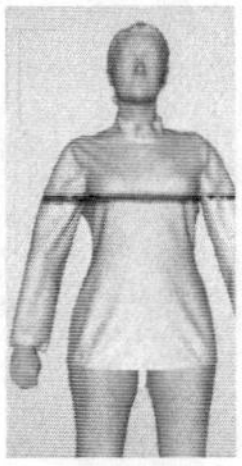	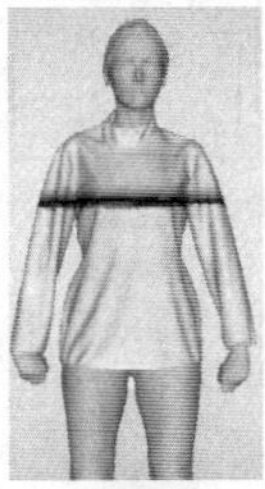	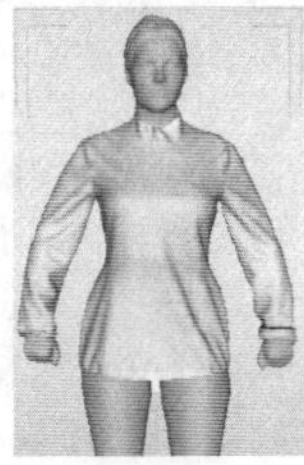	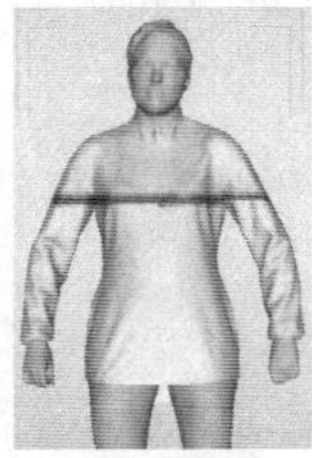	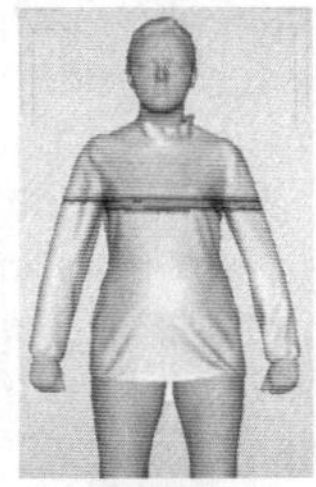	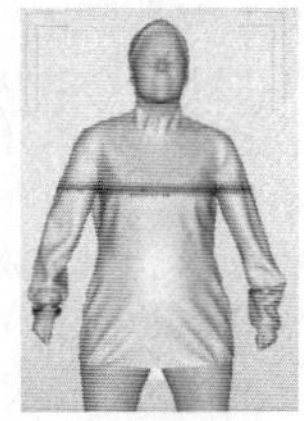
Medial cut of jacket (white)	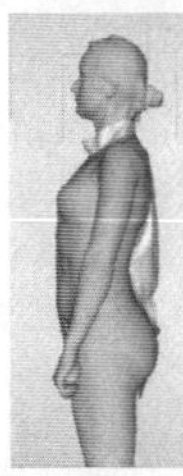	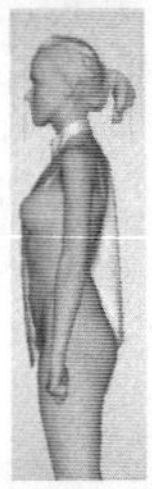	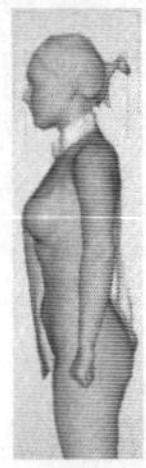	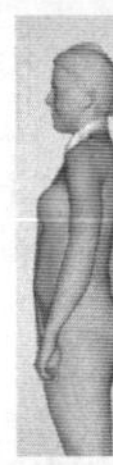	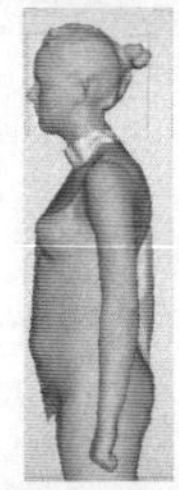	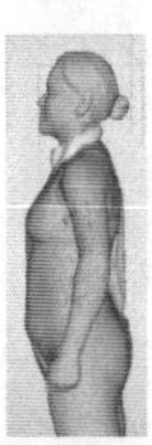
Transverse (bust level)	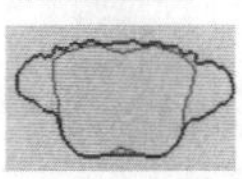	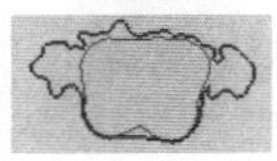		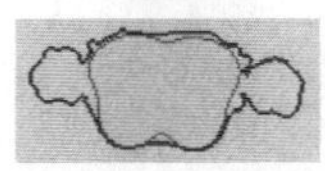	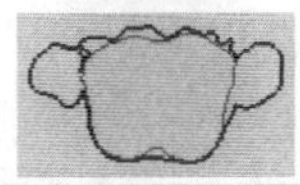	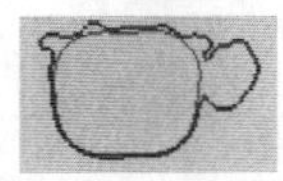

Beside fit and comfort analysis individual peculiarities of a person can be assessed – third person's most prominent bust point is lower than axilla points – therefore transversal analysis shows jacket cut without sleeves, while sixth person is asymmetrical – therefore transverse cut also is not symmetrical.

3.2. System of textile based sensors

The aforementioned methods to be used to identify factors affecting comfort are widely applied and developed so as to allow obtaining results and draw conclusions arising out of them, which are more approximate to the real situation. The proposal as the result of this research is to develop a prototype for a grid system of sensors, which shall be used to experimentally determine to what extent the garment fits to the wearer and the pattern of the garment (in terms of compliance with the specific characteristics of wearer's body) affects wearing comfort and ensures compliance with the relevant functional requirements. Textile based capacitive force sensors shall be integrated into the prototype for the system of sensors to be developed. Furthermore, it is planned to keep developing this idea by integrating heat and moisture sensors.

The capacitive force sensors can be used to determine the pressure between two surfaces. These sensors use the principle as follows: there is a three-dimensional fabric (spacer fabric) arranged between two electrically conductive layers. With the supply of voltage by applying force/ pressure, the distance of both electrically conductive layers is reduced, whereas the electric charge increases, such outcome being expressed as capacity [9], [10]. A set of sensors matrix will be designed in order to increase pressure influence in desired areas and more precise pressure identification. Data will be obtained by cable communication indoors, laboratory tests, or by means of Internet of Things outdoors.

The pressure caused as a result of the interaction between the human body and the clothing, and/or additional elements is, to a great extent, attributable to the fit of clothing, the characterizing indicators being the comfort, freedom of movement, weight, firmness of the fabric, etc. The skin is extremely sensitive to different mechanical stimuli, and the studies show that if the clothing pressure on the human body exceeds 30 mmHg, a person ought to feel discomfort [11]. Increased clothing pressure may indicate to incorrect fit of clothing and reduced performance.

4. Discussion and conclusions

Non-fitting or ill-fitting functional clothing can reduce and/or lose ones' functionality – for example, the insulation [12] between the top and the lining of the garment is tightened/flattened losing the air interlayer, thus the product's heating ability is significantly reduced, discomfort could be also caused by the limited movement ability.

The research will boost further the use of design systems in apparel industry and trade, and reduce the number of trial and error experiments by producers when selecting garment sizes for special target groups. The approach supports the Sustainable Development Goals, i.e. – the need for sustainable development - provides a comprehensive approach bringing together economic and environmental considerations in ways that mutually reinforce each other. As functional clothing is expensive and material consuming process – the improvement and enhancement of the design process and adjusting the design to end user is to be boosted within this research result.

The result of this research will improve the development and operational process in functional clothing industry. The result of the research will show solutions on multi task and multidisciplinary approach – including benefits for textile producers, pattern makers and end users. In the terms of 3D prototyping and ill-fitting detection, research gives advices on improvement of patterns. Research tasks performed within it give opportunity to make conclusions regarding causes of pattern alterations, their depiction in computer simulation, measuring possibilities and thereby the prevention of the causes of such alterations thus easing the work of pattern maker. In addition, smart solutions of ill-fitting detection are suggested.

References

[1] Song G 2011 Improving comfort in clothing (Woodhead) ISBN 1845695399 p 496
[2] EN 13402-2:2014 (standard) Size designation of clothes – Part 3: Body measurement and intervals
[3] Lu, Y., Song, G., Li, J. A novel approach for fit analysis of thermal protective clothing using three-dimensional body scanning. Applied Ergonomics. 2014, 45, 1439 – 1446.
[4] Gupta, D., Zakaria, N. Anthropometry, Apparel Sizing and Design. Cambridge: Elsevier Science & Technology, 2014. 368 p. ISBN 978-0-85-709681-4
[5] Fan, J., Yu, W., Hunter, L. Clothing appearance and fit: Science and technology. Cambridge: Woodhead Publishing, 2004. 239 p. ISBN 1 85573 745 0
[6] Wang, F., Kuklane, K., Gao, C., Holmer, I. Effect of temperature difference between manikin and wet fabric skin surfaces on clothing evaporative resistance: how much error is there? Int J Biometeorol. 2012, 56, 177 – 182. Available from: doi: 10.1007/s00484-011-0411-z
[7] Fu, M., Yu, T., Zhang, H., Arens, E., et.al. A model of heat and moisture transfer through clothing integated with the UC Berkeley comfort model. Building and Environment. 2014, 80, 96 – 104. Available from: doi: 10.1016/j.buildenv.2014.05.028
[8] Li, Y. Computer-aided clothing ergonomic design for thermal comfort. Sigurnost. 2011, 53 (1), 29 – 41.
[9] Giovanelli, D., Farella, E. Force Sensing Resistor and Evaluation pf Technology for Wearable Body Pressure Sensing. Journal of Sensors. 2016, 13 p.
[10] Holleczek, T., Ruegg, A., Harms, H., Troster, G. Textile Pressure Sensors for Sports Applications. In: IEEE Sensors 2010, 732 – 737.
[11] Kamalha, E., Zeng, Y., Mwasiagi, J.I., Kyatuheire, S. The comfort dimension; a review of perception in clothing. Journal of Sensory Studies. 2013, 28, 423 – 444. ISSN 0887-8250.
[12] Williams,J.,T.2009 Textiles for Cold Weather Apparel (Woodhead) ISBN:9781845694111p.432

IOP Conf. Series: Materials Science and Engineering **459** (2019) 012078 doi:10.1088/1757-899X/459/1/012078

Sizing of clothing appropriate for overweight and obese children: methodology stages and the preliminary results

R de Campos[1,2], M A F Carvalho[2] and R S Boldt[2]
1Federal Institute of Rio Grande do Sul, Fashion Department, Campus Erechim, Erechim, RS, Brazil
2University of Minho, 2C2T-Center for Textile Science and Technology, Campus de Azurém, Guimarães, Portugal
raquel.campos@erechim.ifrs.edu.br

Abstract: The article describes the statistical analysis on the main components and the preliminary results for the development of a table of measurements, focused on the overweight and obesity of children from the north of Portugal, as part of an ongoing PhD research in Textile Engineering at the University of Minho, Portugal. In this sense, for the construction of a sample table of measurements, the studies of Huyssteen [1], Silveira [2], and Gill [3], recommend anthropometric studies, health databases and statistical analyzes.

1. Introduction

In Portugal, as in other countries, manufacturers of children's clothing are not prepared to offer clothing suitable for overweight and obese children [4]. A limited supply of clothing that gives them comfort and safety leads this population to wear clothing aimed at other age groups [5]. The use of larger sizes does not take into account their bodies and specificities [6], [7], [8], [9]. For Kim [9], there is a need to segment the tables of measurements of the child population by lean, medium and obese children, in order to increase the coverage of the age range. According to Boguslawska-Baczek [10], it is necessary to find the best way to develop the tables of measurements aiming at designing comfortable and suitable clothing for the body of overweight and obese children [9], [11].

The tables of measurements are the basis for the development of the clothes in series [3], [11], [12]. They have emerged to facilitate commercial practice, provide information to consumers, entrepreneurs and retailers [1], reduce costs and ensure product quality [2]. The tables used for the modelling of garments vary between countries, norms, brands, market segments and fabrics [1], [2], [10], [13], [14], [15]. They are grouped by male, female and child segment [10] from the average population measurements [5], [7], [10], [13], [15], [16], [17], and determine a minimum number of sizes for a maximum number of people [1]. This has an impact on individuals who are far from the average [1], [2], [17], since the tables do not consider their constraints [13], [14], [15], as is the case of overweight and obese children.

As regards the children's segment, the measurement tables show a disparity between body measurements and marks [5] due to the great variation of biotypes between countries [1], [12], [18]. They present the categories: baby, infant and pre-adolescent [5] and can be divided by sex or be unisex. Sizes are represented by age in single numbers (8, 9, 10, 11) or double (8, 10, 12, 14), letters (XS, S, M, L) or words (small, medium and large) [1], [3]. Although many companies use tables formed on the basis of age, this classification is not correct, since it represents poorly the corporal dimensions, given that children of the same age can present different heights [1], [9], [10].

2. Materials and methods

For the development of tables of measurements suitable for clothing of overweight and obese children of the north of Portugal, from the studies of Huyssteen [1], Silveira [2], and Gill [3], it is recommended: current anthropometric studies, selection of the necessary measurements for clothing and statistical analysis.

IOP Conf. Series: Materials Science and Engineering **459** (2019) 012078 doi:10.1088/1757-899X/459/1/012078

2.1 Sample

It involved 816 children between 2 and 12 years old (434 boys and 381 girls). It is important to note that not all of these children belong to the target population of the study. Data from children not in the target group and children aged 2 to 4 years were not considered. The final sample consisted of 205 overweight and obese children, located in 3 cities in the north of Portugal, aged between 5 and 12 years, of both sexes.

2.2 Data collection and Data processing

The anthropometric data collection lasted 6 months, beginning in June 2016 and ending in November 2016. 3D body scanning technology (KBI) was used along with manual collection. After the data collection started the processing of the data. The processing of data went through different software and phases. To guarantee the quality of the data, the outliers (scan errors, recorded body measurements very large or very small) were excluded according to the definition presented in [19].

2.3 Measurements relevant to the study

The definition of the number of measurements required varies among researchers, standards or countries. According to [19], there is no exact number of measurements, being selected according to the relevance and the segment to which the product will be destined, however, the ideal is to select the measures of height and circumference, covering the dimensions of the trunk and extremities, and that can be easily located. Gill [3] presents several researchers that used 33 to 4 measurements. Gill [3] also says that many measurements are difficult to work with, and cover fewer people, which hampers statistical techniques.

3. Results and conclusions

The data were analyzed using SPSS 21 software (Statistical Package for the Social Sciences). In a first phase, was carried out the statistical analysis of the principal components (PCA), which is a technique of multivariate exploratory analysis associated with dimensionality reduction. Its purpose is to transform a set of correlated variables into a set of independent variables [20].

Of the 110 body measurements pre-established by the KBI system at this stage, 26 important variables for children's clothing were considered. However, the factorial structure obtained was not interpretable, and did not result in valid conclusions in the context of this investigation. According to several authors, children vary in growth according to age, gender, hereditary influences and environmental factors, changing rapidly [1], [21]. In the present study, the relationship between the male and female bodies differentiates between the sexes [1], [22].

After an interview with eight professionals working with modelling, it was decided to withdraw some initial variables. Thus, the PCA was again considered, taking into account the following 18 variables: height, bust circumference, waist circumference, hip circumference, abdominal circumference, thigh circumference, knee circumference, calf circumference, ankle circumference, biceps circumference, elbow circumference, wrist circumference, sleeve shirt, bust height, waist height, hip height, thigh height, and knee height. Subsequently, are presented all the steps and justifications that led to the best factorial solution for both sexes.

3.1 Assumptions Verification

Bartlett's sphericity test and the Kaiser-Meyer-Olkin (KMO) measure of sample adequacy are two statistical procedures that allow testing the validity of the factorial analysis, evaluating the quality of the correlations between the variables [23]. The test was performed for both subsamples. The KMO values obtained – 0.919 for the male gender and 0.923 for the female – indicate a very good adjustment of the factorial model to the data in both sexes [24]. Therefore, there is a very good suitability of the subsamples for the application of the factorial analysis. As for Bartlett's sphericity test, it was obtained as test value $p = 0.00 < 0.05$, so, it is rejected the hypothesis that there are no correlations between the variables. The adequacy of the factorial analysis method is confirmed.

3.2 Extraction of factors

Regarding the number of components to be retained in the analysis, there are four most commonly used criteria [23]. For a number of variables below 30, Pestana and Gageiro [24] advise the application of the Kaiser criterion. Therefore, the Kaiser criterion was applied, that is, choosing components with a value greater than 1, the first three main components should be retained for both male and female genders. It corresponds to a percentage of explained variance of 82% for the male gender and 83% for the female gender, presented in table 1.

Table 1. Total Variance Explained

	Male						Female					
Comp.	Value	% Variance	% Cumulative	Value	% Variance	% Cumulative	Value	% Variance	% Cumulative	Value	% Variance	% Cumulative
1	**11,30**	66,49	66,49	7,20	42,36	42,36	**12,34**	68,53	68,53	12,34	68,53	68,53
2	**1,62**	9,54	76,03	5,65	33,25	75,61	**1,49**	8,26	76,80	1,49	8,26	76,80
3	**1,05**	6,19	82,21	1,12	6,61	**82,21**	**1,15**	6,39	83,19	1,15	6,39	**83,19**
4	,60	3,53	85,74				,72	3,98	87,17			
5	,54	3,19	88,93				,46	2,54	89,71			
6	,47	2,77	91,70				,42	2,34	92,05			
7	,33	1,97	93,67				,40	2,20	94,24			
8	,30	1,74	95,41				,23	1,26	95,50			
9	,19	1,14	96,55				,22	1,21	96,71			
10	,17	,98	97,53				,18	1,03	97,74			
11	,14	,82	98,36				,12	,67	98,40			
12	,10	,57	98,92				,09	,51	98,91			
13	,07	,41	99,33				,07	,39	99,30			
14	,03	,19	99,52				,04	,23	99,54			
15	,03	,18	99,70				,03	,17	99,71			
16	,03	,17	99,87				,02	,14	99,84			
17	,02	,13	100,00				,02	,09	99,93			

Table 2. Communalities

	Extraction Male		Extraction Female
Height	,927	Height	,942
Bust height	,962	Bust height	,967
Waist height	,950	Waist height	,956
Hip height	,916	Hip height	,961
Thigh height	,935	Thigh height	,951
Knee height	,917	Knee height	,896
Sleeve shirt	,575	Sleeve shirt	,684
Bust circumference	,811	Bust circumference	,754
Waist circumference	,843	Waist circumference	,774
Abdominal circumference	,913	Abdominal circumference	,871
Hip circumference	,899	Hip circumference	,938
Thigh circumference	,759	Thigh circumference	,836
Knee circumference	,760	Knee circumference	,754
Biceps circumference	,615	Biceps circumference	,694
Elbow circumference	,733	Elbow circumference	,747
Ankle circumference	,752	Ankle circumference	,682
Wrist circumference	,709	Wrist circumference	,853

As can be seen in the calculation of commonalities in table 2, the commonalities indicate the proportion of the main component of each variable. Variables with high values indicate that they are well represented by the components.

3.3 Analysis and profiling of the main components
The matrix of components presents the loadings (or factorial loads). The loadings represent the correlations between each variable and its main component [25]. According to Pestana and Gageiro [24], the loadings must be very large or very small. Still according to these authors, as well as Larose [26], loadings with an absolute value greater than or equal to 0.5 should be considered significant. Field [27] chooses values in modules greater than or equal to 0.4. It is observed that most of the variables have a high loading on the first factor.

Table 3 Component matrix (male)

	Component		
	1	2	3
Height	,921	-,279	-,037
Bust height	,934	-,296	-,055
Waist height	,925	-,290	-,101
Hip height	,897	-,312	-,120
Thigh height	,899	-,334	-,125
Knee height	,879	-,359	-,127
Sleeve shirt	,720	-,237	,016
Bust circumference	,848	,206	,223
Waist circumference	,758	,496	,152
Abdominal circumference	,878	,377	-,013
Hip circumference	,926	,190	-,081
Thigh circumference	,843	,138	-,173
Knee circumference	,862	,125	,041
Biceps circumference	,675	,220	,334
Elbow circumference	,734	,437	,067
Ankle circumference	,528	-,152	,670
Wrist circumference	,423	,464	-,561

Table 4 Component matrix (female)

	Component		
	1	2	3
Height	,945	-,199	-,102
Bust height	,941	-,254	-,126
Waist height	,947	-,204	-,131
Hip height	,950	-,225	-,094
Thigh height	,919	-,293	-,141
Knee height	,902	-,263	-,117
Sleeve shirt	,775	-,239	-,160
Bust circumference	,842	,181	-,107
Waist circumference	,763	,427	,096
Abdominal circumference	,880	,261	,168
Hip circumference	,959	-,078	,109
Thigh circumference	,897	-,159	,076
Knee circumference	,844	,132	,155
Biceps circumference	,639	,531	-,050
Elbow circumference	,695	,339	,386
Ankle circumference	,525	,501	-,394

In order to obtain a more simplified and interpretable structure, there are different rotation methods that seek to eliminate intermediate loadings [24]. In fact, the first component extracted represents a general factor [26]. An orthogonal rotation method – Varimax – was used. This method minimizes the number of variables with high loadings [26]. In the matrixes of components of the rotated solution, presented in tables 5 and 6, the component to which the variable contributes with greater weight was selected. For the male gender, the first component is associated with the variables of height (thigh, bust, knee, hip and waist height, sleeve, height), while the second component is associated with the variables of circumference (waist, abdomen, elbow, bust, hip, knee, biceps and thigh). The third component is associated with the circumferences of the ankle and the bust, which correspond to the smallest circumference measurements of the lower and upper part, respectively. According to [22], the perimeters of the ankle, wrist and neck reveal a complex pattern. In the case of the female gender, there are some nuances compared to those described for the male gender. Thus, the first component is also associated with height (thigh, bust, knee, hip and waist height, height, sleeve), however, it is also associated with the variables of circumference of the thigh and the hip. This association is justified and explained by [22], who report that the distribution of adipose tissue is different between males (it is more abdominal) and females (it is distributed more in the hip and thigh), so the measurements have a greater agreement between the participants of the male gender. The second component is associated with the variables of circumference of biceps, waist, ankle, knee, elbow, calf, bust and abdominal. The third component is associated with the circumference of the wrist. Some variables, such as the female knee circumference, although presenting a lower factorial loading, remained in the model because they were considered relevant.

Table 5 Rotated component Matrix (male)

	Component		
	1	2	3
Thigh height	,916		
Knee height	,916		
Waist height	,903		
Bust height	,902		
Hip height	,899		
Height	,878		
Sleeve shirt	,687		
Waist circumference		,886	
Abdominal circumference		,838	
Elbow circumference		,807	
Bust circumference		,752	
Hip circumference		,717	
Knee circumference		,659	
Biceps circumference		,677	
Hip circumference		,604	
Ankle circumference			,695
Wrist circumference			-,674

Table 6 Rotated component Matrix (female)

	Component		
	1	2	3
Thigh height	,927		
Bust height	,919		
Hip height	,901		
Waist height	,897		
Knee height	,890		
Height	,885		
Sleeve shirt	,788		
Thigh Circumference	,776		
Hip circumference	,772		
Biceps circumference		,800	
Waist circumference		,790	
Abdominal circumference		,723	
Ankle circumference		,701	
Elbow circumference		,686	
Bust circumference		,628	
Knee circumference		,596	
Wrist circumference			,895

4. Future perspectives

Although the 3D scanning system has contributed to the garment industry, anthropometric studies are time-consuming and data processing is required in order to obtain a valid and representative sample. Such studies are of supreme importance to the clothing industry in order to support the design of clothing appropriate to specific populations. Based on the identification of overweight and obese children, the corporal measurements relevant to the development of children's clothing were analyzed through the statistical method of principal components. In a subsequent step, the analysis of clusters will be performed to formulate the tables for this segment. This will contribute to improve the comfort and usability of children's clothing, as well as their self-esteem.

Acknowledgments

This work is financed by Federal Institute of Rio Grande do Sul, Brazil and FEDER funds through the Competitive Factors Operational Program (COMPETE) POCI-01-0145-FEDER-007136 and by national funds through FCT-Portuguese Foundation for Science and Technology, under the project UID/CTM/000264.

References

[1] Huyssteen S V 2006 Development of standardised sizing systems for the South African children's wear *PhD thesis* University of Stellenbosch

[2] Silveira I 2008 Usabilidade do vestuário: factores técnicos-funcionais *Moda Palavra - periódico* Ano 1 no jan-jul pp. 21–39

[3] Gill S A March 2015 Review of research and innovation in garment sizing, prototyping and fitting *Textile Progress* Vol **47** no 1

[4] Campos R, Carvalho M, Lopes H and Xu B 2017 Anthropometric data collection of Portuguese children with

overweight and obesity *In: Proceedings of the 8th International Conference on Applied Human Factors and Ergonomics* Los Angeles

[5] Kang Y, Choi H S and Do W H A 2001 Study of the apparel sizing of children's wear. An analysis of the size increments utilized in children's wear based on an anthropometric survey *J Korean Home Eco Assoc Eng* **2** (1)

[6] Betti M U 2014 Beleza sem medidas: corpo, género e consumo no mercado de moda plus size *Dissertação de mestrado em Antropologia Social* Universidade de São Paulo

[7] Nafiu O and Burke C 2013 Large Clothing Size in Children Is Associated with High Body Mass Index and Clustering of Medical Comorbidities *ISRN Obesity Volume*

[8] Shin S J H and Istook C L 2008 Sizing Systems for Children's Wear in the United States *Texas Tech University* Lubbock TX USA

[9] Kim H and Damhorst M L 2013 Gauging Concerns with Fit and Size of Garments among Young Consumers in Online Shopping *JTATM* Vol **8** Issue 3 Winter

[10] Boguslawska-Baczek M 2013 Analysis of the contemporary problema of garment size *Conference: The 7th Textile Science 2013 conference At Liberec* Czech Republic

[11] Liu K, Zengc X, Bruniauxc P, Wangb J, Kamalhac E and Taoc X 2017 Fit evaluation of virtual garment try-on by learning from digital pressure data Knowledge-Based Systems **133** pp. 174–182

[12] Olaru S, Spânachi E, Filipescu E and Salistean A 2013 Virtual Fitting – Innovative Technology for Customize Clothing Design *24th DAAAM International Symposium on Intelligent Manufacturing and Automation*

[13] Boueri J J 2008 Sob Medida: antropometria, projeto e modelagem. Pires D B (Org) *Design de Moda: olhares diversos* São Paulo Estação das Letras e Cores pp. 347–369

[14] Heinrich D, Carvalho M and Barroso M Julho 2008 Ergonomia e Antropometria aplicadas ao vestuário–discussão analítica acerca dos impactos sobre o conforto e a qualidade dos produtos *Diseño en Palermo* Universidad de Palermo

[15] Sabrá F, Rosa R A, Santos C S and Brito L F F 2013 Desenvolvendo uma nova referência de medidas para o vestuário através da tecnologia de escaneamento de corpos 3d

[16] Campos R, Carvalho M A and Xu B July 2018 Anthropometric Data Collection of Portuguese Overweight and Obese Children using a Kinect Body Image system: Preliminary Results *9th International Conference on Applied Human Factors and Ergonomics (AHFE)* USA

[17] Alemany S, Gonzalez J C, Nacher B, Soriano C and Arnaiz C 2010 Anthropometric survey of the Spanish female population aimed at the apparel industry

[18] Capelassi C H 2013 A evolução histórica do dimensionamento do vestuário *Anais*

[19] Murteira 1993 Análise Exploratória de dados – Estatística Descritiva McGraw-Hill Lisboa

[20] Maroco J 2007 Análise estatística com utilização do SPSS (3ª ed.) Lisboa: Edições Sílabo

[21] Zakaria N 2010 The development of body sizing system for school aged children using anthropometric data *PhD thesis* Universiti Teknologi Mara Malaysia

[22] Loeffler-Wirth H, Vogel M, Kirsten T, Glock F, Poulain T, Ko¨rner A 2017 Body typing of children and adolescents using 3D-body scanning *PLoS ONE* **12**(10): e0186881

[23] Pereira A and Patrício T 2013 SPSS: guia prático de utilização: análise de dados para Ciências Sociais e Psicologia 8.ª ed Lisboa: Sílabo ISBN 978-972-618-736-3

[24] Pestana M H and Gageiro J N 2008 Análise de dados para ciências sociais: a complementaridade do SPSS (5ª ed.) Lisboa: Edições Sílabo

[25] Abdi H and Williams L J 2010 Principal component analysis. Wiley Interdisciplinary Reviews: Computational Statistics, **2**(4) pp. 433–459 doi:doi:10.1002/wics.10

[26] Larose D T 2006 Data Mining Methods and Models John Wiley & Sons, Inc

[27] Field A P 2009 Discovering statistics using SPSS: (and sex and drugs and rock 'n' roll) (OKS Print.) Los Angeles [i.e. Thousand Oaks, Calif.: SAGE Publications]

IOP Conf. Series: Materials Science and Engineering **459** (2019) 012079 doi:10.1088/1757-899X/459/1/012079

An investigation of design of twist knot drape clothes

Z Kazlacheva and J Ilieva

Trakia University, Faculty of Technics and Technologies
Graf Ignatiev 38, 8600 Yambol, Bulgaria

e-mail: zlatinka.kazlacheva@trakia-uni.bg

Abstract. A study about pattern making of twist knot drape shows that in the constructional systems there isn't any mathematically founded dependence between the diameter of the knot and the widths of the draped pieces in the place of twist. The paper presents an investigation of connection between the knot diameter and the widths of the draped pieces in the place of twist in design of patterns with twist knot draperies. The investigation has been made only for woven fabrics. The obtained dependence between the knot diameter and the widths of the draped pieces in the place of twist has made the design of patterns with twist knot draperies fully systematized. The systematization gives possibility for easy and correct pattern making of twist knot drape clothes with various sizes of diameter of knot and drapery parts width. The system facilitates the process of fashion design and pattern making and gives possibility of variety of new designs twist knot drape clothes. It is very useful for ready-to-wear industry.

1. Introduction

The twist knot drape is one of types of drapery in ladies' clothing and as the name of this drapery indicates it is designed in result of a twist of two pieces into a knot. A study about pattern making of this type of drapery shows that in the constructional systems [1, 2, 3, 4, 5] there isn't any mathematically founded dependence between the diameter of the knot and the widths of the draped pieces in the place of twist. It is logical that the bigger width of draped parts leads to the bigger diameter of the knot.

The paper presents an investigation of connection between the knot diameter and the widths of the draped pieces in the place of twist in design of patterns with twist knot draperies, id est searching for a mathematical dependence between them. The investigation has been made only for woven fabrics.

2. Pattern Making

Figure 1 presents a design and pattern making of a twist knot lady's dress. The pattern making is inspired by the geometrical system of Tomiko Nakamichi [1]. The main difference of the pattern making, shown in Figure 1, and the construction, presented in [1], is the number of divisions about drapery folds of the pieces. A previous investigation [6] shows that the two parts of the draped piece can be divided only ones. The bigger number of folds is a result of the twisting in the knot. In Figure 1 W is the diameter of the knot and W is the width of the draped pieces in the place of twist.

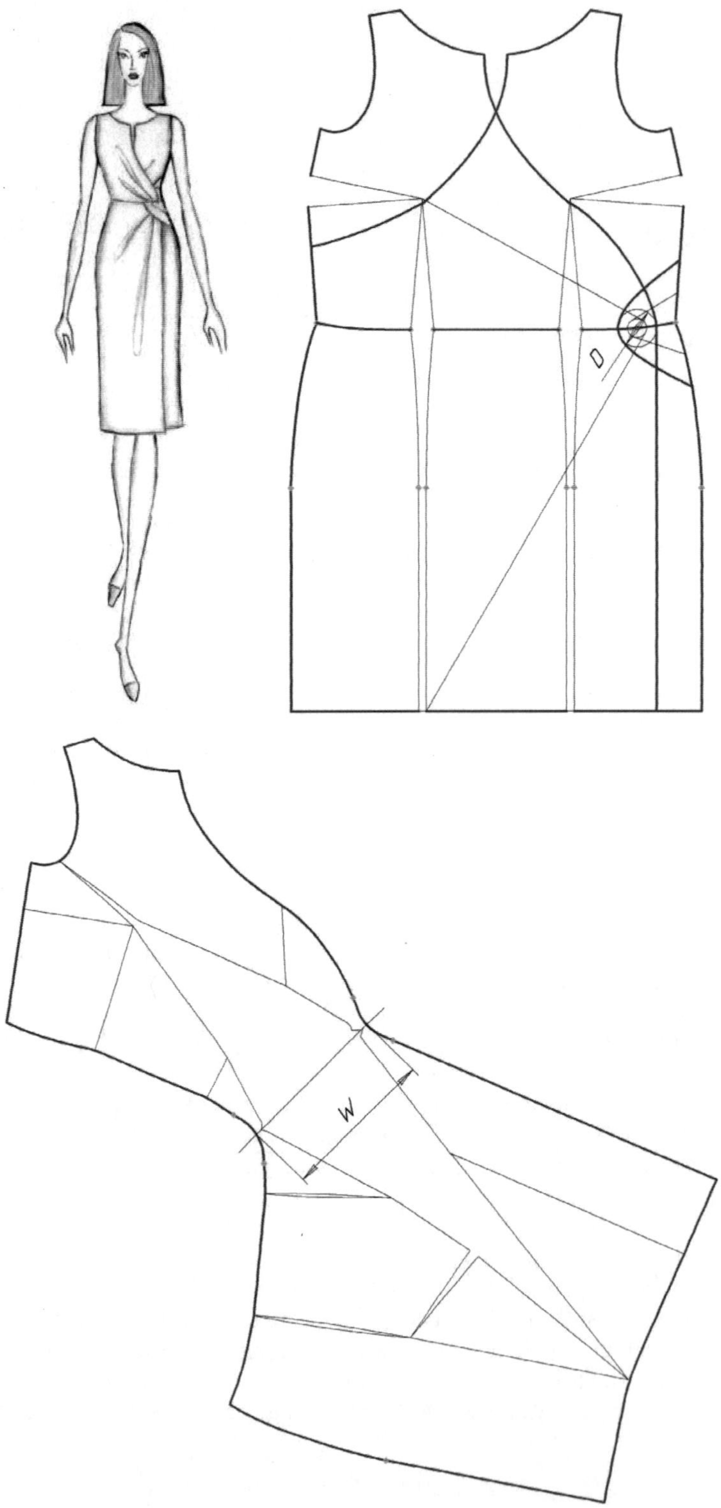

Figure 1. Pattern making of twist knot drape.

3. Methodology

The dependence between the knot diameter and the widths of the draped twisted pieces is searched because it will made the design of patterns with twist knot draperies fully systematized and will lead to design of twist knot drape clothes with various sizes of diameters of knot and drapery parts widths.

For this aim a linear regression is used [7, 8]. The same statistical method leads to easy sizing, which is very important for pattern making, and was successfully used in a previous investigation about pattern making of twisted draperies, presented in [8].

4. Experimental

The fully systemized pattern design has to be suitable for different woven fabrics with good drapability. By this reason four fabrics with good drapability but in different structures are chosen. From the four fabrics double rectangles in skew direction are cut with widths: 32,5 cm (the investigation shows the width bigger than 32,5 cm lead to unnecessary volume), 30, 25, 20 and 16,5 cm (the investigation has shown that for this type of draperies the minimal width has to be 16,5 cm). The double cut pieces are twisted in knots and the diameters of the knot are measured. The twisting and measuring of the knot are presented in Figures 2 and 3. For every combination of material and width five twists in knot and measurements are made.

Figure 2. Twisting of the pieces in knot.

Figure 3. Measurement of twist knot.

After the measurements the dependence between the knot diameter and the widths of the draped pieces in the place of twist has been obtained on the base of a linear regression [7, 8] – formula (1):

$$Y = a + b \cdot X, \quad (1)$$

where Y is the dependent variable; X – independent variable; a – constant; b – coefficient of regressions.

In the presented investigation formula (1) acquires mode (2):

$$W = a + b \cdot D, \quad (2)$$

where W is the widths of the draped pieces in the place of twist; D - the diameter of the knot; a – constant; b – coefficient of regression.

The regression analysis is made with STATISTICA 7.0 [7, 8].

5. Results

The results of regression analysis are: a = -4,4 and b = 5,15, and formula (2) is transformed in dependence (3):

$$W = 5{,}15 \,.\, D - 4{,}4. \qquad (3)$$

The accuracy of the regression model is proved by the values of $p < 0{,}0000$, R-square = 0,9126, and Std. Error of estimate = 1,4429.

Therefore the dependence (3) successfully can be used for determination of the width of the draped pieces in the place of twist – W on the base of the diameter of the knot of twist – D. According to the design idea the turned variant of dependence (3) can be used or determination of the diameter of the knot – D on the base of the draped pieces width – W, formula (4):

$$D = 0{,}2 \,.\, W + 0{,}85. \qquad (4)$$

6. Fashion design application

According to fashion design the twist knot drapery is suitable for casual and formal ladies' clothing in various types of styles [1, 2, 3, 4, 5, 6, 9].

Figure 4 presents a design and pattern making of a lady's dress with a twist knot in the waist area. In fashion design the center of the waist is maybe the most popular place of the knot. For that dress the diameter of the knot D = 7,0 cm is chosen according to the main design idea for rich volume folds of drapery. The width of the draped pieces in the place of twist W = 31,65 cm is determined by dependency (3).

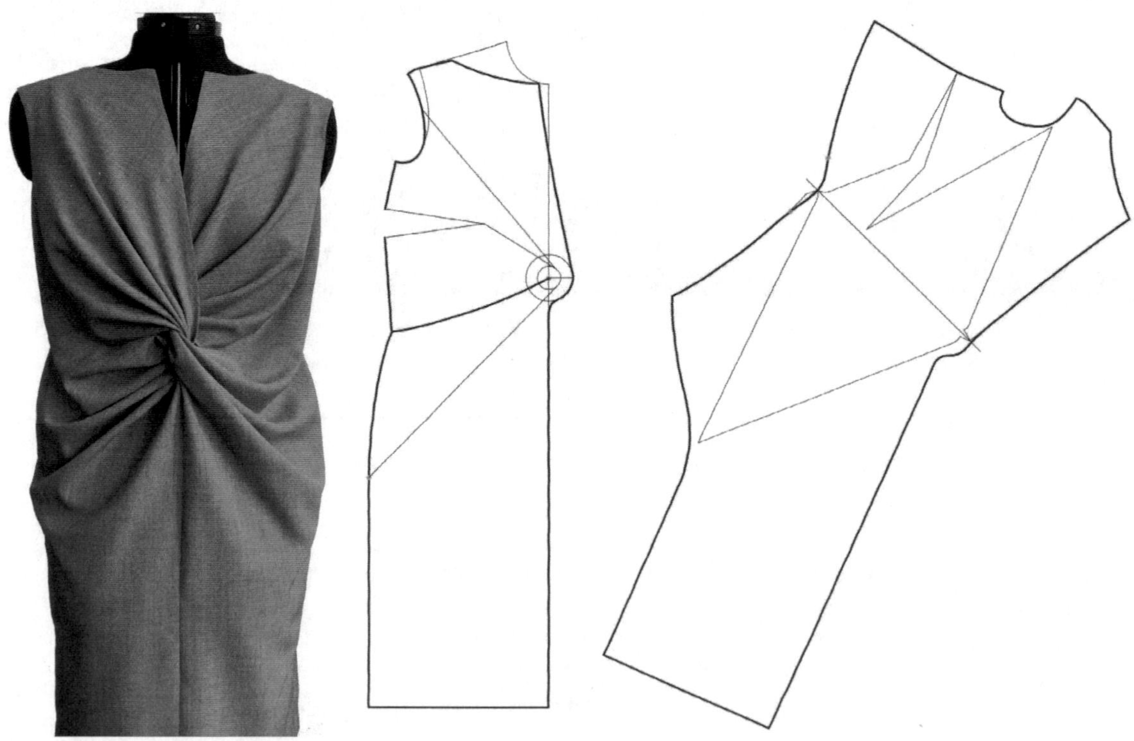

Figure 4. Design and pattern making of a lady's dress with a rich volume twist knot drapery in the center of the front waist area.

Figure 4 center shows the pattern making on the front constructional base. The center of the circle of the knot with diameter D = 7,0 cm is located on the front middle line between the waist and the bust dart apex. The seam of connection of both twisted pieces is under the knot circle by the middle line. The neck opening is formed over the circle of the knot. The seam of drapery fixing is situated between the center of the knot circle and the point of intersection of the waist line and the side seam. The second circle is drawn with the same center of the knot circle, but with diameter which is half of the dimeter of the knot. The lines, which determine the places of additional volume for drapery, are tangents to the smaller circle. They start from the arm hole and bust dart for the top part of the draped piece, and the point of interaction of hip and side seam for the down part of the draped piece.

Figure 4 right present the pattern of the draped piece after additional opening for drapery volume, which sizes the width of the draped pieces in the place of twist W = 31,65 cm. After the additional opening the upper and lower parts of the draped piece are connected each other on a line in diagonal direction (skew direction of the fabric), which determines the place of twisting.

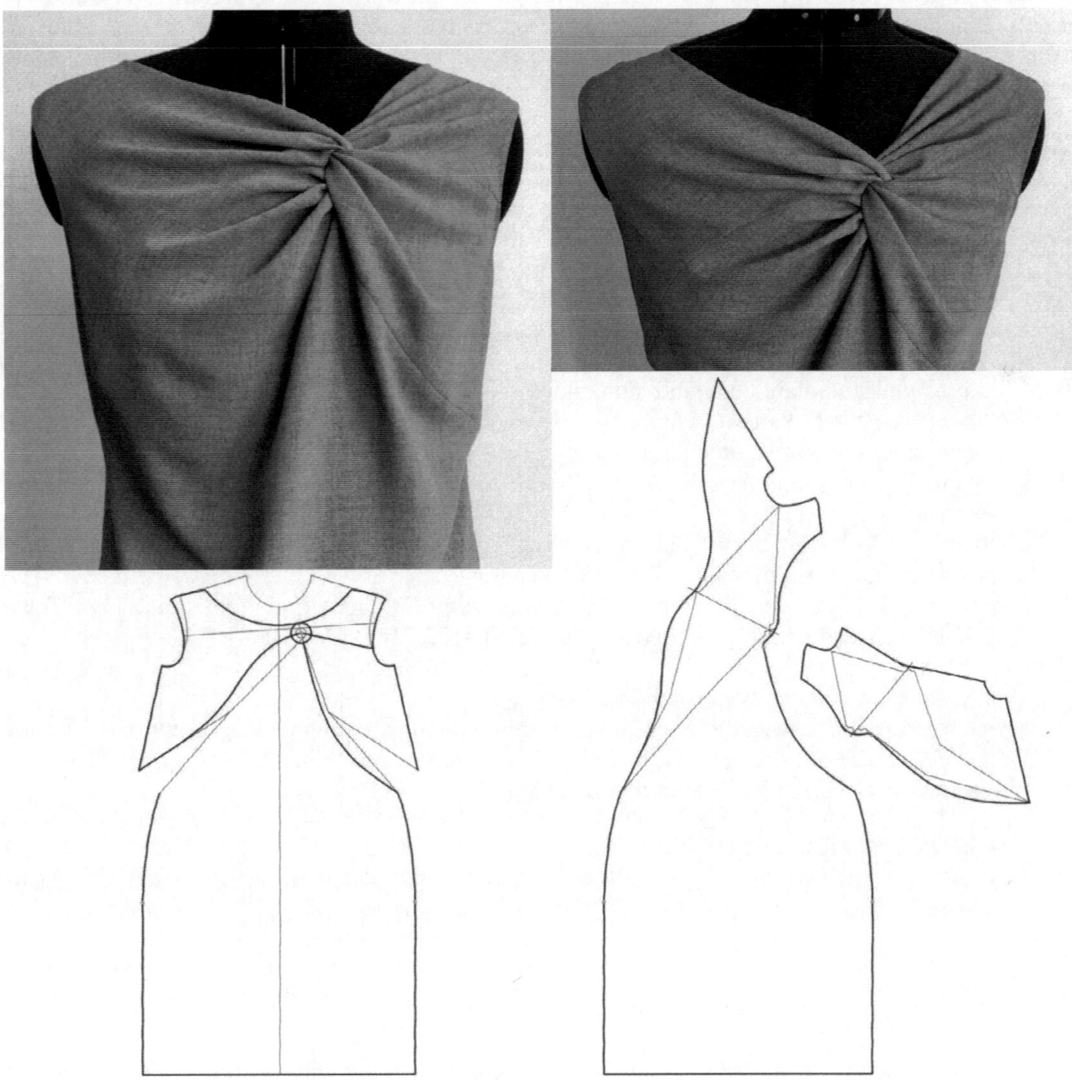

Figure 5. Design and pattern making of a lady's dress with asymmetrically situated in the neckline twist knot.

Figure 5 presents a design and pattern making of a lady's dress with a twist knot, situated asymmetrically at the neckline. The knot divides the neckline in the Golden proportion [10]. For that dress the width of the draped pieces in the place of twist W = 16,5 cm (the smaller size in presented investigation). The diameter of the knot D = 4,15 cm is cm is determined by dependency (4). The pattern making of draped pieces is made similarly to the pattern design of the lady's dress, shown in Figure 4.

In Figures 4 and 5 fashion designs are shown by the experienced models, prepared using presented patterns. According to the materials, lengths, accessories the presented designs can be adapted for casual or formal ladies' dresses in various styles.

7. Conclusion

The obtained dependence between the knot diameter and the widths of the draped pieces in the place of twist has made the design of patterns with twist knot draperies fully systematized. The systemization gives possibility for easy and correct pattern making of twist knot drape clothes with various sizes of diameter of knot and drapery parts width. The system facilitates the process of fashion design and pattern making and gives possibility of variety of new designs twist knot drape clothes. It is very useful for ready-to-wear industry.

8. Acknowledgements

The work is supported by the scientific project 2.FTT/ 2016 'Investigation of the Fashion Design and Pattern making of 3D Elements in Ladies' Clothing' funded by Trakia University and the Fund of the National budget for scientific research in higher education in Bulgaria.

References

[1] Nakamichi T 2010 *Pattern Magic* (London: Laurence King)
[2] McAdam A 2014 Pattern Puzzle - Kimono Twist Dress *Studio Faro* http://www.studiofaro.com/well-suited/kimono-twist-dress
[3] McAdam A 2014 Pattern Puzzle - Jersey Twist Dress *Studio Faro – Well-suited* https://studiofaro-wellsuited.blogspot.com/2014/03/pattern-puzzle-jersey-twist-dress.html
[4] Rissanen T and McQuillan H 2016 *Zero Waste Fashion Design* (London: Bloomsbury Academic)
[5] Chetverikova T 2015 Cadrus: Pattern Making. The Mysterious Knot *Atelie* vol 2015 no 10 http://modanews.ru/journal/atelie/2015/10/cadrus
[6] Kazlacheva Z 2015 Variants of a Lady's Dress with 3D Transformations *Journal of Textile Science and Engineering* vol 5 no 3: 198 doi:10.4172/2165-8064.1000198
[7] Hill T and Lewicki P 2007 *Electronic Statistics Textbook* http://statsoft.com/textbook/stathome.html
[8] Kazlacheva Z and Dineva P 2017 An investigation of pattern making of twisted draperies *ARTTE Applied Researches in Technics, Technologies and Education* (Yambol: Faculty of Technics and Technologies) vol 5 no 2 pp 85-93
[9] Dineva P and Balabanova D 2010 Design of Ladies Cocktail Dresses *Tekstil i Obleklo* (Sofia) vol 58 no 12, pp 369-375
[10] Kazlacheva Z 2017 An investigation of application of the golden ratio and Fibonacci sequence in fashion design and pattern making *IOP Conf. Ser.: Mater. Sci. Eng.* 254 172013

Application of innovative technologies in fashion design education

Z Kazlacheva, V Stoykova, K Georgieva, J Ilieva

Trakia University, Faculty of Technics and Technologies
Graf Ignatiev 38, 8600 Yambol, Bulgaria

zlatinka.kazlacheva@trakia-uni.bg

Abstract. The application of the innovative educational and design technologies offers quality improvement of higher education in fashion design by interactivity, flexibility and dynamics. The paper presents the application of the innovative technologies in education in fashion design subjects, included in the Bachelor and Master programs of Design, Technology and Management in Fashion Industry at Faculty of Technics and Technologies, Trakia University, Bulgaria. According to the examinations results and studying student opinion it can be concluded that application of presented innovative technologies in fashion design education provide easier and more accessible learning of the study material, acquiring more knowledge in a short time, developing the students' creativity, creative and visual thinking and design skills, and generally their application can lead to an increase in the quality of education.

1. Introduction

The fashion design education requires high level of visualization of the study process [1, 2]. The application of the innovative educational and design technologies offers quality improvement of higher education in fashion design by interactivity, flexibility and dynamics.

The use of interactive white board (IWB) in higher education in the fields of engineering (fashion design and engineering, machine engineering, electronics, automatics, computer technology, etc.) is studied in details in previous publications of authors [1, 3]. Every interactive presentation system makes the students' classes closer to digital generation daily life, to their requirements for interactivity, high level of visualization, dynamics and attractiveness in educational process. The managing of objects through stylus or finger similarly to their mobile devises, adding of annotations and notes to slides, run of IWB as traditional white board, saving of the images of IWB in different types of file formats (.pdf, video formats, etc.), real-time sharing of content with remote users are opportunities that are very suitable for effective education at universities today. The interactive projectors are one of optimal decision among various models of interactive presentation systems and it is combine an interactive board and an ultra-short throw projector in one devise and it can transform every traditional white board in IWB.

Add-ons such as CPS, tablets, document cameras can be added to the IWB. The document cameras [1] can to be used as separated devices without integration with IWB, connected to a projector or to a projector and computer, according to the capabilities of the particular model. All objects and actions under the camera's objective are visualized on an IWB or a screen, and by this way the visualization become available to the entire audience. The images can be rotated, captured and added with notes and lines. The ability to record video is also extremely useful. All explanations and actions of the lecturer, drawing and writing can be recorded.

The augmented reality (AR) and virtual reality (VR) are also widely used in the education. Examples of application of AR in different educational fields of natural science, medicine, automation, vocational education and more are presented in [4, 5, 6, 7, 8, 9]. The high level of visualization with the ability to monitor animated 3D objects on all sides and virtual experiments without endangering the students' health, incl. saving of expensive equipment and consumables are just some of the benefits of AR in the learning process. AR is also a means of implementing the so-called blended learning which combines the advantages of traditional (using paper textbooks and manuals) and e-learning, using modern ICT based innovative technologies [1]. It certainly provokes the interest of today's learners and satisfies their needs for a virtual world and modern learning rooms and can therefore motivate them to participate more actively in the learning process. All described abilities of the AR in its application in the learning process can contribute to the better education of the students.

The paper presents the application of the innovative educational and creative technologies in education in Fashion Design subjects, which are included in the Bachelor and Master programs of Design, Technology and Management in Fashion Industry at Faculty of Technics and Technologies of Yambol, Trakia University, Bulgaria.

2. Application of innovative educational technologies in fashion design education

Document cameras, sometimes in combinations with interactive white boards [2], are used in education in subjects in the fields of traditional fashion design and pattern making, especially for visualization of different hand techniques for fashion and textile illustrations, and traditional hand pattern making. Interactive white boards are used in education of work with specialized software as 3D design, CAD systems, and raster and vector graphics apps, especially the suitable and optimal use of drawing and modified tools for specific design and pattern making forms.

The application of IWB and document cameras leads not only to additional visualization. With the help of these educational technologies full or parts of lectures are recorded in videos, which are available for students at the e-learning system, through added links in e-books and e-manuals, or QR codes in traditional paper textbooks, which students can use at any time and from any internet connected device and many times, according to their needs.

Figures 1 and 2 present application of document camera and IWB in fashion design education.

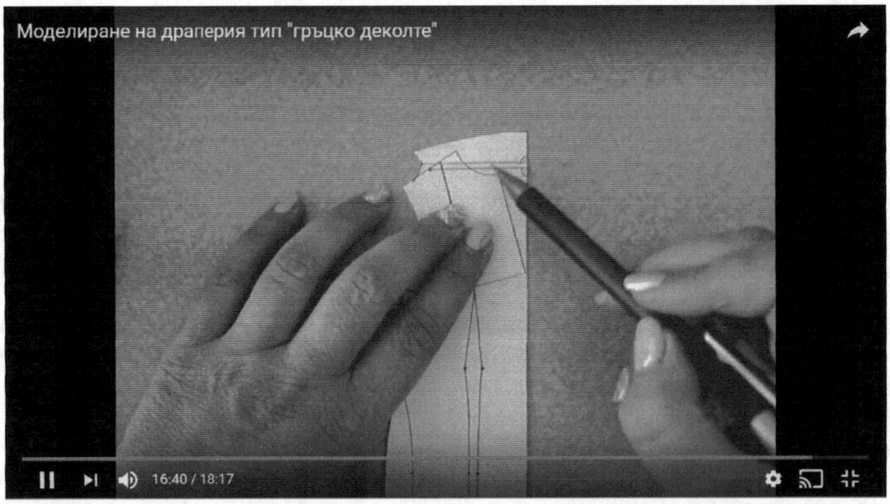

Figure 1. A screen shot of a video lecture about pattern making of ladies' clothing with draperies around front neck opening, made with the help of document camera.

Figure 1 presents a screen shot of a video lecture about a geometric way of pattern making of ladies' clothing with draperies around front neck opening. The presentation of the traditional hand process of pattern making is made with the help of document camera. The video is available online at: https://youtu.be/L39f7N_4x_c (84 views, 2 months ago).

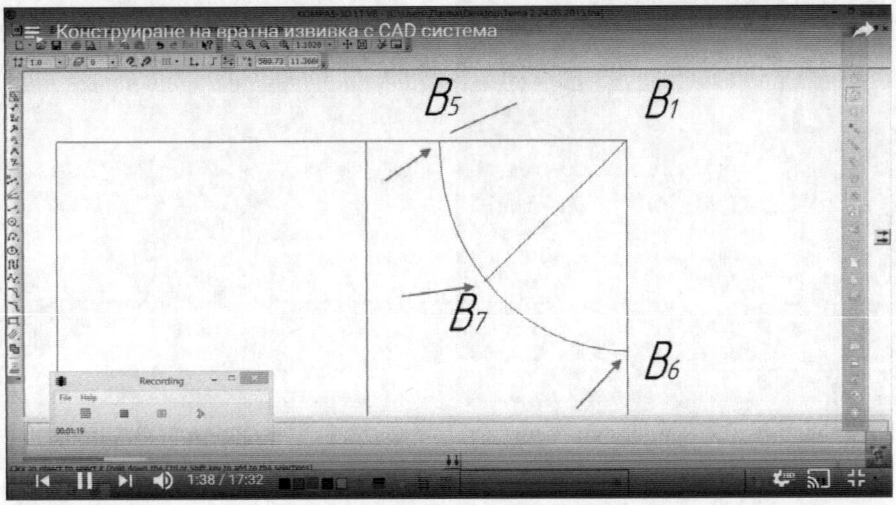

Figure 2. A screen shot of a video about drawing of neckline with a CAD system, which is made using interactive white board.

Figure 2 shows a screen shot of a video about drawing of neckline in constructional base of lady's jacket with a CAD system, which is made using interactive white board. The IWB tools help students in their first meeting with the application of the more interesting drawing tool NURBS in pattern making. The video is available online at: https://youtu.be/qtLGsYeXN5M (634 views, 3 years ago).

The last innovation, which we use in fashion design education, is augmented reality (AR) [4]. Certain themes in textbooks and manuals of fashion illustration, design theory, fashion design and pattern making have been completed with augmented reality markers and the study material is additionally visualized with 3D models which the students can see on the displays of their mobile devises.

Figure 3 presents an example of application of augmented reality in fashion design theory education. The adding 3D dynamic model is a lady's figure with a dress in Y turned silhouette. With the application of AR the basic 12 silhouettes [10, 11] in ladies' clothing have been visualized in every direction and it is easy and successful way for costume forms recognition and differentiation. The augmented reality helps not only with visualization of the type of the silhouette form, AR is very suitable for presentation the different ways of forming of every silhouette. In the case, which is shown in Figure 3, with the help of AR one of the ways for design of Y turned silhouette is presented. That mode is use of one-sided gathers for forming of the volume down Y turned part of the silhouette.

Figure 4 shows a case of use of augmented reality in pattern making education. The adding 3D animated model is a lady's figure with an eight pieces skirt with ruffles. With the use of the AR marker, which is included in the theme for pattern making of ladies' skirts and the other themes in the paper textbook, the students can see the result of pattern making process.

As both photos (Figures 3 and 4) show AR is very suitable for visualization of 3D elements in clothing like gathers in the adding model, presented in Figure 3, and ruffles in the dynamic 3D model shown in Figure 4.

IOP Conf. Series: Materials Science and Engineering **459** (2019) 012080 doi:10.1088/1757-899X/459/1/012080

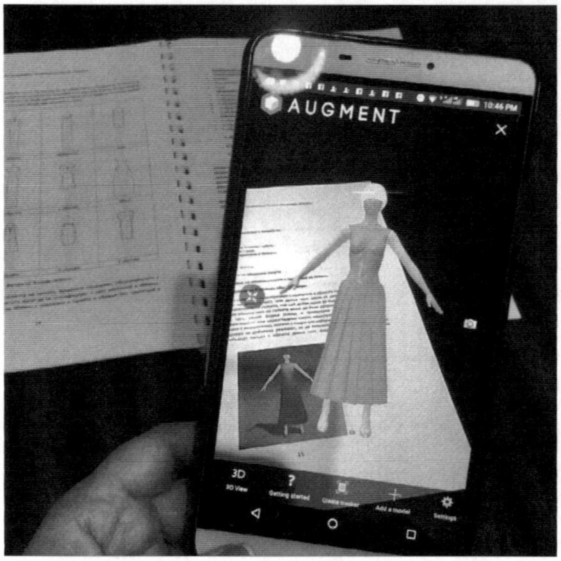

Figure 3. Application of AR in design theory education.

Figure 4. Application of AR in pattern making education.

3. Application of innovative creative technologies in fashion design education

3D printers [12] and 3D markers are used in education of fashion design subjects. With their help the students realized projects of design of decorative elements and small accessories, or parts of bigger accessories, as real models or prototypes. 3D printing and modelling with 3D markers help with 3D forms and models the education in some themes of design theory and fashion design subjects. The priorities of the use of 3D marker compared to 3D printer are not only the lower cost but the easy change of the colors and possibility for realization of more free and artistic design ideas.

Figure 5 presents a decoration for a blouse which is made with 3D marker. This model and other similar creations with 3D marker can be used as real decorative pieces of the garment or as real jewels. Another type of application is as prototypes of a clothing decoration or a bijous, which can realized with embroidery, beads, etc.

Figure 5. Decoration for clothing made with 3D marker.

Our partners from the business help us with digital textile printing for clothing and our students have possibility to realize parts of their project, which they at first have realized with CAD systems, software for raster and vector graphics, or traditional hand way. The students choose the project for printing with software for 3D visualization of clothing. We stimulate our students to use free software and internet applications for design (CAD, raster and vector graphics) and visualization, which give them possibility to create their designs not only in the university classes.

Figure 6 presents a pattern of a lady's dress with textile patterns, which are designed for digital textile printing.

With the help of cooperation with our colleagues from other universities and business we have possibility to present in the process of students' education adaptation of our projects in the software for embroidery, woven and knitted fabrics and decorative bands, and realized designs. For example, the geometrical and floral ornament of the Flower of Life, which is applied in the project for digital printing show in Figure 6, is used in embroidery and woven fabrics digital designs, which are presented in [13] and [14]. Other examples for realization of digital designs of decorative bands, which are presented in [15], are available online at: https://sites.google.com/a/trakia-uni.bg/textile-design/realizirani-proekti.

We use our experience and encourage our students to present their design projects in internet sites and shops about digital printing.

Figure 7 shows a model of lady's garment with deigned textile pattern from an internet shop for digital prints. The model is available online at: https://shopvida.com/products/flowers-933.

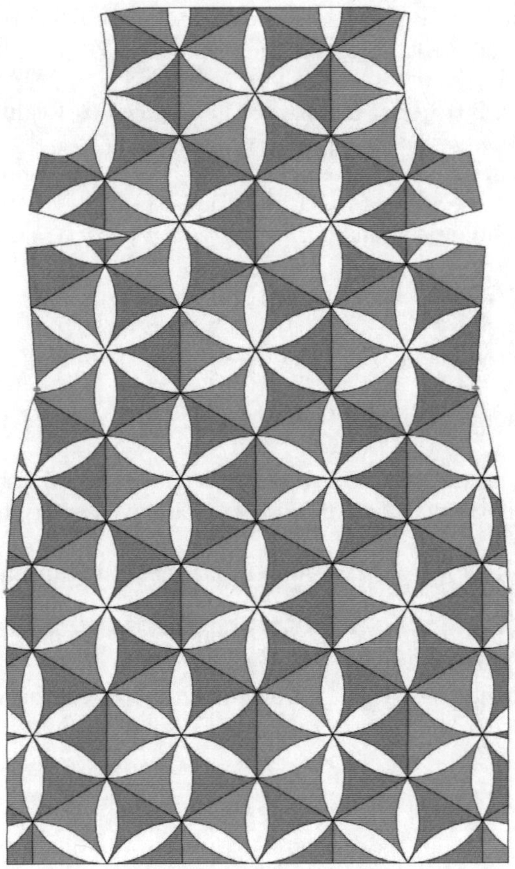

Figure 6. A project of a digital print.

Figure 7. Visualization of a digital printed garment.

IOP Conf. Series: Materials Science and Engineering **459** (2019) 012080 doi:10.1088/1757-899X/459/1/012080

4. Conclusion

According to the examinations results and studying student opinion it can be concluded that application of presented innovative educational and creative technologies in fashion design education provide easier and more accessible learning of the study material, acquiring more knowledge in a short time, developing the students' creativity, creative and visual thinking and design skills, and generally their application can lead to an increase in the quality of education.

References

[1] Stoykova V, Georgieva K, Kazlacheva Z, Ivanova N, Ilieva J 2015 Good practices of students' education in theprogramme of design, technologies *Tekstil i Obleklo* (Sofia) vol **63** no 10 pp 252-258
[2] Wiana W 2018 The Effectiveness of Using Interactive Multimedia in Improving the Concept of Fashion Design and Its Application in The Making of Digital Fashion Design *IOP Conf. Ser.: Mater. Sci. Eng.* 306 012131
[3] Stoykova V, Smrikarov A, Kazlacheva Z, Georgieva K 2012 Some applications of interactive presentation systems in the learning process *Proc. Conf. on e-Learning in Higher Eduvation* (Svishtov: National Centre for Distance Education and "Dimitar Tsenov" Academy of Economics) pp 351-358
[4] Videnov K, Stoykova V, Kazlacheva Z 2018 Application of Augmented Reality in Higher Education *ARTTE Applied Researches in Technics, Technologies and Education* (Yambol: Faculty of Technics and Technologies) vol **6** no 1 pp 1-9
[5] Ivanova G, Aliev Y, Ivanov A 2014 Augmented Reality Textbook for Future Blended Education *Proc. Int. Conf. on e-Learning'14* (Ruse: "Angel Kanchev" University), pp 130-136
[6] Okimoto M L L R, Okimoto P C, Goldbach C E 2015 User Experience in Augmented Reality applied to the Welding Education *Procedia Manufacturing* **3** (2015) pp 6223-6227
[7] Coimbra T, Cardoso T, Mateus A 2015 Augmented Reality: an Enhancer for Higher Education Students in Math's learning? *Procedia Computer Science* **67** (2015) pp 332-339
[8] Calderón R R, Arbesú R S 2015 Augmented Reality in Automation *Procedia Computer Science* **75** (2015) pp 123-128
[9] Hsieh M C, Lee J J 2018 Preliminary Study of VR and AR in Medical and Healthcare *Journal of Nursing and Health Studies* vol **3** no 1 pp 1-5
[10] Kazlacheva Z 2007 Silhouettes in the ladies outer garments *Tekstil i Obleklo* (Sofia) vol **58** no 1 pp 12-15
[11] Kazlacheva Z 2010 Silhouettes of lady's elastomeric clothing *Tekstil i Obleklo* (Sofia) vol **55** no 9 pp 8-11
[12] Verlan V, Irovan M 2018 Applications of 3d printing technologies in the Garment industry *ARTTE Applied Researches in Technics, Technologies and Education* (Yambol: Faculty of Technics and Technologies) vol **6** no 2 pp 85-93
[13] Indrie L, Kazlacheva Z, Ilieva J, Gherghel S 2017 Embroidery - from digital designing to fine art *Industria Textila* (Bucharest) vol 68 no **5** pp 366-369
[14] Indrie L, Diaz-García P, Kazlacheva Z, Montana I, Ilieva J 2018 New technology in producing textile designs and fabrics *Industria Textila* (Bucharest) (in print)
[15] Ilieva J 2015 Fashion design using decorative bands based on the golden and Fibonacci forms *ARTTE Applied Researches in Technics, Technologies and Education* (Yambol: Faculty of Technics and Technologies) vol **3** no 3 pp 265-274

Methodology of the product line matrix formation of the enterprise fabricating clothes

I Klochko[1], L Falko[1] and V Shkuropatskaya[1]
[1] Vladivostok State University of Economics and Service, Vladivostok, Russia

imyak@mail.ru

Abstract. The paper has considered the questions connected with formation of the rational range of the enterprises of clothes production. It has also analyzed the factors influencing the range formation subject to the principles of the rational wardrobe design (a capsular approach and a colour type of one's appearance). The algorithm of the range formation of the enterprises of clothes production in a mass-market segment on the basis of a brand standard subject to the principle of rational wardrobe is developed in the paper. It has created the mathematical model of a rational a product line matrix of the sewing enterprise in a mass-market segment.
Keywords: Product Line Policy, Product Line Matrix, Rational Wardrobe, Capsular Approach, Colour Type of one's Appearance

1. Introduction

Any modern enterprise aims to maximize its profit most often by increasing volume of sales and minimizing the cost price. The prime cost reduction has its limits, but the volumes of realization are limited only to the existing resources of a company and market capacity of consumption.

In this regard the majority of the companies daily face a problem of searching of customers. They search for new markets with the use of various tools: advertisement, price competition, etc. The use of already known tools will not give a desirable result if they do not become an integral component of a certain strategy. This paper considers a product line policy of small and medium business enterprise of clothes production. It is very important to research the development process of a rational product line matrix to implement this policy.

Technological processes transformation in the light industry production of goods has affected all segments of the industry and had the essential impact on the formation of consumer preferences.

The task of producers and sellers of clothes at this stage consists in the development of common integrated algorithms of organization and realization of a product line matrix with the principle of the rational system in basis. In particular, the use of the principle of rationality by a consumer at the creation of his or her own wardrobe, and the reconsideration of a place and role of each element of clothes both when determining an individual capsule and in "person is clothes" system in general is of great interest, where the new forms of the rational interaction appear. On the other hand, the small and medium-sized enterprises producing clothes with chaotically formed range are actively developing under the influence of random factors and often have no common algorithm of the range formation. [1-2, 5, 9, 12, 13].

Despite considerable researches devoted to a problem of the formation of a range now there is no common methodological approach. Methodical tools of the range formation of the sewing industry enterprises are insufficiently developed. On one hand, it would be economically advantageous for the industrial enterprises and, on the other hand, it would satisfy consumers' preferences.

The existence of a stated problem has caused the relevance of scientific researches directed on the improvement of the formation process of a rational commodity range of the enterprises fabricating clothes.

2. Research methodology

The paper has used a system analysis which allowed considering structural and elementary connections in the integrated interaction a producer - a seller - a consumer at the stage of predesign researches of new clothe models from relevance positions and competitiveness. The data obtained on the basis of methods of group questioning and an expert assessment have allocated additional factors of the influence on the definition of the range structure, namely, the necessity for the specification of consumer groups on a color type of appearance. Also the method of the typization and the structurally functional method were used.

3. Research

A product line policy of the enterprise fabricating clothes at the present stage has to consider the factors connected with specific features of a consumer. Appearance of a person defines dimensional characteristics of clothes, a form (silhouette) and color appearance. Features of the identity of a consumer are shown in his or her lifestyle and connected with consumer's style preferences in clothes and a choice of these or those composite and product line decisions. The practice of the formation of a modern person's collection of clothes is connected with a tendency to the rational capsular wardrobe creation. In the course of the design of an industrial collection such approach allows, on one hand, to consider features of production, and on the other hand, to create prerequisites for real customization. [6,13]

Thus, a problem of the formation of a product line policy is closely connected with a problem of the study of the enterprise target audience. An analysis of the target audience gives a chance to define the basic data for the formation of a product line matrix of the enterprise.

A product line matrix is a document in tabular format, in which reflect a company range of products and information with certain characteristics of a range for a certain period of time. The information in a product line matrix is given according to the hierarchical principle, from a category to a unit of a commodity stock account [10].

The present research has offered to consider the following major factors when forming a rational product line matrix of the enterprises fabricating clothes:

- A consumer's lifestyle, a strength of his or her predilection to fashion;
- The segmentation of consumers on style figurative groups to style preferences;
- A level of satisfaction with the individual wardrobe in a consumer group;
- The capsular structure of the client wardrobe and, as a result, the capsule essence of a product line of the enterprise and its continuity in relation to the previous collections;
- The availability in the wardrobe of a consumer of universal products which are coherent between the main capsules;
- A colour type of a person's appearance as one of the criteria of the definition of a collection number coloristic palette of the enterprise product line matrix.

Such approach demands the stable communicative relations between a consumer and a producer. A lifestyle and a individualization of consumer are the elements of an individual image and they form consumer preferences in a concrete target group. In turn, the segmentation of target groups from positions of socioeconomic factors which influence a choice and decision-making on purchase by a consumer, allows connecting the audience's choice to a certain price segment. From this point of view it is expedient to consider preferences of the target audience through a prism of the known brands - standards.

As a result of the research conducted by the authors it is possible to point out the main stages of the formation of a product line matrix of the enterprise:

1. An analysis of analogs (brands-standards):
   - The definition of the target audience.
   - The formation of check-sheets (№ 1 - selection criteria of a suitable analog, № 2 - criteria of an analysis of a collection of a brand - standard).
   - Standard brand choice.
   - The testing/research/analysis of a collection of a brand - standard on commodity groups.
2. An analysis of commodity groups on the basis of the results over the last seasons and the enterprise work on trend watching.
3. The definition of a percentage ratio of commodity categories:
   - The definition of a percentage ratio of commodity categories depending on the results of sales over the last season.
   - The definition of a percentage ratio of commodity categories depending on a color type of consumers' appearance.
   - A comparison of the available data with data of a brand - standard.

4. The formation of a product line matrix.

An analysis of characteristics of a brand-standard corresponds to an analysis stage of analogs in general practice of the design and allows minimizing costs of the enterprise of small and medium business of market researches. An analysis of the sources has showed that at a choice of a brand-standard it is possible to refer to the main criteria of an assessment: Price segment; a consumer's satisfaction/dissatisfaction; loyalty to a brand; the perceived quality; leadership/popularity; the perceived value; the memorable attributes; associations with the organization; awareness on a brand; representation of a brand in the distributive network.

At the same time it is important for an enterprise to consider the results of its own work over the last seasons in which features of its target audience are reflected and which allow to plan, for example, the replenishment of a consumer's wardrobe including purchased clothes/commodity categories acquired by him or her from the last collections.

Each commodity category has its silhouette decisions. An analysis of silhouette decisions helps to trace the actual trends of fashion and to define the strategy of the enterprise in solutions of a model range.

In the offered approach the accounting of a factor "a color type of a consumer's appearance" is new to the formation of a product line matrix in comparison with the existing approaches. The earlier obtained data testify that, for example, in Primorsky Krai of the Russian Federation 54% of all products of an industrial collection have to be designed for a "summer" color type, 28% - for "a winter" color type, 13% - for "a spring" color type, and 5% for "a autumn" color type [11].

It should be noted that the received results are actual for a certain target audience during a concrete period of time for territories of the concrete geographical region.

At the final stage the enterprise independently solves what data to use. However, the carried-out work increases reliability of made decisions. On the basis of the above stated facts the distribution of percentage ratios between commodity categories concerning all collection variety of the enterprise presented in a season is possible. In turn, the advent of coefficients defining a dominating color type of person's appearance and the leading silhouette solutions of a brand-standard allows to receive a product line matrix defining the quantitative indices commodity categories, silhouette decisions and a coloristic decision of a range of the enterprise over the actual period.

From the mathematical point of view, the considered set of characteristics is defined by variables (a commodity group and silhouette decisions), and a group of constant values (a color type of person's appearance) which can be presented in the form of 4 matrixes:

IOP Conf. Series: Materials Science and Engineering **459** (2019) 012081 doi:10.1088/1757-899X/459/1/012081

A constant matrix Y (constant values, the length is constant). A matrix is a vector "color type," consisting of 4 elements: Winter, summer, autumn, spring. For Russia an elements value of a matrix are constant and do not depend on a collection. The values of elements are expressed in percentage.

A matrix of the variable length X (values change, length changes). A matrix is a vector "silhouette" of the variable length. The length of a vector depends on a commodity group, and the values of elements depend on a collection. The values of elements are expressed in percentage.

A matrix of the constant length W (the values change, the length does not change). A matrix is a vector "commodity group" of the constant length. A vector consists of 8 elements: Blouses, trousers and jeans, outerwear, dresses, sweaters and cardigans, t-shirts and tops, shorts and bridges, skirts which values depend on a collection. The values of elements are expressed in percentage.

A matrix of the variable length Z (the values change, the length changes). The quantity of elements of a matrix depends on a commodity group and is a product of elements' quantity of matrixes X, Y, W. The values of elements of a matrix reflect the quantity of products of a certain color type and a certain silhouette in a commodity group concerning the total number of products of this commodity group in a collection expressed in percentage.

A value of a matrix element Z is calculated by the formula:

where:

$$Z_{ij} = \frac{X_i * Y_j * W_k}{100\% * 100\%} \quad (1)$$

$j \in [1;4]$ $k \in [1;8]$ $i \in [1;n]$ i

A k value depends on a commodity group and is appointed by an expert-estimator. In our research we will accept the following k values:

$$\begin{cases} k = 1 - blouses; \\ k = 2 - trousers\ and\ jeans; \\ k = 3 - outerwear; \\ k = 4 - dresses; \\ k = 5 - sweaters\ and\ cardigans; \\ k = 6 - shirts\ and\ tops; \\ k = 7 - shorts\ and\ bridges; \\ k = 8 - skirts. \end{cases}$$

A choice of *k* value at the calculation of an element of a matrix Z is conducted by an expert, depending on what goods are now calculated.

n – a number of possible silhouettes for this commodity group.

$$\begin{cases} n = 4, \quad if k = 1, \quad k = 3, \quad k = 4, \quad k = 5; \\ n = 3, \quad if k = 2, \quad k = 7, \quad k = 8; \\ n = 2, \quad if k = 6. \end{cases}$$

Let's conduct the calculations for a commodity group "blouses" for an autumn-winter collection.

$$Y = \begin{vmatrix} y_1 \\ y_2 \\ y_3 \\ y_4 \end{vmatrix}, \quad (2)$$

where $y_1 = 28\%$, $y_2 = 54\%$, $y_3 = 5\%$, $y_4 = 13\%$.

$$X = \begin{vmatrix} x_1 \\ x_2 \\ x_3 \\ x_4 \end{vmatrix}, \quad (3)$$

where $x_1 = 69{,}7\%$, $x_2 = 30{,}3\%$, $x_3 = 0\%$, $x_4 = 0\%$.

$$W = \begin{vmatrix} w_1 \\ w_2 \\ w_3 \\ w_4 \\ w_5 \\ w_6 \\ w_7 \\ w_8 \end{vmatrix}, \quad (4)$$

where $w_1 = 8{,}895\%$,
$w_2 = 20{,}25\%$,
$w_3 = 20{,}79\%$,
$w_4 = 17{,}205\%$,
$w_5 = 11{,}48\%$,
$w_6 = 9{,}455\%$,
$w_7 = 2{,}36\%$,
$w_8 = 9{,}565\%$.

As in this example the calculation of a range for a commodity group "blouses" is calculated, the calculations have used a w1 = 8,895% coefficient.

$$Z = \begin{vmatrix} z_{11} & z_{12} & z_{13} & z_{14} \\ z_{21} & z_{22} & z_{23} & z_{24} \\ z_{31} & z_{32} & z_{33} & z_{34} \\ z_{41} & z_{42} & z_{43} & z_{44} \end{vmatrix}, \quad (5)$$

where z11 - a range of blouses of a direct silhouette for "winter" color type;
z12 - a range of blouses of a direct silhouette for "summer" color type
z13 - a range of blouses of a direct silhouette for "spring" color type;
z14 - a range of blouses of a direct silhouette for "autumn" color type;
z21 - a range of blouses of a semi-adjacent silhouette for "winter" color type;
z22 - a range of blouses of a semi-adjacent silhouette for "summer" color type;
z23 - a range of blouses of a semi-adjacent silhouette for "spring" color type;
z24 - a range of blouses of a semi-adjacent silhouette for "autumn" color type;
z31 - a range of blouses of the fitted silhouette for "winter" color type;
z3 2 - a range of blouses of the fitted silhouette for "summer" color type;
z3 3 - a range of blouses of the fitted silhouette for "spring" color type;
z34 - a range of blouses of the fitted silhouette for "autumn" color type;
z41 - a range of blouses of a trapezoid silhouette for "winter" color type;
z42 - a range of blouses of a trapezoid silhouette for "summer" color type;
z43 - a range of blouses of a trapezoid silhouette for "spring" color type;
z44 - a range of blouses of a trapezoid silhouette for "autumn" color type.

$$Z_{11} = \frac{X_1 * Y_1 * W_1}{100\% * 100\%} = \frac{69{,}7\% * 28\% * 8{,}895\%}{100\% * 100\%} = 1{,}74\%$$

Thus, the received formula is the tool for the calculation of a range of the enterprise. It can be used for the calculations manually, or to use for special programs, for example, Excel.

5. Conclusions

A problem of the effective planning of a range is solved by the formation of an effective product line policy of the enterprise. The conducted researches on the formation of the rational structure of a range of the sewing enterprise allowed solving the following problems:

1. The factors influencing the formation of a rational range of the enterprises fabricating clothes on the basis of individual consumer preferences are studied.
2. The mathematical model of the formation of the rational structure of a matrix of a range of the enterprises fabricating clothes is developed.

References

[1] Abalikhina I 2004. Improvement of Tools of the Strategic and Operational Management of a Product Line of the Enterprises on Wholesale Trade: Theses... Candidate of Economic Sciences: 08.00.05 p201.

[2] Aboyudu A 2009 Marketing Approach to the Formation of the Optimal Structure of a Product Line at the Enterprise: Theses... Candidate of Economic Sciences: 08.00.05 p130.

[3] Chuvakova S 2012 The Management of a Shop Product Line: A Practical Guide (Moscow: Publishing and Trading Corporation "Dashkov and Co") p260.

[4] Halyavina M 2012 Product line policy of enterprises/ An analysis of the major sights Polythematic Network *Electronic Scientific Magazine of the Kuban State Agrarian University* 84(10) pp 718-726

[5] Kharkova T 2007 Tools and Methods of the Formation of a Product Line Policy of the Industrial Enterprise: Theses... Candidate of Economic Sciences: 08.00.05 p175

[6] Klochko I 2012 Innovative approach to the formation of a product line policy of the small and medium businesses in a service sector *Modern Problems of Science and Education* 06(2015)

[7] Kotsur M 2009 Competitive product line policy in objectives of the enterprise. *Control Over the Management* 4(33) pp 149-155

[8] Nemkov V 2003 Formation of a Product Line Policy in the Maintenance of Economic Stability of the Industrial Enterprise: Theses... Candidate of Economic Sciences: 08.00.05 p148.

[9] Popov E and Balakirev S.2010 Product line matrix - The basis for the formation of a product line policy (*The Ural State Economic University News* vol 4) (Ekaterinburg: Ural State Economic University) pp 101-110

[10] Shkuropatskaya V, Klochko I and Falco L 2014 Ways to improve a range of the small and medium-sized enterprises fabricating clothes *Modern Problems of Science and Education* 03(2014)

[11] Sirotina L 1999 Improvement of the System of Management by a Product Range of the Sewing Enterprise: Theses... Candidate of Technical Sciences: 08.88.28 p230.

[12] Sysoev S and Buzukova E 2012 Product line management in retail *Category Management* (Saint Petersburg: Peter) p288

[13] Zueva T 2000 Development of Methodological Principles of the Design-Programming of a Range of Industrial Clothes: Theses... Candidate of Technical Sciences: 05.19.04 p208

IOP Conf. Series: Materials Science and Engineering **459** (2019) 012082 doi:10.1088/1757-899X/459/1/012082

Eigenvalue analysis for plain-woven fabric structure using shell element and one smoothing cell in the smoothed finite element method

Q T Nguyen[1], **A J P Gomes**[2] and **F B N Ferreira**[1]

[1]Universidade do Minho, 2C2T - Centro de Ciência e Tecnologia Têxtil, Campus de Azurém, 4800-058 Guimarães, Portugal

[2]Instituto de Telecomunicações and Universidade da Beira Interior, R. Marquês de Ávila e Bolama, 6201-001 Covilhã, Portugal

E-mail: quyenum@gmail.com

Abstract. An efficient four-node quadrilateral (Q4) shell element based on the first-order shear deformation theory of plate (FSDT) and the strain smoothing technique in finite elements (referred as SFEM) was proposed for eigenvalue analysis of plain-woven fabric structure. A one smoothing domain (or cell) integration scheme in SFEM was proposed to evaluate the nodal train fields of Q4 shell elements. The numerical result of eigenvalue analysis, which was in the case of free vibration analysis, approximated to that one implemented in the finite element method (FEM) but gave a higher efficiency in computation in terms of central processing unit (CPU) time and numerical implementation.

1. Introduction and kinematics of shells

Finite element methods (FEM) have been developed and widely used for the analysis of eigenvalues in the context of buckling and natural vibration responses of plate/shell structures. The different stages of development of plate/shell finite elements can be found in the works of Yang and his coworkers. [1] Eigenvalue analysis using thin plate/shell finite element models in textile engineering can be referred to [2, 3]. The strain smoothing technique recently proposed by Liu and his coworkers [4], referred as SFEM, which improved the accuracy and convergence rate of the existing conventional finite element finite element method (FEM) of elastic solid mechanics problems [5-8]. This paper presents an efficient Q4 flat shell element based on the FSDT [9-11] and the cell-based smoothed finite element in SFEM [12, 13] for the eigenvalue analysis of thin and moderately thick plain-woven fabric structure. The summarized formulation of Q4 shell plate/shell elements and the smoothing cell integration scheme are respectively presented in the following sections. The numerical results indicated a better efficiency of numerical computation with accuracy results compared to the same Q4 shell elements without integrating cell-based models.

1.1. Kinematics of shells

Based on the FSDT, the displacement components $\boldsymbol{u} = [u, v, w]$ at a local coordinate system (x, y, z) within problem domain Ω bounded by Γ are defined as follows:

$$
\begin{aligned}
u(x,y,z) &= u_0(x,y) - z\theta_x(x,y) \\
v(x,y,z) &= v_0(x,y) - z\theta_y(x,y) \\
w(x,y,z) &= w_0(x,y)
\end{aligned} \tag{1}
$$

where u_0, v_0 and w_0 be the translation displacements, θ_x and θ_y be the rotations about the yz and xz planes in the Cartesian coordinate system.

In terms of the mid-plane deformations, the strain vector $\boldsymbol{\varepsilon} = \{\varepsilon_x \quad \varepsilon_y \quad \varepsilon_{xy} \quad \varepsilon_{xz} \quad \varepsilon_{yz}\}^T$ can be written using Eq. (1), which gives:

$$
\boldsymbol{\varepsilon} = \begin{Bmatrix} \boldsymbol{\varepsilon}^m \\ 0 \end{Bmatrix} + \begin{Bmatrix} z\boldsymbol{\varepsilon}^b \\ 0 \end{Bmatrix} + \begin{Bmatrix} 0 \\ \boldsymbol{\varepsilon}^s \end{Bmatrix} \tag{2}
$$

where superscripts m, b and s are respectively the membrane, bending and the shear terms. The generalized strain vector $\hat{\boldsymbol{\varepsilon}}$ can be written as

$$
\hat{\boldsymbol{\varepsilon}} = \begin{Bmatrix} \boldsymbol{\varepsilon}^m \\ \boldsymbol{\varepsilon}^b \\ \boldsymbol{\varepsilon}^s \end{Bmatrix}, \tag{3}
$$

in which

$$
\boldsymbol{\varepsilon}^m = \begin{Bmatrix} \dfrac{\partial u_0}{\partial x} \\ \dfrac{\partial v_0}{\partial x} \\ \dfrac{\partial u_0}{\partial x} + \dfrac{\partial v_0}{\partial x} \end{Bmatrix}, \boldsymbol{\varepsilon}^b = \begin{Bmatrix} \dfrac{\partial \theta_x}{\partial x} \\ \dfrac{\partial \theta_y}{\partial y} \\ \dfrac{\partial \theta_x}{\partial x} + \dfrac{\partial \theta_y}{\partial y} \end{Bmatrix}, \boldsymbol{\varepsilon}^s = \begin{Bmatrix} \dfrac{\partial w_0}{\partial x} - \theta_x \\ \dfrac{\partial w_0}{\partial y} - \theta_y \end{Bmatrix}. \tag{4}
$$

The constitutive equations can be expressed as

$$
\hat{\boldsymbol{\sigma}} = \hat{\boldsymbol{D}}\hat{\boldsymbol{\varepsilon}}, \tag{5}
$$

where

$$
\hat{\boldsymbol{\sigma}} = \begin{Bmatrix} \hat{\boldsymbol{N}} \\ \hat{\boldsymbol{M}} \\ \hat{\boldsymbol{Q}} \end{Bmatrix}, \hat{\boldsymbol{D}} = \begin{bmatrix} \boldsymbol{D}^m & \boldsymbol{0} & \boldsymbol{0} \\ \boldsymbol{0} & \boldsymbol{D}^b & \boldsymbol{0} \\ \boldsymbol{0} & \boldsymbol{0} & \boldsymbol{D}^s \end{bmatrix}, \tag{6}
$$

in which $\hat{\boldsymbol{N}} = \{N_x \quad N_y \quad N_{xy}\}^T$ is the vector of membrane force, $\hat{\boldsymbol{M}} = \{M_x \quad M_y \quad M_{xy}\}^T$ is the vector of bending moment, $\hat{\boldsymbol{Q}} = \{Q_x \quad Q_y\}^T$ is the vector of transverse shear force and $\boldsymbol{D}$ are stiffness constitutive coefficients matrix, which defined as

$$
\begin{aligned}
\boldsymbol{D}^m &= \int_{-\frac{h}{2}}^{\frac{h}{2}} \begin{bmatrix} \dfrac{E_{warp}}{1 - v_{warp}v_{weft}} & \dfrac{v_{weft}E_{warp}}{1 - v_{weft}v_{warp}} & 0 \\ \dfrac{v_{warp}E_{weft}}{1 - v_{warp}v_{weft}} & \dfrac{\cdot E_2}{1 - v_{weft}v_{warp}} & 0 \\ 0 & 0 & G \end{bmatrix} dz, \\
\boldsymbol{D}^b &= \int_{-\frac{h}{2}}^{\frac{h}{2}} z^2 \begin{bmatrix} B_{warp} & 0 & 0 \\ 0 & B_{weft} & 0 \\ 0 & 0 & H \end{bmatrix} dz, \boldsymbol{D}^s = \int_{-\frac{h}{2}}^{\frac{h}{2}} \frac{5}{6} G \begin{bmatrix} 1 & 0 \\ 0 & 1 \end{bmatrix} dz.
\end{aligned} \tag{7}
$$

where h is the thickness of the plate or shell, G is shear modulus, E is Young's modulus and v is Poisson's ratio corresponding to the warp and weft direction of yarns, B and H stand for flexural moduli and torsional rigidity.

The problem domain Ω is discretized into a set of four-node quadrilateral flat shell elements Ω^e with boundary Γ^e, the generalized mid-plane displacement vector $\hat{\boldsymbol{u}}$ can be then defined as

$$\hat{\boldsymbol{u}} = \sum_{I=1}^{4} \begin{bmatrix} N_I(\mathbf{x}) & 0 & 0 & 0 & 0 \\ 0 & N_I(\mathbf{x}) & 0 & 0 & 0 \\ 0 & 0 & N_I(\mathbf{x}) & 0 & 0 \\ 0 & 0 & 0 & N_I(\mathbf{x}) & 0 \\ 0 & 0 & 0 & 0 & N_I(\mathbf{x}) \end{bmatrix} \begin{Bmatrix} u_I \\ v_I \\ w_I \\ \theta_{xI} \\ \theta_{yI} \end{Bmatrix} = \sum_{I=1}^{4} \boldsymbol{N}_I(\mathbf{x}) \boldsymbol{d}_I, \tag{8}$$

where $N_I(\mathbf{x})$ is the basis function associated to node I of a four-node quadrilateral shell element.

The eigenvalue equation for free vibration analysis can be expressed through the direct application of variational principles and using Eq. (1 to 9), which is given as

$$(\boldsymbol{K} - \omega^2 \boldsymbol{M})\boldsymbol{d} = 0, \tag{9}$$

where ω is the natural frequency, $\boldsymbol{K}$ and $\boldsymbol{M}$ are the global stiffness matrix and the global mass matrix, respectively, which given in [12, 14].

The shear locking phenomenon may appear due to incorrect transverse forces under bending, or in the case of the thickness of the plate tends to zero. To overcome the shear locking phenomena, the approximation of the shear strain $\boldsymbol{\varepsilon}^s$ in Eq. (4) can be formulated with MITC4 as in [9].

1.2. A one smoothing cell integration scheme in SFEM

This technique avoids evaluating derivatives of mesh-free shape functions at nodes and therefore eliminates defective modes. The major techniques used in smoothed finite element methods appear summarized in references [13, 15, 16].

The smoothing operation performed over an element Ω^e bounded by Γ^e, this can be referred as smoothing domain (or cell) Ω_k^s, which is expressed as

$$\bar{\nabla}\boldsymbol{u}(x_k) = \int_{\Omega_i^s} \nabla\boldsymbol{u}(x)\Phi(x - x_k)d\Omega, \tag{10}$$

which has to satisfy the basic conditions:

$$\Phi \geq 0 \; and \int_{\Omega_i^s} \Phi d\Omega = 1, \Phi(x - x_k) = \begin{cases} \frac{1}{A_k^s}, & x \in \Omega_k^s \\ 0, & x \notin \Omega_k^s \end{cases}, \tag{11}$$

where $\nabla\boldsymbol{u}(\mathbf{x})$ is equivalently $\boldsymbol{\varepsilon}(\mathbf{x})$, and Φ is a smoothing or weight function in Ω_k^s, $A_k = \int_{\Omega_k^s} d\Omega$ is the area of smoothing domain.

The strain in smoothing domain Ω_k^s can be further assumed to be a constant and equals $\bar{\boldsymbol{\varepsilon}}(\mathrm{x}_k)$, which gives:

$$\bar{\boldsymbol{\varepsilon}}_k = \bar{\boldsymbol{\varepsilon}}_k(x) = \bar{\boldsymbol{\varepsilon}}(x_k) = \frac{1}{A_k^s} \int_{\Omega_k^s} \boldsymbol{\varepsilon}(x)d\Omega. \tag{12}$$

By substituting smoothing function Φ into Eq. (1), the smoothed gradient of displacement can be defined as

$$\bar{\boldsymbol{\varepsilon}}(\mathrm{x}_k) = \frac{1}{A_k^s} \int_{\Gamma_k^s} \boldsymbol{n}(\mathbf{x}).\hat{\boldsymbol{u}}(\mathbf{x})d\Gamma, \tag{13}$$

where $\boldsymbol{n}(\mathrm{x})$ is the outward unit normal matrix containing the components of the outward unit normal vector to the boundary Γ_k^s.

Eq. (10 to 13) can apply to evaluate the membrane and bending trains of a Q4 shell element being formulated in the previous section using the following shape functions:

$$N_1 = (1, 0, 0, 0), N_2 = (0, 1, 0, 0), N_3 = (0, 0, 1, 0), N_4 = (0, 0, 0, 1). \tag{14}$$

IOP Conf. Series: Materials Science and Engineering **459** (2019) 012082 doi:10.1088/1757-899X/459/1/012082

2. Numerical implementation and results

Mechanical and physical parameters of a plain-woven fabric sample were measured with Kawabata evaluation system for fabrics (KES-FB) [17, 18] and derived as: elastic modulus [gf/cm], $E_1 = 3823.7993$, $E_2 = 14092.4464$ and $E_{12} = 6896.5517$, Poisson's ratio $v_1 = 0.0211$ and $v_2 = 0.0778$, bending rigidity $[gf.cm^2/cm]$ of $B_1 = 0.1237$, $B_2 = 0.1333$ and $B_{12} = 0.0880$, transverse shear modulus $[gf.cm^2]$ of $G = 217.3100$.

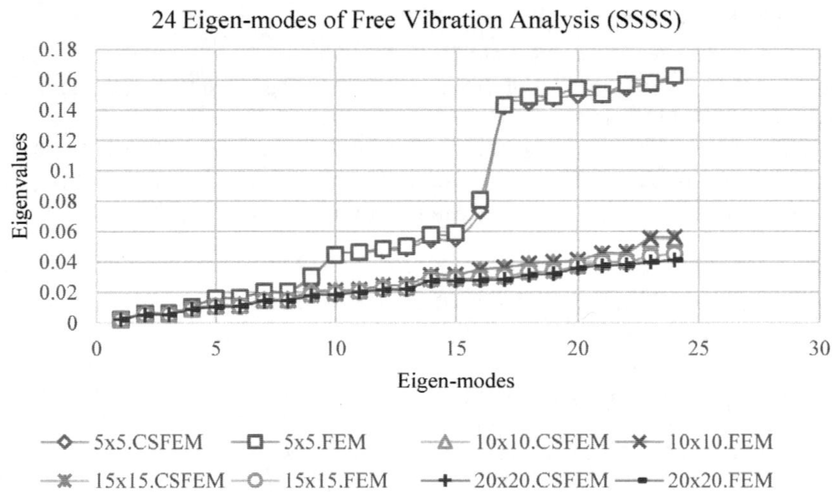

Figure 1. Eigenvalues of free vibration analysis of a SSSS plain-woven fabric sheet with different mesh density subjected natural frequency using 1 smoothing cell

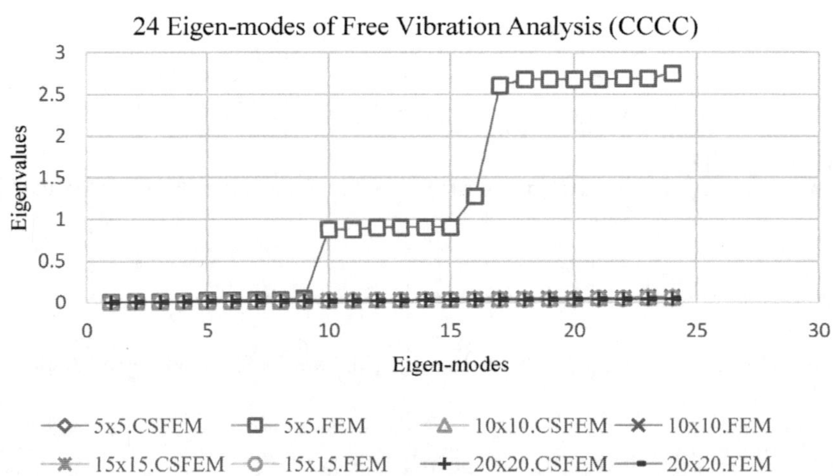

Figure 2. Eigenvalues of free vibration analysis of a CCCC plain-woven fabric sheet with different mesh density subjected natural frequency using 1 smoothing cell

The span-to-thickness ratio l/h of a square woven fabric sample was taken to be 23.5849 in the numerical example. Two types of boundary conditions, simply supported (S) and clamped (C) edges applied for meshes including 5x5, 10x10, 15x15 and 20x20. Both the Q4 shell elements with and without being integrated smoothing cells were programmed.

Figures 1 and 2 indicated that the eigenvalues of 24 eigen-modes extracted from free vibration analysis computed by CSFEM model is approximate and coincided with that one of FEM on the same boundary conditions and mesh configurations. The linear shape functions in SFEM model are constants as presented in Eq. 14. However, the corresponding shape functions of a Q4 element in FEM are bilinear shape functions interpolated in natural coordinates (ξ, η, ζ), that are needed to be transformed into Cartesian coordinates (x, y, z) to evaluate the strain gradient matrices. Thus, the strain smoothing technique reduces the numerical implementation and computation time in terms of the central processing unit time.

3. Conclusions
The one smoothing domain integration scheme in SFEM for a Q4 element, which based on the FSDT, gave an effective computation in terms of CPU time for the eigenvalue analysis, e.g. in the case of vibration behavior of plain-woven fabric structure. One smoothing cell for membrane and bending strain energy integration achieved accurate results and also reduced tasks in numerical implementation compared with FEM.

Acknowledgments
The authors wish to express their acknowledgment to FCT funding from FCT – Foundation for Science and Technology within the scope of the project "PEST UID/CTM/00264; POCI-01-0145-FEDER-007136".

References
[1] Yang H T Y, Saigal S, Masud A and Kapania R K 2000 A survey of recent shell finite elements *International Journal for Numerical Methods in Engineering* **47** 101-27
[2] Oñate E and Kröplin B 2006 *Textile Composites and Inflatable Structures*: Springer Netherlands)
[3] Oñate E and Kröplin B 2010 *Textile Composites and Inflatable Structures II*: Springer Netherlands)
[4] Chen J-S, Wu C-T, Yoon S and You Y 2001 A stabilized conforming nodal integration for Galerkin mesh-free methods *Int. J. Numer. Meth. Engng. International Journal for Numerical Methods in Engineering* **50** 435-66
[5] Liu G R, Zeng W and Nguyen-Xuan H 2013 Generalized stochastic cell-based smoothed finite element method (GS_CS-FEM) for solid mechanics *Finite Elements in Analysis and Design* **63** 51-61
[6] Yue J, Liu G-R, Li M and Niu R 2018 A cell-based smoothed finite element method for multi-body contact analysis using linear complementarity formulation *International Journal of Solids and Structures*
[7] Feng S Z and Li A M 2017 Analysis of thermal and mechanical response in functionally graded cylinder using cell-based smoothed radial point interpolation method *Aerospace Science and Technology* **65** 46-53
[8] Tootoonchi A, Khoshghalb A, Liu G R and Khalili N 2016 A cell-based smoothed point interpolation method for flow-deformation analysis of saturated porous media *Computers and Geotechnics* **75** 159-73
[9] Bathe K-J and Dvorkin E N 1985 A four-node plate bending element based on Mindlin/Reissner plate theory and a mixed interpolation *International Journal for Numerical Methods in Engineering* **21** 367-83
[10] Duan H and Ma J 2018 Continuous finite element methods for Reissner-Mindlin plate problem *Acta Mathematica Scientia* **38** 450-70
[11] Wu F, Zeng W, Yao L Y and Liu G R 2018 A generalized probabilistic edge-based smoothed finite element method for elastostatic analysis of Reissner–Mindlin plates *Applied Mathematical Modelling* **53** 333-52

IOP Conf. Series: Materials Science and Engineering **459** (2019) 012082 doi:10.1088/1757-899X/459/1/012082

[12] Nguyen-Thanh N, Rabczuk T, Nguyen-Xuan H and Bordas S P A 2008 A smoothed finite element method for shell analysis *Computer Methods in Applied Mechanics and Engineering* **198** 165-77
[13] Liu G-R and Nguyen-Thoi T 2010 *Smoothed finite element methods*: Taylor and Francis Group, LLC)
[14] Nguyen-Van H, Mai-Duy N, Karunasena W and Tran-Cong T 2011 Buckling and vibration analysis of laminated composite plate/shell structures via a smoothed quadrilateral flat shell element with in-plane rotations *Computers & Structures* **89** 612-25
[15] Liu G, Dai K and Nguyen T 2007 A Smoothed Finite Element Method for Mechanics Problems *Computational Mechanics* **39** 859-77
[16] Liu G R, Nguyen T T, Dai K Y and Lam K Y 2007 Theoretical aspects of the smoothed finite element method (SFEM) *International Journal for Numerical Methods in Engineering* **71** 902-30
[17] Hu J 2008 *Fabric Testing*: Woodhead Publishing Ltd.)
[18] Fan J, Yu W and Hunter L 2004 *Clothing appearance and fit: Science and technology*: CRC)

IOP Conf. Series: Materials Science and Engineering **459** (2019) 012083 doi:10.1088/1757-899X/459/1/012083

Design of smart garments for sports and rehabilitation

A Paiva, F Ferreira, A Catarino, M Carvalho and H Carvalho

University of Minho, School of Engineering, Centre for Textile Science and Technology (2C2T), Campus de Azurém, 4800-058 Guimarães, Portugal

dmpaiva.s@gmail.com

Abstract. Physical exercise has proved benefits for general health [1] and can reduce the number of sports injuries to one third [2]. However, an athlete that has been injured during sports practice may omit this out of fear of discrimination, and worsen the injury in the weight room, during strength training [5]. Monitoring physiological status of an athlete or rehabilitation patients during training may thus help the person to get an earlier intervention, preventing injuries from getting worse. With this in mind, we propose a set of compression garments – shirt and leggings – with textile sensors to continually monitor heart and muscle activity, breathing rate and temperature. This paper reports the design of the garments and production of the shirt, which comprised a 3-lead ECG system, sEMG (Surface Electromiography) electrodes and a breathing sensor. The ECG (Electrocardiography) system was tested and presented some good results, in particular for very even movements, but the system still needs to be improved, in order to get a better signal, when it comes to movements with a considerable amplitude.

1. Introduction

Studies have shown that physical activity has positive effects in the prevention and treatment of several diseases, such as cardiovascular disease, diabetes, cancer, hypertension, obesity, osteoporosis, and depression [1], and it can contribute to prevent sports injuries, reducing the incidence of sports injuries to one third [2]. Strength training exercises, for instance, is a common practice between athletes, to improve muscle strength and power [3] and is commonly prescribed for health and fitness, athletic training, and prevention or rehabilitation of muscle and bone injuries [4].

One of the issues related with athletes' injuries is early detection. An injury may only be spotted in the weight room, but it may not have occured there. This can happen because an athlete may get injured in the field, and not tell anyone, fearing prejudice, and the injury only worsens in the weight room [5].

In order to increase the quality of training for athletes, rehabilitation patients or other pratictioners, preventing injuries and other health issues, or to avoid them to get worse, it is important to monitor the person while exercising. A system that provides EMG data in real time may help detecting injuries before they get worse, allowing early intervention, or evaluating the quality of exercise of a patient during rehabilitation.

Previously [6], a prototype of a seamless long sleeve t-shirt for cardiac and muscle activity monitoring was reported. The shirt had 4 ECG electrodes and two sEMG (surface electromyography) electrodes located on the biceps. The ECG system performed well, but moving the arms had a negative impact on the signal's response, since two of the electrodes were too close to the armpit. The sEMG electrodes didn't provide any signal.

Lage et al. [7] have presented a prototype of a sleeveless knitted shirt with knitted electrodes to monitor heart rate, as well as a knitted moisture sensor. The shirt was produced in a seamless knitting machine and the sensors are knitted with a silver coated polyamide yarn. For the heart rate measuring, two electrodes are located bellow the chest. The authors found that when the subject is quiet, the system captures a good ECG signal, but when he starts to move the arms, the signal presents motions artifacts and some heart beats are missed.

Manero et al. [8] designed running leggings for muscle activity monitoring, during training. The garment has three pairs of embroidered electrodes to detect EMG signals from the quadriceps, namely the *Vastus Medialis*, *Vastus Lateralis* and *Rectus Femoris*. The system seems to be reliable, according to the performance tests, being able to determine muscle fatigue.

Some products have even made its appearance in the market and the interest for smart clothing has been rising. Athos® has a set of compression garments (shirt and shorts) that has printed sensors with conductive rubber to give information about heart rate, breathing rate and muscle activity [9]. Myontec developed cycling shorts that use textile electrodes to monitor muscle activity of the quadriceps and hamstring muscles [10]. Hexoskin uses a combination of textile and conventional sensors in a shirt to provide information about heart rate and variability, breathing rate, oximetry and activity parameters [11].

Incorporating textile electrodes (or other sensors) presents several challenges in terms of signal reliability. In order to improve the performance of the textile electrodes used in ECG and sEMG, some authors suggest that the electrodes should be covered by a waterproof membrane [12]; should be stretched, but stabilized at the same time [13]; and the structure of the textile substrate should be dense [14] and smooth [15]. Compression has also shown positive results in the quality of the acquired signal [6], since it presses the electrodes against the skin, improving the contact, and holds the electrodes in a steady position.

In this paper, the design and development of a set of sportswear garments – long-sleeve t-shirt and leggings – that integrates textiles sensors for real-time monitoring of the person's activity and physiological data is proposed.

2. Materials and Methods

A set of compression shirt and shorts was designed. The set incorporates ECG, sEMG, breathing and temperature sensors, to continually evaluate the athlete or patient's physiological data. The long sleeve knitted shirt has textile sensors to monitor heart and muscle activity, breathing rate, and skin temperature.

To select the muscles, the work of Delavier [16] was used as a reference. The author describes all strength exercises usually performed in the gym and highlights the muscles that are being worked in each one. By analyzing his work, we conclude that the muscles that are more subjected to effort are the *deltoids*, *pectorals*, *biceps brachii*, *trapezius* and the *triceps* (for the arms), as well as the *quadriceps* (all its portions), *biceps femoris* and *gastrocnemius* (for the legs). The *brachioradialis* muscles was also included, because it is a large muscle of the arm.

The shirt also includes a 3-lead ECG (which can use 3 or 4 electrodes) that, aside from measuring heart beat and intermittent arrhythmias, is a cost-effective way of detecting atrial fibrillation, thus contributing to the prevention of stroke [17].

Breathing rate and temperature monitoring was also included, since they are important parameters to be measured during rehabilitation [18] and the possibility of measuring these vital signs with e-textiles has been reported.

Figure 1. Sketch of the garments

At this first stage, only the shirt was produced, in a seamless knitting machine, which produces the knitted fabric in the form of a tube that can have the size of the body, thus eliminating the need of side seams, and uses jacquard technology to knit patterns in the same substrate. The fabric produced is a polyamide/elastane pique knitted fabric. For the knitted conductive elements (electrodes and conductive tracks), a polyamide/silver yarn – Elitex 110/f34 – was used. The electrodes form a 3D structure (figure 2), reported in [19], while the tracks have the same structure as the rest of the garment. When a jacquard pattern is produced in this kind of machine, the conductive yarn is either cut or loose in the areas that are not being knitted, which means that two tracks can come in contact with each other, which is not desired. So, the undesired yarns need to be cut before insulating them. To insulate the tracks (inside and outside), TPU film was used – Bremis Exoflex. It also decreases the fabric's elasticity, thus creating some compression.

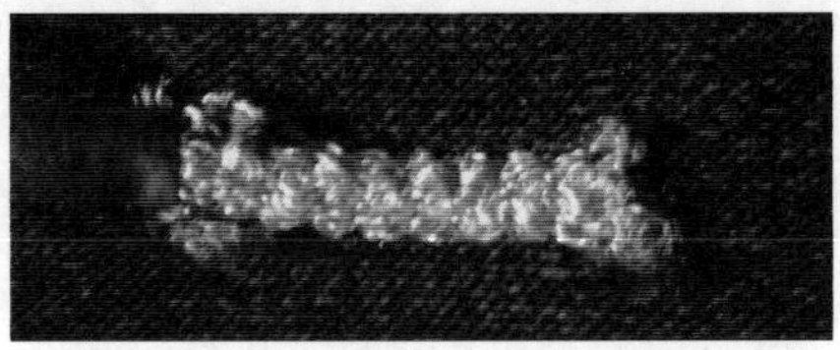

Figure 2. Knitted electrodes

To accurately position the embedded electrodes in the knitted fabric position and study the garment's patterns, a mock-up version of the garment was made and the positions were marked with the shirt dressed. This information was used to digitally design the pattern and produce it in the seamless machine. Figure 4 shows the garments patterns and jacquard pattern in the desired position.

Preliminary tests were executed using a *Shimmer3 ExG* device that can perform a 4-lead ECG (LA-RA, LL-RA, LL-LA and Vx-WCT) or measure EMG signals in two muscles, at the same time. At this stage, only the ECG tests were performed, with the subject standing still and performing *high-pulley curls*. We wanted an exercise that would imply the movement of the arms, to study the effect of these movements on the ECG signal. All tests were carried out with dry and wet electrodes, for comparison, and with and without pressure applied over the electrodes. To create additional pressure, a stretch band was used.

3. Results and Discussion

It figure 3, it is possible to observe the prototype of the knitted shirt. Figure 3a shows the outside of the shirt, where almost all conductive parts are protected by a black TPU layer, except for a small area in the chest, where conductive wires will be sewed to connect to an external device. In the inside of the shirt (figure 3b), only the electrodes are visible, as well a rectangle below the chest that represents the breathing sensor (it will not be considered in these experiments).

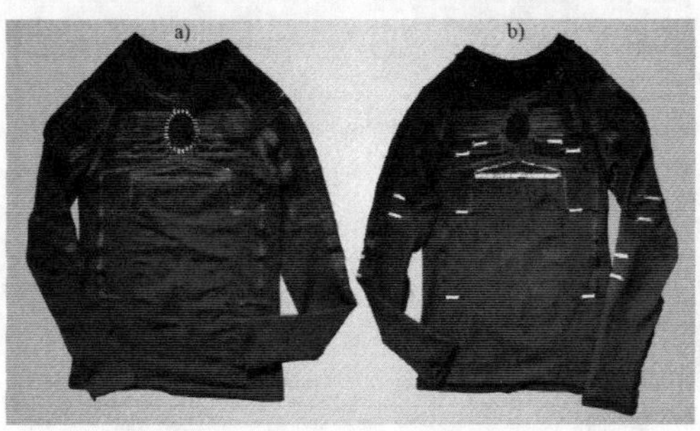

Figure 3. Shirt's prototype: a) exterior and b) interior

When dressing, one needs to manually position the electrodes over the desired muscle, which may require previous training. On the other hand, the TPU inside layer prevents the shirt from sliding, thus keeping the electrodes at the same place. That's especially true for the sleeves, because TPU covers a wider area, unlike the belly region, where only a small portion of this material was used to protect the ECG electrodes (LL and RL).

Figures 4 and 5 show the ECG signal obtained from lead LA-RA with the subject standing still. A full ECG wave is more or less visible in all tests, but there is more noise when the electrodes are dry (Figure 4), which was expected and follows the results from several studies made by other research teams. In both cases (dry and wet), pressure has a positive effect, since it decreases noise, thus showing a cleaner ECG wave.

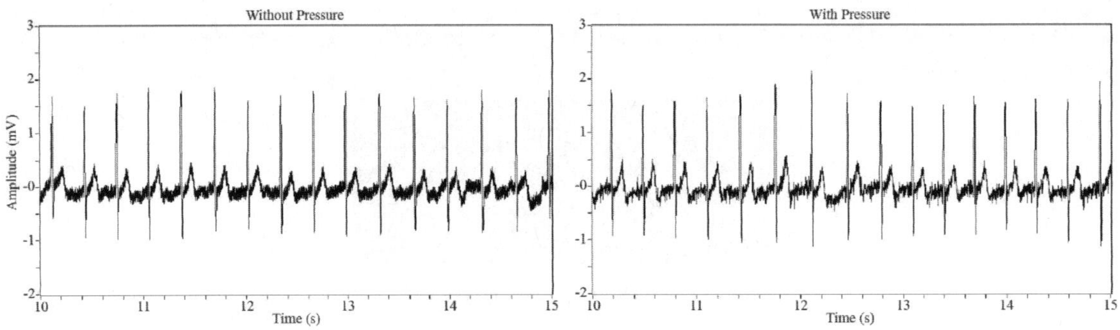

Figure 4. ECG signals in static position, with dry electrodes

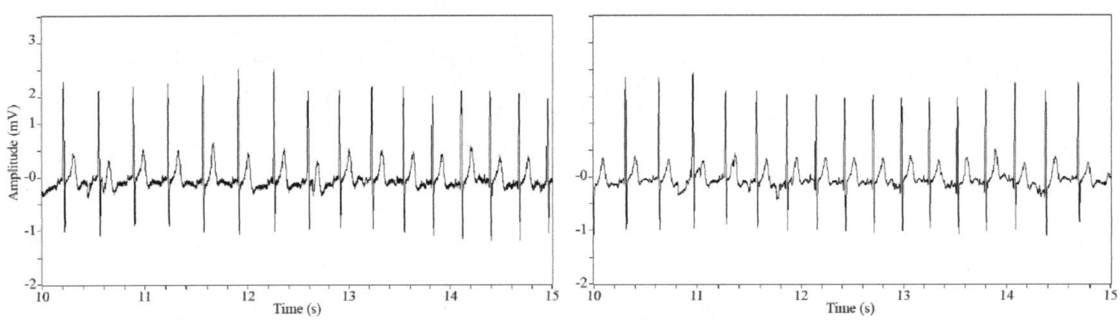

Figure 5. ECG signals in static position, with wet electrodes

When performing front presses, the results get worse (figures 6 and 7). When not wearing the stretch band, the whole range of exercise can be observed (arms moving down – arms down – arms moving up – arms up). When the arms are moving, some oscillations are visible and when the arms are up, there is considerable noise. That happens because when the arms lift, the shirt is stretched and the electrodes loose contact with the skin. The area around the armpit can thus be a tricky area to put textile electrodes.

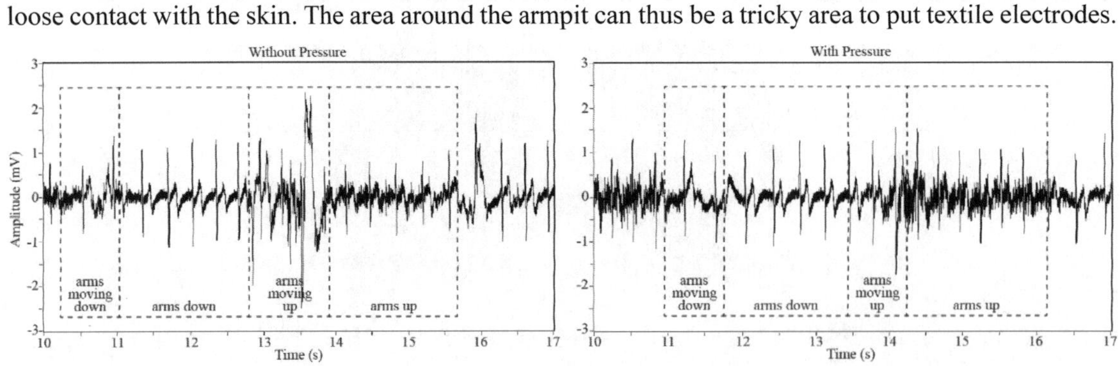

Figure 6. ECG signal while performing front presses, with dry electrodes

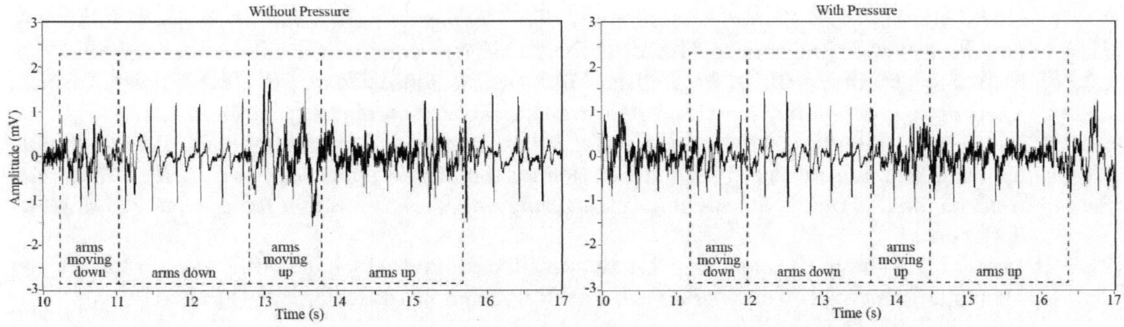

Figure 7. ECG signal while performing front presses, with wet electrodes

The measurement system used seems to be very susceptible to ambience noise. The tests had to be carried out a few meters from the laptop, otherwise noise would overlap the ECG signal.

4. Conclusion

In this paper, ECG measurement using the developed shirt is reported. In the future, tests will be carried out using the EMG electrodes. New experiments need to be prepared until a full clean ECG wave is obtained. This may include finding a more suitable spot to place the electrodes or finding a way to keep them in place, even when moving the arms, without compromising comfort. Alternatively, the movements of the arms can be studied in order to determine the best places to include electrodes.

The TPU film helps keeping the electrodes in place due to the compression created and the rubber like surface that keeps the fabric from sliding over the skin. These features can (and shall) be explored in future experiments.

To carry out the experiments, a *Shimmer3 ExG* device was used, but a device that follows the aesthetic and functional requirements of the garments shall be designed. This is an important objective if we want to shift from a laboratory setting to a user-friendly product that can be used by a person in real scenarios.

Acknowledgements

This work is financed by Project "Deus ex Machina", NORTE-01-0145-FEDER-000026, funded by CCDRN, through Sistema de Apoio à Investigação Cientifica e Tecnológica (Projetos Estruturados I&D&I) of Programa Operacional Regional do Norte, from Portugal 2020 and by FEDER funds through the Competitivity Factors Operational Programme - COMPETE and by national funds through FCT – Foundation for Science and Technology within the scope of the project POCI-01-0145-FEDER-007136.

References

[1] O'Shea S D, Taylor N F and Paratz J 2004 Periferal muscle strength training in COPD *CHEST* **126** pp 903-14

[2] Lanersen J B, Bertelson D M and Anderson L B 2014 The effectivenss of exercise interventions to prevent sports injuries: a systematic review and meta-analysis of randomized controlled trials *Journal of Sports Medicine* **48** pp 871-7

[3] Grimby G and Thomeé R 2003 Strength and endurance *Rehabilitation of Sports Injuries: Scientific Basis* ed W E Frontera (Oxford: Blackwell Science Lda.) pp 258-73

[4] Carpinalli R N and Otto R M 1998 Strength training: single versus multiple sets *Sports Medicine* **2** pp 73-84

[5] Kraemer W J and Dziados J 2002 Medical aspects and administrative concerns in strength training *Handbook of Sports Medicine and Science: Strength Training for Sport* ed W J Kraemer and K Häkkinen (Oxford: Blackwell Science Lda.) pp 163-75

[6] Paiva A, Catarino A, Carvalho H, Postolache O, Postolache G and Ferreira F 2018 Design of a long sleeve t-shirt with ECG and EMG for athletes and rehabilitation patients *Regional Helix*

2018: International Conference on Innovation, Engineering and Entrepreneurship Guimarães

[7] Lage J, Catarino A P, Carvalho H and Rocha A M 2015 Smart shirt with embedded vital sign moisture sensing *SPWID 2015: The First International Conference on Smart Portable, Wearable, Implantable and Disability-oriented Devices and Systems* Lisbon

[8] Manero R B R, Shafti A, Michael J, Grewal J, Fernandez J L R and Althofer K 2016 Wearable embroidered muscle activity sensing device for the human upper leg *2016 IEEE 38th Annual International Conference on the Engineering in Medicine and Biology Society (EMBC)* Orlando

[9] Berger J 2015 The most interesting wearble in fitness: with Athos, has the personal trainer met its match? [Internet] New York: Gear Patrol[updated 2014 September 19; cited 2018 August 10] Available from: http://www.gearpatrol.com

[10] Pesola A 2016 Reduced muscle Activity, sedentary time and cardio-metabolic benefits: effectiveness of a one-year family-based cluster randomized controlled trial *PhD thesis* University of Jyväskylä

[11] Carre Technologies Inc. 2013 Getting started guide for iPhone and OS X users Montreal

[12] Wijesiriwardana R and Mukhopadhyay S 2004 Fiber-meshed transducers based real time wearable physiological monitoring system *IEEE 8th International Symposium on Wearable Computers* pp 40-7

[13] Rattfäll L, Lindén M, Hult P, Berglin L and Ask P 2007 Electrical characteristic of conductive yarns and textile electrodes for medical applications *Medical and Biological Engineering and Computing* **45** 1251-7

[14] Paul G M, Cao F, Torah R, Yang K, Beedy S and Tudor J 2014 A smart textile based face EMG and EOG computer interface *IEEE sensors journal* **14** pp 393-400

[15] Barker R L 2002 From fabric hand to thermal confort: the evolving role of objective measuments in explaining human comfort response to textiles *International Journal of Clothing Science and Technology* **14** 181-200

[16] Delavier F 2006 Strength Training Anatomy (Canada: Montreal: Human Kinetics)

[17] Kristensen A N, Jeyam B, Riahi S and Jenson M B The use of portable three-lead ECG monitor to detect atrial fibrillation in general practice *Scandinavian Journal of Primary Health Care* **34** pp 304-8

[18] Schimtz T J 2007 Vital signs *Physical Rehabilitation* ed S B O'Sullivan and T J Schmitz (Philadelfia: D. A. Davis) pp 82-120

[19] Catarino A, Rocha A M, Carvalho H and Ferreira F 2017 Textile based electrodes for ECG and EMG measurements *ITMC 2017 - International Conference on Intelligent Textiles and Mass Customisation* Ghent Belgium

Virtual method of predicting the accuracy of pattern blocks

X Peng[1] and V E Kuzmichev[2]

Department of Clothing Design, Textile Institute, Ivanovo State Polytechnic University, 21, Sheremetev Ave., Ivanovo, Russian Federation

E-mail: 1347507286@qq.com, wkd37@list.ru

Abstract. Main purpose of this study is to develop the virtual method of predicting the misfit based on new obtaining relations between pattern blocks and body features. In our research, digital twin of female body in CLO3D was used. 3D-2D pattern block flattening technology was used to obtain the body prototype of avatar. The ease value of main structural parts for blouse pattern blocks and the proportions of ease value were obtained by respectively overlapping the blouse pattern blocks with the body prototype. Pattern blocks were analyzed and evaluated in terms of the proportions of ease value. New method can be used to analyze and predict the quality of blouse pattern block effectively and improve the efficiency of virtual garment design.

Keyword: virtual try-on, digital twin, blouse, pattern block, misfit, predicting, accuracy, avatar, flattening technology

1. Introduction

With the improvement of people's living standard, people are no longer satisfied with the warmth of clothes, but more concerned with the comfort and fit of clothes. The fit and comfort of clothing depend on the accuracy and quality of the pattern block. At the same time, the structural design of clothing pattern block is an important part of the whole process of clothing design, and the structural design of the pattern block requires professional experience and skills of the patternmaker, and it takes a lot of time to complete the drafting of the pattern block. Therefore, the technical level of the patternmaker greatly determines the correctness of the pattern block and further affects the comfort of the clothing[1]. And then, with the development of 3D virtual-reality technology, there are many manual of pattern block making and some of them aren't good for the customization in virtual reality. In the well-known virtual-reality technologies such as CLO3D, Lectra3D, OptiTex, and V-Stitcher 3D are shown own structural problems, such as shoulder slope is not consistent with human body, unreasonable distribution of ease to bust girth and waist girth, and so on[2]. These problems not only will cause clothing misfit for the customers, but also will cause learning problems for many garment design amateurs or junior garment pattern-makers.

The aim of this research was to create a new method of detecting the blouse pattern block by finding the new relations of sizes of blouse pattern block, on one side, and measurements of avatar, on the other hand.

2. Methods of research

2.1 Blouse pattern blocks

In our research in order to lay the foundation for detecting the pattern block 122 blouse pattern blocks were collected from Chinese manuals and journals. All pattern blocks were drafted in ET CAD system and were classified as X, H, A style[3]. Then blouse patterns were subdivided into slim, regular, and loose types in according to fit. Thus, the blouse pattern block database was obtained.

2.2 Avatar for the experiment

Body type was 160/84A (girth, cm: bust 84, waist 68, girth 90, neck 37, shoulder width 32.6cm, back length 36.2 cm, distance between SNP and SP 9.7cm and between FNP and front waist 33 cm). It were used for obtaining of the avatar by 3D CLO virtual try-on software.

2.3. Prototype of avatar using 3D to 2D pattern block flattening technology

In order to obtain the bodies prototype, we used 3D pattern block making technology in CLO3D software. The 3D avatar surfaces were flattened into 2D garment patterns. The general scheme of the flattening process is: (1) As figure 1 shows, according to the features of the avatar and the shape of garment prototype, the lines of front and back were drawn on the avatar directly. (2) As figure 2 shows, the surface of the garment prototype based on avatar was obtained by using the flattening tool, and the 2D pattern block of the garment prototype after flattening was obtained at the same time[4].

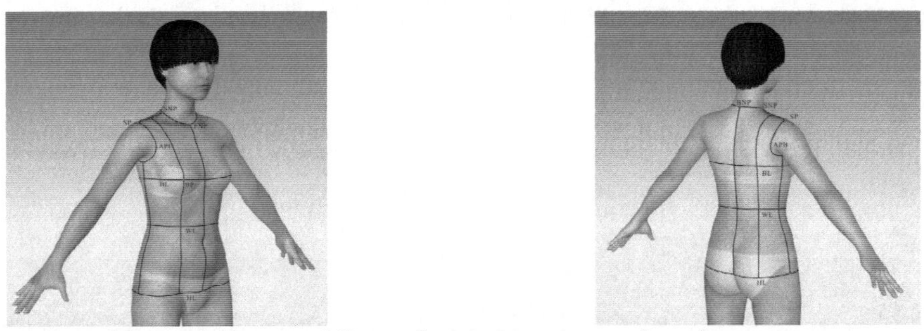

Figure 1. Structural lines of garment prototype based on avatar

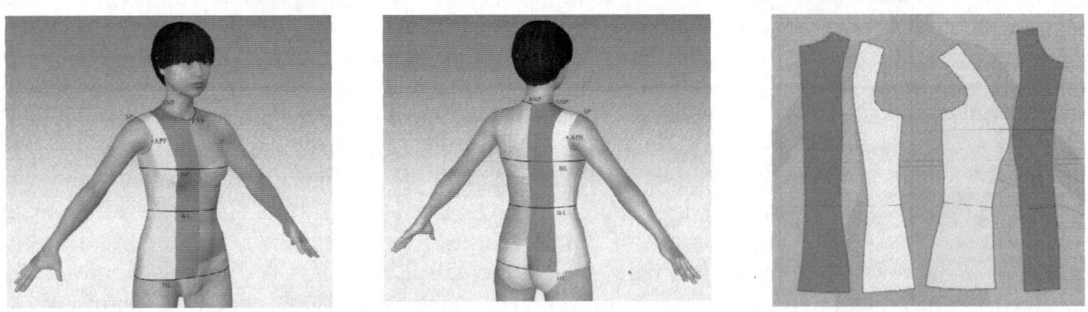

(a) (b)

Figure 2. Garment prototype based on avatar (a) and flattened 2D pattern blocks (b)

In order to ensure the accuracy of the flattened 2D pattern block, the lengths of the key structural curves of the 2D pattern blocks after flattening and the main structural curves of the 3D clothing surfaces and the size of each important structural part were compared respectively, such as the lengths of armhole curves, the length of bust line, and so on. And the test results indicated that the 3D-2D flattening results are precise, which can be used in blouse pattern block checking[5].

2.4. Predicting the accuracy of blouse pattern blocks

As figure 3 shows, 122 blouse pattern blocks were respectively compared with the garment prototype based on avatar. Secondly, the ease amount of the key parts of the blouse pattern block were obtained by measuring the difference value between the important parts of the blouse pattern blocks and the garment prototype. Thirdly, the proportions between the ease allowance of each important part of blouse pattern blocks with good fit were obtained by using SPSS data analysis software, and the interval of the ease amount in each important part were also obtained at the same time. Based on this, a "avatar-blouse" prediction system for the accuracy of blouse pattern was established to predict the accuracy of pattern blocks.

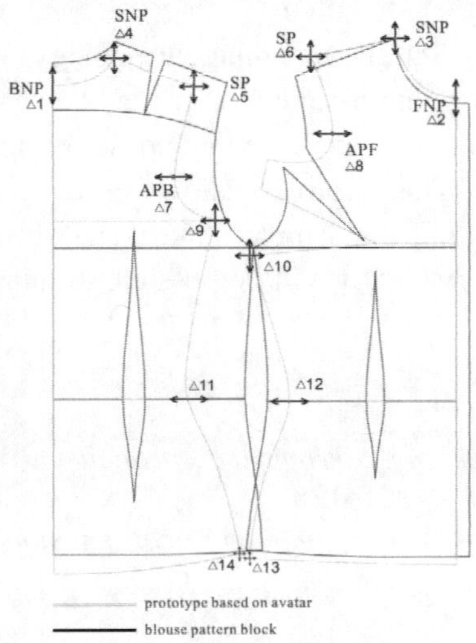

Figure 3. The scheme of overlapping blouse pattern blocks and garment prototype

3. Discussion and results

As figures 4 and 5 show, the blouse pattern blocks were compared with the body prototype respectively by using the method showed in the Figure 3. As shown in figure 4, the ease allowance of each important part of blouse pattern blocks were obtained by overlapping blouse pattern blocks with the garment prototype. Then, the proportions of the ease allowance for each key structural part were obtained by computation. And then by comparing the proportions of the ease allowance with the proportions in the "avatar-blouse" prediction system we have established, we found that the proportions of the ease allowance for the blouse pattern block are rational, and the ease allowance of each key structural part is within the range of each allowance. According to the analysis results, we predicted that the blouse pattern blocks are fit. At the same time, as shown in figure 5, the blouse pattern blocks were analyzed by the same way, and we found that the proportions of the ease allowance in shoulder line, neck line and the back part are too large and not consistent with the proportions in the "avatar-blouse" prediction system. Then we predicted that the blouse pattern blocks are misfit. We further used CLO3D virtual try-on software to check the prediction results. By observing the effect of virtual try-on, we found that the shoulder slope of the blouse pattern blocks do not in accordance with the human body, and the ease allowance of neckline is too large, as shown in figure 5. Thus, the rationality and validity of the method for predicting the accuracy of blouse pattern block are further verified.

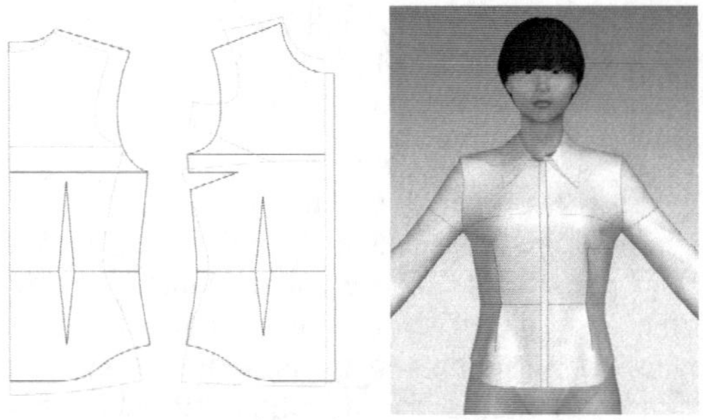

Figure 4. Examples of pattern blocks with good fit and their presentation in avatar

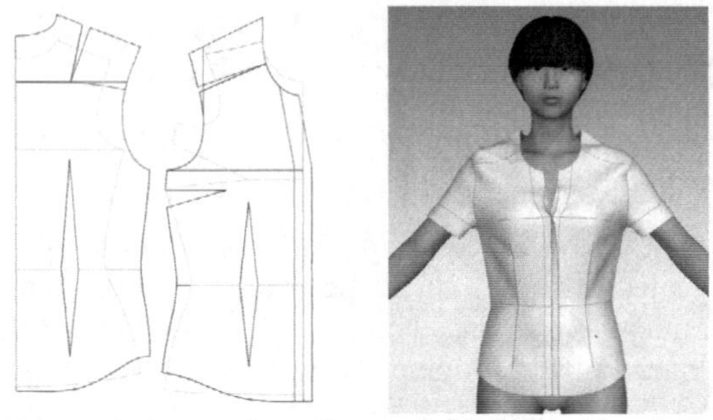

Figure 5. Examples of pattern blocks with misfit and their presentation in avatar

4. Conclusions and future research

In this study, the body prototype was obtained by using 3D-2D flattening technology. And we obtained the ease allowance of blouse pattern block by overlapping pattern blocks and body prototype. And then we developed the prediction system of "avatar-blouse" to evaluate the blouse pattern blocks based on the proportions and intervals of ease allowance. This approach allows us to predict the accuracy and quality of blouse pattern blocks without making the sample. And then, we can alter misfit problems of the pattern block in terms of prediction results. It will be possible to improve the garment production efficiency and the virtual technology.

References

[1] Kyoungok K, Tsuyoshi O and Masayuki T 2017 *Effect of patternmaker's proficiency on the creation of clothing* vol 17, (Poland: AUTEX RES J) p 120-128

[2] Kuzmichev V, Aleksei M, Evgenii S and Mariia M 2017 *Computer reconstruction of 19th century trousers* vol 29, (England: International Journal of Clothing Science and Technology) p 594-606

[3] Zhang W B 2010 *Pattern making for fashion design* (Beijing: China Textile and Apparel Press) pp 68-70

[4] Yang Y C and Zhang W Y 2007 *Prototype garment pattern flattening based on individual 3D virtual dummy* vol 19, (England: International Journal of Clothing Science and Technology) p 334-348

[5] Kaixuan L, Jianping W, Chun Z and Yan H 2016 *Development of upper cycling clothes using 3D-to-2D flattening technology and evaluation of dynamic wear comfort from the aspect of clothing pressure* vol 28, (England: International Journal of Clothing Science and Technology) p736-739

Study on the body girth dynamic size for wetsuit ease design

X Wu[1, 2, 3] and V E Kuzmichev[1, 2, 3]
[1]Fashion of School, Wuhan Textile University, Fangzhi Road, Wuhan, Hubei, China
[2]Ivanovo State Polytechnic University, Textile Institute, 21, Sheremetev Av., Ivanovo, Russian Federation
[3]Wuhan Textile and Clothing Digital Engineering Technology Research Center
xzwu@wtu.edu.cn, wkd37@list.ru

Abstract: The aim of this exploration is to research on how the body measurements change under deformation during diving. We measured 31 body measurements of 57 females in six postures including four postures on the land and two postures under the water. We compared the body measurements, especially focused on the differences between standing and laying prone postures, and check the pattern and pressure with virtual method. It will help designers to find the reasonable dynamic ease and apply it to the pattern block design of wetsuit.

1. Introduction

The divers work under water with different dynamic postures [1], so, skin deformations should be taken into consideration for ease allowance calculating [2]. If the wetsuit is too loose, the water will invade diver' insulation system, if the wetsuit is too tight, the diver meet the difficulties to swim, because the compression wetsuit will be influenced [3].To minimize the flowing water between the skin and clothing during diving process, which mainly depends on diving suit's pattern structure, fabric insulation performance and special craft [4], a dimensional change of body in dynamic positions is a complex issue that cannot be considered only from the aspect of basic anthropometric measurement, but the body morphology parameters must be taken into consideration [5].Some researches about wetsuits only focused on the body standing postures [6].

Our research focuses on the girth under dynamic changes and the corresponding ease values on laying prone postures underwater. Besides, 3D software technology is used to simulate the wearing effect of virtual wetsuit based on 2D pattern block.

2. Experiment steps

1) Four materials M1…M4 are as shown in Table 1. Firstly, we used non-elastic material to test the maximum decrease values ΔG of 11 body girths, 15 testers, to simulate the maximum squeeze change of hydraulic pressure; and used four kinds of materials (M1…M4) in warp and weft directions respectively to stretch different body girths (11 measurements, each part are measured in front, side, back) until to maximum elongation E_M and maximum possible pressure P for each tester, as shown in Figure 1.

Table 1. Experimental materials

No.	Side layers	Middle layer, %	Thickness, mm	Level
M1	Nylon	100% CR	2.85±0.07	High-end

M2	Polyester	SCR(30%CR+70%SBR)	2.80±0.07	Common
M3	Polyester	SBR	2.70±0.03	Low-end
M4	Nylon	SCR(30%CR+70%SBR)	2.99±0.09	Common

*CR is the material with high level of performance, softness, thermal protection, and elasticity; SCR is the common material, which is mixed in proportion by CR and SBR; SBR is material with low level composed of styrene-butadiene rubber.

2) Then we used body surface drawing line method to record body size changes of 31 body sizes of 57 female bodies when postures (based on some typically diving actions) change. The subjects completed two standing and two laying prone postures on the floor and two laying postures under water. We used next abbreviations P0, P1, P2, P3, P4, and P5 to describe each posture, as shown in Figure 2.
Besides, we chose the Chinese body type A as example, and used software CLO to change and adjust the sizes of virtual twin. We joined the body front and profile together of "P1&P0, P2&P0, P3&P0, P4&P0, P5&P0" to compare the outline contour.

- Posture0 – natural standing postures on the floor;
- Posture1 – standing and hands up on the floor;
- Posture2 – laying prone on the floor;
- Posture3 – laying prone and legs curl the floor;
- Posture4 – the same as "Posture2" but under water;
- Posture5 – the same as "Posture3" but under water;

Finally, we applied basic zero ease in wetsuit pattern to obtain a new pattern block.

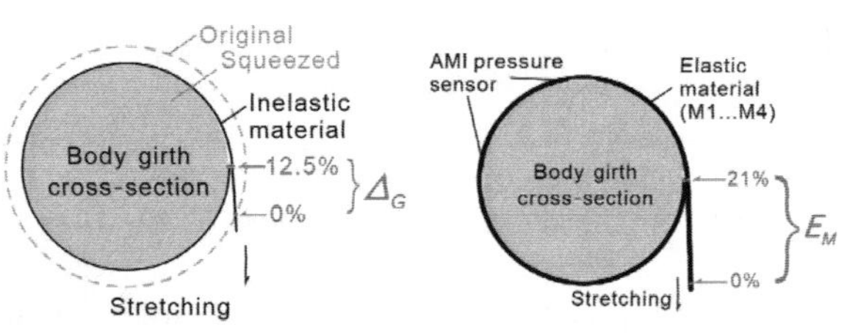

Figure 1. Dynamic postures

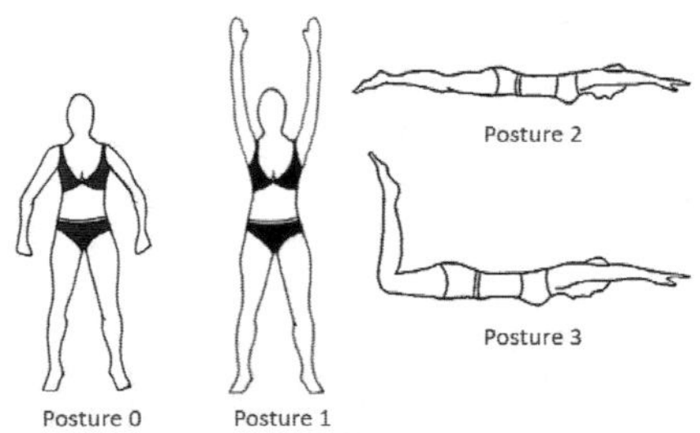

Figure 2. Dynamic postures

3. Results

3.1. Fabric elongation for design ease

We obtained the E_M (average ease for wetsuit affected by the maximum acceptable materials elongation on body) and Δ_G (the body girths maximum decrease values, which can be used to calculate the average material elongation when the body girth is decreased per 1%). We tested the average values of maximum elongation of materials applied to 11 body girths, and found that the E_M is from 13.1 to 18.2%. For materials *M1…M4*, the average pressure value is 2.05 kPa and the average elongation of material is 15.7%, as shown in Table 2.

Table 2. Maximum elongation of materials and pressure

Max average values	*M1*	*M2*	*M3*	*M4*	*Mean*
E_M in warp %	19.5	10.8	9.5	12.7	*13.1 ± 4.4*
E_M in weft, %	21.2	19.7	11.5	20.4	*18.2 ± 4.5*
P, kPa	1.90	1.97	2.39	1.92	2.05 ± 0.23

3.2. Dynamic size change of human body

The differences of *BG*, *WG*, *HG*, *TG*, and *SL* are shown in Figure 3. The maximum difference of BG_F, WG_F, HG_F (and BG_B, WG_B, HG_B) took place between the postures P5&P0. In the circumstance of hands up (set in the same breathing state), the bust line moves upward, the side seam moves forward, the abdomen bulge, the hip, and thighs changed a little.

As shown in Table 3, we obtained the dynamic changes of the human body: *BG* is 0, to *WG* is -1.3%, *HG* is -1.1%, *TG* is -0.2%, and *SL* is 4.4%.

As for wetsuit pattern design, the minimum negative values are applied to fit the human body and satisfy the dynamic change, but, the positive values (such as BG_B) are not considered as additional ease.

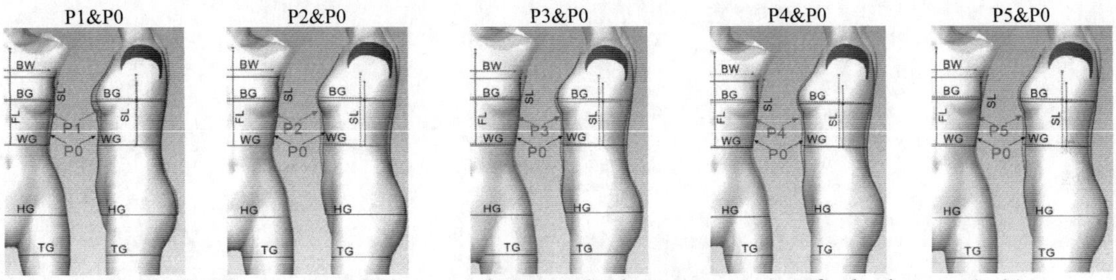

Figure 3. The average differences between body measurements for body type A, %

Table 3. The average differences between body girths for body type A, %

The average differences of body girths

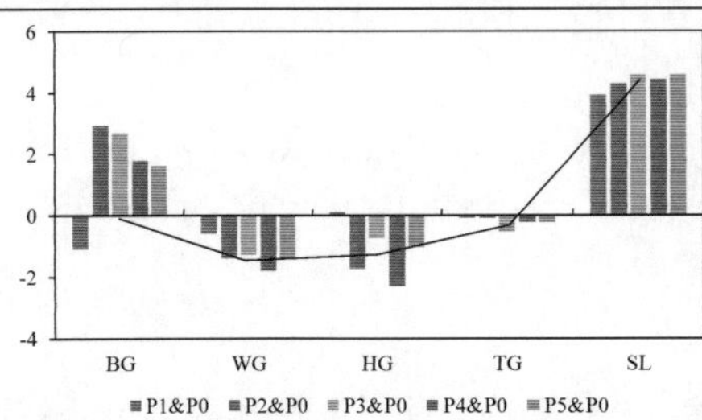

	The average differences of girths front and back part		
	BG_F, BG_B	WG_F, WG_B	HG_F, HG_B
P1&P0	-2.2/1.1	-0.6/0	0.2/-0.1
P2&P0	-3.8/6.7	-1.2/-0.2	0.7/-2.4
P3&P0	-4.0/6.7	-1.2/-0.1	-0.2/-0.5
P4&P0	-4.6/6.4	-1.4/-0.4	0.9/-3.1
P5&P0	-4.9/6.5	-1.3/-0.1	1.0/-2.0

3.3. Pattern adjustment and design

We applied the minimum design ease value to original basic pattern (for keep tight under dynamic conditions), then, an adjusted pattern is designed – New basic pattern *1*. - The ratio we obtained based on previous experiments – the material elongation/the body girth decrease, $R_C = (E_M / \Delta_G)$, Δ_G is dynamic girth changes as the ratio of compression capability of material, which can be used to calculate the average elongation when the body girth is decreased per 1%, $R_C = -2.45 \pm 0.80$. Therefore, multiply R_C the minimum design ease in pattern block *BG* is 0, *WG* is -3.19%, *HG* is -2.70%, *TG* is -0.49%, but *SL* is 4.4%.

In response to this change, we used the 2D graphics software Richpeace CAD, to make careful adjustments to the basic pattern, and redesigned the structural style. As Figure 4 shows, the difference between two patterns is obvious.

Figure 5(a) shows, new design "new pattern *2*" is based on the "new basic pattern *1*"but with raglan sleeve, cutting lines and side piece. The new cutting lines are designed according to body morphology and diving postures characteristics with different colors.

Figure 5(b&c) shows the virtual try-on effect with static standing posture, the corresponding 2D "new basic pattern *1*" and "new pattern *2*" on body type A (*BG*=81.5, *WG*=65.7, *HG*=90.4, cm), both patterns with minimum design ease we mentioned.

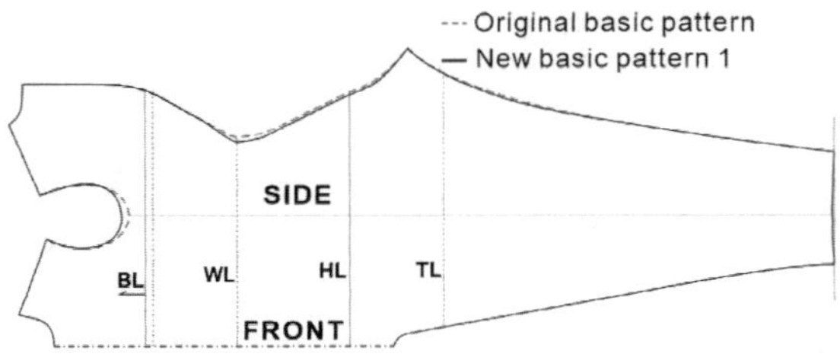

Figure 4. Original basic pattern with zero ease (in dot line) and new basic patterns *1* (in solid line)

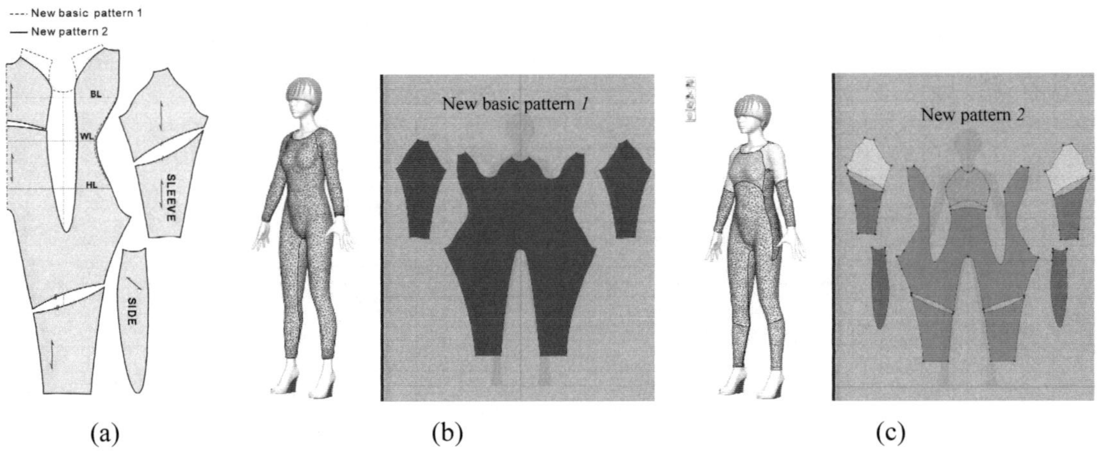

Figure 5. Patterns and virtual try-on: (a) – new basic pattern *1* (in dot line) and new pattern *2* (in solid line); (b) – dressing effect with static standing posture and the corresponding new basic pattern *1*; (c) – new pattern *2*

3.4. Simulation test

In order to check new patterns and simulate the real material properties and the real wearing effect, we adjusted the key data of virtual fabrics properties, marked the key points (174) in CLO software, as shown in Figure 6(a). We take the fabric *M2* as an example.

As shown in Figure 6(b). We cannot make an accurate objective conclusion from the wetsuit appearance, the virtual pressure values can explain the fit degree. In total, the range between the maximum and minimum virtual pressure of basic patterns is significantly, the values at the waist and arm are smaller relatively. The fitting degree of original basic pattern is not enough, but, the new basic pattern *1* with the minimum design ease value is tighter than original basic pattern at the girths of waist, hip and thigh, the pressure values change from bust to thigh is small.The new pattern *2* has relatively higher and comfortable pressure value, which is much tighter than the former 2 patterns.

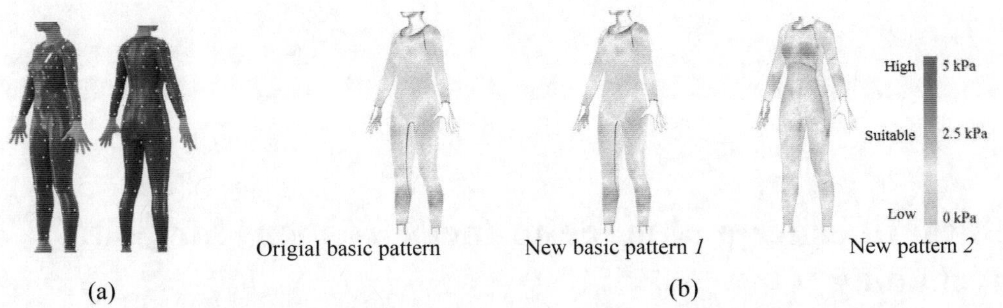

Figure 6. Virtual try-on: (a) – try-on and measure pressure points; (b) – three patterns

The basic patterns have few change when we add the minimum design ease value, because we stretched four kinds of fabrics to the max length of one direction, the other direction shrink less than 3%. We only simulate seamless and more seams patterns (new basic pattern *1* and new pattern *2*), the surface fit degree can be shown through the deformation/elongation. However, the new basic pattern *1* without some cutting design shows that the wetsuit is not fit enough and the ability to resist motion deformation/elongation is poor, as shown in Figure 7.

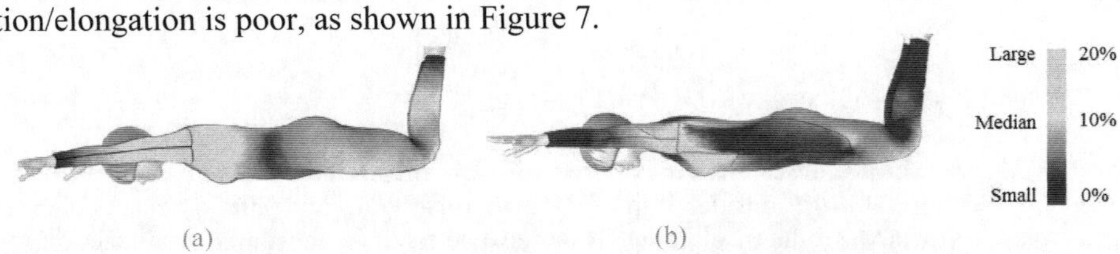

Figure 7. Fabric deformation/elongation: (a) – New basic pattern *1*; (b) – New pattern *2*

4. Conclusion

According to this experiment, those results can predict how dynamic postures influence the body sizes, which can be connected with the pressure on soft tissue, tensile property of materials and parameters of pattern block, and further applied to solve the problems under wetsuit pattern block design.

Under this study with different dynamic postures, we established the relations between their size changes, which provides the possibilities for complex studies of body measurements in dynamic conditions.

Through the use of 2D and 3D technology, we can modify the old pattern and design new style wetsuits, and even simulate some real properties such as the contact performance after adjusting 3D parameters. After continuous adjustment, the new pattern has a better try-on performance and the final pattern can be directly used for factory efficiency production inspection, omitting the repeated actual manufacture modifying works.

References

[1] Nakahashi M 1998 An Effect of A Compressed Region on A Lower Leg on The Peripheral Skin Blood Flow *Journal of the Japan Research Association for textile end-uses* **vol 39** pp 64–69

[2] Xu W and Weiping B 2013 Research on skin surface hardness variation of the adult males with different body fat rate *Chinese Journal of Aerospace Medicine* pp 3

[3] Xinzhou W and Jinsong D 2015 Dynamics analysis of submersible sports and design essentials of wetsuits *Journal of Wuhan Textile University* **vol 5** pp 29–33

[4] Jinghua G and Ping N 2006 Characteristics and development trend of diving suit *Personal Protective Equipment in China* **vol 2** pp 45–47

[5] Petrak S and & Naglic M 2017 Dynamic anthropometry-defining protocols for automatic body measurement *Tekstilec* **vol 4** pp 254–262

[6] Naglic M and Petrak S 2017 Analysis of dynamics and fit of diving suits *In IOP Conference Series: Materials Science and Engineering* **vol 254** p 152007.

IOP Conf. Series: Materials Science and Engineering **459** (2019) 012086 doi:10.1088/1757-899X/459/1/012086

Hierarchical fit criteria of made-to-measure men shirt with virtual technology

J Q Yan[1] and V E Kuzmichev[1,2]

[1]Ivanovo State Polytechnic University, Department of Garment Design, No.21 Sheremetev Avenue, Ivanovo, 153000, Russian Federation
[2]Wuhan Textile University, Fashion Faculty, 1000 Talent Plan Laboratory, No.1 Sunshine Avenue, Jiangxia District, Wuhan, 430200, Hubei province, China

naclplus7@163.com, wkd37@list.ru

Abstract. In this research we explored the new hierarchical fit criteria of made-to-measure (MTM) men shirt in consideration of the male torso types and shirt styles. As the most pivotal factor of MTM products, clothing fit directly determines consumer's satisfaction towards the clothing. To improve the fit of MTM men shirt with an optimized customization process, 118 male bodies were scanned to attain the conventional and novel anthropometric measurements. Meanwhile, the torso type were classified into five distinct groups in accordance with proportions of anteroposterior girths of different levels. Based on the specific groups, the acceptable eases distribution of the pattern were investigated to form the good-fitted men shirts of different styles in virtual try-on platform. With this methodology, the new fit criteria were proposed especially for improving the mass-production of customized men shirt.

1. Introduction

As the prevalent trends in fashion market, the mass customization which are produced in a highly-digitalized process has been developed[1], facilitated with new technologies such as 3D body scanning, 3D virtual try-on, artificial intelligent, etc. Novel investigations of new body measurements[2], detailed morphological features[3] have been conducted with 3D body scanner, the customization process has been also accelerated by observing the virtual try-on effects of pattern blocks instead of making real samples. Even though, precisely predicting and evaluating the fit and appearance of MTM garments before mass-production is still a tough problem for pattern makers and designers.

To make real and virtual good-fitted customized clothing, a more systematic and efficient procedure should adopted with pattern-oriented anthropometric indexes, individualized pattern blocks and optional clothing styles, etc. This research is aimed to propose the hierarchical fit criteria of MTM men shirt and corresponding pattern block drafting method concerning the male torso shapes and shirt styles with virtual technology. Figure 1 shows the methodology of attaining hierarchical fit criteria of virtual system "body-shirt".

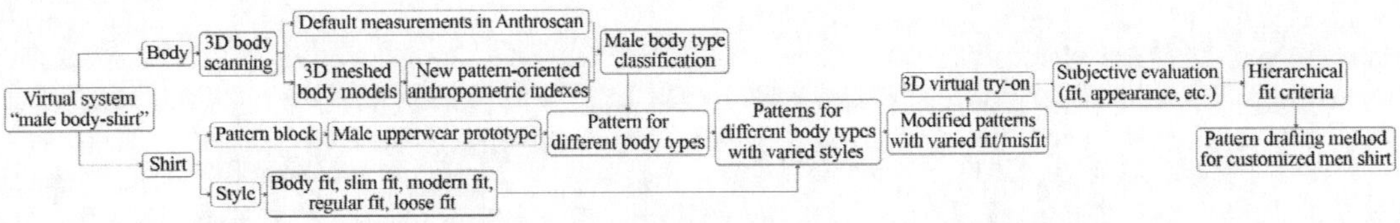

Figure 1. The methodology of investigating the hierarchical fit criteria.

2. Methods

In the virtual environment, the "male body-shirt" was employed as a system considering that the fit and the clothing appearance was formed under the synergy effects of body morphology, pattern blocks, shirt styles, etc. To accomplish the experiment, the following primary software or devices were leveraged: 3D body scanner VITUS Smart XXL for capturing 3D body data, scanning image processing software Anthroscan for generating scanned body models and default measurements, ET CAD 2012 for drafting the 2D pattern blocks, Rhinoceros for marking the feature points and measuring complementary indexes, CLO 3D for 3D virtual simulation, SPSS for data statistics.

2.1. Attainment of pattern-oriented anthropometric indexes

According to the Chinese standard sizing systems for garments (GB/T 1335.1-2008)[4], the essential body measurements utilized for men's wear are height, bust girth, waist girth, hip girth, arm length, neck girth, etc., which is far from adequacy for MTM garments. Thus, 118 male bodies without physiological defects at the age of 18 to 30 were recruited to accomplish the 3D body scanning.

Conventionally, the pattern was sketched according to the predefined linear equations of body measurements. Without considering the specific anthropometric features, the misfit of clothing will appear. Thus, the new indexes revealing morphology for individualized pattern should be proposed with the scanned bodies. Firstly, influential levels were determined as the horizontal plane of SP(shoulder point), bust, waist, the most prominent point of belly, the most concave point of back waist, and hip. Then, the height of these levels were measured as H_{SP}, H_B, H_W, H_C, H_P, H_H, respectively. As the proportionality of body being the main factor influencing the clothing fit[5], anteroposterior girths of different levels were secondly measured. The vertical plane of SP from sagittal view were applied as the bound of front and back torso because of the consistent bound of front and back pieces of pattern. The anterior girths were marked as G_{FB}, G_{FW}, G_{FP}, G_{FC}, G_{FH}, respectively, and the posterior girths were marked as G_{BB}, G_{BW}, G_{BP}, G_{BC}, G_{BH}, respectively. For practicability, the longitudinal distance between SP and other levels were calculated based on the heights as matrix D:

$$D = [D_1, D_2, D_3, D_4, D_5] = [H_{SP} - H_B, H_{SP} - H_W, H_{SP} - H_C, H_{SP} - H_P, H_{SP} - H_H] \quad (1)$$

Where: D1, D2, D3, D4, D5 are the difference between H_{SP} and H_B, H_W, H_P, H_C, H_H, respectively. Similarly, the anterior and posterior girths were marked as matrix GF and GB:

$$GF = [G_{FB}, G_{FW}, G_{FC}, G_{FP}, G_{FH}] \quad (2)$$

$$GB = [G_{BB}, G_{BW}, G_{BC}, G_{BP}, G_{BH}] \quad (3)$$

As Figure 2 shows, longitudinal indexes and latitudinal indexes were measured in both Anthroscan and Rhinoceros.

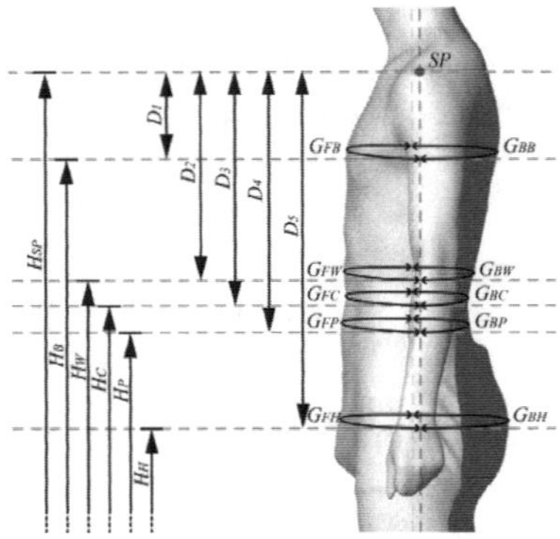

Figure 2. Longitudinal and latitudinal measurements in Anthroscan and Rhinoceros.

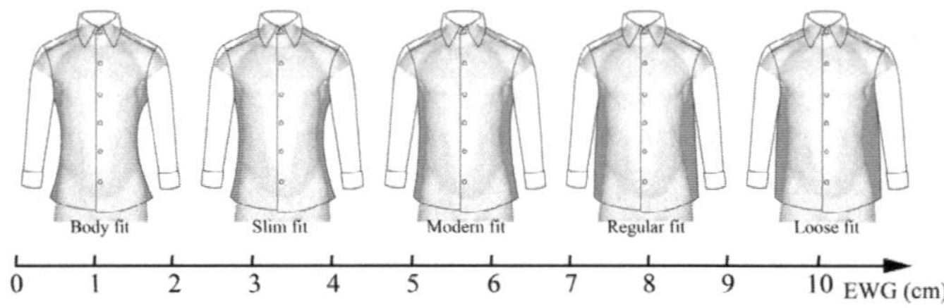

Figure 3. Five shirt styles with different eases to half waist girth (E_{WG}).

2.2. *Classification of male torsos*

For mass customized clothing, it's time-consuming and intricate to adapt one prototype or original block to individual measurements. To simplify the customization, the more representative male torso shape should be investigated to propose the alternative pattern for certain body. To express the proportion of torso, the Matrix P was marked as:

$$P = [P_B, P_W, P_C, P_P, P_H] = GF - GB = [G_{FB} - G_{BB}, G_{FW} - G_{BW}, G_{FC} - G_{BC}, G_{FP} - G_{BP}, G_{FH} - G_{BH}] \quad (4)$$

Where: PB, PW, PP, PC, PH are the difference of front and back girth on the level of bust, waist, the most prominent point of belly, the most concave point of back waist, hip, respectively.

The male torso was classified by K-means clustering method according to the anteroposterior proportional indicator P. With the certain groups of torso, fit criteria could be established for individuals with different morphology.

2.3. Hierarchical fit criteria based on ease parameters

On account of that varied clothing styles have various acceptable ease ranges[5], shirts in five styles as shown in Figure 3 were investigated synchronously to comprehensively reflect the fit and the appearance variants. Hierarchical ease values of different body locations contribute to various shirt styles, the eases to waist girth (E_{WG}) of five styles were primarily explored in this study.

This cession were conducted in virtual try-on software CLO 3D. To eliminate the influence of other fit indicators, the patterns were sketched exactly based on body measurements D, GF, GB. The shoulder and neck area were also modified into balance with corresponding measurements[6]. By adopting the gradient proportion of front E_{WG} (marked as E_{WGF}) and back E_{WG} (marked as E_{WGB}) with interval of 0.5 cm, the fit and appearance of shirt varied. With subjective evaluation of fit, the exact criteria to get good fit and desired styles were concluded with certain values of E_{WGF} and E_{WGB}.

3. Results and discussion

3.1. Male torso types

K-means clustering method was utilized to classified the torso types with cluster number was defined from 2 to 10 gradually. The results shows 5 clusters led to comparatively lower MSE(mean square error). Table 1 shows the average P of different clusters. The associated sagittal views of different body types varied accordingly as Figure 4 shows. The first type has the flat front and concave back contour, the second one has contractive belly and flat back contour with forward thrust shoulder, the third one has the prominent front and concave back contour with forward thrust shoulder, the forth one has the most concave back with backward thrust shoulder, and the fifth one has the flat front and moderately flat back contour. Bodies of different clusters differentiate from the others, hereby the individualized pattern and fit criteria varied enormously with the torso type changes.

Table 1. The average P of different clusters.

	P_B (cm)	P_W (cm)	P_P (cm)	P_C (cm)	P_H (cm)
Cluster 1	3.8	13.1	14.8	14.3	8.2
Cluster 2	- 10.4	- 10.4	- 8.7	- 12.1	- 16.6
Cluster 3	- 6.9	- 0.4	0.3	- 0.4	- 4.8
Cluster 4	9.2	22.8	25.4	24.4	17.6
Cluster 5	- 0.9	6.9	7.7	6.3	0.6

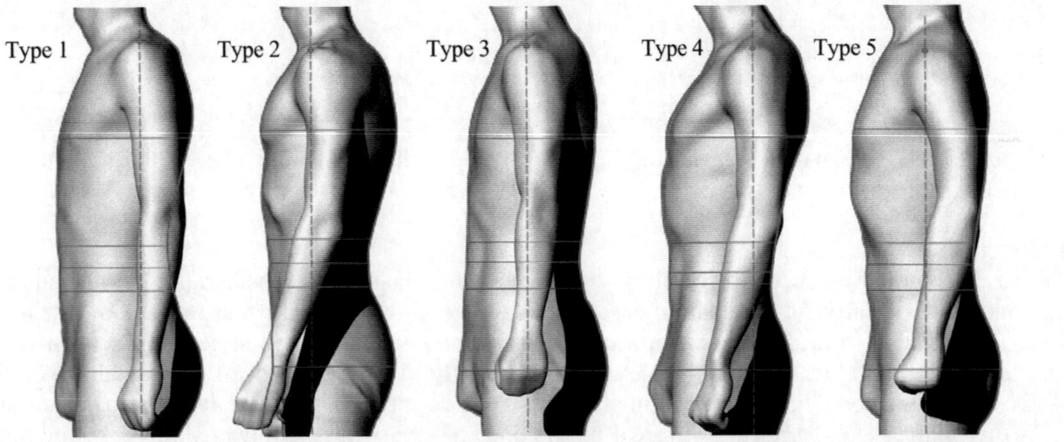

Figure 4. The sagittal views of five types of male torsos.

3.2. The establishment of hierarchical fit criteria

To taking the body of fifth type as an example, the pattern of ready-to-wear men shirt was modified by D, GF, GB, etc. with the ease distributions on bust and hip level were determined. For different styles mentioned before, E_{WGF} was gradually increased from 0 cm to E_{WG}, and E_{WGB} varied accordingly as equation:

$$E_{WGB} = E_{WG} - E_{WGF} \tag{5}$$

E_{CGF} (ease to the half front girth at the level of concave back waist), E_{CGB} (ease to the half back girth at the level of concave back waist), E_{PGF} (ease to the half front girth at the level of prominent belly), E_{PGB} (ease to the half back girth at the level of prominent belly) were modified accordingly with E_{WG} to maintain good fit. The patterns with varied ease allowances were sketched as Figure 5.

The patterns of various styles and distribution of anteroposterior E_{WG} were simulated in the 3D virtual try-on platform CLO 3D. After subjective evaluation of the fit and appearance of virtual clothing (including side-seam, folds and referential lines of different levels, etc.), the best and acceptable ease distribution to attain desired styles were proposed. The best ease distribution revealed best fit with the vertical side-seam, the horizontal referential lines and the minimum folds, and the acceptable one signified slightly skew side-seam and referential lines with few folds, while the unacceptable one represent obvious misfit with the shirt. Figure 6 shows distribution of E_{WGF} and E_{WGB} for attaining the best and acceptable fit for different shirt styles. Obviously, The difference between E_{WGF} and E_{WGB} should be ranged from 2cm to 4cm to maintain good fit. Moreover, the bigger the E_{WG} is, the acceptable range of E_{WGF} and E_{WGB} is wider. And the limit of the acceptable range is the best value ± 2cm, distinct misfit will appear when the eases are out of the range. To achieve the good fit of shirt with certain style for different body type, the fit criteria concerning the distribution of E_{WG} should be applied to balance the proportionality of body and clothing.

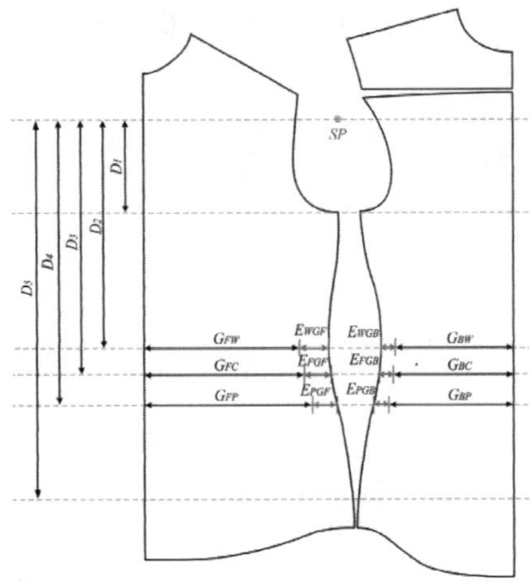

Figure 5. The pattern drafted by D, GF, GB and varied ease distributions.

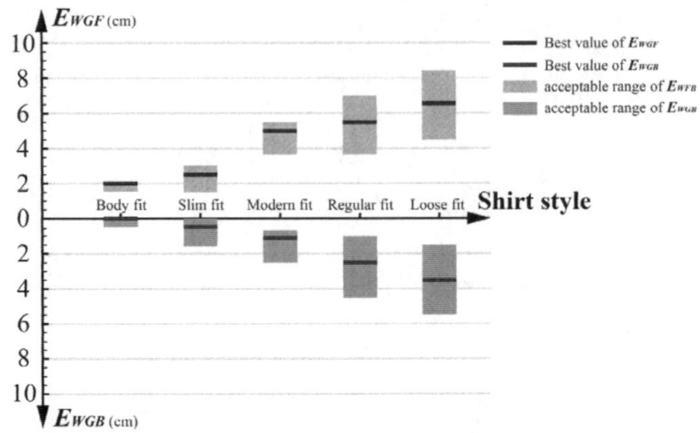

Figure 6. The ease distribution of waist level to attain the best and acceptable fit for different shirt styles.

4. Conclusion

This research proposed the hierarchical fit criteria of MTM men shirt in consideration of the male torso type and shirt styles. With the advised ease ranges for good-fitted shirt, pattern maker or intellectual CAD will be able to draft the customized patterns accurately in desired style and fit for a certain consumer. The new classification of male torso will be helpful to other kinds of individualized clothes. Moreover, the methodology of establishing fit criteria can be applied to other male and female garments.

In the future, the experiments with real bodies and samples will be conducted to verify the results based on the virtual technology. Additionally, more fit indicators will be investigated to acquire the comprehensive and systematic fit criteria for men shirt.

References

[1] Cho Y, Okada N, Park H, Takatera M, Inui S and Shimizu Y 2005 An interactive body model for individual pattern making Int Jnl of Clothing Sci & Tech 17 91–9

[2] Chan A P, Fan J and Yu W M 2005 Prediction of men's shirt pattern based on 3D body measurements Int Jnl of Clothing Sci & Tech 17 100–8

[3] Hu P, Li D, Wu G, Komura T, Zhang D and Zhong Y 2018 Personalized 3D mannequin reconstruction based on 3D scanning Int Jnl of Clothing Sci & Tech 30 159–74

[4] General Administration of Quality Supervision, Inspection and Quarantine of the People's Republic of China 2008 National Standard of People's Republic of China: standard sizing systems for garments – men (China: Standard Press of China) pp 1-36.

[5] Surikova O V, Kuzmichev V E and Surikova G I 2017 Improvment of Clothes Fit for Different Female Bodies Autex Research Journal 17 71–9

[6] Yan J Q, Zhang S C, Kuzmichev V E and Adolphe D C 2017 New database for improving virtual system "body-dress" IOP Conf. Ser.: Mater. Sci. Eng. 254 172029–7

Calculation of the body measurements after analyzing the historical pattern block

S C Zhang[1] and V E Kuzmichev[1,2]

[1]Ivanovo State Polytechnic University, Department of Garment Design, 21 Sheremetev Av., Ivanovo, 153000, Russia

[2]Wuhan Textile University, Fashion Faculty, No.1 Sunshine Avenue, Jiangxia District, Wuhan, 430200, Hubei Province, China

wkd37@list.ru

Abstract. The trends of contemporary fashion design are considerably referring to the historical costumes. Regarding the relevant records of pattern block drafting methods of historical costumes, although the patterns and their concrete indexes are accessible, the anthropometric features of the corresponding historical wearer are uncharted. The purpose of this research is to employ the 2D CAD and 3D virtual try-on technologies to reconstruct the historical costumes to calculate and assess the anthropometric measurements of the historical wearers, validating by the results of virtual system "historical costume – body".

1. Introduction

In the contemporary fashion market and industry, the historical European costume has been one of the most essential inspiration of designing. Especially on the runways of yearly fashion weeks in Paris, Milan, New York, London, the relevant elements and features of historical costumes have been appeared in many notable brands' designs[1].

To optimize the research of historical costume, many authors have applied the new virtual try-on technology to their field, which was proved feasible. However, most researchers investigated only the clothing itself without considering the vital impact of body in the virtual environment[2][3]. Thus, to precisely reconstruct the historical clothing and improve the efficiency of construction, the primary factor is the establishment of identical avatar which is consistent with the historical male body. Namely, in the virtual system "historical costume-body", the appearance, morphology, and anthropometric indicators of the reconstructed virtual avatar should be the same as the ones of the cotemporaneous European body[4]. In this exploration, the virtual male body of 1740s was reconstructed through adaption of body height matchable with the historical pattern. The obtained avatar was manifested as the duplicated historical body in accordance with the measurements from Russian typical figure of men dimensional signs for fashion design system[5].

2. Experiment

2.1. Men's costume

Men's full-skirted coat in 1740s was chosen for reconstruction. Figure 1 (a) shows this coat which is made in the small-patterned figured, extra width at the sides had to be arranged in pleats. These fanned out over the hips below the waist where the side seams ended[6]. Waistcoats were worn for undress from the middle of the century[7]. This costume made of silks, brocades and damasks, such beautiful fabric that decoration became unnecessary. Enhancing by the observations of corresponding inside underwear, fabrics and ingredients, craftsmanship, the reconstruction of these costumes in 3D CAD were proposed with the adapted historical pattern blocks[8].

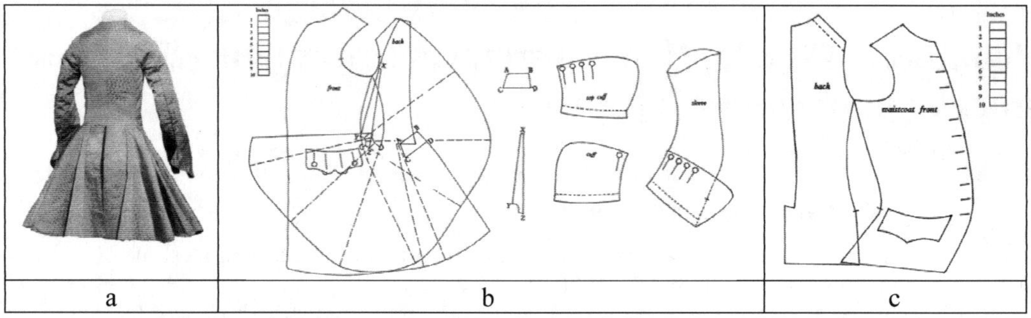

Figure 1. Men's full-skirted coat in 1740s (a), pattern block of coat (b), pattern block of waistcoat (c)

2.2. *Adapting the height of avatar in Clo3D*

After sketching the historical pattern blocks as equal scale, the two kinds of historical clothes were reconstructed in 3D CAD software Clo 3D, with the adjusted avatar height in accordance with the patterns.

Taking the 1740s' clothing as an example (Figure 2.), the concrete method was:

1. Mark the landmarks of BNP (back neck point), waist line(WL) and waistband line(WBL) on the back pattern.
2. Measure the distances from BNP to WL, from BNP to WBL and from WL to WBL on the pattern, respectively recorded as L1, L2 and L3.
3. After virtual historical clothing was dressed on the avatar in Clo 3D, measure the distance from BNP to WL, from BNP to WBL and from WL to WBL along the clothing surface, respectively recorded as L1', L2' and L3'.
4. Adjust the avatar height to ensure L1=L1', L2=L2' and L3=L3' respectively, and record the heights as H1=180.5cm, H2=183.5cm and H3=180.5cm.
5. Calculate the average value H_{mean} =181.5cm of these heights, which would be the originally applied male height of the historical pattern. Moreover, with the pupillary distance of the man in the photo measured, the scaling method could be also utilized to calculate the height with peculiar ratio.

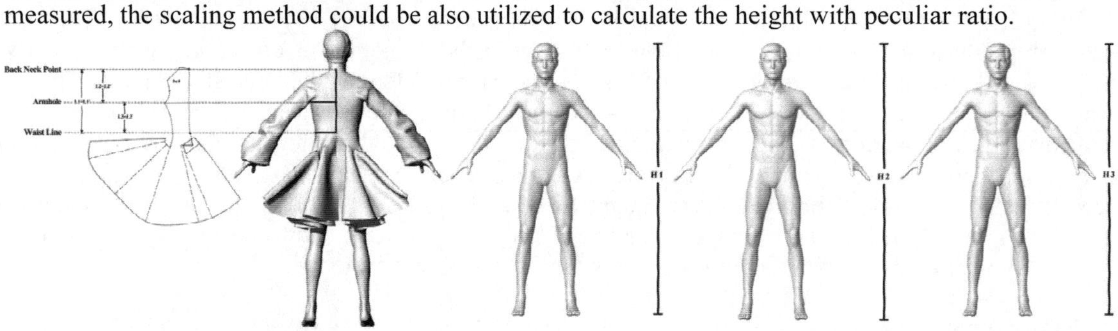

Figure 2. Virtual system "body-coat", adapting the height of avatar during virtual try-on

2.3. *Calculating the bust and waist girth of avatar*

Reconstructed full-skirted coat in 1740s in Clo3D software with the height of European male avatar was 181.5cm. Figure 3a showed upper part of its 3D image. Additionally, the cross-sections on bust level, waist level and the overlapping images of body and costumes were obtained in Figure 3b. The bust girth and waist girth of historical wearer were calculated by the formula (1) and (2) respectively, where L1 represents the circumference of the coat at bust level, as the length of green line (Figure 3b); a represents the distance between the front edge of coat at bust level, as the length of red line (Figure 3); t represents the thickness of coat and underwear. Measured L_1 is 97.61cm, a is 3.78cm, t is 0.5cm in Clo 3D software separately, then we can get the bust girth is 98.25cm, and calculate the waist girth is 74.35cm by same way.

$$L_1 = (\text{Bust Girth-a}) + 2\pi t_{(\text{waistcoat + coat})} \quad (1)$$

$$L_2 = (\text{Waist Girth-b}) + 2\pi t_{(\text{waistcoat + coat})} \quad (2)$$

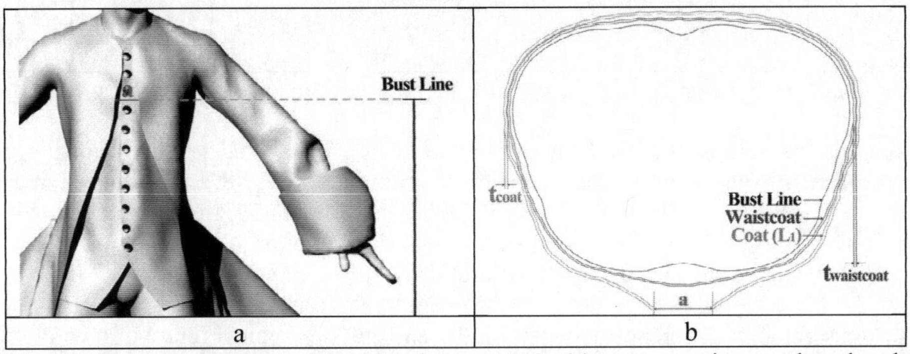

Figure 3. 3D image of Men's full-skirted coat, 1740s (a), cross-sections on bust level (b)

3. Results and discussion

Avatar with the height of 181.5cm that reconstructed in Clo 3D software was imported into Rhinoceros software, 49 anthropometrical data were measured and recorded according to the methods of body measuring which demonstrated in Russian typical figure of men dimensional signs for fashion design system. Figure 4 shows the schemes of bodies measuring in this system.

By using the formula (1) and (2), bust girth and waist girth of virtual model were both obtained. It can be confirmed that this avatar's morphological characteristics conforms to the table of No.4 in Russian typical figure of men dimensional signs for fashion design system, that is, the bust girth between 88-104cm. After combing with the height of avatar, we can get the other 46 measurements.

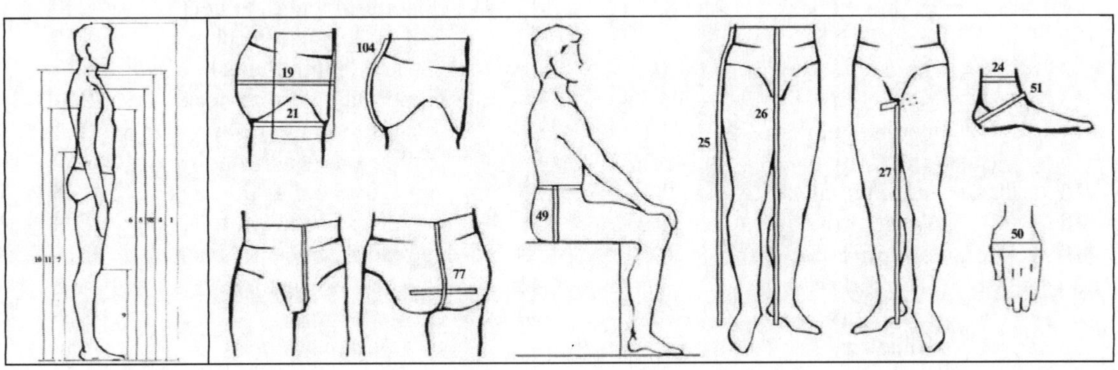

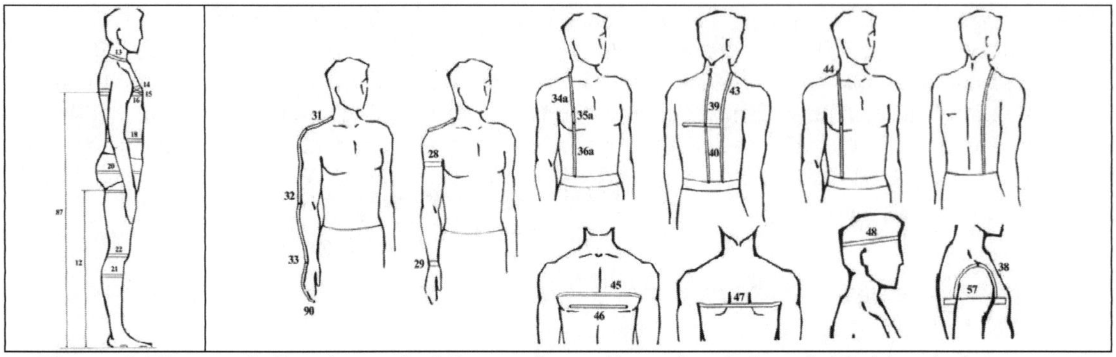

Figure 4. Diagram in Russian typical figure of men dimensional signs for fashion design system

Table 1 contains the anthropometrical measurements which were used in dimensional sign and the difference between 49 sets of dimensions in pairs (the numbers 1,27, 98…26 are following Figure 4). It can be found that 5 pairs are identical and little difference among 39 groups but they still belong to same male body shape standard. There are also 5 sets of data with large variance, mainly concentrated on the length of avatar's arm, such as 32nd, 33rd and 90th.

Table 1. Measurements difference between Clo3D and Russian typical figure of men dimensional signs for fashion design system

No.	Measurement	Difference	No.	Measurement	Difference
1	Height	0.50	27	Crotch height	0.20
98	Neck height front	0.93	104	Waist to lower hip	-0.39
4	Neck height side	0.20	77	Crotch length	-0.70
5	Shoulder point height	-0.06	49	Distance crotch to waist	-0.10
6	BP point height	-3.54	28	Upper arm girth	1.12
7	Waist height	0	29	Wrist girth	0.33
9	Knee height	0	30	Hand girth	0.10
10	Neck height back	0.2	31	Shoulder width	0.35
11	Armpit height back	-0.58	32	Upper arm length to neck	5.40
87	Breast height	0	33	Arm length to neck	4.56
12	Thigh height	0.1	90	Hand length to neck	5.77
13	Neck girth	0.58	34a	Upper bust level to neck	-0.30
14	Upper bust girth	2.18	35a	Bust point to neck	-0.39
15	Bust girth (horizontal)	2.55	36a	Waist to neck front	0.50
16	Under bust circumference	0.40	38	Armhole girth	0.38
18	Waist girth	-0.35	39	Neck to back width	0.04
19	Hip girth (with abdomen)	0.60	40	Neck to waist center back	0.19
20	Hip girth	0.56	43	Neck to waist back	0.08
21	Thigh girth	0.80	44	36a+43	0.58
22	Knee girth	0.74	45	Width armpits	0.80
23	Calf girth	0.80	46	Bust points width	0.60
24	Ankle girth	0.06	47	Back width (armpit level)	0.06
51	Foot girth	0.20	57	Arm diameter (armpit level)	0.37
25	Side seam 3D waistband	0	48	Head circumference	0.23
26	Waistband front height	0			

4. Conclusion

Through the observation of historical costume, the reconstructions in virtual environment were proposed. The way and the algorithm how to get the original male body measurements from the historical pattern in accordance with the pattern block parameters, fabric thickness, air gap, etc. were developed. Thus, the detailed anthropometrical database and morphological features were attained accordingly. While 5 existing measurements of reconstructed avatar in CLO 3D were different with the Russian National Standard. In the future, the avatar will be formed in other software like Optitex, MakeHuman, etc. and make comparisons. The changeable morphology of virtual avatar will be considered when the corset is worn. This research proposed and validated the method of obtaining body measurements from pattern blocks and picture of historical costumes and virtual try-on technologies. This approach will be helpful to accurately exhibit the features of historical costumes for researchers of historical costumes and digital museum.

References

[1] Kuzmichev V E, Moskvin A, Surzhenko E and Moskvina M 2016 Computer reconstruction of 19th century trousers *International Journal of Clothing Science and Technology* **29(4)** pp 594-606

[2] Martin K and Ko H 2009 Virtual Historical Costume Across Cultures and Disciplines In: *Proc. of the 15th Int. Conf. on Virtual Systems and Multimedia* 9-12 September 2009 Vienna Austria pp104-108

[3] Kang Y, Wu S, Ko Y, Kang Z Y, Kim N and Ko H 2013 Digital production of traditional costume *In: Proc. of the 1st Int. Conf. on Digital Fshion London Digital Fashion* 2013 Secretariat 16-17 May 2013 pp 329-335

[4] Kuzmichev V E, Moskvin A and Moskvina M 2018 Virtual reconstruction of historical men's suit *Autex Research Journal* vol.**18** No.3

[5] Central Scientific Research Institute of Scheinon Industry 2005 *Typical Figure of Men Dimensional Signs for Fashion Design* (Moscow)

[6] Norah W 1964 *The Cut of Men's Clothes* (New York: Theatre Art Books)

[7] Vincent W D F 1898 *The Cutter's Practical Guide to Cutting and Making all Kinds of Waistcoats* （London: The J. Williamson company limited）

[8] Vincent W D F 1898 *The Cutter's Practical Guide to Cutting and Making Shirts, Undergarments, Collars, and Specialite Clothing for Various Occupations* (London:The J.Williamson company limited)

The Omni-channel Concept: a Case Study about the Union of Sales Channels in Portugal

M J Abreu and D Miranda
University of Minho, 2C2T-Center for Textile Science and Technology, Campus de Azurém, Guimaraes, Portugal

josi@det.uminho.pt

Abstract: This article contributes to a better understanding of the omni-channel concept in fashion retail, more specifically, in the Portuguese footwear segment and understand if the current challenge of brands is based on the union of all sales channels in order to gain, to locate and to provide a perfect experience to their customers, having regard to the online world and the offline world. The methodology employed was the case study of two Portuguese footwear brands. The collection of data was obtained through two distinct and complementary tools: the semi-structured interview and indirect observation of how brands are moving in their sales channels. The data collected have concluded that there is a big discrepancy between the integration of brands' sales channels, one brand implements more strategies that aims to unify the channels compared to the other.

1. Introduction and state-of-the-art

The technology evolution, in particular, mobile devices (smartphones and tablets), social media/networks and related software with these applications (applications for mobile payments, coupons, digital brochures, location services, among others) have provided new approaches, opportunities and strategies extremely creative for retailers implement in their organizations, creating new points of contact with the consumer. Agis refer that one of the three concepts in Retail 3.0 is the omni-channel, which, according to this author, consist in the integration of the online world with to offline world [1].

Before the globalization, operated the uni-channel concept when the retailer had a unique contact point with the consumer: the physical store [2]. Thus, observing the rise of multi-channel concept to the omni-channel concept that consists in the union of all channels on unique and exclusive means of interaction with the consumer, where they live an exclusive experience with the brand [2,3], allowing access and getting in touch with her through the middle that their want.

Omni-channel should be considered the most advantageous way of extracting the best of each channel to achieve the best for business, for the brand and its consumer [1].

The omni-channel, comparing with the multi-channel, operates in synergy so the distinctions between borders tend to disappear, transforming the world in an environment without walls [4,5].
The figure 1 shows the evolution of concepts presented previously.

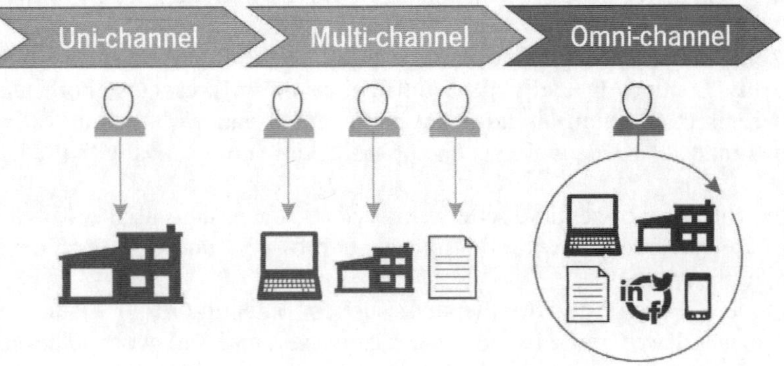

Figure 1. The omni-channel evolution

All the technology advancement and the necessity of companies to readjust their strategies torn necessary the study of their sales channels to analyze the way which these are worked and the activities that they implement.

In the Portuguese retail world, the omni-channel concept is very recent and is taking its first steps. Studies that carried out on this topic are scarce. Even though, there are some efforts in the implementation of the concept, this still have a long and arduous way to go

Sandra Coelho [6] did a study of six national companies in different sectors of retail about the challenges of implementing the omni-channel strategy and concluded that the view of the organizations are the biggest obstacles consisting in the coordination and coherence between channels, the shopping experience and the management of human and technological resources. The companies also made reference to three benefits in implementing this strategy as the increased proximity to customers, increasing brand visibility, particularly at international level and the design of future strategies. National companies are aware of the changing needs, invest in resources and many of them began to explore and to adhere to the point of sale online, focusing on the needs of their customers.

With this topic in mind, the present article have the objective to explore the omni-channel concept and their application on fashion retail, more specifically, in the Portuguese footwear segment and understand if the current challenge of brands is based on the union of all sales channels in order to gain, to locate and to provide a perfect experience to their customers, having regard to the online world and the offline world. Thus, it became necessary to raise the following issues:

RQ1. The omni-channel concept is implemented by the Portuguese footwear brands?

RQ2. Which are the strategies of the omni-channel concept that the brands implement in their channel to create connections between them?

2. Methodology

The methodology employed was the case study of two Portuguese footwear brands. The collection of data was obtained through two distinct and complementary tools: the semi-structured interview and indirect observation of how brands are moving in their sales channels.

The footwear segment was selected because Portugal present a high rate of export of footwear, is a sector that has been growing in the last five years and contribute for the economic growth of the country [7]. Thus, the definition of the "cases" that constitutes the Case Study was obtained through probabilistic samples, once that their constitution is always intentional in this type of method, this is, it meets one criterion pre set [8]. So, the criteria in brands selection were:

• National brands in the footwear segment;

• Implemented in the online world and the offline world so, both brands need to have physical and virtual stores.

After searching some brands having regard to its activity in the online and offline world the choice fell on Eureka and Fly London. In addition to fulfill the criteria presented, both are indisputable leaders in the footwear sector. Eureka is an avant-garde brand and with an innovative design. Fly London is one of the brands of Kyaia which is one of the leading companies with the largest turnover in 2014 [7].

This article, opted for using of semi-structured interviews, once, the initial order of the questions can be modified and restructured; allowing the introduction of new questions or hides the questions that are less relevant to the study [9].

Thus, the development was a script with the key issues for obtaining responses to the theme under study; its elaboration was based on selected brands and the empirical work. The semi-structured interview was recorded through a single mobile device and performed the responsible working directly with online and offline channels.

The analysis of data was made by having regard to the interpretative data search and a reflection of both the views of respondents of each brand. The analysis of the data presentation was performed through objective narratives [8].

The semi-structured interview was held to director of the marketing department of Eureka, and to head of the brand Fly London. The digital platforms of the brands were consulted for get information that has not been possible to collect in the interview. In this way, it was observed information's on brands' website: www.eureka.com, www.flylondon.com, www.kyaia.com and in social media. Some information was also obtained through the exchange of email.

3. Results and discussion

The data proved that Fly London is a reference brand of quality and export level, on the other hand, Eureka is distinguished for being a differentiating brand and avant-garde, investing in quality and innovation of its products. In the next table you can see the general data of the studied labels (Table 1)

Table 1: General data of brands

General data	Fly London	Eureka
Companies	Kyaia, Fortunado O. Frederico, Lda	Alberto Sousa, Lda
Foundation	1993	1986
Segment	Man and Lady	Man and Lady
Products	Footwear, Handbags/Wallets, Accessories	Footwear, Handbags/Wallets, Accessories
Number of employees	620	350
Main markets	United Kingdom, United States of America, Canada and almost all countries of Europe	Portugal and some European Countries

The table 2 presents a summary of the number of sales channels that each brand has at national level and their communication channel. It is important to stress that the brands in study offer products high-touch (footwear, handbags/wallets, accessories), this is, products with sensorial elements that the consumer value and, in most of the time, they wish to touch and experience [10], by this, the brands must contain physical spaces for provide the direct contact between the consumer and the product and, at the same time, to be present on online platforms and learn to communicate their story and provide all the necessary information.

Regarding the sales channels, Eureka is directed to the national market with 23 physical stores in Portugal, one flagship store in Oporto, three outlet stores and the website in www.eurekashoes.com for some European countries.

On the other hand, in Portugal Fly London has less power with two physical stores, one in Oporto and other in Lisbon, focusing in sale of product by the, approximately, 65 multi-brand stores in all the country and the website www.flylondon.com.

In communication channels online Eureka is implemented in seven channels and Fly London only in three channels. In the communication channels offline both brands are present in same number of channels. Also, Eureka offers a means of communication, Eureka Lovers that involves national personalities of various artistic areas that use the Eureka products and disclosed by different media. These data can be observed in table 2.

Table 2: Channels of Brand Fly London and Eureka

Channels of Brand	Fly London	Eureka
Sales Channels		
Physical stores in Portugal	2	23
Flagship Store	-	1
Multi-brand stores in Portugal	65	-
Outlet stores	-	3
Official website	Yes	Yes
Communication Channels		
Online communication channels	Facebook Instagram Newsletter	Facebook Pinterest Youtube Vimeo Instagram Snapchat Newsletter
Offline communication channels	Magazines/Fashion journal Advertising International fairs	Magazines/Fashion journal Advertising Fashion editorials
Other means of communication	-	Eureka Lovers

The table 3 presents the characteristics and the strategies of the omni-channel concept that the brands Eureka and Fly London should implement to unify their sale channels.
This analysis allows comparing the two brands. In this way, Eureka implements thirteen of fifteen omni-channel criteria present in the table 3, while Fly London only implement five of fifteen omni-channel criteria.

Table 3: Activities and characteristics of the omni-channel in Portugal Fly London e Eureka Shoes

Activities and characteristics of the omni-channel in Portugal	Fly London	Eureka
Distribution and logistics		
Click-and-collect (order and pay online and collect in the physical store)	No	Yes
Buy in physical store and receive the product at home	No	Yes
Return of the product in the physical store of orders made on the website	No	Yes
Free delivered on physical store	No	Yes
Free home delivered	Yes	Yes
Auxiliary tools for sale		
Use of mobile technologies on the part of the shop assistants in physical store (e.g. tablets)	No	Yes
Use of technologies within the physical store (e.g. iterative screens)	No	Yes
Possibility to see the extract from the purchase of a product	No	Yes
Online chat to assist consumers in purchasing process	Yes	Yes

Foment the consumer's experience and integration between channel		
Price consistency across all channels	Yes	Yes
Location of the physical store on the brand website	Yes	Yes
Online information about the availability of the product in physical store	No	No
Detailed information about the product (e.g. product characteristics, colors available, sizes, among others)	Yes	Yes
Promote the use of mobile devices to consumers in the physical store (e.g. mobile applications, apps, QR Code)	No	No
Data collection and consumer's behavior		
Getting online statistics about consumer's behavior on the website	No	Yes

4. Conclusions and future work

The data collected from interviews, concluded that, between both brands, exists a big difference concerning the integration of the sale channels. Eureka implements more strategies conceiving unification of both channels in comparison to Fly London.

Fly London's representative, manifest that the brand does not implement omni-channel strategy and the scientific data obtained clearly support that fact. Therefore, it is verified that the brand only implements the multi-channel strategy, which means that the strategies and content in the online and offline channels are developed in an independent and separate way.

In contrast, Eureka, is making good progresses in order to be consider a brand that implements the omni-channel strategy, due to the fact that is developing strategies that aim to unify the sale channels. It is possible to verify that the brand knows its customers and that is making efforts to apply the necessary technology in both channels, which in turn results in a wider activity between them and contributes to the unification. However, Eureka still needs to overcome many barriers, in particular on the mobile channels, in order to achieve unification.

Another aspect that can be noticed is that both brands do not develop specific strategies that drive the use of mobile devices, meaning that, they do not consider these devices to be a very important and fundamental tool to break the walls between the online and offline channels. It is important to consider that the mobile channel is one of the greatest pushers of the omni-channel and is considered to be an integration factor between the physical and virtual channels.

During the development of the present article and principally after its conclusion it was accomplished that the omni-channel subject needs further study. On an academic level there are some limitations, especially if the focus is the strategy in the selling channels. The present theme encompasses many different fields of study, like marketing, management, logistic, distribution, and technology, among others. It would be of great value if there were a clean definition of the right strategy to adopt for a particular channel with the goal to establish a perfect relationship of how to achieve this union.

Another important study would be the characterization of what is necessary to do for a brand to achieve the omni-channel, in other words, what it needs to develop or implement in order to be considered an omni-channel marketer.

It would be also interesting to perform this study on another national fashion segment that go beyond footwear retail with the goal to understand if exist or not the knowledge about the omni-channel concept by another brands and by the consumers it selves.

Through this study, there was an analysis of Eureka's consumer and Fly London's consumer, it was observed that they are not aware of the omni-channel strategy. This study will be published in a future publication.

Acknowledgments

This work was supported by FEDER funds through the Competitivity Factors Operational Program - COMPETE and by national funds through FCT – Foundation for Science and Technology within the scope of the project POCI-01-0145-FEDER-007136.

IOP Conf. Series: Materials Science and Engineering **459** (2019) 012088 doi:10.1088/1757-899X/459/1/012088

References

[1] Agis, D., 2012. *Retail 3.0: Fututo físico e virtual.* s.l.:ATP.
[2] Harris, E., 2012. A Look At Omni-Channel Retailing. [Online] Available at: http://www.innovativeretailtechnologies.com/doc/a-look-at-omni-channel-retailing-0001
[3] Rigby, D., 2011. The Future of Shopping. *Harvard Business Review.*
[4] Brynjolfsson, E., Hu and Mohammad S. R, Y. J. & S. R, M., 2013. Competing in the Age of Omnichannel Retailing. *Summer 2013 Research Feature.*
[5] Verhoef, P. C., Kannanb, P. & Inm, J. J., 2015. From Multi-Channel Retailing to Omni-channel
[6] Coelho, S. C. P., 2015. Desafios do omnichannel na aplicação às empresas nacionais, s.l.: s.n.
[7] APICCAPS, 2015. World Footwear Yearbook 2015. Porto: s.n.
[8] Coutinho, C. P., 2013. Metodologias de Investigação em Ciências Sociais e Humanas: Teoria e Prática. 2ª ed. Coimbra: Almedina.
[9] Robson, C., 2002. Real world research: a resource for social scientists and practitionerresearchers. 2ª ed. s.l.:Oxford: Blackwell Publishers Inc.
[10] Bell, D. R. B., Gallino, S. & Moreno, A., 2014. How to win in an Omni-channel world. MIT Sloan Management Review, pp. 45-54.

IOP Conf. Series: Materials Science and Engineering **459** (2019) 012089 doi:10.1088/1757-899X/459/1/012089

"Do as I say, not as I do" – a systematic literature review on the attitude-behaviour gap towards sustainable consumption of Generation Y

J P Bernardes[1], **F Ferreira**[1], **A D Marques**[1] **and M Nogueira**[2]
[1]Universidade do Minho, Escola de Engenharia, Departamento Têxtil, Campus de Azurém, Guimarães, Portugal.
[2] IPAM - Universidade Europeia, Rua Manuel Pinto Azevedo 748, Porto, Portugal,
joaopedro_bernardes11@hotmail.com

Abstract. This paper contributes to a consolidation of the field by narrowing down the gap between sustainable consumption "attitudes/intentions" and actual consumption "behaviour". The adopted methodology is based on a systematic literature review with an initial sample of more than 1000 publications, this paper considers the following inclusion and exclusion criteria throughout the review: research scope, conceptualisation of the terms (e.g. "Sustainable consumption"; "Ethical consumption"; "Green consumption"; "Green consumerism"; "Responsible consumption"), research design (e.g. quantitative, qualitative experimental studies), segments of consumers (Generation Y) and time frame. A qualitative research approach will be adopted in this paper, to safeguard the inclusion of methodologically diverse studies. Additionally, because the mentioned concepts have evolved over time and raised so many different interpretations, a qualitative approach is considered more pertinent. The focus of the review is to build up coherence regarding the sustainable consumption attitude- behavior gap. The paper concludes with a discussion of challenges and recommendations for future work in the field, namely, the need to encourage and educate consumers to live a more sustainable lifestyle and to design and implement responsible marketing tools to communicate the value of sustainable consumption.

1. Introduction

A consequence of the growing number of conceptual and empirical studies in sustainable consumption is the need to adopt systematic approaches to evaluate and assemble research outcomes that provide a balanced and objective summary regarding a particular topic. The paper reports on the adoption of such approach - systematic literature review - to the published studies relevant to topics within the sustainable consumption domain. So far, research on sustainable consumption is scattered across different disciplines and lacks integration, giving raise to many different definitions of the concept. Research into the responsible, ethical, sustainable, green and socio-political aspects of consumption has grown considerably since the 1990s. The most consistent finding within this literature has been inconsistency between what people say and what they actually do - the so-called "attitude- behaviour" and "intention-behaviour" gaps [1-5].

For instance, in a study of the Millennial generation, [4] observed that even though consumers have a very positive attitude towards sustainability and green products, this attitude is not reflected in actual behaviour. That way, it is of true importance, the understanding of consumers and what influences them. Every consumer is an important part of the fashion system, thus creating a valuable influence in the pursuit of sustainability in the fashion industry. There are several aspects to consider when analysing fashion consumers and sustainability: the consumer knowledge about sustainability, consumer behaviour and consumption habits and feelings associated with sustainable consumption. In fact, the attitude and the behaviour of consumers are in line with the rapid production. Encouraged by low prices and influenced by marketing campaigns and trend changes, consumers tend to speed their fashion consumption [6]. These younger consumers, called the generation Y (Millennials), are defined as a group of people born between 1980 and 2000 [7] and maintain a positive attitude in relation to sustainability in general. However, there is a clear contradiction between how Millennials think about sustainability and what they do when it comes to a sustainable consumption [8]. According to [9], one possible justification for the existence of this gap lies in the consumers' lack of awareness of how their consumption can contribute to the environmental and social problems. Consumers also tend to shift the blame of the current sustainability issues to other institutions like businesses, education and society or even refuse to accept their contribution in the problem itself. There are several efforts required in the areas of innovation, recycling and conservation that could develop a more sustainable consumption way of life on a global scale [10]. A positive attitude is the key to a positive behaviour, therefore it is of true importance that consumers are informed and educated about sustainable consumption and how their behaviour can influence the pursuit of sustainability [11-13].

2. Methodology

This paper reports on a work-in-progress endeavour aiming at developing a comprehensive review of current knowledge in this area since the 1980s. The study adopts the systematic literature review methodology to clarify diverse and confusing theoretical frameworks regarding the so-called attitude–behaviour gap of millennials in respect to sustainable consumption. Although originally developed and employed in medical and health care studies, SLR are gradually gaining momentum in the social sciences, particularly, in management and business fields [14, 15]. The reason for taking this particular methodology is that this field lacks clarity and consensus. By systematically reviewing the literature, we are able to synthetize existing literature and empirical evidence in a transparent and replicable way, identify areas where knowledge is still scarce and point out future research questions to academics, practitioners and policy makers [16, 17]. This paper reports on the first stage of this SLR. From a range of 1079 papers in total, we identified 128 papers that need to be thoroughly revised. Given that this paper reports on a work-in-progress, it limits itself to the analysis of 26 studies focused on four key concepts: sustainable/green/ethical consumption, attitude-behaviour gap among millennials, fashion consumption and finally, the characterization of Generation Y/Millennials.

2.1. Data collection: Studies were identified by searching three electronic databases – Web of Science, Scopus and Google Scholar – using a combination of key words, search strings (AND and OR) and root terms, e.g. *Sustain* Consum** to include all studies using the words sustainable and sustainability, consumers, consumer, consumption, consumerism, etc. Table 1 presents the array of combinations searched and reproduced in the three databases, from more complex strings to simpler combinations.

Table 1. Key words for data collection.

Green Consum* OR Ethic Consum* OR Sustain* Consum*	AND	Generation Y OR Millennial	AND	Attitude-behavi*r gap	AND	Fashion OR Apparel

2.2. Study selection and data extraction: Criteria to include and exclude studies were defined and applied to the set of 1079 papers: studies published after 1990 in order to grasp the millennials attitudes only, written in English, empirically based on millennials consumption behaviour, and initially, focused on fashion or apparel purchases only. As this criteria was too narrowing, we opened the search to other types of products and services in order to grasp millennials attitudes towards green, sustainable or ethical consumption, in general. After discarding duplicates, within and across databases (196 in total), 983 titles and abstracts were first extracted to a data sheet and thoroughly reviewed and analysed. This process resulted in 128 potentially relevant articles to include in the final SLR, from which 26 are reported in this paper. The steps of data selection and extraction are reported in Figure 1.

IOP Conf. Series: Materials Science and Engineering **459** (2019) 012089 doi:10.1088/1757-899X/459/1/012089

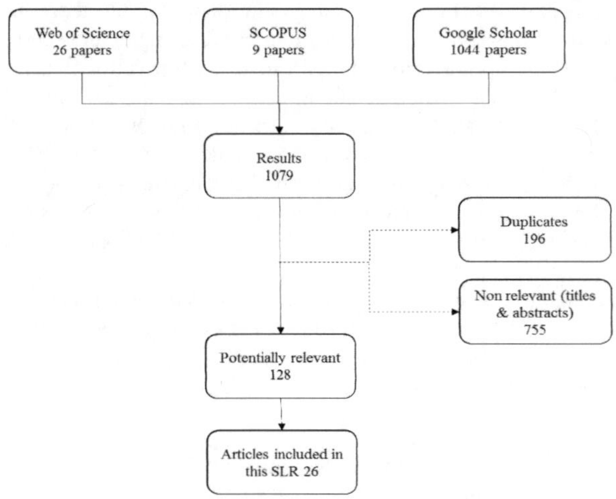

Figure 1. Systematic Literature Review steps

During the selection process, studies focusing only on the supply side of green, ethic or sustainable products/services were excluded, given that the focus of this study is on the demand side, particularly from millennials. Hence studies dealing companies' sustainability strategies or corporate social responsibility or corporate green strategies, were discarded. Studies dealing with green marketing or green supply chains, in general, were also excluded. For each study, data from full texts were extracted, including author(s), publication year, title, key words, research questions, methodology and main results. These were subject to thematic analysis, giving place to four main themes discussed in the following section.

3. Analysis and discussion: key concepts defined

3.1. Sustainable/green consumer behaviour: A large body of literature deals with green, or sustainable, consumer behaviour. These terms are used interchangeably in most of the articles reviewed. This concept is interpreted as an improving pace of consumption that thrives to minimize the depletion of natural resources for future generations, by changing consumers' habits in their purchasing, use and recycling behaviour [12]. It involves the study of the various procedures that individuals, groups or organisations adopt to buy, use and dispose products/services/experiences to meet their needs and the impact of these behaviour on society. Research on the topic started in the 60s by characterizing green and conscious consumers in terms of their socio-demographic features and personality traits. Later on the concept of consumers' pro-environmental attitudes and values was brought to the debate to explain the differences between an eco-centric/biospheric value orientation and an anthropocentric/altruistic value orientation. While the former translates consumers that seek to protect Nature, the latter translates consumers who seek to protect Nature to provide a clean environment for humankind [11]. Attached to green or sustainable consumption, studies also discuss consumers' doubts regarding sustainable brands' credibility and greenwashing that is companies positioning themselves as "green" although they do not behave in a sustainable manner [18].

3.2. Attitude-Behaviour gap in green/sustainable consumer behaviour: Given than presently the estimated market share of green products is less than 4% worldwide [28, 29], researchers call for the need to understand consumers' motivations, concerns and values and design strategies to stimulate green consumption and promote environmental consciousness. As emphasized in the literature, it is difficult to influence behaviour without first impacting attitudes and values, particularly when claims and attitudes may not always translate into actual behaviour [19]. Various attitude-behaviour models have been developed to explain pro-environmental consumer behaviour, namely, the theory of reasoned action [21], the extended theory of planned behaviour [22], the theory of underlying personal and cultural values orientations [23]. The common denominator in what green purchasing is concern, is that consumers who claim themselves as pro-environment try to purchase only green products [24, 25] and are truly committed to engage in efforts to solve environmental problems claiming their willingness to contribute personally to their solution [26]. The concept of self-identity is also suggested as a good predictor of green/sustainable consumption behaviour and an empowering approach to explain the attitude-behaviour gap of Gen-Y consumers [27].

3.3. Fashion industry's impact on sustainability: Fashion industry is defined in the literature as the set of "companies and individuals involved in the creation, production, promotion, and sale of items that (a) have novel and specific aesthetic and functional properties, (b) trigger psychological reactions related to desire and need, and (c) are adopted by a group of people for a limited amount of time" [12] . For reasons that will be here discussed, fashion apparel consumption have been accused of being unsustainable and detrimental to the natural environment in the sense that entails renewable and non-renewable natural resources depletion, emits dangerous substances into the air, water and land through toxic chemicals used in the production of fibres and textiles and creates unmanageable quantities of postconsumer solid waste sent to landfills [12, 30]. This is because fast fashion apparel, often referred to as throwaway fashion, on one hand, is generally not designed to last long, and on the other hand, due to its lower prices and fast turnaround of fashion trends, encourages consumers to buy more than necessary and to dispose their clothes after only wearing for a few times [31, 32] ultimately ending up in landfills instead of being recycled, re-used or re-purposed [33]. Paradoxically while consumers are becoming more environmentally and socially conscious, fast fashion has led to contradictory behaviour, because the literature shows that most consumers disregard sustainability when buying clothes [11, 31]. In fact, studies show that consumers find it difficult to recognize sustainable brands and are often not willing to pay premium prices for green products or buy unfashionable apparel [11].

3.4. Generation Y or Millennials' behaviour towards sustainable fashion consumption: According to [34, 35] Generation Y or Millennials, accounting for more than 25% of the world's population, are people born between 1980 and 1995-2000. They are characterized by being civic-minded, socially conscious individuals and consumers, searching for quality goods, willing to pay more for brand names and willing to pay extra for sustainable offerings [34]. The majority (77 %) claims to be concerned about the environment [24], for the fact that they were born during an era of environmental consciousness and active calls for action which, in turn, translates into a generation who is more apt to adopt pro-environmental behaviours [5]. Additionally, millennials are capable to influence their parents spending habits, which make them an attractive segment for marketers. However, the literature also shows that Gen-Y consumers exhibit disparities in their consumption behaviour making it difficult to marketers to comprehend their attitudes [27, 37]. One important feature of Gen-y is that they tend to convey their identity through clothing and thrive to be seen as having different styles and appearances from their peers to show their uniqueness and self-image [27, 35, 36]. Reasons attached to self-esteem are also present in the literature, to justify millennials willingness to spend large amounts of money on clothing [27]. Hence, apparel retailers acknowledge the importance of researching their purchasing habits, but face challenges in trying to segment them accurately. Moreover, as the fashion market deals with changing consumer tastes and styles, understanding the basic psychology of apparel consumption became utterly important [27, 37].

4. Conclusions

This study presented the results of a work-in-progress analysis and synthesis of the body of knowledge so far produced about green/sustainable consumption among Gen-Y consumers. Based on the systematic literature review methodology, the purpose was to uncover areas of inquiry that may help marketers to be innovative to attractive this segment of consumers, already considered as the most influential consumer group in the world [5]. Our review shed some light on the complexity of understanding concepts that are intertwined, namely, sustainable/green consumer behaviour, and attitude-behaviour gap in green/sustainable consumer behaviour, fashion industry's impact on sustainability and Generation Y or Millennials' behaviour towards sustainable fashion consumption. The literature is unanimous on the need to engage in in-depth research about this particularly segment and acknowledges that few researchers have examined Gen Ys' attitudes and behavioural intentions toward green issues [27, 34] which makes the contribution of this endeavour clear and pertinent.

We conclude this paper by collating opportunities for future inquiry in three different clusters. Firstly, this study provides clear features of Gen-Y consumers' purchase intention for green/sustainable fashion apparels, highlighting brand, style, price, and social identity as the most influential factors affecting their behaviour. However, knowledge should be deepened on how this segment develops their commitment towards pro-environmental purchasing behaviours and how their self-identity that is, their values, beliefs and education influence their interpretations of symbolic and functional attributes of the green/sustainable product or service [27, 36]. Secondly, further research should investigate the effects of education on green/sustainable consumption and propose consumer educational programs that will encourage fashion consumers to change their behaviour toward reducing clothing consumption and waste production and to contribute to more sustainable practices integrated into their everyday lives [11, 12, 37]. Thirdly, in terms of implications for practitioners concerned with their green marketing strategies, this study unveils the need to stay connected with Gen-Y consumers using their own formats of communication, such as social media. As millennials are technology-oriented and highly exposed to social influence, tending to attach more importance to environmental messages that are communicated within their social networks [5], companies are advised to communicate their sustainability initiatives in posts, blogs, press releases to encourage consumers to choose sustainable apparel collections [18, 24]. Moreover, as consumers claim to lack knowledge about companies offering sustainable apparel brands and state that they would like to learn more, marketing efforts, campaigns, and promotional activities should be placed on accurate communication messages promoting the benefits of green products and explaining the reasons why they are normally more expensive, due to its enhanced quality and limited availability when compared to non-sustainable products [18, 24, 37]. Hopefully this work-in-progress will stimulate future research on the topic, by integrating existing studies and providing a comprehensive set of compelling questions that needs further inquiry.

Acknowledgments

"This work is financed by FEDER funds through the Competitivity Factors Operational Programme - COMPETE and by national funds through FCT – Foundation for Science and Technology within the scope of the project POCI-01-0145-FEDER-007136".

References

[1] Belz F M and Peattie K 2009 *Sustainability marketing: A global perspective*. Chichester: Wiley.

[2] Bray J, Johns N and Kilburn D 2011 An exploratory study into the factors impeding ethical consumption *Journal of Business Ethics* **98** (4) p 597–608.

[3] Carrigan M and Attalla A 2001 The myth of the ethical consumer - Do ethics matter in purchase behaviour? *Journal of Consumer Marketing,* **18** (7) p 560–578.

[4] Bernardes JP, Ferreira F, Marques A D and Nogueira M 2018 Generation Y's sustainability attitude-behaviour gap Conf. proceedings of CIMODE – 4th International conference on fashion and design Madrid.

[5] Muposhi A, Dhurup M and Surujla, J 2015 The green dilemma: Reflections of a Generation Y consumer cohort on green purchase behaviour. *The Journal for Transdisciplinary Research in Southern Africa* **11** (4).

[6] Birtwistle G and Moore C M 2006 Fashion adoption in the UK: a replication study. Paper presented at the Anzmac Conference, Brisbane, CA.

[7] Zemke R 2001 Here come the Millennials. *Training Magazine* **38** (7) p 44-49.

[8] Cui Y, Trent E, Sullivan P and Matiru G 2003 Cause-related marketing: How Generation Y responds. *International Journal of Retail & Distribution Management* **31** (6) 310-320.

[9] Connolly J and Prothero A 2003 Sustainable consumption: Consumption, consumers and the commodity discourse. *Consumption, Markets and Culture* **6** (4) p 275-291.

[10] Fisk G 2006 Envisioning a Future for Macromarketing. *Journal of Macromarketing* **26** (2) p 214-218.

[11] Gurtner S and Soyez K 2016 How to catch the generation Y: Identifying consumers of ecological innovations among youngsters. *Technological Forecasting and Social Change* **106** p 101–107

[12] Ruppert-Stroescu M, Lehew M, Connell K, and Armstrong C 2015. Creativity and Sustainable Fashion Apparel Consumption. *Clothing and Textiles Research Journal* **33** (3) p 167–182.

[13] Jang Y J, Kim W G and Bonn M A 2011 Generation Y consumers' selection attributes and behavioral intentions concerning green restaurants. *Int. Journal of Hospitality Management* **30** (4) p 803–811.

[14] Nill A and Schibrowsky J A 2007 Research on marketing ethics: A systematic review of the literature. *Journal of Macromarketing* **27** (3) p 256-273.

[15] Thorpe R, Holt R, Macpherson A and Pittaway L 2005 Using knowledge within small and medium-sized firms: A systematic review of the evidence. *International Journal of Management Reviews* **7** (4) p 257-281.

[16] Pittaway L, Robertson M, Munir K, Denyer D and Neely A 2004 Networking and innovation: a systematic review of the evidence. *International Journal of Management Reviews* **5** (6) p 137-168.

[17] Tranfield D, Denyer D and Smart P 2003 Towards a methodology for developing evidence-informed management knowledge by means of systematic review *British Journal of Management* (14).

[18] De Lenne O and Vandenbosch L 2017 Media and sustainable apparel buying intention. *Journal of Fashion Marketing and Management: An International Journal*, **21** (4).

[19] Kanonuhwa M and Chimucheka T 2014 Green Marketing and Purchase Behaviour of Generation Y-Consumers. *Mediterranean Journal of Social Sciences* **5** (20) 2785.

[20] Kaiser F G, Wölfing S and Fuhrer U 1999 Environmental attitude and ecological behaviour. *Journal of environmental psychology* **19** (1) p 1-19.

[21] Mannetti L, Pierro A and Livi S 2004 Recycling: Planned and self-expressive behaviour. *Journal of environmental psychology* **24** (2) p 227-236.

[22] Kalafatis S P, Pollard M, East R and Tsogas M H 1999 Green marketing and Ajzen's theory of planned behaviour: a cross-market examination. *Journal of consumer marketing* **16** (5) p 441-460.

[23] Soyez K 2012 How national cultural values affect pro-environmental consumer behavior. *International Marketing Review* **29** (6) p 623-646.

[24] Karavasilis G, Nerantzaki D, Pantelidis, P, Paschaloudis, D and Vrana V 2015 What generation y in Greece thinks about green hotels *World Journal of Entrepreneurship, Management and Sustainable Development* **11** (4) p 268-280.

[25] Laroche M, Bergeron J and Barbaro-Forleo G 2001 Targeting consumers who are willing to pay more for environmentally friendly products *Journal of consumer marketing* **18** (6) p 503-520.

[26] Dunlap R and Robert E 2002 *Environmental concern: conceptual and measurement issues*. In Dunlap and Michelson eds Handbook of Environmental Sociology. Westport, CT Greenwood Press

[27] Valaei N and Nikhashemi S R 2017 Generation Y consumers' buying behaviour in fashion apparel industry: a moderation analysis *Journal of Fashion Marketing and Management: An International Journal* **21** (4) p 523-543.

[28] Ritter A, Borchardt M, Vaccaro G, Pereira G and Almeida F 2015 Motivations for promoting the consumption of green products in an emerging country: exploring attitudes of Brazilian consumers. *Journal of Cleaner Production* (106) p 507-520.

[29] Severo E, Guimarães J and Dorion E 2018 Cleaner production, social responsibility and eco-innovation: generations' perception for a sustainable future *Journal of Cleaner Production* (186) p 91-103.

[30] Chen H L and Burns L D 2006 Environmental analysis of textile products. *Clothing and Textiles Research Journal* (24) p 248–261.

[31] Presley A and Meade L 2018 The Business Case for Sustainability: An Application to Slow Fashion Supply Chains. *IEEE Engineering Management Review* **46** (2).

[32] Pookulangara S and Shephard A 2013 Slow fashion movement: understanding consumer perceptions—an exploratory study *Journal of Retailing and Consumer Services* **20** (2) p 200–206.

[33] Kozlowski A, Bardecki M and Searcy C 2012 Environmental impacts in the fashion industry *The Journal of Corporate Citizenship* **45** p 17–37.

[34] Tang C and Lam D 2017 The role of extraversion and agreeableness traits on Gen Y's attitudes and willingness to pay for green hotels. *Int. Journal of Contemporary Hospitality Management* **29** (1) p 607-623.

[35] Naderi I and Van Steenburg E 2018 Me first, then the environment: young Millennials as green consumers. *Young Consumers* **19** (3)

[36] Piacentini M and Mailer G 2004 Symbolic consumption in teenagers' clothing choices. *Journal of Consumer Behaviour: An International Research Review* **3** (3) p 251-262.

[37] Jung S and Jin B 2016 From quantity to quality: understanding slow fashion consumers for sustainability and consumer education. *International journal of consumer studies* **40** (4) p 410-421.

IOP Conf. Series: Materials Science and Engineering **459** (2019) 012090 doi:10.1088/1757-899X/459/1/012090

Millennials: is 'green' your colour?

J P Bernardes[1], F Ferreira[1], A D Marques[1] and M Nogueira[2]
[1]Universidade do Minho, Escola de Engenharia, Departamento Têxtil, Campus de Azurém, Guimarães, Portugal.
[2] IPAM - Universidade Europeia, Rua Manuel Pinto Azevedo 748, Porto, Portugal, joaopedro_bernardes11@hotmail.com

Abstract. Millennials are defined as the group of people born between 1980 and 2000 and maintain a positive attitude in relation to sustainability. However, there is a contradiction between how Millennials think and act when it comes to sustainable consumption. It is interesting to determine which factors can influence their sustainable attitude and behaviour, and those influences can be personal and situational. Six different factors were selected from the literature review and were subject of analysis. The research method will be through three different focus groups with Millennial participants. After the analysis of the focus groups it was possible to understand that Millennials show a very positive attitude towards sustainability but this attitude is not translated in 'green' footwear purchasing behaviour.

1. Introduction

The ecological footprint that humans leave on Earth is a complex concept and it embodies the space that each of us takes up on the planet, how does it become necessary to produce to satisfy the individual consumption, the amount that each individual consumes of energy and the waste produced by each person in a certain space of time [1]. It is necessary to use the natural resources of our planet with enough awareness, so that future generations are not affected by the current lack of sustainable planning [2]. Associated to a sustainable development is the concept of sustainable product. This concept can be misleading because the impacts that any product has on social and ecological environment depends not only of its production, but also of its use by consumers [3]. In fact, the attitude and behaviour of consumers are in line with the rapid production. Encouraged by low prices and heavily influenced by marketing campaigns and constant changes of trends, consumers tend to speed up their fashion consumption [4]. Consumption of footwear in Portugal increased in 2016 to 57 million pairs, the highest figure in the last six years. In 2017 Portugal produced 82 million pairs of footwear and is an industry that now accounts for 3.4% of value added in Portuguese manufacturing industry [5]. Experts blame fast fashion for the dramatic expansion of our wardrobes [6]. Consumers are an important part of the fashion system and can create a valuable influence in the pursuit of sustainability in the fashion industry. [4]. The younger consumers, the generation Y (Millennials), is defined as a group of people born between 1980 and 2000 [7] and maintain a positive attitude in relation to sustainability in general. However, there is a clear contradiction between how Millennials think about sustainability and what they do when it comes to a sustainable consumption [8].

Since only studying the 'green' attitudes will not produce conclusive results on the behaviour of young consumers to buy sustainable products, it is interesting to determine what factors influence the attitude-behaviour relationship. Several studies suggest that there is a difference of attitude of 'green' purchasing behaviour due to the complex nature of personal and situational influences [9]. Personal factors are internal influences. The attitude is a personal factor important in the purchase of sustainable products but doesn't explain why the young

IOP Conf. Series: Materials Science and Engineering **459** (2019) 012090 doi:10.1088/1757-899X/459/1/012090

consumers don't buy this type of products. Other examples are awareness, trust, priorities, emotion and control [9]. Situational influences are the external influences that a person cannot control, but which affect the relation attitude-behaviour. Examples are: time, opportunity, money or the ability to perform the desired behaviour [10].

2. State of Art

Millennials' purchasing power is rising and will be at its peak between the period of 2020 and 2050 [11]. The methodology adopted in this paper is qualitative through the use of three focus group, all participants still with no established purchasing power, but they will be the adult consumers of the next 30 years. So, for companies its necessary to understand them now, in order to know how to establish lasting connections in the near future. In order to understand Millennials, some influencing factors were selected and will be further subject of analysis throughout this paper, and they are:

2.1. Consumption habits

Habits are behavioural routines that are repeated regularly and tend to develop subconsciously. A habit is a fixed form of thinking, acting or feeling and is determined by the repetition of a prior experience. When a behaviour is often performed, can conduct an immediate behaviour under the control of a usual process [12].

2.2. Economic availability

The author [9] perceives a relationship between income and ecological footprint. Consumers with higher incomes have a negative footprint, while consumers with lower income have a positive footprint. The price of 'green' products depending on the economic availability is the most common barrier for consumers not buying 'green' products or services. However, the amount of expendable money also seems to be a personal factor since it is related to the willingness to pay a certain price [13].

2.3. Physical availability

The physical availability of 'green' products in relation to the time that a consumer has or is willing to spend on shopping, influences the amount of 'green' purchases 'made. Physical availability is related to the ease or difficulty of obtaining or consume a certain product. Consumers want to find sustainable products in regular stores, alongside conventional products [14].

2.4. Personal Benefits

Personal benefits refer to the efforts that a person has to do to satisfy their needs and desires. Consumers will assess whether the expected behaviour corresponds with their own personal concerns. There is the tendency to decide in favor of their own interests, which is thus related to the fact that consumers often fail to estimate the actual impact of buying 'green' in their lives [15]. A consumer will buy a 'green' product, when in addition to the environmental benefit, they realize some benefit individual direct in acquiring the product [16].

2.5. Consumer Consciousness

This factor is defined as the amount of time that was spent in the processing of information about 'green' and sustainable products. It is important that consumers are aware of the fact that buying 'green' can be a criterion of purchase and should also pay attention to the communication of organizations about products of this type [17]. Consciousness holds a key role in the acquisition of sustainable products and is a bonus when making a 'green' purchase. When consumers are not aware of the existence of these types of products, they will never be able to form strong attitudes and intentions regarding the purchase of sustainable products. So, the level of consciousness influences the relationship between 'green' attitudes and purchase intentions [18].

2.6. Personal perceived importance
The personal perceived importance is a type of attitude that explains whether consumers consider sustainability issues important to them, or as a problem of society as a whole. So, the self-involvement level with the environment determines the amount of 'green' purchases consumers do. Consumers can have a very positive attitude in relation to sustainability, but at the same time feel that the preservation of the environment isn't their responsibility [19].

As will be explained in the next section of this paper (3. Methodology), this paper will analyze each of the influencing factors stated above through the focus group methodology [20].

3. Methodology
The main goal of this paper is to study the Generation Y's perceptions regarding sustainability and their 'green' consumption habits of footwear in Portugal. The research method will be through the analysis of three focus group only with a total of 30 Millennial participants. The analysis of the focus groups allows a better understanding between the gap of what Millennials think about sustainability and how they perceive it, with their actual consumption habits regarding footwear. In order to design a solid focus group, the influencing factors selected from the literature review will be further subject of analysis on the focus groups. This paper will analyze each of the influencing factors allowing a better understanding of the Generation Y's sustainability perceptions and consumption habits in the footwear industry in Portugal.

4. Results and Discussion
4.1. Millennials perceptions on sustainability and 'green' products
This first section of the discussion is focused in understanding Millennials' thoughts on sustainability and 'green' products in general. After these first three questions is the analysis and results to each of the influencing factors selected from the literature review. All of the questions below were asked in the three focus groups.

<u>What do you think sustainability means?</u>
Some of the participants only mentioned the waste that a person leaves behind but the majority agreed that "Sustainability is all about balance. Its using the resources in a way that satisfies our needs and at the same time it doesn't harm the future generations". Every participant seemed to understand the concept of sustainability concerning the environmental aspect, but lacked to mention the social and financial pillars inherent to the concept. A more complex and complete definition of Sustainability was explained to them in order for them to fully understand the following questions.

<u>Who do you consider the main responsible for sustainability? How/Why?</u>
Every participant agreed that the main responsible are the Humans. They justified by saying that "We are the main users of its natural resources and the only ones with the ability to manage them." But they also felt like the responsibility isn't all on the consumer, stating that "our buying habits and behaviours are a reflection of the education that has been given to us by society". They explained that they don't feel educated enough by brands and companies: "The impacts of every purchase needs to be explained to consumers, so that we can make an informed and aware decision and know what the real repercussions of our actions are. Brands should worry about educating consumers more than selling to consumers".

<u>What are your thoughts on 'green' products? Why?</u>
Every participant described 'green' products as a product with higher quality but also higher prices. And "due to these higher prices, we opt for the easiest and more convenient choice and don't pay attention to a more sustainable alternative". The long-term personal benefits from consuming 'green' are overweighed by the financial aspect. And also "'When we think of green' products, what comes to mind are always non-appealing products. There is still a lot to explore in terms of production and design. This stigma has to change and we need to be reeducated".

IOP Conf. Series: Materials Science and Engineering **459** (2019) 012090 doi:10.1088/1757-899X/459/1/012090

Do you buy 'green' products? What are the benefits you find in doing so?
Answers were divided in this question. Some of the participants said "Yes" because they choose to buy fruits and vegetables directly in farms or local producers but the majority answered negatively. Regarding the benefits from consuming 'green', Millennials answered that "there are a lot of health related long-term benefits in consuming 'green' products, but right now we don't have the financial power to do so".

4.2. Millennials' influencing factors in purchasing 'green' footwear

The next section of the discussion is the analysis of each of the influencing factors regarding the consumption of 'green' footwear:

4.2.1. Consumption habits

Do you have the habit of buying sustainable footwear? Why/Why Not?
Every participant responded negatively with the opinion that footwear brands should offer a quality guarantee for their products, with maintenance and repair services due to the materials, therefore justifying the higher prices. Also, it would be very important if brands explained and educated consumers on why a certain model is more durable and sustainable (materials, production and process). One big barrier from buying 'green' footwear comes from the lack of communication between consumers and brands. They are not aware if footwear brands have or haven't got 'green' lines, and they believe this is the brand's fault and not only theirs.

Would you be able to start buying 'green' footwear?
"Yes" was the answer from everyone. They justified by saying that "it lasts longer, has higher quality and you're helping not only yourself but the environment". The main barrier that is still keeping them from acting on their positive attitude towards sustainable consumption are the higher prices.

4.2.2. Economic Availability

Do you think 'green' footwear is worth the extra financial effort? Why/Why Not?
All responded positively but only if the price isn't much higher than 'normal' footwear: "Nowadays prices between regular and 'green' products are very different. A slightly higher price seems fair because of the sustainable materials inherent in producing a 'green' shoe. We know it will last longer, it's an investment, but not if the prices are as high as they are now".

4.2.3. Physical Availability

Can you identify any footwear brand with a 'green' line at stores or online?
This factor also had unanimity because no participant was able to name a sustainable footwear brand and therefore they don't have the ability to identify a 'green' footwear model. They said that "We can't spot a 'green' shoe so we don't really know if we've seen one before".

4.2.4. Personal Benefits

What are the personal benefits you believe there are in consuming 'green' footwear? Why?
To this question there were two important statements, shared by two participants but with general agreement by everyone were: "Health. No only ours but everyone else's, because in consuming 'green' we are reducing our ecological footprint thus helping the planet"; "Space for all. Because the less we consume, the less its produced to satisfy the individual consumption. And with less production, the less waste and less space we take in our planet".

4.2.5. Consumer Consciousness

When it comes to producing footwear, what are the materials you think are the most eco-friendly? And the least?
Participants felt uncertain with this topic so only a few answered this question. Their answers to the most eco-friendly materials were "Cotton", "Cork" and "Synthetic leather". Regarding the least they responded "Leather". It is clear that they don't search for this type of information and therefore have very limited knowledge in sustainable materials.

IOP Conf. Series: Materials Science and Engineering **459** (2019) 012090 doi:10.1088/1757-899X/459/1/012090

Before buying footwear, do you see if it's made with sustainable materials? Why/Why Not?
To this question none of the participants answered positively. This is due to their lack of research on the matter and therefore they can't detect a sustainable shoe from a regular one.

Before buying footwear what are the aspects that can influence more your purchasing decision?
Every participant agreed that the design is the aspect that influences the most when they are buying footwear. Following the design comes the price and then the comfort. The materials were not even mentioned.

4.2.6. Personal Perceived Importance

How important is eco-friendly footwear to you and to the pursuit of sustainability?
They feel that it is "Very important. The term 'friendly' says it all. Sustainable products can improve not only our quality in living but also everyone else's lives". Another participant added that "We know we need to consume in a more sustainable way, but in order to do that we need products that appeal to us. Specially if its fashion! It's not only what the materials are, but how the whole design of the shoe is. We won't buy a product we don't like just because of its impacts of the environment. It has to be a win-win situation and companies still fail to understand that". One student even said that "If I have two identical shoes to choose from, one being sustainable and the other one not, with identical prices, I would opt for the sustainable model without a doubt". Millennials feel that they are very limited choices in sustainable fashion and that, allied to the high price, are the main barriers that prevent them from living a 'greener' lifestyle.

5. Conclusions

After selecting and analyzing the influencing factors that can positive or negatively impact the gap between Millennials' attitude and behaviour concerning sustainability and sustainable fashion, some interesting conclusions are now possible to obtain. Portuguese Millennials are fully aware of their role in society and therefore know that they should consume 'green' and that by doing so they are contributing positively to the environment and improving their lives as well. However, Millennials don't spend too much time researching for sustainable alternatives and have very little knowledge in the matter. In their opinion, the blame for this lack of education is not only theirs but also the brands'. Discussing with them was possible to see that no participant was able to identify a sustainable footwear brand, but showed great interest in buying a 'green' pair of shoes if only there were more options to choose from and the prices are not much higher than 'regular' footwear. It was possible to note that the extra financial effort that is inherent in buying sustainable footwear doesn't justify the personal benefits of being 'green'. Millennials want to be 'green', but they can't afford to be 'green'. They also felt that, even though they don't buy 'green', they can be if they consume in a smart way. One of the participants even added that "consumers don't need to have all of their pairs of shoes made with sustainable materials, but if they are taken care of they can last longer. Maintenance is key in achieving a more sustainable lifestyle too. It's all about balance".

After analyzing the factors above it is possible to conclude that even though Portuguese Millennials have a very positive attitude towards sustainability and 'green' products, this attitude is not reflected in actual behaviour. The main limitation of this study is that the results from the analyzed sample are 100% Portuguese, therefore it is not possible to extrapolate these results to other countries. Even though it is the same generation worldwide, each culture has its own impacts on the Millennials education, and therefore, consumption habits. As future research it would be very interesting to add 'Marketing Efforts' as an influencing factor by analyzing the marketing strategies done by sustainable footwear brands and how they engage with the Millennial generation. Also, it would be interesting to study different ways to increase consumer awareness in what comes to buying 'green', having the 'Marketing Efforts' factor as the starting point.

Acknowledgments
"This work is financed by FEDER funds through the Competitivity Factors Operational Programme - COMPETE and by national funds through FCT – Foundation for Science and Technology within the scope of the project POCI-01-0145-FEDER-007136".

References

[1] Kiperstok A 2005 Sustentabilidade ambiental: produção e consumo; I congresso internacional de cooperação universidade-indústria UNINDU Brasil

[2] Canciglieri J R, Cardoso R and Pereira S 2011 Uma Visão Tecnológica sobre o Desenvolvimento de Produtos e a Sustentabilidade. Congresso Brasileiro de Gestão de Desenvolvimento de Produto, Porto Alegre.

[3] Stegall N 2006 Designing for Sustainability: A Philosophy for Ecologically Intentional Design. Design issues **22** p 96.

[4] Birtwistle G and Moore C M 2006 Fashion adoption in the UK: a replication study. Paper presented at the Anzmac Conference, Brisbane, CA.

[5] APPICAPS 2017 Retrieved on June 2018. Available at: https://www.apiccaps.pt/publications/plano-estrategico/116.html/

[6] Dirksen K 2008 Fashion guide I: Slow Fashion. Retrieved on June 2018. Available at: http://faircompanies.com/news/view/fashion-guide-i-slow-fashion

[7] Zemke R 2001 Here come the Millennials. *Training Magazine* **38** (7) p 44-49.

[8] Cui Y, Trent E, Sullivan P and Matiru G 2003 Cause-related marketing: How Generation Y responds. *International Journal of Retail & Distribution Management* **31** (6) 310-320.

[9] Csutora M 2012 One more awareness gap? The behaviour–impact gap problem. *Journal of consumer policy* p 1- 19.

[10] Ajzen I 2012 The theory of planned behavior. In P A M Lange, A W Kruglanski and E T Higgins (Eds.), *Handbook of theories of social psychology* **1** p 438-459 London UK.

[11] Forbes 2014 Retrieved on June 2018. Available at: The 10 social and tech trends that could shape the next decade. Available at: http://www.forbes.com/sites/sarwantsingh/2014/05/12/the-top-10-mega-trends-of-the-decade/2/#47cbd7e55eae.

[12] Ajzen I, Czasch C and Flood M G 2009 From Intentions to Behavior: Implementation Intention, Commitment, and Conscientiousness. *Journal of Applied Social Psychology*, **39** (6) p 1356-1372.

[13] Bray J, Johns N and Kilburn D 2011 An exploratory study into the factors impeding ethical consumption. *Journal of Business Ethics* **98** (4) p 597–608.

[14] Vermeir I and Verbeke W 2006 Sustainable food consumption: Exploring the consumer "attitude–behavioral intention" gap. *Journal of Agricultural and Environmental Ethics*, **19** (2) 169-194.

[15] Rokka J and Uusitalo J 2008 Preference for green packaging in consumer product choices – do consumers care? *International Journal of Consumer Studies* **32** p 516–25.

[16] Nottage A 2008 The green agenda gets personal. Marketing July p 33–5.

[17] Jones S and Eden C 1981 OR in the community. *Journal of the Operational Research Society* p 335-345.

[18] Silva C and Souza T M 2012 Desenvolvimento de Sistema para Dimensionamento do Aquecedor Solar Popular de Água - *Revista Sodebras* **7** (3) January.

[19] Laroche M, Bergeron J and Barbaro-Forleo G 2001 Targeting consumers who are willing to pay more for environmentally friendly products. *Journal of consumer marketing* **18** (6) p 503-520.

[20] Krueger R and Casey M 2009 Focus Groups A Practical Guide for Applied Research, 4th edition SAGE Publ. Thousand Oaks CA *Harvard Business Review* **37** p 117– 124.

IOP Conf. Series: Materials Science and Engineering **459** (2019) 012091 doi:10.1088/1757-899X/459/1/012091

Pattern Design Methods for Non-Conventional Bodies

R Boldt, M Carvalho, F Ferreira and H Carvalho

University of Minho, School of Engineering, Centre for Textile Science and Technology (2C2T), Campus de Azurem, 4800-058 Guimaraes, Portugal

rachelsagerb@gmail.com

Abstract. The main goal of the study presented in this paper was to compare three different methodologies of pattern design methods. The case study was based on an individual body shape with a severe deformity in the spine with additional different measurements of the bust, waist and hip girths when compared with common standard measurement tables. This methodology comparison aims primarily to identify its performance regarding garment Fit and performance, in particularly, its potential related to electronic sensor integration for vital signals monitoring. Different performance results were obtained, with the best results obtained with the flat pattern block extracted from the 3D digital model surface methodology, followed by the flat pattern block using the studied individual personal measurements and the worst results obtained with the flat pattern block developed from a typical measurements table used by the industry and fashion schools.

1. Introduction

The moment to wear clothes can be considered decisive for the sale of clothing products. During this experiment the consumer evaluates different attributes, aesthetic and technical, from the interaction of his body with the product. According to some authors, Fit performance is considered one of the most important attribute in the purchase decision [1-3]. It is considered that "fit is the ability to be the right shape and size" [4]. It is also directly connected to the anatomy, body volumes [5] and the clothing ability to cover the body [6]. However, the ineffectiveness of clothing adjustment to the body, lead consumers to dissatisfaction and consequently to refute the clothes [2, 7].

The ergonomic factors are the ones with the greatest impact in the flat pattern design process, fundamental for the performance of the garments. Similarly, the physical constitution of the user and his postural dynamics, interferes in the garment design process and consequently in the performance of the garment [8].

Focusing on the range of products with functional attributes, like smart wearable clothes for vital signals monitoring, the right fit to the wearer's body plays a fundamental role in the performance of the integrated electronic system. For this range of products, the correct position of the sensors, at specific points along the user's body, is fundamental for the monitarization process to be performed as expected [9].

In industrial clothing production are commonly used traditional pattern-making methodologies, also referred as flat pattern design. These are based on the transfer of body measurements (3D) to a flat surface (2D) [5,10]. In traditional methodologies, the measurements transferred are based on anthropometric tables. These measurement tables are normally built from data collection of a given sample, in order to serve a segment of the population or a target audience [1, 10, 11]. The flat pattern blocks still have the function of acting as guides for the stylization of different models, corroborating to the success of fit in a collection [1].

The traditional development of flat pattern blocks methods is very dependent on the accuracy of the measurements of the anthropometric tables. Therefore, several studies of data collection are being done worldwide in order to understand and contemplate the largest number of individuals. However, despite the proposed measurement tables consider a large population sample, they often fail to satisfy the continuous changing body of the population. Furthermore, the individuals with unconventional characteristics are excluded and marginalized from industrial consumption [5, 12]. Similar difficulties occur during anthropometric standard identification in high ethnic diversity populations [13].

The use of technological tools such as three-dimensional digitization supported by 3D body scanners and virtual prototyping enabled by 3D CAD software provides relevant advances related to the efficiency of ergonomic validation of projects in several product segments [1, 5, 10, 14, 15]. These tools have been available in the market for more than a decade, however, they have regular evolutions and expansion of possibilities of use, expecting continuous research advances in this area.

These technologies act as external agents that present great potential for the expansion of the methodological possibilities of developing flat pattern blocks. They help and broaden the understanding of the users' individual needs, as well as contribute to the definition of design decisions during the process of developing new products.

The present paper aims to explore the contribution of these technological pattern design methodologies, comparing its development efficiency in creating clothes to people whit non-standard bodies.

2. Methodology

The research procedure of this study falls within the exploratory field of qualitative character. The techniques adopted were; direct documentary research with a bibliographical survey and experimental research, with direct documentation of the data obtained from the development of pattern design blocks through three different methodologies: (1) flat pattern blocks developed using the "Esmod" methodology with the measurements of the "Alvanon" European mannequins (size 40 and 48); (2) flat pattern blocks developed using the "Esmod" methodology with the studied individual personal measurements; and (3) flat pattern blocks extracted from the digitized 3D body surface of the studied individual. This study aims to compare different working procedures and the fit and performance efficiency of each one, in the individual context, when attending a population target with a non-standard body shape.

3. Results and discussion

The experiment consisted in comparing garment fit performance from three different flat pattern design methodologies. Additionally, aims to observe the aesthetics and the efficiency of each method in the ability to reference points in function of the body design, as a future goal, to predict the location of sensors intended for smart wearable clothing.

The experiment studied the interaction of three variables: Body shape, textile materials and flat patterns. The first two variables are dependent and the last independent [16]. Because it represents a preliminary study related to the new technological possibilities of clothing development, it was decided to virtually simulate the body shape, the textile material and the prototype fit evaluation. The software Clo 3d version 4.1 was used in this study.

The variable "Body shape" simulates the female body shape with severe deformity in the spine and girth dimensions of bust, waist and hip, different from the standard measurements imposed by the anthropometric measurements table used in this study. The body shape simulation was done from the stylization of the standard "Avatar" available in the 3D Clo software. New dimensions of bust, waist and hip girths were defined for the used “Avatar” and the displacement of the spine was defined to enhance the differences to a standard body, producing an unique 3D body image.

IOP Conf. Series: Materials Science and Engineering **459** (2019) 012091 doi:10.1088/1757-899X/459/1/012091

For the textile material variable, a highly stable fabric was used in order to minimize the deformation of the flat pattern, and thus confirm its ability to translate the body size and shape to the two-dimensional plane. In this experiment was used the virtual simulation of 100% cotton twill. The simulation was based on real fabric samples and simulated in the software, according to the system manufacturer's instructions. Additionally it was applied a plaid print. The print was designed to aid during clothes fit visualization on the body and consequently its potential referencing.

The first sample garment was developed from the flat pattern blocks developed using the "Esmod" methodology with the measurements of the "Alvanon" European mannequins (size 40 and 48); The body shape and the measurements used were defined ty the anthropometric study carried by Alvanon for the European population, using its mannequins sizes 40 and 48. Two scenarios were presented: Method 1 - Meet the dimensions of the bust; and hip girths ignoring the waist girth; (referring to size 40) or Method 2 - Meet the waist girth dimension ignoring the bust and hip girths (referring to size 48). It was decided to evaluate both methods. The 2D drawings were directly designed on a 2D Cad System and subsequently simulated in a 3D environment. The results of the simulation wear are presented in Figures 1 and 2.

It was possible to infer that Method 1 had too much trouble in adapting itself to the body shape under study. It was possible to observe too much tension in the abdominal area. Because of that, this garment cannot be dressed.

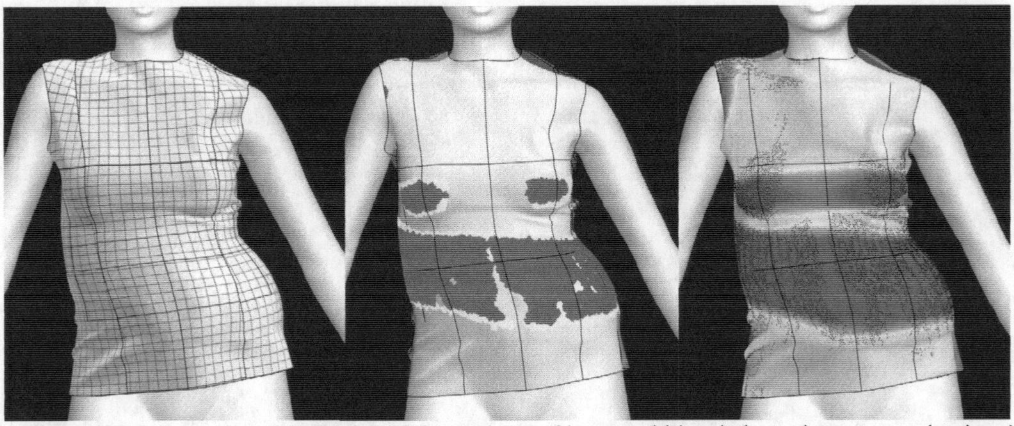

Figure 1. Methodology (1) – Method 1 - Dimensions of bust and hip girths as in mannequin size 40, ignoring the waist girth.

In Method 2, using the measurements table for size 48, we can observe that the garment is wearable, as it does not present too much compression on any part of the body. However, because of the large measurements of size 48, the garment was too baggy in the bust and hip regions, resulting in the sensors inability to operate in this garment, due to the lack of contact between the sensors and the body.

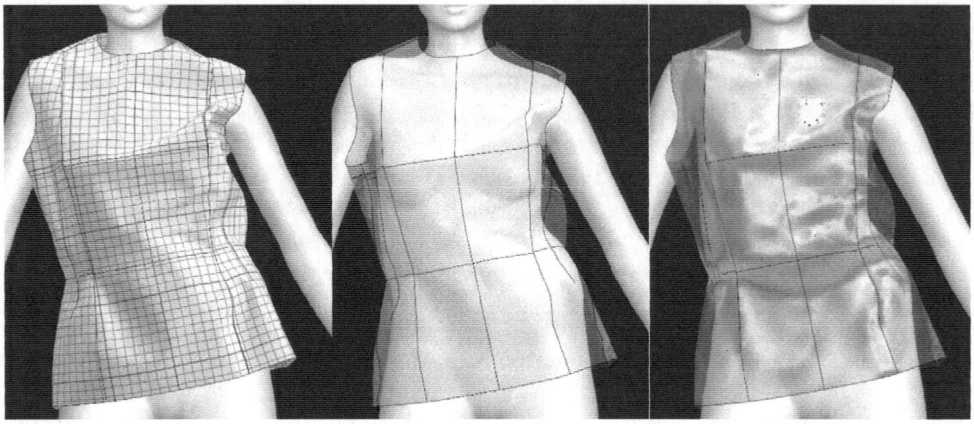

Figure 2. Methodology (1) – Method 2 - Dimensions of waist girth as in mannequin size 48, ignoring the bust and hip girths.

The second sample garment was obtained from the flat pattern blocks developed using the "Esmod" methodology, using with the waist, hip and bust girths measurements according to the individual under study personal measurements. In this experiment were maintained the same procedures of drawing and simulation as in the previous ones. The results of the simulation wear are presented in Figure 3.

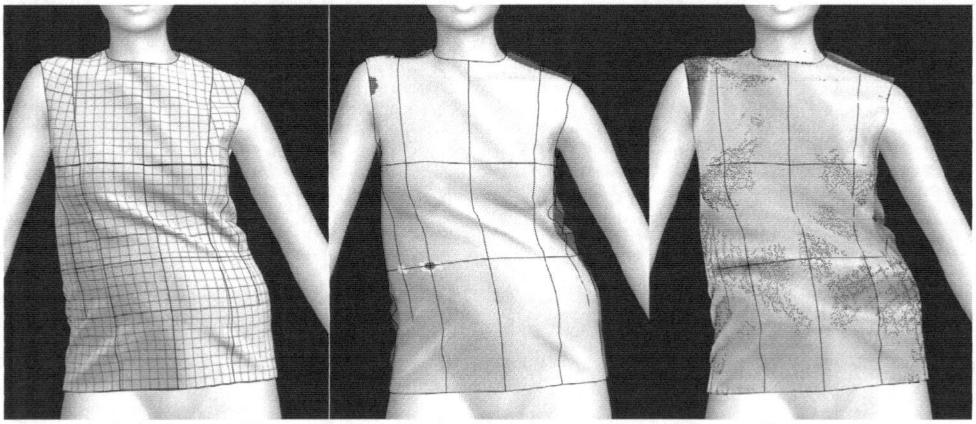

Figure 3. Methodology (2) - "Esmod" methodology with the studied individual personal measurements

It was possible to infer that a better fit result was obtained with this methodology. The virtual garment presented a more harmonious appearance and a better distribution of pressure, when compared with the previous methodologies. However, by ignoring the asymmetry of the user, the fit was compromised, because was too baggy and twisted. This condition will possibly interfere with the location and operation of wearable sensors.

In the third proposal was used the Software Clo 3D version 4.1 for the garment development, with the goal of creating a flat pattern block from the individual's body shape surface under study. The procedure adopted started by importing the 3D body model. Then, using the tools "line (avatar)" which acts on the delimitation of the 3D surface contour to convert into a 2D image, using the software "flatten" tool. The same stylistic orientation of the "Esmod" methodology was maintained. The contours were defined vertically in order to maintain an aesthetic standard for comparison. The results of the simulation wear are presented in Figure 4.

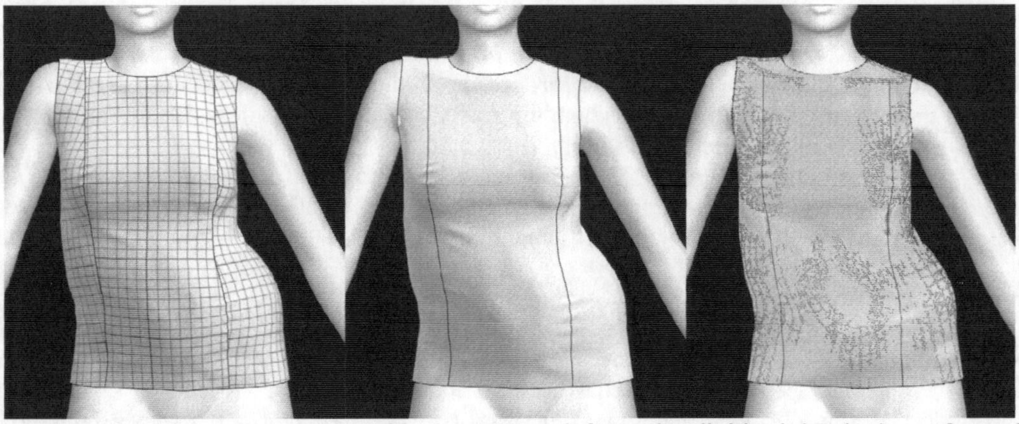

Figure 4. Methodology (3) - Pattern blocks extracted from the digitized 3D body surface of the studied individual

It was possible to infer that this third methodology performed best results regarding fit. The design of the flat patterns when dressed appeared more harmonious and balanced. It respected the particularities of the individual's body shape, resulting in a garment with six unique and not symmetrical flat patterns. We can observe that the garment when dressed, does not present baggy, twisted nor shows excessive compression. We can perceive a much more stable interaction with the body, and consequently, has a greater possibility to assure greater accuracy with wearable sensors. However, further stylisations resulting from the flat patterns block of this methodology are complex, still requiring as reference the traditional styling methodologies. Therefore, further studies in this area are required.

4. Conclusion

This study analysed three different methodologies for the construction of pattern design for the production of clothing for non-symmetrical body, aiming integration of sensors for vital signals monitoring. These methodologies are: (1) flat pattern blocks developed using the "Esmod" methodology with the measurements of the "Alvanon" European mannequins (size 40 and 48); (2) flat pattern blocks developed using the "Esmod" methodology with the studied individual personal measurements; and (3) flat pattern blocks extracted from the digitized 3D body surface of the studied individual. The purpose of this comparison was to verify the results ability in relation to Fit and the capacity of the garment to incorporate wearable sensors when considering a user with individual and non-standardized body characteristics.

It was possible to observe different performances in the three different methodologies, resulting from different interpretations of the same volumes and measurements. The first methodology was less efficient as it ignored the measurements of the individual studied. Resulting in a garment too tight, which made dressing impossible, when size 40 was used, and too loose, when size 48 was used. These results allowed us to conclude that it would be impossible to incorporate the sensors in a garment produced with this methodology and assure their positioning in the places where they are required for vital signals monitoring. Additionally, the aesthetics was very poor in both simulations. The other two methodologies reached a better fit result when compared with the first methodologies. However, they showed differences in fit efficiency and in the ability to pre-set points that meet specific body parts of the individual under study.

The personalised flat pattern block showed better compression distribution, but still has a low performance in relation to the Fit in a non-symmetrical body. As a consequence, it was possible to observe a baggy and twisted garment when clothes are worn. On the other hand, the digital extraction of the flat patterns block methodology presented a better performance regarding Fit. It respected the particularities of the drawing of the individual's body, presenting himself more balanced and stable. In contrast, it presents challenges for future stylizations, as there are no geometric points of orientation in the layout of the base, turning stylization difficult through traditional techniques.

The definition of the ease value and the properties of the materials used are essential for the location of the sensors and the proper functioning of the smart wearables. These variables were simplified in this study but require further research.

This paper presents the initial results from the research surrounding the technological possibilities of pattern design for clothing development, in particularly, its potential related to electronic sensor integration for vital signals monitoring in non-standard bodies.

Acknowledgments
This work is financed by Project "Deus ex Machina", NORTE-01-0145-FEDER-000026, funded by CCDRN, through Sistema de Apoio à Investigação Cientifica e Tecnológica (Projetos Estruturados I&D&I) of Programa Operacional Regional do Norte, from Portugal 2020 and by FEDER funds through the Competitive Factors Operational Program (COMPETE) POCI-01-0145-FEDER-007136 and by national funds through FCT-Portuguese Foundation for Science and Technology, under the project UID/CTM/000264.

References

[1] F. Sabra, *Modelagem, tecnologia em produção de vestuário*, 1st ed. São Paulo, 2009.
[2] O. Rahman and H. Navarro, "Fashion Design for Short Male Consumers Fashion Design for Short Male Consumers," *Des. J.*, vol. **6925**, pp. S2679–S2688, 2017.
[3] E. Frederico, R. R. Torres, G. C. Rodriguez, and B. M. Silva, "Satisfação Com O Consumo De Vestuário Feminino De Tamanhos Especiais," *Gestão Reg.*, vol. **31**, no. 93, 2015.
[4] Oxford, "The Oxford Dictionary." Oxford University Press, 2002.
[5] J. Fan, W. Yu, and L. Hunter, *Clothing appearence and fit: Science and Technology*. A: Woodhead Limited and CRC Press, 2004.
[6] J. Workman and E. Lentz, "Measurement Specifications for Manufacturers' Prototype Bodies," *Cloth. Text. Res. J.*, vol. **18**, no. 4, pp. 251–259, 2000.
[7] L. K. Abraham, "Consumers ' conceptualization of apparel attributes and apparel quality," Iowa Srate University, 1992.
[8] A. Saltzman, *El Cuerpo diseñado: sobre la forma em el proyecto de la vestimenta.* Bueno Aires: Piados, 2004.
[9] A. Paiva, D. Vieira, J. Cunha, H. Carvalho, and B. Providência, "Design of a smart garment for cycling," *Lect. Notes Electr. Eng.*, vol. **505**, no. i, pp. 229–235, 2019.
[10] D. E. Treptow, *Inventando moda: planejamento de coleção*, 3°. Brusque, 2005.
[11] L. da Rosa, "Vestuário Industrializado: Uso da Ergonomia nas Fases de Gerência de Produto, Criação, Modelagem e Prototipagem," PUC Rio, 2011.
[12] Z. Stjepanovic, A. Cupar, S. Jevsnik, T. K. Stjepanovic, and A. Rudolf, "Construction of adapted garments for people with scoliosis using virtual prototyping and CASP method," *Ind. Textila*, vol. **67**, no. 2, pp. 141–148, 2016.
[13] D. G. e S. Andrade, "Calças perfeitas, Investigação experimental sobre modelagem de calças adequada à tipologia física da mulher negróide," University of Lisboa, 2013.
[14] Y. Hong, X. Zeng, P. Bruniaux, and K. Liu, "Interactive virtual try-on based three-dimensional garment block design for disabled people of scoliosis type," *Text. Res. J.*, vol. **87**, no. 10, pp. 1261–1274, 2017.
[15] G. A. Pires, L. C. Paschoarelli, L. M. Pereira, and F. A. P. Scacchetti, "Protótipos Físicos e Virtuais (CAD 3 D): uma Pesquisa Descritiva sobre o Processo de Construção de uma Saia Godê Physical and Virtual Prototyping (CAD 3D): a Descriptive Research on the Process of Building a Flared Skirt," *Des. Tecnol.*, vol. **11**, 2016.
[16] M. J. V. Freixo, *Metodologia Científica : Fundamentos, métodos e técnicas*, 3rd ed. Lisboa, 2011.

Fashion brands strategies and contemporaneous consumer behavior on social media in Portugal

M P Carizani[1] and A D Marques[1]
[1]Universidade do Minho, Escola de Engenharia, Departamento Têxtil, Campus de Azurém, Guimarães, Portugal.

marinacarizani@hotmail.com

Abstract. This study presents two main analysis focused on Portuguese fashion brands strategies and contemporaneous consumer behavior. The first analysis consisted on the case studied and observation of three Portuguese brands (Patrick de Pádua, Luís Carvalho and Nair Xavier) aiming to understand their behavior on social media. During 4 months, Facebook and Instagram were monitored aiming to understand the strategy of each brand and the consequent results. The second analysis was a online survey with 436 valid answers focused on the understanding of Portuguese fashion consumer behavior. The case studies analysis allowed to understand that to have more followers and thus produce more interactivity on a continuous basis with their public, the brands have to create contents consistent with their markets and publics, with the brand language and also with the brand image. The results obtained on the survey allow to conclude that the number participants that share opinions about fashion products and services are still limited. It was also possible to conclude that the majority of the survey participants know several Portuguese brands, but usually don't buy their products.

1. Introduction

Nowadays, people connect, interact and share experiences with other people and brands in just seconds. A vast number of recent tools allow people to interact in the online world without any influence of geographical and demographic boundaries [1]. These aspects trigger an exponential growing of social media users, reaching large numbers worldwide. According to Simon Kemp's study [2], between January 2017 and January 2018, worldwide internet users and active social networking users grew 7% and 13% respectively. In the 90's people started to become conscious about their acts of consumption, not only in terms of the exhaustion of natural resources by the over-high consumption, but also of the pollution that these acts and all their daily activity caused in the conservation of the nature, and consequently of those resources [3]. This concern began to reflect on people consumption habits and towards this new context Paiva and Proença [3] report that in the last decades, particularly from the beginning of the 21st century, a new consumer with a "greener" behavior has emerged. This new consumer is called "green consumer", "citizen consumer" or "conscious consumer".

Carvalhal [4] reported that the factors previously used to define the target public, such as age, gender and social class no longer define the target public neither their behavior. The contemporaneous consumer has more information, access to this information, purchasing power and especially a high critical sense. The contemporaneous consumer does not admit any type of product and each brand needs to have one or more differentiating factors that stand out from the infinite existing options. These aspects lead to a process of personalization in which, an evolution occurs from a child's stage of identification to a state with a superior awareness and differentiation [4]. Accordingly, it is possible to observe this change through the growth of homonymous brands of young designers with greater consciousness of their role in the world, which fit the contemporary consumer behavior by implementing their values on their brands and products. All of these aspects indicate that fast fashion networks and social media approaches have to reposition themselves in the market and work differently on product development lines. The current fast, degradable and unsustainable attitude no longer interest the general public as it was before.

This new attitude introduces a high level of complexity in terms of market segmentation which demands substantial changes on the analyzes of generations and the range age studied to make them able to analyze and to know better this new consumer and, consequently, to help the brands and companies to transmit with more relevance their contents [5]. In fact, this generation of new consumers is transversal to age, gender, nationality, religion and social condition and its distinctive feature is the ability to be informed and conscious about all the above-mentioned aspects and even other.

The outcome of these types of analysis request from companies a capability to adopt a research approach more focused on the human being. The contemporaneous, independently of the role (buyer or consumer) has inherent expectations that no longer agree with a passive consume. He is expecting pleasant experiences and assume a position in which the corporations are accountable for the impact of the products on his body, culture and environment. This revolution generates a wide-ranging transformation in the dynamics between sellers, service providers and customer [6].

This study aims to understand the abovementioned aspect in Portugal, i.e. understand the behaviour/attitude Portuguese brands on social media and also to understand if Portuguese consumer is aligned with the contemporaneous consumer.

2. Methodology Research

In this study two main analysis were performed. The first consisted in three case studies aiming to understand the behavior of three Portuguese brands on social media and its effects. The second analysis consisted on the understanding if the Portuguese participants behavior is aligned with the contemporaneous consumer.

2.1. Patrick de Pádua, Luís Carvalho and Nair Xavier brands: cases study

The first step of this study consisted in the brands selection. The ModaLisboa fashion show is one of the most important fashion events in Portugal and several new designers/brands have the opportunity every year to show their work. To select the brands to be investigated in this work, the brands had to have shown its collection on a recent catwalk ModaLisboa and have also the following requirements:

. Portuguese origin;
. Women's and/or men's wears;
. Be present on social networks Facebook and Instagram;
. Produce a minimum of two collections and a maximum of four collections per year;
. No fast fashion neither slow fashion.

Based on the defined requirements the following three brands were selected: Patrick de Pádua, Luís Carvalho and Nair Xavier. The cases study were carried out with longitudinal analysis of their social networks Facebook and Instagram. This quantitative longitudinal analysis aims to analyse and understand each brand evolution/progress during a period of 4 months and also identify their behavior, patterns and changes.

2.2. Social media behavior, fashion consciousness and Portuguese fashion brands: online survey

The second step of this study aim to understand the social media, the fashion consciousness and the behavior of Portuguese residents about Portuguese fashion brands. The main objective was to understand if in Portugal the consumers attitude is aligned with the Carvalhal and other authors opinion about the contemporaneous consumer. For that purpose, an online survey was used to collect data about the following aspects:

. Behavior of Portuguese consumers in social media;
. Behavior of Portuguese consumers about fashion products;

. Sensibility of Portuguese consumers regarding the issues associated to the fashion industry worldwide;
. Knowledge and consume of Portuguese brands.

The online survey collected responses from 18th to 8th June 2018 on the Google Docs platform. This survey had 30 questions and the participants voluntarily answer to the survey with standardized questions [7]. The survey was sent by e-mail to professors of the University of Minho and shared in the social networks. The survey was available online for 21 days via a link on the Google Docs platform with 24-hour access.

3. Results and discussion

3.1. Patrick de Pádua, Luís Carvalho and Nair Xavier brands: cases study

From January 14th to May 20th, 2018, was carried out a longitudinal analysis of Facebook and Instagram profile of Luis Carvalho, Nair Xavier and Patrick de Padua brads. Figure 1 shows the results obtained through the analysis of the profile of the brands with respect to Facebook. It was analyzed and collected data about the following points: number of followers, number of publications, reactions and likes of followers, number of shares, number of comments, number of replies of the brands in comments and number of brand reactions in comments.

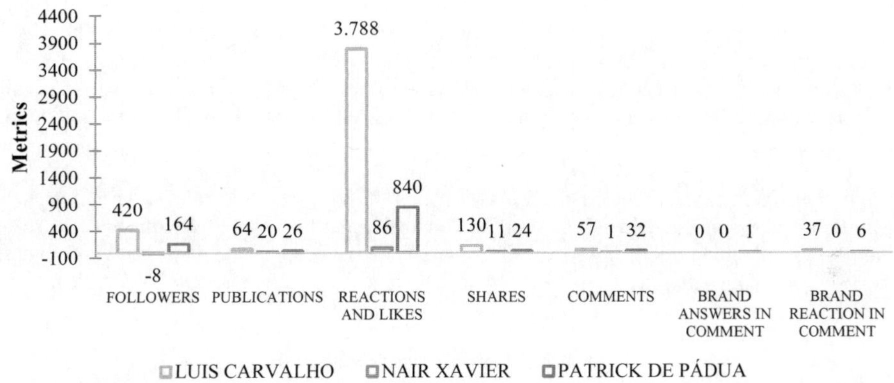

Figure 1. Facebook longitudinal analysis results (from 14th January to 20th May of 2018)

As can be seen in Figure 1, the brand that had the highest increase of the followers were Luis Carvalho, followed by Patrick de Padua. Nair Xavier brand lost a total of 8 followers during the analysis time. When analysing Figure 1, can be seen that the increase of followers is related to the number of contents and interactions of each brands. In fact, Luis Carvalho brand had more publications (64), more reactions and likes (3788), more shares (130), more comments (57), and more reaction in comments (37) when compared to the other brand's metrics. In this sense, it can be comprehended that as high was the capability of the brand to chase the public and as more effective was the interaction to the followers and as high was the number of new followers

Additionally, a longitudinal analysis of the Instagram profile was also carried out. Figure 2 shows the obtained results in which the following metrics were collected: number of followers, number of publications, number of likes and number of comments.

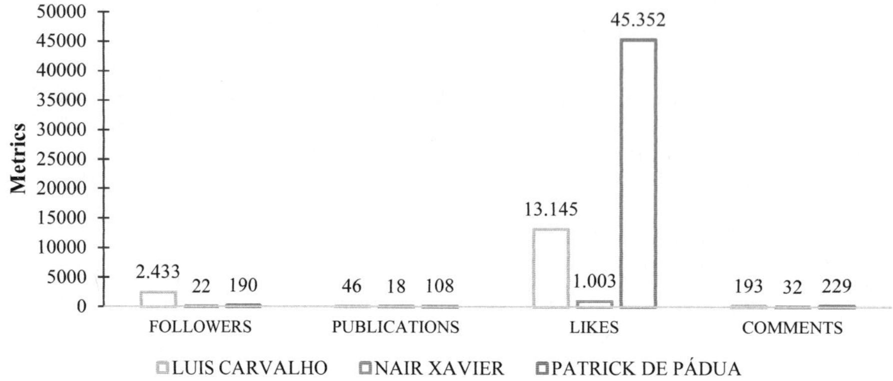

Figure 2. Instagram longitudinal analysis results (from 14th January to 20th May of 2018)

By analyzing Figure 2, some important aspects can be observed and discussed. Firstly, it can be seen that similarly to the tendency identified on Facebook analysis, Luis Carvalho was the brand with the highest increase of followers (2433) followed by Patrick Pádua (190) and, finally, Nair Xavier with 22 new followers. Secondly, it is also possible to perceive that the highest number of publications (108) and likes (45352) was obtained for Patrick Pádua brand. However, these results do not lead to a significant increase of the followers when compared to Luis Carvalho brand and actually, the new follower's number was 190 when compared to 2433 obtained on Luis Carvalho brand.

3.2. Social media behavior, fashion consciousness and Portuguese fashion brands: online survey
The online survey on consumer behavior in Portugal regarding Social Media and conscious consumption collected a total of 436 valid answers. The majority of the answers were obtained from people with ages ranging from 19 to 29, in which 75% were female and 25% males.

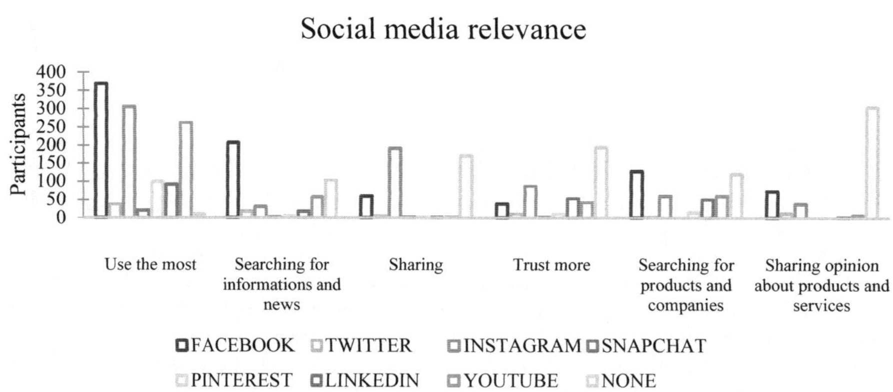

Figure 3. Survey participant's answers regarding the differences on the social media relevance.

Figure 3 resumes the results obtained in regarding to the main aspects related to each social media. It can be observed that, in Portugal, Facebook is the most used social network (85%), indicating that Facebook has the higher time of utilization. Moreover, Facebook is the social media selected by the majority of the participants to find information and news. Regarding the sharing behavior it was observed that 44% of the participants share their daily life using Instagram, 39% of the participants don't share their life on social media and 14% answered that use Facebook for that purpose. Additionally, it was also observed that 45% of the participants do not trust in any social media here suggested.

When analysing the behavior of the participants regarding the searching for products and companies it can be observed that excluding Snapchat, all the social media are used for searching products and companies. Finally, the majority of the participants (70%) do not use the social media to share opinion about products and services. When asked about the origin of the brands, 239 participants confess to know only on certain occasions about the origin of the brands, they consume, 111 don't know about the origin, 86 affirm to know and 11 admit not knowing the origin of the brands.

Another question of the survey was about the consumption of Portuguese fashion brands. The obtained answers were the following: 220 of the participants claim to consume Portuguese brands, 152 declare that they do not know if they consumer or not because of the abovementioned aspect and 64 confess that they are not consumers of Portuguese fashion brands.

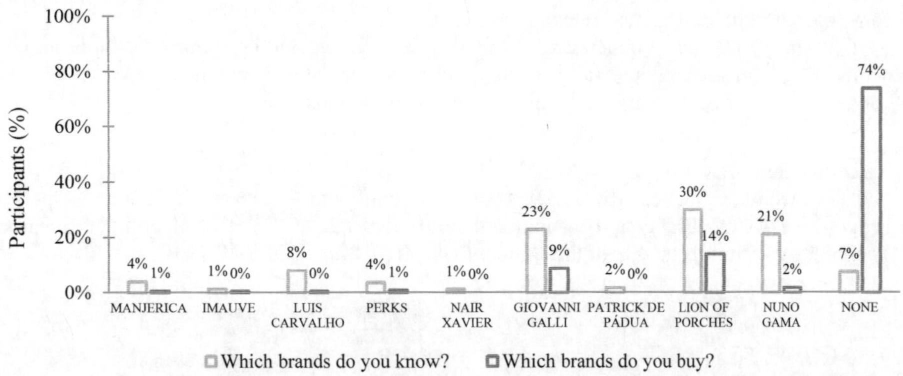

Figure 4. Survey results regarding the Portuguese brands.

Aiming to better understand if the participants know and consume Portuguese fashion brands, the following nine brands from Portugal were selected: Manjerica, Imauve, Perks, Nair Xavier, Giovanni Galli, Lion of Porshes, Nuno Gama and also Luis Carvalho, Nair Xavier e Patrick de Pádua. After considering these 9 brands, two questions were applied as can be seen in Figure 4. It can be understood that the four most well-known brands are Lion of Porches (30%), followed by Giovanni Galli (23%), Nuno Gama with 30% of participants and finally Luis Carvalho with 8%. Similarly, when asked about "Which brands do you buy?", it is important to highlight that even knowing the majority of the nine brands selected, 74% of the participants don't assume that they don't buy these fashion brands products. Only 14% of the participants buy on Lion of Porches and 11% on Giovanni Galli.

4. Conclusions

Social networks users are increasing more and more users and this phenomenon indicates that the more a brand knows who its target audience is, the better it will be able to generate content and interact with its users. In Portugal, according to the analysis performed in this study, it can be concluded that the number of Facebook users is higher when compared to other social networks and also was comprehended that a significant number of people is using Instagram. Despite the higher use of Facebook when comparing to the other social media, it was verified that the level of user-brand interaction is higher on Instagram. The Nair Xavier analysis indicated that Instagram generate a higher interaction user-brand compared to Facebook even with a lower number of publications.

The fashion brands that want to connect directly with their consumers in order to generate value, recognition and interaction should have a higher assertiveness when investing their resources Facebook and Instragram social networks. It is important to emphasize that, as seen on the social media analysis for Luis Carvalho, Nair Xavier and Patrick de Pádua brands, it is imperative for a brand to be not only concerned with the number of likes in its publications. To have more followers and thus generate more interactivity on a continuous basis, is necessary to create content that is consistent with the audience target, with the brand language and also with the brand image. When the communication is not planned and has not a well-defined targeted, it does not lead to an increase of followers even with high number of likes in several publications.

Sharing is one of the main trends of social network behavior worldwide. Through the analysis of the consumer behavior in Portugal it can be concluded that the users are increasing the share of their daily lives and their opinions about products and services on social networks. In contrast, it can be concluded that the number of participants that share opinions about products and services is still very limited due to the apprehension to trust in social networks. However, this apprehension is not visible when considering the sharing of users daily lives in social networks.

In addition, it was also possible to verify that the majority of the survey participant know several brands of Portuguese origin, but do not consume their products. In this regard, it is possible to conclude that in order to generate superior knowledge and relevance in the market, Portuguese brands must invest in social media campaigns aiming to get attention of the contemporary consumer through current and relevant contents.

This study allowed to understand the Portuguese brands and consumer's attitude in a global perspective considering several aspects related to the social media behaviors as well as contemporaneous attitude about fashion products and brands.

Acknowledgments

"This work is financed by FEDER funds through the Competitivity Factors Operational Programme - COMPETE and by national funds through FCT – Foundation for Science and Technology within the scope of the project POCI-01-0145-FEDER-007136".

References

[1] Kotler P, Kartajaya H, Setiawan I 2017 *Marketing 4.0* (Lisbon: Actual Editora)
[2] Kemp S 2018 *Digital In 2018 - Essential insights into social media, mobile, and ecommerc use around the world.* Available at: https://digitalreport.wearesocial.com/. pp. 50-79
[3] Paiva T, Proença R 2011 *Marketing Verde* (Lisboa: Actual Editora) pp. 14-120
[4] Carvalhal A 2016 *Moda Com Propósito - Manifesto pela grande virada* (São Paulo: Editora Paralela)
[5] Morace F 2011 *Consumo Autoral - As gerações como empresas criativas.* 2nd Edition (São Paulo: Estação das Letras e Cores Editora Ltda) pp 9-20
[6] Brown T 2010 *Design Thinking - Uma Metodologia Poderosa Para Decretar o Fim Das Velhas Ideias* (Rio de Janeiro: Elsevier) pp. 104-189
[7] Saunders M, Lewis P, Thornhill A 2008 *Research Methods for Business Students* 5th Edition. (Harlow: Pearson Education Limited)

IOP Conf. Series: Materials Science and Engineering **459** (2019) 012093 doi:10.1088/1757-899X/459/1/012093

Overview of textbooks published for Hungarian fashion design students

E Csanák

Senior Lecturer, Óbuda University Sándor Rejtő Faculty of Light Industry and Environmental Engineering Institute of Product Design, H-1034 Budapest, Doberdó 6

csanak.edit@rkk.uni-obuda.hu

Abstract. In this article, the author introduces two textbooks published for the Hungarian fashion design students. Textbooks *Collection development* and *Design of fashion accessories* were published by the Hungarian National Office for Vocational Training and Adult Education in 2015. The two textbooks created based on the curriculum of same name subjects were issued with a goal to create competitive, high-quality literature for the Hungarian young professionals, in their mother tongue. The overview introduces the contents of the books, presents the some content editing issues, and article announces the opinion of the peer review and introduces some sales and library statistic. The textbooks created for college, BA and MA students, and possibly will match the interest of the professionals and fashion enthusiasts too.

1. Importance of competitive literature in training of young fashion professionals

Fashion surrounds us! The World of Fashion has always been a glamorous profession, attacking thousands of young people. Every year a new generation of designers is looking for their place in this world, thus: entering an accredited institution has been always the culmination of the dreams and the fulfillment of the hopes for the talented. [1]

Fashion is a complex market with cruel rules. Development of a fashion collection created not for its own advantage, but for the benefit of the company, and designed in the spirit of sustainable development requires talented and skilled designer, who is familiar with the marketing of fashion goods. [2] Improvement of a competitive course curriculum associated with valuable literature, is a strategic goal of all national training programs and is also a goal for accredited educational institutions.

2. General intentions of the publishing office, target audience and classification of the textbooks

The existence of valuable literature in the mother tongue of the student is essential in training. Specific fashion literature is usually published in English. Literature, which deals with the process of fashion design, and development of a fashion collection has never been released in Hungarian. Since fashion design education has a significant institutional background in Hungary, the conjecture of the publisher general intention of the National Office was to fulfill this gap, creating two valuable, classified and approved textbooks in Hungarian. This last aspect was exceptionally important, however the publisher's hopes were if arousing the interest of the art-loving audience too, the proposal will definitely result in unique and successful book market products.

Two textbooks have been proposed to be published: *Collection development* and *Design of fashion accessories*. Both books were hired to write Dr. Edit Csanák, the Author of this text.

2.1. General intention of the publishing office

Complex books are rare in the book market, and notable literature is rarely accessible in the mother tongue of the student. This circumstance constitutes a serious problem during the studies. Obtainable university scripts that focus on a specific matter of the profession are usually not available in the free sale, specific texts obtainable via electronic libraries are as well accessible only to a limited number of readers. Therefore, the overall aim to create textbooks that extensively deal (if possible) with almost all the theoretical and practical issues of fashion design, hopefully an attractive, comprehensible and valuable literature has been perceived to result, describing the content on up-to-date practical examples.

Figure 1. Covers of books *Collection development* (left) and *Design of fashion accessories* (right)

2.2. The target audience of the books

The publishing house's intention was that the books will be used as first in college-level teaching[1], as well as on the high-educational (BA and MA) courses. Since specific task has been to write popularly literature, the two books can be offered to all the readers interested in Fashion design.

2.3. Classification of the books

Books *Collection development* (Fig. 1 – left) and *Design of fashion accessories* (Fig.1 – right) were peer reviewed and pedagogical peer reviewed in more steps. They were officially classified and approved as textbooks for Higher Education use in 2014.

3. Proposal development methodology, and contents of the textbooks

Texts that cover more aspects of the design, responding to from the base to specific professional questions, providing information necessary for successful acting on the international fashion scene are not too common. Since precedent for the books did not exist the contents have been edited in more steps and were developed upon the curricula of the same name subjects, based on prominent contemporary fashion literature [3] [4], bearing in mind that the final result has to be as highly professional as easily readable.

[1] This type of education is called OKJ (National Qualification Register = Accredited Educational Program) in Hungary. The courses last two years, and give higher education degree. They are organized by accredited state- and private schools.

3.1. Contents of textbook 'Collection development'

Contents cover all the theoretical and practical questions of collection design. The text is diluted with short stories, and stimulating practical exercises. The subjects can be easily tracked to students and teachers due to the logical table of contents, and the connection with the curricula of a fashion design course. Book has 260 pages and contains 297 color illustrations. Chapters and titles of the book are as follows:

1. **The fundamentals of fashion design:** The concept of fashion, fields and areas. Mechanism, societal aspects, and influence. The vertical principle of spread vs. horizontal.
2. **The fashion industry:** Economic importance. Fields and categories of the Fashion World.
3. **Fashion trends:** The cultural and social background of the phenomena, concept, and features. International committees of style and materials fairs. Types of trend forecasting.
4. **Definition and concept of the collection:** Types of collections (based on consumer demand, concept, the manner of the performance, product type, client need, etc.) Seasonal collection: concept, function. The concept of 'wardrobe essentials'. (Fig. 2)

Figure 2. Illustration No. 72 in the book *Collection development* – Wardrobe essentials

5. **Designing by type of collection:** Types of collections. Aspects of designing: womenswear collection and menswear collections (basic types of the products). Clothing worn by both sexes. Baby clothes. Underwear.
6. **Designing of a multi-product brand collection:** The importance of fashion accessories in the collection and aspects of design. Knitwear and footwear. Fur and lather in the collection.
7. **Designing of sportswear collection:** History of sportswear, product types, advanced materials. Influence of sportswear on fashion. Streetwear, casual, outdoor, and denim.
8. **Special categories:** Workwear, protective clothing, corporate clothing, uniforms, wedding wear; types, aspects of design, fabric- and manufacturing features.
9. **Product design process:** 4 stages of product development. Garment, as a product. The lifespan of textile products. The Garment Supply Chain.
10. **Framing of the concept and establishment of the requirements:** Sources of research. Experimentation. Conceptual solutions: storyboard, sketch-book, and mood-board. Trend information in practice. Requirements, market demands.
11. **The design process:** The creative process. Requirements, and differences of different type of fashion drawing: sketch, fashion drawing, fashion illustration, fashion flat.
12. **The process, conditions and aspects of prototyping:** The sampling process. Selection and procurement of raw materials. Small accessories: buttons, zippers, coat, labels, etc. Decorative elements, graphic design. Management of the collection. Development of tech-packs.
13. **Promotion of the collection:** Types of fashion photography. Preparation for photo-shooting. Types of advertising material. Methods of brand promotion, social media, online marketing.
14. **Final thoughts:** Tasks of the Fashion Designer. The lifespan of the product.

IOP Conf. Series: Materials Science and Engineering **459** (2019) 012093 doi:10.1088/1757-899X/459/1/012093

3.2. Contents of the textbook 'Design of fashion accessories'

This book – as a *'little sister'* of the previous –, is more complex and multifaceted; no similar book can be either mentioned in the international literature. The thematic is in part related to contents of the textbook *'Collection development',* and is principally based on the curriculum of the similar name subjects of the Higher Education institutions, and is based relevant foreign and Hungarian literature.

During the content editing, the correct classification of fashion details was very important, since more fashion accessories are very differently systematized in the professional literature, since the materials and technology applied for their fabrication are assorted, and each type has heterogeneous sorts and varieties! Thus, maybe the greatest value of this book – in addition to the qualitative professional content diluted with historical materials, exciting short stories, and reviews of certain moments of the fashion history, when some accessories were of special, unusual significance – is the clear classification of the add-ons.

Book has 174 pages and contains 204 color illustrations. Chapters and titles of the textbook *'Design of fashion accessories'* – without the ambition to comprehensively mention all – are the following:

1. **About the fashion accessories in general:** Influence of the fashion on the accessories. Importance of fashion accessories for the fashion market. Fashion detail, as a product. The effect of the accessories on the outlook, and their significance in the shaping of clothing. (Fig. 3)

Figure 3. Illustration No. 2 in the book *Design of fashion accessories* – The contribution of fashion details to the overall image

2. **Anatomy of fashion accessories:** The importance of fashion accessories in formation of clothing. (Fig. 3) Sorts and types.
3. **Garment accessories:** Leather accessories. Rods and sticks. Umbrellas and sun umbrellas. Fans. Handkerchiefs.
4. **Fashion details:** Types of footwear. Scarfs, scarves, eyebrows. Ties. Hats, caps, turbans, etc. Gloves and muffins. Socks and stockings.
5. **Jewelry, watches and glasses**
6. **Design process and product design:** Accessory collections based on trend, style, and market demand. Stages of the work: information and inspiration, application of the latest trends, establishment of the criteria. Sketches, designs; 2D, 3D virtual modeling. The prototyping process.
7. **Special materials and specific technologies**
8. **The prototyping process:** The process of making a prototype of leather goods. Making fashionable textiles details of. The process of producing a prototype of shoes.
9. **The fundamentals of fashion jewelry and goldsmith**
10. **Fashion accessory presentation:** Scenes and types of presentation, methods of market introduction.
11. **Final thoughts:** Tasks of Accessory Designer. Product lifespan according to sustainability.

IOP Conf. Series: Materials Science and Engineering **459** (2019) 012093 doi:10.1088/1757-899X/459/1/012093

4. Editing challenges: Text editing issues, visual elements and graphic design

The graphic design, the typography and the page layout, the overall format of a book is as important for the reader, as its content. Visual image is vital in case of fashion design, because readers interested in such field are aesthetically and visually sensitive people; they want to see beautiful images, and attractive design.

4.1. Editing of the content: layout issues

Maintaining the official format of the accredited textbooks was a criterion. In order to it, the Office has given a template appropriate for preparation of a textbook to the author. All the contents were firstly edited in this template, which contained detail instructions according to the general formatting rules, the rhythm of the text, the content of the entire chapter, the format of the figures (illustrations).

The template has given suggestions to the author, for instance how make the text more readable by inserting short stories and curiosities. Each chapter is to be completed with feeds, and the answers to the textual ones should be at the end of the books.

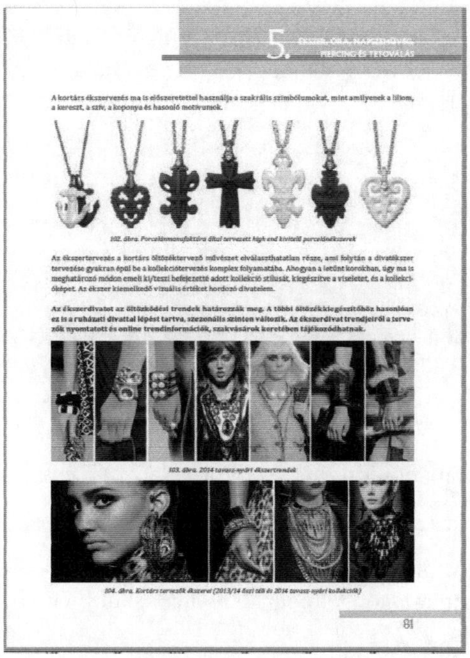
5.

A kortárs ékszervezés ma is előszeretettel használja a szakrális szimbólumokat, mint amilyenek a liliom, a kereszt, a szív, a koponya és hasonló motívumok.

Az ékszertervezés a kortárs öltözéktervező művészet elválaszthatatlan része, ami folytán a divatékszer tervezése gyakran épül be a kollekciótervezés komplex folyamatába. Ahogyan a letűnt korokban, úgy ma is meghatározó módon emeli ki/teszi befejezetté adott kollekció stílusát, kiegészítve a viseletet, és a kollekcióképet. Az ékszer kiemelkedő vizuális értéket hordozó divatelem.

Az ékszerdivatot az öltözködési trendek határozzák meg. A többi öltözékkiegészítőhöz hasonlóan ez is a ruházati divattal lépést tartva, szezonális szinten változik. Az ékszerdivat trendjeiről a tervezők nyomtatott és online trendinformációk, szakvásárok keretében tájékozódhatnak.

81

Figure 4. Sample page layout – Pg. 81 from the book *Design of fashion accessories*

After the revision of the draft, in order to maintain the aesthetical level the author edited the content again, wrapping the text and placing the images. The page layout should been created according to a regular format of the series of textbook of the publishing office. However, soon became clear, that the standard page layout could not be accessed and cannot be one-in-one adjusted to this specific task; usage of the template didn't resulted the desired layout.

4.2. Image editing issues

The two books contain altogether 501 color illustrations by licensed usage of the campaign- and product images of 21 designers and 42 companies. Over 280 graphic montages have been edited in various vector graphics and raster graphics editors, and further 10 diagrams and graphs give answer to specific issues. Further 160 pictures have been canceled, changed and re-conceptualized during the work. All the illustrations have been created by the author.

Images are typically wide, breaking the text over and above, contributing to aesthetic page layout (Fig. 4). The rhythm of the images supports text memorization and helps the learning process.

5. Summary of the results, and analysis of the statistic

Peer rewievers of the books were emphasizing the value and quality, highlighting their overall contribution to the training: *'A very exhaustive, detailed analysis that encompasses almost every field of expertise. Topics integrated into tasks, are to deepen the gained abilities, encouraging students to take new ones. Quotations that capture entire chapter, and typically brilliant examples and outstanding images, make the book a pleasant read. Interesting and readable short stories enable versatile student orientation, and providing great support for the teachers.'* (Judit Bányász, Textile designer, reviewer of a textbook '*Collection development*'). *'Outstanding Book! The writing style is dynamic, the structure is logical, the illustrations are interesting. The book reads with pleasure. Author of the book not only with the text but also with the pictures helps students to acquire the necessary competencies. Tasks are interesting, issues are coherent to the text of the textbook. With its look and content, the book is in line with the age of potential readers, who are already willing to become fashion designers when reading the introduction.'* (Gabriella Dr. Parajné Toth, the pedagogical peer reviewer of the textbook *'Design of Fashion Accessories'*)

Both books have been declared as textbooks year 2015, and their national distribution is provided by the Library Supply Non-Profit Ltd (KELLO). Statistic statement (Hungarian National Office for Vocational Training and Adult Education, January 2018) reports on about nearly 170 pieces sold. (Fig. 5)

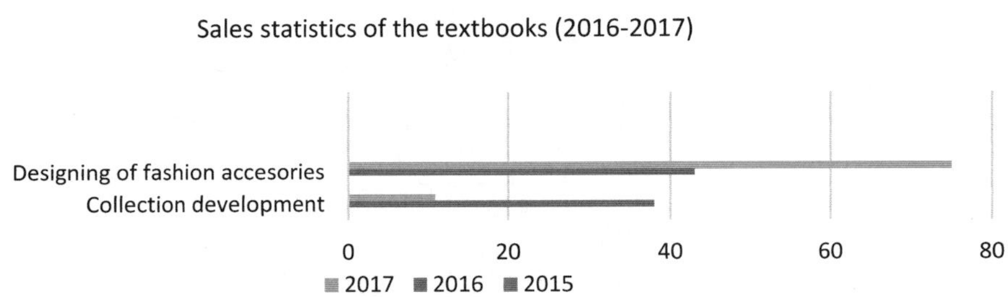

Figure 5. Diagram of sales statistics of the books *Collection development* and *Design of fashion accessories*

Statistic statement of the University Library of Óbuda University reports about rising tendency of borrowing; both books are popular among students, and constantly are on the waiting list...

6. Conclusion

High-quality professionally literature helps the education of domestic Fashion Intelligence. The two Hungarian textbooks '*Collection development*' and '*Design of Fashion Accessories*', introduced in this article, were published for college and university students by the National Office for Vocational Training and Adult Education (NSZFH, Budapest), year 2015. The article attempted to draw attention to the significance of high-quality professionally literature in the education, providing insights into the chapters of book, and highlighting the complexity of writing and editing of this kind of writing.

References

[1] E Csanák, Kollekcióalakítás, Budapest: NSZFH, 2015
[2] E Csanák, Öltözék kiegészítők tervezése, Budapest: NSZFH, 2015
[3] R Sorger, J Udale, Choosing a concept - The Fundamentals of Fashion Design, AVA Publishing, 2006
[4] A Gwilt, A Practical Guide to Sustainable Fashion (Basics Fashion Design), New York: Fairchild Books, 2014

IOP Conf. Series: Materials Science and Engineering **459** (2019) 012094 doi:10.1088/1757-899X/459/1/012094

Sustainable entrepreneurship in the reuse of textile waste: H Sarah Trading case study in Portugal.

C Jordão[1], A C Broega[1], R Puppim[1,2] and A D Marques[1]

[1]University of Minho, Centre for Textile Science and Technology, Campus de Azurém 4800-058, Portugal

[2]Instituto Federal de Goiás, Avenida Vereador Vagner da Silva Ferreira, quadra 1, lote 1-A, Parque Itatiaia, Aparecida de Goiânia, Brasil

carijordao@hotmail.com

Abstract: The current business models are experiencing changes guided by the solution for serious global problems, related to the indiscriminate use of the planet resources, as well as the unrestrained production of waste and disposals. Thus, companies are being stimulated to rethink their market strategies, based on sustainable entrepreneurship theme, aimed at enhancing economic enterprise development and at the same time having environmental and social purposes. Faced with this new reality, this article aims to analyze how the structures of the new business models, with a focus on sustainable entrepreneurship, for the management of textile wastes, are being designed. So, a qualitative research was carried out, through a case study of a Portuguese company - H Sarah Trading - that works with waste management. The research results point to the construction of business structures that are moving towards sustainable development, balanced with economic profitability, and environmental and social benefits.

1. Introduction
Sustainable entrepreneurship has occupied space in the creation of new companies that seek to adapt their products and services, in order to require for a new and more conscious consumer that is already emerging. Its growth is also justified by the need to reach a balance between financial profitability and economic development, within justice and social welfare, environmental conservation and the rational use of natural resources [1].

This way, many companies begin to make use of sustainable entrepreneurship, as an agent of social and economic transformation, to find solutions to the negative impacts that are generated by the production processes. In the textile sector, the issues related to the management of its discards become even more evident, due to the relevance of this segment in the world economy.

In Europe, for example, 5.8 million tons of textile waste is discarded every year, 75% are sent to landfills or for incineration and only 25% to recycling [2]. Specifically, in Portugal, 5% of the 4607 tons of urban solid waste consists of garments or household textiles, ending up in landfills or are incinerated [3].

The present context motivates us to understand the changes that are being implemented by the textile and clothing sector in search for new waste management systems that can incorporate the concept of sustainability to minimize ecological and social crises, aiming at the maintenance and perpetuation resources for future generations [4].

Thus, this article aims to conduct an exploratory study to analyze the structure of business models, focusing on sustainable entrepreneurship, to minimize the impacts caused by textile waste, possibly minimized with reuse and/or recycling.

Therefore, the company H Sarah Trading, which operates with waste management and it is based in Portugal, was chosen as the unit of analysis for the case study of this research. The selection was based on the differentials of its corporate structure, the immense network of established partnerships and the benefits delivered to its stakeholders.

Based on data collected and analyzed, it is expected that the investigation can become relevant in the development of knowledge for future investigations on the business models that are managing textile waste, guided by sustainable entrepreneurship as well as for the reflection of new paths searching for a more fair and sustainable society.

2. Background

Sustainable entrepreneurship is a recent theme of studies, and still in process of conceptual construction. This subject contains a nascent flow of research that finds an intersection between sustainable management and entrepreneurship [5]. Despite presenting several definitions, sustainable entrepreneurship can be understood as the discovery and exploitation of economic opportunities that come from states of market imbalances and that can initiate the transformation of a sector towards an environmentally and socially sustainable state [6].

The environmental aspect aims to reducing the exploitation of valuable resources for the environment, reducing the conception and use of harmful substances, minimizing the environmental pollution and the production of waste. On the social area, sustainability can encompass the promotion of social integration, and the development of safer and fairer societies, with respect to human rights, diversity and equality, as well guaranteeing the main means of subsistence [7].

The discovery, development, and exploitation of opportunities are related to a generic process and the activities of sustainable entrepreneurship. The economic, social and ecological benefits generated refer to the triple bottom line, which must be kept in balance.

The delineation of these trends leads to the construction of different models of sustainable entrepreneurship, which are defined with a typology that considers three indicators: the niche of sustainability explored, the role of sustainability and the use of corporate social responsibility. As for the *niche* of sustainability exploited, this could be environmental or social (although hybrid business also exists). The role of sustainability is related to what this concept represents for the company, in other words, whether it means to achieve its end or the purpose of the organizational culture. Social responsibility is related to the company's awareness of assuming its active role in generating positive impacts on society [8].

Thus, sustainable entrepreneurial orientation is identified as an adaptive, innovative and externally oriented capacity that refers to the organizational predisposition to accept proactive, innovative and risk-taking processes, practices and behaviors, towards the achievement of sustainable development [9].

3. Methodology

This study was developed with a first phase through a literature review, in current indexed studies and books, which investigate the theme of sustainable entrepreneurship. In the second moment, a qualitative-exploratory research was carried out, through a case study, of H Sarah Trading Ltda.

This company has been operating in the Portuguese market since 2006 and is holder of a license to operate in waste management. It presents itself as a liaison between the citizen and the textile recovery cycle. Its activity aims to reduce the quantities of textile waste destined to landfills by reusing/recycling these materials, contributing to the awakening of a sustainable awareness among citizens [10].

The data collection was done through a semi-structured interview, *in loco*, at the unit in Braga-Portugal. The research was applied in July 2018 by the researchers of this study, intervening the two project managers of the company: Liliana Neves Cruz and Daniela Filipa Pereira. In addition, the technical materials available, as reports, videos, and the company website were consulted.

4. Results and Discussion – Case Study

The data on the H Sarah Trading case study were collected with the prior definition of parameters in order to systematize the analysis procedure and was divided according to the following items: i) analysis of the business model; ii) identification the environmental and social benefits of the business; iii) the relevance of the business to the partners' network; and iv) communication strategies for visibility and brand building.

4.1 Business Model Analysis

The H Sarah was established in 2006, but the waste management activities began in 2010. The company holds the Waste Management Operator's Permit and has an integrated system of selective screening of textiles, giving new meanings to the material that is collected. This material includes used clothes, footwear, and toys, which are deposited in more than 3 thousand containers spread throughout the national territory of Portugal, in public or private areas.

Another collection channel is through partnerships, based on protocols with public and private entities that receive garment donations and assign them to H Sarah to undergo hygiene and selection processes and then return according to the specific needs of the organizations.

Today, there are 3 poles that work as transfer warehouses, in the cities of Lisbon, Braga and Seia, the last one being the center of sorting and headquarters of the company. The collected materials are sent to one of these centers, where the initial screening is carried out in 3 categories: shoes, toys, and clothes. After this initial separation, the materials are sent to the screening unit of Seia, which has a crew of 80 people.

The used clothes undergo a triage according to the quality of the pieces and depending on this, it takes different destinations. Pieces that are classified as good quality, in the new or semi-new state, are intended for donation. Thus, they are referrals to the partner institutions, after having gone through the service of sorting and cleaning and separation according to the demand of each institution, thus meeting the social objectives of poor families.

After these sorting processes, the pieces go to a process of hygiene in a fumigation machine, which disinfects the pieces by vapors, which reduces the energy consumption, compared with a traditional washing process.

The second level with the usable parts, that do not meet the quality of donation, are valued for export to African countries, but from this point, the company has no further intervention in the screening of materials.

The third level quality parts, that not suitable for use, are directed to recycling and shipped for export, mainly to Pakistan and India. In this case, according to information from the company managers, they do not fall into the status of waste, since they have already undergone transformation processes.

The second and third level of parts represents the core business of the company, the main area of operation of H Sarah Trading. It is the business generated by these two types of materials, that the company can remain economically active and competitive in the waste management market.

It was observed that the business model under analysis is gaining strength by the construction of partnerships, which allow the company to capillarity of waste collection points, through protocols. Some of these agreements, according to the interviewees' testimony, are established by the exchange of services or financial compensation, reverted to benefit to the public attended by the partner entities.

4.2 Identification of the environmental and social benefits of the business

The business model developed by H Sarah Trading presents environmental benefits by helping to minimize the problem of the disposal of garments as urban solid waste. By proposing environmentally correct solutions to the destination of these materials, H Sarah ends up reducing the environmental impact of pollution and minimizing the extraction of natural resources. Currently, the company collects an average of 350 tons of waste per month. In 2017, 79,63% of total waste was destined for reuse (valorization and donation), 7,56% for textile recycling, 1,24% for recycling of other materials (such as plastic and cardboard) and 11,58% are not used and thus are directed to landfills.

About social benefits, one can observe the company's commitment to actions that promote its active participation in society in the sense of developing cooperation bridges and promoting corporate responsibility. A social contribution observed, refers to the service of sorting, sanitizing and re-routing of donations according to the needs of each partner institution. Through this action, the institutions receive the donations back according to their specific demands, to facilitate processes.

In addition, it was observed the company's interest in contributing as protagonist or partner in social actions. With the project "Sarah goes to school", for example, the company aims to sensitize children and young people to the practice of environmental education and correct recycling and reuse through playful activities of manual labor with textile.

The association with the ABAE - Blue Flag Association of Europe, which coordinates the "Eco Schools" program in Portugal, contributes to reinforcing social actions in the field of textile recycling. Thus, H Sarah entered as a partner in the project "Used clothes are not finished" which now has more than 200 schools enrolled. The objectives of the project are to inform and raise awareness about the importance of textile reuse and recycling, to promote the reuse of end-of-use clothing and to provide the community with delivery sites with the correct referral of this waste. The themes of this project involve issues of cultural valorization and the year 2018, had as a motto "Our Forest in Tissue" that invites the students to personalize discarded pieces of fabric inspired by icons of the forests of Portugal.

Another example of H Sarah's social contribution is the partnership with the Dorcas Project, which was born in 2013 in Cartaxo-Portugal. The project consists of volunteers who, in addition to other social activities, make dresses, (as an example of figure 1), from shirts, sheets, and other donated garments, which are sent to girls in need mainly from African countries.

According to the information gathered, the structure of the analyzed business model follows the principles of sustainable entrepreneurship. It can remain economically active in the market, generating more than 120 jobs, contributing to boost the national economy, while at the same time, it brings environmental and social benefits to partners and society.

4.3 Relevance of the business to the network of partners

Based on the benefits that have been mapped out, it can be said that H Sarah Trading turns out to be a relevant business for its stakeholders. Currently, the company has a network of almost 50 municipalities and parishes and more than 20 public and private institutions with which it maintains different partnership protocols.

For municipalities and parishes, H Sarah contributes to improving the management of the collection of municipal solid waste, clothing, shoes, and toys, by making collection points and disposal service environmentally correct for reuse or recycling. This fact minimizes costs of municipal funds with the management of this waste and mainly reduce impacts with the environmental pollution of the materials that would be directed to landfills.

Regarding public institutions, H Sarah has relevance for facilitating the work of sorting donations, because the clothes that are received by these entities end up not meeting their demands of size or gender, for example. Thus, with the partnership created, H Sarah offers the service of sorting and disinfection or even financial consideration, or in foodstuffs, used in favor of the actions with the poor public served by these entities.

Private institutions also benefit from the partnership because they end up adding a positive brand image in the minds of their consumers by supporting sustainable causes and providing them with points for the correct collection of textiles.

4.4 Communication Strategies for visibility and brand building

The relationship between sustainability and economic development appears to be an antagonistic structure and thus, many business models, guided by sustainable entrepreneurship, end up being criticized and generate doubts about the ethical business question. This also happens because many companies use the banner of sustainability only to achieve their economic goals and not a true business value.

IOP Conf. Series: Materials Science and Engineering **459** (2019) 012094 doi:10.1088/1757-899X/459/1/012094

In this sense, communication becomes a key tool to build a relationship of transparency with all the public involved, having a notable role in convincing consumers of the importance of sustainable consumption.

Consumers also expect more concrete information on how their behavior, for example, can affect waste reduction. The transparency of the production process is crucial to gaining consumer confidence. All communication channels should be widely used, and web paper and social media channels contribute to reaching a wider audience with an interest in sustainability issues [11,12].

According to Liliana Cruz, there is a concern to present in a transparent way, for the different publics of the Organization, all the activities, and processes that are developed by the Company. For the partners are presented monthly and quarterly reports with information of quantities and destination of waste. Already for society in general, the company is intensifying work on social platforms with the dissemination of videos of different stages of the waste management process, to demystify all activities that are developed by the Company.

The design of the containers presents differentials itself because they are designed in "t-shirt shape" and have the possibility of personalizing the visual identity according to the needs of the partners, as an example of figure 2. They are also explanatory with respect to the type of material, which can be collected and leave contacts for further information on waste disposal. In this way, they end up attracting the visual attention of the people while being a point of contact of the brand with the public.

Figure 1. Dorcas Project's dress

Figure 2. H Sarah's waste container

5. Conclusions

The research shows that today's societies are experiencing changes in their production and consumption systems to ensure the survival of future generations and reduce environmental impacts. This fact exposes the need for radical changes with the adoption of sustainable postures, both in people's daily lives and in productive systems, which need to find innovative and disruptive solutions to their waste.

The productive chain of the textile and clothing industry continues to follow these trends and starts to create industries and services focused on sustainable entrepreneurship. The case study carried out in Portugal with waste management, company H Sarah Trading, represents a business model that is based on the Triple Bottom Line [13], by adding environmental and social benefits to the development the Organization.

The structure of the business is being consolidated mainly by the creation of collaborative strategies through a network of partnerships that makes the business more financially viable to deal with complex environmental problems, caused by the harmful impact of incorrect disposal of post-consumer textile waste.

The use of containers in the public spaces of the cities, to collect the textile discards, is presented as a strategy to engage with consumers to attract attention and awaken sustainable awareness. The results of the projects implemented with different stakeholders of the Organization strengthen the partnership network and gives visibility to the Company's positioning.

It should be emphasized that the communication strategies becomes fundamental to establish a sustainable business transparency relationship and design a positive image of the Brand for different communication channels, especially social platforms with greater reach.

Thus, it can be concluded that sustainable entrepreneurship has consolidated itself as a path with no return, to companies and emerges as a promising alternative for the textile sector to manage the impacts of its discards in a creative, innovative and financially viable way. In addition, it is emphasized that through entrepreneurship, companies can create new markets for sustainable products and services, influencing their productive chain that becomes more sensitive to environmental and social issues.

Acknowledgements
"This work is supported by FEDER funds through the Competitiveness Factors Operational Programme - COMPETE and by national funds through FCT – Foundation for Science and Technology within the scope of the project POCI-01-0145-FEDER-007136".

References

[1] Sachs I 2008 *Desenvolvimento includente, sustentável e sustentado* (São Paulo: Garamond)
[2] Eurostat Statistc Explained. Homepage, http://ec.europa.eu/statistics-explained/index.php/Environment, (accessed 2017/12/20)
[3] APA-Agência Portuguesa do Ambiente. Homepage, https://www.apambiente.pt/, (accessed 2018/5/17)
[4] Vezzoli C 2010 *Design de Sistemas para a Sustentabilidade* (Salvador: Edufba)
[5] Belz F M and Binder J K 2017 Sustainable Entrepreneurship: A Convergent Process Model *Business Strategy and the Environment,* **26** (1), pp.1–17. https://doi.org/10.1002/bse.1887
[6] Hockerts K and Wüstenhagen R 2010 Greening Goliaths versus emerging Davids Theorizing about the role of incumbents and new entrants in sustainable entrepreneurship *Journal of Business Venturing*, **25** (5), pp. 481–492. https://doi.org/10.1016/J.JBUSVENT.2009.07.005
[7] Kraus S, Burtscher J, Vallaster C and Angerer M 2018 Sustainable entrepreneurship orientation: A reflection on status-quo research on factors facilitating responsible managerial practices *Sustainability* , **10** (2) https://doi.org/10.3390/su10020444
[8] Borges C 2014 Empreendedorismo sustentável e o processo de criação de empresas. In: Borges, Cândido (Org.*) Empreendedorismo Sustentável* (São Paulo: Saraiva)
[9] Criado G A, Cervera-Taulet A and Iniesta-Bonillo M A 2017 Sustainable Entreprurial Orientation: A Business Strategic Approach for Sustainable Development *Sustainability*, **9** (9), 1667 https://doi.org/10.3390/su9091667
[10] H Sarah Homepage, http://www.sarah-trading.com/ (accessed 2018/02/10)
[11] Vehmas K, Raudaskoski A, Heikkilä P, Harlin A and Mensonen A 2018 Consumer attitudes and communication in circular fashion *Journal of Fashion Marketing and Management* https://doi.org/10.1108/JFMM-08-2017-0079
[12] Han S L C, Henninger C E, Apeagyei P and Tyler D 2017 Determining effective sustainable fashion communication strategies, In *Sustainability in Fashion* (Cham: Palgrave Macmillan), pp. 127-149
[13] Elkington J 2012 *Canibais de garfo e faca* (edição histórica) (São Paulo: M. Books)

IOP Conf. Series: Materials Science and Engineering **459** (2019) 012095 doi:10.1088/1757-899X/459/1/012095

Exploring "Everyday Experience Methods" as a means to assess the (social-)psychological comfort of clothing

L L Matté[1,2], A C Broega[1] and M E B Pinto[3]
[1]Centre for Textile Science and Technology, Department of Textile Engineering, University of Minho, Guimarães, Portugal
[2]Department of Fashion Design, Federal University of Technology - Paraná, Apucarana - PR, Brazil
[3]Department of Psychology and Psychoanalysis, Londrina State University, Londrina - PR, Brazil

E-mail: livialauramatte@yahoo.com.br

Abstract
One of the main challenges for the Science of Clothing Comfort, is to uncover the intricacies and complexities of the psychological aspects of the clothing comfort perception. The concept and the parameters of this comfort dimension are still in the process of maturing, and the methods and tools to assess this component are still underdeveloped. One of the main drawbacks of the research on the psychological comfort of clothing resides in the dependency of the context and situational information, to elaborate a psychological comfort or discomfort perception. In the realm of the psychological or social-psychological comfort, it's primordial that the context variables are considered. This pilot study aims at examining the feasibility of assessing the perception of (social-)psychological comfort of clothing, using an everyday experience method. For this purpose, the PACO application (pacoapp.com) was the chosen platform.

1. (Social-)Psychological Comfort

At the interface between the human body and its surrounding environment, clothing plays a determinant role in the subjective perception of a user's comfort status. In the scope of clothing, comfort includes the physical, physiological and social-psychological balance between an individual, their clothing and their environment [1]. This balance is fundamental when it comes to the person-clothing relationship and the particularities that involve the use of a product that interacts continuously and dynamically with the body and is often compared to a second skin [2, 3]. Psychological comfort, or social-psychological comfort, is a process of hedonic judgment, whereby the brain forms a subjective perception of sensory stimuli, influenced by many factors. Due to its subjective nature, and diversity of influential parameters, psychological comfort is, according to Slater, the hardest comfort component to evaluate [4].

The choice of clothing, intentionally or unintentionally, communicates as an "open letter"[5], so that feeling of psychologically comfortable will largely depend on the reactions and public responses received by the user. This observation points out the importance of comfort assessments in real situations, as well as reveal the need for interdisciplinary studies to understand psychological comfort. The apparent diversity and complexity of the set of parameters to be considered in the process of evaluating the psychological comfort of garments, warns of the indispensability of seeking creative ways and combining methods and contributions from other fields to collaborate in this task.

IOP Conf. Series: Materials Science and Engineering **459** (2019) 012095 doi:10.1088/1757-899X/459/1/012095

To the best of our knowledge, there are, up to this date, no methodological proposals or instruments dedicated to assessing the psychological comfort of clothing. We believe, that the complexity of variables, and the problem with ecological validity in assessing such parameters in laboratory settings, are probably the reasons why this dimension is underdeveloped. A starting point for a more concrete assessment of socio-psychological comfort may be following the path of the Everyday Experience Methods.

2. Everyday Experience Methods

Everyday Experience Methods refer to a paradigm for studying social-psychological phenomena as they occur in the ebb and flow of everyday life. This heading proposed by Reis, Gable and Maniaci, contemplates a myriad of procedures and measures that aim 'to capture life as it is lived'. There are many different techniques, by which researchers can study daily experiences, and although the names of these methods vary (experience sampling, diary methods, ambulatory assessment, ecological momentary assessment, diary methods, intensive longitudinal designs) in sum, they allow researchers to study experience, behavior, environments, and even, physiology of individuals, self-reported or non-self-reported, in their natural settings, in (close to) real time, and on repeated occasions [6–9].

This range of methods started to flourish in the early 1970s, being the Experience Sampling Method (ESM), conceived by Mihaly Csikszentmihalyi and colleagues, the most prominent one. The ESM, "was developed as an attempt to adapt the scientific method to the understanding of human behavior and human experience instead of the reverse; that is, to understanding only those aspects of human experience that were amenable to experimental laboratory investigation." The ESM intends to study what people do, feel, and think during their daily lives [6,10–12]. In ESM protocols, when receiving a signal (usually via pagers or smartphones) the participant must respond as soon as possible. Typically, in ESM studies, participants receive eight daily notifications at random times. The questions can have both objective and subjective nature, but there are always the basic questions for characterization: where the person is, with who is he/she, and what is he/she doing at the moment of receiving the signal. The sampling of data can assume three forms: *time or interval-contingent*, when participants record their experiences at predetermined intervals (e.g. at the end of the day or every four hours); *signal-contingent,* where participants describe their activity at the moment a signal (with random or fixed schedules) is delivered; or *event-contingent,* which requires a report every time certain events occur, for example, the act of dressing, which was the selected event for the experiment to be described in the next topic [10].

A fundamental benefit of the Everyday Experience Methods is that, they permit the examination of reported experiences within its natural, spontaneous context. Kaiser, emphasizes the importance of context when studying clothing, asserting that "clothes do not exist in limbo; instead, they are embedded in context - the more complete framework or social circumstances of daily life." [13].

Also, by sampling experience the moment it occurs, it is possible to avoid the potential distortions associated with the use of retrospective diaries. The act of dressing and the decision-making process to decide which clothes to wear is an ordinary, mundane one, so, it is primal that the collection of data occurs at least, close to real time, avoiding the reliance on recollection since these events are not exactly memorable. It is clear that the method is not free of limitations, a considerable disadvantage, for example, is the demand it imposes on the interviewees because of the commitment to participate in an experiment for several consecutive days, besides, the experiences with random signal-contingent can also be inconvenient in certain situations. [8, 10, 11]

3. Method

This pilot study was developed as part of a Ph.D. research on the social-psychological comfort of clothing. At this phase, we aimed at examining the feasibility of assessing the perception of psychological comfort of clothing, using an everyday experience method [10, 12, 14]. For this purpose, the PACO application (pacoapp.com) was the chosen platform.

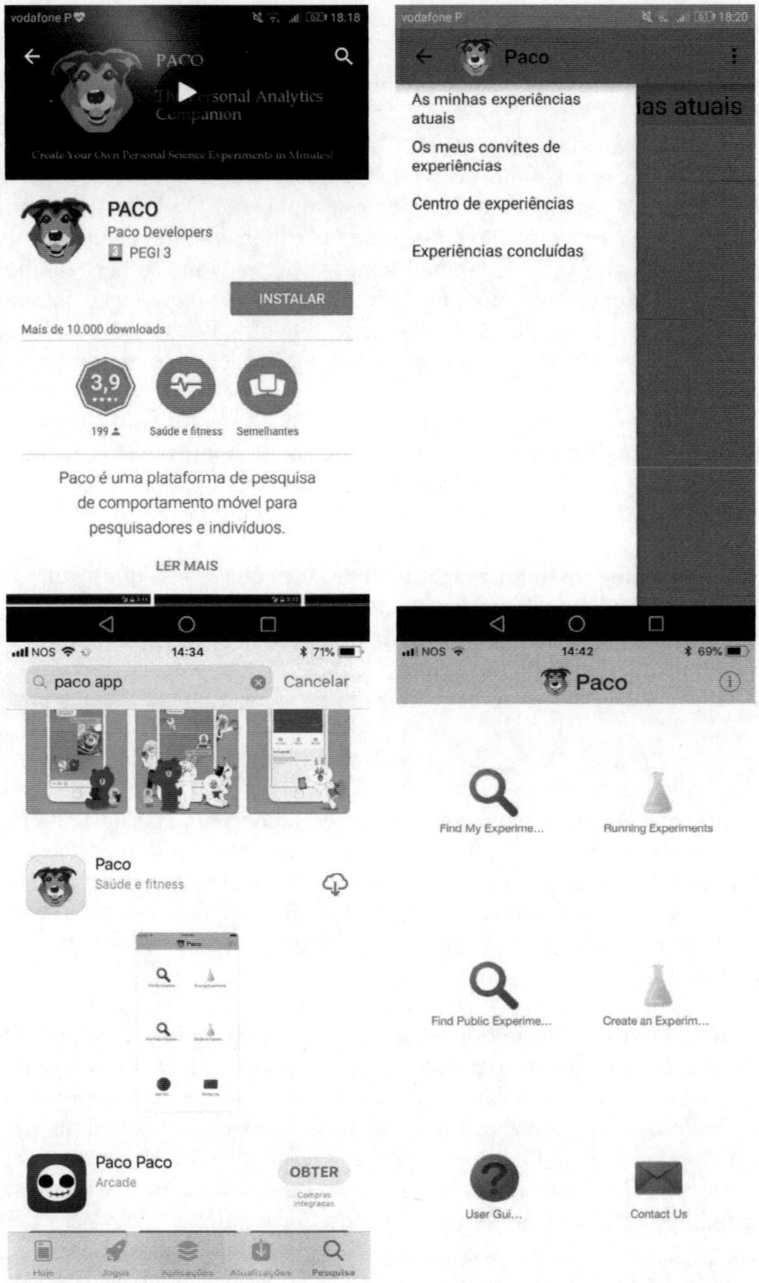

Figure 1. The interface of PacoApp for Android (top) and IOS (bottom).

PACO (Personal Analytics Companion) is an open source ESM (Experience Sampling Method) [12] that enables the scaling of qualitative research through quantitative, computational techniques [15]. The platform provides customizable software to design and deliver remote research instruments to mobile devices. The figure 1 shows two sample screenshots of the application interface for Android and IOS.

A group of clothing comfort and fashion design experts were invited to take part in the pilot study. They received guidelines for downloading and installing the Paco App and a protocol to participating

IOP Conf. Series: Materials Science and Engineering **459** (2019) 012095 doi:10.1088/1757-899X/459/1/012095

in the trial. For seven consecutive days, requests were sent to participants to fill out a small form (7 questions) immediately after the dressing time (in the morning) and at the end of the day (5 questions), which classifies the adopted model as "event sampling", since there are two specific "events" during the day that require responses.

Although the procedure was defined as event-contingent sampling, the researchers decided that it was important to send notifications, both visual and sonorous, as a reminder. After each notification (visual and sound), participants completed the open questions and the Likert scale ratings, directly in the application, concerning current context (location, weather, situation, and company) as well as satisfaction with clothing, self-esteem, social appropriateness and other components of user's momentary subjective experience with clothing. The design of the experience also included a request to take a picture (selfie) that showed the outfit choice for the day. Each report took approximately 1–2 min to complete. The answers were time-stamped, allowing for accurate recording of the time that the responses were received.

The purpose was to explore the potentialities and limitations regarding the use of this specific App as a viable tool to instrumentalize the assessment of the social and psychological aspects of clothing comfort. Some of the issues to be addressed from this phase of testing were:

-Test the App and evaluate whether the record of information fed in by the participants would be reliable.

-Explore the available possibilities to include the questions that were essential for the assessment of the psychological comfort of clothing

-Test whether participants would receive daily notifications as programmed by the researchers.

-Identify the difficulties faced by the volunteers, both, in the process of installing the App and at the moment of the answering, also evaluating the suitability of the scales that were chosen for the survey.

4. Results

Despite the application had already been tested by the researchers (in android), the platform had some problems in processing the images and had also a delayed reception of the answers. So, in this phase, it was essential to ensure that the App worked properly. It was also important to look for alternatives to make Paco perform well for users of IOS and, because the App was not fully compatible with this operating system, the researchers had to find ways to adapt the questionnaires to circumvent the limitations of the platform.

Technically, the App performed very well, and the participants declared that taking part in this study was enjoyable. This experimentation phase also brought relevant outputs that helped us to improve the questionnaire for the next phase of the research, when applying the event-sampling method and inquiries using Paco to a larger audience. Some of the interesting findings during the testing of the App were that a) we noticed that the lack of specific instructions (in the protocol) to make the selfies photography would be a problem, as we observed that some of the images did not show the clothing as a whole, (most of them, just above the waist); b) it was possible to adjust the questionnaire, expanding the options of some questions (for example: For what occasion did you get dressed?) and modifying some of the scales (From a simple 5 point Likert scale to a smiley scale); c) the time interval within which it was possible to respond to the questions needed to be shortened as it was noted that in some occasions, the participants only responded much later than they were supposed, since the application allowed them not to respond immediately after the notifications; d) although some may have found it fun to answer the questions and take the pictures every day, we are aware that the commitment to respond twice a day and also the disposition to photograph its own outfits, even in the "not so good days" are significant obstacles.

The overall evaluation of the platform was positive, and the main advantages of the method, besides portability, were that it obtains reports about people's experience as it occurs, not relying on memory and reconstruction, and the most important, it enables the correlation between the perception of psychological comfort of clothing and social contexts/situations.

IOP Conf. Series: Materials Science and Engineering **459** (2019) 012095 doi:10.1088/1757-899X/459/1/012095

5. Final remarks

The need for a deeper understanding of psychological comfort presents itself as a still vast field of research, especially regarding new and creative methods for evaluating this comfort component. With this paper, we sought to explore and present the methodological richness and innovative to be found in social-psychological research and, hopefully, to present ideas for expanding the methodological diversity employed in the Comfort Science.

Everyday Experience Methods (diary-methods, experience sampling method, event sampling, ecological momentary assessment, and all those studies that use systematic measures of momentary experience and behavior), offer more than just another methodological alternative; the possibility to understand and assess the aspects of daily activity in real-time, including momentary physiological sampling, will allow researchers to evaluate comfort in a broader sense.

As pointed by Reis et. al, experience methods are meant to complement, not substitute for, experimentation [10], that way we believe that even the other dimensions of clothing comfort research such as ergonomic, sensorial and thermophysiological, can benefit from methods that include everyday experience protocols.

Acknowledgments

This work is supported by FEDER funds through the Competitivity Factors Operational Programme - COMPETE and by national funds through FCT – Foundation for Science and Technology within the scope of the project POCI-01-0145-FEDER-007136. The first author would also like to gratefully acknowledge the support from the Federal University of Technology, especially, the Fashion Design Department and the Office of Research and Graduate Studies.

References

[1] Branson D H and Sweeney M 1991 Conceptualization and measurement of clothing comfort: Toward a metatheory *Critical linkages in textiles and clothing subject matter: Theory, method and practice* ed S B Kaiser and M L Damhorst (Monumet, CO) pp 94–105

[2] Hosseini Ravandi S A and Valizadeh M 2011 Properties of fibers and fabrics that contribute to human comfort *Improving Comfort in Clothing* pp 61–78

[3] Sanches M C de F, Ortuño B H and Martins S R M 2015 Fashion design : the project of the intangible *Procedia Manuf.* **3** 2311–7

[4] Slater K 1985 *Human Comfort* (Springfield: Charles C. Thomas)

[5] ECO U 1989 O hábito fala pelo monge. *Psicologia do vestir* pp 3–20

[6] Mehl M R and Conner T S 2012 *Handbook of research methods for studying daily life* (New York, NY.: Guilford Press)

[7] Bolger N and Laurenceau J-P 2013 *Intensive Longitudinal Methods: An Introduction to Diary and Experience Sampling Research* (New York: The Guilford Press)

[8] Hektner J M, Csikszentmihalyi M and Schmidt J a. 2007 *Experience sampling method: Measuring the quality of everyday life.*

[9] Schwartz J E, Neale J, Marco C, Shiffman S S and Stone A A 1999 Does trait coping exist? A momentary assessment approach to the evaluation of traits *J. Pers. Soc. Psychol.* **77** 360–9

[10] Reis H T, Gable S L and Maniaci M R 2014 Methods for Studying Everyday Experience in Its Natural Context *Handbook of Research Methods in Social and Personality Psychology* ed H T Reis and C M Judd (New York: Cambridge University Press) pp 373–403

[11] Bolger N, Davis A and Rafaeli E 2003 Diary Methods: Capturing Life as it is Lived *Annu. Rev. Psychol.* **54** 579–616

[12] Larson R and Csikszentmihalyi M 2014 Flow and the Foundations of Positive Psychology pp 21–35

[13] Kaiser S B 1993 Linking the social psychology of dress to culture: a contextual perspective *Social science aspects of dress: New directions* (International Textile and Apparel Association) pp 39–47

[14] Mehl M R, Conner T S and Csikszentmihalyi M 2012 *Handbook of Research Methods for Studying Daily Life* (New York, NY.: Guilford Press)

[15] Evans B 2016 Paco - Applying Computational Methods to Scale Qualitative Methods *EPIC2016 - Pathmaking* pp 348–368

IOP Conf. Series: Materials Science and Engineering **459** (2019) 012096 doi:10.1088/1757-899X/459/1/012096

Design thinking as necessary constituent of creative industry

O Savchenko

National Technical University "Kharkiv Polytechnic Institute", Kharkiv, Ukraine

savchenko.khpi@gmail.com

Abstract. Creative management is implemented at the pre-design and design stages of the innovation cycle and considers the innovative product as a complex structural. Accordingly, the management associated with the creation of a new system of knowledge, skills and abilities must be in a certain way divided into a number of constituent elements. It should also be noted that creative management should be aimed at the full disclosure of creative abilities of people. The design thinking activates the development of knowledge, skills and abilities, because through the use of a qualitatively different methodical device raises the creative potential to a higher level.
Keywords: creativity, design thinking, innovative product, creative management, knowledge

1. Introduction

Creative management has now announced itself as a real-life knowledge area. Moreover, it becomes universally accepted. Creative management is closely linked to the innovative management that operates with the intellectual product.

It is of considerable interest to study the experience of the world in monitoring innovation indicators, as this process is very dynamic under the influence of new trends of development: globalization, formation of knowledge economy, open innovation. To continue with, it was proposed to compare the activities with a set of different value indexes (Global Innovation Index, for Ukraine 64 ranking position from 30 world countries; Patent activity, 27 position). The high position of Ukraine in the ranking in terms of Global Creativity Index, 45th ranking position from 20 countries.

Creative industries represent the sector of the global economy with the most dynamic development. Creative industries represent the sector of the global economy with the most dynamic development. Creative industries represent for 3.4% of global GDP. It is important to note that some young people employed in the creative economy account for 25% of the world's population. At the same time, the economic impact of creative industries on the overall economic development of the country is small. The potential of the creative sector is not used effectively.

Experience shows that the use of creative-heuristic techniques in scientific and technical creativity can increase the efficiency of the work of scientific and engineering teams several times. However, this rich experience is hardly ever used to improve managerial decision-making in the economic and organizational spheres.

Thus, we can say, that innovative management, on the one hand, creative - on the other, have the same object of consideration, but study it in one case (innovative) from the outside, as ready, in another (creative management) - from the internal, as the process of creation.

Creative management is closely linked with the development of such areas of management as personnel management, strategic management, innovation management, "self-management", organizational culture of the enterprise, marketing. Desing thinking is not a new concept, but one that in recent years has entered a growing range of industries in the world. For an innovator, design thinking is an approach that minimizes uncertainty and the risk of innovation. Focusing on rapid iteration of the idea and active involvement of the client in an open dialogue throughout the process, the company is better positioned to understand the root cause that really needs a solution, as well as the product or service functions that they need to provide to meet this need. The activity of future managers is related to work with people, it is aimed at strengthening and better utilizing the creativity of the collective (creative management), as well as the fact that the staff has activated its own intellectual abilities. There

is a certain contradiction between the constant growth of knowledge and the low creative level of students.
Ukrainian industries that could be successful on the European market would be IT products, food, light industries and agricultural products. First of all, enterprises need to get information. According to international experience, the implementation of the concept of a creative economy opens up new horizons of socio-economic progress. The creative industries, which make up its material foundation, are becoming a dynamic and highly profitable sector of the global economy. They depend on the realization of creative capital.
Most of the resources used in the process of creating innovations and their commercialization are classified as specific and interspecific. The key interspecific resource is the human capital of workers engaged in innovation. The team principle of organization of work is often characteristic of innovation activity.

The intellectual property market is one of the most dynamic markets in the information economy. Assessment of intellectual property can be designated as the process of establishing (in monetary terms) the utility of the results of intellectual activity, the individualization of products, work or services performed. The training of organizations and the ways in which it occurs are discussed in the light of the different types of knowledge that are used by human resources. Communication and dissemination of information are identified as critical elements of this process. Creating a learning environment becomes a central task for an enterprise that wants to manage its own knowledge base.

The concept of "absorbing capacity" establishes a link between information and knowledge outside the company and training, organizational structures within the enterprise. This approach connects the enterprise's ability to recognize the significance of new, external information and its ability to assimilate new information with existing similar knowledge (embodied in human capital), and finally apply joint knowledge for commercial purposes.

Accordenly, national policy should invigorate educational instititions notably in the realm of technical skills, vocational qualification and entrepreneurial talent.

The purpose of the article is to study the impact of the ability to apply forms of a design thinking, creative approach to management activities of business processes, present design thinking as a strategy for product innovation and business innovation.

2. Materials and methods

We now present our review of the literature of this area, starting with a description of the nature of design thinking, its characteristics, and processes.

A number of researchers state that the development of intellectual - creative potential of youth is on the lower boundary of the average norm. At the same time, it was determined that high indicators of creative potential of a person and personal qualities of future managers influence the efficiency of management activity and their self-realization.

A shot definition of creativity. ….The process by which something new comes into existence and a longer one: “a process of being sensitive to problems, deficiencies, gaps in knowledge and missing elements, disharmonies and so on; identifying the difficulty, searching for solutions, making guesses, or formulating hypotheses about the deficiencies: testing and retesting them, and finally communicating the results” [Jalan and Kieiner, 1995].

The nature of design thinking in many areas, knowledge is generated and accumulated through action (i. e. something done and evaluate the results). That is, knowledge is used to produce work, and work is evaluated to obtain knowledge. Creative people tend to work in two different ways: either as finders or as makers [Owen, 2007]. Finders demonstrate their creativity through discovery. They are driven to understand and to find explanations for phenomena not well understood. Makers are equally creative, but they are forced to synthesize what they know in new designs, arrangements, patterns, compositions and concepts. Given the fundamental technological differences between how quest and manufacturers think and work, other factors can equally identify differences in the professional fields and, therefore, help determine the nature of design thinking. One such factor is the content with which a field of works. A conceptual map can be drawn to represent the four stages of creative process (Figure 1).

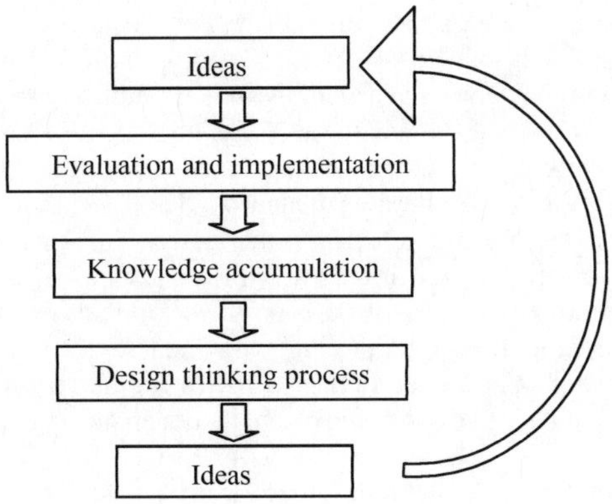

Figure 1. Stages of creative process

Ideation. Make sense of everything that heard, generate tons of ideas, identify opportunities for design, and test and refine your solutions.

Incubation. Design is an iterative activity thatonly has broad guidelines but no fixed process. What's more important is that critical insights, sensitivity to consumer needs and beautiful solutions come from the creative chaos encouraged by an open design process. A common focus on communication, creativity, analytical thinking, reliability and repeatable processes.

Implementation. To bring solution to life. Will figure out how to get idea to market and how to maximize its impact in the world protection of property rights); costs of production; costs of promoution; losses from administrative barriers.

The main feature of design thinking, as opposed to analytical thinking, is not a critical analysis, but a creative process in which sometimes the most unexpected ideas lead to a better solution of the problem.

Problem definition is first and most important stage of design thinking, because if you incorrectly determine what the problem is, then the solution that you come to in the process will not be the solution to the problem that needs to be solved. After you have identified the problem, you also need to determine who your end user is (whose problem do you solve?) And what result do you want to achieve (what is the successful result of the project?).

Design thinking involves an analytical component that includes such areas as science, for example. It is more analytical when using the process, and their content is more symbolic than the real in this subject, as a rule, is abstracted in its analyzes. The second component is symbolic, which includes a wide cover of symbolic content and synthetic processes.

Design thinking from this perspective is about borrowing designers ways of thinking and working in order to promote innovation in businesses, as well as balance the analyticalbusiness mindset with the intuitive and creative design mindset.

Iterative and repetitive often from different points of view, the creative process is not linear and should never be seen as one. And it is very social, with some of the best ideas arising from empathy, interaction and exchange. This should occur because workplaces play an integral role in ensuring the creative process and providing employees with the support, inspiration and adaptive conditions necessary for interaction and innovation.

The reseach focused on the difference in creative management from its traditional forms. Non-standard creative approach; develop theories of cooperation, harmonization; team work; flexibility is closely related to creative performance; creative performance has its original not in innate cognitive abilities but in the motivation of the creative person. Characteristics of traditional management: management of the rules, develop theories of struggle, individual and group work, strict order of performance.

Next, we present literature regarding expertise, expert versus design thinkers, and expertise in design. Many articles in the design thinking literature were identified and described design thinking, design cognition, design behavior, design studying, design reasoning. Many authors have studied the nature of essence and the various processes that underlie the process of design thinking [Owen, 2007; Stempfle & Badke-Schaube, 2002].

Then we present our design model of thinking. The first step is to meet customers. This phase is devoted to exploring on the fly, opening yourself to creativity and confidence that, as long as you remain grounded in the desires of the people you engage, your ideas will evolve into the right solutions. It is very important to use the provided situation. Opportunity is a situation in which changes in technology or economic, political, social and demographic conditions create the potential for creating something new.

Researchers who are interested in measuring and supporting design thinking have great opportunities to conduct a wide range of experimental studies that can lead to important findings. For instance, researchers may examine the effects of the design thinking process on various learning outcomes. They can also investigate the effects of different tasks and their complexity relative to enhancing design thinking skills, which in turn are assumed to increase students' learning outcomes. It would also be interesting to know if design thinking skills mediate the learning process. In other words, design thinking skill may serve as a mediator that clarifies the nature of the relationship between an independent variable (e.g., problemsolving skill) and a dependent variable (e.g., math test scores) [Shute and Torres, 2012].

The design thinking is that designers and designers' work processes can help systematically extract, teach, learn and apply these human-centered techniques in solving problems in a creative and innovative way in our designs, in our businesses, in our lives and in our countries. The invention is a technological solution that becomes an innovation only when it enters into social practices.

Innovation is closely linked to the social context and that it must be rooted in the understanding of the needs of specific groups of people.

The design thinking activates to development of knowledge, skills and abilities, because through the use of a qualitatively different methodical device raises the creative potential to a higher level.

Tim Brown also emphasizes that Design Thinking techniques and strategies of design belong at every level of a business [Tim Brown, 1986].

"Design thinking begins with skills designers have learned over many decades in their quest to match human needs with available technical resources within the practical constraints of business. By integrating what is desirable from a human point of view with what is technologically feasible and economically viable, designers have been able to create the products we enjoy today. Design thinking takes the next step, which is to put these tools into the hands of people who may have never thought of themselves as designers and apply them to a vastly greater range of problems" [Tim Brown, 1986].

Tim Brown, CEO and president of IDEO, is also a proponent of using design thinking in business but he defines design thinking as the"human-centered approach that designers use in the development process". He believes that "drawingfrom the designer's toolkit allows businesses to better integrate the needs of people, the possibilities oftechnology, and the requirements for business success".

Bill Moggridge, co-founder of IDEO, has shown "The 'Design Thinking' label is not a myth. It is a description of the application of well-tried design process to new challenges and opportunities, used by people from both design and non-design backgrounds. I welcome the recognition of the term and hope that its use continues to expand and be more universally understood, so that eventually every leader knows how to use design and design thinking for innovation and better results." [Dagnino-Pastore, Farina, 1999].

Finally, we discuss the findings from the literature, showing the importance of design thinking and providing suggestions for future research.

The research by [Johansson-Sköldberg, Woodilla, Cetinkaya, 2013] has given more information with regard to the outline a distinction between 'designerly thinking' and 'design thinking.' The authors suggest that'designerly thinking' refers to the practical skills and innate competencies of a designer and is "rootedin the academic field of design." On the other hand, 'design thinking' refers to the use of designpractices and competencies by disciplines outside of design.

Creative thinking techniques: Brainstorming (heuristic abilities), than people coming together with the sole purpose of generating ideas; logical generating activities.

Thus, Design Thinking is an iterative process in which knowledge is constantly being questioned and acquired so it can help us redefine a problem in an attempt to identify alternative strategies and solutions that might not be instantly apparent with our initial level of understanding. Design Thinking is often referred to as 'outside the box thinking', as designers are attempting to develop new ways of thinking that do not abide by the dominant or more common problem-solving methods – just like artists do. At the heart of Design Thinking is the intention to improve products by analyzing how users interact with them and investigating the conditions in which they operate. Design Thinking offers us a means of digging that bit deeper to uncover ways of improving user experiences.

Design mastery is found in the practical skills and innate competencies of a designer. Designers aretrained to frame problems and reason through abductive thinking [Dorst, 2011].

In general, thinking constrictions may be presented:

- a way of looking at the world,
- worrying of what other people may say,
- that's a daft idea' and 'people will laugh at it',
- there is only way of doing this….,
- seeing things the way they have always been,
- the process of killing ideas before allowing them to form,
- information as a liability.

The rules of creativity entreprener:Ragnar Siil, business-manager partner of Creativity Lab (Estonia) and coordinator of expert group for development cultural and creative industries of European Commission said: «Cultural creative – key to successes of business on creative industries».

Main goals of creative management [Pererva, P., Savchenko, O. ,2013]:

- assessment of the creative potential of a person – a specialist who wants to be involved in solving of creative problems,
- creating a creative atmosphere in a team that has worked for a long time and has its own traditions, informal leaders, the distribution of role functions that has justified itself,
- acquisition of skills and skills in the formation of creative thinking,
- studying methods of solving non- standard tasks,
- ability to apply forms of a creative approach to management activities,
- creative management, promotes the development of skills of non-trivial creative decisions in business.

Creative activity is the key to an innovative economy; and in the present conditions, without any communication and atmosphere of creativity, no innovations are possible. In the practice of spatial development, these are the same key concepts as demography and economics. It is worth noting that, in our opinion, the development of creative industries and creative economy in Ukraine should not be entirely the responsibility of the state; her task is to create a comfortable environment for creativity (the number of enterprises in the creative sector is growing, but they lack the conditions for work) and, if possible, not to interfere at all in the activity of the creators (who, in need of freedom in their activities, and only under its conditions can produce creative goods and services).

3. Results

Thus, the program of formation of thinking of future marketers, engineers, economists should include the following areas: - the acquisition of skills and abilities in the formation of creative thinking; - the study of methods for solving non-standard problems; designers are not effective in conveying that design skills, a more serious problem may be that adopting design thinking and misusing tools and methods threaten the future of design and reliability of designers. There is a need to better convey the value of design excellence through methods in design thinking, but additional methods must also be developed that can combine empathy, analytical and intuitive, deductive, inductive and inductive approaches that effectively manage the innovation process. The analysis that has been carried out allows concluding that the design thinking to help design better products, services, processes, strategies, spaces, architecture, and experiences. Design thinking helps develop practical and innovative solutions for decision making. It is a human-focused, prototype-driven, innovative design process.

Creative management, promotes the development of skills of non-trivial creative decisions in business. In creative management, you can distinguish the following main goals: assessment of the creative potential of a person, a specialist who wants to be involved in solving creative problems, creating a creative atmosphere in a team that has worked for a long time and has its own traditions, informal leaders, the distribution of role functions that has justified itself.

It is necessary to increase the turnover of the country, namely, creative products.

Design thinking is not only for designers but also for creative employees, freelancers, business leaders, students of all specialties. It's for anyone who seeks to infuse an approach to innovation that is powerful, effective and broadly accessible, one that can be integrated into every level of an organization, product, or service so as to drive new alternatives for businesses and society.

References

[1] Amitabh Jalan, Brian H. Kleiner, (1995) // New developments in developing creativity, Managerial Psychology, Vol. 10 Issue: 8, pp.20-23, https://doi.org/10.1108/02683949510100750

[2] Razzouk, R., Shute, V. (2012) /Review of Educational Research, What Is Design Thinking and Why Is It Important? AERA: Florida State University, Vol. 82, № 3. - pp. 330–348. http://rer.aera.net330

[3] Stickdorn, M., Schneider, J. (2011) This is Service Design Thinking: Basics - Tools – Cases // The Netherlands, Amsterdam, 2011. – pp.150.

[4] Brown, Tim. (2014) About IDEO. Available// http://www.ideo.com/about/
Johansson-Sköldberg U. Woodilla J. and Çetinkaya M. (2013) Design Thinking: Past, Present and Possible Futures // Creativity and Innovation Management, 22.2. – pp. 121-146.

[5] Dorst K. (2011) The core of 'design thinking' and its application// Design Studies, 32.6. – pp. 521-532.

[6] Communication and Information Resource Centre for Administrations, Businesses and Citizens. https://circabc.europa.eu/sd/a/357eb96f-e51c-4a6a-bc3c-ea7ef147fdec/20150630.xlsx (Accessed: 11 July 2015).

[7] Skavronska, I. (2016) Osoblivosti rozvitku creativity industriy in Ukraïni Materiali II Vseukraïnskoï NAUKOVO-praktichnoï internet-konferentsiï of mizhnarodnoyu participation // Aktualni problemi upravlinnya sotsialno-ekonomichnimi systems, Lutsk, / PBB Lutsk NTU, Part 5,-pp.218–222. www.economosvita.lntu.edu.ua

[8] Pererva, P., Savchenko, O. (2013) Innovation development as the factor of structural convergence / Strategy of innovation development of economy, P. 2. Monography / Ed. by. Pererva, P., Savchenko, O. - Generous manor plus, - pp. 8-20.

[9] http://uacrisis.org/ua/28293-creativite-l-entreprise-pourrait-sauver-leconomie-de-lukraine-particulier-exportations-experts

IOP Conf. Series: Materials Science and Engineering **459** (2019) 012097 doi:10.1088/1757-899X/459/1/012097

How to communicate a new fashion brand through social media and public relations' perspective.

P Scheide[1] and A D Marques[1]
[1]University of Minho, School of Engineering, Department of Textile Engineering, Campus of Azurém, Guimarães, Portugal

Email: adinis@det.uminho.pt

Abstract. Fashion communication has been undergoing a transformation and there is a drop in tradicional marketing investments which indicates new modalities of brand's communication. The purpose of this study is to discuss the branding strategy relevance through social media and public relations' perspective, aiming to understand how digital marketing platforms have been used and integrated by new fashion brands. This paper presents an ongoing Master's research including data collection method through a qualitative content analysis, aiming to find some features can contributed to improve their market position.

1. Introduction

The fashion industry is highly competitive and there is a retail tendency of mergers and acquisitions by large companies. In recent years, the transition from a mass market to a niche-segmented market forced companies to know their target audiences, track their behaviours, understand their tastes and desires, and also seek for the best way to deliver customized services and products [1].

Nowadays, brands are facing serious challenges and it is even more difficult to consolidate a new brand in the market. Fashion communication has been undergoing a transformation, there is a decrease in tradicional marketing investments which indicates new modalities of brand's communication [1]. There is no doubt that the digital methods need to be incorporated into the marketing strategy of a fashion brand to achieve both sales and recognition results. Unlimited internet access, fast transmission, content flexibility and low-cost consumer mobility point toward a new era for companies that want to increase their brand and drive sales. [2]

In addition to the new investments in interactivity and connectivity, we can also observe an increasing search for strategies that value the continuous relationship between brands and consumers. According to Cietta [3], Italian fast fashion companies are innovating in communication, putting the brand in evidence and giving more visibility through unusual ways of approaching consumers. Instead of publishing campaigns in magazines and newspapers, they choose to invest in public relations and do some events in fashion spots and nightclubs, for example. Therefore, brands and consumers can communicate with each other without any restriction of time and place. As a result, the traditional process of one-way communication changes to a two-way communication in a scenario where brands can benefit their exposure to fortify relationships with their customers and, consenquently, increase their sales [4].

This research discusses the branding strategy relevance through social media and public relations' perspective in order to understand how new digital marketing platforms have been used and integrated by fashion industry professionals. These practices have already been adopted and used by many companies as sources for gathering relevant information, establishing brand awareness and creating

IOP Conf. Series: Materials Science and Engineering **459** (2019) 012097 doi:10.1088/1757-899X/459/1/012097

relationships with their prospective customers. On top of that, influencing a huge number of individual acceptance and the usage of products and services, with minimal effort attempts required to influence them [5]. Kim A. and Ko E. [6] said that even luxury brands have resorted to using digital platforms to survive, increasing competition within the market.

Carvalhal [7] believes that consumers like brands for what they represent. The more meaningful is a brand, stronger and more inspiring it will be. Building a fashion brand is similar than building a meaning. Hence, to create a new fashion brand, it is necessary first to create its identity. Brands with a stronger meaning are more easily differentiated in the market, consequently, it will result in a greater growth potential than brands which differ only by the products characteristics. Brand identity is what brands transmit to the consumers, which provide them to understand the essence and expression of a brand. [8].

Talking about public relations investments, Tungate [9] pointed out that celebrity sells and gives brands a well-defined personality for a minimum of effort. Golshorkhi, mencioned at Tungate [9], believes that the connection between fashion brands and the media is based on relationships where fashion PR professionals are always working to stimulate friendships with journalists.

2. Research Methodology

For this investigation, a qualitative methodology was chosen more appropriate as the researchers are investigating a small sample. Mark Saunders *et al* [9] stated that the classification of the types of study that are most used in research methods are exploratory, descriptive and explanatory, however each of the embodiments may have more than one purpose.

Qualitative data include texts as well as images, movies, audio-recordings, cultural artefacts, and more. It is an analysis more interpretive instead of statistical. The goal of qualitative research is to find correlations and create casual models that can be generalized [10]. A case study is an important research strategy when the researcher wants to comprehend the complex social phenomenon and allows him to focus on a case and maintain a holistic perspective [11].

In the scope of this study, the question to be analysed is the attempt to understand the factors of differentiation in the launch of new fashion brands. Then, the focus of the research is to collect data and observe the branding and communication strategies that were relevant to positioning of three fashion brands recently launched in the Portuguese market.

The participant observation, where the researcher interacts with the study object, causes the same to be identified with the brand culture, making the research process highly valued, due to the permanence and an in-depth knowledge of it [12].

The expected results in these case studies include some answers to relevant questions: which is the best way to position a new brand in the market, how is the impact of branding on consumer behaviour, how social networks could be used as a fashion communication channel, how social media guidelines can be used to increase global recognition of the brands, how the public relations strategy, bloggers and celebrities can influence consumers, among others. The non-probability sample technique selected as most appropriate for this study was the purposive sampling. According to Saunders *et al* [10], the purposive or judgemental sampling enables us to use our judgement to select cases that will best enable to answer the research questions and to meet the objectives. This form of sample is often used when working with very small samples such as in case study research.

In order to achieve this purpose, we first will do a literature review, second a case study with data collection method, qualitative content analysis and semi-structured interviews, and finally cross the information about the three brands aiming to find some features that contributed to their brand position. The semi-structured interview commences with a set of interview themes but is prepared to vary the order in which questions are asked and to ask new questions in the context of the research situation [9].

3. Case Studies and Discussion

This research focused on data collecting from available information on brand's websites and their social networks. The remaining information will subsequently be obtained throughout the research process over interviews with brand's Marketing Managers.

Considering the topic and objectives proposed, a search for new Portuguese fashion brands was made in digital press such as magazines, newspapers, online blogs, social networks and fashion & lifestyle bloggers. After a brief observation of thirty new fashion brands, eight brands were chosen for sample analysis (Table 1).

Table 1: Brands General Information for Sample Analysis (update on August 15th 2018)

Brand Name	*Founded*	*Brand Concept*	*Facebook Followers*	*Instagram Followers*	*Press Interest*	*Blogger Interest*	*Relevance Research*
Cinco Store	2014	Independent design, minimalist	18.758	66.100	Low	High	Poor
Isto	2017	Independence, transparency, organic, ethical	2.613	4.780	High	Moderate	Excelent
Kaoa Shop	2013	Well designed products, finest clothing	113.399	257.000	Low	Moderate	Poor
Misci	2017	Luxury in tradition, handmade details, ethical	12.488	1.604	High	Low	Somewhat
Naz	2017	Sustainable, minimalist, conscious, democratic	5.699	4.313	High	Low	Somewhat
Perff Studio	2017	Seansonless, inovation, functional fashion, confident	2.292	2.612	High	Moderate	Excelent
Sienna	2016	Passion, proudly made in Portugal, clothes to live	8.490	14.700	Moderate	High	Excelent
Imauve	2016	Practical, sophisticated minimalism, essential pieces	2.778	2.430	High	Low	Somewhat

These brands went through a direct observation analysis regarding the relevance of the brand concept, website and online store, social networks, press interest in the brand and endorsement by bloggers.

Therefore, the choice of this small sample was based on the purposive sampling using the self-judgment and also the criteria mentioned above. After this analysis, three brands were evaluated as excellent by the researchers, so Isto, Perff Studio and Sienna were selected to be the case studies of this research. It was decided to choose brands from different segments, with a high press and bloggers interested, founded after 2016, with a different brand concept, concerned about the digital brand communication (Figure 1).

Figure 1: Brand's Visual Communication – Instagram Data Collection
Source: https://www.instagram.com/siennainspo, https://www.instagram.com/isto.pt, https://www.instagram.com/perffstudio

3.1. Sienna

Sienna was born and raised between two dressmakers and it is proudly made in Portugal. The brand segment is casual women wear with a design minimalist and very feminine.

Sienna has online store and it is present in the social networks Facebook, Instagram and Pinterest, but the last one will not be collected in this research. On Instagram, they usually share inspire and lifestyle content, as show in figure 1. The brand is only 2 years old, however is always be mentioned in the most important Portuguese websites and blogs about fashion and lifestyle. In addition, a lot of Portuguese bloggers love the brand and they are addict on their products. We can frequently find inspire pictures on the fashionists Instagram's account. According with some of these bloggers, Sienna is one of the most fashionable Portuguese casual brand.

As mentioned in websites and Portuguese digital magazines, the success of Sienna cames from the acceptance of the bloggers, so this type of investments in public relations was extremely important to its grown.

3.2. Isto.

Isto is an organic men's brand which was born to prove that premium menswear can be awesome, affordable and timeless. It is also a platform for communication through its own digital magazine aiming to be a community of consumers whose feedback is the ultimate reward and driving force.

The naming Isto means "this" in Portuguese and stands for the brand concept: independent, superb, transparent and organic. Actually is a new way of looking at fashion. And it is exactly this point of view which probably fascinated the press. Because nowadays the brands which are ethical, concerned about people, industry transparency and mainly who wants to make the difference in the world are really important to the media.

Isto has an online store and social networks Facebook e Instagram. The brand is freshly launched, but we can already see relationships with bloggers and digital influencers. Its content is more transparent and simple, where can be find bloggers wearing the brand, lookbook and also pictures inside the office and industry.

3.3. Perff Studio

Perff Studio was born from high fashion and technical innovation. The brand position is athleisure wear, which represents a new movement in the sports and fashion sector that benefits people's lives from mental wellbeing to everyday living in stylish comfort. Perff is a contemporary fusion of hybrid performance fashion which enables people who enjoy working out and love the comfort and benefits of sportswear.

The differentiate brand concept and the high investments in communication allowed them to conquer the press interest. In the last one and a half year, they got digital reports in great Portuguese websites. The brand has some connections with influencers and sometimes send special gifts for bloggers, in order to increase the brand awareness and the followers. In the recent months, they started a relationship with a press agency aiming to contribute to better impact the press. They did a press day where fashion opinion makers attended and got to know this new Portuguese brand.

Perff has online store and it is present in the social networks Facebook and Instagram where they perfectly show the brand concept through creative images and great content, how can be seen on the figure 1.

3.4. Discussion

Due to the small sample and by reason of the investigation is still in progress, no final conclusions can be drawn by now. Besides, it was possible to perceive that findings point to a possible influence from social media networks and public relations investments to enhance the brand equity and brand awareness.

As previous literature discussed, new marketing strategies offer can establish loyal customer relationships with a new brand and affect consumers' individual perceptions of the product, pictures, videos, etc. Furthermore, as seen in the data collection, activity in social media increases brand exposure and brand awareness, and therefore influences brand equity. Additionally, investing in public relations strategies can also contribute to increase the level of brand awareness and improve the communication with customers, ensuring more reliability through the digital influencers. This discussion section will further discuss the results found in the course of the case studies investigation.

4. Preliminary Conclusions

New fashion brands have to deal with mature markets, strong competitors, high market share, established and notorious brands and to generate brand awareness in potential markets. On the other hand, there are always market opportunities in some niche markets, further to innovative and differentiate products or quality services, even if the brand is new.

Keeping in mind that the transition from a mass market to a segmented market contributed to change the consumer's habits, it is important to emphasize that we – as fashion communication researchers – need to constantly study how to achieve an effective brand strategy to surprise this consumer, if we want to build a successful and valuable brand.

This paper describes part of an ongoing Master's Thesis research study aiming to contribute to the subject of Branding and Fashion Consumers. What we can conclude by now is that the consumers are using even more new digital channels to interact with their love brands, so the social networks and digital public relations are quite important to a new fashion brand.

Although there were certain limitations considering it is still an ongoing research, what it is intend to achieve at the end of this investigation:

- To identify the newest trends in fashion communication;
- To comprehend the strategies used by brands to impact the consumer in the digital environment, especially on social networks Facebook and Instagram;
- To analyse the impact of bloggers and influencers on accepting a new brand;
- To realize if a differentiated brand concept has a competitive advantage in market positioning;
- To understand if there is a link between the information gathered in different brands in order to identify similarities in the strategies used that contribute to the brand success.

Acknowledgments
"This work is financed by FEDER funds through the Competitivity Factors Operational Programme - COMPETE and by national funds through FCT – Foundation for Science and Technology within the scope of the project POCI-01-0145-FEDER-007136".

References

[1] Kartajaya H, Setiawan I, Kotler P, 2017 *Marketing 4.0 – do Tradicional ao Digital.* (São Paulo: Sextante)
[2] Chiele, P R, 2016 *Luxury Branding: construindo marcas de luxo* (Porto Alegre: Editora do Autor)
[3] Cietta, E, 2010 *A revolução do fast fashion: estratégias e modelos organizativos para competir nas indústrias híbridas (*São Paulo: Estação das Letras e Cores)
[4] Kim, A J and Ko, E 2012 Do social media marketing activities enhance customer equity? An empirical study of luxury fashion brand. *Journal of Business Research*, **65**, pp.1480-1486.
[5] Subramani, M R and Rajagopalan B 2003 Knowledge-Sharing and Influence in Online Social Networks via Viral Marketing. *Communications of the ACM*, **46**, Issue 12, pp.300-307
[6] Kim, A J and Ko, E 2010 Impacts of Luxury Fashion Brand's Social Media Marketing on Customer Relationship and Purchase Intention. *Journal of Global Fashion Marketing: Bridging Fashion and Marketing*, **1**, Issue 2, pp.164-171
[7] Carvalhal, A, 2016 *Moda com propósito.* (São Paulo: Paralela).
[8] Randall, G, 2000 *Branding: a practical guide to planning your strategy,* 2nd edition (London: Kogan)
[9] Tungate, M, 2005. *Fashion Brands: Branding style from Armani to Zara.* (London: Kogan Page)
[10] Saunders M, Lewis P and Thornhill A, 2009 *Research Methods for Business Students* (London: Financial Times Prentice Hall)
[11] Yin, R K 2009 *Case Study Research: Design and Methods* (California: Sage Publications Inc)
[12] Kuckartz, U, 2014 *Qualitative Text Analysis: a guide to methods, practice & using software* (London: Sage Publications Ltd)

AUTHOR INDEX

AUTHOR INDEX

AUTHOR INDEX

Institute of Physics
Dirac House, Temple Back
Bristol BS1 6BE UK

ISSN: 1757-8981
ISBN 978-1-5108-8005-4

CURRAN ASSOCIATES INC.
proceedings.com